Mate Choice

Mate Choice

THE EVOLUTION OF SEXUAL DECISION MAKING FROM MICROBES TO HUMANS

Gil G. Rosenthal

PRINCETON UNIVERSITY PRESS
PRINCETON AND OXFORD

Published by Princeton University Press, 41 William Street, Princeton, New Jersey 08540

In the United Kingdom: Princeton University Press, 6 Oxford Street, Woodstock, Oxfordshire OX20 1TR

press.princeton.edu

Jacket image: Swindern's Love birds, *Psittacula swinderniana*. Courtesy of Alamy

ISBN 978-0-691-15067-3

Library of Congress Control Number 2017940516

British Library Cataloging-in-Publication Data is available

This book has been composed in Minion Pro with Scala Sans Display

Printed on acid-free paper ∞

Printed in the United States of America

10 9 8 7 6 5 4 3 2 1

CONTENTS

PREFACE

A cherished bit of apocrypha about Marvin Minsky, a founder of artificial intelligence:

About twenty years ago, he assigned a graduate student a seemingly tractable problem for a summer project: connect a camera to a computer and make the computer describe what it sees […] *Despite machine vision's enormous progress, that problem has still not been solved.*

—Hurlbert & Poggio (1988, p. 218)

In the summer of 2010 I thought I could write a book on mate choice in a mere two years. My research program on the topic was broad and aspired to be integrative in the tradition of the scientists I trained with: Peter Marler, Chris Evans, and Mike Ryan. I wanted to figure out *why* living things are sexually stimulated by some things but not others, and *how* one individual's object of disgust is another's object of desire. In more clinical terms, my long-term goal was (and remains) to identify the mechanistic underpinnings of variation in mating preference and use those underpinnings to test explicit hypotheses about evolutionary patterns. This meant I needed to straddle sensory ecology, animal behavior, and evolutionary genetics. Being broad inevitably means having to sacrifice some depth; for me, part of the trade-off was not being able to fully immerse myself in any one of these disciplines, and another was a sometimes myopic focus on small freshwater fishes. With this book, I wanted to both give myself a comprehensive education on what is known about mate choice, and share that education with anyone interested. I also hoped to shake up a field that yielded so many exciting ideas and so much interesting data over the past 30 years yet nevertheless has managed to spend that time in a conceptual rut. Over the past three decades, advances in machine vision have brought us self-driving cars and automated face recognition. The immense corpus of work on mate choice over that same time period has not yielded comparable disruptions to the thinking that was popular in the 1980s. I thought a comprehensive work that encompassed the whole sweep of mate-choice research would spark connections among fields and study systems.

For this book to be genuinely integrative, and not just a collection of superficial reviews, I needed to be on intimate terms with a massive literature that ranged from the protein structure of odorant receptors, to the response of discrete brain regions to attractive faces, to the neurobiology of foot fetishes. I came into this endeavor knowing that my expertise was an inch deep and possibly less than a mile wide. I felt comfortable taking this on as a traveler's guide rather than a technical manual, designed to get people to draw novel connections with fields they didn't normally think about. As the months progressed I developed a stronger appreciation both for the need for a book that cut across taxonomic and methodological silos and for the magnitude of the scholarly effort required to do those silos justice. I wanted to make sure I had a sufficient enough understanding of each of the subdisciplines to be able to make a cogent critical analysis of their relationships to the broader framework of mate choice. Sifting through the literature led down a Borgesian labyrinth of exponentially expanding citations.

A delightful labyrinth, however—there is so much work out there that has been ignored by the scholarly world I know, and so much to discover that is illuminating or just amazing. In the amazing lie the pitfalls, though. Hurlbert and Poggio (1988) are frequently cited as authoritative sources for the Marvin Minsky story, but the secondary sources omit the "apocrypha" bit. I may have been fool enough to think that I could synthesize mate choice in two years (while maintaining a funded and aspirationally integrative experimental research program; the U.S. National Science Foundation nurtured many of the ideas herein), but Minsky surely knew better than to think he could solve machine vision in a summer. Yet the "cherished bits of apocrypha" are, quite literally, infectious: appealing and hard to source, they propagate in the literature and take root with secondary citations.

The mate-choice literature is full of poorly sourced "cherished bits," which I have done my best to elucidate, surely committing Marvin Minskys of my own along the way. The scholarly due diligence has been overwhelming, even as I have striven to keep up with the exploding literature. About a third of the studies I cite here have been published since I started this book. Fortunately, some of these have included broad and synthetic surveys of several key areas, which have proven invaluable in shaping my thinking and in summarizing important topics. Many of these are featured in the "Additional reading" section at the end of each chapter.

I am especially grateful to the students and postdocs in my lab, who have been patient with my book-writing solipsism and have provided an endless source of energy and inspiration. The early stages of the book were heavily influenced by discussions of drafts with graduate students Chuck Carlson,

Rongfeng Cui, Sarah Flanagan, David Garcia, Brad Johnson, Dara Orbach, Kim Paczolt, Emily Rose, Michelle Ramsey, Grace Smarsh, Victoria Smith, and Mattie Squire. Among these I am particularly grateful to Pablo Delclós and Dan Powell, who stuck through all the way to the final drafts and provided incisive comments on multiple versions of each chapter. Lauritz Dieckman and Nick Ratterman deserve thanks for their insightful help gathering the relevant literature. Connie Woodman, Andrew Anderson, Mateo García, Stephen Bovio, Megan Exnicios, Gastón Jofre Rodríguez, Chris Holland, and E. V. Voltura came in with fresh eyes to help clarify and refine the later drafts. This book was also greatly improved by detailed comments on various chapters from Christian Bautista Hernández, Mark Kirkpatrick, Carlos Passos, Molly Schumer, Machteld Verzijden, and Bob Wong. Maria Servedio provided incisive comments on multiple drafts of the later chapters.

For helpful discussion and criticism I am indebted to Suzanne Alonzo, Peter Andolfatto, Ricardo Azevedo, Spencer Behmer, Chris Blazier, Dan Blumstein, Felix Breden, Rob Brooks, Sabrina Burmeister, James Cai, Ginger Carney, Sergio Castellano, Iain Couzin, Charles Criscione, Zach Culumber, Molly Cummings, Jenny Gumm, Heidi Fisher, Rosemary Grant, Osvaldo Hernández Pérez, Hans Hofmann, Kim Hughes, Spencer Ingley, Mike Jennions, Adam Jones, Alex Jordan, Hanna Kokko, Topi Lehtonen, Bruce Lyon, Constantino Macías Garcia, Tami Mendelson, Bill Murphy, Gail Patricelli, Steve Phelps, Andrea Pilastro, Jonathan Pruitt, Margaret Ptacek, David Reznick, Dan Rubenstein, Paul Samollow, Manfred Schartl, Ingo Schlupp, Giovanna Serena, John Swaddle, Greg Sword, Bettina Tassino, Michael Tobler, Joe Travis, Thor Veen, Mary Wicksten, Ashley Ward, Alastair Wilson, Harold Zakon, and Marlene Zuk. I am especially grateful to John Endler, Rick Prum and an anonymous reviewer for their candid and constructive comments, which greatly improved the final manuscript.

All the line drawings in this book are by Matt Stephens, to whom I also owe a debt for a quarter-century of friendship, scientific insight, and Sierra Madre escapades. I am thankful to Lauren Bucca, Dimitri Karetnikov, and Brigitte Pelner, at Princeton University Press for their assistance with production and to Karen Verde for copyediting. This book would not have happened without the tireless support and encouragement of my editor, Alison Kalett.

My parents' improbable mate choices had personal consequences. My mother, Maghy Spampinato Rosenthal, taught me to love writing. My father, Howard Rosenthal, taught me to love science and gave me his invaluable and detailed perspective on the entire manuscript. My daughters Carmen and Jamila, themselves the product of unlikely choices, were respectively a toddler and a zygote when I started this book. They have been a constant reminder of the important things in life, and as insightful enthusiasts of aesthetics and animals respectively they have helped me finish it.

This book is dedicated to Rhonda Struminger, not only for proving that imprudent mating decisions can yield offspring of tremendous Quality, but for her unflagging support—moral, intellectual, and logistical—over these last years as this book has come to fruition.

Finally, this whole book owes a debt to Mike Ryan, who initially encouraged me to propose this book and whose integrative perspective on mate choice has heavily influenced mine. I have striven to meet his standards of writing and scholarship, and this book is a tribute to his insight and guidance.

PART I

Mechanisms

CHAPTER 1

Mate Choice and Mating Preferences

AN OVERVIEW

1.1 INTRODUCTION

Hiking in the eucalyptus woods of northern Australia, we might come upon an odd structure with a promenade of shells and bones leading up to a curving, symmetric arch. We might reasonably speculate that we have stumbled upon an indigenous ceremonial site, or perhaps a contemporary art installation (fig. 1.1a). Diving off Japan's Okinawa Prefecture, we come upon a similar structure—an "alien crop circle" in the popular media (fig. 1.1b). We are astonished when we discover that the architects were a male great bowerbird (*Ptilonorhynchus nuchalis*) and a male pufferfish (*Torquigener* sp.), and that these structures only function in the context of courtship and mating. As amazed as we are by the structures' builders, we should be awestruck by the aesthetics of the females they were built to impress. How intricate their aesthetics, how exacting their desires, must be in order to drive males to such cognitive and physical extremes? Why do females even bother to choose males on the basis of these structures, rather than simply mating at random?

Mating is an expensive, risky, and intimate interaction, and over an individual's lifetime one expends time and energy on facilitating some matings, and time and energy on avoiding others. Who a **chooser**[1] mates with and who she pairs with will affect how long she lives and how many healthy children and grandchildren she has. Mate choice determines which sperm fuse with which eggs, and therefore ultimately shapes how lineages split apart or merge together. It can drive the evolution of elaborate traits that hinder critical tasks like finding food and avoiding predators, in direct opposition to **natural selection**. The role of mate choice in both **reproductive isolation** among species and in **sexual selection** made it a key concept in Darwin's *Origin of Species* (1859). There was widespread skepticism over his conjecture that mating **preferences**—a "taste for the beautiful," in Darwin's

[1] Boldfaced terms are defined in the glossary.

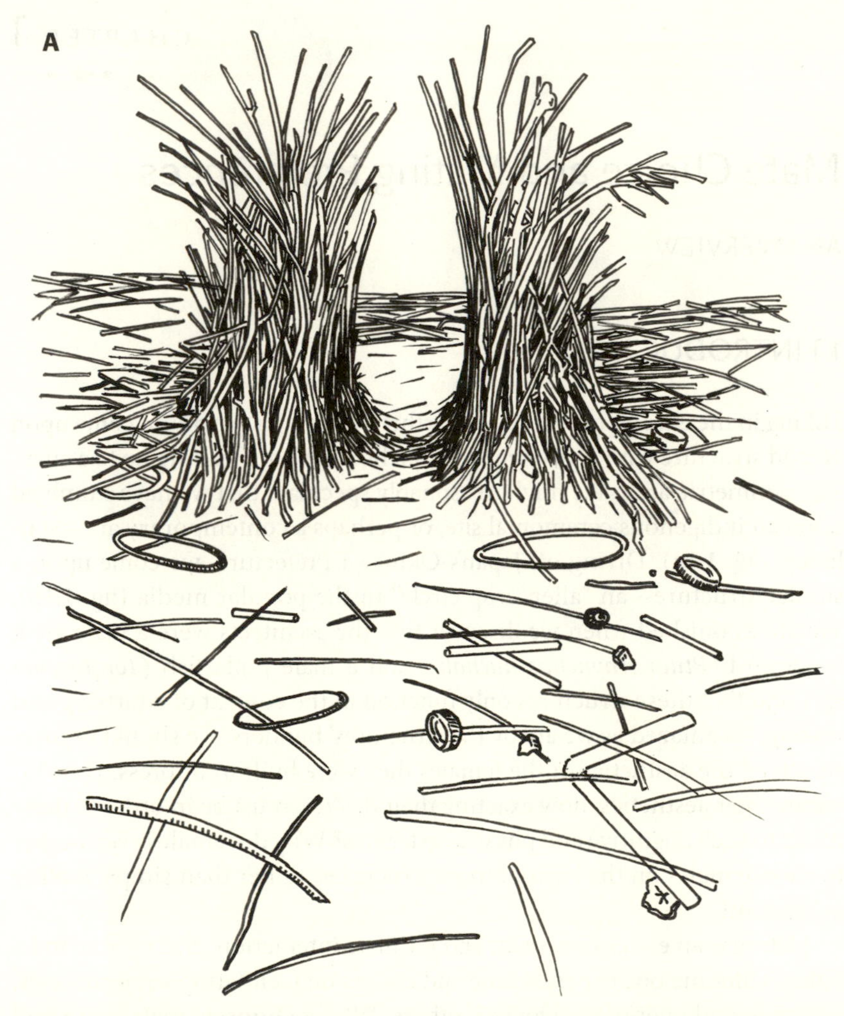

Figure 1.1. (a) Bower of a satin bowerbird, *Ptilonorhynchus violaceus*, Queen Mary Falls, Queensland, Australia. Drawing from photo by Gail Patricelli.

memorable phrasing—could explain the seeming paradox of so much exuberant scent, texture, and sound in nature. Accordingly, he devoted the bulk of his next major work, *The Descent of Man, and Selection in Relation to Sex* (1871), to making the case for the central evolutionary role of sexual selection, particularly via mate choice.

Almost a century and a half later, mate choice continues to present a unique problem in evolutionary theory. Like predators coevolving with their prey, or hosts with their parasites, those courting and those choosing form

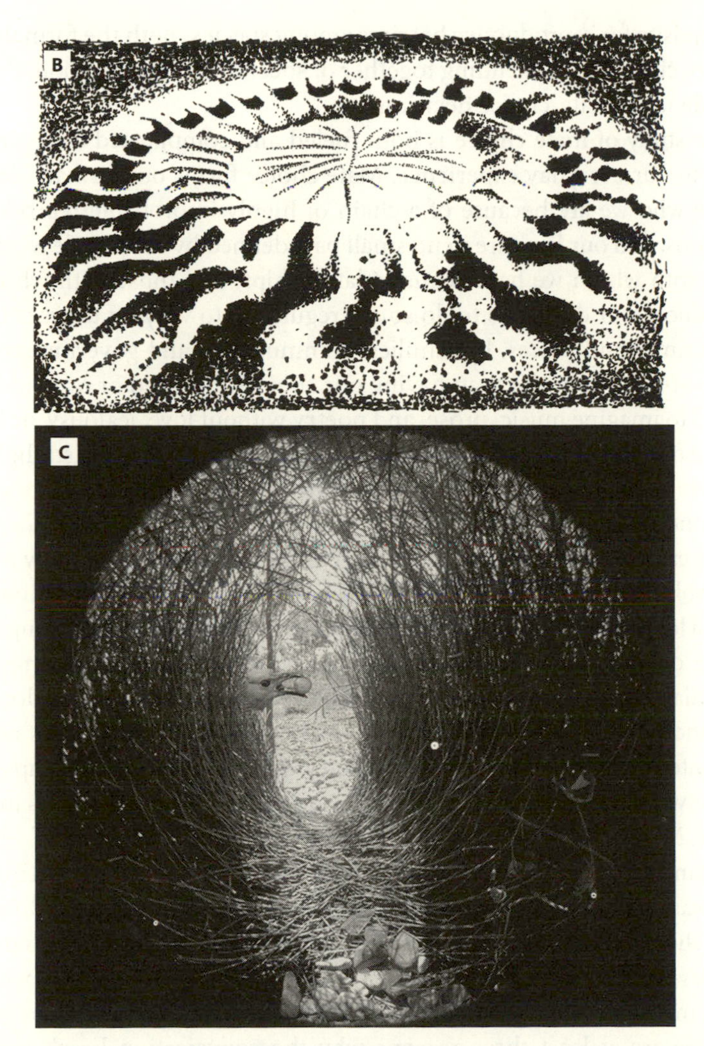

(b) Bower of a pufferfish, *Torquigener* sp., off Okinawa prefecture, Japan (Kawase et al. 2013). (c) Composite image of a displaying male great bowerbird (*P. nuchalis*) and bower as it appears to a choosing female, © 2017 John Endler.

a feedback loop, where **chooser** decisions can select for particular **courter** behavior and vice versa. In the case of mate choice, however, the same genome influences the behavior of both actors, and the interests of both are partly aligned and partly in conflict (Arnqvist & Rowe 2005). This kind of dynamic can lead to rapid evolution of elaboration of signals and choices within a species, which can lead to marked diversity of such signals and choices between species. Such divergent mate-choice patterns are often a prerequisite for reproductive barriers among species. Both the formation of new species, and the blending together of species via hybridization, depend on individual mate-choice decisions.

The study of mate choice is both fueled and complicated by its importance to our everyday experience. Mate choice forms our human identity: we are who we are because of a chain of highly improbable reproductive decisions, and our lives are in no small part defined by the people we desire, those with whom we have sexual relationships, and those with whom we reproduce. Our decisions to do so are regulated, to varying degrees in different times and places, by families, communities, and governments; few things are more painful than having our choices thwarted or overridden. It is hard to imagine music, prose, and poetry without love, jealousy, or heartache. And when we court each other and choose each other by starlight, we do so to the soundtrack of crickets and frogs doing the same. Mate choice surrounds us.

It is easy to make the case that mate choice is important, but how it actually evolves and how it actually works remains essentially mysterious. We are at a loss to explain much of the beauty in the world, from birdsong to the palette of colors on a coral reef, because we know that these things arise from mate choice, but we are still striving to understand how. We don't understand why choosers pay attention to so many different things or how they integrate information into a unitary decision to mate. Perhaps most visibly, we still fail to agree on the importance of **adaptive** processes in mate choice. My first scholarly exposure to mate choice was in the fall of 1993, in a freshman seminar on "Sex and Evolution" led by Jae Choe. At the time, the field was consumed by a debate about the extent to which an individual's mate-choice decisions impact the "genetic quality" of her offspring. Two decades later, we remain mired in, and limited by, the argument of whether or not mate choice is optimally designed to pick mates bearing "good genes."

There are at least three reasons why the conversation hasn't changed much over a generation. The first reason is that work on mate choice is hard to do; the core of mate-choice research involves inferring and predicting mating decisions indirectly and/or over long timescales. This is because mate choice as a phenotype is inherently slippery; we're usually measuring behavioral decisions, which are inherently contingent on the stimuli presented, and can only be measured indirectly. We can readily measure the spectral reflectance of the components of a bower and calculate how they catch the sunlight over the course of a day, but it's much more challenging to measure how these components influence the likelihood that a female will mate with the male who produced it. The next chapter deals with the technical challenges of measuring mate choice.

The second reason is the Balkanization of our approaches to studying mate choice. Those who study humans are generally associated with entirely different disciplines (anthropology and social and evolutionary psychology) than the majority of their colleagues working on non-humans (biology and its subfields, as well as comparative psychology). Biologists, moreover, are further subdivided into quantitative geneticists, behavioral ecologists, ethologists, and behavioral neuroscientists. The massive literature on mate choice is a mixed blessing, since it makes it difficult for any individual to have in-depth knowledge of more than one of these areas. A major goal of this book is to bring these fields together toward a synthetic understanding of mate choice.

The third and perhaps principal reason for the field's slow progress is that we have always thought about mate choice primarily in terms of its functional *consequences*. Starting with Darwin (*The Descent of Man, and Selection in Relation to Sex*, 1871) and sexual psychologist Havelock Ellis (*Sexual Selection in Man*, 1905), and continuing on to the present (Andersson, *Sexual Selection*, 1994; Eberhard, *Female Control: Sexual Selection by Female Choice*, 1996; Arnqvist & Rowe, *Sexual Conflict*, 2005), the focus has not been on mate choice as an intricate psychological and behavioral process in its own right, but on mate choice as an agent of sexual selection. Evolutionary models sometimes rely on fanciful assumptions about mechanisms; conversely, empirical studies of mechanism frequently assume optimal design. Conversely, to the extent that those who study mate-choice mechanisms think about **fitness** consequences, they often assume these mechanisms are systems optimally designed to maximize the benefits of mate choice to choosers, rather than systems cobbled together from available genetic variation that sometimes lead choosers astray. It is tempting to think of choosers as actuaries, evaluating expected lifetime fitness, and taxonomists, recognizing **conspecifics** and **heterospecifics**, and executing each of these tasks both perfectly and separately. Yet relatively little attention is paid to how mate choice actually works, although this is crucial to understanding both how it evolves and how it imposes selection. How does a female bowerbird actually experience her choices (fig. 1.1c)? Our focus on courter traits, rather than chooser preferences, has produced some stumbling blocks for evolutionary theory: one important example is that the predictions of most sexual selection models depend entirely on whether the net direct benefits of mate choice are positive or negative, yet we seldom measure this directly.

What is **total selection on mate choice**, and how does it affect the way preferences and sampling strategies evolve?

The standard approach in the mate-choice literature is to begin by reviewing theoretical and conceptual models, then discussing empirical evidence in light of the theory. Inspired by Darwin's inductive approach in the *Origin* and the *Descent*, I have attempted to turn this approach upside down and interpret theory in light of what we know about how mate choice works. Accordingly, I focus this first section of the book on natural history—a broad description of the mechanisms, ontogeny, and phenotypic expression of mating preferences and mate choice. I have deliberately chosen my language to minimize a priori assumptions about any adaptive functions of choosing particular mates over others. In the second section, I use this perspective to address how mate choice evolves and acts as an agent of selection, and how it generates fitness consequences for individuals and evolutionary consequences for populations and species.

Part of the challenge of studying mate choice arises from the enormous scope of mate choice as a phenomenon. The contemporary literature on mate choice is immense. Choice can range from the simplicity of a single-celled protozoan exchanging genes only with another individual emitting a particular signaling molecule, through the protracted mutual courtship of humans and other vertebrates. What these vastly different mechanisms have in common is that they impose variation in the mating success of the *individuals being chosen*—courters. Preferred courters will, by definition, have an advantage over unpreferred ones (but see Long et al. 2009 for a counterexample).

By contrast, the magnitude and direction of mate choice's benefits to *choosers* is hugely variable, and while sophisticated mate-choice mechanisms offer more opportunity for nuanced **evaluation** and comparison, they also offer greater entry for subversion and deceit. Some of the mechanisms involved in mate choice, like the tuning of peripheral sensory receptors, are universal among organisms. Other mechanisms, such as selective attention to particular traits, are highly labile among species, within species, and even within individuals. A recurring theme of this book is the importance of the processes promoting and maintaining within-population variation in mate choice.

Like the mechanisms used for mate choice, the ecological theater of mate choice spans the full range of natural history. Mate choice occurs in everything from parrots that grow up with both parents, to parrotfish that are cast off into the plankton as fertilized eggs. There is mate choice among anglerfish in the deep ocean that encounter mates so rarely that when they do, males permanently attach to females, their circulatory systems fusing together; and there is mate choice among crickets in noisy choruses, surrounded by thousands of courters and choosers. Both the mechanisms constituting mate choice and the selective pressures shaping it are thus as variable as can be among taxa. The diversity of mate-choice mechanisms provides the potential for wonderful natural experiments, but these are again limited by the difficulties inherent in measuring choice. One person can easily go to museums and measure a morphological trait in a hundred species. An individual research lab working on mate choice can manage at most a handful of species with similar maintenance needs. Accordingly, taxonomic clustering adds to the intellectual Balkanization of mate choice. Social-context effects on mating (chapter 6) are one example. Nearly all studies of sexual reward come from one rodent species, and nearly all studies of mate-choice **copying** come from one fish family and one bird species. With different model systems come different constraints as to what we can measure, different traditions of what's important to think about, and different networks of researchers. This book attempts to bring these approaches together and survey mate choice across taxa (including humans), striving to avoid being too biased by my own inordinate fondness for livebearing fishes.

Attempts to fit mate choice between two covers are, perhaps sensibly, few and far between. Since Bateson's (1983) eponymous edited volume on mate choice, we have gained considerable ground in understanding sensory, perceptual, and cognitive mechanisms, and in understanding the evolutionary causes and consequences of choice. Across fields, the literature on mate choice has exploded. The aim of this book is to present a conceptually unified approach to thinking about what Darwin termed the "taste for the beautiful." It is not intended to be an exhaustive review of the literature, particularly since any such effort would be both redundant with the Internet and obsolete by the time of publication. I have attempted to synthesize the work of hundreds of people, but the papers I cite are probably biased by my taxonomic and geographic parochialism (Wong & Kokko 2005). I have tried to abide by the late Stephen Jay Gould's (1994, p. 164) maxim that erroneous ideas are useful, since they can invigorate science by stimulating new avenues of thought, while misleading facts are corrosive. Therefore, I have endeavored to be meticulous in terms of characterizing my sources, but have allowed myself some qualified speculation in the hopes of generating new conversations about mate choice. Nevertheless, I am a tourist to many of the subdisciplines involved, and although I have made an effort to have each chapter read by at least one expert colleague, the book surely retains mistakes and misconceptions that are entirely my own.

In this chapter, I begin by describing a basic framework for thinking about mate choice and mate preferences, and then provide an outline for how the book attempts to address key questions about how they work, how they evolve, and how they act simultaneously as targets and agents of selection.

1.2 WHAT IS MATE CHOICE?

> It is possible therefore that the emotional reactions aroused by different individuals of the opposite sex will, as in man, be not all alike, and at the least that individuals of either sex will be less easily induced to pair with some partners than with others. With plants an analogous means of discrimination seems to exist in the differential growth rate of different kinds of pollen in penetrating the same style.
>
> —Fisher (1930, p. 143)

One of the challenges of learning cell biology or neuroanatomy is the sheer amount of new vocabulary it entails. Students are overwhelmed by trying to keep the anterior cingulate cortex straight from the torus semicircularis. Animal behavior, by contrast, tends to assign specialized meaning to ordinary terms: while this makes them more accessible to a broad audience, it can lead to semantic confusion and anthropomorphism. Further, different authors use myriad terms to describe comparable processes. Edward (2015) provides a comprehensive review of the terminology surrounding mate choice and mating preferences. This book is biased toward multicellular animals with neurally mediated behavior, to which some authors prefer to restrict the term "choice." But neither sensory perception nor neural processing are required for a mechanism that discriminates among potential mates. Most contemporary scientists (Edward 2015; Kokko et al. 2003; Servedio & Bürger 2014) use variations on Halliday's (1983, p. 4) definition, which extends mate choice to a broad range of mechanisms in even the simplest creatures:

> Mate choice can be operationally defined as any pattern of behavior, shown by members of one sex, that leads to their being more likely to mate with certain members of the opposite sex than with others.

The focus of this book is on neurally mediated behavior in animals, but chemically or morphologically mediated, non-neural mate choice is the only option for plants (Burley & Willson 1983) and microorganisms (e.g., Lin 2009) and plays an important role in postmating choice in animals (chapter 7). If we change "patterns of behavior" to the more general "phenotypes," Halliday's definition encompasses mate choice in each of these systems.

Halliday's definition of mate choice is restricted to opposite-sex interactions. A growing number of studies (reviewed in Bagemihl 1999; Bailey & Zuk 2009b; Poiani 2010) have highlighted the ubiquity of same-sex sexual behavior across the animal kingdom; and more generally, an individual's sexual partners (homo- or heterosexual) can influence that individual's fitness independent of whether gametes are exchanged (chapter 14). Further, as I will argue below, "mating" is not a discrete event. I therefore suggest a modification of Halliday's (1983) definition to take into account the possibility of same-sex partners: **mate choice** can be defined as any aspect of an animal's phenotype that leads to its being more likely to engage in sexual activity with certain individuals than with others.

1.3 CHOOSERS AND COURTERS, NOT FEMALES AND MALES

Many authors use "female" and "male" as convenient shorthand for the individuals choosing and courting. This is because the distribution of potential mates is usually male-skewed, meaning there are many more males available to mate at a given time than females. Accordingly, females tend to be choosier than males. A growing body of work, however, emphasizes the role of male mate choice not only in sex-role-reversed species like pipefish and jacanas, but also in species where strong female choice is present as well. And sexual selection can act strongly on females (reviewed in Clutton-Brock 2009; chapter 8). Accordingly, I use the terms *chooser* and *courter* throughout this book. In addition to being sex- (and gender-) neutral, these terms refer to behavioral roles rather than to permanent aspects of the phenotype. Ecological circumstances like the availability of suitable territories can determine which sex chooses. In monogamous or hermaphroditic animals, the roles of courter and chooser can even reverse over the course of a single social interaction. It is therefore more productive to think about mate choice in terms of the roles that individuals are playing over the course of a given interaction.

1.4. MATE CHOICE IS DISTINCT FROM SEXUAL SELECTION

Mate choice has *ab ovo* been cast as the handmaiden of sexual selection:

> I cannot here enter on the necessary details; but if man can in a short time give beauty and an elegant carriage to his bantams, according to his standard of beauty, I can see no good reason to doubt that female birds, by selecting, during thousands of generations, the most melodious or beautiful males, according to their standard of beauty, might produce a marked effect. (Darwin 1859)

In the *Origin* and, more extensively, in the *Descent of Man, and Selection in Relation to Sex* (1871), mate choice plays an important, novel, but ultimately supporting role in sexual selection. The superstar is the evolution of intricate songs and elaborate plumage in birds, which Darwin regarded as self-evidently "beautiful." Darwin's proposed agent of sexual selection was the strikingly un-Darwinian "taste for the beautiful": slippery, anthropomorphic, mystical, and tautological. Part of the reason this hypothesis was rejected by Darwin's contemporaries was because, uncharacteristically, he treated this taste for the beautiful as axiomatic and failed to provide a satisfying evolutionary explanation for its origin or maintenance (Cronin 1991; Milam 2010).

Darwin's Victorian contemporaries were skeptical that animals, particularly female animals, could be capable of making aesthetic distinctions. This skepticism was of course exacerbated by the conventional sexism of Darwin's time, but Darwin didn't help his case by treating female cognitive inferiority as a self-evident fact of nature, and by throwing up his hands at the complexity of mate choice. As Milam (2010) points out in her excellent social history of the science of mate choice, Darwin's implication of "love or jealousy" as agents of choice suggested that sophisticated cognitive mechanisms might be involved; indeed, Darwin (1871) argued that mate choice would be restricted to species with "powers of the mind [that] manifestly depend on the development of the brain."

The subsequent century and a half of research on mate choice and mating preferences has largely continued Darwin's approach of viewing preferences in light of their role as agents of selection, even as it has partly demystified mate choice and demonstrated its operation in very simple biological systems. The foundational theory of R. A. Fisher (1930) focused on courter traits as the driver of preference evolution; the hypotheses of Zahavi (1975) further posited that these traits provided information to choosers about the genetic makeup of their offspring. The trait-centered perspectives of these evolutionary biologists were mirrored by the work of mid-twentieth-century **ethologists**, who mainly viewed mating interactions through the lens of courters (invariably males) "priming" the sexual receptivity of their mates. This obviated the need to think about any kind of agency on the part of choosers. Yet some courters are better than others at "priming," which allows us to think about sexual receptivity as a mechanism of mate choice (chapter 6); indeed, in the Australian redback spider, insufficiently stimulated females devour their suitors before mating occurs (Stoltz & Andrade 2010). Courters' efforts to induce choosers to mate, and choosers' responses to such efforts, set the stage for **sexual conflict** (chapter 15).

As will become clear throughout this book, mate-choice decisions can be adaptive, non-adaptive, or **maladaptive**, and mating preferences do not map cleanly to the traits that courters express in the real world: choosers often prefer combinations of traits that are unavailable in real courters. Mate-choice mechanisms are subject to multiple selective forces independent of the mating outcomes they shape. Conversely, strong preferences based on individual compatibility do not generate sexual selection. As Bateson (1983, p. ix) wrote in his preface to *Mate Choice*:

> When the term "mate selection" is used for what animals do, it can quickly lead to unconscious punning and the assumption that a preference for a particular kind of mate necessarily has implications for sexual selection. As will become plain, the assumption is false.

1.5 PREFERENCE AND ANTIPATHY UNDERLIE REALIZED MATE CHOICES

1.5.1 Preference

To understand what Bateson meant about the study of mate choice not being about sexual selection, it is important to distinguish between "mate choice" and "mate preference." These are sometimes thought synonymous, but they can be very different things. A **preference** is a chooser's internal representation of courter traits that predisposes her to mate with some phenotypes over others (Heisler et al. 1987; Jennions & Petrie 1997). Cotton and colleagues (2006a) separate "preference" into preference functions (fig. 1.2, see page 16), which correspond to the preceding definition, and the process of sampling and deciding among mates. For clarity, I restrict "preference" to the narrower definition. In chapter 6, I will return to the interdependence of preferences with mate sampling and mating decisions.

Preference by definition means that choosers are influenced differently by different stimuli; it is an inherently comparative process, even though it is convenient to think about "absolute" preferences in some contexts (chapter 2). Preference can be applied to a ranking of individuals (Harry prefers Sally over Marie) or of discrete characters (female túngara frogs, *Physalaemus pustulosus*, have a preference for calls followed by short, high-energy bursts called "chucks" over those without), or to a continuous function (female swordtails, *Xiphophorus hellerii*, have a **directional** preference function for longer tails on males). Univariate and multivariate preference functions are discussed below. Crucially, preferences don't have to be realized into choices; indeed, choosers often have strong preferences for trait values, or combinations of traits, that are unavailable in courters (Fisher et al. 2009). Such "hidden" preferences are universal and have the potential, once revealed by novel courter traits, to induce rapid and permanent evolutionary change (chapter 13).

1.5.2 Antipathy

We tend to think of mate choice in terms of which individuals or traits are most preferred, rather than which are rejected by choosers. Often, choosers expend more energy avoiding rejected courters than seeking out preferred ones. Even when rejection is relatively cost-free, mate choice invariably involves rejecting many more mates than one accepts; indeed, choosers may forgo mating altogether if no acceptable options are available (or, notably in flowering plants, to self-fertilize rather than mate with a different genetic individual; Burley & Willson 1983). Most preference envelopes—the space of acceptable mates—are very narrow, in the sense that there are almost always infinitely more ways that a courter can be unacceptable than acceptable. And, as Darwin recognized, choosers may often select not the most enticing available mate, but the least repulsive:

> [T]he female, though comparatively passive, generally exerts some choice and accepts one male in preference to others. Or she may accept, as appearances would sometimes lead us to believe, not the male which is the most attractive to her, but the one which is the least distasteful. (1871, p. 273)

Darwin returns frequently to this point and uses the term **antipathy** to describe when a particular individual, category, or trait value is less attractive than others. The term "antipathy" has not seen wide use in the recent literature, but as I will argue throughout this book, choosers' rejection of unattractive courters is generally more important than their acceptance of attractive ones, because of the downside risk of inappropriate matings (Clemens et al. 2014). The literature on sexual conflict (chapter 15) often refers to **resistance** (Holland & Rice 1998). For example, in female seaweed flies, *Coelopa frigida*, a female's mate choice is manifested mostly by physical rebuffs and signals of rejection (Blyth & Gilburn 2011). Preferred males are the ones that encounter the least resistance. In birds and mammals particularly, researchers frequently measure both negative (**aversive**) and positive (**appetitive** or **proceptive**) behaviors when characterizing preferences. For example, Forstmeier and colleagues (2004) assigned positive and negative preference scores based on the observation of appetitive and aversive behaviors, respectively. The most preferred courter is thus not only the individual who elicited the most positive responses, but also the least negative responses; the most beautiful is also the least distasteful. This may not matter much to courters, but the subjective value of mating encounters—whether choosers perceive them as positive or negative experiences—can affect a chooser's future mating decisions, making them more averse or more responsive to particular courter phenotypes (chapter 6).

Mate choice, then, is the phenotypic manifestation of preference and antipathy. Realized choices are constrained by the availability of courter phenotypes, by courter actions toward choosers, and by the way that choosers sample potential mates. Mate sampling and preference functions are intertwined, in that the sampling experience can change chooser preferences (chapter 6).

1.6 PREFERENCE FUNCTIONS

1.6.1 Overview

Chooser preferences are internal representation of the properties of courter stimuli. We also apply the term "preference" rather loosely to behavioral or other measures that vary according to courter stimulus; accordingly, a common way of representing preference is to plot a chooser's response—our choice of assay—as a function of courter trait values. This is a **preference function** (Wagner 1998; fig. 1.2), in which the mating response varies with the value of a trait. Preference functions are convenient concepts in studies of sexual selection via mate choice; in the idealized case where they perfectly predict realized mate choice and are the same for all choosers in a population, they represent **fitness functions** for chooser traits (chapter 14).

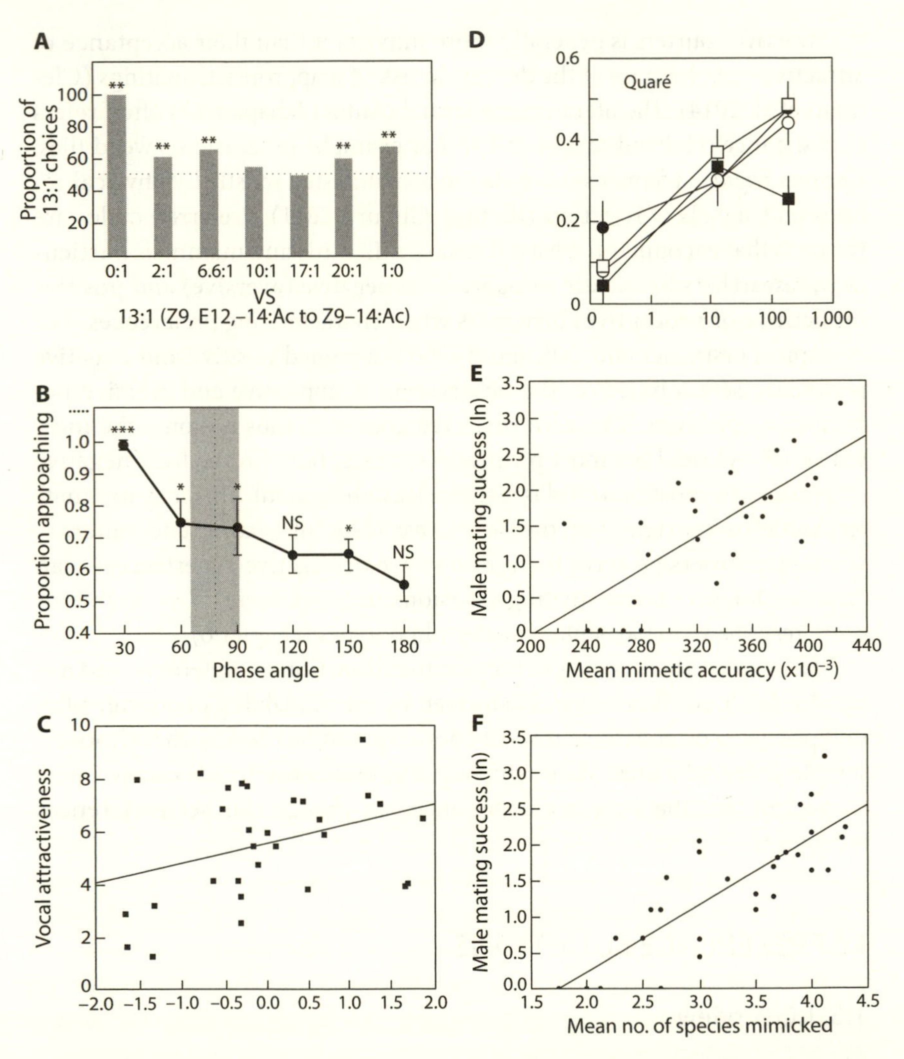

Figure 1.2. Preference functions. (a) Proportion of male almond moths orienting to the species-typical female pheromone blend when presented as an alternative to the ratio on the Y-axis (Allison & Cardé 2008); (b) approach probability of female midwife toads to the temporally leading call in a pair of stimuli presented with varying degree of temporal offset. At 30 degrees of phase angle, the leading call immediately precedes the following call; at 180 degrees, they are exactly antiphonal (Bosch & Márquez 2002); (c) Mean attractiveness rating of female voices by British men as a function of higher harmonic frequencies (Collins & Missing 2003); (d) Proportion of a male's courtship displays that elicit a "glide" sexual response from female guppies from the Quare drainage, Trinidad, as a function of dietary carotenoid concentration (ppm) and male and female population of origin (circles/squares—males from low/high carotenoid-availability steams; filled/unfilled—females from low/high streams) (Grether 2000); and mating success of male satin bowerbirds as a function of (e) the accuracy and (f) taxonomic diversity of their courtship vocalizations (Coleman et al. 2007).

The conceptual usefulness of preference functions is limited, however, when we start to consider the comparative nature of preference, whereby a stimulus's attractiveness is contingent on comparisons to other traits (chapter 6).

Preference functions represent a measure of mate choice in relation to continuous or ordinal variation in a courter trait. Measuring preference functions, by definition, requires sampling chooser responses to multiple trait values, and requires that chooser response be expressed as a continuous or ordinal variable (Wagner 1998). For empiricists, the response is typically represented as the frequency or duration of a particular behavior associated with mating (chapter 2) or as the proportion of individuals in a sample choosing a particular trait. Preference functions can either represent absolute responses to a stimulus (fig. 1.2c–f) or relative preferences between or among stimuli (fig. 1.2a–b).

Preference functions vary considerably among individuals in the same population or species (Jennions & Petrie 1997; chapter 9), which can have fundamental consequences for sexual selection (chapter 15) and speciation (chapter 16). To understand among-individual variation in mating preferences, we need to characterize preference functions for distinct individuals (Wagner 1998), which poses several challenges—notably that assaying preferences inevitably changes individual experience. Preference assays and repeated testing of individuals are discussed in the next chapter.

The **shape** (or "form"; Cotton et al. 2006) of a preference function falls into a few broad categories (Ritchie 1996; Edward 2015): **unimodal**, with choosers preferring an optimal trait value (fig. 1.3a); more rarely, **bimodal** or multimodal, where choosers prefer two or more distinct trait values (fig. 1.3b); or **directional**, where preferences increase (or decrease) monotonically with trait value (fig 1.3c). While some preferences appear to be directional within the range of current courter phenotypes, any preference will be limited, at the very least, by minimum **thresholds** for detection at the low end of trait values, and by sensory receptor saturation or cognitive constraints at the high end (fig. 1.3d; chapter 3). Most trait values we actually measure are also physically constrained to be positive (fig. 1.3d).

Preferences may also be **categorical**, with choosers attending only to traits within a certain range of values, but distinguishing little within that range (fig. 1.3e; chapter 4). Note that categorical preferences are analogous to Edward's (2015) "threshold" preferences; I use the term *categorical* for consistency with the cognition literature (chapter 4). As noted above, continuous preference functions will always have maximum and minimum thresholds, although threshold values may be unobserved or unattainable in actual courters.

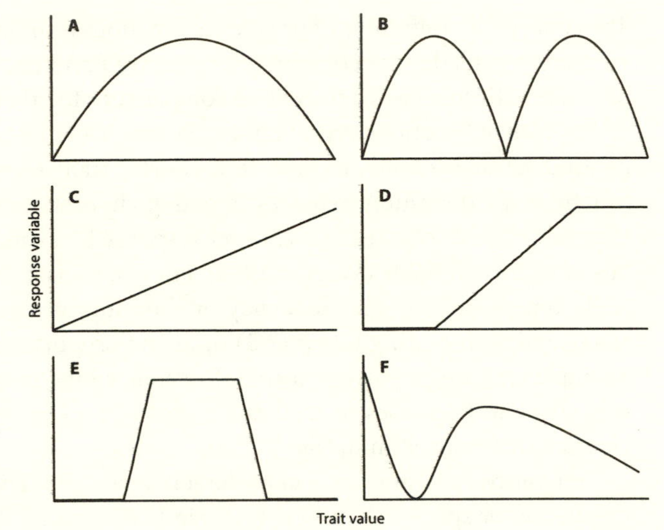

Figure 1.3. Shapes of preference functions: (a) unimodal; (b) bimodal; (c) directional; (d) sigmoid; (e) categorical; (f) complex.

Finally, preferences may be complex (fig. 1.3f). There are few examples of complex (or bimodal) preferences in mate choice, although it should be noted that we rarely have the statistical power to fit complex functions to preference measures. Mori's (1970) "**uncanny valley,**" however, is a familiar reminder that preferences can be highly nonlinear. Mori, a roboticist, was pessimistic about efforts to make robots more humanlike in morphology and behavior, because too-humanoid robots would be less appealing to humans (think *Star Wars'* creepy C3P0 versus the adorable R2D2). People, Mori argued, exhibit maximal disgust toward stimuli that are slightly dissimilar from healthy humans, notably corpses, resulting in a preference function like the one in figure 1.3f. Beyond popular culture, there has been limited work on the uncanny valley. Matsuda and colleagues (2012), however, showed that infants prefer faces of their mothers over those of unfamiliar women, while intermediate faces between the two women lie in a zone of reduced attractiveness. Karl MacDorman and colleagues have conducted a detailed series of studies of the cognitive mechanisms underlying the uncanny valley in adults (e.g., MacDorman & Chattopadhyay 2016, and references therein). If preferences are largely driven by low-level sensory responses (chapter 3), they will generally be directional or unimodal in shape (Ryan & Keddy-Hector 1992), whereas integration of multiple cues may result in categorical or complex preferences (chapters 4 and 5).

It is useful to extract summary measures from preference functions that correspond to biologically meaningful properties of chooser behavior. These parameters are referred to by different names by different authors; for example, the term "strength of preference" has been variously used to discuss each of the first three properties described below: peak preference, responsiveness, and choosiness. For simplicity's sake, I use these terms throughout this book when characterizing preferences.

1.6.2 Peak preference

The **peak preference** (fig. 1.4a) is the trait value that elicits the maximum response from choosers. Peak preference corresponds to the "ideal point" in political science and economics (Poole 2005). If preferences are multimodal or complex, it makes sense to identify local maxima; that is, multiple peaks within a preference function.

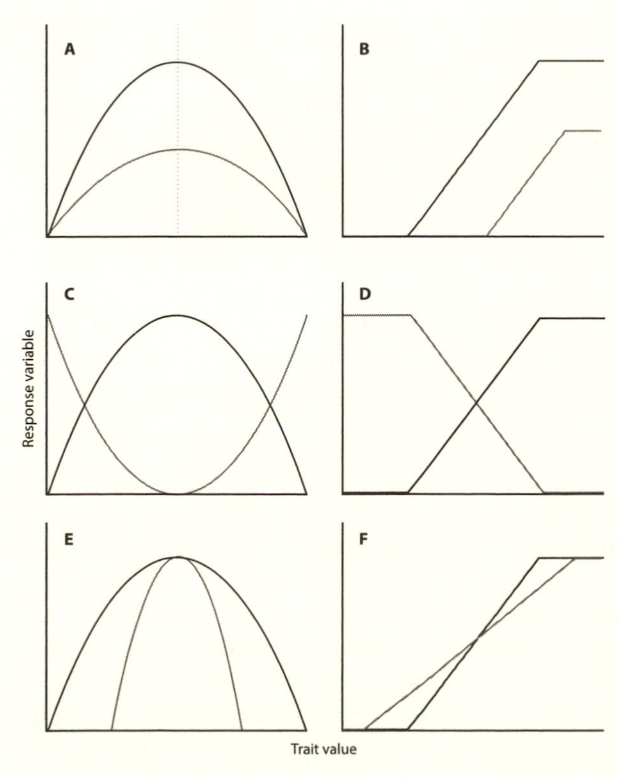

Figure 1.4. Properties of preference functions. Dashed line in (a) indicates peak preference. Black and gray preference functions in (a) and (b) differ in responsiveness; preference functions in (c) and (d) differ in valence; in (e) and (f), in choosiness.

1.6.3 Valence

The term **valence** has not seen wide use in the literature, but genotype, social experience, and environmental effects can all act to flip the direction of preference—that is, preference versus antipathy. For example, female Gouldian finches (*Erythrura gouldiae*) show Z-linked reversals in preference for male color morph (Pryke 2010) and male zebra finches (*Taeniopygia guttata*) show experience-dependent reversals in preference for bill color (fig. 1.4c–d). Both guppies (*Poecilia reticulata*, Endler & Houde 1995) and great bowerbirds (*P. nuchalis*) show preferences for some colors and antipathy for others. Variation in the direction of preference is important to distinguish from variation in preference function or preference shape, since shifts in preference direction can lead to rapid evolution and diversification of courter traits. Valence is explored in detail in chapter 5.

1.6.4 Responsiveness

Responsiveness (fig. 1.4a–b; Brooks & Endler 2001) is the chooser response averaged over the distribution of a courter trait in a specified reference population, whether based on natural variation or on the values determined by the experimenter. Responsiveness corresponds to the motivation to mate, which in turn often depends on physiological receptivity (chapter 6). Responsiveness is particularly important in the context of sexual conflict (chapter 15), since it can evolve in response to courter signals that impose costs on choosers. Resistance (Holland & Rice 1998) is essentially the inverse of responsiveness (Rosenthal & Servedio 1999); a chooser with the gray preference function in figure 1.4a has lower responsiveness and higher resistance to courter traits than a chooser with the black preference function.

1.6.4 Choosiness

A crucial but more complicated property of preference functions is **choosiness**, conceptually defined as "time or effort that [a chooser] is prepared to invest in making a choice" (Brooks & Endler 2001). This definition implies that a choosier individual is going to devote more resources to sampling among mates, and is therefore going to base mate choice off a more accurate estimate of courter trait distributions (Janetos 1980; chapter 6). However, it is useful to decouple choosiness from mate sampling, since two individuals can sample the same number of mates and nevertheless differ in choosiness.

Reinhold and Schielzeth (2015) define choosiness as "change of mating propensity with changes in trait values" and provide an extensive discussion of the underlying statistical issues. Put simply, we can quantify choosiness by measuring how concentrated chooser responses are in particular areas of trait space. For unimodal, Gaussian preference functions, choosiness can be quantified as the standard deviation of response about the peak preference (Gray & Cade 1999; fig. 1.4e); more generally, unimodal choosiness is the maximum slope of the cumulative distribution function (Reinhold & Schielzeth 2015). For directional preference functions, choosiness is expressed as the maximum slope of the linear preference function (Pomiankowski 1987; fig. 1.4f) or, for saturating functions, as the slope of the logistic regression (Reinhold & Schielzeth 2015). Models of preference evolution tend to focus on choosiness for directional preference functions and on peak preference for unimodal functions, and to ignore more complex preference functions altogether (chapter 14).

Responsiveness and choosiness are intertwined. At low levels of responsiveness—for example when an individual is physiologically unreceptive—all mates will be rejected. Similarly, at high levels of responsiveness, choosers may mate with any available mate. Choosiness in both of these cases will therefore be zero, but the consequences will be maximally different: an unresponsive chooser mates with nobody, thereby increasing sexual selection by eliminating a chooser from the mating pool. A hyperresponsive chooser mates with anybody, weakening sexual selection and promoting interspecific hybridization. Further, choosiness is where we perhaps see the biggest disconnect between measured preference functions and realized choices. This is because two choosers may be equally selective, but may differ in their observed choosiness because they encounter a different distribution of courter traits from which to make choices. I will return to the relationship between choosiness, responsiveness, and the effective distribution of courter signals (Edward 2015) in chapter 6.

1.6.6 A concordance of preference-function properties and behavioral mechanisms

It is useful to think about how these rather abstract properties of preference functions map on to actual behavioral mechanisms in choosers (table 1.1), since it is these mechanisms that evolve and modulate social and environmental effects on preferences. As noted above, preference involves appetitive or proceptive behaviors toward a sexual stimulus, while antipathy involves

Table 1.1. Properties of preference functions and their corresponding behavioral mechanisms

Functional property	*Behavioral mechanisms*
Preference	Appetitive behavior Proceptive behavior
Antipathy	Aversive behavior
Peak preference	Sensory tuning Perceptual tuning
Valence	Hedonic assignment
Responsiveness	Sensory tuning Perceptual tuning Motivation Receptivity
Choosiness	Sensory tuning Perceptual tuning Motivation Receptivity Mate sampling Comparative evaluation

aversive behavior (chapter 2). Peak preference is often a function of the tuning of the sensory periphery (chapter 3) or of downstream perceptual mechanisms (chapter 4). Valence depends on the hedonic assignment of subjective value to stimuli (chapter 5), and responsiveness depends on physiological receptivity and motivation to mate (chapter 6). Choosiness is more complicated, since it is intertwined with responsiveness as noted above, and is additionally dependent on mate sampling and mechanisms for comparative evaluation (chapter 6), and with the breadth of sensory or perceptual tuning.

1.6.7 Real preference functions are highly dimensional

Even the simplest preference is a so-called function-valued trait (Stinchcombe & Kirkpatrick 2012; McGuigan et al. 2008b; Rodriguez et al. 2013) where the chooser's phenotype depends on the value of a stimulus. Preferences are, however, highly multidimensional, with choosers attending to multiple interacting traits, which means that we are invariably measuring a subset of the multivariate space within which preferences operate (chapter 4).

These multivariate preference functions acquire even more dimensions when we consider that they are not hard-wired properties of individuals.

Mate-choice mechanisms can be modulated in a host of ways, starting with their sensory underpinnings (chapter 3); for example, the retinal pigments of brown trout (*Salmo trutta*) change according to season (Muntz & Mouat 1984), and the ears of female cricket frogs (*Acris crepitans*) are tuned to lower frequencies with increasing age (Keddy-Hector et al. 1992). The signaling environment changes the conspicuousness of signals and therefore their assessment by choosers (chapter 3), and choosers vary in their motivation to mate and therefore their choosiness (chapter 6). Even when tested on the same sets of stimuli under identical conditions, choosers are often highly inconsistent in their responses (chapter 9). Preferences vary according to a host of ecological factors including nutritional condition, predation risk, and parasite infection (chapter 11), and as a function of social interactions (chapter 12). All of these factors can modulate, eliminate, or even reverse preferences. If we want to characterize an individual's "preference phenotype," we need to take into account how preferences vary according to history and circumstance.

Further, preferences are heavily dependent on the set of traits being compared and on how choosers are comparing them. Chooser responses to a series of traits presented in isolation may be very different from those to a series of traits encountered by comparing among courters. Interactions among multiple traits within the same courter can have unpredictable effects on preference functions for individual traits, and the multivariate distribution of courter traits is frequently misaligned with that for preferences. I will return to each of these issues over the course of this book.

1.7 STAGES OF MATE CHOICE

Mate choice is not a discrete event, but rather a process with multiple, distinct stages starting well before mating and, in systems with parental care, continuing throughout the lifetime of the chooser. The importance of each of these stages varies with the natural history of the organism in question. Each stage typically involves different processes of sensation, perception, and evaluation, different mechanisms for exercising choice, and different risks and rewards for choosers and courters. It is important to note that these stages encompass social mating and thereby do not require any attempt to produce offspring; choosers can behave differently toward courters or courters' offspring depending on their sexual relationship, with consequences for fitness (chapter 8).

Mate choice doesn't even require two individuals to meet. For example, male collembolans *Orchesella cincta* deposit spermatophores. Females can pick up only one spermatophore, and the spermatophores of males exposed to rivals are more attractive (Zizzari et al. 2013). Choosers are therefore making a decision based entirely on the "extended phenotype" (Dawkins 1983) of courters.

The convention traditionally has been to divide mate choice into "premating" and "postmating" stages. Before mating, choosers behave differently toward different courters as a function of the latter's signals and cues. After mating, choosers can exert mate choice by skewing fertilization in favor of some courters more than others, and/or by differentially allocating resources to the offspring of attractive versus unattractive courters.

The pre/postmating dichotomy is perhaps insufficient, because it fails to capture important mate-choice processes surrounding mating itself. There is also a big difference between choosers merely evaluating signals produced by courters and choosers engaging in intimate activity. Even before copulation or gamete release, physical contact increases the likelihood of pathogen transmission or physical injury. Individuals perform courtship behavior around and during mating, and fertilization bias and subsequent investment can be influenced by these behaviors (Eberhard 1996; chapter 7). It is also the case that our understanding of mate-choice mechanisms overwhelmingly comes from work on the earlier stages of choice: it is much easier to study the pre-contact stages of mate choice, for example through experimental manipulation and playback of signals, than intimate interactions where touch and contact chemical cues might be involved. I suggest that a third category, the "peri-mating" stage of choice, is useful in distinguishing interactions before and after intimate contact.

1.7.1 Premating

Premating choice involves the detection and evaluation of courter signals, and is by far the best-understood stage of mate choice. Premating choice can be performed at minimal cost to the chooser (chapter 6), and can be readily performed by both males and females. An **advertisement** (e.g., nest decorations, long-range visual displays, frog calls, or birdsong) can be generated by the courter without attending to a specific chooser, and the chooser can evaluate it without directly interacting with the courter.

In part because advertisements can be manipulated or synthesized (chapter 2), experimental studies of mate choice have overwhelmingly focused on premating processes. And because nearly all mechanistic studies have been on sensory reception, we have a rich picture of how choosers detect advertisements, and how variation in detection affects mate choice.

Detection of advertisements—distinguishing them from background noise—is obviously required for mate choice. In order for a courter to be a candidate for mate choice, a chooser has to know that it exists. Some workers see detection as a separate process from mate choice. Parker (1983) makes a distinction between "passive attraction" and "active mate choice." In the former, choosers are simply attending to the courters who provide the greatest stimulation, and thereby the greatest probability of being detected. In the latter, they are discriminating among readily detected mates. This distinction is not very valuable, because the properties of "passive attraction"—for example, sensitivity to particular wavelengths of light, acoustic frequencies, or volatile molecules—can evolve and be modulated by experience and environmental input just like "active choice" can.

On the other hand, some scholars vastly overestimate the importance of detection, equating detectability to preference and preference to choice (chapter 3). As I will argue in chapter 7, much of the unexplained variation in mating outcomes relative to mating preferences lies in the later stages of mate choice. These stages are intimate and interactive and therefore more resistant to experimental study (but see chapter 7), but may in fact account for most of the variation in realized mate choice. Detection is just the necessary first step in choosing a mate, and should not be considered in isolation from the rest of the process.

Having detected a courter's signal, the chooser then evaluates the courter. Based on that evaluation, she may respond by behaving in a way that facilitates or inhibits subsequent stages. For example, male satin bowerbirds and their kin are noted for producing elaborate courtship structures (bowers; fig. 1.1). Females evaluate the bowers first with males absent, then return to a subset of these bowers to observe courtship displays when males are on the nest. Finally, females return a third time to copulate with a fortunate minority of males (Coleman et al. 2004). In general, mate choice proceeds through increasingly close-range and individually directed phases of courtship, and blurs into the peri-mating phase.

1.7.2 Peri-mating

Peri-mating choice includes activities soon before, during, and soon after mating when partners are in close physical contact. I use "peri-mating" (Fedina & Lewis 2008) rather than the more common "pericopulatory" (e.g., South & Lewis 2012), since the former encompasses external fertilizers that do not copulate, like broadcast-spawning fishes and flowering plants. In these species, but even more so in species with internal fertilization, mating is physically intimate and presents choosers with potential risks ranging from injury to transmission of sexually transmitted diseases. The opportunity for peri-mating (and postmating) choice is heavily skewed toward the sex with control over fertilization and provisioning, usually the female (but see, e.g., Paczolt & Jones 2010). There are ample opportunities for choosers to express preferences immediately before, during, and after mating, notably by permitting, denying, prolonging, or terminating copulation, or by physically attacking and even consuming courters (chapter 7). Just as detection has been labeled "passive choice" (Parker 1983), some workers have described peri-mating behavior as distinct from mate choice, in the context of courters "priming" choosers for mating by inducing physiological receptivity (e.g., von Schilcher 1976; Riede 1983). While courters can certainly manipulate choosers into mating, receptivity constitutes an important mate-choice mechanism, since different stimuli are better or worse at inducing receptivity (Eberhard 1996).

1.7.3 Postmating

Postmating choice is the subject of a large body of correlational and experimental work (chapter 7). Choosers can make sexual decisions after mating; first, through differentially biasing fertilization success, and second, through the resources that are invested into offspring after fertilization; a chooser behaves differently toward its offspring from different courters. The postmating phase gives a chooser ultimate control over a courter's fate. At the one extreme, she can devote all of her **residual reproductive value** to a single mating with a single courter, allowing only his sperm to fertilize her eggs and devoting herself to caring for their offspring at the expense of any future broods. At the other, she can kill him and eschew his sperm (Andrade & Kasumovic 2005).

1.8 MATE CHOICE AS A PROBLEM IN ANIMAL COMMUNICATION

Mate choice is about communication: a receiver, the chooser, must interpret a signal from the courter (Ryan 1990). This is true whether we are referring to a female fruit fly flying away from the song of a male, to hormone receptor proteins in the reproductive tract binding to signal molecules in the seminal fluid, or to the awkward give-and-take of first-date conversation. This can be a highly dynamic process, where chooser behavior is likely to influence subsequent actions by the courter. All mate choice involves the sensation, perception, and evaluation of courter signals, and their integration into a behavioral decision. The first part of this book details how these steps operate.

Sensation is by far the best understood step in how choosers interpret signals, and is the focus of chapter 3. Sensation involves the conversion of stimuli in the environment into internal neural and chemical responses. Sensation is a sine qua non in mate choice. If a stimulus can't be sensed by a chooser, there is no way that it can influence her preference. Mechanistic studies of mate choice have overwhelmingly focused on the relationship between the **sensory periphery** (the structures like eardrums, olfactory epithelia, and photoreceptor arrays that transduce environmental information into internal information) and **detectability** (the probability of detecting a courter signal against background noise). A sexual stimulus is identified as such from a much broader pool of stimuli in a noisy environment; before it can have an effect on mate choice, it must be discernible from the background noise. As detailed in chapter 3, there has been a great deal of work on mate choice in the context of how sensory mechanisms respond to courter signals, and how this response depends on the sensory environment.

In some microbial systems, sensation is the sole determinant of mate choice; mating can only occur between individuals emitting a highly specific agonist and individuals expressing a highly specific odorant receptor. This is the case in the unicellular fungus *Cryptococcus neoformans*, where individuals respond only to pheromones from their own molecular "mating type," and both the pheromone and the response are associated with genetic variation at a single mating-type locus (Lin 2009, 2010; fig. 1.5). It should be noted, however, that even this most elementary molecular response is context-dependent (chapter 11), with factors like light and nutrient abundance modulating reproductive decisions (Lin 2009, 2010). Even in *Cryptococcus*, mate choice involves the integration of information across multiple sensory inputs.

In most animals, however, mate choice involves **perception**, whereby multiple sensory inputs, often from multiple sensory **modalities**, are integrated into a neural representation that a chooser can then use to make decisions (Levine 2000). We know that choosers attend to numerous traits during mate choice (Partan & Marler 2005; Candolin 2003), but there has been relatively little mechanistic work on how perceptual integration influences mating decisions (but see Griffith & Ejima 2009). This topic is explored in chapter 4.

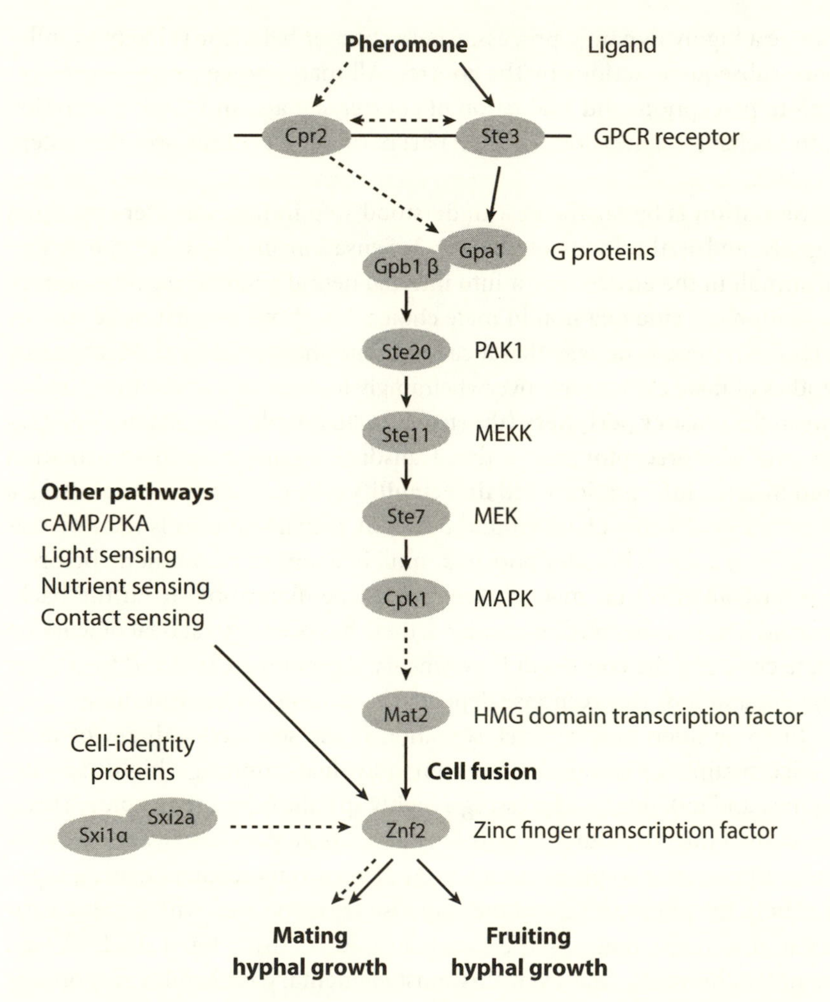

Figure 1.5. Signal transduction pathway leading to mating in the unicellular yeast *Cryptococcus neoformans*. Chemoreceptors respond to exogenous pheromones that signal "mating type," triggering a signaling cascade that leads to cell fusion and meiosis. Environmental variables can modulate the expression of transcription factors leading to mating. After Lin (2009); Lin et al. (2010).

Perceptual representations of courters, or of particular features of courters, are assigned a valence, a positive or negative assignment of subjective value. This evaluation (chapter 5; Musch & Klauer 2003) depends heavily on the courter's motivational state and on the pool of potential mates (chapter 6). Evaluation, like sensation, is by definition part of mate choice. In *Cryptococcus*, evaluation is performed by the activity of the transcription factor TF2, which integrates input from a pheromone signal with information about the environment to produce a decision whether or not to exchange genes (fig. 1.5).

Evaluation may be highly variable even among individuals with evolutionarily conserved sensory and perceptual mechanisms, as in members of the two sexes within one species. In a similar vein, the same trait is often attractive to choosers in one species and repulsive to choosers in a closely related species. For example, green swordtail females show a preference for males with a colorful "sword" ornament on the caudal fin (Basolo 1990a), while sheepshead swordtails show antipathy for the same trait (Wong & Rosenthal 2006). The traits are sensed and perceived by females in both species, but evaluated differently. Therefore, the same stimulus induces choosers to mate in some cases and to reject in others. Evaluation is discussed in chapter 5.

1.9 PROSPECTUS

The first section of this book describes the natural history of mating preferences and mate choice. In chapter 2, I provide an overview of how empiricists measure mate choice. Chapter 3 focuses on the importance of sensation and sensory modality in shaping mate choice, drawing on the substantial literature on the sensory ecology of mate choice. Chapter 4 discusses behavioral and evolutionary-genetic studies of chooser preferences for multiple courter traits in the light of the psychological literature on multimodal and multitrait perception. Similarly, chapter 5 uses the scant behavioral data on evaluative processes and mate choice as a point of departure to discuss the general neural mechanisms that are likely to be involved in the process of assigning subjective value to mates. Chapter 6 addresses the cognitive and ecological constraints on how choosers sample and decide among multiple candidates; chapter 7 discusses mechanisms of mate choice once mating has occurred, and chapter 8 discusses the dynamics of systems where social and/or reproductive mate choice is mutual.

The middle of the book focuses on variation in mate choice and mating preferences. Mate choice is primarily interesting because it operates so differently among species, among individuals, and even within individuals. Chapter 9 provides an overview of variation in mating preferences, focusing on the repeatability of mate choice. Chapter 10 covers genetic variation in preference, along with a brief overview of genomic approaches to elucidating the genetic basis of mate choice. Chapter 11 addresses how preferences are shaped by the physical environment and the ecological community, including predators, parasites, and nutrition. Finally, chapter 12 considers the role of the social environment in shaping preferences, from parental effects during early development through mate-choice copying during an interaction with a courter.

The final portion of the book concerns origin, evolution, and consequences of mate choice. In chapter 13, I describe how mating preferences can arise in a context unrelated to mating, due to selection in another context or due to basic constraints on organismal function. Chapter 14 focuses on how natural selection can act directly on mating preferences as a result of costs or benefits of choosers mating with particular courters. In chapter 15, I address the core focus of contemporary mate-choice research, namely the theoretical and empirical literature concerning the co-evolution of chooser preferences and courter traits. Chapter 16 reviews the ample literature on mate choice and speciation, as well as the more novel topic of mate choice and genetic exchange among species. Chapter 17 specifically deals with mate choice and human evolution. Human examples are cited throughout the book, but the distinct corpus on mate choice from evolutionary psychology, and some of the peculiarities of what we find attractive, both require special treatment. Finally, chapter 18 concludes the book with suggestions toward a synthetic theory of mate choice.

1.10 ADDITIONAL READING

Andersson, M. 1994. *Sexual Selection.* Princeton, NJ: Princeton University Press.

Bateson, P. (ed.) 1983. *Mate Choice.* Cambridge: Cambridge University Press.

Darwin, C. 1871. *The Descent of Man, and Selection in Relation to Sex.* London: John Murray.

Milam, E. L. 2010. *Looking for a Few Good Males: Female Choice in Evolutionary Biology*, Baltimore, MD: Johns Hopkins University Press.

CHAPTER 2

Measuring Preferences and Choices

2.1. INTRODUCTION

How do we actually study mate choice and mating preferences? If all we are interested in is contemporary sexual selection—how courter traits influence fitness—the answer is relatively straightforward. As Arnold (1983, p. 71) pointed out in an influential early review on sexual selection, "one need not identify the agent of sexual selection (e.g., rival males versus discriminating females) in order to measure its impact." One need only measure realized mating outcomes (section 2.2). We can simply observe a set of focal choosers in a natural or experimental population, and either determine how they behave differently toward individual courters, or use genetic markers to assign parentage to the choosers' offspring. We can then estimate the relationship between courter phenotypes and our measure of choice. If the aim is purely to measure how sexual selection currently acts on courter traits, the formal selection theory of quantitative genetics (Brodie et al. 1995; Schluter 1988) provides a robust statistical methodology for measuring how individual traits are associated with mating success.

Arnold prefaced his review with a caveat about this approach:

> The aim is merely to characterise sexual selection by its statistical effects on phenotypic characters within a generation. This, of course, tells us nothing about how selection actually worked in the past, nor does it enable us to extrapolate into the future. (1983, p. 67)

In order to do both of these things, it helps greatly to try to understand what is going on inside the heads and bodies of choosers, which is considerably more challenging than measuring mating outcomes. We need to measure not only the mate choices of choosers—how choosers discriminate among actual mates—but also the underlying preferences: choosers' internal representation of courter traits. There are several reasons that mating outcomes don't tell us much about mating preferences.

First, the courters that are available to mate will be a non-random subset of the courters available in the population. This is primarily because some courters will be better than others at making themselves available, whether by aggressively excluding other courters or by occupying times and places where they're likely to encounter choosers. In many species, females prefer to mate with subdominant males, but these males are prevented from mating by more aggressive rivals (Wong & Candolin 2005). Mating outcomes therefore paint an incomplete and skewed picture of underlying preferences (Postma et al. 2006).

Second, another complication with merely measuring realized mating outcomes is that a chooser's evaluation of a courter can be dependent on her current and previous encounters with other courters. Chapter 6 discusses how a chooser's mating decisions are influenced by her interactions with the pool of courters that are currently available. In general, choosers become less choosy if circumstances make it difficult to sample multiple courters. And further, the attractiveness of one courter is dependent on the characteristics of other candidates (Bateson & Healy 2005; chapter 6). Mating decisions can be greatly influenced by a chooser's social and ecological environment throughout her lifetime (chapters 11 and 12), and understanding these influences is critical to understanding how ecological and social circumstances can influence mate choice and therefore mating outcomes.

Third, the current pool of courters represents only a small part of the total range of phenotypes that might elicit a mating response. Choosers often have "hidden" preferences that don't exist in available courters, and novel signals are often more attractive than anything produced by the current pool of courters. Indeed, mating signals often evolve in response to latent chooser preferences (chapter 13). Hidden preferences can also exist for novel *combinations* of traits that are unavailable in courters: for example, female swordtail fish *Xiphophorus birchmanni* prefer males with large bodies and small dorsal fins, but males with large bodies have large dorsal fins (Fisher et al. 2009). The scope of possible chooser decisions is therefore almost always much larger and more highly dimensional than the space occupied by courter traits. We can therefore gain only limited insight into preferences by restricting ourselves to the phenotypic combinations expressed by a contemporary population of courters.

Finally, we are often specifically interested in testing responses to courter traits that lie outside the range of variation available to chooser populations. For example, choosers may reliably fail to mate with heterospecifics, but we may wish to know which trait combinations or which signaling environments are required to maintain **reproductive isolation** between two species.

To address these limitations, we turn to manipulative studies of preference. Here, we present focal choosers with controlled variation in signals and measure behavioral, physiological, or neural responses which we hope are an accurate proxy for how choosers would make mating decisions with respect to the corresponding properties of real courters.

In this chapter, I begin by discussing how mating outcomes are measured, then present a conceptual framework for thinking about how preferences are structured. I then discuss the options for empirically measuring mating preferences, and the pitfalls associated with each approach.

2.2 MEASURING MATE CHOICE USING MATING OUTCOMES

Studies of mating outcomes range from behavioral observations of social affiliation in the wild, to laboratory experiments where individual choosers and courters are paired in isolation. When we measure mating outcomes, we are usually interested in how choosers allocate resources (time, energy, gametes) among courters, and how this differential allocation of resources affects courter fitness. One of the most important things to know about mate choice is how it influences the **reproductive success** of both courters and choosers. This most commonly involves counting the number of offspring using direct observation and/or molecular parentage analysis (see Jones et al. 2010 for a review of parentage methods as applied to sexual selection studies). This approach has the advantage that it can be performed in virtually any system, including wild populations, which in some cases can be tracked over time to generate multi-generation pedigrees and long-term measures of sexual selection (Dunn et al. 2012). Molecular parentage analysis further enables us to partition measures of reproductive success between the sexes; in species that bear live young, for example, the identity of one parent is often known to the researcher and the other can be inferred by genotyping (e.g., Culumber et al. 2014). This allows us to compare the difference in the variance in reproductive success between males and females, an important component of sexual selection (chapter 15).

Importantly, direct measures of reproductive success allow one to directly measure sexual selection, as I discuss in chapter 15. Further, direct measures of courter reproductive success can shed light on the validity of behavioral assays of preference; for example, patterns of siring success by distinct, Y-linked morphs of *Xiphophorus nigrensis* correspond to relative preferences by females in experimental studies (Ryan, Pease, & Morris 1992). Finally, by chronicling actual outcomes of mating in wild populations, direct measures can provide estimates of the frequency of hybridization between species or incipient species, which is critical to understanding the role of mate choice in speciation and genetic exchange.

Parentage analysis can also be used to decouple the social and genetic effects of mating. For example, studies of nesting passerine birds show that **social mate** (chapter 8), rather than genetic ancestry, accounts for most of the variance in number of offspring that survive to fledge (Hadfield et al. 2006; Whittingham & Dunn 2005). This means that in these species, choice of (and choice by) a social partner is more important to fitness than any purely genetic benefit of mate choice (chapter 15).

There are several major limitations to what we can glean from direct measures of reproductive success. It is difficult to attribute variation in courter reproductive success to variation in chooser decisions. It is even more difficult to use reproductive success alone to identify the stages of mate choice (pre-, peri-, or postmating) that are most important to sexual selection. Measuring reproductive success also entangles intersexual selection via mate choice with intrasexual competition. Aggressive exclusion of some courters by others can act in opposition to premating choice (Wong & Candolin 2005), while sperm competition (Birkhead & Moller 1998) can do the same for postmating choice (chapter 7). Further, differences in embryo viability and survival (as might occur with interspecific hybrids or with close relatives) can make it difficult to differentiate between natural selection against offspring and sexual selection against less-preferred mates. These problems can be somewhat mitigated by incorporating measures of chooser behavior (like latency to copulate, Grillet et al. 2006) into statistical models that estimate the effect of chooser behavior on variance in courter reproductive success.

Second, merely counting offspring provides an incomplete picture of how mate choice affects the fitness of choosers and courters. This is because courters and choosers can provide benefits to (or impose costs on) each other and each other's offspring long after offspring are born. For example, female blue tits spend more time feeding their young when mated to males with bright ultraviolet feathers (Limbourg et al. 2004), and, as discussed previously, parental care by social mates can be the dominant influence on offspring fitness. There can also be indirect genetic costs or benefits of mating with particular courters, who may produce offspring of varying viability or attractiveness (chapter 14).

Finally, as mentioned earlier in this chapter, parentage analysis constrains the preference space being studied to that occupied by currently expressed courter phenotypes: parentage can tell us nothing about hidden preferences for unavailable traits or trait combinations. We are limited by whatever variation happens to be present in the courters we sample. We therefore need to be able to measure chooser preferences independent of what happens to be present in contemporary courters. The remainder of this chapter addresses how we measure preferences.

2.3 EMPIRICAL ASSAYS OF PREFERENCE: WHERE TO BEGIN

An exciting thing about mate choice is that univariate and multivariate preference functions do not coincide with the distribution of courter ornaments; this property is what generates preferences for exaggerated features and novel trait combinations. A review by Ryan and Cummings (2013) compiled 22 studies showing chooser preferences for traits outside the range of courter variation. As discussed in chapter 13, while much of this "hidden" preference space may arise from a deep history of natural and sexual selection on courters and choosers, it may more commonly involve other forces shaping the cognitive, perceptual, and sensory systems recruited for choosing.

Preference space is also *operationally* hidden when we fail to consider how animals might be communicating. When designing mate choice experiments, it is crucial to consider all possible modes of communication to avoid missing key signals. As detailed in the next chapter, all electrically and nearly all chemically mediated mate choice is hidden to our direct experience, as are decisions based on acoustic frequencies and wavelengths of light outside our range of sensitivity. Puzzling patterns of chooser responses to a trait may arise from our failure to properly consider what choosers actually attend to. It is therefore imperative that we have some idea of the ***Umwelt***—the perceptual universe (von Uexküll 1909)—of our study organism. This is true both with regard to multiple sensory modalities, and to the range of sensitivity that animals can perceive; for example, we have found only quite recently that choosers in birds, butterflies, and fish attend to the ultraviolet range when choosing mates (chapter 3).

2.4 MEASURES OF PREFERENCE

How do we actually measure preferences? We are interested in preferences because they underlie mate choices, and what we would really like to know is a chooser's probability of favoring or spurning a courter (via mating, fertilization, or preferential treatment) given that the courter expresses a certain combination of traits. In other words, we would very much like to know how often the same individual, given the same experience and the same conditions, would mate with a particular courter given a large number of opportunities. We cannot rewind time and replay life's tape again and again, so we are faced with an inherent trade-off. If we measure preferences based on actual mate choices (assuming we can appropriately control for courter behavior, section 2.2 above), we can often measure them only once per individual. This is because mating changes individual experience and motivation (chapter 6), both of which are likely to influence subsequent measurements. Further, using only realized matings limits our ability to use playbacks (section 2.5) to manipulate signal variation.

On the other hand, we can use a more indirect assay of preference, like a behavior that is predictive of mating. In this case we can measure responses to artificial stimuli, and we can test the same chooser repeatedly, but we are making the assumption, often not explicitly tested, that our response measure is predictive of the mating (or postmating) outcome if a chooser were to encounter a particular stimulus on a courter in nature. Further, although the impact of exposure to signals is likely to be less than that of mating, any experience with courter stimuli is liable to result in subsequent changes in chooser behavior.

A possible solution to this dilemma is to use multiple instances of the same genetic individual. By using inbred lines of fruit flies (*Drosophila* spp.) and other laboratory model organisms (Chenoweth & Blows 2006; e.g., Ratterman et al. 2014), one can tightly control the experiences of genetically identical individuals, for example by testing each individual on only one stimulus. Unfortunately, phenotypic variation can actually be greater in inbred lines than in outbred lines because of inbreeding's effects on developmental stability (Møller & Swaddle 1997; Falconer & Mackay 1996), which makes it harder to accurately estimate preference phenotypes.

2.4.1 Direct measures of mating

In addition to directly measuring mating outcomes (section 2.2) we can measure quantitative properties of mating itself. For example, in fruit flies (*Drosophila*), latency to copulate is widely used as a measure of female preference (Grillet et al. 2006), and copulation duration, which determines the amount of ejaculate transferred, is used as a measure of male preference (Gilchrist & Partridge 2000). In species where mating is brief or involves little or no physical contact, like fish, quantity of gametes released (Alonzo 2009) or copulation attempts (Gabor & Aspbury 2008) can be used. These measures have the advantage that they are directly measuring investment in mating, but the disadvantage is that, for internally fertilizing species they can only be used for interactions between live individuals; further, the effects of mating often make it impossible to conduct repeated trials on the same individual.

2.4.2 Behavioral measures of preference

Studies using artificial or abstract stimuli use behavioral, physiological, and, more recently, **gene expression** measures that offer reasonable proxies for preference. The major critique of such indirect measures is that they are not necessarily predictive of whether a chooser would actually mate with, and ultimately reproduce with, a courter expressing a particular phenotype. However, some chooser behaviors are potentially better predictors of courter *fertilization* success than mating itself, since choosers can make fertilization and allocation decisions after mating (chapter 7).

Further, an expanded definition of mate choice can encompass non-reproductive sexual relationships, and between individuals of the same sex. When the focus is on preferences in the context of *social* mates rather than *reproductive* mates, direct measures of behavior are the only way to measure mating preferences. Choice of social mates involves performance of certain activities (sexual acts, reciprocal grooming, foraging) more with some individuals than with others (chapter 8). Direct measures of behavior are therefore, by definition, the only appropriate metric for measuring the non-genetic consequences of mating.

Preferences before mating are best measured with so-called proceptive behaviors (Beach 1976). These behaviors are uniquely and specifically expressed in a sexual context, and often involve the mechanical or physiological facilitation of gamete transfer. Accordingly, they are generally good predictors of copulation and therefore of premating choice and pair-bonding in natural interactions. Examples include copulation solicitation displays in birds and lordosis in mice (fig. 2.1). In some cases, proceptive behaviors are as simple as movement toward a point source. This is the case with so-called **phonotaxis** experiments in frogs and toads, in which two or more sounds are played back from speakers some distance apart, and choosers are operationally defined as expressing a preference when they come within a predefined area adjacent to one of the speakers. Females only exhibit phonotaxis

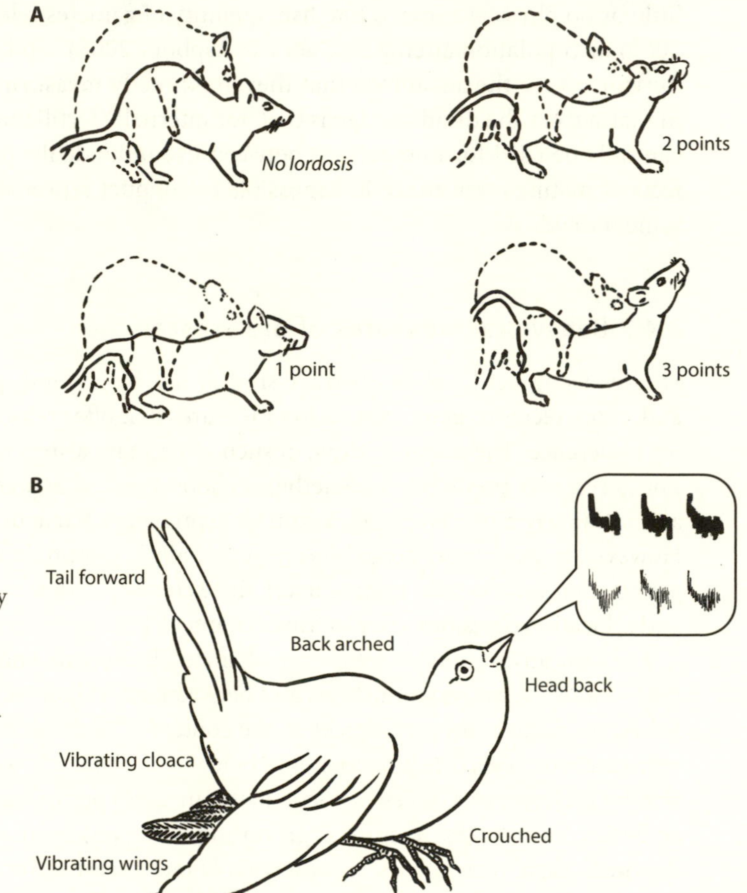

Figure 2.1. (a) Hardy & deBold's (1971) point system for scoring lordosis in female rats; (b) copulation solicitation display in female canaries includes song and postural cues (Amy et al. 2015).

when they are sexually receptive (Lynch et al. 2005). In orthopteran insects, a powerful tool for assaying phonotaxis is the so-called Kramer treadmill, or spherical locomotor compensator, where a subject is placed on a rolling ball with stimuli played back from different orientations. Electronic sensors then track the magnitude and direction of movement (e.g., Verburgt et al. 2008).

In other cases, researchers are forced to use less specific assays to infer preferences. Time in proximity to a stimulus, or **association time**, is widely used in studies of fish (e.g., Fisher et al. 2009). Changes in gross locomotor activity are also often interpreted as proxies for preferences; the number of times a subject hops from perch to perch is a standard preference assay in passerine birds (e.g., McGlothlin et al. 2004). A major advantage of these behaviors is that they are very easy to define operationally (section 2.8) and can be scored automatically by machine-vision systems (section 2.9). They remain controversial, however, because they do not specifically measure sexual responses. For example, association time of female poeciliid fish can be used to measure both a mating preference for male phenotypes (Wong & Rosenthal 2006) and a nonsexual preference for female shoalmates (Wong & Rosenthal 2005). Nevertheless, evidence suggests that association time in poeciliids correlates well with proceptive behavior (Cummings & Mollaghan 2006), the likelihood of reproduction (Walling et al. 2010), and with realized mating patterns in the wild (Ryan, Pease, Morris, et al. 1992; Culumber et al. 2014). David and Cézilly (2011) pointed out that assays that depend on locomotor activity, like perch changes and association time, may be confounded by personality differences among choosers; exploratory behavior before testing explained several important measures of preference (fig. 2.2).

Some of the most powerful assays, thus far used largely in birds, involve direct measures of choosers engaging courter stimuli. Gaze-tracking

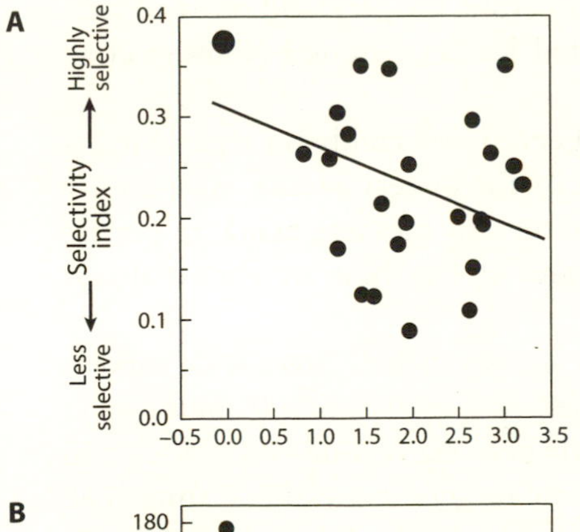

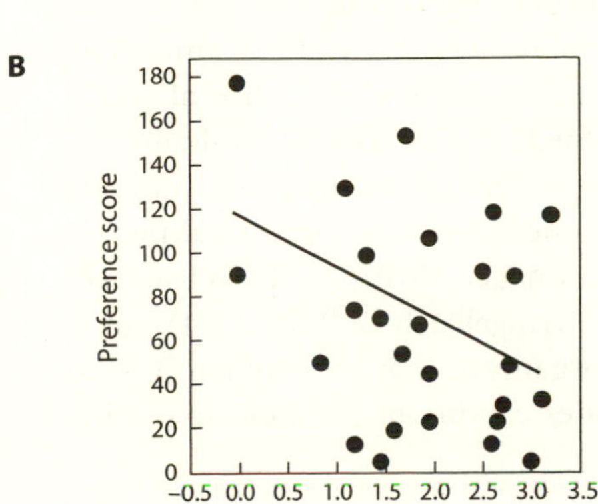

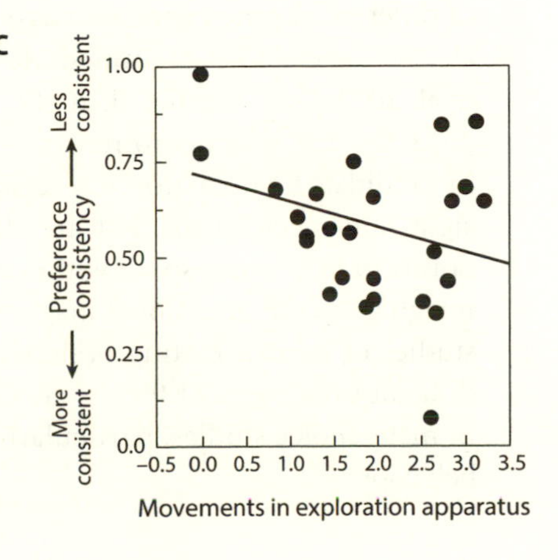

Figure 2.2. Movement in an exploratory task confounds preference measures. Female zebra finches (*Taeniopygia guttata*) that were more active when exploring a four-chamber choice apparatus were subsequently (A) less selective or choosy, and showed weaker preferences for preferred males, but (B) were more consistent in their preferences. After David & Cézilly (2011).

(Yorzinski et al. 2013) may represent a powerful new way to study the attentional processes involved in mate choice (chapter 4). Operant responses, where choosers are rewarded with stimuli based on pecking a key, have been instrumental in studying song preferences in zebra finches (Riebel 2000; Holveck & Riebel 2007).

Assays specific to humans, like surveys, interviews, and mining Internet searches, are introduced where relevant in later chapters.

2.4.3 Neural, physiological, and genomic measures of preference

Choosers may be differentially stimulated by courter traits without exhibiting any outward behavioral differences. Further, characterizing the underlying sensory and neural substrates requires measures of physiological responses below the whole-organism level. Physiological measures of chooser response can reveal mating biases in addition to, or in the absence of, behavioral assays. For example, *plethysmography*, used to measure changes in blood flow to the sex organs, was used in a recent study to measure physiological arousal of humans to erotic imagery, and revealed substantial disagreement with self-reported levels of arousal (Chivers et al. 2010; chapter 18).

Electrophysiological recordings of nerve fibers and brain regions in response to sexual stimuli are invaluable for understanding mate-choice mechanisms, since they can be used to directly measure neural responses across stimulus gradients (chapters 3 and 4), but these are almost always terminal procedures.

Terminal assays of brain or sensory activity include brain-wide surveys of differential gene expression associated with mate choice (Cummings et al. 2008) and studies of **immediate-early gene expression** (IEGs; Sockman et al. 2002; Desjardins et al. 2010), candidate gene expression (Wong et al. 2012; fig. 2.3), and neurotransmitter concentrations (Pawlisch et al. 2012) in candidate neuroanatomical regions. Such studies provide invaluable insight into mechanisms, but are clearly unsuitable for mapping multiple points in preference space within the same individual. Functional neuroimaging (mostly functional magnetic resonance, fMRI) has been useful in studies of human mating preferences (Kringelbach & Rolls 2004) but is difficult to implement for the smaller organisms that are commonly used in mate-choice studies, particularly under conditions favorable to mating behavior.

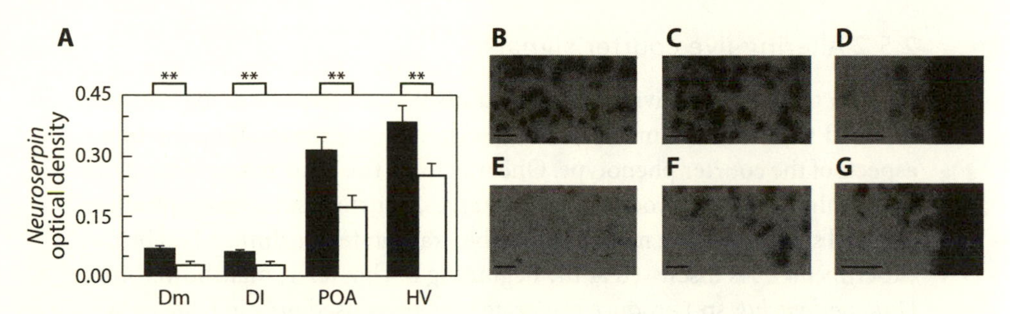

Figure 2.3. Expression of the *neuroserpin* gene in forebrain reward (Dm) and social behavior regions (DI and POA) and in the hypothalamus (Hv) of female swordtails *Xiphophorus nigrensis* immediately following dichotomous choice trials. **, p < 0.01; *, p < 0.05. B–D and F–G are representative images of Dm, Dl, and POA from high- (top) and low-preference females. After Wong et al. (2012). Images courtesy of Molly Cummings.

Finally, physiological, neural, and genomic responses can be used to study chooser preferences expressed after mating. For example, Gil and colleagues (2004) used assays of unfertilized egg hormone titers to show that female canaries increase egg testosterone after exposure to preferred songs, and Otti and colleagues (2014) recently measured transcriptomic responses to ejaculate variation in bedbugs (chapter 7).

2.5 STIMULI USED IN PREFERENCE ASSAYS

2.5.1 Live courters

At one extreme of the gamut of stimuli that we can use to understand mate choice are live courters and choosers commingled together (e.g., Culumber et al. 2014). Such assays by themselves tell us little about preferences, because chooser behavior is confounded by the behavior of courters toward one another (Wong & Candolin 2005). Further, the behavior of choosers is contingent on that of other choosers (mate copying; chapter 12) meaning that individual choosers in a group do not constitute independent data points.

Live courters can be confined from one another and from choosers, presented sequentially, or choosers can be provided spaces that courters can't access (Coria-Ávila et al. 2005). Here, chooser preferences may be obscured by mutual mate choice, and preference behavior may depend on the amount of effort the courter puts in (chapter 8). Nevertheless, interactions with live courters (or recently dead ones, e.g., Marcillac et al. 2005) are often the only way to measure peri- and postmating choice.

2.5.2 Altering live courter signals

In order to identify individual courter traits that are important to choosers, we need to be able to manipulate these traits while controlling for other aspects of the courter phenotype. One way to do this is by manipulating the signals that courters produce. For examples, courters sometimes produce external structures, like nests, that can be exaggerated or diminished by the experimenter. As discussed at the beginning of chapter 1, male bowerbirds (*Ptilonorhynchus* sp.) produce elaborate structures used only in mate attraction and courtship, called bowers. By manipulating the quantity and arrangement of decorations on bowers, researchers have gained insight into how bower decorations interact with courtship displays to influence mating preferences (Patricelli et al. 2003) and how the orientation and position of objects within bowers determine their visual conspicuousness (Kelley & Endler 2012a, 2012b). Courter signals can also be manipulated by abstracting them from other aspects of the phenotype. This is a common approach to studying preferences for olfactory signals: for example, Fisher and colleagues (2003) evaluated female preferences for familiar and unfamiliar males in pygmy lorises (*Nycticebus pygmaeus*) using wooden dowels scented with male urine. When signals are thus abstracted from courters, this manipulation is analogous to playback (next section).

Surgical alterations provide another opportunity for manipulating live courters. Perhaps the least subtle example of this is the widespread practice in *Drosophila* mate choice of using decapitated courters as a control for studies of pheromone preference (e.g., Marcillac et al. 2005). In spiders *Leucauge mariana*, Aisenberg and colleagues (2015) manipulated both putative courtship structures on the chelicerae of males, and sensory structures on the chelicerae of females, to test the hypothesis that cheliceral courtship was associated with differential mating success. In vertebrates, Andersson's (1982) classic study showing preferences for extreme tail length in widowbirds did so by clipping male tail feathers or elongating them by gluing on extensions. In fish, elongations of the caudal fin can be similarly shortened by amputation (Rosenthal et al. 2002) or exaggerated with plastic extensions (Basolo 1990b).

Patterns on the body surface can also be altered by experimenters. In fish, dark melanin pigments can be removed by applying dry ice to the skin (Morris et al. 1995), or added with tattoo ink (Rosenqvist & Johannson 1995); in birds, multiple studies have applied sunscreen to remove the ultraviolet component of feather reflectance (Limbourg et al. 2004). Burley and Szymanski (1998) added colored leg bands to male zebra finches to test female preferences for novel stimuli (chapter 11), while ten Cate and colleagues (2006) painted the beaks of males and females to test for the effects of early learning on preferences (chapter 12).

A less invasive approach to manipulating the signals of live courters involves altering the signaling environment, as has been done in multiple fish species (Seehausen & van Alphen 1998; Maan et al. 2006; Cummings et al. 2003). By using color filters to manipulate incident light on courters, Kingston and colleagues (2003) were able to decouple color from morphology in evaluating the mating preferences of female pygmy swordtails (*Xiphophorus pygmaeus*) for distinct male genetic morphs. Adding humic acid to a test aquarium abolished chooser preferences for species-typical chemical cues in another swordtail, *X. birchmanni* (Fisher et al. 2006).

While manipulations of live courters can generate arrays of stimuli that provide powerful tests of preference, biological constraints and ethical concerns limit the range of manipulations that can be performed. Further, even the least invasive manipulations of live courters have the potential to produce confounding effects on courter behavior; for example, male pygmy swordtails may behave differently toward females under yellow light than under full illumination (Kingston et al. 2003). To avoid these concerns, we turn to playback.

2.5.3 Playback

Playback is defined by McGregor and colleagues (1992) as "the technique of rebroadcasting natural or synthetic signals to animals and observing their response." Playback provides an analytical approach whereby an experimenter can quantitatively manipulate specific signal components while holding others constant. This approach eliminates the confounding effects of courter behavior, and, crucially, allows for the unveiling of hidden parts of preference space. As I will discuss below, however, playback carries its own set of problems.

While the term "playback" is typically used today to refer to the electronic reproduction of a temporally patterned signal, the term encompasses presentation of sounds, chemicals, static dummies, or robots. Acoustic (Gibson 1989), chemical (Teale et al. 1994), and visual (Clark et al. 1994) cues can all be presented in the field.

Playback stimuli can be drawn from a natural repertoire, as in most studies of responses to birdsong (Searcy and Marler 1981). Isolation of chemical signals from courters and presentation to choosers (see above) is equivalent to presentation of recorded song or video. Recorded stimuli can also be manipulated to produce novel stimuli (Rosenthal & Evans 1998; Witte et al. 2001). Playback can involve the simultaneous presentation of signals in multiple modalities (e.g., Rosenthal et al. 2004) and can be interactive, with chooser behavior determining courter signaling behavior (Otter et al. 1999; Butkowski et al. 2011; Gierszewski et al. 2016; reviewed in King 2015).

A particularly powerful approach is to synthesize stimuli *de novo*. This allows the experimenter to measure preferences outside the range of natural variation in courter traits. Synthetic stimuli have been used widely in acoustic (Phelps et al. 2001), chemical (Ruther et al. 2009), and visual preference studies (Woo et al. 2011).

Playbacks offer a degree of control and precision that is unavailable from observational studies or direct manipulation of live courters; by their very nature, therefore, playbacks are prone to a number of potential pitfalls that may limit the external validity of experimental results (Chouinard-Thuly et al. 2016; Powell & Rosenthal 2016). Two main issues have been the focus of attention: **signal fidelity**, whereby playbacks fail to represent signals appropriately (Fleishman et al. 1998; D'eath and Dawkins 1996), and **pseudoreplication**, whereby playbacks fail to adequately sample multivariate signal variation.

2.5.4 Signal fidelity

Signal fidelity, which involves deficiencies in the physical characteristics of stimuli, is specific to the sensory modalities being presented and the perceptual biology of the organisms being tested (chapter 3). The basic problem is that we typically want to give choosers representations that approximate what they would experience in nature, but are constrained from doing so when we set up experiments. In some cases this is because we are only presenting a subset of the stimuli that choosers attend to in mate choice. For example, female subjects in song playback studies in birds only exhibit preferences if treated with estrogen (Searcy 1992), which increases sexual motivation (chapter 9), likely because they are not experiencing the interactive visual and olfactory elements of courtship.

Stimuli can also be deficient because of the physical characteristics of the output medium. This is a particular problem with video playback of visual images. This is because video technology is designed to produce a reduced representation of visual scenes specific to the human visual system. Zeil (2000) points out that important features like size and distance can be misrepresented by a two-dimensional stimulus, since key cues like motion parallax are unavailable in flat video images.

Further, video monitors represent color via linear combinations of red, green, and blue light tuned to the sensitivities of the three classes of human cone photoreceptors (chapter 3): an image appears yellow to us because an area of the screen is emitting red and green light, but no blue light. For nonhuman animals with three classes of cones within the human-visible range, it is possible in principle to adjust the color balance appropriately, but this approach does not work for the countless animals that have more than three photoreceptors or are sensitive in the ultraviolet. Another complication of this correction is that it becomes difficult to compare responses across animals with different visual sensitivities, or even within the same animal across seasons (Muntz & Mouat 1984).

A final point is that courter signals and chooser preferences are highly multidimensional (chapter 4). Especially with synthetic stimuli, there is a risk that traits key to mate choice are misrepresented or absent. With video animations in particular, artefactual spatiotemporal cues may slide courter signals into Mori's (1970) "uncanny valley" (chapter 1; Rosenthal 2007). It is therefore important to validate synthetic stimuli against live or recorded references (Fisher et al. 2009; Wilczynski et al. 1995; Moravec et al. 2010).

Nevertheless, video stimuli continue to be invaluable for the manipulation of visual signals (Woo & Rieucau 2011). Despite the limitations of visual fidelity, preferences measured in video playback studies have been consistent with those measured in parallel studies manipulating live individuals (Kingston et al. 2003; Rosenthal et al. 2002).

2.5.5 Pseudoreplication

Pseudoreplication is an important concept in the design and interpretation of behavioral experiments and deserves special consideration in the context of preference assays using playback. Pseudoreplication was coined by Hurlbert (1984, p. 190) as "testing for treatment effects with an error term inappropriate to the hypothesis being considered." Put more simply, it is the use of a sample size in a statistical test that is inappropriate to the hypothesis being tested (McGregor 2000), as illustrated with the following hypothetical example. A researcher seeks to test the hypothesis that female golden-cheeked warblers prefer the songs of conspecific males to those of the sympatric black-capped vireo. She uses a single recording of a male warbler song and a single recording of a vireo song. She then quantifies preference by alternating playback of the two songs to 40 female warblers in the field and measuring the number of approaches to the speaker during playback of each song. Finally, she uses a paired t-test (N = 40) to test the one-tailed null hypothesis that female golden-cheeked warblers do not prefer conspecific over heterospecific song.

In the 1980s, numerous studies on mate choice (and other social interactions) in birds employed a similar design, using playback of a small number of natural signals. Kroodsma (1989) was perhaps the first to point out that such a design did not compare the attractiveness of male warblers to that of male vireos; it compared the attractiveness of subject #1, who happens to be a warbler, to that of subject #2, who happens to be a vireo. With reference to the narrow hypothesis that subject #1 is more attractive than subject #2, the appropriate sample size is the 40 females tested. With reference to the researcher's original hypothesis of conspecific song preference, the sample size shrinks to a single data point: the relative attractiveness of the two individuals being tested. Without sampling responses to more males, we don't know if choosers prefer conspecific songs in general, or if this particular male's song just happens to be more attractive than that one's.

Without adequately sampling responses to multiple exemplars, it is impossible to discern whether differences in response are due to differences in stimulus classes, or due to idiosyncratic differences among individuals. This is a particular concern since multiple courter traits can interact to determine preferences (chapter 4). This problem can be addressed by using multiple natural exemplars and performing appropriate statistical analyses. Synthetic stimuli, which are generated from specified parameters, offer the opportunity to eliminate this idiosyncratic variation. Parameters can be modeled on data sampled from multiple natural signals. Even synthetic stimuli, however, leave open the possibility that responses depend on interactions between a manipulated parameter and a parameter that is arbitrarily fixed. For example, the attractiveness of a repeated acoustic mating signal might depend on an interaction between pulse rate and dominant frequency. Simply holding dominant frequency constant and varying pulse rate would provide an incomplete picture of how sexual selection via mate choice is acting on the signal. In other words, this gives us only a low-dimensional slice of a complex preference function. Studies using synthetic stimuli based on population data should nevertheless strive to include multiple exemplars.

Another problem, by no means specific to playback experiments, is that the identity and experience of choosers is almost always pseudoreplicated.

In the warbler example above, the wild population is subject to similar social and ecological conditions that can drive preference variation, and choosers can behaviorally interfere with one another.

2.6 REPEATABILITY OF PREFERENCES

A key question in understanding preferences is the extent to which they are invariant properties of individuals. Permanent differences between individuals in preferences arise as a result of genetic and environmental causes (chapter 9). Variation *within* individuals, besides that due to measurement error, places an upper bound on the **heritability** of preferences and therefore on their potential to evolve. In quantitative genetics, **repeatability** is the fraction of phenotypic variance (for example in choosiness over a set of stimuli) that is due to among-individual differences rather than within-individual differences (Lynch & Walsh 1998; Wolak et al. 2012; chapter 9). Estimating repeatability requires testing the same individual at least twice on the same set of stimuli. Nakagawa and Schielzeth (2010) provide methods for estimating repeatability for discrete and continuous variables.

Estimates of repeatability should be taken with a grain of salt, and as Dohm (2002) points out, repeatability as actually measured maps only loosely on to heritability (chapter 9). It is important to note that experimental measures of repeatability test choosers in quick succession under identical conditions, and therefore fail to capture ontogenetic or environmental components of within-individual variation. On the other hand, any presentation of a stimulus to a chooser changes its experience, and therefore its response to subsequent stimuli. This underscores the importance of controlling for order effects (section 2.8.3 below), but it also compromises estimates of repeatability. Inconsistency between one day's response and the next may be an artifact of a chooser's habituating to a familiar stimulus (Jordan & Brooks 2012).

What we would really like to know is whether, if we replayed life's tape, the same chooser would make the same decisions over and over again. What we can actually measure is a chooser's set of sequential decisions, each one likely to be contingent on the last; errors in how we measure these will also inevitably result in lower repeatability estimates. As discussed in chapter 9, there is broad variation in empirical estimates of chooser repeatability, and it is often weak or undetectable.

2.7 SEQUENTIAL VERSUS SIMULTANEOUS ASSAYS

Preference is comparative by definition. At one extreme, choosers on **leks**, like female sage grouse, are simultaneously faced with multiple courter displays (Höglund & Alatalo 1995). At the other, choosers at low densities, like solitary carnivores, may rarely interact with more than one courter at a time. Both the costs of mate sampling and the cognitive constraints involved vary with the distribution of courter signals and with the natural history of how courters and choosers interact, and these can have dispositive effects on preference (chapter 6).

Experimental measures of preference strive to quantify the decisions choosers would make in nature. They typically involve either **sequential choice tests** (also called single-choice tests or, somewhat misleadingly, "no-choice" tests), or **simultaneous choice tests** where two or sometimes more stimuli are presented to choosers at the same time. Single-choice tests operationally include tests where a putative sexual stimulus is presented simultaneously with a nonsexual stimulus like white noise or a juvenile shoalmate.

The contrast between sequential and simultaneous tests is discussed at length in a recent meta-analysis by Dougherty & Shuker (2014) and associated commentaries. Choosiness is as a rule lower in a no-choice context (e.g., Hoikkala & Aspi 1993; Doherty 1985; fig. 2.4). Dougherty and Shuker's (2014) analysis showed this effect is driven by females; males are no less choosy in sequential assays (but see commentary by Kokko & Jennions 2015). The effect is particularly pronounced when choosers are attending to within-species signal variation rather than between-species differences (fig. 2.5). This suggests that the effect is sensitive to the relative costs of mate acceptance and rejection; for females, but generally not for males, mating with a heterospecific carries greater downside risk than not mating at all (chapter 16).

Ryan and Taylor (2015) called attention to the importance of different cognitive mechanisms that may be recruited for no-choice versus comparative tasks, and for the simultaneous and sequential integration of information about different courters. Responses to stimuli are also not necessarily **transitive** across simultaneous comparisons; that is, a preference for A over B and B over C does not necessarily mean that choosers prefer A over C (Kirkpatrick et al. 2006; chapter 6).

Similar problems occur when scaling from two choices to three or more choices. Raffa and colleagues (2002) showed that the sample size required to detect a significant preference in choice experiments was an increasing

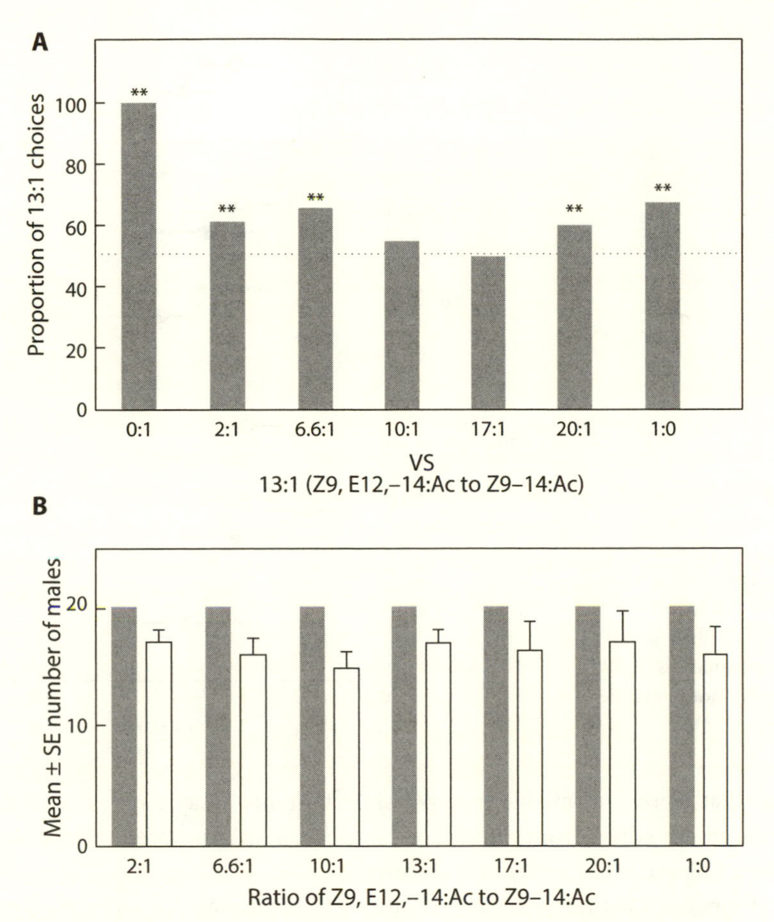

Figure 2.4. Responses of male almond moths (*Cadra cautella*) to a gradient of pheromone blends in a wind tunnel. (A) Proportion of individuals orienting to olfactory sources in simultaneous choice assays versus a standard 13:1 ratio. *, $p < 0.01$. (B) In no-choice assays, the number of individuals initiating upwind flight (hatched bars) and contacting sources (open bars) is independent of pheromone ratio. After Allison & Cardé (2008).

function of the number of options. A third, less-preferred stimulus can also influence the relative attractiveness of two more attractive stimuli (the decoy effect). I will return to the costs and cognitive biology of comparative evaluation in chapter 6.

Shackleton and colleagues (2005) made a methodological point that no-choice one-on-one interactions with individual courters could provide more sensitive assays of preference than simultaneous-choice tests with confined individuals, as they allowed access to a full repertoire of chemical and tactile

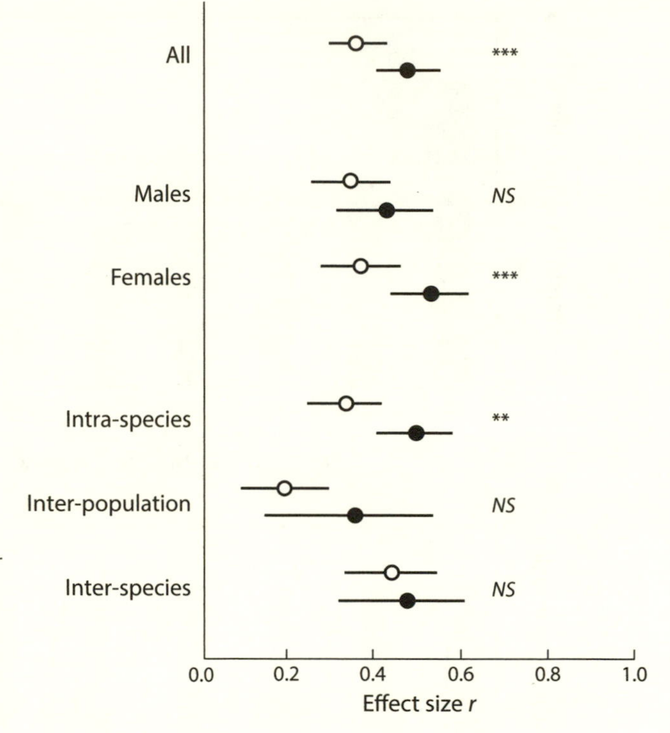

Figure 2.5. Mean strength of preference for no-choice (white diamonds) and two-choice trials (black diamonds) in a meta-analysis of 38 studies by Dougherty & Shuker (2014).

cues that were unavailable in most simultaneous assays. However, interactive assays suffer from the difficulty of disentangling preferences of the interacting partners, as discussed above.

Wagner (1998) further pointed out that simultaneous-choice tests conflate variation in preference with mate-sampling behavior: choosers that spend more time with the less-preferred stimulus may simply be doing so because they take more time to assess each stimulus. Simultaneous-choice tests can also amplify preferences by highlighting minor differences between two stimuli (Wagner et al. 1995; Wagner 1998; chapter 6). Single choices, however, can produce a **ceiling effect** (Martin & Bateson 1993) whereby choosers respond equally to a range of courter signals if no others are currently available. For example, in association-time assays in poeciliid fishes, choosers will almost always associate with another fish if it is presented in isolation.

It is also important to note that simultaneous and no-choice tests are often used for answering different questions. Specifically, no-choice assays test **recognition**: whether a courter signal is within an envelope that can elicit proceptive behavior from a chooser. Simultaneous assays test **discrim-**

ination, whether the chooser exhibits differential behavior among a set of alternatives. I return to the relationship between discrimination and recognition in chapter 6.

Taken together, these issues suggest that we should be sensitive to the nature of the comparison when designing and interpreting assays of preference. Depending on the natural history of the organism and the hypotheses being tested, single or comparative assays may be more appropriate. Given sufficient time and resources, it is a good idea to evaluate multiple assay designs and multiple comparisons when estimating preferences (Dougherty & Shuker 2015; Ryan & Taylor 2015).

2.8 OTHER CONCERNS WITH STUDY DESIGN AND INTERPRETATION

Anyone planning to embark on, or to interpret, behavioral experiments should read Martin and Bateson's (1993) primer on studying animal behavior. Tests of mating preference are particularly sensitive to lapses in good experimental protocol, many of which are all too apparent in the literature. The key issues are outlined below.

2.8.1 Standardized measurement

Ideally, data on preferences should be collected blind with respect to the stimulus presented to the chooser, to avoid possible bias on the part of the experimenter. This can be accomplished, for example, by scoring videotapes of phonotaxis with the sound turned off, or by using motion-tracking algorithms (see section 2.9 below). Videotaping and tracking also permit long-term archiving of raw data. Often, blind scoring is impossible, as when observing natural interactions between courters and choosers. In these cases, it is particularly important to have rigorous **operational definitions** of behavior explicitly specifying what an animal has to be doing in order to be scored as expressing the behavior. Measures of proximity, defined by the position of the chooser in space, are particularly easy to score operationally. A good operational definition is robust to the individual scoring it and to so-called observer drift, when the same person's definition of a behavior changes over time. Martin and Bateson (1993) provide a measure for calculating inter-observer reliability that can be used to estimate the robustness of operational definitions.

2.8.2 Positive and negative controls

As will become apparent in the latter half of this book, the absence of chooser preferences is just as important to trait evolution, speciation, and hybridization as their presence. Negative results in preference trials are resistant to interpretation, though, because the failure of choosers to show a preference for two stimuli could be due to a situational lack of motivation to mate (chapter 9) or due to problems with signal fidelity (section 2.5.3). It is therefore important to include appropriate positive controls in any study purporting to interpret negative results as absence of preference. For example, studies of call recognition by female túngara frogs pair a heterospecific call against a burst of white noise in a simultaneous-choice assay. Females that fail to respond to either stimulus could be doing so because they fail to recognize the heterospecific call as a mating signal, or because they are not motivated to mate (a so-called **floor effect**). To distinguish between these alternatives, experimenters "sandwich" each recognition test with a trial that includes conspecific calls to which females are a priori expected to respond (Ryan & Rand 1999). The "sandwich" trials provide a positive control for chooser motivation. Similarly, using pairs of live courters provides a positive control for signal fidelity in studies using video playback (Rosenthal et al. 2002). A less common problem is that choosers fail to show a preference because they exhibit an elevated sexual response across all stimuli (a ceiling effect), in which case negative controls (stimuli like white noise, which are *a priori* expected to not elicit a sexual response) are in order.

2.8.3 Controlling for order effects

As discussed in section 2.6, every presentation of a stimulus to a subject has the potential to change her response to subsequent stimuli. Further, chooser behavior can show ontogenetic or seasonal changes over the course of an experiment, and can even change depending on time of day (Toomey & McGraw 2012). Finally, as noted in section 2.8.1, the way observers score behavior can change over the course of an experiment. For these reasons, it is imperative to control for **order effects** any time one is doing explicit comparisons of data collected at different time points, like sequential choice tests or successively presented simultaneous-choice tests, by systematically or randomly varying the order of stimulus presentation across subjects and groups.

2.8.4 Controlling for side biases

Side biases deserve special mention because of the prevalence of simultaneous-choice assays of preference (section 2.7). The two sides of a testing arena can differ in lighting, acoustic properties, residual olfactory cues, and the characteristics of playback equipment. It is standard practice to alternate the sides of the arena on which stimuli are presented, within and/or among subjects. While this controls for confounding effects, a side bias nevertheless introduces noise in the data when sides are alternated, and therefore increases the chance of false rejection of the null. One can test for consistent side biases within and across choosers by testing responses to identical stimuli on either side (Baugh et al. 2008).

2.9 SYNTHESIS: MEASURING MATE CHOICE AND MATING PREFERENCES IN THE TWENTY-FIRST CENTURY

2.9.1 The siren song of big data

In this chapter, I have attempted to lay out best practices for mate-choice research, as well as make prospective mate-choice biologists aware of the ongoing controversies in how we measure preferences. There are some key ways in which technology could be recruited toward more meaningful and sophisticated assays of behavior, and there are some novel ways we could think about measuring mate choice in general.

For better or for worse, the empirical measure of mate choice has largely escaped the big-data revolution of the 2000s. For the better, because animal behavior has never been more hypothesis-driven than it is today, and this stands in contrast to "discovery"-driven data-mining approaches that sometimes fall prey to the exuberant pan-adaptationism of an earlier age; for the worse, because our picture of how mating preferences are structured remains limited by our ability to collect rich data sets. There are some key ways in which technology could be recruited toward more meaningful and sophisticated assays of behavior.

Almost two decades after Wagner's (1998) seminal paper on measuring preferences, there are still relatively few studies that measure even univariate preference functions in a given environment, let alone that explore multivariate preference space or how preference functions respond to social and ecological conditions. Many studies of preference, including the bulk of my own work, still compare chooser responses to just two courter states (e.g., conspecific versus heterospecific, or presence/absence of an ornament). In order to begin to map multivariate preferences, experimenters need to measure responses to multiple trait values under multiple conditions. Further, given that responses to one courter trait are often contingent on the value of one or more additional traits, we need to make an effort to sample multidimensional preference space more thoroughly. This is an incredibly burdensome task, particularly if we are to conduct a dense sampling of trait values, and one that is rendered more complicated by the comparative nature of preference (Ryan & Rand 2003; chapter 6; Blows et al. 2003; Gerhardt & Brooks 2009).

In order to sample preference space more thoroughly, we can become more efficient at collecting large amounts of data. Machine-vision and acoustic recognition systems allow for automated processing of behavioral responses (Ahern et al. 2009; Iyengar et al. 2012), which can operate in tandem with automated systems for stimulus playback (Butkowski et al. 2011). Automation can greatly increase the number of trials that can be run per unit time; one is still limited, however, by the number of tests that can be performed on any single individual.

Interactive playback (e.g., Butkowski et al. 2011; Otter et al. 1999; King 2015) could perhaps be combined with heuristic optimization to intensively sample stimulus space. This is only feasible using synthetic stimuli that can be automatically modified in response to chooser behavior, and choosers whose preferences are robust to order effects (section 2.8.3). In the context of measuring mating preferences, *heuristic optimization* might work as follows. A chooser is initially presented with two synthetic courter signals, each bearing an arbitrary combination of traits. If she prefers courter A over courter B, courter B is discarded and a novel courter, C, is generated that is an exaggerated version of courter A relative to B. This kind of "hill-climbing," replicated within and across choosers, could serve to map preference functions without the need for saturated sampling across multivariate space. The limitation of this approach is that, given that chooser repeatability is often quite low in univariate tests, fluctuations in chooser response could drive the hill-climbing algorithm in idiosyncratic directions. Further, the approach assumes that preferences are transitive across comparisons, which need not be true (Kirkpatrick et al. 2006). Nevertheless, heuristic optimization could be a powerful tool for exploring hidden preference space. Novel attractive and unattractive phenotypes suggested by heuristic optimization could be subsequently presented to choosers in conventional preference trials.

Finally, tracking software now permits characterization of position, orientation, and gaze of hundreds of individuals (Gallup et al. 2012). In an open-field or perhaps even natural setting, this enables multivariate analyses of the behavior of multiple, individually identified courters and choosers. This may allow researchers to disentangle chooser preferences from courter behavior, and to quantify the effect of mate copying on preference (chapter 12).

2.9.2 Examining our experimental paradigms

As the following chapters should make clear, mating decisions are exquisitely sensitive to a chooser's prior history, recent experience, and immediate context. We are improving as a community insofar as standardizing, systematically varying, or at least reporting these factors in published studies, but much of our work involves taking animals that have impoverished social experience (chapter 12) and testing them individually, even in the case of highly social animals. Although individual trials are much more tractable and powerful for statistical analysis, putting choosers in these highly contrived situations may introduce artifacts that incorrectly characterize preferences. Similarly, the physical environment—factors like ambient light, water chemistry, and visual background—matters a great deal both to stimulus perception (chapter 3) and to chooser behavior (chapter 11). Many of these key variables go unreported and likely unmeasured in many studies.

2.9.3 Taxon sampling and research bias

An alien reading this book might imagine Earth's skies teeming with zebra finches and corn borers, her waters seething with sticklebacks and swordtails, and her nights enlivened by the sounds of túngara frogs and field crickets. Some species and taxonomic groups are notably overrepresented. Our taxon bias is probably the most obvious manifestation of how we decide what aspects of mate choice to study based on what's feasible and what's already been established. Kokko and Jennions (2015) point out that we do not generally attempt to study mate choice where we think it's not operating. As mentioned in chapter 1, there are inherent difficulties in experimentally studying behavior across a large number of species and doing systematic comparative surveys. Different species are also suited for asking different questions about mate choice; we will hear a great deal about guppies and bowerbirds in early chapters, and about zebra finches and spiders in later ones.

At the same time, there is a strong argument to be made for the advantages of studying a system intensively. The systems above give us insights into mate choice because we can integrate studies encompassing—among others—psychology, neuroscience, physiology, and population genetics into an increasingly rich picture of how individuals make mating decisions and of the evolutionary consequences of those decisions.

Whenever possible, it is desirable to at least study both preferences and mating outcomes in the same system. Preferences are inconsequential unless they can be meaningfully related to real or potential mate choices, and mate choice is impossible to understand without characterizing the underlying preferences. As I will argue in the remainder of this book, understanding the mechanisms involved in preference is critical to understanding how mate choice evolves and how it operates as an agent of selection. These mechanisms generate rules and strategies that determine how a chooser integrates myriad streams of external information into its reproductive decisions.

How these streams are perceived, evaluated, and weighted is in no small part a function of natural history: the sensory modalities used for navigation and social interaction, the rate at which choosers encounter courters, the physical costs of mating, and so forth. The dark side of hypothesis-driven research is manifest in a tendency to gloss over natural history in an effort to test theoretical predictions, which in turn can be compromised by unrealistic assumptions about mechanisms. A particular model system is thereby touted as well suited for answering a particular theoretical problem. But ultimately, the organism poses the questions to be answered.

2.10 ADDITIONAL READING

Dougherty, L. R., & Shuker, D. M. 2014. The effect of experimental design on the measurement of mate choice: a meta-analysis. *Behavioral Ecology* 26, 311–319.

Martin, P., & Bateson, P. 2007. *Measuring Behaviour: An Introductory Guide.* 3rd ed. Cambridge, UK: Cambridge University Press. 186pp.

Wagner, W. E. 1998. Measuring female mating preferences. *Animal Behaviour,* 55, 1029–1042.

CHAPTER 3

The First Steps in Mate Choice

PREFERENCE FUNCTIONS AND SENSORY TRANSDUCTION

3.1 INTRODUCTION

Before a chooser can express a preference for any aspect of a courter's phenotype, he or she has to be able to detect it, a process called sensation (Levine 2000; fig. 3.1). Sensory biology places an upper bound on the breadth and complexity of preference space—the preference envelope—since anything outside a chooser's range of sensitivity is invisible, inaudible, and odorless: undetectable.

Detection involves a response by the sensory periphery to a stimulus. **Receptor** cells in the periphery transduce a physical property of the environment (photic energy, molecular displacement, molecular structure) into the neurochemical information that choosers see, hear, and smell. Mate choice can involve any sensory modality, and choosers often attend to courter characteristics using every modality they possess.

Sensory biology is of fundamental importance to mate choice and to its evolutionary effects. Greater sensory stimulation often translates into stronger preference. The vast majority of the studies reviewed at the time by Ryan and Keddy-Hector (1992) found directional preferences for greater sensory stimulation. For the most part, choosers prefer signals that are brighter, louder, and smellier, and courter traits can evolve simply because they elicit more stimulation in choosers (Ryan 1990; Ryan, Fox et al. 1990). This **sensory exploitation** is a special case of **preexisting biases** favoring some courters over others (chapter 13); at the very least, a signal that elicits more stimulation is going to be easier to detect at a distance or against background noise.

Even the simplest possible sensory process, a response based on odorants binding to a single type of odorant receptor (section 3.3 below), encapsulates much of the richness of mate choice: the expression of a single sensory receptor gene can generate a preference function that responds differentially to different courters (producing different quantities and types of molecules), that is multivariate with respect to courter traits (intensity and

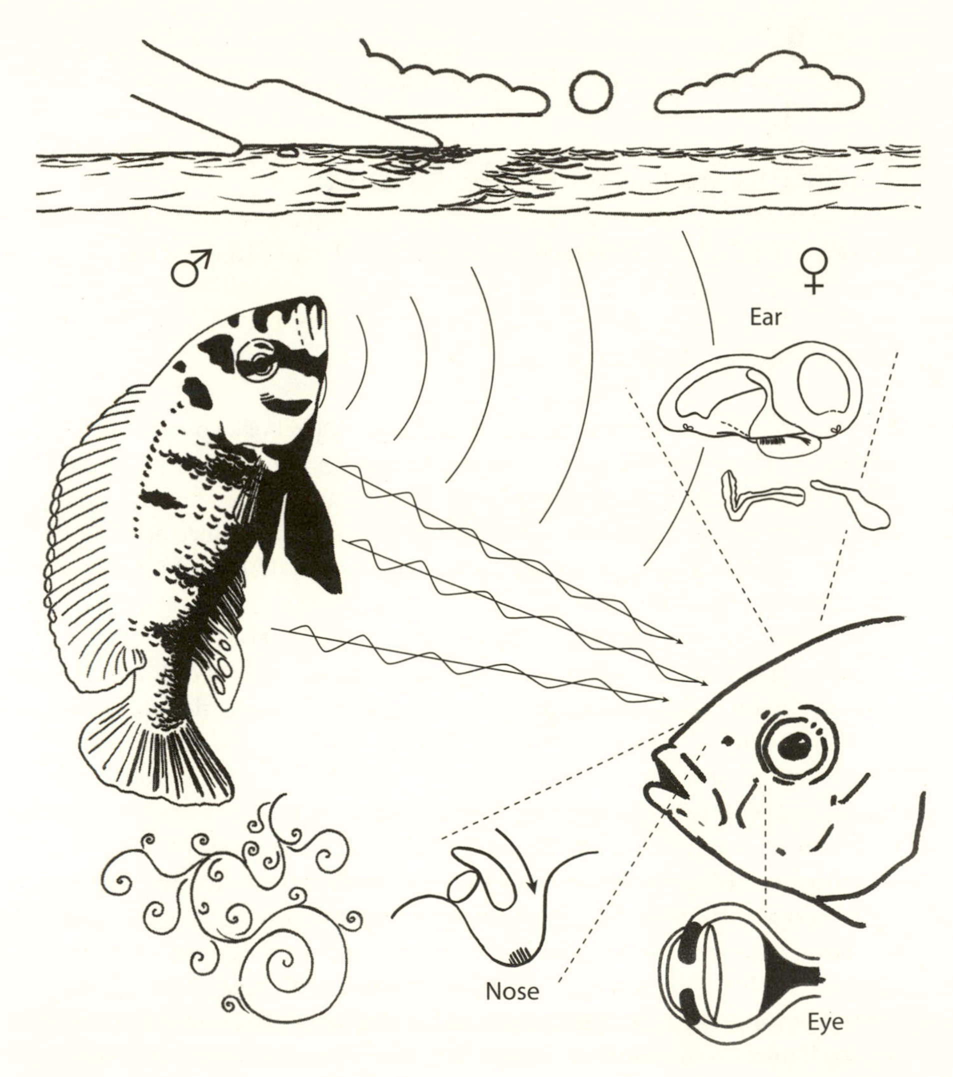

Figure 3.1. Reception and transduction in three modalities for a hypothetical vertebrate. (Left) Chooser signals are attenuated and degraded by the environment before reaching the ears, eyes, and nose. (Right) Sensory structures transduce environmental cues into neurochemical structures: hearing: (A) inner ear and (B) hair cell; vision: (C) retinal cells and (D) phototransduction; olfaction: (E) olfactory rosette, (F) odorant receptor neurons, and (G) odorant reception. Teleost ear drawing after Meyer et al. (2012); chemoreceptors after Kaupp (2010); phototransduction from Larhammar et al. (2009); vertebrate hair cell after Dumont & Gillespie (2003); nose drawn from a live haplochromine cichlid; all others after Bradbury & Vehrencamp (2011).

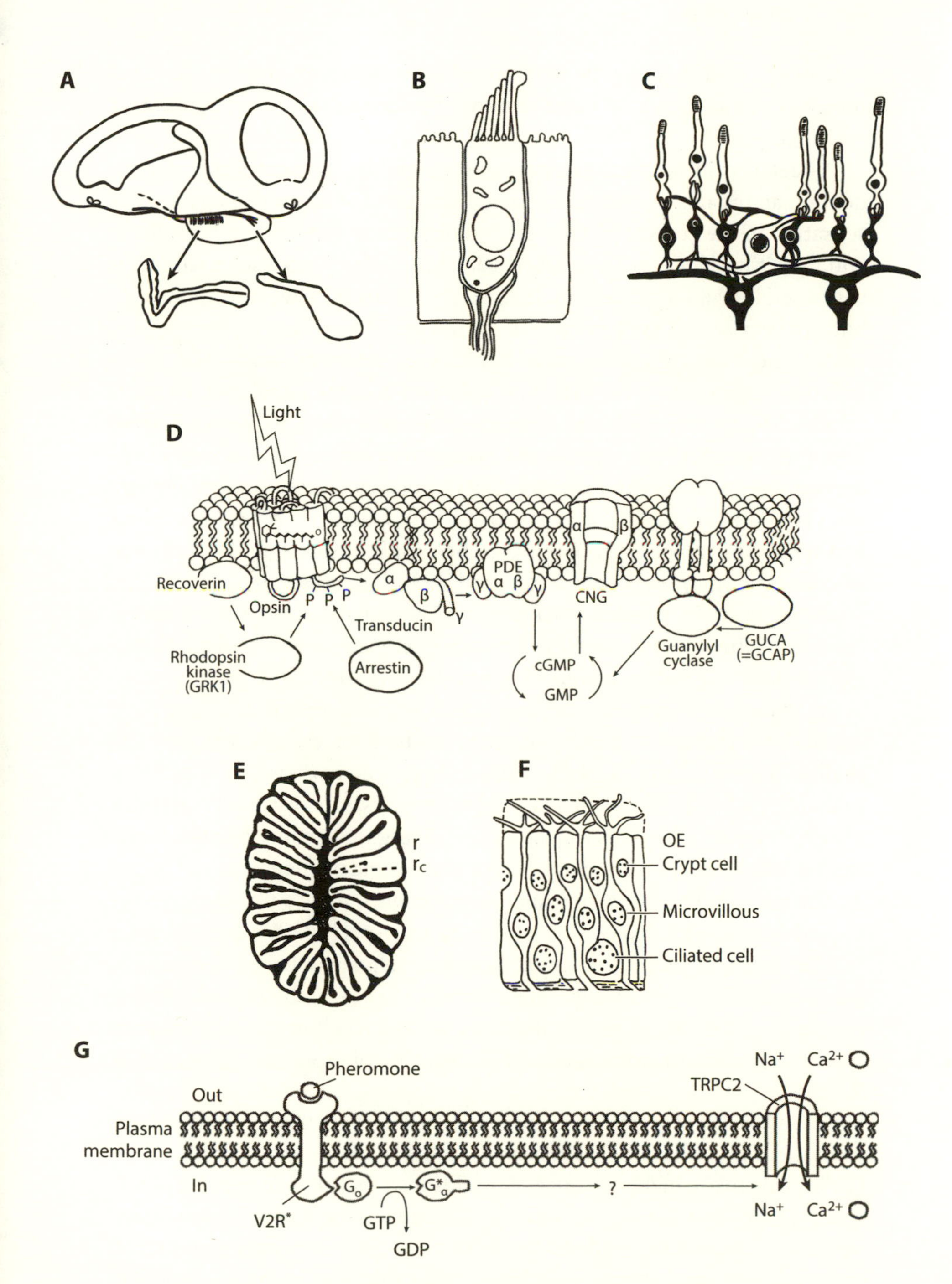
A
B
C
D
Light
Recoverin
Opsin
P P P
Transducin
α
β
γ
PDE
α β
γ γ
CNG
α β
Rhodopsin kinase (GRK1)
Arrestin
cGMP
GMP
Guanylyl cyclase
GUCA (=GCAP)
E
r
rc
F
OE
Crypt cell
Microvillous
Ciliated cell
G
Pheromone
Out
Plasma membrane
In
V2R*
Go
G*α
GTP
GDP
?
TRPC2
Na+
Ca2+
Na+
Ca2+

specificity), and that may be maximized by nonexistent courters (a novel molecule could elicit a stronger response by serving as a better ligand to a given receptor).

As detailed in the next section, sensory biology is inextricably dependent on an organism's natural history. The field of **sensory ecology** studies sensation in the context of how it helps organisms survive and reproduce, and how it itself acts as an agent of selection. Color vision, for example, is very useful for a forest-dwelling animal that feeds on fruit during the day, but not very useful for a nocturnal carnivore. Differences in the physical environment—even among microhabitats centimeters apart—also favor different sensory abilities. Therefore, different courter signals will be maximally detectable in different sensory environments. This **sensory drive** hypothesis predicts that signals will evolve to optimize detectability to the appropriate receivers (Endler 1992, 1993). For this reason, sensory biology can play a central role in the formation of new species via so-called **ecological speciation**, because chooser preferences and courter signals will diverge in concert with divergence in ecological niches. Recently diverged species flocks of freshwater fishes (sticklebacks, Boughman 2001; Lake Victoria cichlids, Hofmann et al. 2009) provide elegant examples of ecological speciation involving both receivers and signalers. In both cases, females occupying shallower or deeper water have different visual sensitivities, and male signals have evolved to maximize detection by females, with red males in shallow water and blue or black males in deep water where red light is unavailable. As I will address further in later chapters, sensory drive represents a special case of mate-choice mechanisms diverging in concert with ecological factors. Mate choice and speciation is the topic of chapter 16.

The interaction of sensation and environment can also play a role in the breakdown of mate choice as a barrier to interspecific gene flow (chapter 16). For example, recently diverged blue and red pairs of Lake Victoria cichlids hybridize in turbid water where signals cannot be distinguished (Seehausen et al. 1997) and swordtails hybridize as a result of organic pollution interfering with chemoreception of conspecific pheromones (Fisher et al. 2006).

Sensation can also influence macroevolutionary patterns, like the rate of speciation. Frogs have two "eardrums" in the inner ear; the *basilar papilla* (BP) is sensitive to higher-frequency sounds than the *amphibian papilla* (AP). Ryan's (1986) hypothesis predicted that the range of acoustic frequency sensitivity in female frogs and toads would determine the range of potential male call types and therefore facilitate speciation via premating isolation: the more morphologically complex the AP, the broader the range of sensi-

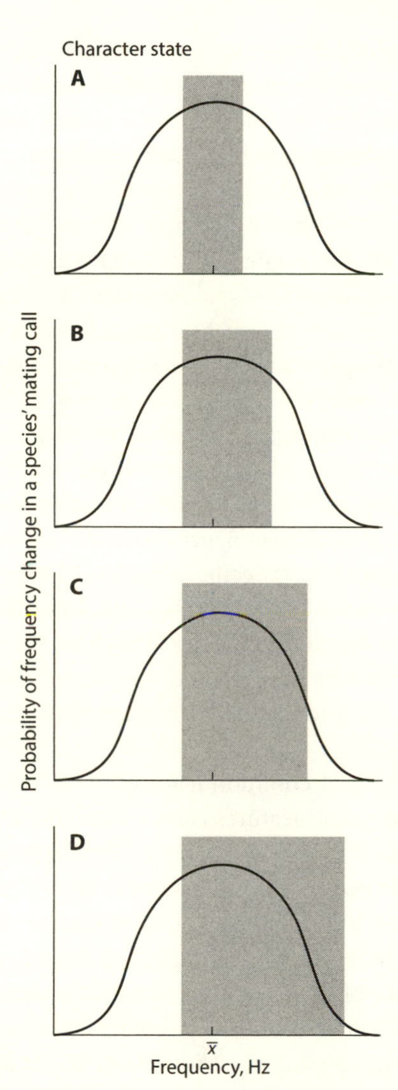

Figure 3.2. (after Ryan, 1986). Frequency-sensitivity range increases the opportunity for signal divergence. The solid curve represents the probability density function of the effects of a random mutation changing the current dominant frequency (x). From top to bottom, frequency sensitivity spans a greater range of possible frequencies.

tivity to acoustic frequencies, and therefore the greater the opportunity for calls to diverge (fig. 3.2). Phylogenetic analysis confirmed this prediction: rates of speciation were indeed higher in lineages with broader frequency-sensitivity ranges in the AP (Ryan 1986). However, Richards (2006) showed that across anurans, most of the frequency variation in calls across species was in the range of the other acoustic transducer, the BP. Richards (2006) suggested that the evolution of the complex AP could have evolved to perform for other auditory tasks, like detecting prey and avoiding predators.

While Richards's (2006) analysis rejected the hypothesis that frequency sensitivity of the AP predicted call diversity, a more broadly tuned AP could have allowed the basilar papilla to become specialized for communication.

Despite the central importance of sensation to how preferences are structured, it is important to recognize that it is only the first step in a sophisticated process that involves perceptual integration (chapter 4) and assignment of subjective value (chapter 5). We might predict that diversity in visual signals is associated with variation at the visual periphery, but this is not the case. Despite the exuberant diversity of visual signals in bowerbirds, there is little functional variation in opsin structure or in retinal photoabsorbance (see section 3.4 below) among bowerbirds, or even among birds in general (Coyle et al. 2012). Sensory tuning, as measured by cone opsin gene expression, accounted for only 3.36% of the variation in behavioral color preferences within a population of bluefin killifish, *Lucania goodei* (Fuller et al. 2010). As detailed further in chapter 5, gross detectability may conflict with the ability to discriminate intraspecific variation (Delhey et al. 2013) or with the direction of preference (Wong & Rosenthal 2006). We may assign too much importance to sensory processes (Dawkins & Guilford 1995) because we can readily quantify peripheral tuning as well as the structure and expression of sensory receptor genes involved in mate choice (Horth 2007; Springer et al. 2011).

In this chapter, I outline the important common features of all sensory systems (section 3.2). All of these common features can be used to explain chooser features downstream of sensation, through perception to the motor output of behavior. These shared features are what's most important in terms of our understanding of mate choice, but what draws our attention about mate choice is the diversity of ways in which it's accomplished: female gray treefrogs (*Hyla arborea*) scan a group of males on a rainy night, looking at and listening to an array of suitors (Gomez et al. 2011); female flour beetles feel their partner stroking their elytra during sex (Edvarsson & Arnqvist 2000); male bedbugs (*Cimex lectularius*) taste the hemolymph of their mates for a rival's sperm (Siva-Jothy & Stutt 2003). I focus on the particulars of how sensory systems work in each of the principal modalities (sections 3.3–3.6); for more detail on sensory processing of communication signals across the animal kingdom, see Bradbury and Vehrencamp (2011). I conclude by addressing the relationship between sensitivity, sensory constraints, and mating preference (section 3.7).

3.2 COMMON FEATURES OF SENSORY SYSTEMS

3.2.1 Basic functions of sensation

The task of all sensory systems, from chemoreception in jellyfish to vision in raptors, is to transduce energy in the environment—radiant energy, kinetic energy, or chemical potential—into electrochemical signals within an animal's nervous system (fig. 3.1). Doing so usefully requires being able to detect important signals while ignoring irrelevant ones (section 3.2.5 below). These basic requirements give rise to several ubiquitous properties of sensory systems that are important to mate choice. Many of these same properties apply across scales: for example, we can talk about thresholds in both the electrophysiological responses of sensory receptor cells, and in behavioral measures of detection which may incorporate downstream neural processing beyond the sensory periphery (chapter 4).

3.2.2 Thresholds

All sensory tasks have a threshold—the point at which a stimulus can just be detected. These thresholds form a lower bound for which stimulus values can elicit a response from choosers (chapter 2). Below a lower threshold, a stimulus cannot be distinguished from background and internal noise. The term threshold can lead to "unconscious punning" *sensu* Bateson (1983) in that it conjures a step function characteristic of a categorical preference function (chapter 4). This is emphatically not the case; preference functions flatten out at extreme stimulus values no matter what.

Thresholds are useful to sensory biologists in terms of mapping sensitivities; for example, the threshold **amplitude** at which an acoustic signal can be detected is dependent on **frequency** in all auditory systems; in visual systems, the threshold intensity depends on the wavelength and area of a stimulus (Levine 2000). Since signal intensity decreases with increasing distance, sensory thresholds influence the range over which a chooser may detect a signal, and are therefore an important parameter in determining the distance over which a signal may be detected.

Sensory thresholds can vary markedly both within and among individuals, in ways that are relevant to mate choice. In threespine sticklebacks, for example, the threshold sensitivity of females to longer (red) wavelengths of light is correlated with the degree of red coloration in their fathers (Rick et al. 2011), consistent with a genetic correlation between traits and preferences (chapter 15). Within individuals, sensory thresholds can change over ontogeny, as seen in humans' declining acoustic sensitivity with age (Brant & Fozard 1990). Sexual receptivity can also increase threshold sensitivity; in humans, olfactory thresholds are lowest during ovulation and highest during menstruation (Navarrete-Palacios et al. 2003). Much of the within- and among-individual variation in choosiness and receptivity described in later chapters may be accounted for by shifting sensory thresholds.

At extreme values, all sensory receptors **saturate**—that is, they reach an upper threshold beyond which an increase in stimulus intensity fails to increase response; increased stimulation becomes impossible. Courter signals probably reach saturation intensity in a minority of cases, involving close-range acoustic, chemical, or tactile cues.

Sensory thresholds play an important role in determining which courters are chosen. All else equal, courters that produce signals that are detected at a longer distance are more likely to be chosen as mates (Forrest 1994). Parker (1983) and later workers (e.g., Bailey et al. 1990) have distinguished such "passive choice," where mate choice is merely a function of detection, from "active choice" based on the evaluation of multiple detected stimuli. This distinction may not be entirely useful. Sensory thresholds represent an effective mechanism for choosers to discriminate among potential mates, and they can and do vary in response to proximate cues or as a result of selection. Raising a threshold in this context increases choosiness, since a smaller subset of courters will be detected and therefore eligible for consideration as potential mates. The distinction between passive choice and active choice may therefore be artificial (Dawkins & Guilford 1996; chapter 6).

3.2.3 Nonlinear responses and Weber's law

Sensory systems have to deal with a vast range of signal intensities. For example, there are a billion times as many photons striking the eye at a given moment on a sunny day as there are on a moonless night (Rieke & Rudd 2009). Accordingly, sensory systems respond nonlinearly to increasing stimulation. Above threshold, it is convenient to speak of **just noticeable differences** (JNDs)—what is the smallest difference in intensity between two signals that an observer can detect? (The JND is distinct from the **just meaningful difference** required to elicit a preference, which is another way to think about choosiness over narrow ranges of chooser trait values.)

In the 1830s, E. H. Weber showed that the JND increases with stimulus intensity. **Weber's Law** (also known as the Weber-Fechner law in honor of his contemporary Gustav Fechner) hypothesized that receivers were attending to the ratio between stimulus intensities, rather than to their absolute value, in making JND judgments (Akre & Johnsen 2014):

$$\text{Sensory stimulation} = k \log I$$

where k is a scaling constant and I is stimulus intensity. The logarithmic relationship means that the more intense a stimulus, the weaker the marginal effect of a given increase in signal intensity on sensory stimulation (Dehaene 2003). Weber's law holds for important psychophysical variables like brightness and loudness (fig. 3.3). The decelerating scaling of discrimination (so-called **nonlinear compression**) with increasing trait value propagates to properties like human perception of time and number, measured in terms of both behavior and neural activity (Akre & Johnsen 2014). This has major consequences for mate choice because as signal values increase, it may become more and more difficult for courters to provide increasing stimulation to choosers (Cohen 1984).

Ratio comparison does not fully explain the scaling of JND for all stimuli, nor even necessarily across a broad range of values for a given stimulus (Akre & Johnsen 2014). For example, people's assessment of length and flash rate are accelerating functions of stimulus value. Stevens's (1957) **power law** is:

$$\text{Sensory stimulation} = kI^a$$

where k is a scaling constant and I is stimulus intensity as before, and a is an exponent specific to modality and stimulus characteristics. The power law

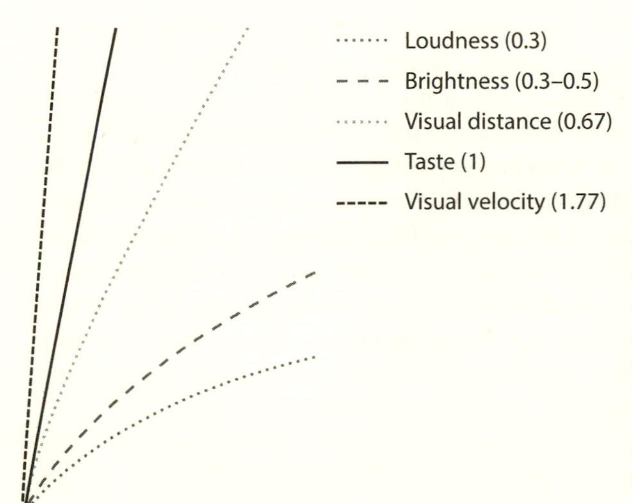

Figure 3.3. Power functions (arbitrary units) for humans' assessment of the relationship between the measured difference and the perceived difference between two stimuli based on exponents given in Stevens (1957). Perceptual responses can be decelerating or accelerating functions of stimulus size.

provides a more general description of sensory responses, with the exponent determining whether increasing stimulus values result in an accelerating or decelerating response (fig. 3.3). If adaptive models of trait-preference coevolution are correct (chapter 15), decelerating functions should be the norm rather than the exception for preferred traits, and it will be interesting to see whether this holds true in more systems.

Choosers often attend to variation in the number of stimulus units; for example, song syllables in oscine passerines, call pulses in frogs, and spatial pattern elements in fish. Weber's law describes numerosity—perception of number—in diverse nonhuman animals and in human infants, whereby subjects attend to the ratio of two numbers rather than the absolute difference (Buckingham et al. 2007; Nieder & Miller 2003). Akre and colleagues (2011) asked whether nonlinear compression could account for the scaling of mating preferences in túngara frogs, where females prefer male calls followed by a larger number of broadband "chucks." As predicted by nonlinear compression, females' responses in binary-choice tests were predicted by the ratio of chucks between paired stimuli. The greater the ratio of chucks in one stimulus to those in the other, the more females chose the greater stimulus; this relationship held independent of the actual magnitude (number of chucks) of the difference (fig. 3.4). Remarkably, the relationship was the same for the bat predators of túngara frogs, reinforcing the point that nonlinear compression of numerical information is broadly shared. As with the above examples, there is a nonlinear relationship between sensory stimulation and preference, which has its roots in a widespread mechanism for making numerical judgments.

Nonlinear compression means that it may pay courters to produce traits along another axis of signal variation (chapter 4) thereby resulting in more complex, multidimensional signals (Rowe 1999; Akre et al. 2011). It also means that lower stimulus values are more discriminable and that, all else equal, choosiness is greater when comparing stimuli of smaller magnitude. For example, Bush and colleagues (2002) found that the unimodal female preference for pulse rate in the treefrog *Hyla versicolor* fell off more sharply toward lower values. Weber's law therefore makes general predictions about how courters and choosers should diverge in response to selection against hybridization, with shifts toward lower magnitude proving more effective (Akre & Johnsen 2014).

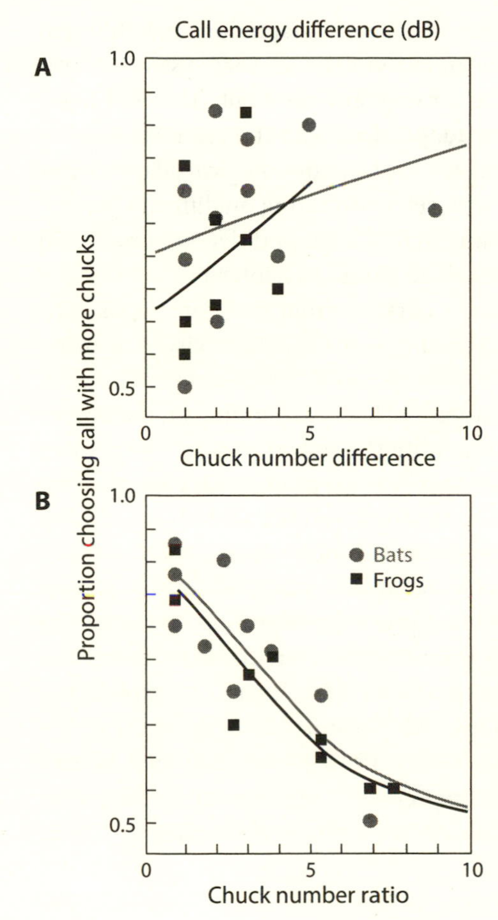

Figure 3.4. Weber's law provides a good fit for responses of both female túngara frogs and their bat predators to pairs of stimuli differing in the number of acoustic ornaments ("chucks") added to a species-typical call. (A) proportion choosing the call with more chucks as a function of difference in chuck number; (B) proportion choosing chucks as a function of chuck ratio.

3.2.4 Physiological adaptation and habituation

Signal detection poses different challenges in different environments. The sensory systems of choosers adapt evolutionarily in response to selection on mate choice (chapter 13) but more generally in response to selection on sensory functions associated with the organism's ecology. For example, planktivorous fish are sensitive to the ultraviolet light scattered by small particles in the water column. These small particles are food for plankton-feeders, but merely visual noise for fish that forage on larger items; accordingly, piscivorous fish have filters on their corneas that block ultraviolet light (Siebeck & Marshall 2007). UV blocking in piscivores is an evolutionary adaptation.

Sensory biologists, by contrast, use the word "adaptation" in a different sense, to mean the process whereby an individual's sensory system accommodates short-term changes in the sensory environment. This will hereafter be referred to as **physiological adaptation**, with the unmodified term referring to evolutionary adaptation. Of course, physiological adaptation is itself a phenotype that can evolve, with selection acting on different genetic variants that accommodate environmental change in different ways. I will return to the evolution of function-valued traits in chapter 11.

When we step into the light from a dark environment, our pupils constrict and we switch from using our *rod cells* for low-light vision to using our *cone cells* for bright-light vision. This means that a chooser's sensitivity to particular courter stimuli can change markedly depending on the immediate context, and can therefore complicate the task of parsing one courter from another. As I will discuss in chapter 6, **perceptual constancy** presents a challenge for choosers making comparisons, and the underlying mechanisms can be recruited as mating preferences (Endler et al. 2014).

Further, sensory systems can physiologically adapt to courter signals themselves. **Habituation** is a universal property of sensory systems: the same stimulus presented again and again will yield a reduced response. Release from habituation has been proposed as a mechanism to explain the widespread preference of female songbirds for more complex vocal repertoires in singing males (chapter 13; Searcy 1992; Ryan 1998) with behavioral, electrophysiological, and gene-expression responses attenuating after repeated song playback (Dong & Clayton 2009; but see Sockman et al. 2002, 2005).

3.2.5 Tuning curves and signal detection theory

Sensory systems typically exhibit open-ended, power- or Weber's-law responses to changes in quantity, like light intensity, sound pressure, or odorant concentration. Yet when it comes to changes in **quality**,[1] like color, tone, or chemical structure, sensory systems exhibit tuning curves (fig. 3.5), which are usually unimodal. This means that different wavelengths, frequencies, or chemical mixtures from courters will elicit different sensory responses on

[1] Here and throughout this book, I use the word "quality" according to the dictionary definition of "an essential or distinctive characteristic, property, or attribute," rather than the alternative definition "character with respect to fineness, or grade of excellence." This latter definition seems to have infected much of the mate-choice literature, without addressing what determines fineness or excellence. I return to the Q-word in the latter half of this book.

the part of receivers. We can measure tuning curves for receptor molecules, sensory neurons, or downstream neural structures. The peak of a tuning curve can predict peak preference (e.g., Nocke 1972; Kostarakos et al. 2008; Ryan et al. 1992). Tuning curves can therefore shape preference functions (fig. 3.6), although receiver sensitivity may yield antipathy rather than preference (chapter 5).

Tuning curves can range from very choosy, or *selective* (narrow) to very **permissive** (broad). The breadth of tuning curves is sometimes determined by inherent constraints; for example, the absorbance spectrum of visual pigments is consistent enough that spectral absorbance curves (fig. 3.5a) can be robustly predicted by knowing the peak absorbance; the breadth of the curve can be modulated by switching between vitamin A1 and vitamin A2 as the *chromophore*, the molecule that absorbs the photon and triggers a change in protein structure (Govardovskii et al. 2000; for more detail see section 3.4 below). The olfactory system contains both narrow and broadly tuned inputs from odorant receptors, with the former responding to specific odorants and the latter involved in ensemble coding of multiple odors (Christensen & White 2000). The permissiveness—the breadth—of a tuning curve should correlate with choosiness, although this has received little attention (but see Domingue et al. 2007; Vedenina & Pollack 2012).

Tuning curves determine an inherent trade-off in responding to signals, which applies equally whether receivers are discriminating them from background noise or, say, to signals produced by a closely related species. This problem was formalized as *signal detection theory* by electrical engineers in the mid-twentieth century, and applied to animal communication by, among others, Haven Wiley (1983, 2006, 2017). As illustrated in figure 3.7, a selective (narrowly tuned) receiver runs the risk of ignoring potential mating opportunities (analogous to type II error in statistics) while a permissive one runs the risk of responding to inappropriate stimuli (type I error). Signal detection theory applies to every phase of mate choice, from detecting stimuli at the sensory periphery to resource allocation after mating, and provides a useful framework for understanding the trade-offs underlying the evolution of mating decisions. As I will cover in chapter 16, choosers are typically more selective when they occur in sympatry with closely related species and there is a fitness cost to hybridization. A common response to selection against mating with heterospecifics (or other inappropriate courters) is **peak shift** displacement, whereby sensitivity is displaced toward more extreme values of the preferred signal (Ryan & Cummings 2013; chapters 6 and 12).

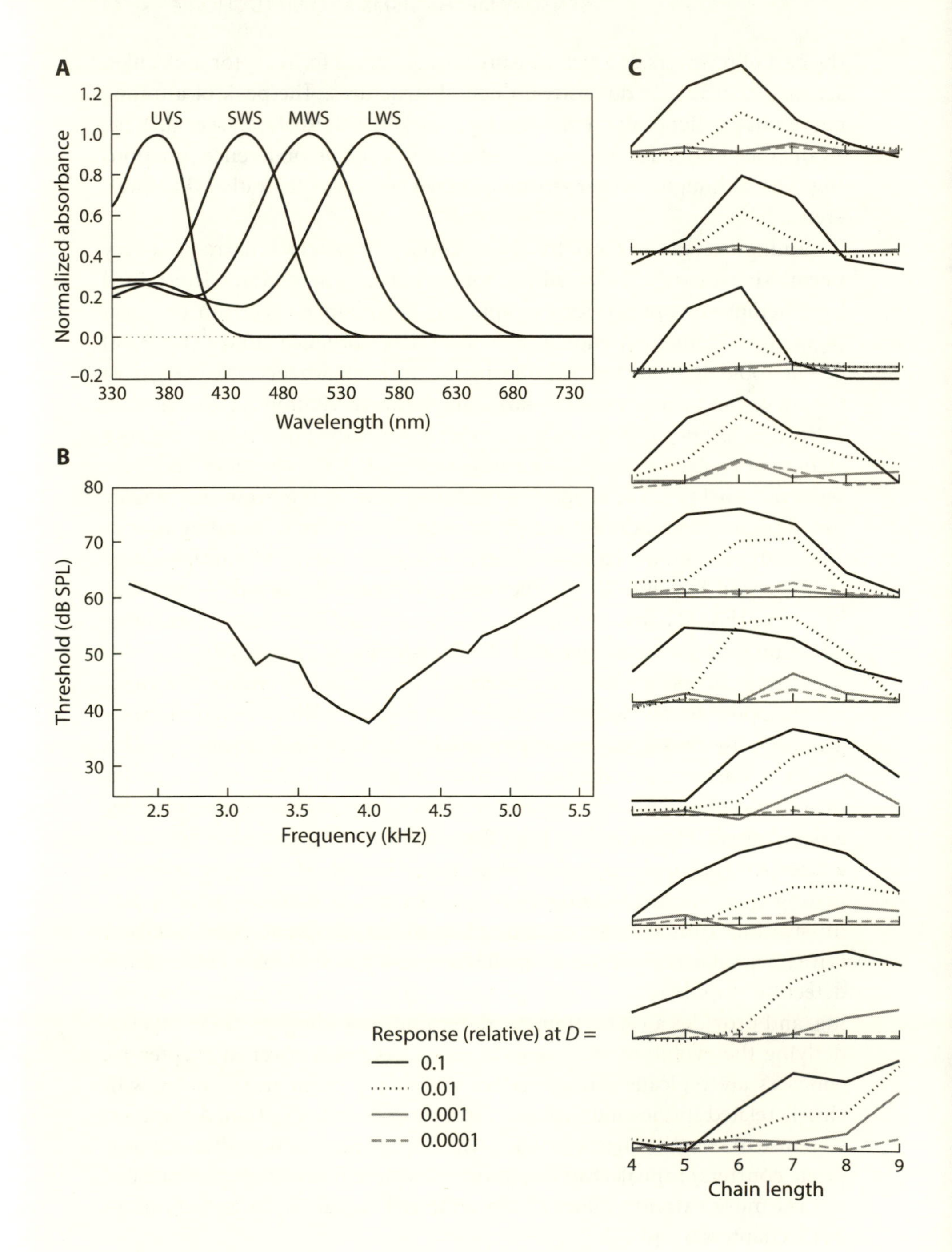

A
UVS
SWS
MWS
LWS
Normalized absorbance
1.2
1.0
0.8
0.6
0.4
0.2
0.0
−0.2
330
380
430
480
530
580
630
680
730
Wavelength (nm)
B
Threshold (dB SPL)
80
70
60
50
40
30
2.5
3.0
3.5
4.0
4.5
5.0
5.5
Frequency (kHz)
C
Response (relative) at D =
0.1
0.01
0.001
0.0001
4
5
6
7
8
9
Chain length

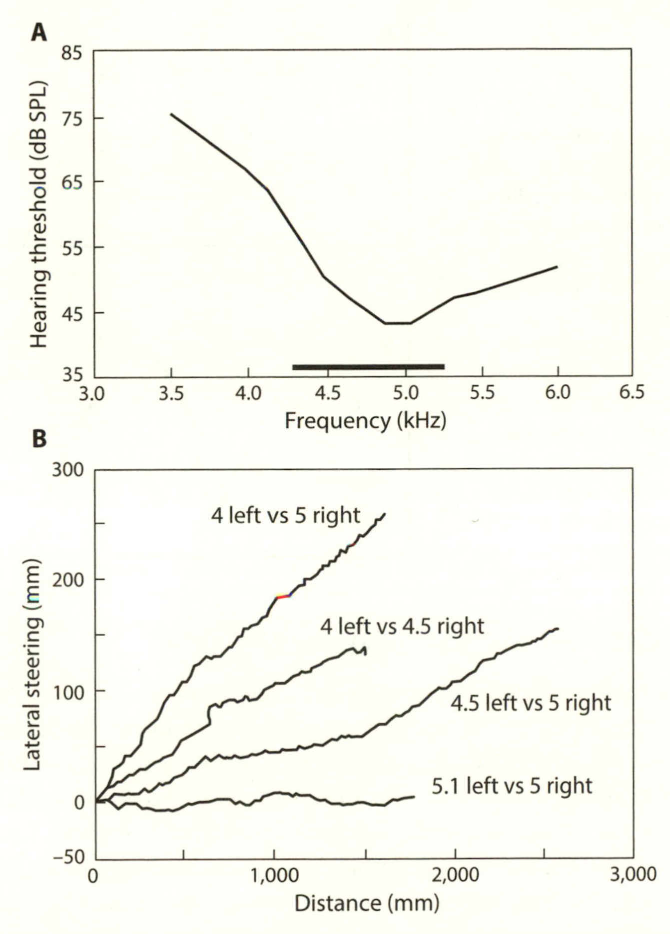

Figure 3.5. (*opposite*) Tuning curves across sensory modalities. (A) Absorbance spectra of visual pigments in the retina of blue tit (*Parus caeruleus*; Hart 2001). (B) Sensitivity (threshold value required for response) of an auditory neuron mediating phonotaxis in the cricket *Paroecanthus podagrosus* (Schmidt et al. 2011); (C) change in optical density of individual glomeruli in the rat olfactory bulb in response to aliphatic aldehydes with carbon chains of different length (Meister & Bonhoeffer 2001).

Figure 3.6. (*above*) (A) Threshold response levels of auditory interneuron (AN-1) of female field crickets *Gryllus campestris* to synthetic male calls varying in carrier frequency. Dark line shows mean response and gray lines show individual chooser responses. Black bar indicates male call range. (B) Direction of phonotaxis toward binary stimuli varying in carrier frequency (kHz). 5-Hz calls are strongly preferred over lower-frequency calls. After Kostarakos et al. (2008).

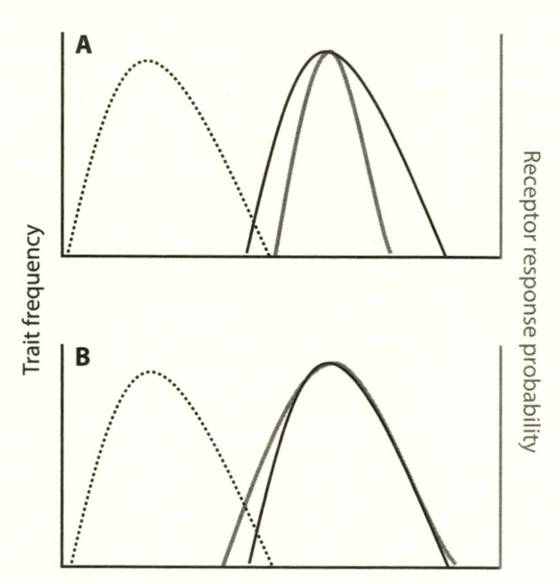

Figure 3.7. Tuning curves and signal detection theory. Dashed and solid black curves represent the frequency distribution of costly and beneficial stimuli (e.g., heterospecific and conspecific), respectively. (A) A selective (narrow) curve ensures rejection of inappropriate stimuli but also rejects appropriate ones (type II error). (B) A permissive (broad) curve ensures acceptance of appropriate stimuli but also accepts appropriate ones (type I error).

3.2.6 Detecting signals in noisy environments

The most basic challenge in sensation is discriminating a stimulus of interest from everything else in an often complex environment. The greater the signal-to-noise ratio (SNR) in a courter signal, the more likely a chooser is to detect it, and therefore the lower the cost to that chooser of evaluating it (assuming it's not outweighed by risky attention from predators; Rosenthal et al. 2001, 2002). Environmental noise is a major driver of communication systems, with both signalers and receivers adapting—physiologically and evolutionarily—in response. Signal-to-noise ratio constrains the sensory modalities that choosers can use to evaluate courters, and is therefore a major driver of macroevolutionary differences in sexual communication. The specificity of sensory systems to filter out noise (Witte et al. 2005), or sensory systems exhibiting peak shift (chapters 6 and 12) away from a source of noise (ten Cate et al. 2006), can be a source of selection on courters and a driver of speciation (chapter 15). By disrupting chooser sensory processes, human-induced noise can weaken sexual selection (Wong et al. 2007) and permit hybridization (Seehausen et al. 1998; Fisher et al. 2006; reviewed in Rosenthal & Stuart-Fox 2012).

Accounting for noise is key to understanding the behavior of choosers in nature. Hirtenlehner and Römer (2014) showed that minimum detectable differences along three axes of preference (chirp rate, amplitude, and carrier frequency) were markedly worse outdoors than in a laboratory anechoic chamber. Choosiness is (probably as a rule) weaker in noisier environments, so laboratory experiments may provide high-biased estimates in this regard. Wollerman (1999) experimentally added noise to playback stimuli, determining that female chorus frogs *Hyla ebraccata* detect differences between courters only above a signal-to-noise ratio of +1.5dB. Based on distributions of calling males in the field, she then argued that females were unlikely to be evaluating more than one male at a time at a given location. Noise can therefore modulate the costs of mate choice.

3.2.7 Receptor genes and evolution

A big part of the appeal of studying the sensory substrate of mate choice is that we can determine the structural genetic basis of sensory variation by studying sensory receptor genes. A basic understanding of **structural** and **regulatory** evolution of DNA sequences is helpful to understanding the molecular evolution of sensory receptors, or other genes involved in mate choice. Structural evolution involves changes in DNA sequences that are transcribed and ultimately translated into proteins; structural changes therefore yield changes in amino acid structure and ultimately in protein function. It is possible to directly link DNA sequence to function by experimentally replacing amino acids and measuring how the protein's activity changes. Only some DNA changes result in amino acid changes, though. A useful parameter when studying structural evolution is the so-called dN/dS ratio, which is the rate of change at "non-synonymous" sites that produce an amino acid difference, divided by that for synonymous sites that don't result in amino acid changes. Values of $dN/dS > 1$ are consistent with **diversifying selection** on an allele, since we see more amino acid substitutions than expected under a neutral model; values < 1 are consistent with **purifying selection** (or stabilizing selection) against functional change (Graur & Li 2000).

Regulatory evolution involves changes in the sequence of regions of DNA that regulate the expression of genes. Regulatory elements range from promoters upstream of a coding region, to stretches of misnamed "junk DNA" that can modulate transcriptional efficiency. Analogous to dN/dS, we can use techniques like comparing the rate of interspecies divergence to interspecific polymorphism (Andolfatto 2005) to estimate signatures of selection on regulatory regions. It is harder to develop specific predictions relating structural changes in regulatory sequence to functional changes in gene expression, but most phenotypic changes probably involve changes in gene regulation rather than protein function (Mackay et al. 2009).

Receptor genetics in the context of mate choice suffers from a typological view of choosers, whereby variation within and between choosers is given little attention (see chapter 9). This is problematic for two reasons. First, as with mate-choice mechanisms in general, there is likely to be meaningful within-species variation in sensory mechanisms. For example, opsin structure and expression predict mate preference in guppies (Sandkam et al. 2015a) and vary more among populations of guppies than they do among species in the group (Sandkam et al. 2015b). Second, the expression of sensory receptor genes, like virtually all downstream aspects of chooser behavior, is modifiable by social and environmental experience at a variety of timescales (e.g., Fuller et al. 2005; Hofmann et al. 2010). Sensitivity is thus a function of both a chooser's developmental history and her shorter-term interactions with the environment, a point that I will return to in later chapters.

We can detect the phenotypic consequences of regulatory changes, and of environmental inputs, by measuring **gene expression**. Two species may share an identical palette of sensory genes, but their sensory periphery may differ markedly as a result of expression differences between the two species (Carleton & Kocher 2001). The interpretation of gene-expression studies, including contemporary RNASeq studies, requires us to keep a few caveats in mind. For any gene, expression varies dramatically within an organism over time and space. Many genes, such as the combination of genes expressed during the sensitive period for song learning in passerine birds, are only expressed during a narrow window of development. The hemoglobin gene, for example, is in the genomic DNA sequence of every cell within the body, but is only expressed in red blood cells and mesangial cells (Nishi et al. 2008). Gene expression also can vary dramatically as a consequence of epigenetic effects and environmental influences, which will be discussed in later chapters. This means that environment and developmental history have to be very carefully standardized when making inferences about genetic differences underlying differences in gene expression. It is also important to keep in mind that while we often use gene expression as a proxy for protein concentration, there may be nonlinear relationships between gene expression and both the function and abundance of proteins (Zhou et al. 2013). Despite these caveats, gene expression provides a useful assay of neural activity (chapter 2) and sheds light on neurogenetic mechanisms underlying female preference (e.g., Wong et al. 2012).

3.3 CHEMORECEPTION

The oldest and most widespread sensory modality is chemoreception, which includes smell and taste (gustation). Chemoreception involves responding to molecules based on their chemical structure and concentration. Mate choice associated with chemical cues is near-universal among animals, with anuran amphibians constituting a notable exception. Even in birds, long thought to be largely visual and auditory creatures, recent evidence suggests that courters produce complex chemical signals encoding individual identity, and that choosers attend to these signals (Leclaire et al. 2012). Since choosers can exhibit differential responses to courters purely on the basis of the affinity of chemosignals to odorant receptor proteins, chemically based mate choice is possible both in microorganisms and in the gametes of multicellular organisms. In broadcast-spawning marine invertebrates, for example, interactions between "chemoattractants" released by eggs and odorant receptors on sperm cells mediate both intra- and interspecific mate choice (Evans et al. 2012). Spatiotemporal sampling of chemical cues is also important to navigational tasks like locating individual mates (Jacobs 2012; chapter 6).

The first step in chemoreception is to solubilize chemical cues in an aqueous medium—in vertebrates, a layer of mucus—so that they can contact odorant receptor neurons (ORNs; fig. 3.1). The thickness and chemical composition of the fluid surrounding olfactory receptor cells can influence olfactory sensitivity to specific odorants (Ache & Restrepo 2000). In particular, so-called odorant binding proteins can solubilize hydrophobic (non-water soluble) compounds, thus extending the range of compounds to which organisms are sensitive (Pelosi 1994; Hekmat-Scafe et al. 2000).

Dissolved odorant molecules then bind to odorant receptor proteins on odorant receptor neurons. In vertebrates, binding to receptors triggers a G-protein-coupled cascade, which activates a neural response; in insects, the receptor is a two-molecule complex that includes a receptor and an ion channel opened by binding to the odorant (Kaupp 2010). In vertebrates, each ORN expresses a single type of odorant receptor protein (Ache & Restrepo 2000) while in at least some invertebrates, there can be multiple receptor classes in a single olfactory cell (Bargmann & Kaplan 1998). Mice express about 1,100 functional receptor genes in olfactory tissues, compared with only 100–300 in humans (Mombaerts 2004). This is still one or two orders of magnitude greater than the number of receptor genes expressed in other sensory modalities, and sets the stage for highly specific responses to particular odorants at the molecular level. Psychophysical measures indicate that humans can discriminate about 3 trillion odors, six and seven orders of magnitude higher than the colors and tones we can discriminate, respectively (Bushdid et al. 2014), with no vocabulary beyond simile to describe them. Because odorants bind directly to the protein products of receptor genes, sensitivity to a specific odorant can thus be modified by changing the structure or expression of an individual gene, while minimizing **pleiotropic** effects on responses to other signals.

The diversity and specificity of odorant receptors mean that there are genes in choosers for which single-allele differences can account for dramatic variation in mate choice. Male moths (order Lepidoptera) make some of the simplest reproductive decisions. For male silkmoths, *Bombyx mori*, the detection of the female pheromone bombykol is both necessary and sufficient to elicit sexual behavior. A single, narrowly tuned receptor, *BmOR1*, is responsible for detecting bombykol. Sakurai and colleagues (2011) replaced *BmOR1* with a gene for a receptor tuned to pheromones of the diamondback moth *Plutella xylostella*. As predicted, males directed sexual behavior at *Plutella* cues and were unresponsive to those of conspecifics. In *Heliothis* moths, a small cluster of the genome with four odorant receptor genes arrayed in tandem is responsible for species-specific differences in male response to female pheromones, and therefore for sexual isolation between two closely related species (Gould et al. 2010). In E-strain *Ostrinia nubilalis* (European corn borers), about 3% of males are permissive, responding to cues of *O. furnacalis* as well as those of conspecific females. This permissive response is due to broader tuning of the receptor neurons responsible for pheromone detection (Domingue et al. 2007).

As will become apparent shortly, in other modalities it is much less likely that individual receptor genes play such a strong role in preference variation. In olfaction, however, individual odorant receptors can therefore mediate sexual isolation and among-individual preference variation. Simple receptor-ligand systems may play an important role in vertebrates as well. A notable and widely studied example is the class I ligand proteins of the major histocompatibility complex (MHC). These small proteins play a role in the immune system's recognition of antigens, and selection maintains a diverse repertoire of tandem MHC alleles in natural populations, including humans. When expressed on the cell surface, MHC or related urinary proteins constitute chemical cues that receivers, including humans, attend to (Leinders-Zufall et al. 2004; Sturm et al. 2013; Milinski et al. 2013). Mate choice based on MHC or other small, rapidly evolving proteins may be a widespread mechanism of compatibility-based mate choice in vertebrates,

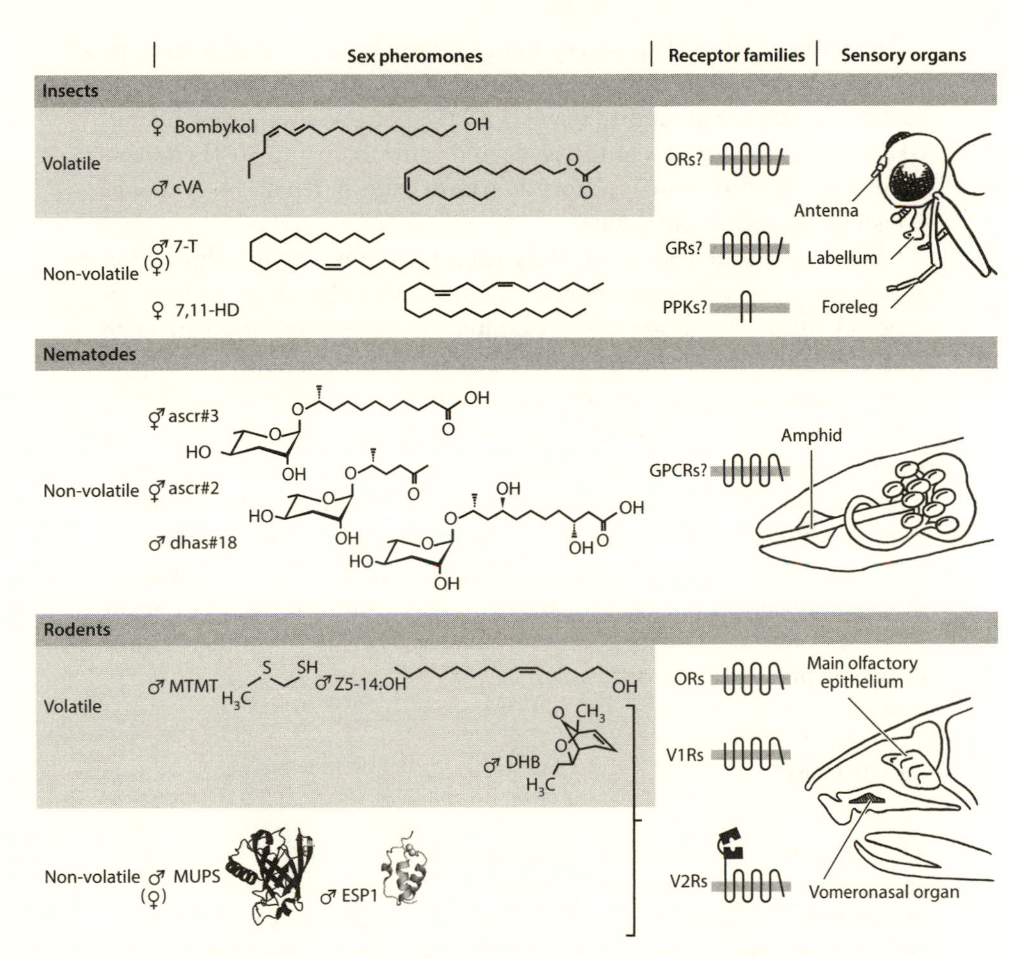

Figure 3.8. Diversity of sensory mechanisms in olfaction. Distinct signals and distinct receptor molecules are associated with different anatomical structures in insects, nematodes, and rodents. After Gomez-Diaz & Benton (2013).

although empirical evidence is mixed (Edwards & Hedrick 1998; Penn 2002; chapter 14).

The modularization of chemical communication is further facilitated by the fact that in many insects and vertebrates, distinct neuroanatomical systems exist for processing social and non-social information (fig. 3.8). In insects, pheromone receptor neurons project to the macroglomerular complex, while other odorant receptor neurons project to the main antennal lobe. In mammals, pheromone-sensitive neurons in the vomeronasal organ project to the accessory olfactory bulb, while other ORNs project to the main olfactory bulb (Christensen & White 2000).

The **active space** of chemoreception can range widely, and chemoreception is likely recruited in all stages of mate choice: from long-distance attraction to signalers in moths (David et al. 1983); to close-range evaluation MHC conjugates secreted in the urine and saliva in mammals (Leinders-Zufall et al. 2004); to peri-copulatory arousal of males by female pheromones (Amstislavskaya & Popova 2004).

Chemosignals are also most likely used in postmating discrimination among sperm or offspring between multiple sires (Sanchez-Andrade & Kendrick 2009). There are fewer examples of taste being used in mate choice, but female decorated crickets do discriminate among male spermatophores based on their amino acid composition (Gershman et al. 2012).

Chemoreceptor distributions can vary widely among individuals and over ontogeny, and in response to environmental input. Ontogenetic changes in response to external cues are exemplified by coho salmon (*Oncorhynchus kisutch*), which upregulate expression of chemoreceptors sensitive to odorants they experienced as juveniles in their home stream. Sensitivity to natal odors presumably helps salmon home in on the stream of their birth (Nevitt et al. 1994). Social influences on chemical sensitivity may be a major force in shaping preferences (chapter 12).

3.4 VISION

Perhaps because of humans' own bias toward vision and visual metaphors, empirical models of mate choice and mating preferences have heavily emphasized vision (fig. 3.1). Nearly all studies of visual mate choice have focused on vertebrates, arthropods, and mollusks with image-forming eyes that produce spatially explicit visual scenes. Vision almost always relies on the contrast between patterns of light emanating from a natural source, usually the sun, and reflected off the courter. Visual systems can detect variations in time, space, intensity, and wavelength. Spatiotemporal patterns (Rosenthal 2007) are encoded beyond the photoreceptor cells performing sensory transduction, and are discussed in chapter 4. Polarization sensitivity, which is encoded at the periphery in arthropods and some fishes (Pignatelli et al. 2011), has been shown to mediate mating decisions in butterflies (Sweeney et al. 2003) and swordtail fish (Calabrese et al. 2014)

Vision usually involves forming an image by focusing light onto an array of photoreceptors, a task accomplished by the so-called ocular media. In vertebrates, light is filtered through the cornea, then focused by the lens before passing through sensory interneurons and striking the retina. In insects, light passes through individual **ommatidia**—components of a compound eye—before striking the retina. The ocular media—the lens and cornea—can each serve as spectral filters, selectively admitting only certain wavelengths of light; in some insects and vertebrates, oil droplets in photoreceptor cells are also color filters. The size and optics of the ocular media also determine the **spatial acuity** of the retina—how fine-grained the objects are that an animal can see. In nocturnal, crepuscular, or deep-water animals, the back of the eye is coated with a reflective *tapetum* that increases photon capture while reducing spatial acuity (Ollivier et al. 2004).

Like olfaction, vision also involves transduction of an external signal by a G-protein coupled receptor protein, in this case an **opsin** (fig. 3.1). Mutations in opsin genes can dramatically affect sensitivity. For example, five amino acid substitutions are responsible for the sensitivity difference between human red and green cone pigments, which enables us to see in color (Yokoyama & Radlwimmer 1998). The opsin protein is covalently bound to a *chromophore* molecule. The chromophore is usually vitamin A1, but some species have mixes of vitamin A1 and A2 (Govardovskii et al. 2000). The A1/A2 ratio can change both wavelength sensitivity and the breadth of the tuning curve for the photoreceptor. The A1/A2 ratio can change over the course of a season in concert with changes in environmental light conditions (Muntz and Mouat 1984). The chromophore absorbs a photon and changes its chemical structure, which in turn induces a structural change in the opsin and a chemical cascade that results in a neural response.

In addition to the A1/A2 ratio, the response of photoreceptors to light intensity and wavelength is determined by opsin protein structure. Most vertebrates and arthropods express a small number of different opsin genes in the retina at one time. Some, like rod opsin (rhodopsin), are expressed in the rod cells used in low-light and peripheral vision (*scotopic vision*), which involve perceiving **brightness contrast** across regions of the visual field. Cone opsins are, in vertebrates, expressed in the cone cells involved in *photopic* vision under brightly lit conditions. Combinations of different cone opsins are used in color vision, and receivers perceive a **chromatic** signal by comparing inputs between cone types (chapter 4). Color vision is therefore shaped by the diversity of opsin proteins within a chooser's retina. Mammals express two or three cone opsins, and other diurnally active vertebrates and arthropods, notably poeciliid fishes and stomatopod crustaceans, express on the order of a dozen more. Changes in the abundance of different cone opsin classes (e.g., red, green, and blue) will change wavelength sensitivity. Finally, visual sensitivity at the periphery can be altered via changes in the amino acid structure of individual opsins.

There is considerable geographic variation in both opsin expression (Fuller et al. 2004) and structure (Dann et al. 2004) as a function of visual microhabitat. The visual periphery can also change markedly as a result of experience. Rearing in different light environments can induce both morphological changes in the retina (Kröger et al. 1999) and changes in opsin gene expression (Fuller, Carleton et al. 2005; Hofmann et al. 2010).

Most of the information in a visual scene is captured in spatiotemporal variation in brightness (Rosenthal 2007). Nevertheless, color plays a widespread and important role in mate choice (Amundsen & Forsgren 2001; Kodric-Brown & Nicoletto 2001; Baldwin & Johnsen 2009; Limbourg et al. 2004; Brooks & Endler 2001a; Cole & Endler 2015). In Lake Victoria cichlids, color sensitivity appears to play a central role in speciation: selection favors divergent sensitivity to red and blue along a light gradient, which in turn leads to divergent mating preferences for red and blue courters in different microhabitats (Maan et al. 2006). When color information is blocked by environmental disturbance (eutrophication), individuals hybridize with heterospecifics (Seehausen et al. 1997).

3.5 HEARING

Hearing, or perception of far-field acoustic cues, is restricted to arthropods and jawed vertebrates, but plays a fundamental role in mate choice, and most of our understanding of mate-choice mechanisms comes from acoustic studies of passerine birds, anuran amphibians, and orthopteran insects. Hearing allows choosers to attend to variation in the intensity, frequency, and temporal patterning of sound. Perception of near-field signals, including substrate-borne vibrations (e.g., treehoppers, Rodríguez et al. 2006; jumping spiders, Elias et al. 2005; naked mole rats, Rado et al. 1998) and close-range acoustic signals (e.g., malarial mosquitoes, Charlwood & Jones 1979) is critical to mate choice in some taxa, and for the purposes of this section shares many similarities with hearing.

The function of auditory systems depends heavily on the physics of peripheral filters that interact with sound before sensory transduction. A sound signal can only be detected if it moves one part of a chooser's body relative to another. For vertebrate ears, a critical step in this process is *impedance matching*—for example, if a sound is traveling through air and strikes the body of the receiver, most of the sound's energy will be reflected back into the air. Impedance matching is accomplished by transducing sounds among materials of different densities before it interacts with sensory neurons. Peripheral structures also act to amplify and filter sounds before they reach receptor cells. This has led to a bewildering diversity of mechanisms for coupling external signals to the ear, including the external pinna in mammals, Weberian ossicles (discovered by Ernst Weber, mentioned above) in otophysan fish, and otoliths (ear stones) in most vertebrates. Near-field reception involves sensory hairs or antennae that bend in response to pressure differences in the medium (Bradbury & Vehrencamp 2011).

In vertebrates, one or more external membranes in the outer ear vibrate in response to sound. This movement is amplified and mechanically coupled with the movement of fluid in the inner ear via a series of bones in the middle ear. Fluid motion in the inner ear then moves one or more inner ear membranes, which causes the response of specialized mechanoreceptors on the membrane called hair cells. Displacement of the hair cell triggers a G-protein-coupled cascade as with olfactory and photic receptor neurons. The spatial arrangement of hair cells on the membrane is often *tonotopic*, like a piano keyboard, with different parts of the membrane stimulated by different frequency ranges. The peripheral response thus encodes both frequency and intensity (amplitude). Choosers attend to both of these parameters in mate choice; preference for lower-frequency, higher-amplitude calls is widespread (Ryan & Keddy-Hector 1992; Ritschard et al. 2010). Differential frequency sensitivity among sensory receptors permits ears to perform *Fourier analysis* decomposing complex signals into constituent series of sine waves differing in amplitude, frequency, and phase (Bradbury & Vehrencamp 2011).

In contrast with vision and especially olfaction, variation in hearing among individuals and species is generally due to differences in peripheral anatomy than to molecular variation in sensory receptors. A classic study by Narins and Capranica (1976) measured auditory nerve responses of the basilar papilla in male and female coquí (*Eleutherodactylus coqui*), the national frog of Puerto Rico. Coquí are so named because males produce a two-note call: the *co* is used in intrasexual aggression, while the *quí* is an advertisement call to females. The former matches the auditory sensitivity of the male amphibian papilla, while the courtship component matches the peak sensitivity of the female amphibian papilla (Narins & Capranica 1980; fig. 3.9a).

The frequency sensitivity in the ear is driven in large part by the biomechanics of the structures to which sensory cells are mechanically connected. In both vertebrates and insects, recent work has shown that active feedback from sensory neurons acts as an amplifier for low-frequency sounds. In

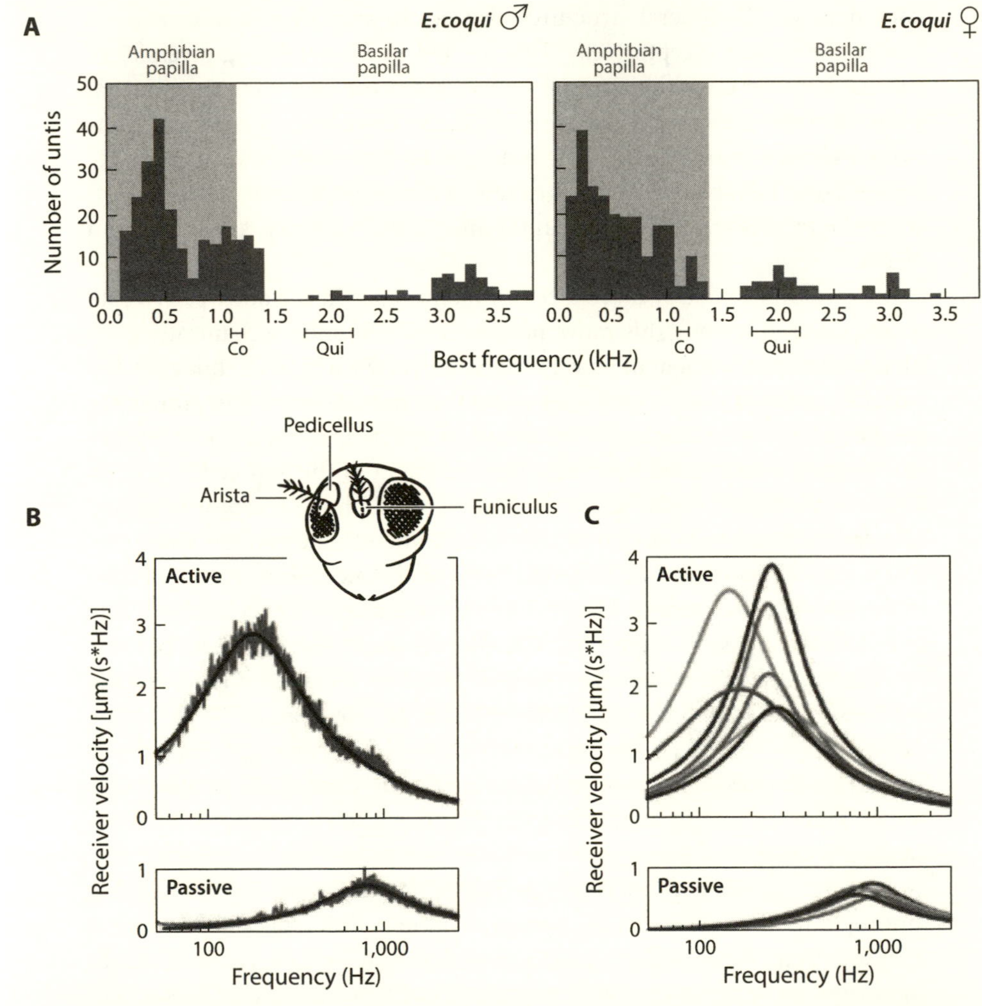

Figure 3.9. (A) Distribution of best excitatory frequencies for auditory nerve recordings in male (left) and female (right) coqui treefrogs. Shading indicates sensitivity regions corresponding to amphibian and basilar papillae. After Narins & Capranica (1976). (B) Free mechanical fluctuations of active (top) and anesthetized (bottom) antennal ears of *Drosophila teissieri* fit a simple harmonic oscillator model (black line). (C) Species-specific variation in active and passive (anesthetized) tuning of antennal ears across nine *Drosophila* species. After Riabinina et al. (2011).

Drosophila, this active feedback is responsible for interspecific variation in frequency tuning. Genes that drive this coupling of sensory cells with auditory membranes may be important to functional variation among individuals and species (Riabinina et al. 2011; fig. 3.9b).

Sexual dimorphism in hearing is often a special case of size-dependent variation (chapter 9) in auditory sensitivity. Size dependence takes two forms: first, auditory morphology covaries with size. For example, the basilar papillae of larger cricket frogs are thicker, which yields a slightly lower frequency range in larger individuals (Keddy-Hector et al. 1992). Second, spatial localization of sounds—a critical element of mate searching (chapter 6)—requires comparing phase or intensity at two different points on the body—usually the left and right ear. The higher an acoustic frequency, the greater the interaural distance required to detect it. Body size—which largely determines interaural separation—thus places a major constraint on signal localization. Similar constraints affect the production of mating signals; for these reasons, body size, peak frequency sensitivity, and peak call frequency are tightly correlated within and among species (Fletcher 2004).

A further source of variation in hearing, for ectotherms, is temperature, which affects both the frequency response of the auditory periphery and the frequency characteristics of auditory signals (Gerhardt & Mudry 1980). This presents a challenge for the mechanisms involved in mate choice, and a potential source of variation in preferences (chapter 11). The tympanal membrane of the ear of female tree crickets exhibits a permissive frequency-response spectrum that allows it to sense male calls at a wide range of temperatures (Mhatre et al. 2011); this means, of course, that frequency selectivity at the periphery is unlikely to constitute an effective means of mate discrimination by itself.

3.6. OTHER MODALITIES

Electroreception, the perception of changes in voltage, is present in many organisms, but the use of electrical signals in mate choice is restricted to mormyrid and gymnotiform freshwater fishes. Males use modified muscle tissue to produce frequency- and voltage-modulated electrical signals during courtship and aggressive interactions, and males and females detect these signals via specialized ion channels. The fact that signals are produced and received via elegantly simple mechanisms makes electrocommunication a highly tractable model for integrating sensory biology with mate choice (Feulner et al. 2009).

Touch, or somatosensory sensation, is by definition only expressed at close range and is the key modality in peri-mating choice (chapter 7; Eberhard 1994). It is likely much more important to mate choice than we might think, bound as we are by our visual and acoustic biases. Tactile signals are ubiquitously phylogenetically widespread; for example, males in many spider species perform rhythmic movements of the genitalia during copulation, which are not required for sperm transfer and likely function in tactile stimulation of a female *in copula* (Huber 2005). In unisexual whiptail lizards (*Cnemidophorus uniparens*), one female will perform male-like "doughnut" courtship behavior, constricting and biting her partner. Latency to ovulation is lower in females in the presence of a male-like partner, suggesting that physical contact plays a direct role in stimulating ovarian development (Crews et al. 1986; fig. 3.10). The importance of touch to human pre- and peri-mating courtship is intuitive, and survey respondents self-report tactile stimuli to be important to sexual arousal (Herz & Cahill 1997). By contrast with olfaction, vision, and hearing, however, there has been very little work done on somatosensory mechanisms in the context of mate choice. This is largely because of the difficulty of doing manipulative experiments with tactile stimuli (but see chapter 7 for some clever counterexamples).

Figure 3.10. In parthenogenetic whiptail lizards, touch stimuli (the "doughnut pose") induce ovulation (Crews et al. 1986). Photo by David Crews.

3.7. SENSORY CONSTRAINTS ON MATING PREFERENCES

3.7.1 The environment determines the sensory envelope

An animal's ***Umwelt***, or sensory world (von Uexküll 1909), is shaped by natural selection. The acoustic sensitivity of bats to very high frequencies, for example, is largely driven by their use of echolocation to find prey, as is the sophisticated response of their moth prey to those same signals (Windmill et al. 2006). In a given environment, there will be a sensory envelope that optimizes finding food, avoiding predators, and communicating with others. Unsurprisingly, therefore, a huge amount of the variation in sensory systems is driven by the overall features of the sensory environment. For example, in fishes, ultraviolet (UV) sensitivity may facilitate foraging in species and ontogenetic stages that eat plankton, but may present a liability for adults preying on larger organisms (Cummings et al. 2003).

Pine forests scatter low-frequency sounds. Accordingly, resident cricket frogs have their peak acoustic sensitivity shifted to higher frequencies than conspecifics in oak/juniper savannas (Wilczynski et al. 1992). Among surfperches off the California coast, photic wavelength sensitivity is strongly correlated with the distribution of available light in different microhabitats (Cummings & Partridge 2001). In animals inhabiting lightless environments like caves, spatially explicit vision is absent altogether (Yamamoto & Jeffery 2000).

Overwhelmingly, the sensory envelope—and therefore the maximum possible extent of a chooser's preference function—is determined by these ecological constraints. This point is evident before signals even reach the chooser, since the transmission of courter signals depends heavily on how they interact with the environment, and since the environment acts differently on signals in different modalities. More subtly, different modalities are under different selection pressures in each taxon. In a tropical pond, for example, acoustic signals are attenuated and degraded due to the way they reflect off the substrate and the water surface; electric signals can range farther or closer depending on water conductivity; chemical signals are at the mercy of currents and water chemistry; and visual signals depend on the distribution of sunlight as filtered through the water column, and filtered again after it reflects off the courter and travels to the chooser. Across modalities, therefore, signals vary in how they depend on the environment (Ryan & Cummings 2013; Rosenthal & Ryan 2000). In chapter 11, I discuss how variation in the sensory environment affects the expression of mating preferences.

3.7.2 Intrinsic differences among modalities

In addition to differences in how they interact with the environment, the structural differences among sensory mechanisms for different modalities mean that each modality is intrinsically constrained in a unique way with respect to how preferences can be expressed. At one extreme, olfactory sensitivity can be modulated by changes in the expression or structure of single genes (similarly, in electrocommunication, changes to the kinetics of ion channels can also change sensitivity to courter signals). The high dimensionality of chemoreception means that communication can be entirely **private** within a species; chemosignals may pass between courters and choosers sharing distinctive odorant receptors, without being eavesdropped upon by prey or predators. These properties mean that mate preferences can evolve at the olfactory periphery relatively free of constraints imposed by selection in other contexts. A change in the expression or structure of a single gene can determine detection of an odorant without disrupting the rest of the system—meaning that sensitivity to individual odorants need not be highly constrained by natural selection. Bargmann (2006) argued that olfaction should evolve more rapidly than other modalities: "the olfactory system, like the immune system, tracks a moving world of cues generated by other organisms, and must constantly generate, test, and discard receptor genes and coding strategies over evolutionary time." The physics of chemical communication also mean that chemoreception can function at every stage of mate choice, from detection of courters at a distance to intimate interactions during and after mating.

At the other extreme, visual systems are constrained by broadly shared physical constraints. Because different types of photoreceptors interact to shape color perception and visual sensitivity to broad classes of environmental stimuli, evolution of peripheral sensitivity in the context of mating is constrained by visual function with respect to other tasks. With vision, the opportunity for private communication is limited for two reasons: first, physical constraints restrict usable wavelengths to a narrow band of the electromagnetic spectrum (Levine 2000), and second, all biological material changes the distribution of incident light in some way.

In contrast with olfaction, vision provides choosers with spatially and temporally explicit representations of courter stimuli. Spatial resolution is limited by eye size and therefore by body size; smaller eyes have lower spatial resolution, which limits choosers' ability to attend to detailed patterns. Because vision relies on line of sight, visual displays are usually directed at a single receiver or at small groups of receivers (Rosenthal 2007), and are, in most natural habitats, restricted to medium-range interactions.

Since factors like habitat structure and body size impose tight constraints on the visual periphery, changes in visual preferences at the periphery are most likely to reflect selection on visual performance more broadly, outside the context of mating. Visual preference evolution should therefore often involve changes at higher levels of processing (chapters 4 and 5). The most commonly studied candidate mechanism for variation visual sensitivity, variation in the structure and expression of retinal opsin genes, would be particularly ill-suited for modulating specific responses to sexual signals. If selection favored antipathy toward red spots on males by females in guppies, loss of the long-wavelength-sensitive opsin proteins would be countered by natural selection favoring red sensitivity in the context of foraging (Rodd et al. 2002). Further, females would no longer be able to discriminate between red spots and black spots on males. Rather than losing sensitivity to certain wavelengths, choosers may be expected to evolve antipathies toward particular visual traits (Coleman et al. 2004; Morris 1998a; Wong & Rosenthal 2006) that can be accomplished without compromising sensitivity (chapter 5).

Hearing occupies a somewhat intermediate position in terms of constraints on preference evolution. On the one hand, the structure of auditory systems offers little scope for sensory receptor genes to evolve in the context of mating preferences as odorant receptor proteins would. The complex anatomy of the auditory periphery, however, means that auditory sensitivity can be partitioned into different functional modules; for example, anatomically distinct inner ear membranes can be tuned separately to acoustic frequencies produced by potential mates and to more general acoustic tasks (Richards 2006), analogous to separate olfactory systems for social and nonsocial functions (section 3.3).

Compared to vision, different species may inhabit vastly different acoustic *Umwelts* (Rosenthal & Ryan 2000). Among mammals alone, sensitivity ranges from "infrasonic" in elephants (Heffner & Heffner 1980), through "sonic" in humans, to "ultrasonic" in bats, cetaceans (Liu et al. 2010), and many insects (Jang & Greenfield 1996). This variation may afford courters and choosers a relative degree of privacy with respect to predators and prey.

Acoustic signals can be broadcast to a wide range of choosers in the early stages of mate choice; singing whales interact at a scale of kilometers (Tyack & Whitehead 1983). At the other end of the spectrum, acoustic communication can play a role in very intimate interactions around mating, as in the vocalizations of female rodents during copulation (Thomas & Barfield 1985).

3.7.3. Taxon-specific constraints on sensory modalities

Taxon- and modality-specific constraints on the sensory periphery may also influence our thinking about the importance of peripheral processes in mate choice. Female túngara frogs use their ears primarily to locate suitable mates. They use their eyes to find food and avoid predators. By contrast, female guppies use their eyes not only to locate suitable mates, but to find food and avoid predators. Mike Ryan's "sensory exploitation" hypothesis, which places chooser sensory mechanisms at the center of the evolution of courter signals, was inspired by his work on mate choice for acoustic signals in túngara frogs, where female hearing in the context of mate choice may be relatively free of constraint; John Endler's "sensory drive" hypothesis, driven by his work on mate choice for visual signals in guppies, places the environment as the driver of sensory function and therefore signal structure.

The two most influential frameworks for thinking about the role of sensory biology in sexual selection are therefore intertwined with the natural history of the systems used to develop them. The intellectual differences among proponents of various evolutionary explanations for preferences may therefore be driven in part by the biology of the empirical systems we each focus on.

3.7.4 Sensory systems can be "fooled"

In the previous chapter, we saw how video works by combining red, green, and blue light to give humans the illusion of color. This is because color is represented as a combination of inputs to red, blue, and green photoreceptors. Multiple types of stimuli, or metamers, can therefore be used to elicit the same sensory response. Similarly, multiple classes of molecules may elicit the same response from an odorant receptor, as with artificial sweeteners.

This provides an opportunity for courters to use multiple avenues to elicit the same sensory response, and presents a challenge to the maintenance of honest signals in mate choice (chapter 15). For example, female swordtail fish (*Xiphophorus hellerii*) exhibit an effectively open-ended preference for male size, which can be accounted for by increased stimulation of a greater number of retinal photoreceptors. Males can elicit this preference either by increasing body mass or by growing a caudal appendage called a sword, which is less costly to produce (Basolo 1998b). In some lineages, females retain the preference for body size but have secondarily lost, or even reversed, the preference for swords (Rosenthal et al. 2002; Wong & Rosenthal 2006). "Fooling" of receivers can occur at multiple levels of processing, and extends to such remarkable phenomena as sexual deceit in orchids (Gaskett et al. 2008). As I will return to later, this sets the stage for the evolution of more sophisticated discrimination mechanisms in choosers (Macías Garcia & Ramirez 2005).

3.8 SYNTHESIS

Sensory systems determine the envelope of stimuli that choosers can respond to. The simplest sensory process can generate chooser preferences among mates that are multivariate with respect to courter traits and that are maximized by courter traits outside the natural range of variation. Universal, low-level properties of sensory systems favor chooser preferences for novelty; while they facilitate preferences for extreme courter traits, power-law effects mean that they also dampen the effectiveness of extreme signals.

Depending on context and modalities, there is sometimes—but by no means always—opportunity for preferences to diverge at the sensory periphery rather than at higher levels of processing. Few if any studies have explicitly quantified the relative contribution of the sensory periphery to mate choice. There is mixed evidence as to how preference covaries peripheral sensitivity across scales (e.g., Richards 2006; Fuller et al. 2010; Sandkam et al. 2015a). And a growing series of studies (e.g., Macías Garcia & Ramirez 2005; Wong & Rosenthal 2006) show that preferences can evolve via explicitly non-sensory processes, as manifested in antipathy toward more detectable stimuli. In the following chapters, I discuss how downstream processes—integration of information from multiple sensory channels and the cognitive processes of evaluation and memory—are likely to influence mate choice.

All other things being equal, more physical energy from the courter translates into greater detectability. This has allowed sensory mechanisms to be coopted into **adaptationist** arguments about sexual selection, since courters with more resources will elicit greater sensory stimulation in choosers—and therefore, the assumption goes, greater preference. But as the following two chapters will argue, this assumption is often invalid.

3.9 ADDITIONAL READING

Bradbury, J. W., and S. L. Vehrencamp. 2011. *Principles of Animal Communication*. 2nd ed. Sunderland, MA: Sinauer.

Endler, J. A. 1993. "Some general comments on the evolution and design of animal communication systems." *Philosophical Transactions of the Royal Society of London B: Biological Sciences,* 340: 215–225.
Gomez-Diaz, Carolina, and Richard Benton. 2013. "The joy of sex pheromones." *EMBO Reports* 14:874–883.

CHAPTER 4

Beyond the Periphery

PERCEPTION, COGNITION, AND MULTIVARIATE PREFERENCES

4.1 INTRODUCTION

4.1.1. From univariate to multivariate preferences

The preceding chapter focused on the very first—and most thoroughly studied—step in a chooser's interaction with a courter: transducing energy in the environment into electrical and chemical signals within an organism. The way scientists think about mate choice has focused on this kind of univariate process, in which a single property of a courter maps neatly onto a single preference function in a chooser.

Yet universally, choosers attend to multiple features of courters when making mating decisions. Even the simplest chemosensory response involves attending to two aspects of the incoming chemosignal—specificity and concentration. Mate-choice interactions are often highly complex, recruiting more than one sensory modality (Candolin 2003). Consider the bowerbird from chapter 1: females evaluate a physical structure built by males, as well as male vocalizations, motor displays, and plumage (Patricelli et al. 2003). Or fruit flies, where females evaluate several different cuticular hydrocarbons, volatile molecules on the surface of the male's exoskeleton, in addition to a visual, tactile, and acoustic display produced by vibrating and extending the wings (Dickson 2008; Ferveur 2010). Multidimensional and often multimodal preferences are the rule.

Nonlinear interactions among preferences for multiple traits are the norm rather than the exception (section 4.2; Rowe 1999; Partan & Marler 2005). As Stevens (2014) points out, we can be misled about communication if we fail to consider how signal components interact to influence preference. An illustrative example is presented in figure 4.1. To a human observer, the horizontal line appears longer in the topmost drawing in figure 4.1a. When we remove the flanking lines (fig. 4.1b) we accurately perceive the lines as the same size. This is the Müller-Lyer illusion (Pressey 1967), whereby the orientation of the flanking marks acts to exaggerate or diminished the

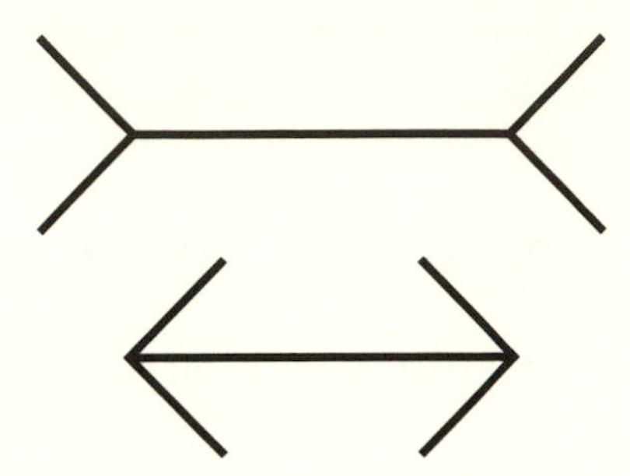

Figure 4.1. The Muller-Lyon illusion illustrates the hazards of incompletely sampling preference space. The two horizontal lines are the same size.

apparent size of the target line. It is easy to see how this illusion could affect our interpretation of how preferences act on traits; without taking this interaction among stimuli into account, we might reach an erroneous conclusion about how mating preference depends on the length of a horizontal stripe in courters. The Müller-Lyer illusion has been documented in a variety of nonhuman animals, and may be a driving force behind posture and positioning in courtship (Kelley & Kelley 2014).

Mate choice thus involves both the processing of complex information within sensory modalities, and the integration of information across modalities. The magnitude of this task becomes apparent when we take into account that choosers are attending to courters' extended phenotype, which can include courtship structures like bowers and the characteristics of a courter's territory (Balmford 1991). When we measure chooser responses to courter variation, we sometimes do so without carefully considering what choosers are actually attending to. This is especially the case when we test preferences "for" courter properties like social status, familiarity, and health. We test choosers on their responses to different categories of courters, but we are actually measuring responses to multivariate differences between potential mates, and we often remain agnostic about how those differences are salient to receivers.

It is useful to think about choosers having multivariate preference functions which vary according to a combination of two or more courter traits (chapter 1). Multivariate preference functions occupy multidimensional **perceptual space** (Ryan & Rand 2003; Ryan & Cummings 2013). As discussed below, choosers often exhibit **configural** preferences for specific combinations of traits, and these combinations can interact in nonlinear ways to determine attractiveness. Further, preferences are often categorical in nature, with abrupt transitions from preferred to unpreferred over a small interval in trait values. Taken together, these observations suggest that we should not expect multivariate preference functions to be straightforward extensions of the smooth univariate preference functions often assumed by theoretical models.

4.1.2 Visualizing and measuring higher-dimensional preferences

Visualization presents a challenge when we think about preferences involving more than two dimensions. A standard approach is to use a variable-collapsing algorithm like multidimensional scaling or principal components to capture the important axes of covariance among courter traits, and then map preferences onto those axes (fig. 4.2a–b). Multivariate preferences do not, however, map neatly onto the distribution of traits in space. Figures 4.2a–b show the same sets of túngara frog calls, with more strongly shaded points eliciting a stronger preference. The difference between the two panels is that in figure 4.2a, calls are arranged in two-dimensional space according to their similarity with regards to multiple acoustic measures, whereas in figure 4.2b, they are mapped according to their perception by choosers, based on similarity of responses.

An alternative variable-reducing scheme is to use canonical rotation to dimensions of preference that constitute the major axes of selection, i.e., the axes through multidimensional space that maximize choosiness (fig. 4.2c–f). This approach does, however, assume that preference functions along these axes have linear or quadratic shapes, meaning that complexity in preference functions may be hidden. Nevertheless, these approaches can sometimes reveal complex preferences for disjunct points in multivariate space (fig. 4.2f).

Measuring multivariate preferences is an arduous task. Imagine testing preferences along just two different dimensions. Five values along each axis are the bare minimum for detecting anything more complex than a linear preference function. The experimental assay thus requires $5 \times 5 = 25$ parameter combinations to be tested, which even for a small sample size of 20—well below the lower bound of statistical power to detect a meaningful effect (Rosenthal & Servedio 1999)—means that there are 500 tests to be done. Further, if one is attempting to characterize preference functions in a single individual, one is measuring responses in an older, more experienced individual at test #25 than at test #1. We must therefore be economical in what we choose to measure; accordingly, our estimates of preference variation are constrained by our a priori ideas about which traits are worth assaying for preference.

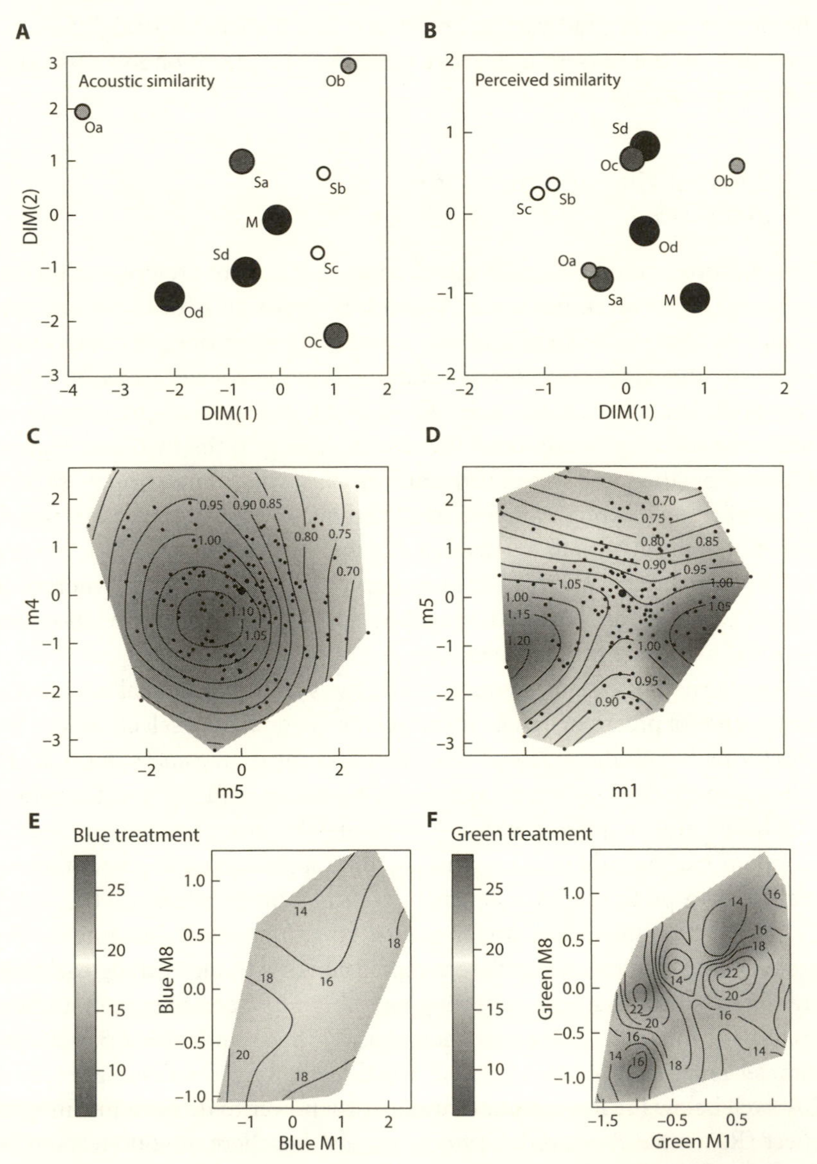

Figure 4.2. Slices of multivariate preference functions. Attractiveness (more strongly shaded = more phonotactic responses) of individual male calls to female túngara frogs in (A) multivariate acoustic space; and (B) in perceptual space based on similarity of responses (Ryan & Rand 2003). (C) Preferences of female gray treefrogs along two axes of stabilizing selection (unimodal preferences, m5 and m4) and (D) m5 and an axis of disruptive selection (bimodal preference, m1). White dots represent synthetic calls used and large black dot indicates standard reference call used in simultaneous choice tests (Gerhardt & Brooks 2009). (E) Markedly different preferences of female guppies along multivariate trait axes under blue light and (F) under green light (Cole & Endler 2015).

Approaches to sampling multidimensional preference space take two forms: measuring responses across a spectrum of naturally occurring courter variation (the multivariate trait distribution; fig. 4.2a–b, e–f), and measuring responses across a multivariate parameter space defined by the experimenter (fig. 4.2c–d; Brooks et al. 2005; Oh & Shaw 2013). Using natural variation is sufficient if we are primarily interested in preference functions as selection surfaces on contemporary courter trait distributions. For example, Hohenlohe and Arnold (2010) analyzed a large data set on traits and preference in three independent lineages (*Drosophila*, *Desmognathus* salamanders, and the Malawi cichlids *Pseudotropheus*). They estimated that variation in both traits and preferences collapsed into only two major dimensions. It is not clear, however, how much of this is due to selection from preferences acting to canalize trait variation, or to structural and developmental constraints limiting the extent to which courter traits can address chooser preferences. If we want to understand trade-offs between courter signals (chapter 15) or preferences for novel trait combinations (chapter 16), however, we have to assay chooser preferences for stimuli not represented by the current trait distribution of courters. In chapter 15, I will return to how preference mechanisms map on to variation in courter traits, and in turn what this mapping means for how preferences are realized in natural populations.

4.1.3. A chooser's-eye-and-nose perspective on multivariate preferences: the case of *Xiphophorus*

Focusing on multiple display traits, as opposed to multiple receiver mechanisms, has constrained our thinking about mate choice, since courter traits can be infinitely and arbitrarily subdivided in ways that don't match the way they are sensed and processed by choosers. A chooser-focused perspective might be helpful in guiding how we measure multivariate preferences. Table 4.1 summarizes the preferences that, as of this writing, have been documented in the swordtails and platyfishes (Teleostei: genus *Xiphophorus*). These studies have measured female preference by varying a particular feature of males and measuring preference, typically in simultaneous choice assays (chapter 2). From these experiments, we know that female *Xiphophorus* typically prefer live conspecifics over live heterospecifics; when they do so, they attend to both chemical and visual cues produced by males (table 4.1). Within visual cues, they attend to temporal features, spatial features, and chromatic features. Within spatial features, they attend to the magnitude

Table 4.1. Summary of pairwise stimuli manipulated in behavioral studies of female *Xiphophorus*

Stimulus type	*Reference*
Live, interacting courters	
Conspecific vs. heterospecific (open-field)	Clark et al. 1954
Conspecific vs. heterospecific (both visual and olfactory cues)	Ryan & Wagner 1987; Willis et al. 2011, 2012
Small, noncourting vs. large, courting males	Ryan et al. 1990; Tudor & Morris 2011; Ramsey et al. 2010; Wong et al. 2012
Conspecific vs. sympatric heterospecific (only visual cues)	Hankison & Morris 2003
Conspecific vs. bigger, allopatric heterospecific (only visual cues)	Rosenthal & Ryan 2011
Allopatric vs. sympatric heterospecifics (only visual cues)	Rosenthal & Ryan 2011
Spotted caudal vs. no spotted caudal	Fernandez & Morris 2009
Courtship rate	Wong et al. 2011
Courter chemical cues	
Conspecific vs. heterospecific	Crapon de Caprona & Ryan 1990; Hankison & Morris 2003; McLennan & Ryan 1996, 1999; Verzijden et al. 2012; Verzijden & Rosenthal 2011
Allopatric vs. sympatric heterospecifics	Fisher & Rosenthal 2010; Rosenthal & Ryan 2011; Hankison & Morris 2003
Poorly fed vs. well-fed males	Fisher & Rosenthal 2006a, 2006b
Females vs. males	Rosenthal et al. 2011
Experimental manipulations of courter visual traits	
Conspecific vs. heterospecific	Fisher et al. 2006; Verzijden & Rosenthal 2011; Butkowski et al. 2011
Dorsal fin (small vs. large)	Fisher et al. 2009; Fisher & Rosenthal 2007; MacLaren & Daniska 2008; MacLaren et al. 2011; MacLaren & Fontaine 2013; Robinson et al. 2011
Size (small vs. large)	Rosenthal & Evans 1998; Fisher et al. 2009; MacLaren & Daniska 2008; MacLaren et al. 2011; Lyons et al. 2013; Rosenthal & Ryan 2010
Courting vs. non-courting	Rosenthal et al. 1996

Table 4.1. (*continued*)

Stimulus type	*Reference*
High-contrast versus low-contrast sword	Trainor & Basolo 2006
Bars vs. no bars	Morris 1998
Symmetric bars vs. asymmetric bars	Morris 1996; Morris et al. 2006; Tudor & Morris 2009; Lyons et al. 2013;
More bars vs. less bars	Morris 1995
Sword ornament vs. no sword	Basolo 1990a, 1990b; Rosenthal & Evans 1998; Rosenthal & Ryan 2010
Long sword vs. short sword	Basolo 1990a, 1990b, 1995; Rosenthal & Evans 1998; Rosenthal et al. 2002; Walling et al. 2008, 2010
Disembodied sword vs. swordless male	Rosenthal & Evans 1998
Blue vs. yellow body	Kingston et al. 2003
Orientation of polarized light	Calabrese et al. 2014
More saturated vs. less saturated sword	Rosenthal et al. 2002; Cummings et al. 2003
UV vs. non-UV sword	Cummings et al. 2003
Sword in front vs. sword in back	Haines & Gould 1994
Cut crescent vs. no cut crescent	Culumber & Rosenthal 2013
Proximal vs. distal sword	Trainor & Basolo 2006
Body depth	Rosenthal & Ryan 2010

of the body, dorsal fins, vertical bars, and swords. Within swords, they attend both to the length of the ornament and to the contrast pattern between pigmented areas within the sword (Basolo & Trainor 2002). Black and long-wavelength pigmentation are produced by melanophore cells and erythrophores, respectively, which contain pigment granules called melanosomes and pterinosomes, which in turn contain melanin and pterin pigment molecules (Matsumoto 1965); in the closely related guppy, females attend to variation in ratios of these pigment molecules (Grether et al. 2004; Deere et al. 2012).

From these kinds of studies, we can draw some inferences about how male traits should respond to sexual selection via female preferences, but focusing on male signals misses much of what is going on with female preferences. First of all, there is of course no such thing as a "pterin detector" within the perceptual system of females; more generally, the mechanisms used to produce signals are often entirely distinct from those used to detect them. Even in animals like songbirds, where there is some overlap between production and perception of courtship signals, distinct neural mechanisms are recruited to produce and to interpret display traits. As I will revisit later in this book, this means that multidimensional trait space is invariably misaligned with respect to multivariate preference space.

Second, there is a network of integration rules among preferences that determines how choosers respond to particular combinations of stimuli (section 4.4). These rules can only be discerned by manipulating multiple traits simultaneously. We can deepen our understanding of how mating preferences work in *Xiphophorus* by reinterpreting behavioral data from the perspective of female preference mechanisms rather than male traits. We do this by analyzing multidimensional preferences in terms of their constituent components, including the distinct peripheral structures and sensory receptors used to transduce courter signals (chapter 3), and by identifying the rules that govern how females integrate information from multiple sensory channels. We can then determine how these processes map onto courter traits.

How can we rephrase the multivariate sexual communication of swordtails in terms of female preferences rather than male traits? A primary question concerns the dimensionality of preference: do all these preference experiments tell us that females have 31 different preferences, each subject to its own set of internal and external constraints, and each capable of evolving on an independent trajectory? Alternatively, does manipulating each of these traits simply reveal a special case of a single, broad female preference for overall sensory stimulation? Much of the data in table 4.1 suggests such a minimalist model of preferences in *Xiphophorus*, as follows:

> Mating preferences in these fish are determined by gross visual and chemical stimulation. The former is determined by maximizing contrast among visual photoreceptors, as produced by high color saturation and lots of moving edges. The second is determined by the concentration of a highly species-specific odorant. In each case, more is better.

The above is essentially the scenario that the late Chris Evans and I advocated when we showed that female *X. hellerii* showed equivalent responses to males with swords versus swordless males of the same total length (Rosenthal & Evans 1998). Increasing apparent size acted to increase gross visual stimulation to the retina, which in turn predicted chooser preference (see also MacLaren 2017). The preference for swords had previously been found in species that diverged from *X. hellerii* prior to the evolution of the ornament (Basolo 1990b; Basolo 1998a), and we argued that a novel ornament could evolve in response to a permissive, low-level preference for large apparent body size.

The model begins to unravel, however, when we see that swords are neither universally attractive, nor the target of a low-level preference for more gross stimulation. Haines and Gould (1994) showed that choosers were sensitive to the orientation and position of the sword. Female *X. nigrensis* failed to prefer conspecific males with longer swords, and even showed antipathy for a high-contrast, colorful sword like that of *X. hellerii* (Rosenthal et al. 2002). In *X. birchmanni*, a species where males have secondarily lost the sword, females exhibited strong antipathy for swords added on to males of their own species (Wong & Rosenthal 2006). Crucially, females in both species retained ancestral preferences for larger male bodies (Fisher et al. 2009; Ryan et al. 1990b). These experiments show that choosers can express an antipathy (or preference) for swords that is perceptually distinct and evolutionarily independent from the preference for large size; the loss of preference for swords, combined with the retention of preferences for large size, indicates that sensory stimulation is insufficient to explain preference.

Similarly, preference for flanking vertical bars flips from preference (Morris et al. 1995) to antipathy (Morris 1998a) among females in different *Xiphophorus* species. Females prefer symmetric bars (Morris 1998b) or, in older females of at least one species, asymmetric bars (Morris et al. 2006). Since it is impossible to view the lateral surfaces of both sides of a fish simultaneously, this requires a sophisticated cognitive comparison between distinct retinal representations (see chapter 5). The preference for more symmetric (or asymmetric) males, like the antipathy for swords, also cannot be explained as a function of low-level sensory stimulation.

Comparing preferences among species and individuals therefore reveals that visual preferences are multidimensional. Further insights come from manipulating multiple traits in the same experiment. For example, female *X. birchmanni* are not very choosy among available males, even though males vary in body size by a factor of three. Fisher and colleagues (2009) used computer animations to isolate male traits independently, and found that females preferred males with larger bodies but shorter dorsal fins. Among conspecific males, larger males have larger dorsal fins. The direction of

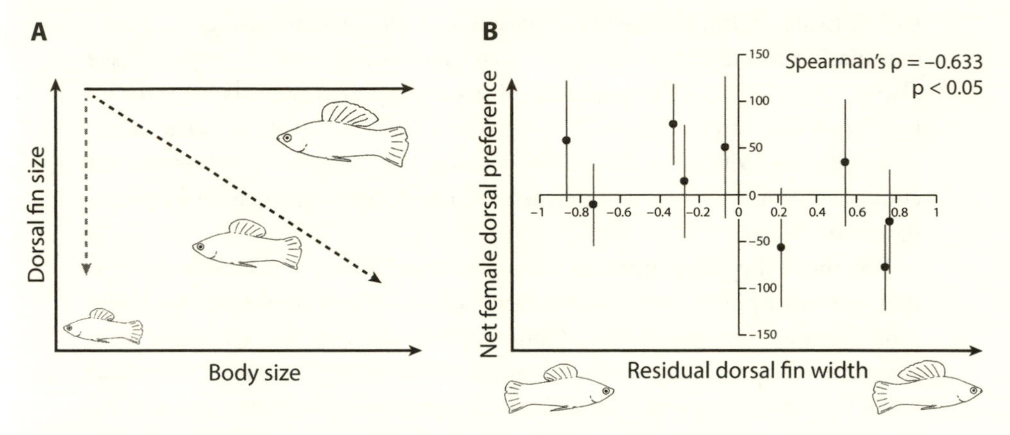

Figure 4.3. Female preference for multiple traits in *Xiphophorus birchmanni* is misaligned with male trait variation. (a) Larger males have bigger dorsal fins, but females prefer males with large bodies (thick, solid arrow) and small dorsal fins (light gray, dashed arrow) resulting in a net preference (dashed arrow) oriented perpendicular to the main axis of male trait variation. (b) As predicted by single-choice experiments, females prefer males with proportionately small dorsal fins.

female preference is thus effectively perpendicular to the prevailing axis of male trait variation (fig 4.3a). In other words, females prefer big males with small dorsal fins, but they're stuck with big males with big dorsal fins or small males with small dorsal fins. Indeed, among naturally occurring male phenotypes, females preferred males with the lowest residual of dorsal fin on body size (fig. 4.3b; Fisher et al. 2009). This result illustrates the misalignment of male traits and female preferences (chapter 15).

The two known modalities of mate choice interact nonlinearly to determine mate preference. In *Xiphophorus*, females typically have strong preferences for chemical cues of conspecifics, whereas preferences for visual traits are much less predictable. When chemical cues of conspecifics are paired with visual cues of heterospecifics, or vice versa, chemical preferences override, i.e., are **hierarchical,** over visual preferences (Crapon de Caprona & Ryan 1990; Hankison & Morris 2003).

This reframed narrative about *Xiphophorus* illustrates three general points, which I will expand on in the second half of this chapter. First, choosers attend to many different aspects of courter phenotypes, especially when we hold all else constant. Second, manipulating more than one variable shows that preferences interact non-additively; in other words, the attractiveness of one trait is dependent on another trait. Third, chooser preferences and courter traits do not map on to each other neatly in multivariate space.

4.2 MECHANISMS OF PERCEPTUAL INTEGRATION

4.2.1 Overview

In the previous section, we saw that the mapping between traits and preferences can be messy. Independent of how traits are generated by courters, however, they produce multiple streams of sensory information, which then have to be integrated by choosers. In fruit flies, a workhorse for neuroethology, we know a great deal about how stimuli are sensed and about the early levels of perceptual processing. By contrast, we know very little about how different streams of sensory information are integrated into a mating decision (fig. 4.4; Dickson 2008). Functional neuroanatomical studies are beginning to shed light on the anatomical loci of preference integration (chapter 12). In male Gouldian finches, visual mate choice is strongly lateralized, like many other cognitive tasks; occluding the right eye deprives the left hemisphere of the brain and inhibits mate choice, whereas occluding the left eye produces a similar response to when both eyes are available (Templeton et al. 2012).

In this section, I focus on case studies in vision, audition, and olfaction where multiple mechanisms interact to determine receiver response. I begin by focusing on systems involving simple stimulus combinations and small numbers of neurons, which illustrate how complex, nonlinear properties

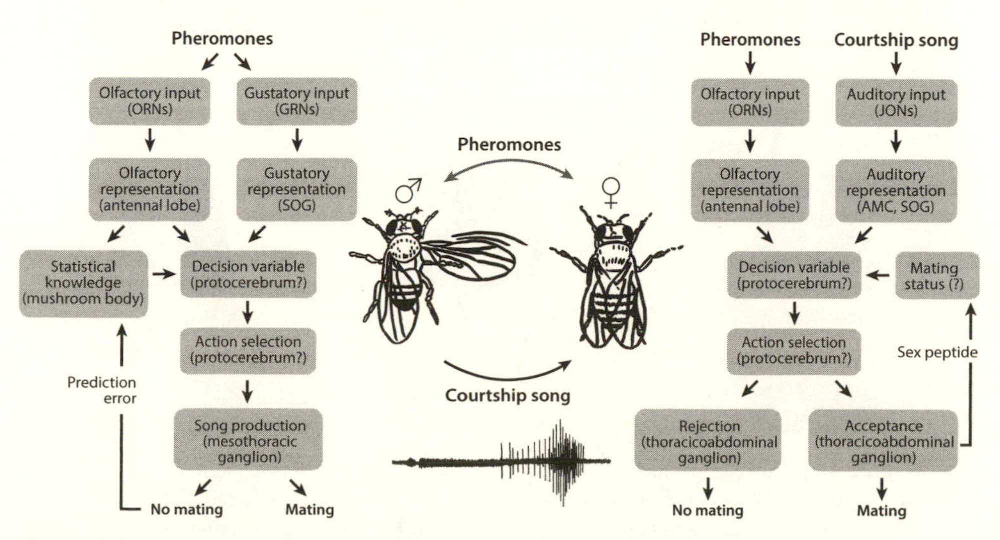

Figure 4.4. Integration of multiple sexual signals into mating decisions in male and female *Drosophila*. After Dickson (2008).

can emerge in even the most basic mechanisms underlying mate choice. I then turn to multimodal preferences, where choosers integrate information from more than one sensory channel, before discussing general properties of how multiple components of preference are integrated into unitary mate choice.

4.2.2 Color perception in vertebrates

Choosers across taxa attend to color—specifically, to the hue of a trait on courters (reviewed in Cummings & Ryan 2014). For example, female limnetic sticklebacks *Gasterosteus aculeatus* attend to male throat color. In most populations, females prefer red throats. They attend to both the brightness contrast and the chromatic contrast between the red throat patch and the background skin color (Milinski & Bakker 1992; Baube et al. 1995). That is, they prefer courters displaying radiance spectra with most of their energy concentrated around 650 nm (fig. 4.5a) over spectra that are yellower (fig. 4.5b) or pinker, i.e., less saturated (fig. 4.5c).

Radiance spectra are distributions of photons, as a function of wavelength, reaching the chooser's retina. In order for choosers to have a preference for red, they must be able to evaluate "redness" efficiently enough to also be able to perceive spatial and temporal variation in the signal (Rosenthal 2007). This means sampling the continuous, one-dimensional color spectrum with a minimal number of different types of receptor cells. The

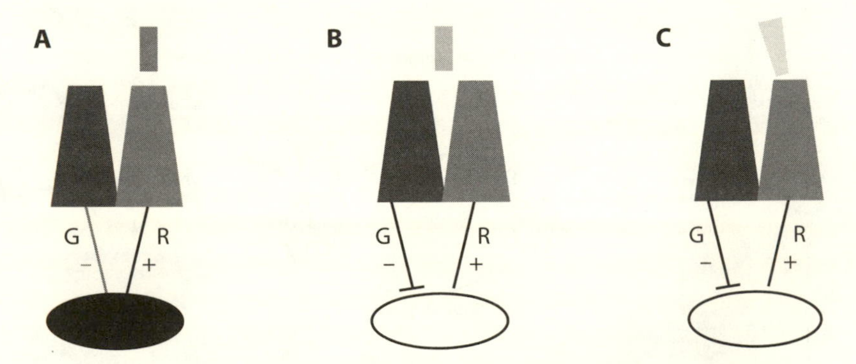

Figure 4.5. A simple mechanism for preference for redder courters explained by R+G− opponency channels. (a) Red light (darker gray) stimulates only red cone photoreceptors, which activates R+G− cell; (b) and (c) orange (lighter) and pink (lightest) light stimulate both red and green photoreceptors, resulting in both excitatory and inhibitory inputs to the opponency channel.

color experienced by most humans is transduced by three cone cell types (red, green, and blue). We'll ignore the blue cone for now and focus on the red and green cone cells, whose distinct response curves depend on five amino acids within cone opsin genes (Yokoyama 2000). To simplify somewhat, these two sensory receptors connect to a sensory interneuron called an R+G- cell, which operates as a Boolean operator, firing only in the presence of input from the red cone cell AND the absence of input from the green cell. This comparison is called the chromatic signal (Twig et al. 2003). The different courter spectra in figure 4.5 elicit different responses from this system, with the red signal eliciting the maximal response. Analogous systems exist for comparing olfactory (Meister & Bonhoeffer 2001) and auditory inputs (Gerhardt & Hobel 2005).

This minimal system for comparing sensory responses between two receptors already contains many of the important properties of multidimensional preferences. First, the response of the system is determined by the logical rule governing the integration of the two inputs. In the above example, the rule is that response is an AND function: preference is maximized when the red input is high AND the green input is low. There are also green detectors (R-G+), as well as luminosity ("brightness") detectors that act as logical OR cells, responding positively to input from any type of cone cell (fig. 4.6; Jacobs 1981).

Second, the interactions between sensory inputs produce nonlinear outputs: the chromatic signal depends on an interaction between the sensitivity curve of the green (medium-wavelength-sensitive) receptor and the sensitivity curve of the red (long-wavelength-sensitive) receptor. In evolutionary-genetic terms, this means there are epistatic interactions between the underlying sensory receptor genes. This interaction is illustrated in figure 10.8 below.

Third, the mapping between courter traits and chooser preferences is complex even at this early stage of processing. At a single point in time, a courter signal is a function of light intensity over the range of wavelengths that the chooser can detect, reflected off a three-dimensional surface. This function is generated by an interaction between morphological structures, pigment reflectances, and behavioral decisions on the part of both courter and chooser. This signal is then transduced by a courter using a two-dimensional projection onto an array of receptors sensitive to different wavelength ranges. There is no congruence between the mechanisms that a courter uses to present an iridescent blue patch in a certain orientation relative to the sun and the chooser's eye, and the mechanisms that the chooser uses to interpret the signal.

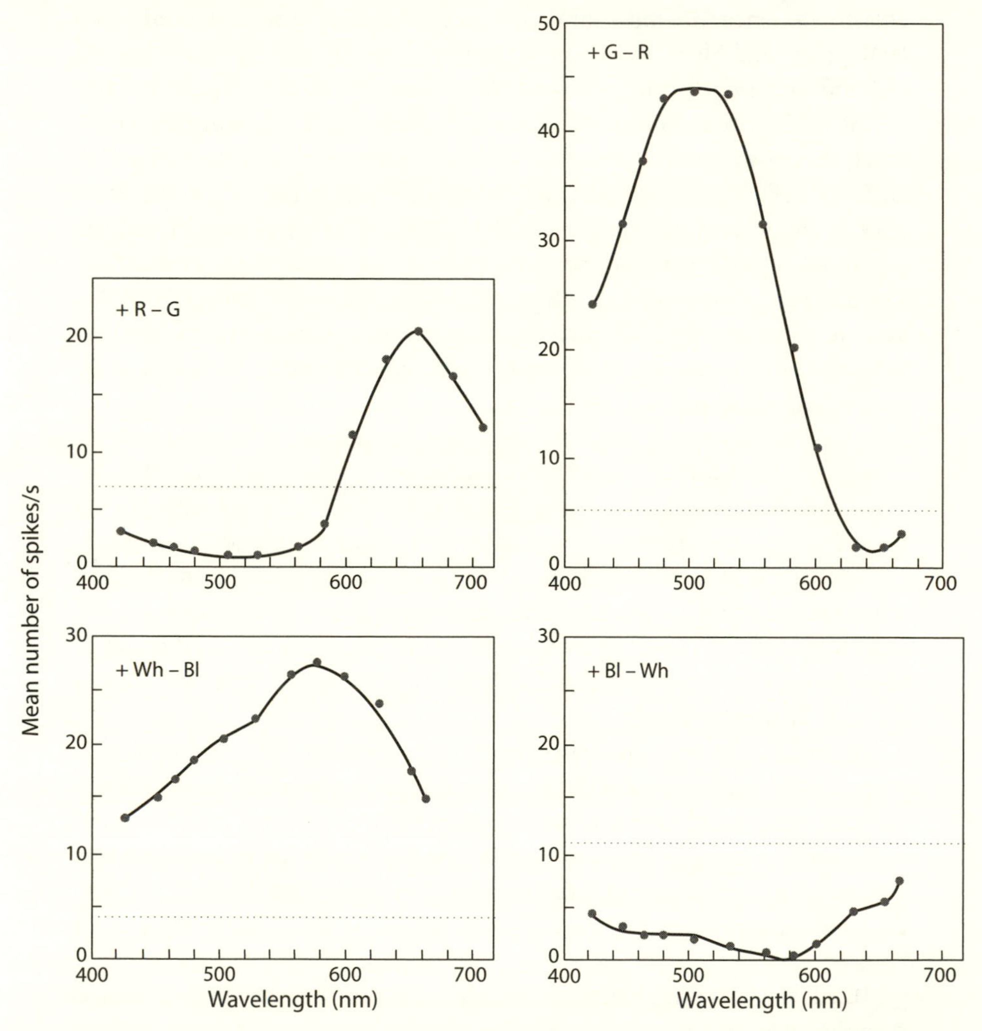

Figure 4.6. Color-opponent systems in rhesus macaques. Cells in the lateral geniculate nucleus show specific responses as a function of wavelength. Top, chromatic system with antagonistic inputs from medium- and long-wavelength cones; bottom, achromatic (luminosity) system with additive inputs from the two cone types (after Jacobs 1981; adapted from Bradbury & Vehrencamp 2011).

Fourth, it follows from the above point that more complex processing does not necessarily prevent choosers from being fooled. Just as different types of courter traits can elicit the same level of gross stimulation at the periphery (chapter 3), so can multiple combinations of traits elicit the same response. As we saw in chapter 2, multiple combinations of color and brightness (so-called metamers) can produce the same chromatic signal in the retina. This is part of how humans are fooled into displaying sexual and affiliative responses toward their LCD screens.

4.2.3. Auditory integration in female frogs

Female anurans (frogs and toads) will typically exhibit phonotaxis (chapter 2) only toward the calls of males of their own species. The distinctive features of the call of any given frog are multivariate functions of frequency and intensity over time. Features are parsed by selective neurons, starting in the periphery with hair cells rendered frequency-selective by their position in the inner ear (chapter 3). Inputs from these cells are integrated in the auditory thalamus and torus semicircularis of the brain. Some thalamic neurons in the brain function as AND gates, responding maximally when two specific frequencies are present (Mudry & Capranica 1987); these are analogous to the R+G+ cells in the retina, above. Rose and Capranica (1984) identified populations of neurons in the torus semicircularis of toads that are sensitive to different temporal properties of stimuli: high-pass, low-pass, and band-pass, which respectively filter out low frequencies, high frequencies, and sounds outside a frequency range; and band-suppression, which show decreased responses to specific pulse rates. Thus, feature-detecting neurons respond to combinatorial features of complex stimuli, and these responses in turn drive motor output—the phonotactic response to appropriate signals (Hoke et al. 2007).

Peripheral neurons can also integrate information across a perceptual spectrum. Gerhardt and Höbel (2005) showed that adding an intermediate frequency between two spectral peaks reduced the attractiveness of the call of male treefrogs *Hyla cinerea*. They argued that this response was consistent with *tone-on-tone suppression*, the widespread property of auditory neurons to show a suppressed response when an inhibitory frequency is presented at equal or greater amplitude than an excitatory frequency. This is yet another perceptual bias that can shape preferences (chapter 13).

4.2.4 Configural perception of odorant mixtures in insects

Mate choice involving olfactory communication almost always involves the integration of sensory responses to multiple odorants (with silkmoths a notable exception of an animal that responds to a single compound; Sakurai et al. 2011). In insects, receivers often exhibit configural preferences for particular ratios of pheromone components, rather than attending to the absolute concentration of individual compounds (reviewed in Smadja & Butlin 2009; Weddle et al. 2013). A human analog for configural preferences is the way we perceive perfumes; we may perceive low concentrations of perfume as attractive, but are repelled by its components of civet-gland extract and ambergris (or their synthetic analogs). Meyer and Galizia (2012) used non-sexual cues in honeybees to show that this configural perception (or *synthetic coding*) is mediated by "configural neurons" in the antennal lobe that respond only to combinations of multiple inputs from distinct classes of olfactory receptor neurons (chapter 3). The antennal lobe also contains "elemental neurons" that respond to individual odorants rather than mixtures.

The macroglomerular complex of the antennal lobe of male codling moths (*Cydia pomonella*) presents a remarkable extension of choosers integrating multiple signals, in that it responds to a configural blend of both courter signals and habitat cues (fig. 4.7). Chooser response thus depends on females not only producing a specific odorant combination, but also doing so in a specific habitat (Trona et al. 2013). This provides an elegantly simple mechanistic basis for environment-dependent preferences (chapter 11). Across taxa, receivers can switch between elemental and configural

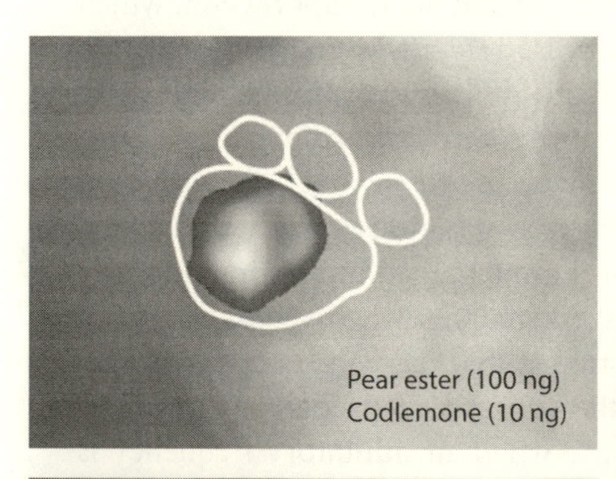

Figure 4.7. Calcium imaging of the antennal lobe of male codling moth stimulated with sex pheromone (codlemone) and/or a plant volatile (pear ester). Note enhanced reaction to both odorants in tandem. After Trona et al. (2013). Photos by Federica Trona.

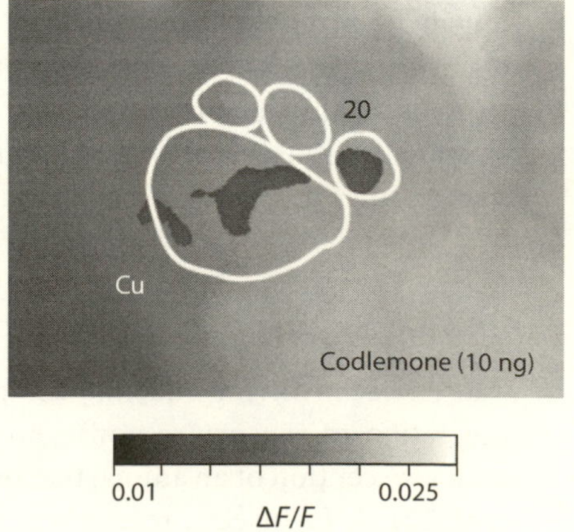

percepts depending on odorant concentrations, previous experience, and motivation, through mechanisms that are poorly understood (Jacobs 2012). Such flexibility underscores the importance of higher-level cognitive processes, experience, and motivation in the processes underlying mate choice, as discussed in subsequent chapters.

4.2.5 Multisensory integration in birds

The preceding examples have shown that complex, context-dependent preferences can emerge from interactions among a handful of neurons responding to one stimulus in one sensory modality. Yet, in animals as diverse as flies (Billeter & Levine 2013), frogs (Rosenthal et al. 2004), and fish (Crapon de Caprona & Ryan 1990) choosers attend to cues in more than one sensory modality. Multimodal courtship is perhaps most striking in birds, where courters have evolved choreographed "song and dance" combinations (Dalziell et al. 2013; Williams 2001; fig. 4.8). There has been limited mechanistic work done on the integration of auditory and visual cues in mate choice, but an immediate-early gene (IEG) expression study showed that in zebra finches, where females attend to both auditory and visual courtship cues from males, the *visual* cues of courtship alone are sufficient to elicit activity from female *auditory* forebrain regions (Avey et al. 2005). This is consistent with findings in other contexts, like spatial localization, that show that multimodal processing involves the cross-activation of sensory-specific brain regions (Giard & Peronnet 1999).

4.3 CATEGORICAL PERCEPTION

Categorical perception (fig 4.9a) deserves special mention because it is a form of perceptual processing that is widely distributed among animals and that defines our everyday human experience. The way we communicate with one another and our internal representation of the world starts with the names of things. We understand each name as its own entity and this allows us to divide up the world accordingly. We are able to name things because categorical perception underlies human speech recognition, in which continuous variation in acoustic signals is partitioned by receivers into distinct phonemes. Classic experiments by Liberman (1970) showed that acoustic intermediates between phonemes (for example, "ba" and "da") are perceived by listeners as either one or the other, with a sharp transition between the two (fig. 4.9b).

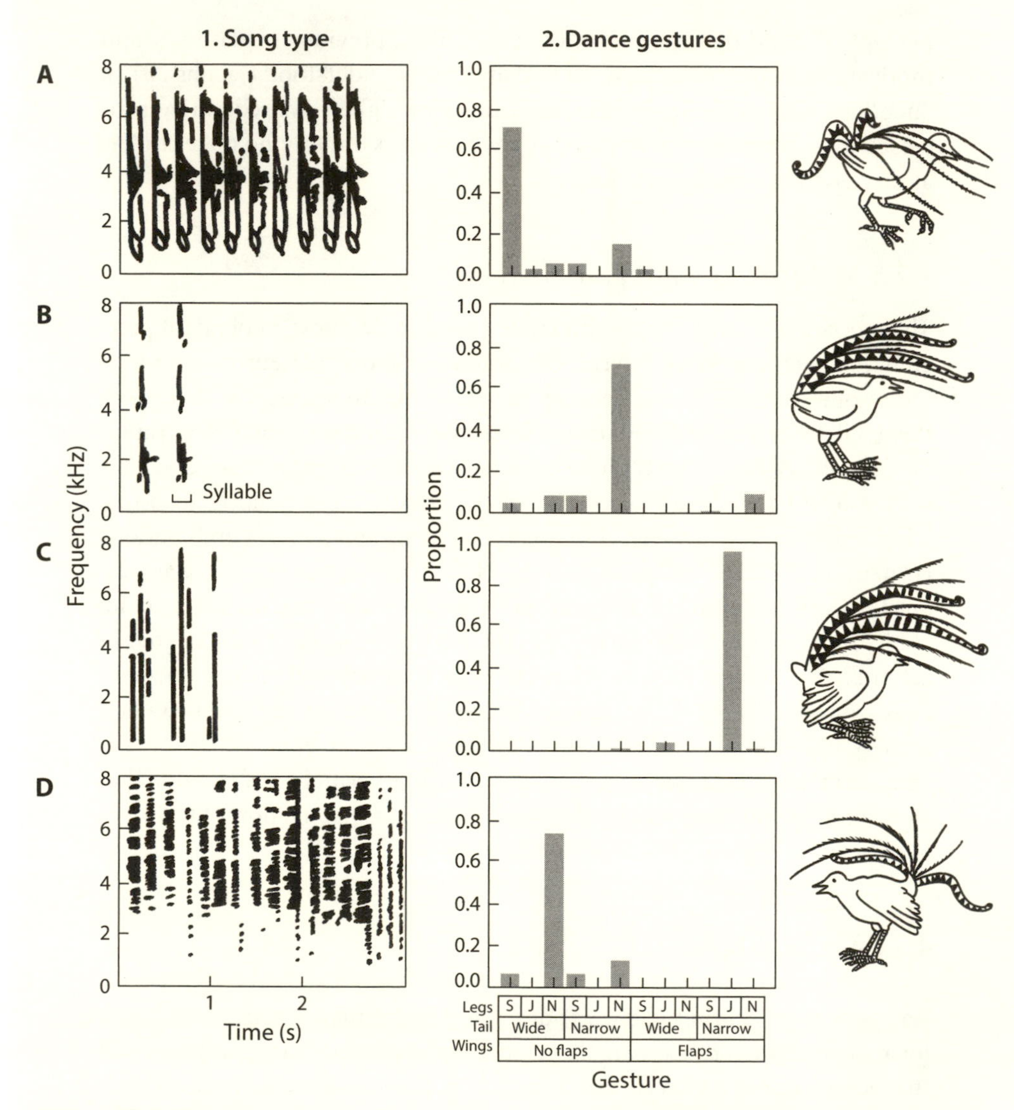

Figure 4.8. Choreographed audiovisual displays in male superb lyrebirds, *Menura novaehollandiae*. Distinct song types (left) are predictably accompanied by distinct postural combinations ("steps," "jumps," or none for the legs), flaring of the tail, and/or extension of the wings. Sonograms and charts after Dalziell et al. (2013); drawings after Mulder & Hall (2013).

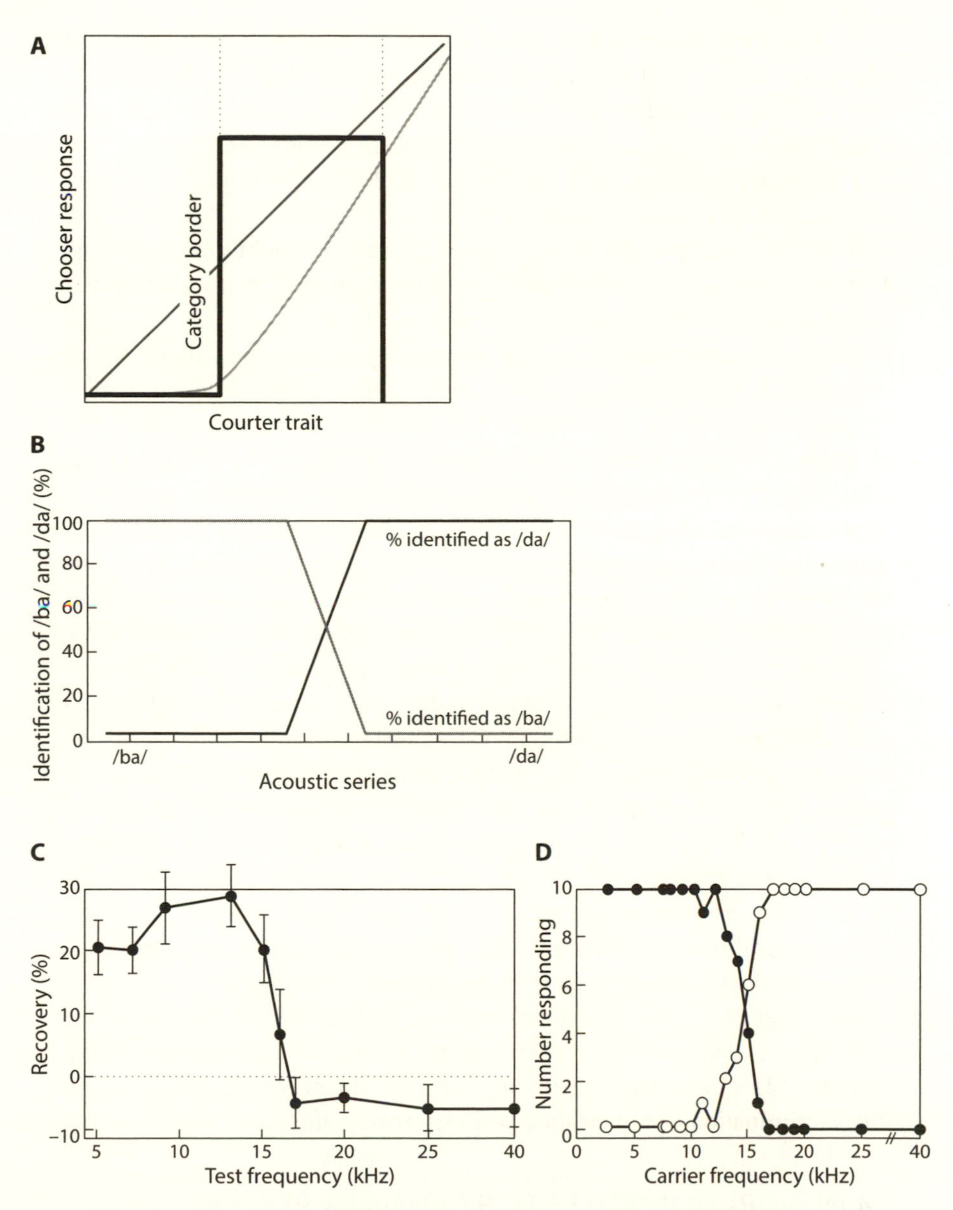

Figure 4.9. (A) Categorical perception (thick line) along a trait axis. Linear (dark gray) and threshold (light gray) responses shown for comparison. (B) The *ba-da* continuum in humans (after Liberman 1970). When "ba" and "da" syllables are acoustically morphed into one another, there is a sharp break at which they are identified as one or the other. (C) Discrimination in crickets. Female crickets do not distinguish between the 20Hz stimulus and other stimuli above 16 Hz. (D) Labeling. Females make an abrupt switch between attraction to calls (black circles) and repulsion (white circles) at around the same frequency. After Wyttenbach et al. (1996).

A categorical response to a signal differs from a continuous response (dark gray line in fig. 4.9a) in that choosers attend to signal differences on either side of a boundary (discriminating, fig. 4.9c), while discriminating much less between stimuli differing by the same amount on the same side of a boundary (labeling, fig. 4.9d; Ryan et al. 2009). It is important to emphasize that this second condition is what discriminates categorical perception from a simpler threshold response (chapter 3; light gray line in fig. 4.9a), in which signals are ignored below a critical level of stimulation but which elicit an increasing response above threshold.

There is reason to believe that categorical perception is widespread in mate choice. Rhesus macaques, chinchillas, and of course humans exhibit categorical responses to continuous variation in human phonemes (reviewed in Harnad 1987). In túngara frogs (Baugh et al. 2008), and crickets (Wyttenbach et al. 1996; fig. 4.9c–d), choosers or competitors show categorical responses to natural or modified courter advertisement calls. Categorical responses to distinct song elements (Nelson and Marler 1989) are a prerequisite for the preference of some female songbirds for an increasing number of distinct song "syllables" (Catchpole 1987; but see Byers & Kroodsma 2009).

While most animal studies have focused on acoustic signals, there is evidence for categorical perception of spatial orientation in both rats (Maki et al. 2001) and rhesus macaques (Wakita 2004). Given the ubiquity of vertical bars and horizontal stripes in animal signals (Rosenthal 2007), categorical perception of spatial patterns may play an unappreciated role in mate choice. By contrast, categorical perception of color may be unique to humans and contingent on language; nonhuman primates do not show categorical responses to color variation (Davidoff & Fagot 2010).

Categorical perception means that continuous variation in courter signals could elicit discontinuous responses from choosers and abrupt transitions in preference, and thereby drive the punctuated evolution of courter traits (Baugh et al. 2008), particularly if exacerbated by mate choice copying (chapter 12). In chapter 16, I discuss the role of categorical perception in the discrimination of conspecifics from heterospecifics.

4.4 INTEGRATION RULES FOR COMPLEX PREFERENCES

4.4.1 Overview

The mechanisms governing different aspects of preference must ultimately coalesce into a unitary evaluation of a courter. Therefore, an inherent property of mate choice is the set of rules that govern how these preferences interact. We can discern these rules experimentally by manipulating the values of two or more traits in combination and assaying chooser preference.

A convenient way of thinking about interactions between preference mechanisms is by linking them with logical (Boolean) operators (Castellano et al. 2012). In some cases, like the thalamic gates in frogs that serve to apply an AND function, we can point to neural circuits that act as Boolean operators; in others, they are at the least useful metaphors for organizing patterns of responses. Two traits may each be *sufficient, but not necessary* to elicit a mating response. This is equivalent to a logical OR operation, since a response to either trait will elicit a preference. By contrast, either trait by itself may be *necessary, but not sufficient*; preference is a logical AND depending on the values of both traits. Finally, there may be a hierarchical relationship between the two traits, whereby one is *necessary and sufficient* and the other is neither *necessary nor sufficient*, but enhances preference if the first trait is present.

These divisions allow us to classify integration rules into five categories inspired by Partan and Marler's (1999) and Partan's (2004) classification of multimodal signals (fig. 4.10).

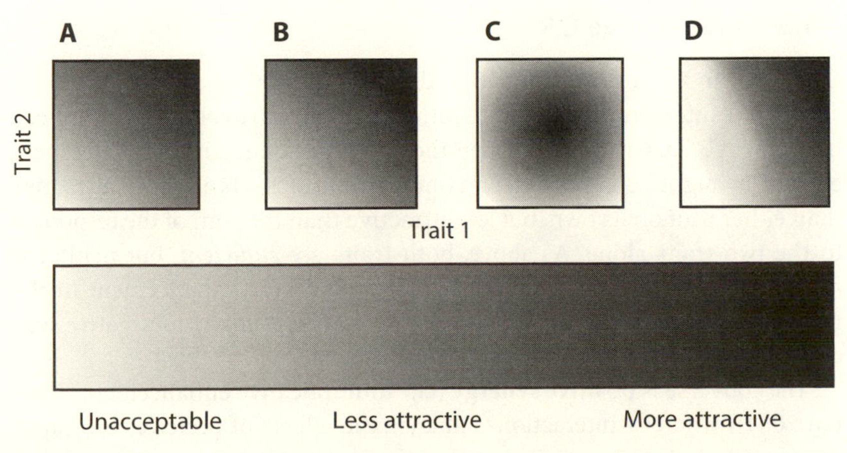

Figure 4.10. Interaction rules among preferences: (A) additive OR; (B) multiplicative OR (positive synergy shown); (C) configural AND; (D) hierarchical modulation.

4.4.2 Additive OR

Within OR rules, two preference mechanisms may integrate **additively**, with chooser preference a weighted sum of the response to each variable (Castellano et al. 2012; "summation," Partan 2004; fig 4.10a). Choosers can therefore show preferences if either or both trait variables lie within a certain range, and the value of each trait makes an independent contribution to preference. Brightness detectors in the retina are a simple example of additive integration, since they respond in proportion to input from any or all upstream photoreceptors. For example, equivalent increments of sword length or of body size are equally effective in eliciting preferences in female *X. hellerii* (Rosenthal & Evans 1998). Additive preferences for different courter traits could arise from traits addressing the same sensory mechanism, as originally surmised for *X. hellerii*, but additivity can also be a result of linear weights being assigned to inputs at higher levels of processing (Castellano et al. 2012). For example, preference mechanisms may also be additive in the closely related *X. birchmanni*; here, however, larger bodies are more attractive but swords are less attractive (Wong & Rosenthal 2006; Fisher et al. 2009). It remains to be tested whether an increase in sword length, which makes a male less attractive, can be compensated by a scalar increase in body size. For additive integration, each signal component is *sufficient, but not necessary* to elicit a preference. In other words, a courter could be equally attractive with a preferred value of one trait and a less-preferred value of the other trait, or vice versa.

4.4.3 Multiplicative OR

OR rules can also be **multiplicative**; different preference mechanisms have synergistic interactions with one another. **Negative synergy** (cf. "minor enhancement"; Partan 2004) is when there is redundancy in the preferences elicited by signal components: a combination of traits is more attractive than either trait on its own, but less attractive than the sum of the responses to the two traits alone. As above, both traits are *sufficient*, but neither is *necessary*. There are few explicit examples of minor enhancement in the mate-choice literature, perhaps because ornaments subject to negative synergy would be unprofitable for signalers.

The converse is **positive synergy** (cf. "multiplicative enhancement"; Partan 2004), whereby interactions enhance the effects of preference components, such that the two traits together elicit a stronger response than either would in isolation (fig. 4.10b). Each component is, again, *sufficient*, but not

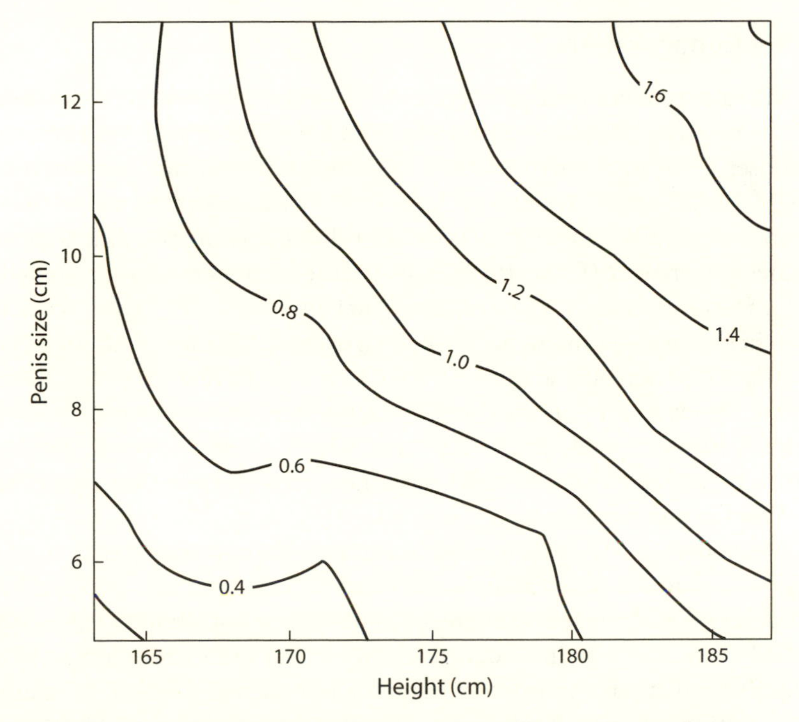

Figure 4.11. Bivariate, multiplicative preference function for male penis size and height (controlled for shoulder-to-hip ratio) in heterosexual women. 1 = mean attractiveness. After Mautz et al. (2013).

necessary. Positive synergy is common in mate choice. Female Cope's gray treefrogs, *Hyla chrysoscelis*, prefer artificial calls that maximize the product of call rate and call duration, traits that are negatively correlated within choosers (Ward et al. 2013; chapter 15). In humans, women prefer both taller men, and men with larger penises, but the effect of penis size on attractiveness is stronger in taller men (Mautz et al. 2013; fig. 4.11).

Weber's law suggests that positive synergy (along with hierarchical modulation, see below) should be common. This is because, for courters, Weber's law imposes diminishing marginal returns on increasing ornamentation. Adding a novel trait may therefore elicit a stronger preference than increasing the value of the original trait (Akre & Johnsen 2014). This is also true if the first trait elicits a categorical response, such that increasing trait value would not produce a concomitant response from choosers (i.e., a smaller and a larger trait on the same side of the category boundary). In both of these cases, adding a novel trait might make a courter more attractive than increasing the magnitude of the original trait.

4.4.4 Configural AND

AND rules produce configural preferences, whereby specific combinations of the two trait values are required to elicit a positive or negative response; chooser behavior is based on the *relationship* between two or more traits (fig. 4.10c). Configural preferences require the nonlinear integration of more than one stimulus input: it is the relationship between stimuli that determines preferences (Rowe 1999). This is true of preferences for complex odor mixtures, acoustic phrases, and visual displays. Face (*figura*) recognition in humans is perhaps the best known example (Maurer et al. 2002), but configural preferences are ubiquitous, from insect preferences for pheromone blends to the "uncanny valley" in humans. In other words, courter attractiveness depends on trait 2 being within a narrow range of values given trait 1, and vice-versa: choosers show preference for the whole, but not for the individual parts.

Without thinking about it, most of the time when we manipulate courter signals we are changing multiple variables from the chooser's perspectives. Choosers prefer courters with precise combinations of stimulus values. So-called sex recognition, individual recognition, and species recognition often require sophisticated integration of multiple context-dependent streams of information. As I will discuss in later chapters, these configural preferences are key to the role of mate choice in reproductive isolation and to the diversification of courter signals across species.

There has been limited neuroethological work on the mechanistic basis of configural preferences, with some notable exceptions in songbirds. Woolley and Doupe (2008) identified two distinct brain regions in the auditory telencephalon of female zebra finches that process preferences for complex traits. Males produce "directed" songs to females, and "undirected" songs alone, which vary subtly in their acoustic properties. Females prefer the directed song of their mate over his undirected song, largely on the basis of a highly configural trait: within-song variability in a particular type of song syllable. Compared to females that hear their mate's undirected songs, females that hear their mate's directed song have higher IEG expression in the caudomedial mesopallium (CMM) of the auditory telencephalon. Females also show behavioral preferences for the directed songs of their mates over those of unfamiliar males, but the CMM is equally active when females are presented with unfamiliar directed songs. Meanwhile, the caudomedial nidopallium (NCM) responds to familiarity but not to whether the song is directed or undirected. Two distinct anatomical regions are therefore recruited to evaluate signals along subtle multivariate axes of variation. Intriguingly, a previous study in starlings had suggested that these structures in turn contain distinct systems (associated with distinct IEG markers) for processing novel songs depending on the acoustic features of previously experienced stimuli (Sockman et al. 2005; chapter 12).

4.4.5 Hierarchical modulation

Hierarchical modulation (fig. 4.10d; "modulation"; Partan & Marler 1999) is another widespread integration rule. For example, female túngara frogs will only mate with a male if he produces a species-typical "whine" call; adding a "chuck" after the whine increases the call's attractiveness. While a chuck can "rescue" a marginally attractive whine, a chuck by itself fails to elicit a response (Phelps et al. 2006). Similarly, female gray treefrogs *Hyla versicolor* prefer calls of longer duration over calls of shorter duration, but adding an animation of vocal sac inflation makes short-duration calls as attractive as long-duration ones (Reichert & Höbel 2015). Addressing the first mechanism is both *necessary and sufficient* to elicit a preference, with the second mechanism (neither *necessary nor sufficient*) acting to modulate the first. In other words, courters need to express trait 1 within a certain range in order to be recognized as mates, but high values of trait 2 can increase a courter's attractiveness. Hierarchical effects are the norm when choice happens sequentially, since signals produced at later stages will not be evaluated if choosers reject courters based on earlier signals (Uy & Safran 2013). Castellano (2009a) points out that multiplicative rules (above) could produce similar responses to hierarchical modulation (for example prioritizing species-typical cues over other cues of attractiveness). Hierarchical modulation is also related to configural AND preferences, because traits in combination elicit a better response, although one of the traits is not necessary to elicit a response.

4.4.6 Integration rules: summary

Integration rules are perhaps an unappreciated axis of preference variation, particularly since environmental noise can alter the costs and benefits of attending to particular traits (Bro-Jørgensen 2010). Integration rules may represent an important and unappreciated avenue for diversification of preferences: indeed, among wolf spiders, combining varying visual and vibrational cues yields markedly different results across species (Hebets & Papaj 2005; fig. 4.12). Configural and hierarchically modulated preferences, where

Species	Male foreleg morphology	Courtship	Visual	Vibration	Visual + vibration	Source
S. rovneri		Unimodal	.37	.79	.89	Scheffer et al., 1996
S. duplex		Unimodal	.00	.89	.89	Hebets & Uetz, 1999
S. uetzi		Bimodal	.25	.71	.50	Hebets & Uetz, 1999
S. stridulans		Bimodal	.50	.63	.63	Hebets & Uetz, 1999
S. crassipes		Bimodal	.50	.50	.75	Hebets & Uetz, 1999
S. ocreata		Bimodal	.64	.69	.93	Scheffer et al., 1996

Figure 4.12. Variation in integration rules for vibratory and visual displays in *Schizocosa* wolf spiders. Response patterns are consistent with negative synergy in *rovneri, duplex, stridulans,* and *ocreata,* and configural preferences in *uetzi* and *crassipes. Numbers indicate proportion of females responding to listed display components.* After Hebets & Papaj (2005).

it is the nonlinear interaction of trait values that determines preference, are probably the norm, and by their very nature involve epistatic interactions among associated genes. Elucidating these rules, and gaining a better understanding of their underlying mechanisms and phylogenetic distribution, should prove invaluable in understanding mate choice.

4.5 SYNTHESIS: COMPLEX PREFERENCES AS INTEGRATED PHENOTYPES

4.5.1 Illusions

Just as with lower-level sensory processes, the network of interactions that makes up configural preferences can be fooled. Túngara frogs attend primarily to male auditory signals, but the synchronized inflation of the vocal sac enhances preference and likely facilitates localization (Rosenthal et al. 2004). Taylor and Ryan (2013) showed that a large silent interval between the whine and the chuck in túngara frogs, which normally would eliminate the chuck's enhancement of attractiveness through hierarchical modulation, could be "perceptually rescued" by a signal in a different modality, the visual stimulus of an inflating vocal sac. Kelley and Endler (2012a) showed that courting great bowerbird males create a forced perspective illusion by creating a texture gradient (larger to smaller) of objects they collected in the environment. Reversing the gradient led males to restore the original structure; notably, males reverted the experimental manipulation even for experimentally improved illusions, leading to stable individual differences in the effectiveness of the illusion (Kelley & Endler 2012b). Illusions may be important to mate assessment (chapter 6); Kelley and Kelley (2014) catalog a number of other illusions that affect the perceived size and conspicuousness of ornaments and may therefore play an important role in preference.

4.5.2 Attention

Perceptual attention plays a fundamental role in determining whether a chooser evaluates one courter rather than another (chapter 6), and in determining how choosers evaluate multiple traits within each courter. Attention is important to how integration rules operate; some signal components may function mainly to elicit attention from a receiver (Partan 2013). If attentional components function simply to elicit attention, they should be the target of permissive preferences that modulate, say, a configural preference for another set of traits.

Attention should also switch between modalities depending on environmental challenges to sensation (Bro-Jørgensen 2010). For example, female wolf spiders have logical OR preferences for male substrate-borne vibrations, whose propagation depends heavily on the physics of the substrate, and male visual displays, whose propagation depends on line-of-sight. By attending to either stimulus, choosers can evaluate courters in heterogeneous environments (Uetz et al. 2013). The role of attention in mate choice is an exciting new area of research, with gaze-tracking studies in humans (Fromberger et al. 2012) and peacocks (Yorzinski et al. 2013) providing a promising way forward.

4.5.3 Correlations among preferences

As seen in the preceding sections, preferences can involve a complex network of interactions among perceptual networks within an individual. An important evolutionary question is the extent to which these interactions are constrained by one another. In some cases, multiple aspects of mating preferences address different male traits and are uncorrelated with one another (fig. 4.13a), can respond independently to selection (Brooks & Couldridge

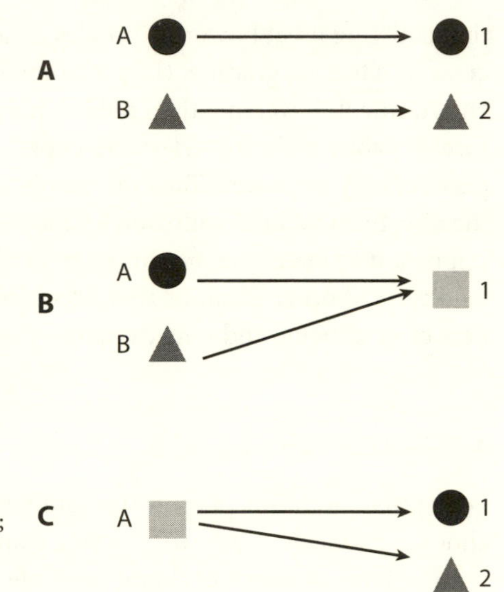

Figure 4.13. Possible relationships among preferences and traits. (a) Independent preferences (1 and 2) for multiple traits (A and B); (b) the same preference mechanisms, or correlated mechanisms, addressing different traits; (c) uncorrelated preferences for the same trait.

1999; Tomlinson & O'Donald 1989), and can evolve independently of one another (Wong & Rosenthal 2006).

Correlated preferences (fig. 4.13b), in which an individual chooser's preference with respect to one set of stimuli is correlated with its preference with respect to another, can arise in a number of ways. In the extreme, the same perceptual mechanisms may be stimulated by different courter signals: an individual's preference for red pterin coloration is the same as her preference for red carotenoid coloration. The same genes may influence seemingly distinct tasks, even in different modalities (Frenzel et al. 2012). **Physical linkage** (as with moth pheromone receptor genes; Gould et al. 2010) or statistical correlations among underlying genetic loci can also generate correlations between functionally independent preferences (chapter 10). Conversely, multiple preference mechanisms may address the same courter trait (fig. 4.13c). Choosers assess both chromatic and brightness contrast, for example, in responding to variation in carotenoid pigment. There are also trade-offs inherent in the design of any perceptual system, for example between spatial acuity and photic sensitivity in vision (chapter 3).

4.5.4 Evolutionary ecology and the mechanisms of preference integration

Mating preferences recruit multiple sensory, perceptual, and cognitive processes. Even when choosers seem to attend to fairly simple variations in courter phenotypes, the network of underlying mechanisms may be quite complex. For example, preferences for specific syllabic arrangements of birdsong (Vallet & Kreutzer 1995) require, at a minimum, mechanisms for encoding the frequency and temporal structure of courter signals within and across song syllables. It is the integration of multiple mechanisms that ultimately accounts for such phenomena as the "uncanny valley," whereby intermediate stimuli are perceived as less attractive than two extremes (Mori 1970; Matsuda et al. 2012).

Multivariate preferences involve the assignment of different weights to different sensory inputs, and a spectrum of rules governing how these inputs are integrated into a behavioral response. While the sensory periphery may often be constrained by the broad sensory environment (chapter 3), the weighting and integration of multiple stimuli can evolve, and vary according to environmental context (Cole & Endler 2015; chapter 11) and between the sexes. For example, women give much more weight to odor cues when rating a partner's attractiveness than men do (Herz & Cahill 1997).

If we are interested in the evolution of complex preferences, a very basic question is whether the marginal net benefit of evaluating an additional trait is positive or negative. Theoretical models (e.g., Pomiankowski & Iwasa 1993, 1998) assume that there is at best no cost to assessing multiple courter signals. Rowe (1999), however, pointed out the vast evidence from the psychology literature to the effect that multicomponent signals can be more easily detected and remembered, particularly if they covary in a redundant fashion. The marginal costs of multicomponent assessment may be a driver of covariance among courter traits.

Compared to the vast literature on the sensory basis of mate choice, scant attention has been paid to the evolutionary ecology of higher-order processing in mate choice (but see Ryan et al. 2009; Kelley & Kelley 2014; Akre & Johnsen 2014). This is perhaps not surprising, since characterizing integration rules requires factorially more behavioral comparisons than measuring univariate preference functions, and much larger sample sizes. Nevertheless, studying the evolution and ecology of stimulus integration is essential to understanding mate choice. Mechanistic constraints, divergent selective pressures, and basic microevolutionary pressures each make it difficult for chooser preferences and courter traits to align in multivariate space. As I will revisit in chapter 15, this both dampens the effect of sexual selection and facilitates the evolution of novel courter traits.

The integration of multiple streams of information implies that they are assigned a **hedonic value**: positive, neutral, or negative. Integration rules therefore incorporate these hedonic assignments into a unitary preference. In the next chapter, I discuss the flexibility and centrality of hedonic value in structuring mating decisions.

4.6 ADDITIONAL READING

Akre, Karin L., & Sönke Johnsen. 2014. Psychophysics and the evolution of behavior. *Trends in Ecology & Evolution* 29:291–300.

Hebets, E. A., and D. R. Papaj. 2005. Complex signal function: developing a framework of testable hypotheses *Behavioral Ecology and Sociobiology* 57:197–214.

Rowe, Candy. 1999. Receiver psychology and the evolution of multicomponent signals. *Animal Behaviour* 58: 921–931.

CHAPTER 5

Aesthetics and Evaluation in Mate Choice

5.1. INTRODUCTION: "A TASTE FOR THE BEAUTIFUL"

> On the whole, birds appear to be the most aesthetic of all animals, excepting of course for man, and they have nearly the same taste for the beautiful as we have. This is shewn by our enjoyment of the singing of birds, and by our women, both civilized and savage, decking their heads with borrowed plumes, and using gems which are hardly more coloured than the naked skins and wattles of certain birds.
>
> —(Darwin 1871, p. 697)

> [Hair] on the face is considered by the North American Indians "as very vulgar," and every hair is carefully eradicated. (...) On the other hand, bearded races admire and greatly value their beards. (...) We thus see how widely the different races of man differ in their taste for the beautiful. (p. 888)

> We may admit that taste is fluctuating, but it is not quite arbitrary. (p. 814)

The previous two chapters focused on how choosers integrate input from multiple courter stimuli. This chapter deals with evaluation, whereby different stimulus combinations elicit different hedonic values: in the context of mate choice, they increase or decrease the probability of a mating decision by a chooser. Recognition, usually invoked in the context of "conspecific mate recognition" or "partner recognition," is a special case of evaluation in which positive hedonic value is assigned to certain classes of individuals or certain stimulus envelopes (chapter 2). Evaluation means that a given courter will respond positively to some stimuli and negatively to others.

Darwin invokes choosers' "taste for the beautiful" as a primary agent of sexual selection, and relies on the words *beauty* and *beautiful* throughout the *Descent of Man*. For the purposes of thinking about mate choice, we can be satisfied with a circular definition of beauty: a courter signal is "beautiful" insofar as it makes a chooser more likely to mate with, or invest in, a courter bearing that signal.

But Darwin is alluding to something deeper insofar as what makes something beautiful to a chooser, as the quotes above suggest. He often pairs "beauty" with "taste," which conveys something fleeting and subjective, but Darwin's own "taste" was clearly awakened in his rhapsodic descriptions of avian courtship displays, suggesting a more universal appeal. Darwin's ambiguity here reflects a millennial debate about the nature of beauty, which can be simplified into a continuum between a Platonic or objectivist view, whereby certain stimuli are inherently pleasurable to any chooser—"not quite arbitrary," and a Sophist or subjectivist view, whereby the hedonic value of a stimulus depends on the internal, "fluctuating" properties of the chooser (Reber et al. 2004).

Scholars of mate choice have overwhelmingly taken the objectivist view, leading to a kind of search for a universal aesthetics of mate choice with a heavy emphasis on sensory stimulation (chapter 3). In this view, beauty is also Truth, or at least Quality, with a sort of universal aesthetics that favors more viable courters. Yet, as Darwin noted, the same stimulus can elicit polar-opposite responses in different human populations, and even at different times within populations. Clearly, then, the extreme objectivist view of beauty is incorrect in the context of mate choice. As will become clear in upcoming chapters, the taste for the beautiful is indeed fluctuating both within and among individuals.

In the previous chapter, I described how sensory inputs from various stimulus features can combine in non-intuitive ways to determine preference. A central property of this process is that responses are not simply increasing functions of peripheral stimulation. In this chapter, I focus on how some stimuli are hedonically marked as attractive, while others are marked as unattractive. Perceptually salient stimuli can elicit positive, neutral, or negative responses—indeed, the same stimulus can elicit very different responses among receivers who can all detect it equally well.

Importantly, evaluative responses go both ways. As Darwin noted, a chooser may often choose not the "most attractive" courter but the "least distasteful." A "preference" for A over B may reflect not only an attraction to A, but an aversion to B, and the two are the result of very different underlying mechanisms. We must therefore consider not only the "taste for the beautiful" but also the "aversion to the ugly." The aversion to the ugly may be the more important of the two mechanisms, because of the downside risk of mating with inappropriate mates (chapter 14). For example, Clemens and colleagues (2014) found that female grasshoppers *Chorthippus biguttulus* assigned far greater negative weight to unattractive call components than the positive weight they assigned to attractive components.

Finally, I suggest that preference evolution is at least as likely to involve evaluative mechanisms as opposed to the sensory periphery, perhaps leading to abrupt discontinuities in preferences among choosers.

5.2 UNIVERSALS OF BEAUTY?

5.2.1 Symmetry

Is there a shared "taste for the beautiful"? The notion that certain configural relationships are inherently and universally more beautiful than others is an old one: Aristotle wrote that "the chief forms of beauty are order and symmetry and definiteness." Much like the dogma that bigger, brighter, and louder signals indicate better mates, the convergence of philosophical objectivism with "good genes" thinking has led to a search for transcendent properties of courter signals that indicate "quality." Chief among these is bilateral symmetry. The original argument, made by Møller (1990) and others, was that fluctuating asymmetry—deviation from a symmetrical average—is an indicator of developmental instability, such that more symmetrical courters are in better condition and have better genes. Accordingly, there should be broadly shared chooser preferences for symmetry: Beauty is Truth. A slew of studies in the 1990s seemed to provide support for this notion, documenting correlations in courters between symmetry and measures of fitness, and showing that choosers preferred to mate with more symmetrical individuals.

In the late 1990s, support for fluctuating asymmetry as a universal target of mate choice began to waver, and detailed meta-analyses by Palmer (2000) and Clarke (1998) called into question the "good genes" aspect of this story (chapter 15), with the putative relationship between asymmetry and fitness-related traits in courters (let alone fitness benefits for choosers) appearing to be largely an artifact of selective reporting and confirmation bias. Further, experimenters often found effects only when asymmetries were extended to extremes outside the range of natural variation among courters. Empirical studies began to accumulate, showing no preference for symmetry or preferences for asymmetry (Tomkins & Simmons 1998). The heritability of symmetry and other putative developmental instabilities is also weakly heritable and therefore an unlikely source of indirect fitness benefits (Polak 2008). These issues—biased reporting of relationships between traits and fitness, the conflation of courter fitness with chooser fitness, and exaggerated manipulation of putative fitness-related traits—pose a general problem when interpreting studies aiming to test "good genes" models of mate choice evolution (chapter 15).

Given the weak evidence for adaptive advantages of symmetry preferences, are there mechanistic constraints that might explain widespread preferences for symmetric traits? Enquist and Arak (1994) showed that preferences for symmetry could emerge from easier recognition of self-similar images when viewed in different orientations and positions (Kirkpatrick & Rosenthal 1994). This may therefore be an instance of correlated traits (in this case left and right halves of a visual stimulus) simply being more detectable (chapter 4; Bullock & Cliff 1997). The greater detectability of symmetry may have a simple mechanistic basis in Weber's law (chapter 3) as explained by Shettleworth (1999), as follows. In animals like fish, any information about symmetry must come from a sequential comparison of the number of elements on two sides of the courter's body. But a sequential presentation of symmetrically distributed elements will always produce greater sensory stimulation than one of the same number of asymmetric distributed elements. If there are n elements displayed on each side of a symmetrical courter, the response predicted from Weber's law is proportional to $\log n + \log n = 2\log n$; a mismatched pair, by contrast, will always elicit a lower response—at most $\log (n - 1) + \log(n+1)$, which is always smaller than $2 \log n$.

Symmetry therefore appears to generate stronger receiver responses as a result of basic constraints on perception. But greater sensory response does not automatically translate into greater preference. Indeed, female paradise whydahs, *Vidua paradisaea*, prefer experimentally manipulated males with more asymmetric tail feathers (Oakes & Barnard 1994). In two species of swordtails, preference for symmetry versus asymmetry appears to reverse itself as a chooser ages, with younger females preferring symmetric signals and older females preferring asymmetric signals (Morris et al. 2006). And while humans are drawn to symmetric faces bearing neutral expressions, symmetric faces showing symmetric emotional expressions fall into the "uncanny valley" (Kowner 1996). While symmetry preference may reflect mere detectability, reversals of symmetry preference do not. Even though symmetry detection may be universal among creatures with image-forming eyes, symmetry isn't beautiful for everyone.

5.2.2 Consonance

An acoustic parallel to symmetry is the search for musical universals. Darwin saw universal beauty in music: "The perception, if not the enjoyment, of musical cadences and of rhythm is probably common to all animals, and no doubt depends on the common physiological nature of their nervous systems." However, recent evidence suggests that one important aspect of music perception, synchronizing to a beat, is restricted to humans and perhaps a subset of mammalian and bird species (Patel 2014).

A second feature of music is the configural relationship of acoustic features to one another. Combinations of frequencies in small-integer harmonic ratios with one another, termed *consonant*, feature prominently in European classical music and contrast with *dissonant* combinations of tones characteristic of jazz, rock and roll, and other musical forms. The mechanistic argument for greater detectability of consonant sounds is that these frequencies are likely to address the same *critical band* in the basilar membrane (Plomp & Levelt 1965; chapter 3); a more recent study suggested that consonant frequencies could elicit synchronized activity from coupled oscillators (Shapira Lots & Stone 2008).

As with symmetry, however, preference for consonance is not universal. Two-month-old babies (Trainor et al. 2002) and newly hatched domestic chicks (Chiandetti & Vallortigara 2011) show spontaneous preferences for consonant tones. Túngara frogs, by contrast, fail to prefer low harmonic ratios, leading Akre and colleagues (2014) to suggest that the harmonic structure of many animal calls might be due to biomechanical constraints on production rather than selection favoring consonance. Starlings (Hulse et al. 1995) and tamarin monkeys (McDermott & Hauser 2004) perceive the difference between consonant and dissonant sounds, but fail to show a preference. Just as with symmetry, consonance is a configural property of acoustic stimuli that many choosers can attend to when evaluating courters, and that is probably more inherently detectable than other combinations of stimuli (Trainor 2008). Yet the *preference* for consonance varies in its attractiveness among species, and, as evidenced by the spectrum of human musical tastes, among and within individuals of the same species.

5.3 DETECTION AND EVALUATION AS DISTINCT COMPONENTS OF MATE CHOICE

The examples above underscore the point that preference is distinct from mere detectability. Our everyday experience is shaped by stimuli that provoke a hedonic response—that is, we evaluate them as positive or negative. It is intuitive that hedonic response is not simply an increasing function of our ability to detect these stimuli. Many things are highly salient but unattractive or aversive: the sound of car alarms, the smell of feces (but see Pynchon 1973; chapter 17), the pain from a stubbed toe. Evaluation is the process between detection and a behavioral response, whereby choosers assign a subjective value of attractiveness to a stimulus. The hedonic value of a stimulus—positive, neutral, or negative—can vary among choosers even if they all detect the stimulus equally.

Despite its obvious importance, evaluation has been given scant treatment in the mate-choice literature relative to the plethora of studies on the sensory periphery. Studies of sexual selection have generally assumed that courter conspicuousness—or detectability—is of a piece with attractiveness to choosers, and with courter "quality" (chapter 14). Yet while there is surely an overall trend for more conspicuous traits to be more attractive (Ryan & Keddy-Hector 1992), there are quite a few examples of less conspicuous courters being more attractive to choosers (Griffith et al. 1999; Wong & Rosenthal 2006; Morris 1998a; Saetre et al. 1997). We can say that a courter signal has positive hedonic value (or "beauty") to a chooser if it increases her probability of mating or investing in the courter's offspring, and vice versa for negative hedonic value. Figure 5.1 shows the relationship between conspicuousness and hedonic value and the resulting consequences for behavioral preferences.

Hedonic value is extremely labile at multiple levels (see section 5.6 below). Among species, female *Xiphophorus hellerii* show increasing preferences for male "sword" ornaments (Basolo 1990), while swords make males less attractive to female *X. birchmanni* (Wong & Rosenthal 2006; fig. 5.2a). Among populations, female house sparrows in most populations prefer males with larger throat badges, while females in one island population prefer males with smaller badges (Griffith et al. 1999). Among individuals, some genotypes of male *Drosophila melanogaster* prefer females with pheromones, some favor females without pheromones, and some show no preference (Pischedda et al. 2014; fig. 5.2b). Such genetic variation in valence can respond to selection: Wilkinson and Reillo (1994) imposed artificial selection on longer and shorter eye span in male stalk-eyed flies, and found that fe-

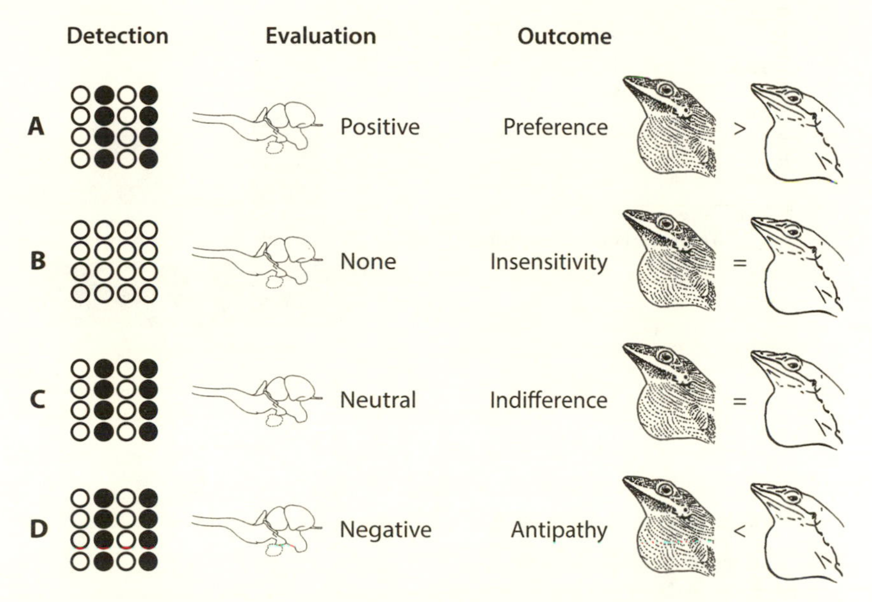

Figure 5.1. Detection and evaluation. In this hypothetical lizard species, males express either orange (stippled) or blue (white) color on the dewlap. The right-hand panel shows chooser preferences as a function of courter color. (a) **preference**: orange pigment preferentially stimulates long-wavelength-sensitive photoreceptor cells in the cone mosaic of theretina of a female. This sensory stimulation is assigned positive hedonic value in the brain and results in a preference for orange. (b) **insensitivity:** choosers lack the sensory capacity to express a preference, for example lack of chromatic vision. Therefore, they choose at random with respect to color. (c) **indifference:** Choosers perceive color differences among courters but do not assign them hedonic value, and therefore also choose at random. (d) **antipathy**: stimulation of long-wavelength cones elicits an aversive response and therefore a preference for blue males.

males in the short-eyed line reversed the wild-type preference (fig. 5.2c). Shifts in valence are therefore a candidate for mechanisms of preference divergence as a consequence of sexual conflict (chapter 15), and that can lead to speciation (Turner & Burrows 1995; chapter 16).

Within individuals, both female satin bowerbirds (Coleman et al. 2004) and female swordtails (Morris et al. 2006; fig. 5.2d) change the sign of their preferences with age, such that the least-preferred trait values when young are the most-preferred when older. Social experience can have particularly dramatic effects on preferences (chapter 12): for example, male zebra finches reverse their preferences for female beak color depending on early experience (ten Cate et al. 2006; fig. 5.2e), and there are widespread examples of choosers reversing their preferences after observing another chooser with a

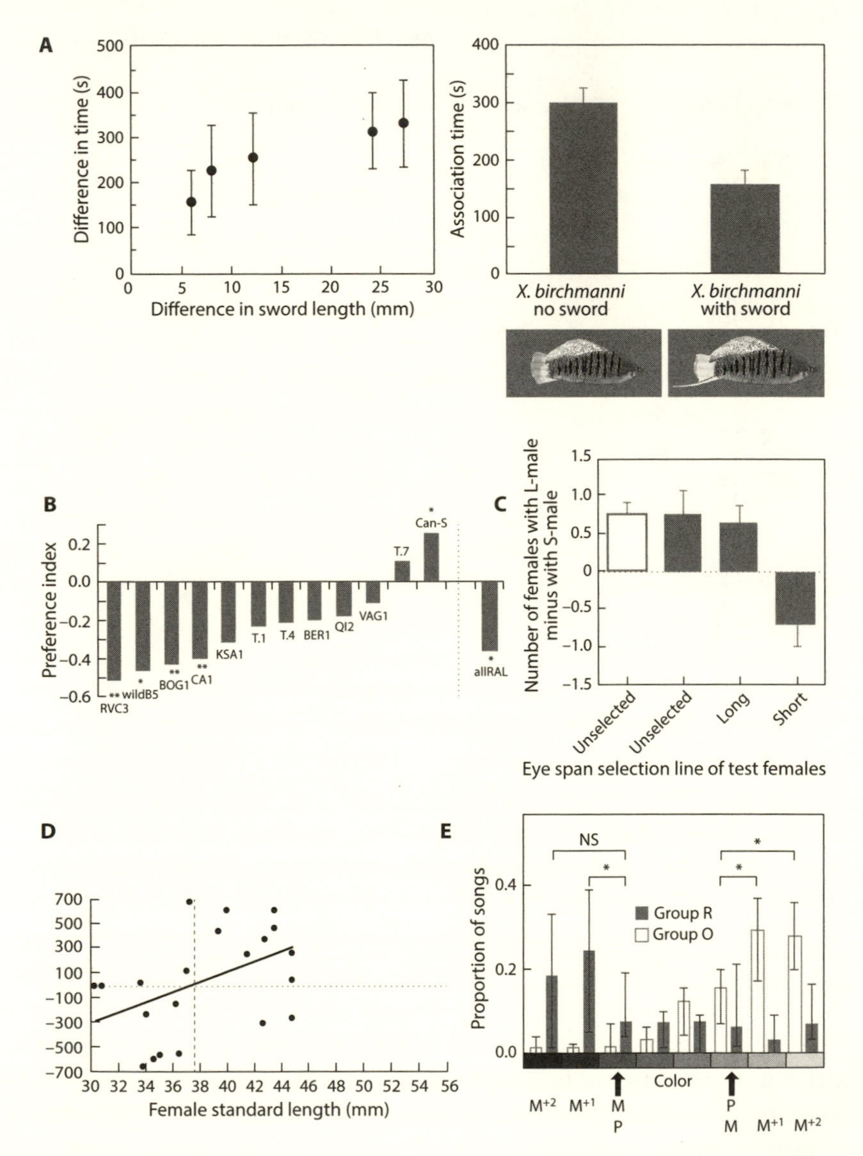

Figure 5.2. Hedonic reversals in mating preferences. (A) Between species: *Xiphophorus hellerii* females prefer males with longer "sword" ornaments (Basolo 1990a), *X. birchmanni* prefer males with no sword (Wong & Rosenthal 2006); (B) Within species: some inbred genotypes of *Drosophila melanogaster* prefer females lacking cuticular hydrocarbons (positive values; Pischedda et al. 2014); (C) Preferences reverse in response to selection; Y axis shows female preference for long-eyespan stalk-eyed flies in unselected lines and lines selected for longer and shorter male stalk length (Wilkinson & Reillo 1994); (D) Preferences reverse with age; positive values show stronger preferences of *X. malinche* for males with asymmetric males (Morris et al. 2006); (E) Preferences reverse with social experience; males reared with red-beaked mothers direct more songs at red-beaked than orange-beaked females, and vice versa. Note peak shift favoring extreme values over the training set (chapter 12; ten Cate et al. 2006).

previously unpreferred courter (mate copying; chapter 12). While evaluation and perception are related (section 5.5), evaluation is variable and can be decoupled from upstream perceptual processes.

How does evaluation determine the relationship between stimulus input and behavioral response? In general, stimuli will have positive hedonic value if they stimulate mechanisms that increase mating probability and/or inhibit mechanisms that decrease mating probability, and vice versa. Attractive traits are attractive because they elicit a positive preference—that is, a sexual response—from choosers (fig. 5.1a). For a stimulus to be attractive, it must generate a peripheral response *and* be propagated by downstream filters *and* be coupled to a behavioral or other process that generates a positive response. Unattractive traits, by contrast, can be unattractive in a multiplicity of ways; they can fail to elicit a sexual response because of insufficient stimulation at the periphery (fig. 5.1b; chapter 3), *or* because they are not admitted by downstream filters (fig. 5.1c), *or* because they elicit a nonsexual or aversive response (fig. 5.1d). This latter mechanism is required for the valence of a courter stimulus to flip—that is, go from positive to negative.

Mating preferences may result from multiple combinations of proximate mechanisms, with different aspects of different courters eliciting positive or negative hedonic responses. It is instructive to think about what mate choice looks like at the two extremes of the continuum. As is often implicitly assumed, preference may be based entirely on differential attraction, with every courter eliciting some sexual response and the most preferred courter being the most attractive; alternatively, and perhaps more commonly, some courters may elicit sexual arousal while others provoke fear or disgust. In some systems where "courter" behavior is entirely coercive, differential response may be based entirely on avoidance, with choosers expending less effort to avoid the least distasteful mates. For example, in the poeciliid fish *Gambusia affinis*, where males lack courtship and obtain matings by coercion, females actively avoid males. Intriguingly, brain expression of neuroserpin, a candidate gene associated with mate choice, is negatively correlated with association with males in *G. affinis*. By contrast, female *X. nigrensis*, a species in which most males court females, are more likely to associate with males, and there is a positive correlation between brain neuroserpin expression and association with males (Cummings 2012). If mate choice is based more on rebuffing coercive advances than on differential proceptive behavior toward attractive males, this should markedly affect the cost of mate choice as a function of the number of mates sampled (chapter 6).

5.4 MECHANISMS OF EVALUATION

5.4.1 Overview

For our purposes, evaluative mechanisms are any structures that assign hedonic value to courter signals—that is, they modify the probability that the signal will result in a mating response. There are many ways in which neural structures can modify the valence of hedonic value, and probably many such structures in the chain connecting courter signals to chooser response. As I discuss below, compared to the sensory periphery they are therefore likely to represent a broader mutational target as well as one less constrained by functions other than mate choice.

5.4.2 Attraction versus indifference

Given that choosers perceive a stimulus, a basic part of the decision-making process is simply whether to attend to it in the context of mate choice. Indifference to a stimulus (fig. 5.1c) can be differentiated from insensitivity (fig. 5.1b) behaviorally by complementing mate-choice studies with conditioned-response trials, in which subjects can be trained to respond to a detected, but non-preferred, stimulus in a non-sexual context. For example, Witte and Klink (2005) found with video-playback experiments that female sailfin mollies (*Poecilia latipinna*) failed to respond to a novel ornament in males. They then conducted an additional study where they presented the ornament in isolation, paired with a food reward. After a training phase, females showed a feeding response when presented with the ornament. This showed that they perceived the stimulus, but were indifferent to it in the context of mate choice.

Indifference—as opposed to insensitivity—is the result of filters downstream of the periphery that prevent courter signals from inducing a sexual response. In túngara frogs, for example, males will respond to a broad range of conspecific and heterospecific calls, while females respond only to conspecifics. Hoke and colleagues (2007, 2008, 2010) showed differential IEG expression in response to conspecific and heterospecific calls in the auditory midbrain, which acts as a sensorimotor interface connecting sensory inputs with motor outputs. The auditory midbrain thus serves as a "gate" governing the relationship between sensation and response (chapter 4).

In songbirds, an important gate is the "high vocal center" (HVC), the primary brain region involved in the production of song by males and its perception by both sexes. Lesioning the HVC of female canaries produces indiscriminate responses to conspecific versus heterospecific songs (Brenowitz 1991) and attractive versus unattractive songs (Del Negro et al. 1998). A correlational study showed that the volume of the HVC and another song nucleus, lMAN, predicted the choosiness of female European starlings (*Sturnus vulgaris*) for differences among conspecific songs (Riters & Teague 2003). In vertebrates, the catecholamines dopamine and norepinephrine may be particularly important in modulating the hedonic marking of stimuli. Pharmacologically increasing brain dopamine titers caused female starlings to lose preferences (Pawlisch & Riters 2010). Importantly, sexual motivation—as measured by proceptive behaviors directed toward stimuli —was not affected by this manipulation (chapter 9). Rather, treated females decreased their sexual response to conspecifics and increased their response to heterospecifics. Similarly, pharmacological inactivation of noradrenergic neurons reduces choosiness for songs but not sexual motivation in estrogen-implanted zebra finches (Vyas et al. 2008; see also Pawlisch et al. 2011). Natural variation in estrogen titers, associated with breeding condition, is associated with marked differences in catecholamine synthesis in brain regions associated with sexual behavior (Riters et al. 2007), suggesting that the hedonic marking of sexual stimuli is an important source of variation in preferences (chapter 9). Multiple mechanisms may therefore serve as filters modulating sexual responses to stimuli, suggesting that hedonic marking may provide a large mutational target for preference evolution. In vertebrates, dopamine and norepinephrine pathways may be the first place to look for hedonic reversals of mating preferences.

5.4.3 Attraction versus antipathy

When choosers dislike a courter trait, it means that they perceive the trait, but it makes them *less* likely to mate with that courter. This means that a stimulus evokes an avoidance response rather than a sexual response. The most familiar example is the **Westermarck effect**, namely, that experience with an individual during childhood elicits sexual disgust—imagining sexual activity with a sibling or parent is highly aversive (Lieberman et al. 2000; chapter 12). It is noteworthy that the disgust is entirely confined to the sexual domain, because other activities with close relatives are positively marked. Sexual disgust—imagining sexual activity with a non-preferred partner—varies across women's menstrual cycle (chapter 9), while disgust in other domains like disease avoidance—imagining things like "stepping on dog poop"—does not. This suggests context-sensitive avoidance of specific sexual stimuli (Fessler & Navarrete 2003). As noted above, the same stimulus can elicit disgust or arousal in different individuals, or even in the same individual at different points in ontogeny. Further, sexual arousal acts to dampen sexual disgust (Borg & de Jong 2012). And numerous studies of cross-fostering and mate copying show that preferences are often completely reversed by appropriate social experience (chapter 12). **Associative learning** can also determine the hedonic marking of arbitrary stimuli; a stimulus associated with a positive reinforcer elicits a positive response and vice versa. The role of associative learning in the development and modification of preferences is discussed in later chapters.

How do choosers "flip" their responses from attraction to disgust? The brain contains structures that assign hedonic value to sensory information. There are multiple cases where individuals routinely flip their subjective evaluation of ecologically meaningful stimuli. For example, the dramatic transition of individual *Schistocerca gregaria* from grasshoppers into plague locusts involves an abrupt switch from avoiding contact with conspecifics, to actively initiating contact. This switch is mediated by an increase in serotonin levels induced by exposure to social stimuli (Anstey et al. 2009). Similarly, it is common for organisms to experience life history switches in the cues they prioritize. Female cotton leafworms *Spodoptera littoralis* show a dramatic shift in olfactory behavior, associated with changes in activation of odor-specific glomeruli in the antennal lobe: before mating, they strongly prefer cues associated with adult food sources over those associated with oviposition sites; after mating, this preference reverses (Saveer et al. 2012). Flips in preference could similarly arise from switches in prioritizing cues produced by different sets of courters.

The above examples show that the same stimulus can flip from attractive to aversive, or vice-versa, as a function of individual experience. Stimuli can also exhibit evolutionary reversals in hedonic value. An evolutionary flip in the taste preference for glucose has occurred in German cockroaches in response to selection via poisoned glucose-containing baits. In wild-type cockroaches, glucose stimulates sugar-gustatory receptor neurons, and toxic alkaloids like caffeine stimulate bitter-gustatory receptor neurons. In glucose-averse populations, glucose stimulates the bitter-gustatory receptor neurons, while other sugars continue to stimulate sugar-sensitive neurons (Wada-Katsumata et al. 2013). The valence switch thus involves glucose being assigned the same hedonic value as environmental toxins. This rapid shift in the valence of a sugar suggests that hedonic value can reverse itself quickly in response to selection. Tait et al. (2016) describe a similar mechanism associated with switching from hawthorn to apple hosts in *Rhagoletis pomonella*.

At least one such reversal in the hedonic value of stimuli has occurred during the diversification of estrildid finches, this time in the context of social interactions. Males in some species are gregarious, occupying mixed-sex flocks, whereas males in other species are territorial and solitary. The medial bed nucleus of the stria terminalis (BSTm) projects to basal forebrain areas including the ventral pallidum, which plays a key role in vertebrate pair-bonding (chapter 8). Vasotocin-immunoreactive (VT-ir) neurons in the BSTm show increased activity in response to social stimuli in gregarious species, but not in solitary species (Goodson & Wang 2006). These neurons may therefore play a major role in assigning or removing positive hedonic value from a stimulus.

In the context of mate choice, there has been little work on naturally occurring hedonic reversals, but mutation studies suggest that small genetic changes can indeed result in reversals of hedonic value in sexual stimuli. Female silkmoths *Bombyx mori* produce pheromone blends consisting of an 11:1 ratio of bombykol to bombykal. Males normally produce a full sexual response in response to bombykol alone, but ignore bombykal. In a mutant strain, the pattern is reversed; males respond fully to bombykal but ignore bombykol. The mutants are characterized by functional disruption of a transcription factor, *Bmacj6*, which is associated with both reduced density of bombykol receptors at the periphery, and projection of bombykal-sensitive neurons into the toroid, a brain region normally associated with positive response to bombykol (Fujii et al. 2011). Again, a single mechanism yields a reversal of preference, this time via changes in both sensitivity and hedonic marking.

Another group of moths, the genus *Ostrinia*, provides a compelling model for dissecting the neural and genetic mechanisms underlying preference reversals. The European corn borer *O. nubilalis* has two distinct strains characterized by female pheromone blends with highly skewed, opposite ratios of E- and Z-isomers of 11-tetradecenyl acetate. Female E-strain females produce E-skewed pheromone blends, and E-strain males respond sexually only to E-skewed blends. Z-strain females produce a Z-cocktail which is necessary for a sexual response in Z males. The neuroanatomy behind this reversal is straightforward: odorant receptor neurons (ORNs) specific to the major pheromone component project into the medial glomerulus, and ORNs specific to the minor component project into the lateral glomerulus (Kárpáti et al. 2008). In *Drosophila*, the activation of an individual glomerulus can mediate attraction to a stimulus (Semmelhack & Wang

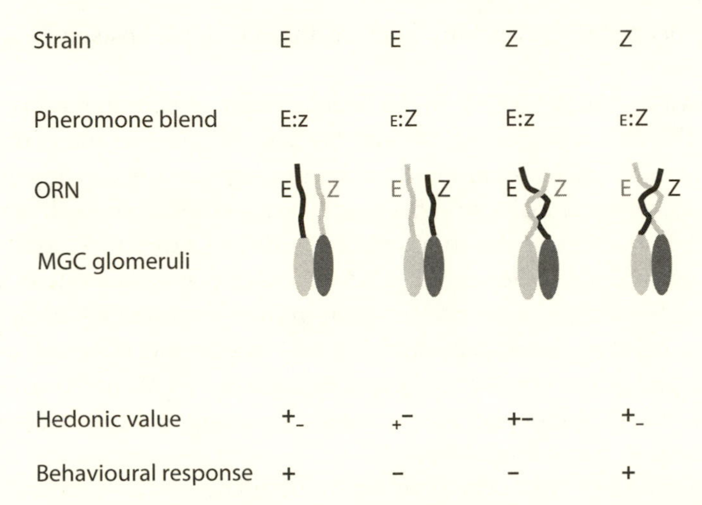

Figure 5.3. Schematic of neuroanatomy underlying hedonic reversals of preference between male *E*- and *Z*-strain corn borers (Kárpáti et al. 2008): axonal projections of olfactory receptor neurons into medial (light gray) and lateral (dark gray) MGC glomeruli. For both strains, neurons responding to the strain's own major pheromone project into the medial glomerulus and neurons responding to the minor pheromone project into the lateral glomerulus.

2009), so a plausible model is that stimulating the medial glomerulus elicits a positive hedonic response and stimulating the lateral glomerulus elicits a negative hedonic response (fig. 5.3). Indeed, the activity of ORNs tuned to heterospecific pheromone components inhibits sexual behavior in male *O. furnacalis*. The corn borer system thus highlights the importance of considering mechanisms of aversion as well as attraction.

Perhaps the most striking example of a hedonic reversal associated with mate choice occurs in rats infected with the protozoan parasite *Toxoplasma gondii*. *Toxoplasma* must be transmitted into a cat in order to reproduce, and infected rats, while similar to uninfected rats on a variety of generalized behavioral measures, are attracted specifically to cat urine—a stimulus that is very salient and very aversive to uninfected rats. House and colleagues (2011) used IEG-based neuroanatomical data to argue that *Toxoplasma* not only attenuates the "innate fear response," but induces competing sexual attraction by stimulating the same pathways associated with positive responses to female odor (fig. 5.4). Again, evaluative mechanisms represent a theater for the evolution and co-option of behavioral responses to sexual stimuli.

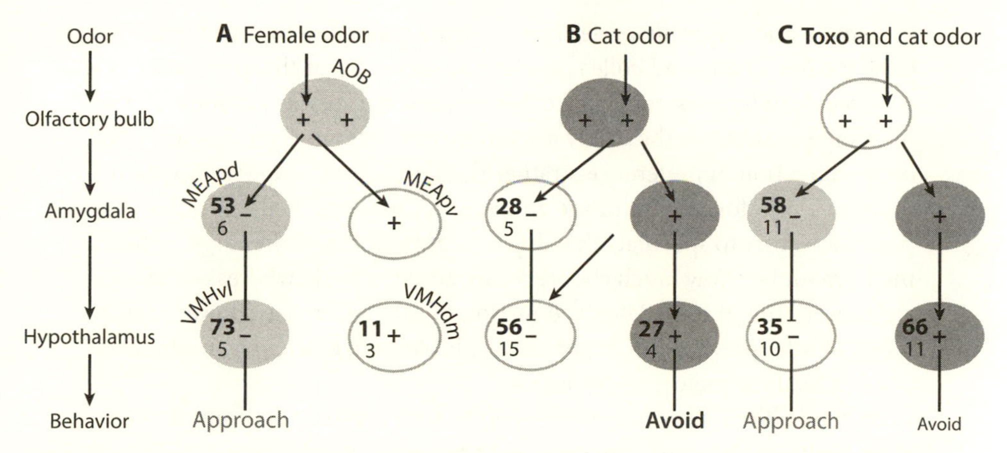

Figure 5.4. *Toxoplasma* parasites hijack sexual arousal in the limbic system of male rats. Numbers indicate raw density of c-Fos immediate early gene, with standard errors in smaller print. AOB, accessory olfactory bulb; MEApd and MEApv, posterodorsal and posteroventral medial amygdala; VMH, ventromedial hypothalamus. (A) female odor elicits approach behavior; (B) in uninfected individuals, cat odor elicits avoidance by stimulating the "defensive" dorsomedial part of the ventromedial hypothalamus; (C) in infected rats, cat odor continues to stimulate defensive pathways, but also stimulates pathways associated with sexual arousal. After House et al. (2011).

5.4.4. Stimulus valence and sexual preference

Within a species, one of the most important axes of preference variation is preference for male stimuli versus female stimuli. Despite what is usually a conserved set of sensory mechanisms between the sexes, there are sharp differences in what choosers find attractive depending on their sex and, in humans, gender[1] and sexual orientation. As in the previous examples of valence reversal in insects, preferences for male versus female stimuli in fruit flies involve differences in axon targeting. *Drosophila melanogaster*, express male- and female-specific splice variants of the *fruitless* gene. Males lacking expression of the male-specific gene exhibit atypical preference behavior, including preferentially courting other males. Kimura and colleagues (2005) showed that *fruitless* induces sexually dimorphic neural circuitry, which changes the valence of sensory inputs with respect to motor response (Kohl et al. 2013).

[1] I follow the American Psychological Association publication manual (2001) and use "sex" when referring to biological distinctions and "gender" when referring specifically to the human social construct.

Female mice typically prefer to mate with males and to investigate male genital odor. Zhang and colleagues (2013) showed that this preference was reversed, with females preferring other females, in mutants lacking either serotonergic neurons or the ability to synthesize serotonin. This systemic effect on the direction of preference—rather than on overall sexual motivation—hints at a role for serotonin metabolism in assigning hedonic value. It is pure conjecture to speculate that dopaminergic and noradrenergic neurons might modulate how much choosers care about stimuli, whereas serotonergic neurons might modulate the direction of this preference. However, these recent studies point to the importance of looking to evaluative mechanisms as a key locus of preference variation.

Systemic differences in neurotransmitter metabolism are unlikely to explain variation in sexual preferences in humans. We are nearly unique in having stable heterosexual and homosexual phenotypes in both sexes (Bailey & Zuk 2009b), which allows researchers to disentangle chooser gender from chooser preferences. A functional magnetic resonance imaging (fMRI) study by Kranz and Ishai (2006) showed greater activity of the thalamus and medial orbitofrontal cortex (OFC) of the brain in heterosexual men and homosexual women shown female faces, and greater activity of those brain regions in heterosexual women and homosexual men shown male faces. Intriguingly, functional studies show that the OFC, which receives inputs from the thalamus, is involved in assigning reward value and subjective pleasantness to stimuli in general (reviewed in Kringelbach 2005). A meta-analysis of fMRI studies further suggested that the medial OFC, which was activated by faces of the preferred sex in Kranz and Ishai's (2006) study, is involved in the evaluation of reinforcers (i.e., stimuli with positive valence), while the lateral OFC evaluates punishers (stimuli with negative valence; Kringelbach & Rolls 2004). The OFC may therefore be important to modulating all three possible evaluative states (preference, indifference, and dislike; fig. 5.2) in response to a conserved set of sensory inputs. There are intriguing parallels between the medial and lateral OFC in humans, the medial and lateral MGC glomeruli of corn borers, and the lateral horn of *D. melanogaster* (Schultzhaus et al. 2017) in assigning hedonic value to stimuli.

5.4.5 Hedonic value and preference functions

For simplicity's sake, this chapter has focused on discrete, binary preferences. But measured and modeled preference functions are continuous and often unimodal, with a single stimulus value, or combination of stimulus values, eliciting the maximal response (chapter 2). Unimodal preferences can sometimes be explained by sensory processes like receptor tuning (chapter 3), but they can also arise when stimuli elicit competing negative and positive hedonic responses. Female *D. melanogaster* prefer to lay their eggs on substrates containing low concentrations of ethanol. This preference depends on the response to alcohol by two distinct populations of gustatory neurons—one associated with aversive stimuli, one with appetitive stimuli. "Appetitive" neurons are activated more strongly at lower concentrations that are beneficial for oviposition, whereas at high (and toxic) concentrations of ethanol, aversive responses predominate (Azanchi et al. 2013). A comparable case in mate choice may be chooser responses to courtship displays, which can often elicit a combination of proceptive and fear responses (Huxley 1966; Coleman et al. 2004; Fisher & Rosenthal 2007). Unimodal or complex preference functions in choosers may thus reflect combinations of hedonic markings.

5.5. EVALUATIVE MECHANISMS AND PERCEPTION ARE RELATED: "BEAUTY IN THE PROCESSING EXPERIENCE"

Throughout this chapter, I have emphasized the distinctness of perceptual mechanisms and evaluative mechanisms. Yet these are of course functionally integrated components contributing to mating decisions, and some of the examples of preference reversals above have been accompanied by changes at the sensory periphery (e.g., Leary et al. 2012; chapter 3). Reber and colleagues (2004) make the compelling argument that hedonic value is influenced by what they term "fluency of processing," or the ease with which stimuli are analyzed and interpreted. In this view, preferences for familiar individuals (chapter 6) are enhanced by facilitating effects of prior experience on perception. Fluency of processing may also act as a source of selection on correlations among multiple courter traits, if some correlations are more familiar or inherently easier to process (chapter 4).

5.6 PLASTICITY AND EVOLVABILITY OF EVALUATIVE MECHANISMS

Stimulus valence can be remarkably labile, both in response to experience and in response to selection. For example, ten Cate and colleagues (2006) found that male zebra finches raised with mothers experimentally painted with red beaks expressed, as adults, attraction to red-beaked females and aversion to yellow-beaked females (fig. 5.2e). Males raised with yellow-beaked mothers exhibited the complementary pattern. Contrasting experiences before maturity can therefore flip the hedonic value of sexual stimuli. Evolutionary flips in preference may be widespread as well. In a classic experiment, Wilkinson and Reillo (1994) imposed 13 generations of selection on male eye-span length in stalk-eyed flies. Lines selected for short eye span showed a preference reversal, with females favoring short-stalked males (fig. 5.2c).

The range of mechanisms associated with hedonic reversals, and the specificity of some of these mechanisms—such as an aversive response to glucose but not to other sugars—suggests that evaluative mechanisms may be relatively free of the broad constraints associated with upstream sensory or perceptual processes, and that they may collectively represent a broad mutational target in the context of preference evolution. Most importantly in the context of preference evolution, evaluative mechanisms can flip—a single mutational step can potentially cause a shift from attraction to aversion toward the same stimulus. Evolutionary theory often assumes that preferences vary continuously, or that zero preference is an inevitable intermediate between aversion and attraction. Reversal of valence facilitates the evolution of assortative mating and sympatric speciation, as shown in an influential model by Turner and Burrows (1995). Later workers have suggested that such reversals are biologically implausible (Ritchie 2007), but numerous examples of behavioral reversals of preference, underlain by simple changes in brain chemistry or neural architecture, suggest that abrupt reversals of preference functions may be an important feature of mate-choice evolution. Beauty is fluctuating indeed; and if not quite arbitrary, variable enough to encompass a near-infinite variety of tastes.

5.7. ADDITIONAL READING

Fujii, T., Fujii, T., Namiki, S., Abe, H., Sakurai, T., Ohnuma, A., ... Shimada, T. (2011). Sex-linked transcription factor involved in a shift of sex-pheromone preference in the silkmoth *Bombyx mori*. *Proceedings of the National Academy of Sciences, 108*(44), 18038–18043.

Kringelbach, M. L. 2005. The human orbitofrontal cortex: linking reward to hedonic experience. *Nature Reviews Neuroscience, 6*, 691–702.

Reber, R., Schwarz, N., & Winkielman, P. 2004. Processing fluency and aesthetic pleasure: is beauty in the perceiver's processing experience? *Personality and Social Psychology Review, 8*(4), 364–382.

CHAPTER 6

From Preferences to Choices

MATE SAMPLING AND MATING DECISIONS

6.1 INTRODUCTION

The previous chapters detailed the internal processes that choosers use to evaluate potential mates. These processes can be represented as multivariate preference functions, whereby a chooser's response is a function of courter trait values. But how do preferences translate into actual mating decisions? Mate choice is by definition a comparative process, in the sense that for mate choice to exist, there have to be differences between courters such that a chooser has to be more likely to mate with some courters relative to others, depending on some difference between them. One can imagine an extreme version of Parker's (1983) "passive" mate choice (chapter 3), where chooser response is determined solely by the probability of detecting a courter signal (fig. 6.1a–b). In this case, the probability of mating is determined only by the chooser's preference function and the distribution of courter traits in the population.

The norm, however, is for choosers to devote time, energy, and cognitive bandwidth to evaluating potential mates. Consider two species of frogs with the same preference function for the peak auditory frequency of male calls. In one, females mate with the first male they detect, so that detectability determines preference as in figure 6.1b. In the other, females sample multiple males and mate with the one whose peak call frequency is closest to their peak preference (fig. 6.1c). In the first instance, the distribution of chosen males as a function of call frequency will resemble the preference function, whereas in the second, it will be a narrower distribution clustered around the peak preference (Jennions & Petrie 1997). Indeed, the bigger the sample evaluated by the female, the narrower the distribution of chosen males.

This second, apparently simple rule—sample *n* males, mate with the male who most closely matches peak preferred frequency—is a daunting cognitive and perceptual task. This is because choosers would need to (1) segment the signals of individual courters from environmental noise and from one another; (2) take into account environmental effects on signal perception,

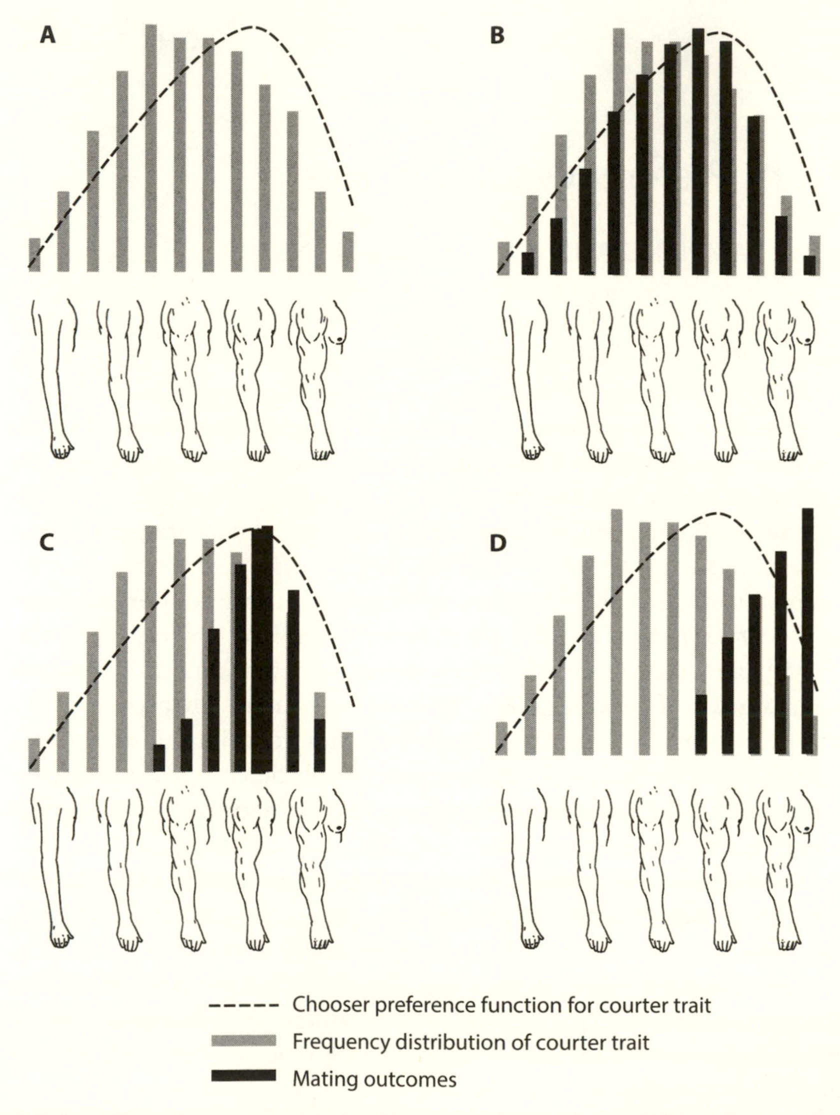

Figure 6.1. Sampling rules, courter trait distributions, and mate choice. (A) Distribution of an arbitrary courter trait and chooser preference function for that trait. (B) *Fixed threshold* rule: choosers mate with the first courter to exceed an internal criterion—in this case, detectability by a perceptual bias that serves as a preference function. Detectability determines the distribution of courters a chooser encounters. If all choosers have the same preference function and this fixed sampling rule, the relative mating success of courters as a function of size is proportional to the preference function. (C) Best-of-*n* rule: choosers sample a predetermined number of courters, mating with the most attractive one. Choosiness is higher and sexual selection on courters is stronger with increasing *n*. (D) Best of *n* applied when preferred (smaller) courters are excluded by intrasexual competition; courter mating success is skewed toward larger males compared to (C).

whereby factors like temperature and habitat-dependent signal degradation change the structure of courter signals as they reach receivers; and (3) remember the spatial location and individual identity of each candidate courter.

The environment also imposes limitations on choice. Depending on the distribution of courter signals and the nature of the environment, sampling an acceptable *n* may be prohibitively costly. Deciding among mates is costly in terms of time, energy, and the risk of predation or parasitism; there is an inherent trade-off between sampling more mates and doing something else. Even if they are small, the cognitive challenges described in the previous paragraph will scale positively with *n*. Further, courters routinely counteract mate choice by interfering with one another or by coercing matings from choosers, which effectively skews the distribution of courters with whom a chooser can mate (fig. 6.1d; section 6.10.3). Small and/or skewed samples will inevitably increase the mismatch between preferences and realized mating outcomes.

Finally, preference functions are dynamically intertwined with mate sampling. The process of choice—short-term interactions with courters—can influence every aspect of preference functions—responsiveness, choosiness, preference shape, and the valence of preference. This adds up to the distribution of mating outcomes generally being broader than chooser preference functions, and to mate sampling representing an important target of selection and one that is exquisitely sensitive to proximate ecological factors (chapter 11) associated with cost and risk. And surprising preferences—with important consequences for sexual selection and genetic exchange—can emerge from the process of comparing different stimuli that cannot be predicted from the properties of individual courters.

This chapter focuses on mate sampling and on how choosers decide among sampled mates. The thinking in this regard has largely focused on the fitness consequences for courters and choosers (e.g., Gibson & Langen 1996). For courters, how does mate sampling affect choosiness and therefore variance in courter fitness, and how do courters exploit sampling and decision mechanisms? For choosers, how does fitness depend on different putative sampling schemes and decision algorithms? Much of this literature flows back to an influential paper by Janetos (1980) that explored the fitness consequences of several hypothetical sampling schemes. These proposed algorithms provide a useful framework for classifying observed variation in mate sampling in animals. In keeping with the organization of this book, I will return to the fitness consequences of mate sampling for choosers and courters in later chapters. Here, I will describe what we know about the heuristic rules that animals use to evaluate finite pools of courters, and about the cognitive constraints underlying mate assessment and comparisons among mates.

6.2 THE BIOLOGICAL CONTEXT OF MATE CHOICE

6.2.1 Mate sampling and mating systems

It is instructive to consider two extremes of mate-encounter scenarios and the challenges they present to choosers. At one extreme, choosers encounter courter signals one at a time—sequentially. In order for mate choice to occur, they must either remember individual courters and where to find them, and/or dynamically adjust their preference functions as a function of experience with courter signals. At the other extreme, choosers are presented with a number of courter signals simultaneously. Here, they must parse sensory inputs into distinct streams corresponding to individual courters and localize them in space in order to make a decision. As a matter of fact, choosers typically will be sampling multiple courters at a time, over an extended period of time—both sequentially and simultaneously. A further challenge is posed by the fact that courters produce multiple signals and that these signals change over the course of courtship. Changes in preferences of individual choosers therefore can involve differential weighting of courter signals. A chooser cannot evaluate, let alone remember, every potential mate. A chooser has a finite pool of courters to choose from, and a finite amount of time, energy, and cognitive resources at her disposal to sample and evaluate mates. And, as we saw in chapter 4, preference space does not necessarily map well to the trait combinations that are in fact available in the real world. Mate sampling is further complicated by the fact that courters constrain choosers. In addition to mutual mate choice, which is covered in chapter 8, courters can coerce matings, and both intrasexual competition among courters and mate guarding can restrict the ability of choosers to sample other potential mates. These constraints mean that a chooser will often end up with a mate whose phenotype deviates substantially from her preference optimum (fig. 6.1d).

The biological context of choice is a function of the **mating system**, a population-level concept that encompasses a broad array of strategies used by choosers and courters in obtaining mates (Kokko et al. 2014). The mechanisms of mate choice play an instrumental role in structuring mating systems, and mating systems in turn shape the evolution of mate choice (Emlen

& Oring 1977; Shuster & Wade 2003; Kokko & Jennions 2008a; chapter 14). In order to begin thinking about mate sampling and mating decisions, it is therefore helpful to consider the array of theaters in which choosers operate.

An important aspect of the mating system has to do with the reproductive life history of choosers (Shuster & Wade 2003; Partridge & Endler 1987). If choosers are extremely **iteroparous**—reproducing multiple times throughout their lives—or if mate choice occurs mainly after mating (chapter 7), individual mating decisions become less important. At the opposite extreme are "one-shot" choosers like male spiders *Argiope bruennichi*, for which a single mating is all they get: their genitals are damaged during copulation and they are prey to sexual cannibalism (Schulte et al. 2010). An individual mating decision is therefore all-important. As I will discuss further in chapter 14, the costs and benefits of each stage of mate choice shape how much choosers invest in each mating decision.

Two properties of the mating system arise from an interaction between the reproductive life history of choosers and the phenotypes of courters, and are particularly relevant to mating decisions. These are the distribution of courters from a chooser's point of view, and the time course over which prospective mates are evaluated. Together, these variables determine the distribution of courter signals that choosers can attend to, and therefore the mechanistic constraints on mating decisions. For example, if courter densities are high and females integrate mating signals within a short time window, evaluating the same kinds of signals produced simultaneously by different courters requires perceptual segmentation of individual signals (Miller & Bee 2012). By contrast, searching among widely dispersed courters, or integrating information about them over a long period of time, requires choosers to either store representations of courters in memory or to dynamically adjust mating preferences (Castellano & Cermelli 2011). Each of these constraints will be addressed in turn below. In chapter 15, I will address the question of why mating systems vary so dramatically among—and often within—species (Shuster & Wade 2003; Shuster 2009).

6.2.2 Effective distribution of courter signals

The effective distribution of courter signals is what the mating scene looks like from the point of view of choosers: how many courters, and how many different types of courters, does a chooser have the opportunity to sample? The cumbersome term "effective distribution of courter signals" is awkward in comparison with "encounter rate," which stands in for the same thing in many models. I avoid "encounter rate" since it leads to unconscious punning (*sensu* Bateson 1983), conjuring up an image of choosers moving from courter to courter and evaluating them in sequence. This excludes the frequent case where choosers must simultaneously perceive and evaluate signals from multiple courters.

The availability of mates to choose from is the most basic requirement of mate choice, and has long been recognized as a determinant of sexual selection (Emlen & Oring 1977; Eshel 1979; Aronsen et al. 2013). Variation in effective courter density is intimately tied to mating system, and accounts for much of the interspecific variation in the strength of sexual selection and the extent to which one sex or the other chooses (Kokko & Rankin 2006). Indeed, reversals in the choosy sex can occur over the course of a breeding season as a consequence of changes in sex ratio (chapter 8). At one extreme are lek-breeding systems or choruses, where choosers can sample multiple courters simultaneously on an aggregation at relatively low cost (Höglund & Alatalo 1995), but are faced with the challenge of parsing multiple signals in a hurry. At the other are so-called resource-based systems where courters are more widely dispersed. For example, in pine engraver beetles (*Ips pini*), choosers encounter and evaluate courters one at a time, rarely revisiting the same males (Reid & Stamps 1997).

While many studies have measured and manipulated density in the context of mate searching, fewer have quantified the effective distribution of courter signals from the chooser's point of view. An exception is a recent study by Deb and Balakrishnan (2014) who used acoustic data from the field combined with psychophysical data on auditory sensitivity to estimate the active space of courters in choruses of tree crickets (*Oecanthus henryi*). Males were spaced far enough apart that only about 15% of females could hear two or more males simultaneously. Females would therefore have to actively move to sample multiple courters. By contrast, Murphy (2012) estimated that female barking treefrogs, *Hyla gratiosa*, sampled the calls of on average four males simultaneously at a given location.

Sensory perception is by no means the only determinant of the subjective density of courters, which is also affected by internal factors including a chooser's sensory abilities to detect courters, and the perceptual and cognitive constraints involved in evaluating and distinguishing among courters. External factors include sources of risk like predation, competition from other choosers, and interference from courters. All of these constrain a chooser's ability to make and execute a decision to mate, whether because they divert attention from mate choice or make the task of choice more costly in terms of energy, lost opportunities, or increased exposure to risk. For example, a chooser in an area frequented by aerial predators might spend less time out in the open interacting with courters, thereby lowering the density of courter signals experienced per unit time. Changes in the subjective density of courters, as a consequence of factors like sensory impairment and predation risk, are perhaps the most important ecological source of variation in realized mating preferences (chapter 11). This variation can be among species or broad taxonomic groups, as mentioned above, or can occur on the scale of individual mating decisions.

6.2.3 Temporal integration of mating decisions

The second point to consider is the time over which mating decisions are integrated. For example, female *Drosophila* will accept or reject a mate based on a few seconds' evaluation of a multimodal display (Ferveur 2010). By contrast, satin bowerbirds (*Ptilonorhynchus violaceus*) evaluate multiple signals over the course of several days. Males build elaborate structures called bowers (somewhat different in structure from those of the great bowerbird in fig. 1.1), which females initially sample when males are absent from their territories. A female then returns to a subset of the sampled bowers when males are present, at which point males produce an elaborate auditory and visual courtship display. Different individual females assign different weights to each of these characteristics, mostly as a function of their age and experience (Coleman et al. 2004; Uy et al. 2001; fig. 6.2).

The pertinent timescale can vary depending on what aspect of mate choice we are looking at. In vertebrates, experience with acoustic signals can trigger the onset of reproductive condition (Lea et al. 2001); for example, Hinde and Steel (1976), showed in canaries that the songs of conspecific males, but not heterospecifics, induced nest-building and proceptive behavior. The distribution of signals can therefore influence chooser behavior on the timescale of a breeding season. At the other end of the spectrum, phonotaxis decisions by gray treefrogs appear to integrate information about conspecific call variation over a course of 1–2 minutes (Schwartz et al. 2004).

Different courter signals are often assessed at different stages (section 6.8), with the marginal cost of sampling to choosers typically increasing as choice progresses. In black grouse *Tetrao tetrix*, females initially survey the males on a lek from the treetops, then fly down to interact with a handful of males multiple times, finally mating with one of the visited males before leaving the lek (Rintamäki et al. 1995). Gibson (1996) showed that visits to

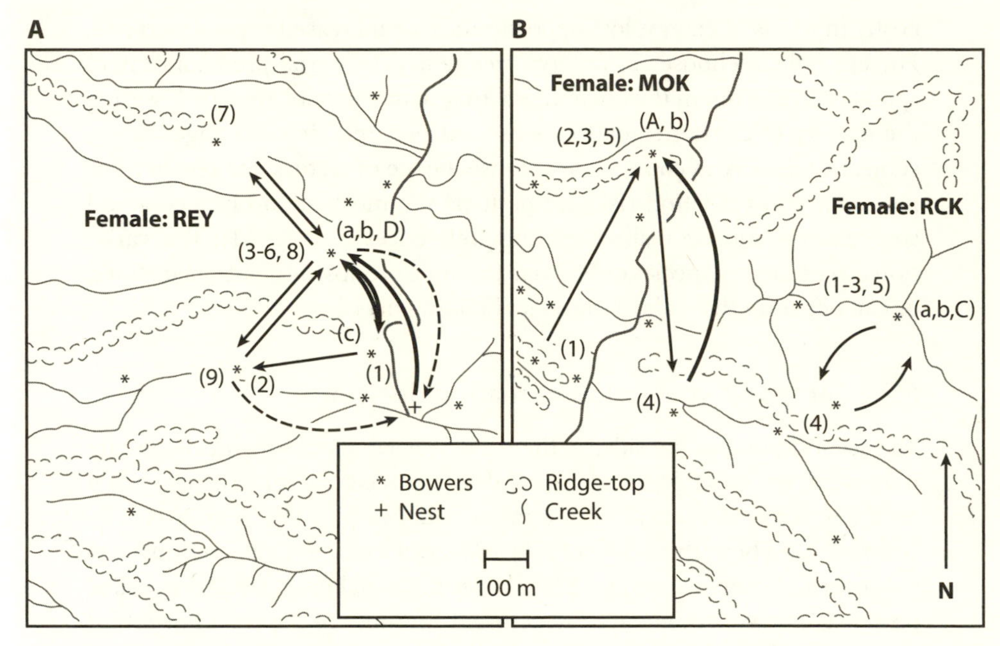

Figure 6.2. Mate-searching patterns of three female satin bowerbirds (after Uy et al. 2001). Numbers indicate early courtship visits, letters indicate later visits, with uppercase letters indicating visits that involved copulation. Thin solid lines indicate earlier visits, thicker lines indicate later visits, and dashed lines indicate movement to the female's nest.

male territories by female sage grouse (*Centrocercus urophasianus*) on leks were determined by the characteristics of males' long-range acoustic broadcasts, while the probability of mating once visited depended on the rate of a stereotyped visual display; together, these yielded a multiplicative preference for males with high display rates and specific acoustic features (fig. 6.3).

6.2.4. Mate sampling and the sensory environment

The sensory context of mate choice has a heavy influence on the effective distribution of courter signals and on the temporal integration of these signals, regardless of the approach choosers use to sample mates. Environmental effects on the signaling environment are covered in chapter 3; a variety of environmental effects, from ambient noise to changes in the transmission medium, can change the size and nature of the distribution of courters from the point of view of choosers.

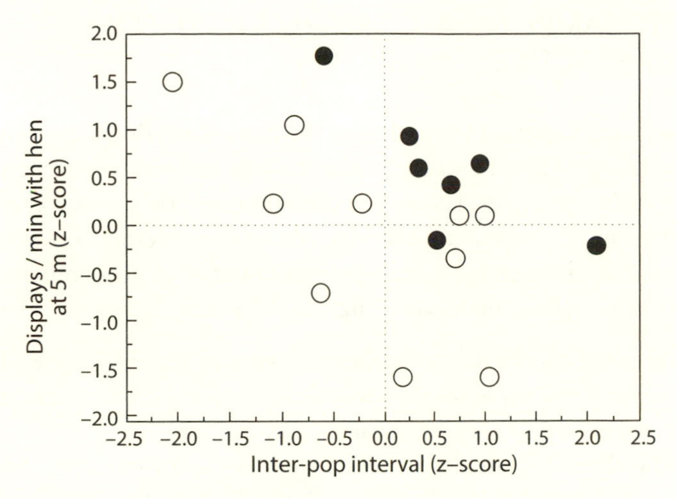

Figure 6.3. Multiplicative interaction between close-range courtship displays as a function of inter-pop interval, a long-range display, on female preferences in lekking sage grouse (*Centrocercus urophasianus*). Filled symbols indicate males that were mated with at least one focal female, open symbols are males that were rejected by all females (Gibson 1996).

Independent of the sampling algorithm used, **perceptual constancy** of choosers represents a formidable challenge when sampling courters. If courters are assessed in different times or places, robust preferences require that choosers have mechanisms to account for environmental effects on signal perception. One such mechanism is **chromatic adaptation**. Recall from chapter 3 that the detection of visual signals depends on incident light reflecting off a courter and passing through a physical medium before striking the eye of a chooser. The physical qualities of the light and of the medium therefore change the distribution of the photons bouncing off the displaying bright-red male and striking the receiver. Chromatic adaptation weights the contribution of different cone inputs inversely in proportion to their overall response to background, which is why we see a green afterimage immediately after staring at a red light source. Endler and colleagues (2014) showed that male great bowerbirds (*Ptilonorhynchus nuchalis*) take advantage of a female's chromatic adaptation (in the short-term, physiological sense) to the reddish walls inside a bower (fig. 1.1), flaunting brightly colored found objects during a courtship display that maximizes visual contrast. Mechanisms for perceptual constancy may play an unappreciated role in linking environmental heterogeneity to variation in mate choice.

6.3 MATE SAMPLING ALGORITHMS IN THEORY AND PRACTICE

In this section, I attempt to synthesize the mate-sampling algorithms proposed by Janetos (1980) and later workers with the empirical data on how choosers make mating decisions. The primary question is whether a chooser's evaluation of a given courter depends on her interactions with other courters, either through previous experience or immediate comparison. We can therefore organize mate-sampling algorithms along two broad axes:

1. **Sequential** algorithms involve the assessment of multiple courters at different times, while **simultaneous** algorithms involve contemporaneous assessments of signals from multiple courters.
2. **Static** algorithms are dependent on choosers alone, including their prior history before mate sampling. **Dynamic** algorithms depend on choosers and on the distribution of courters experienced.

Empirical evidence is consistent with multiple types of rules across taxa, with different algorithms likely recruited within a population and even within the same individual (reviewed in Beckers & Wagner 2011). There is no reason a chooser could not in principle apply multiple rules to different stages of a mating decision, or to different traits at the same stage.

6.4 SEQUENTIAL AND STATIC: FIXED-THRESHOLD RULES

With a "**fixed-threshold**" rule (Janetos 1980), a chooser will accept the first courter she encounters whose attractiveness exceeds a static criterion. This is a type of passive mate choice *sensu* Parker (1983), where a chooser cannot dynamically alter her choice patterns. Fixed-threshold is the simplest rule because it relies on a purely internal standard, and arguably the most parsimonious mechanistically, since it can arise completely from something as simple as the threshold sensitivity of a peripheral neuron. Phelps and colleagues (2006) provide a version of the fixed-threshold rule that incorporates simultaneous assessment, as well as uncertainty in the chooser's evaluation of courter signals: here, attractiveness of a trait is specified by a fixed preference function, plus a positive or negative random error term; choosers mate with the most attractive mate in a comparison set. This rule seems to be operating in decorated crickets *Gryllodes sigillatus*, where twice-mated females show consistent pre- and postmating preferences between their first and second mate, with no previous mate effect on preferences (Ivy & Sakaluk 2007). Phelps and colleagues (2006) showed that this rule predicted both simultaneous and sequential (no-choice) comparisons in túngara frogs (but see Kirkpatrick et al. 2006).

6.5 SEQUENTIAL AND DYNAMIC: ADJUSTABLE THRESHOLDS

6.5.1 Experience-dependent preferences

Under the pure fixed-threshold rule, if no courter meets the criterion value then the chooser simply fails to mate. We know, however, that animals markedly lower their thresholds based on mate availability. An extreme example is the infelicitously named "prison effect" (Bailey & Zuk 2009a), a particularly unfortunate example of anthropomorphic terminology. Here, heterosexual individuals engage in sexual activity with same-sex individuals when the opposite sex is unavailable for a length of time (Van Gossum et al. 2005; Engel et al. 2015). (The sexual dynamics in actual prisons are surely much more complex.) Similarly, lack of access to conspecifics induces choosers to mate with heterospecifics when opposite-sex conspecifics are unavailable (Clark et al. 1954). Even when favored mates are available, choosers may opt to mate with a less-favored courter if risks act to decrease the subjective density of courters. For example, female swordtails *Xiphophorus birchmanni* lose their olfactory preferences for conspecific over heterospecific males if the latter are closer to a shelter that provides refuge from predators (Willis et al. 2012).

Most animals probably incorporate information about the distribution of courter signals into their mate-sampling algorithms. The female frogs in Phelps et al.'s (2006) study were all taken from pairs in the process of mating, which may have standardized perceived mate availability. A refinement of the fixed-threshold algorithm is the **last-chance option**. This is a fixed-threshold strategy, but if no above-criterion courter is encountered, the chooser will mate with the last courter encountered independent of his attractiveness. Beckers and Wagner (2011) argued that female field crickets *Gryllus lineaticeps* exercised a last-chance option, relaxing their preferences for male calls after 24 hours of acoustic isolation.

Last-chance option is a special case of a more general **adjustable threshold**, described by Janetos (1980, p. 109) as a chooser "starting out as very picky indeed, but becoming less so as she runs out of time, until her last chance, when she will mate with any conspecific male." Castellano (2015) provided an approximate Bayesian model of a similar choice process, whereby thresholds are dynamically updated based on prior information.

There is abundant evidence that choosers adjust their preferences dynamically. As I will discuss in chapters 11 and 12, preferences and sampling algorithms can in no small part be shaped by environmental and social influences on chooser phenotypes and on the decisions of individual choosers. In this case, the influence is the distribution of sampled courters experienced by a chooser. Expressed preferences frequently change as a function of sampling history. This starts, in some cases, with the induction of sexual receptivity by courter signals (see above). Once choosers are receptive, they ubiquitously relax their preferences as the effective distribution of courter signals becomes less dense, becoming less choosy as mates become less frequent (e.g., Shelly & Bailey 1992; Willis et al. 2011) and as the breeding season draws to a close (Forsgren et al. 2004; Passos et al. 2014).

The adjustable-threshold algorithm further makes directional predictions about how experience should shape chooser behavior. A standard signal should be more attractive if chooser previously experienced a less attractive signal, and less attractive if they experienced a more attractive signal. A number of studies support this prediction (Collins 1995 and references therein; Bailey & Zuk 2009b; Wagner et al. 2001). For example, Lyons and colleagues (2014) manipulated attractiveness of a courter signal by digitally altering inter-syllable length in songs of Lincoln's sparrows. Females showed stronger preferences for intermediate songs if they had been previously exposed to unattractive songs (fig. 6.4).

In addition to the stimulus values that choosers are exposed to, the variation that choosers experience also influences preference. Under the simplest adjustable-threshold algorithm, choosiness increases with increasing variability in the courter traits experienced. Several studies have supported this prediction (reviewed in Wiegmann et al. 2013). For example, recent experience with variable distributions of female size induced male guppies to preferentially court large females; if males hadn't experienced female size variation, they courted at random with respect to size (Jordan & Brooks 2012). Experiencing a variable distribution therefore increased choosiness. However, when uncertainty is incorporated into chooser evaluations, choosiness can either increase or decrease as a complex function of the costs of choice (Wiegmann et al. 2013).

Sometimes, exposure to a single stimulus can increase choosiness relative to naïve individuals. In sticklebacks, previous experience with either

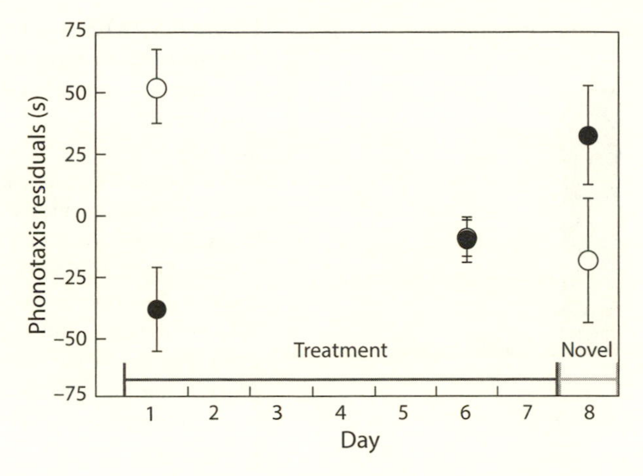

Figure 6.4. Residual phonotaxis time controlling for side bias (mean +/- SE) of female Lincoln's sparrows as a function of social experience. Filled symbols are females that were exposed to unattractive songs, open symbols are females that were exposed to attractive songs. Females had stronger preferences for a novel intermediate song (right) if they had been previously exposed to unattractive song (Lyons et al. 2014).

conspecifics or heterospecifics makes heterospecifics less attractive to choosers; females become choosier after having experienced either stimulus (Kozak et al. 2013). But prior experience does not always increase choosiness. Kárpáti and colleagues (2013) studied the short-term response of male Z-strain European corn borers to shifts in pheromone ratios. When males are presented with a stream of pheromones emanating from a single source, they nearly always respond to a plume of Z-strain pheromone and almost never respond to hybrid (50:50 ratio) and E-strain blends (fig. 6.5a). If the Z-plume was interrupted after males started orienting and was replaced with clean air, they stopped moving toward the source, but continued toward it if it was replaced with a hybrid or heterospecific blend (fig. 6.5b–c). Therefore, previous experience with a species-specific pheromone ratio reduced choosiness, resulting in a permissive response to the pheromone blends of other species relative to naïve controls. Kárpáti and colleagues showed that this pattern was at least partly a consequence of sensory neurons becoming more permissive in response to repeated stimulation. They argued that generalizing to broad sets of stimuli enabled choosers to efficiently "lock on" to courter signals, but noted that this might facilitate hybridization between strains (chapter 16). These results are consistent with Wiegmann and colleagues' (2013) theoretical prediction that uncertainty in assessment can decrease choosiness.

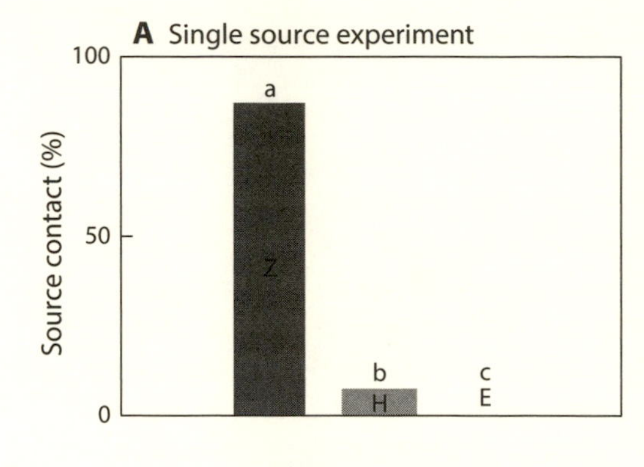

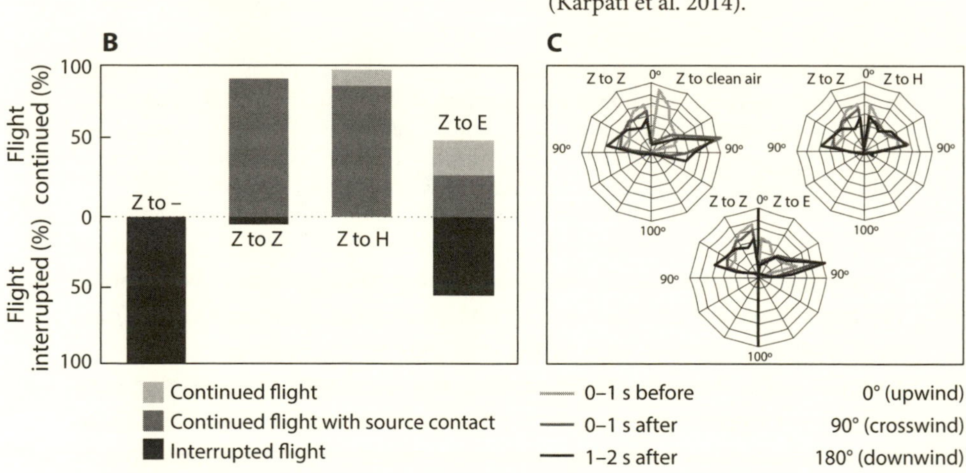

Figure 6.5. (A) Z-strain male moths *Ostrinia nubilalis* show strong preferences for conspecific pheromone blends in single source trials; (B) while discontinuing the pheromone plume results in interrupted flight, choosers continue flying toward the source when it is replaced with a 50:50 "hybrid" blend and with a heterospecific E-strain blend. (C) distribution of track angles 0–1 seconds before (light gray) and 0–1 (middle gray) and 1–2 seconds (dark gray) after the switch. The Z-to-Z swap is provided on the right for comparison (Kárpáti et al. 2014).

6.5.2 Familiarity and the Coolidge effect

Choosers vary with respect to the positive versus negative effect of experienced stimuli on subsequent decisions. In chapter 5, I suggested that the hedonic value of courter stimuli was an underappreciated mechanism for chooser responses to selection on preferences. Here and in later chapters, I will draw attention to the growing number of studies on variability in the hedonic value (chapter 5) of learned preferences. Experience with courters can have both positive and negative effects on preference. Izzo and Gray (2011) found that female field crickets *Gryllus texensis* tested sequentially were more likely to prefer conspecifics after prior experience with conspecifics, and heterospecifics after experiencing heterospecifics. Female *G. rubens*, by contrast, were not affected by previous experience, always preferring heterospecifics.

Another example of between-species differences in the effects of experience comes from two sister species of swordtails, *Xiphophorus malinche* and *X. birchmanni*. Exposure to males has opposite effects on females of the two species: *X. birchmanni* prefer familiar phenotypes, while *X. malinche* avoid them (Verzijden et al. 2012). These effects mirror those seen as a consequence of experience before maturity (Verzijden & Rosenthal 2012; Cui et al. in press), suggesting that the same mechanisms may be involved in specifying preferences or antipathy for familiar phenotypes at different stages of ontogeny (chapter 12). More generally, repeated experience with stimuli can have either positive or negative effects. Different responses to recent and long-term experience may account for preference differences—including asymmetries in conspecific mate preference—between species.

The most basic question we can ask about adjustable preferences, then, is whether prior experience with an individual or trait makes choosers subsequently more or less disposed to favor that trait. Preferences for familiar individuals or phenotypes are widespread, and can be explained by the hedonic value of fluency of processing (chapter 5) or by facilitating detection of a familiar stimulus (Dukas 2002). On the other hand, choosers can often have a preference for novelty (or an aversion to the familiar). A widespread phenomenon is the so-called rare male (rare courter) effect (Partridge & Hill 1984; Partridge 1988), whereby choosers are biased toward the less common phenotype in sequential (Janif et al. 2014) and simultaneous (Royle et al. 2008) tasks. For example, Janif and colleagues (2014) found that men and women rated pictures of bearded men more attractive if they were surrounded by clean-shaven men, and vice versa. As I will revisit later, such preferences for rare or novel stimuli could maintain diversity in courter phenotypes and genotypes.

There are often marked sex differences in the preference for novelty. There are many examples of female choosers preferring to mate with familiar individuals (Fisher et al. 2003), and familiarity preferences are the basis for pair bonding (chapter 8). On the other hand, males often exhibit the so-called **Coolidge effect**, favoring unfamiliar females (Kelley et al. 1999; Joseph et al. 2015).

In a rare study directly addressing between-sex differences in familiarity preference, Tan and colleagues (2013) found dramatic differences between the sexes in *D. melanogaster* in responses to novel versus unfamiliar individuals, and to novel versus unfamiliar phenotypes. Females preferred familiar individuals and familiar phenotypes, whereas males preferred novel phenotypes. Mutants of either sex for a co-receptor essential to olfaction did not show a preference, indicating that olfactory cues are necessary for familiarity-based preference (Tan et al. 2013). The shared olfactory substrate raises the intriguing possibility that males and females vary primarily in the valence assigned to familiar and/or novel stimuli (chapter 5).

6.5.3 Generalization and peak shift

An important question is whether the positive or negative effects of familiarity are specific to the individual courter that a chooser has experienced, or whether choosers exhibit **generalization**, whereby learned responses to one stimulus are extended to similar stimuli. For example, in Kárpáti and colleagues' (2013) experiment, described above, choosers generalized to a broad array of pheromone blends after initial exposure to a species-typical pheromone blend. Qvarnström and colleagues (2004) fitted paired male collared flycatchers (*Ficedula albicollis*) with a novel trait, a red stripe on the white forehead patch. The next year, females with previous experience with red-striped males—but not naïve females—were more likely to mate with red-striped males and produced more offspring with them. Females thus generalized from their previous experiences with individual red-striped males, to all a range of males bearing that trait.

A related phenomenon is **peak shift**, which arises when individuals are trained to discriminate among two stimuli. With peak shift, individuals show maximal responses to exaggerated values of previously experienced stimuli. A study of **discrimination learning** of courter signals by Verzijden and colleagues (2007) illustrates both peak shift and generalization. The authors used food rewards (rather than sexual responses) to train zebra finches to discriminate between synthetic songs; for one stimulus, pecking a key would yield a food reward, while for the other, birds were rewarded if they refrained from pecking. One set of songs varied in syllable number, the other in the internal placement of an odd syllable (fig. 6.6a–b). Subjects showed generalization with respect to syllable number, whereby birds trained to peck for the longer-syllable call pecked for all calls with that number of stimuli or more. For the negative stimulus, they showed peak shift, with more extreme values of the stimulus eliciting a stronger response than the original conditioned stimulus (fig. 6.6c–d). By contrast, while they learned to respond appropriately to the placement of a novel song element, they did not generalize with respect to element placement (fig. 6.6e–f). Positive and negative experiences with courters varying in multiple traits could therefore potentially influence chooser preferences for subsequent courters. Peak shift and generalization in the context of early learning are discussed in chapter 12.

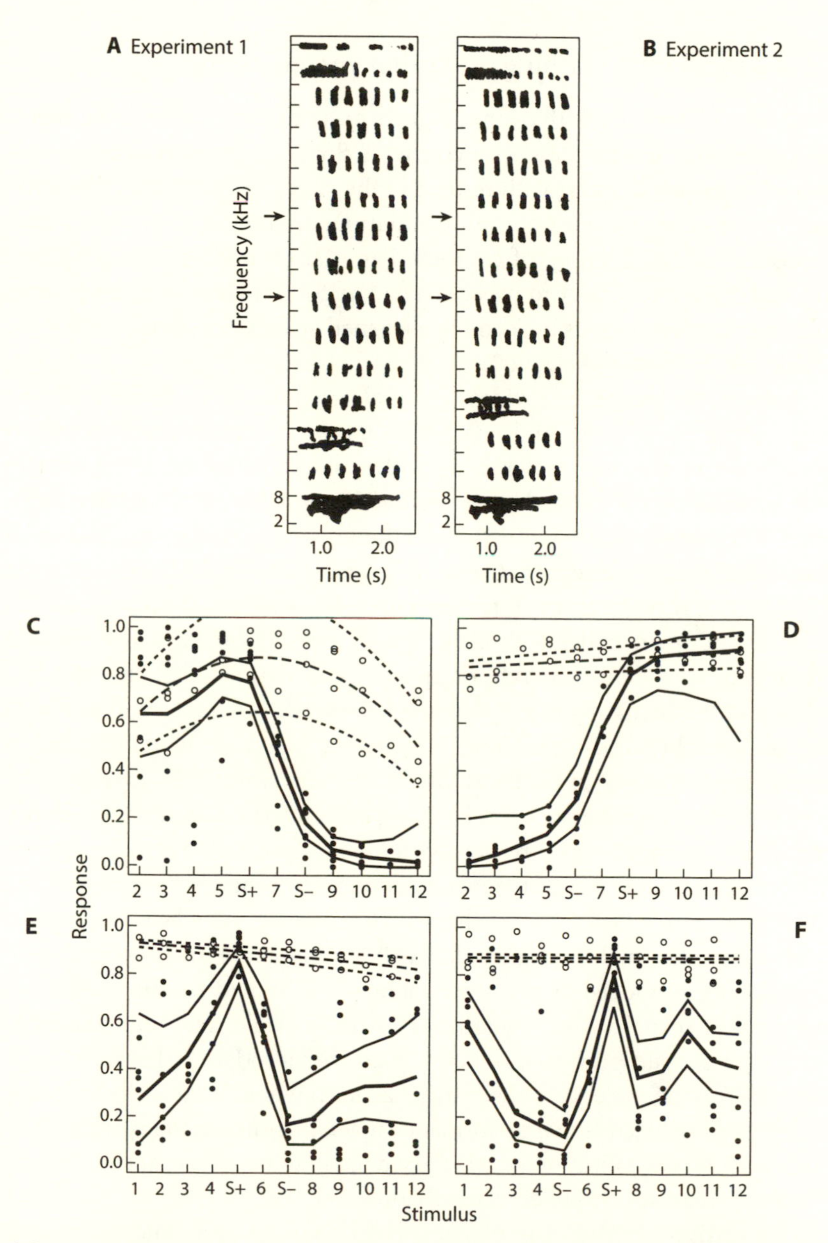

Figure 6.6. (A) Synthetic songs of zebra finches varying in syllable numbers; arrows indicate stimuli used for training in conditioned-response experiments; (B) synthetic songs varying in the placement of an odd element. (C–D) Generalization curves after training on six-syllable song as the positive stimulus and eight-syllable song as the negative stimulus, and vice-versa. Open circles and dashed lines indicate responses of control (untrained) individuals. (E–F) Generalization curves after training on the placement of an odd element. AfterVerzijden et al. (2007). Song spectrograms courtesy Machteld Verzijden.

6.5.4 Associative learning and other mechanisms

How does experience with courter signals exert positive or negative effects on subsequent interactions with courters? Some of the mechanisms discussed in earlier chapters with respect to the salience of individual stimuli, like Weber's law and release from habituation, are expected to impact how choosers evaluate a group of courters. Familiar stimuli may be positively marked if they are easier to process (chapter 5) and choosers may be able to minimize search costs by forming a learned search image of courters (cf. Dukas 2002; Chittka et al. 2009).

Associative learning provides a general mechanism for dynamically incorporating the positive and negative outcome of experience into subsequent mating decisions. Interacting with a signal of an attractive courter may provide **sexual reward**—the positive hedonic value of sexual experience (Kavaliers & Choleris 2013)—and thereby trigger sexual arousal when that signal is re-experienced. Coria-Ávila and colleagues (2005) measured the behavioral preferences of female rats that had learned to associate an arbitrary odor on males with either paced copulation, in which females controlled male copulation rate, or unpaced copulation, where females had to fend off male mating attempts. Females learned to associate odors with sexual reward (paced copulation), suggesting that positive reinforcement could shape chooser preferences. Similarly, experiments by Saleem et al. (2014) proposed that associative learning accounted for learned preference of male *D. melanogaster* for responsive females. Associative learning dependent on sexual reward need not involve actual sexual activity with a chooser. Roberts and colleagues (2014) showed that female mice learned to positively associate a male's volatile odor with a nonvolatile major urinary protein, darcin, produced by the same male. Conditioned sexual reward can even assign sexual valence to arbitrary stimuli. Male rats were trained to associate a somatosensory cue—being fitted with a rodent jacket—with copulation. Trained males exhibited sexual arousal when fitted with the jacket, and showed reduced sexual performance when unclothed (Pfaus et al. 2012). Such learned sexualization of arbitrary cues may be important in shaping diverse human sexual preferences (chapter 17).

Associative learning may play a role in shaping preferences for novel traits, as in the experiment by Qvarnström and colleagues (2004) described above. Their manipulation targeted males on prime habitat that made them more likely to rear successful young. The authors suggest that females' second-year preferences arose by associating red-striped males with positive sexual or social experiences (a successful brood) in the first year.

Conversely, forced mating, harassment, or avoidance by courters may trigger aversive behavior when signals from those courters are re-experienced. Indeed, Dukas (2008) found that male *Drosophila persimilis* were less likely to court heterospecifics if they had recently been rejected by a heterospecific female. Successful copulations and unsuccessful interactions with unreceptive females act as positive and negative reinforcers, respectively, in rats (reviewed in Pfaus et al. 2001).

In addition to associative learning, preference from the familiar can arise through the so-called mere exposure effect, whereby stimuli that are more easily processed are more attractive (Hill 1978; Reber et al. 2004; chapter 5). Habituation (chapter 3) can also increase preferences for familiar courters or repertoires, by attenuating aversive responses to elements of courter displays. Patricelli and colleagues (2002) suggested this mechanism to explain the increasing preference with age of female satin bowerbirds for high-intensity male displays. Preference learning on a longer timescale is the focus of chapter 12.

In other cases, the loss of sensitivity from habituation may cause loss of preference. Dishabituation when presented with novel signals has been invoked to explain preferences for signal complexity, like novel syllabic arrangements in birdsong (Searcy 1992). Sockman and colleagues (2002, 2005), however, suggest that preferences for novelty are canalized by what choosers found attractive: female starlings only respond to novel elements in the context of the long-bout songs they already find attractive.

One can detect a signature of experience in the neuroanatomical loci associated with complex preferences. In chapter 4, I described how activity in the caudomedial mesopallium of the auditory telencephalon (CMM) of starlings was associated with configural properties of courter signals (Sockman et al. 2005). CMM activity shows a neural signature for experience-modulated choosiness. Female starlings are choosier with respect to song length, and the CMM shows a greater differential response, in choosers that have been previously exposed to long songs (reviewed in Sockman 2007).

Empirical evidence thus suggests that the relative attractiveness of a set of courters is not something choosers ascertain by ranking them independently. Rather, a chooser's response to a courter is an emergent property of their relationships with one another, and there is an array of mechanisms that can modulate the effect of experience. Adjustable thresholds mean that the attractiveness of a given courter is path-dependent and can, like simultaneous comparison below, account for departures from **transitive** preference.

6.6 SIMULTANEOUS AND STATIC: COMPARATIVE EVALUATION AND (IN)TRANSITIVITY

Simultaneous comparisons are probably widespread where courter signals are found at high effective densities. Murphy and Gerhardt (2002) used a combination of playback experiments and field observations in barking treefrogs (*Hyla gratiosa*) to show that females were adopting a simultaneous algorithm. Females did not merely choose the first male they detected in a chorus; rather, they moved to a position where they could detect multiple calls simultaneously, then choose among detected options. Females thus had to localize and discriminate among detected males (fig. 6.7).

There are a number of ways that choosers can simultaneously evaluate courters. Dynamic algorithms that require choosers to remember and distinguish multiple individuals are costly, as should become clear in the next section. In humans and other animals, decisions are generally made using heuristic shortcuts that are faster and more computationally efficient (Bateson & Healy 2005; Hutchinson & Gigerenzer 2005). Much of this is accomplished by attending to the *differences* among stimuli rather than trying to parse their absolute attributes. For example, auditory neurons in the forebrain and midbrain of female zebra finches are tuned to configural properties that maximize differential response to acoustic differences among courters (Woolley et al. 2005). This inevitably means that a signal's attractiveness is dependent on its relationship with another signal.

Attending to differences means that courters cannot be assigned an individual "score" for attractiveness (e.g., Phelps et al. 2006). Since choosers attend to multiple traits (chapter 4), attractiveness depends on how these differences are weighted by choosers (Miller & Todd 1998); and the weighting of these differences depends on the phenotypes of other courters (Bateson & Healy 2005). Accordingly, choosers can express complex patterns of preferences. Kirkpatrick and colleagues (2006) performed "round-robin" assays of female túngara frogs' responses to many variations of conspecific male advertisement calls, with each stimulus tested in comparison to another. Each call was attractive above threshold in a no-choice situation, with females showing strong preferences for calls in the stimulus set over white noise or to heterospecific signals. The authors first tested for **transitivity**, which would suggest that choosers assigned an attractiveness value to a call independently of the attributes of other stimuli. In other words, if choosers have a preference for A over B and B over C, they should prefer A over C. There was no evidence for any model consistent with transitivity. Further, Kirkpatrick and colleagues (2006) found no evidence for strict **intransitivity**

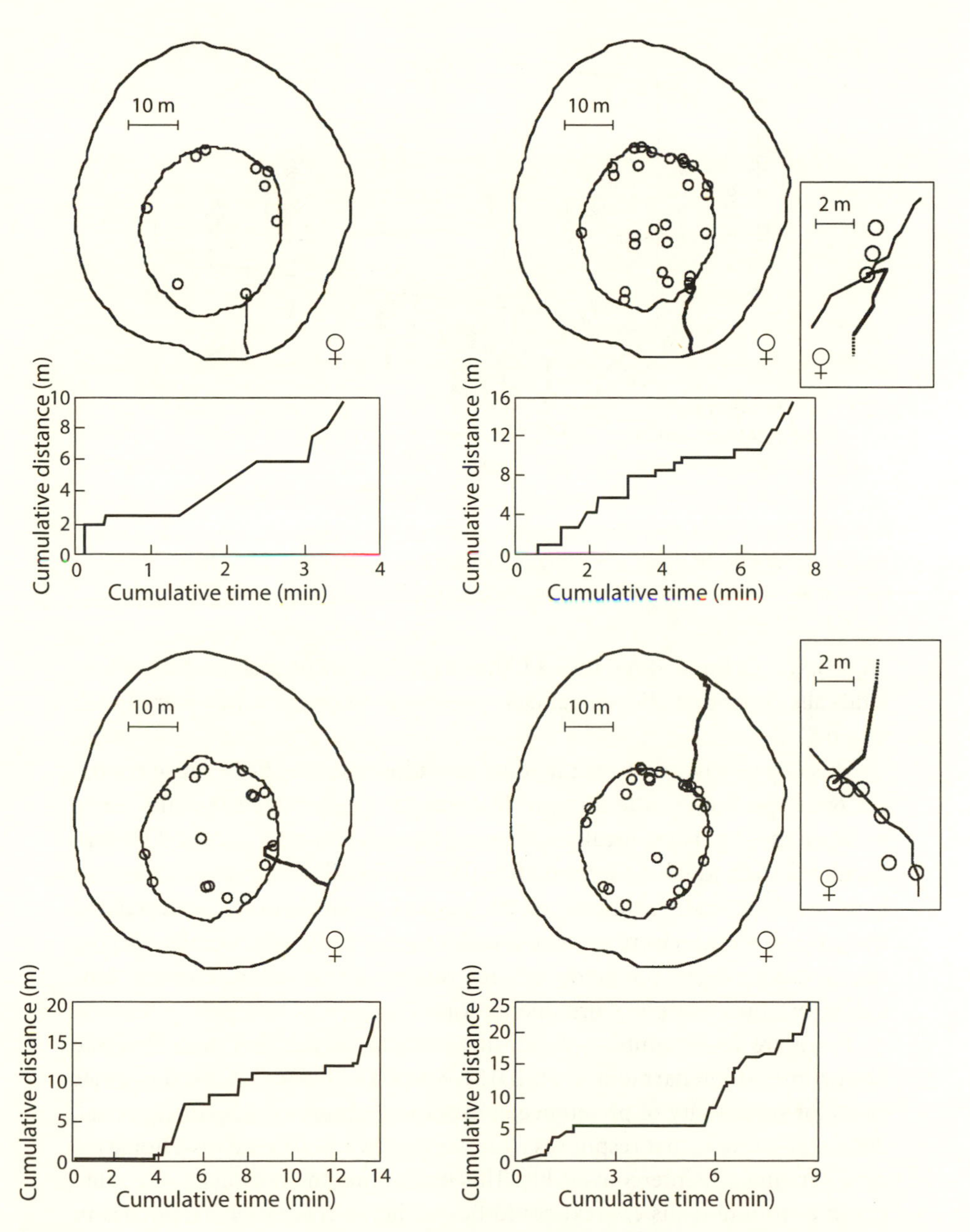

Figure 6.7. Paths taken by female gray treefrogs when sampling choruses of calling males. Outer shape represents the fence surrounding a pond, represented by the inner shape. Females were released at the fence line; thick line shows the path taken by each female. Graphs show cumulative distance moved over the sampling period (Murphy & Gerhardt 2002).

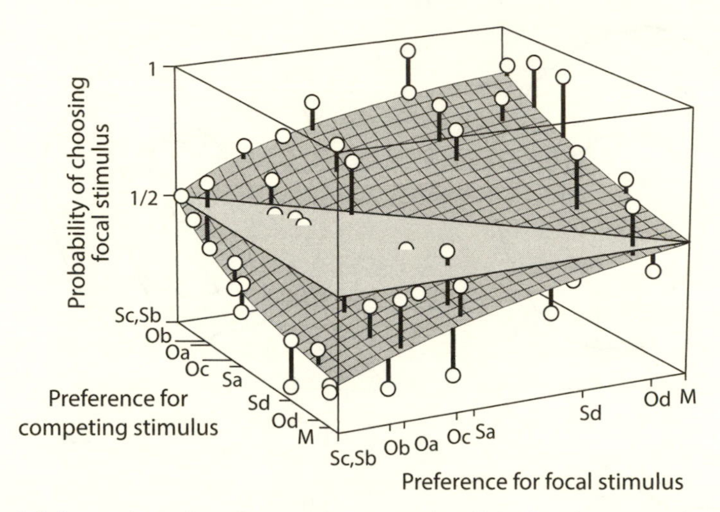

Figure 6.8. "Round robin" preferences for nine male calls by female túngara frogs in simultaneous-choice assays in Kirkpatrick et al. (2006). The curved and hatched surface shows expected results if preferences were transitive (Ryan et al. 2009).

(A > B, B > C, but C < A). Instead, the attractiveness of each of the conspecific calls depended idiosyncratically on which other call it was paired with (fig. 6.8).

This dependence on the individual stimulus pair may be because choosers were specifically attending to features that maximized the difference between those particular calls. For example, Calkins and Burley (2003) manipulated individual male traits in California quail, *Callipepla californica*. Females chose males primarily on the basis of variation in the manipulated trait. When females were presented with choices between naturally varying males, the strongest predictors of preferences were composite traits combining multiple aspects of ornamentation.

There are few examples of transitive choice in the literature (but see Dechaume-Montcharmont et al. 2013). Surprisingly, one of the few other cases for transitivity of preference comes from túngara frogs (Phelps et al. 2006), who found that responses from single- and simultaneous-choice assays were linearly interconvertible. This meant that the responses of choosers in a simultaneous context could be predicted from their responses in a sequential ("single-choice") paradigm, and were therefore independent of comparison stimuli. Kirkpatrick and colleagues (2006) excluded nonresponders from analysis, which may have obscured preference patterns (see below). However, Lynch and colleagues (2006) found that manipulating estrogen titers in females increased the probability of response, but did not affect choosers' discrimination in a simultaneous task. Phelps and colleagues (2006) used synthetic stimuli that manipulated conspecific and heterospecific call components, rather than natural variation which may have provided more possible features for comparison. The decision rules underlying mate choice are therefore still unclear even in one of the most thoroughly studied mate-choice models.

Another consequence of comparative evaluation is that irrelevant options can bias the outcome of choice. A well-known instance is the "decoy effect," where the characteristics of a third, irrelevant stimulus can determine the direction of a chooser's preference between two stimuli (Bateson & Healy 2005; Locatello et al. 2015; Lea & Ryan 2015). Along these lines, female fiddler crabs are more likely to mate with a large-clawed male if he is adjacent to a smaller male than if he is alone or adjacent to a larger male. Callander and colleagues (2013) suggested that this effect might be analogous to the Ebbinghaus illusion in human psychophysics, whereby a visual stimulus is perceived as larger if it is surrounded by stimuli smaller than itself and vice versa (fig. 6.9). The Ebbinghaus illusion is similar to the effects of prior courter attractiveness discussed above under adjustable thresholds (Lyons et al. 2014; fig. 6.4). Courters can dynamically adjust their relationships to other courters in light of their relative attractiveness; for example, Oh and Badyaev (2010) found that less-ornamented male house finches were more likely to switch social groups, and that "socially labile" courters had higher pairing success. Preference for a particular stimulus, or a particular courter, is thus heavily dependent not only on the physical environment (chapter 3) but also on the social environment formed by other courters.

The number of choices and the distribution of variation can also affect the process of comparison. More options and more variety among options can be costlier to evaluate and make it harder to apply simple heuristics, resulting in the so-called paradox of choice (see Kacelnik et al. 2011). In the context of feeding responses in gypsy moth larvae, Raffa et al. (2002) found that preferences were clearly expressed when subjects were given two choices, but that responses became inconsistent and statistically nonsignificant when the number of choices was increased. Lenton and Francesconi (2011) found that increasing the variety of potential mates—when controlling for the number of options—made humans less likely to choose attractive partners (the consensus option) in a time-constrained speed-dating task, and more likely not to choose at all.

Finally, choosers may be attending to group-level properties (Billeter et al. 2012). This may be the case if choosers sample multiple patches of courters

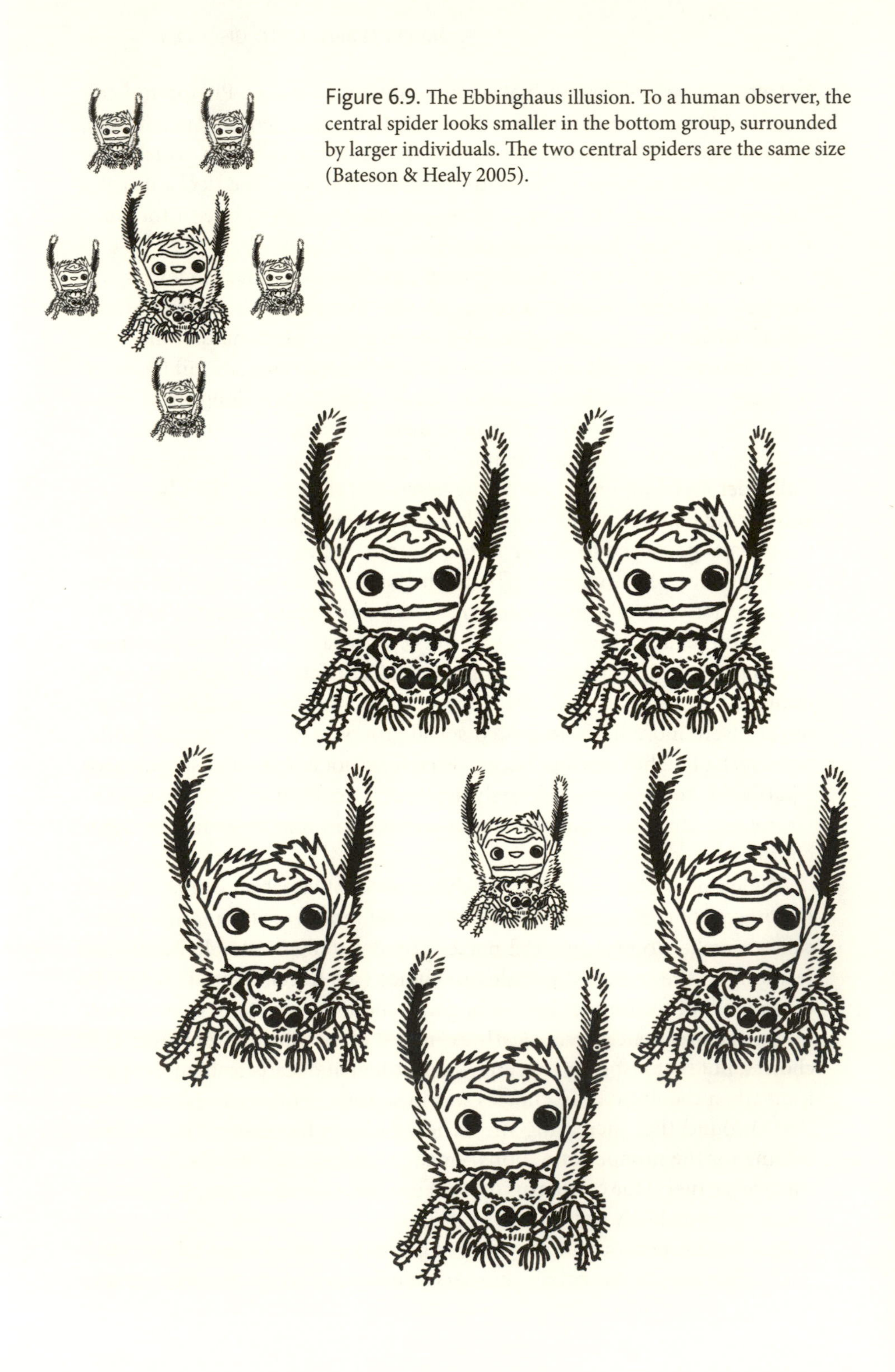

Figure 6.9. The Ebbinghaus illusion. To a human observer, the central spider looks smaller in the bottom group, surrounded by larger individuals. The two central spiders are the same size (Bateson & Healy 2005).

(Hutchinson & Halupka 2004), integrate signals from multiple courters until a threshold is reached (e.g., Stoltz & Andrade 2010), or if interactions between courters themselves generate attractive or aggressive signals. The decision to mate with an individual courter, and that courter's individual mating success, may stem from group as well as individual attributes.

If mate choice depends on relationships among a set of stimuli being tested, rather than the properties of individual stimuli, this raises a concern for how we measure preferences empirically (Calkins & Burley 2003). While manipulating ornaments independently removes confounding effects of correlated traits (chapter 2), we may be artefactually encouraging choosers to focus on a subset of differences they normally don't attend to.

6.7 SIMULTANEOUS AND DYNAMIC: BEST-OF-*N*, COMPARATIVE BAYES, AND RANDOM WALK

Janetos's (1980) best-performing algorithm was "best of *n*," whereby a chooser evaluates a fixed number of courters and selects the most attractive one. In a situation that is actually simultaneous, such as when a chooser receives continuous or repeating signals emitted by multiple courters at the same time, best-of-*n* could be achieved through a relatively simple "behavioral mnemonic" whereby a chooser simply spends more time orienting toward the most attractive chooser in a group, and is therefore subsequently more likely to move toward that stimulus (Akre & Ryan 2010). Even for a behavioral mnemonic, signals must somehow be individually labeled in order for a best-of-*n* strategy to work. How do choosers extract information and associate it with an individual chooser? Localization—pinpointing the source of a signal in space—is a critically important task in mate choice, since a chooser has to be close to a preferred courter in order to mate, and far away from a rejected one to minimize the risk of mating. The basic challenge of localization in at least one sensory modality (Bradbury & Vehrencamp 2011) is one that active choosers have to solve, no matter their decision rules.

If signals overlap in time, as they do in high-density choice situations like choruses and leks, choosers have to solve the so-called cocktail party problem of parsing out distinct signals from individual courters (Miller & Bee 2012). For example, in bushcrickets, *Tettigonia viridissima*, the responses of auditory neurons to a 45db signal are suppressed when the lower-intensity signal is given simultaneously with one at 60db, allowing a chooser to orient and move toward the louder stimulus (Römer & Krusch 2000). A similar phenomenon is the **precedence effect**. Here, two sounds that overlap in time are perceived as a single stimulus, and choosers orient to the first source. Precedence effects are found in mating responses of many anurans (Bosch & Marquez 2002) and insects (Snedden & Greenfield 1998) and are likely to drive the temporal dynamics of choruses.

If choosers or courters are mobile, a behavioral mnemonic is insufficient. Best-of-*n* becomes more of a cognitive challenge, because choosers must not only segment signals from different courters, but also retain distinctive representations of multiple courters in **working memory**. Akre and Ryan (2010) found that female túngara frogs have longer working memory for acoustically more complex calls, and suggested that working memory might be an important source of selection on signal repetition by courters (chapter 15). Working memory may thus place constraints on signal structure as well as on mate sampling.

Choosers must also associate the preferred signal with an individual Wiegmann (1999) showed that female field crickets *Gryllus integer* remember the location of a previously encountered call. If courters as well as choosers are mobile, **individual recognition**—behavior contingent on the unique phenotype of a particular individual—is required for comparison and repeated assessment (e.g., Luttbeg 1996; Castellano & Cermelli 2011) of mobile individuals. The neurogenetic basis of individual recognition in mammals is beginning to be well understood (Winslow & Insel 2004; Choleris et al. 2007) and may represent a mechanism for variation in choosiness. Individual recognition may also prove important in understanding preferences for novelty or "rare males," since individuals with distinctive traits may be easier to recognize and distinguish in memory. This may increase the probability of repeat visits to the "odd man out," and may also select for the repeated evolution of novel traits (chapter 15).

In addition to the cognitive constraints on best-of-*n*, the costs of searching for potential mates can make the strategy prohibitive (Luttbeg 2002; Wiegmann et al. 2010). This is exacerbated by the fact that cognitive constraints and environmental noise could require choosers to make multiple visits to estimate courter attractiveness. Luttbeg's (1996) "comparative Bayes assessment" algorithm incorporated uncertainty about courter attractiveness, which could arise due to sensory constraints in heterogeneous environments, discussed above, or due to internal noise in neuronal coding (Neuhofer et al. 2011). Similar to an adjustable-threshold model, choosers start with a prior estimate of courter trait distributions, which they then update through repeated sampling of multiple individuals. The "random walk" algorithm of Castellano and Cermelli (2011) is an adjustable-threshold model that uses an approach similar to Luttbeg (1996) and Phelps and colleagues (2006) in that choosers acquire unreliable information until a criterion is reached, and mate with the first courter to exceed that criterion.

6.8 SAMPLING MULTIPLE TRAITS

Choosers often attend to courter displays in a predictable order. Studies in multiple systems have shown that choosers often assess courter traits sequentially, with long-range cues serving to filter the pool of males that is subsequently visited; Uy and Safran (2013) review studies from 11 species where choosers attend first to long-range cues in one modality, and then to short-range cues in another (e.g., Rintamäki et al. 1995). The mate-searching algorithms operating in one step may be different from those operating in the next; for example, choosers could use an adjustable threshold to select a small set of courters, which in turn are evaluated with a random-walk algorithm. Applying multiple strategies to multiple stages may be instrumental in maximizing the marginal benefit of choice (chapter 14). Indeed, Luttbeg and colleagues (2001) showed that different stages of the mate-sampling process in sticklebacks were differentially sensitive to manipulated costs. The notion of successive filters in mate sampling is similar to Miller and Todd's (1998) "sequential aspiration" model of human mate choice and Hutchinson and Gigerenzer's (2005) "Take the Best" algorithm. In both of these, prospective mates are progressively winnowed on the basis of traits that are progressively costlier to assess. An interesting problem, which I will return to in chapter 8, arises when choosers face a trade-off when evaluating two sets of traits (e.g., Thünken et al. 2012).

Uy and Safran (2013) also reviewed studies where choosers attend to multimodal courter displays that are concurrently presented; here, choosers are displaying, say, visual and acoustic cues concurrently. In some cases, detectability of one of the cues acts as a low-cost filtering algorithm as with sequential sampling, above. Studies of perceptual attention and mate choice (Yorzinski et al. 2013; Suschinsky et al. 2007) can tell us whether simultaneous evaluation is an accelerated version of sequential sampling, with unacceptable mates filtered out after a quick assessment.

In addition to using cues as successive filters, choosers can also simultaneously integrate information from multiple cues to determine a courter's attractiveness (independently, or in comparison to other courters). Integrating multiple traits may allow choosers to estimate courter attractiveness more precisely and with greater computational efficiency, particularly if traits are correlated among choosers (Rowe 2000; chapter 4), and may be necessary for choosers to individually identify courters (section 6.7). Multiple traits may increase working memory (Akre & Ryan 2010), facilitate perceptual binding of spatially or temporally disjunct aspects of a courter's signal (Taylor & Ryan 2013), or interact to address low-level perceptual mechanisms (Endler et al. 2014). These may increase or decrease the cost of assessment and/or the reliability of signals for choosers.

6.9 RECOGNITION

The mate-choice literature often refers to mate recognition, often in the context of choosers attending to conspecific sexual signals but not to heterospecific signals (Paterson 1985; chapter 16). It is important to note that this term is very distinct from individual recognition, discussed above. Empirically, choosers recognize signals, or combinations of signals, if the signals are preferred over a null control (silence, white noise, blank screen) in no-choice trials (e.g., Kárpáti et al. 2013; fig. 6.5). Recognition is formally a fixed-threshold rule, and defines the boundary below which signals are rejected even if no other courters are available. But as noted above, manipulating the effective distribution of courter signals can have marked effects on threshold responses. As the prison effect attests, though, the envelope of recognition can be broadened *in extremis*. Even baseline recognition is incredibly broad among males in many **polygynous** species, who frequently exhibit sexual behavior toward heterospecifics, dead individuals, and inanimate objects (Moeliker 2001; Gwynne & Rentz 1983). At the other extreme, in many cases the set of recognized signals is nonexistent for individuals outside the window of sexual receptivity.

In the stricter sense, a courter signal is recognized if there are some circumstances under which it elicits proceptive behavior in sexually receptive courters; a signal is not recognized if it fails to elicit a mating response even when hell freezes over. The window of recognition is important because it defines the outer bounds of stimulus space that are relevant to choosers; outside this window, choosers simply fail to attend to stimuli as sexual signals.

6.10 EXECUTING CHOICES

6.10.1 Mating decisions encompass acceptance and rejection

Mating decisions ultimately result in differential behavior by choosers toward courters. We normally think of these in terms of choosers exhibiting proceptive sexual behavior toward chosen individuals and either ignoring those not chosen or exhibiting aversive behavior toward unattractive courters (Kavaliers et al. 2006). In some cases, however, mate choice is almost entirely driven by differential aversive behavior, especially if courters mainly secure matings by coercing and harassing choosers. In these systems, mating decisions are mainly a function of behavioral resistance to mating, with preferred—or least distasteful—courters eliciting the least amount of resistance (e.g., Shuker & Day 2002). I discuss the appropriateness and extent of applying "choice" to coercive situations in chapter 15, in the context of sexual conflict.

The timing of acceptance versus rejection is likely to be asymmetric over the course of a mating decision. Choosers, particularly females, may often reject inappropriate courters very quickly, even before these have completed a signaling bout (Clemens et al. 2014), but may take much longer to commit to a mating (Uy et al. 2001). Behavioral experiments where we play back non-tactile signals may therefore tell us more about which potential mates are filtered out as unacceptable than about which courters actually end up mating. Careful analysis of each stage of choice will be pivotal in understanding the evolution of multiple chooser preferences for multiple courter traits, as well as the evolution of mate-sampling algorithms.

A chooser's mate-choice decision can also go well beyond mating or not mating. Rejection (and for that matter acceptance) can entail physical injury or cannibalism of courters (Houde 1997; Andrade 1996; Stoltz & Andrade 2010). Behavioral mating decisions by choosers can therefore have the direst of fitness consequences for courters. Further, as should become clear in the next chapter, mating is in many ways only the beginning of the decision for choosers, who will allocate differential fertilization and differential resources to the gametes of different courters.

6.10.2 Choosers respond interactively to courters

Choosers also actively engage with courters, inciting courtship and sometimes agonistic interactions between courters. Female canaries attend to vocal agonistic displays between males and prefer the displays of males that overlap their opponents (Leboucher & Pallot 2004). Inducing courters to repeat displays can improve assessment from the point of view of choosers (Luttbeg 1996; Mowles & Ord 2012). These interactions blur the line between Wiley and Poston's (1996) "direct" mate choice governed by female decision rules, and "indirect" mate choice arising from courter responses to choosers. In mutual mate choice (chapter 8), the outcome is a consequence of reciprocal behavioral interactions between individuals.

6.10.3 Courters can override the mating decisions of choosers

Intrasexual competition can prevent choosers from expressing their preferences (Wong & Candolin 2005). Further, "courters," almost always males in this case, can override preferences through forced copulation and the prevention of remating. For example, in stream water striders, *Aquarius remigis*, female mating opportunities, or lack thereof, are largely determined by the aggressiveness of nearby males (Wey et al. 2015).

After forced copulation, a female can still exercise choice with regard to fertilization and investment in progeny. The effect of forced copulation on postmating choice will be discussed in the next chapter.

6.10.4 Choosing nobody: responsiveness is data

If responsiveness is synchronized among choosers, or if choosers are mostly unresponsive, this acts to increase competition among courters and therefore sexual selection (Partridge & Endler 1987; Shuster & Wade 2003). For choosers, sexual responsiveness is one mechanism of expressing preference. Failure to exhibit proceptive behavior may be an advantageous strategy in the absence of suitable mates (chapter 14). Along with selfing in hermaphrodites, choosing not to mate represents a decision that can positively or negatively impact chooser fitness and that has consequences for courter fitness and for genetic exchange among lineages.

But unresponsive individuals—those who choose not to mate with anybody—are largely discarded from our data sets and from our thinking. Unresponsive individuals (usually defined according to some behavioral criterion like failing to visit both sides of a stimulus arena) are excluded from analysis, since failure to respond might be an artifact of husbandry conditions or stimulus deficiencies (chapter 2). When we analyze simultaneous data, we compare the behavior of responsive individuals; and when we model mating decisions, the possible outcomes are typically restricted to choosing one courter or choosing another.

Phelps and colleagues (2006) raised a concern with the widespread practice of modeling mating decisions as forced-choice tests, where it is not possible to withhold a response. In a forced-choice test, the decision inevitably reduces to the difference in attractiveness between two stimuli, rather than depending on their absolute attractiveness. They argued that unforced-choice tests, which offer subjects the option to withhold a response, may provide a better representation of natural variation in response and preference. If nonresponders are included in analyses, however, positive controls are required to control for artifactual variation in responsiveness.

6.11 THE MARGINAL COST OF SAMPLING AND CHOICE

The entire edifice of sexual selection theory is built on largely untested assumptions about the costs and benefits of mate choice (chapter 14). Mate choice is not expected to evolve at all unless the indirect and direct fitness benefits of choosing at least equal the costs. Accordingly, the literature on mate sampling has focused on building algorithms that confer the biggest net payoff to choosers, in terms of the costs of sampling versus the benefit of mating with the most attractive chooser. Whatever these payoffs may be, if a chooser mates with a courter at random, the net benefit will, on average, equal the net benefit of mating with the average available courter. I will return to the costs and benefits of mate choice in chapter 14.

Any mechanism to detect sexual partners is by definition a choice mechanism, since it induces sexual behavior in response to some signals and not others. Much of the process of mate choice—particularly the routine peripheral filtering of unacceptable signals—falls under the category of what Parker (1983) calls "passive" (chapter 3). But mate choice—in this case, under a fixed-threshold sampling rule—occurs even if choosers are maximally lazy and simply mate with the first positively marked chooser they detect, since some courters are more likely to be detected than others. Mate choice can thus occur without any investment in sampling behavior or cognitive architecture. Responsiveness, choosiness, and peak preference can still be modulated by varying the envelope of signals that choosers attend to.

It is also important to keep in mind that *not choosing* can be costly. If choosers were to mate only with the first courter they detect, they would have to reject mating attempts from all subsequent courters, which is energetically expensive (Shuker & Day 2001). Conversely, if choosers were to mate with every male they encountered, they would likely incur survivorship costs (Priest et al. 2008; Franklin et al. 2012). Either way, mate choice could occur in the absence of mate sampling, through differential detection in the first case and through postmating processes in the second.

In chapter 14, I will return to the question of why individuals should sample and which rules they should adopt. Most attention has focused on the evolution of preferences, rather than sampling strategies, but as Lindström and Lehtonen (2013) observed, "*how* females sample mates is more sensitive to external conditions (especially costs) than *which* males they prefer" (italics theirs). Sampling is costly not only in terms of cognitive effort and time (Chittka et al. 2009) but also in terms of risk and energetic expenditure.

Choosers' ability and motivation to sample is shaped by the environment in which sampling takes place. One of the few generalizations we can make about the evolutionary ecology of mate sampling is that increasing search costs, whether by increasing risk or locomotor effort, reduces the number of courters sampled and thereby decreases choosiness (Alatalo et al. 1988; Lindström & Lehtonen 2013). For example, Milinski and Bakker (1992) found that female sticklebacks in still water were choosy, preferring brightly colored over dull males. This preference disappeared when females were forced to swim against a current. Booksmythe and colleagues (2008) found that female fiddler crabs *Uca mjoebergi* failed to express their preferences for a male signal as travel distance increased. Strong preferences expressed under a low-cost situation disappeared when females had to travel to sample males. In humans, geographic propinquity has historically been the major predictor of mating outcomes (Lewontin et al. 1968), an effect only partially attenuated by modern phenomena like online dating (Rosenfeld & Thomas 2012). The relationship between latent preferences and expressed choices is a tenuous one, mediated by the circumstances under which choice occurs.

It is worth noting that the environment for choice is partly comprised of courters and the signals they produce. Courters can either inhibit mate sampling, as through coercion of choosers and aggression toward other courters, or facilitate it, as through redundant and conspicuous signals that make it easier for choosers to detect and discriminate among signals. Courters can thus make it easier or harder for an individual to exercise choice for a given amount of effort.

Many of the sources of variation in mating preferences (chapter 9) manifest themselves via sampling costs. Predation increases the risk of traveling from place to place, attending to courtship displays, or associating with conspicuous courters. Environmental disturbance reduces the active space of courter signals (chapter 3) and therefore the number of courters that can be sampled in a given location. And poor chooser nutritional condition reduces the amount of time and energy that can be diverted to mate choice. When sampling costs are higher than the expected payoff of choosing a preferred courter, choosers should mate at random.

The few empirical studies of the cost of choice suggest that the cost to choosers can range from small to substantial. Byers and colleagues (2005) estimated energy expenditures of female pronghorn antelope (*Antilocapra americana*) using a "sampler" strategy of switching between male-defended "harems" over the course of their estrus period, versus "quiet" females remaining in the same harem. They estimated the cost of choice as being equivalent to about half a day's metabolic expenditure, which they in turn calculated as being 59 times higher than the cost for female sage grouse sampling on leks (Gibson & Bachman 1992). Such empirical measures of the costs and benefits of choice are indispensable for us to be able to make predictive models of how mate choice operates across mating systems.

6.12 SYNTHESIS

We focus a great deal of attention on *preferences*, the internal properties of choosers that bias mating decisions toward certain types of courters over others. This is because we can readily measure their behavioral proxies and some of their mechanistic underpinnings, and also because they are individual phenotypes for which we can in principle measure genotypic effects and response norms to individual variables. But preferences only have meaning insofar as they are expressed through mate sampling and mate choice. Preferences abstracted from choice are convenient operational constructs, but conceptually misleading, for two reasons.

First, mate preferences are dynamically coupled to mate sampling. Preferences underlie sampling decisions, and in turn are shaped by the sampling experience. Second, insofar as we observe effects of environment, experience, or genotype on preferences, these effects are perhaps most often due to their effects on sampling behavior. Any discussion of preference evolution, or of the evolutionary consequences of preferences, must therefore take place within an understanding of how mate sampling operates and how sampling responds to proximate factors. Preferences are only exposed to selection when they are manifested as choices, and mate sampling is the process that translates preferences into choices.

Different forms of sampling recruit different cognitive mechanisms, but in general there may be hard trade-offs between the sophistication of the perceptual mechanisms used to sample courters and the number of courters that must be sampled in order to make a mating decision. More accurate and precise perceptual discrimination among courter signals enables choosers to exhibit steeper preference functions, thereby increasing the ability to assess differences among courter phenotypes. A similar effect is obtained by attending to more dimensions of courter traits. A more complex cognitive apparatus may thus reduce the number of courters that a chooser has to sample.

The breadth of the mechanisms involved in making mating decisions is evident when we think about one axis, density. If courter signals are sparsely distributed, then choosers will have to use fixed-threshold rules unless they can represent the state of previously encountered courters in memory. If courters are experienced simultaneously, choosers will have to attend to differences among courters unless they can remember *and* segment individual courters. The cognitive constraints involved in choosing may canalize the coevolution of courter signals and chooser preferences.

We are just beginning to gain a basic characterization of mate sampling, in terms of documenting inter- and intraspecific variation in sampling rules at different stages of preference, although we can rule out some algorithms by analyzing patterns of observed sampling behavior, and in some cases ascertain strategies through careful experimentation. Dougherty and Shuker's (2014) meta-analysis of 38 studies showed that choosiness was much stronger in simultaneous versus sequential experimental paradigms (chapter 2). This means that choosers generally do not have a fixed threshold for mate choice independent of the courters they experience. This could be because choosers are applying a single adjustable-threshold rule in both kinds of comparisons, or because in the simultaneous tests they are attending to differences among choosers rather than absolute attributes (Ryan & Taylor 2015). Elucidating the rules that connect distributions of courter traits to chooser mating decisions is essential to understanding mate choice in an evolutionary context.

A chooser may accept a mating from a courter, but mating with a chooser is no guarantee of a fitness benefit for a courter. The next chapter covers choices made after mating. In chapter 8, I return to mating decisions, but in the context of mutual assessment and mutual choice.

6.13 ADDITIONAL READING

Bateson, M., & Healy, S. D. 2005. Comparative evaluation and its implications for mate choice. *Trends in Ecology & Evolution,* 20(12), 659–664.

Kirkpatrick, M., Rand, A. S., & Ryan, M. J. (2006). Mate choice rules in animals. *Animal Behaviour,* 71, 1215–1225.

Miller, C. T., and Bee, M. A. 2012. Receiver psychology turns 20: is it time for a broader approach? *Animal Behaviour* 83 (2), 331–343.

CHAPTER 7

Mate Choice During and After Mating

7.1 INTRODUCTION

In the last chapter, I described the decision rules that choosers use to translate preferences into realized matings. An almost definitional assumption when we think about mate choice is that it is these matings that are the endpoint of chooser decisions and of mating success. As described by Eberhard (2009), this point of view characterized Darwin's (1871) approach to sexual selection and predominated for the next century through the genesis of sexual selection theory. An influential paper by Parker (1970) pointed out that courters continue to compete after mating has occurred. Parker focused on **sperm competition**. Just as males compete agonistically for mating opportunities, so do they compete for fertilization, whether through direct interactions between gametes or a host of other mechanisms (reviewed in Birkhead & Møller 1998). What we think of at first blush as "mate choice"—a behavioral decision to engage in sexual activity with a given partner—is usually only the first in what can be a lengthy series of decisions made by choosers, extending through fertilization, the allocation of resources to gametes, and in many species to gestation and parental care.

Chooser mechanisms before mating were long thought secondary to mating outcomes; chooser mechanisms after mating were long thought secondary to fertilization outcomes. However, evidence that reproductive decisions happened during and after mating began to emerge with the renewed interest in mate choice in the mid-1980s (Birkhead & Møller 1993). In a paper that is "seminal" by multiple definitions, Thornhill (1983) showed that postmating processes were the primary mechanism for mate choice in female scorpionflies (*Harpobittacus nigriceps*). Premating choice was rare: in about 3.5% of observations, females would steal a nuptial gift (insect prey) from a male and then mate with another male. In all other cases, acceptance of a nuptial gift occurred in tandem with mating. Females mated longer when consuming a larger nuptial gift (peri-mating choice) and were less likely to remate (postmating choice). Both the size of the gift and the size of the male were positively associated with female **fecundity**. In this species, therefore, peri- and postmating mechanisms expressed in sequence played a bigger role in female preference than premating choice did.

The years following Thornhill's (1983) paper were marked by considerable controversy over the importance of cryptic choice (e.g., Ward 1993; Simmons et al. 1996; Birkhead 1998), largely because of the methodological difficulty of disentangling choice from other processes—notably sperm competition—that could result in differential fertilization. These methodological challenges, and the experimental approaches that have largely neutralized them, are discussed in section 7.3 below. There is no longer any controversy over whether cryptic choice is important to sexual selection and reproductive isolation, but its importance remains relatively unappreciated. The opportunity for cryptic choice occurs whenever choosers mate multiply, which they nearly always do. The focus of this chapter is on peri- and postmating choice, starting with intimate interactions when individuals are mating or about to mate, and continuing through differential fertilization and treatment of offspring. At this point, choosers have made a decision to mate with a courter (chapter 6), but can still exercise choice in a multiplicity of ways that are important to their fitness and dispositive to courter reproductive success. For semantic convenience and consistency with the earlier literature, I will use the term **cryptic choice** to collectively refer to peri- and postmating processes (fig. 7.1). As described in chapter 1, the term "peri-mating" (Fedina & Lewis 2008) refers to intimate interactions when a chooser and a courter (or, in the case of broadcast spawners, their gametes) are in intimate physical contact. Postmating choice occurs after individuals have physically separated.

Peri- and postmating are useful partitions from a behavioral point of view, and have the advantage that they apply equally to reproductive and non-reproductive sexual interactions. When thinking about chooser life histories and the potential for genetic exchange, however, a more important distinction is between **prezygotic** and **postzygotic** processes. As I will argue below, cryptic prezygotic choice is in many ways qualitatively similar to premating choice, since in both cases choosers have the opportunity to sample multiple courters. With postzygotic choice, choosers have already committed to a set of courters and must make decisions about how to allocate resources between them.

Just as choosers need to sample multiple courters to exercise premating choice, they need to mate with multiple courters (i.e., be **promiscuous**) to exercise cryptic choice. This chapter, like the previous ones, assumes that choice is a one-way process, and focuses on females choosing among male sperm or ejaculates. Cryptic choice by males, in the context of strategic

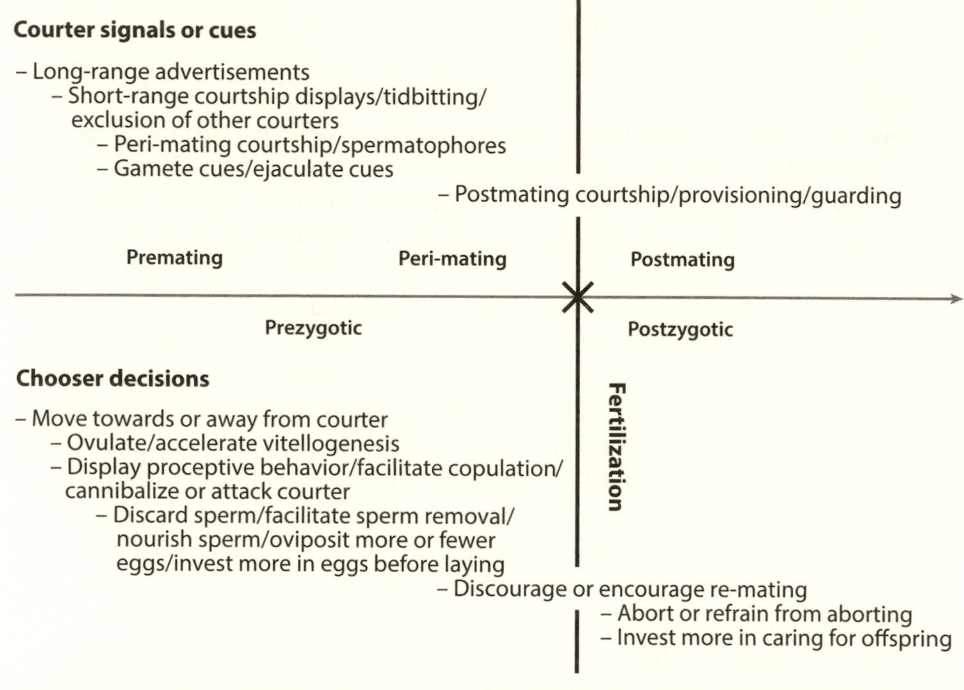

Figure 7.1. Stages of mate choice, revisited. Chooser decisions are in part adapted from Eberhard (2011). Timing of fertilization (peri- or postmating) varies between species, as does the window over which each type of decision operates. Ovulation, for example, can be induced by pre- or peri-mating cues, and remating can occur both before and after fertilization from a previous courter. Abortion and differential care, by definition, occur after fertilization.

ejaculate allocation (Wedell et al. 2002), is discussed in the next chapter. It is very often the case that sperm from multiple males coexist in a female's reproductive tract, or in the environment surrounding eggs of broadcast spawners. Even if mating is sequential, choosers must decide what resources to allocate to a current mate and what resources to save for subsequent matings (Sheldon 2000; Edward 2014).

Mating, therefore, is just the end of the beginning in mate choice. Eberhard (2011) summarizes his encyclopedic review of candidate mechanisms (Eberhard 1996) with 23 "female processes whose selective performance can result in cryptic female choice." The reader is referred to Eberhard (1996, 2011) for details; here I highlight a selection of studies showing evidence for cryptic choice and reinforce the point that most of mate choice is hidden from view.

7.2 REMATING AND CHOICE OF MULTIPLE MATES

The overriding decision in cryptic choice is whether to mate multiply and how often to mate. Just as the number of courters sampled places an upper bound on premating choice, so does the number of mates place an upper bound on cryptic choice. As I explore further in chapter 15, remating rate is a prime theater for sexual conflict in promiscuous animals (Rice 1996; Alonzo & Pizzari 2013). By mating multiply with preferred males, or by mating closer to a window of sexual receptivity (Eberhard 1996), females can increase the odds of fertilization by a male with a preferred phenotype (Simmons 1986, 1987). Multiple mating mitigates the cost of mating with genetically incompatible mates, including close relatives and heterospecifics (chapter 14). There is some evidence that the tendency to remate is modulated by premating preferences. Pitcher and colleagues (2003) showed that female guppies were reluctant to remate unless the second male was more ornamented than the first, in which case the second male also had a disproportionately greater share of paternity. This "trading up" (Pitcher et al. 2003) may be a way to mitigate downward adjustment of a threshold of mate acceptance (chapter 6).

At the same time, it is in a courter's interest to inhibit remating in a chooser he's mated with, and to induce it in a chooser previously mated to another. There is therefore **sexual conflict** (chapter 15) over remating. This is particularly the case since most females exhibit so-called *last-male sperm precedence*, whereby the last male to mate before fertilization sires a disproportionate share of the offspring (Birkhead & Pizzari 2002). Males of internally fertilizing species have an array of approaches to delaying remating, ranging from aggressive mate guarding and injury to female genitalia, to the production of mating plugs that act as physical barriers—chastity belts—to intromission (Eberhard 1996). Males frequently produce substances that act to delay remating. In scorpionflies *Panorpa cognata*, male saliva secretions provided as nuptial gifts contain substances that inhibit remating (Engqvist 2007). In *D. melanogaster*, male ejaculates induce non-receptivity in females, thereby delaying remating, and also induce a female to oviposit more eggs; moreover, ejaculates are toxic, reducing female survivorship (Rice 1996). Females, however, are not passive recipients of this manipulation. In both *D. melanogaster* (Pitnick et al. 2001) and houseflies *Musca domestica* (Andrés & Arnqvist 2001), female responses coevolve with male signals, such that females show an attenuated response to manipulation from males of their own genotype.

Courters can also act to hasten remating by choosers who have previously mated with another. The best-known example of this is the so-called Bruce effect in rodents, whereby exposure to a novel male during early pregnancy induces abortion of previously implanted offspring and renewed sexual receptivity (Bruce 1960). This response appears to be mediated by chemosignals, as the soiled bedding of a novel male is sufficient to elicit the Bruce effect. Again, however, females are not passive, and can actively avoid trigger stimuli. The effect is variable and highly dependent on age, stage of reproduction, and social context (Schwagmeyer 1979; Eberhard 1996). This effect has been studied almost exclusively in laboratory populations, where, among other concerns, it may be exaggerated by close confinement of females with novel males. Mahady and Wolff (2002) found weak evidence for the Bruce effect in wild voles. Recently, however, Roberts and colleagues (2012) found a strong effect in wild gelada monkeys (*Theropithecus gelada*), with 80% of females terminating pregnancies following replacement of the dominant male in a social group. There are few studies of how the Bruce effect covaries with courter phenotype (besides the definitionally context-dependent traits of social dominance and novelty). Male-pregnant pipefish (*Syngnathus scovelli*) disproportionately abort the eggs of unattractive females (Paczolt & Jones 2010).

Choosers can also exert behavioral control over the distribution of courters with whom they mate. Female Tanganyikan cichlids (*Julidochromis transcriptus*) select spawning sites strategically to favor **polyandry**, thereby reaping putative indirect benefits as well as direct benefits of parental care from multiple males. They do so by selecting wedge-shaped structures in the substrate, with the smaller end of the wedge occupied by a subdominant male who would otherwise be competitively excluded by a larger male, who occupies the larger end of the wedge (Kohda et al. 2009; Li et al. 2015).

Interactions over mating rate thus illustrate the point that all stages of mate choice involve the manipulation of choosers by courters, but also involve plastic, evolvable mechanisms that choosers can use to bias responses in favor of one courter or another. As pointed out previously (Rosenthal & Servedio 1999), and as I will return to in chapter 15, some mechanisms of manipulation (like advertisement calls) are easier to ignore than others (like physical damage to the genitalia).

7.3 BIASING FERTILIZATION

7.3.1 Overview

Fertilization decisions have the potential to encapsulate the totality of mate choice: during or after mating, a chooser can devote her entire reproductive output to a single courter, or, at the other extreme, utterly reject all his gametes. Fertilization can happen immediately upon gamete release, as in broadcast spawners; or sperm can be stored for months after insemination. Females in many species have spermathecae, specialized organs for sperm storage, and some of these organs provide nutrients to sustain sperm for long periods of time (Eberhard 1996). In some cases, therefore, a female may choose a male well after the male himself has died (López-Sepulcre et al. 2013).

Fertilization bias plays a particularly important role in reproductive isolation. When choosers are mated to both conspecifics and heterospecifics, fertilizations are often strongly biased towards conspecific (Zeh & Zeh 1997). This phenomenon, known as **conspecific gamete precedence** (and primarily in the context of *conspecific sperm precedence*) is widespread in plants and animals (Gregory & Howard 1994; reviewed in Howard 1999; Lorch & Servedio 2007). I will return to conspecific sperm precedence in chapter 16.

As noted above, the notion that cryptic choice plays a role in fertilization bias initially met with skepticism, in large part because of the methodological challenges involved in disentangling chooser- from courter-driven processes. In chapter 2, I outlined the standard approaches for measuring premating preferences. Observational approaches, including direct measures of realized parentage, suffer from the entanglement of mate choice with intrasexual competition (chapter 6). We therefore turn to experiments in which choosers are presented with a set of courter cues, either sequentially or simultaneously, whereupon we test for differential behavioral (or physiological, or genomic) responses as a function of courter phenotype. As detailed below, the same kind of manipulative approach can be applied to some aspects of cryptic choice, like differential fertilization as a function of peri-copulatory courtship (Eberhard 2011), interactions between sperm and ovarian fluid (Evans et al. 2013), or postmating effects of premating stimuli (Limbourg et al. 2004). For external fertilizers, one can even perform simultaneous choice tests on sperm that are directly analogous to tests on animals (Oliver & Evans 2014).

For internal fertilizers, studying fertilization bias involves measuring skewed parentage toward different courters' sperm simultaneously present in the reproductive tract. A standard approach is to have a female mate with two males in succession; the response assay is $\mathbf{P_2}$, the proportion of fertilizations assigned to the second male. This measure makes it difficult to disentangle cryptic choice from sperm competition (Birkhead 1998). Since sperm are difficult to genotype directly (but see below), P_2 is typically measured with parentage analysis (chapter 2), which carries the additional complication that fertilization bias is conflated with differential embryo mortality or abortion.

P_2 becomes a powerful assay, however, when combined with experimental manipulations of courter signals or chooser responses. For example, in two species, P_2 is significantly higher when females are anesthetized, thereby disrupting sensation and cognitive processing. This indicates a role for chooser behavior during or after mating in biasing paternity (arctiid moths *Utetheisa ornatrix*, LaMunyon & Eisner 1993; red flour beetles, Fedina & Lewis 2004).

It is sometimes possible to directly characterize sperm cells from different sires. Simmons and colleagues (1999) used different radioisotopes to label sperm from different males, which allowed a coarse measure of sperm concentrations in different parts of the female reproductive tract. A relatively new technique, competitive PCR (Bretman et al. 2009; Bussière et al. 2010; Hall et al. 2010), enables one to measure the relative contribution of different males to the population of sperm stored in the reproductive tract (or in a specific area of the female genitalia) prior to fertilization. This approach requires each potential sire to have alleles at a given locus (like a microsatellite marker) that are unique to that particular male. The proportion of stored sperm attributed to that male can then be computed by comparing the sequencer peak area for his allele relative to a standard curve obtained by mixing known quantities of sperm. In field crickets *Gryllus campestris*, competitive PCR reveals that conspecific sperm precedence arises both from biased storage in the spermatheca and from biased fertilization by stored sperm (Tyler et al. 2013; see below). Finally, in genetic model systems, recent advances permit the direct visualization and quantification of gametes in the reproductive tract using transgenic sires producing different colors of fluorescent-protein expressing sperm (e.g., Manier et al. 2010).

7.3.2 The morphology of "female control"

In internal fertilizers, females as a rule do not make it easy for sperm. The theater for much of cryptic choice is an intricate genital apparatus of impressive complexity, characterized by a chemical environment that serves as a choice mechanism (see below) and by morphological structures that play an important role in fertilization decisions. As with premating choice, where the focus is typically on the signals produced by courters rather than the mechanisms choosers use to evaluate them, studies of biased fertilization are generally biased toward male processes: male reproductive structures have received much more attention than female genitalia (see meta-analysis by Ah-King et al. 2014).

This bias toward male structures is unfortunate, given the remarkable intricacy and diversity of female genital morphology (reviewed in Eberhard 1996) and its pivotal role in determining the consequences of cryptic choice. In some cases, morphology provides a primary choice mechanism when chooser behavior is circumvented by forced copulation; for example, the "corkscrew" oviduct of female ducks spirals counterclockwise, while the penis of conspecifics spirals clockwise; experiments with artificial genitalia suggest that the morphology by itself hinders fertilization when copulation is forced (Brennan et al. 2010; fig. 7.2). In other cases, intricate genital morphology hinders direct interference by males attempting to remove or displace a rival's sperm (Eberhard 1996).

Female genitalia provide us with the only straightforward morphological evidence of mate choice, thereby allowing us to test comparative hypotheses about mate-choice evolution using data gleaned from preserved

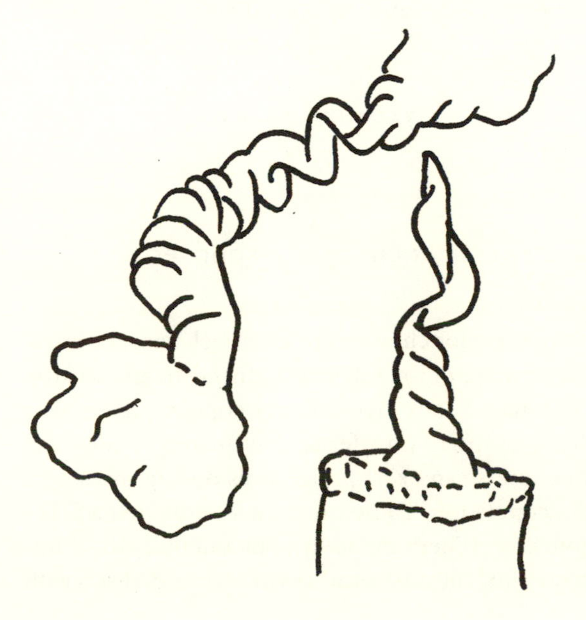

Figure 7.2. Genitalia in male and female ducks (*Anas* sp.). The female oviduct (left) spirals clockwise; the male penis (right), counterclockwise. After Brennan et al. (2010).

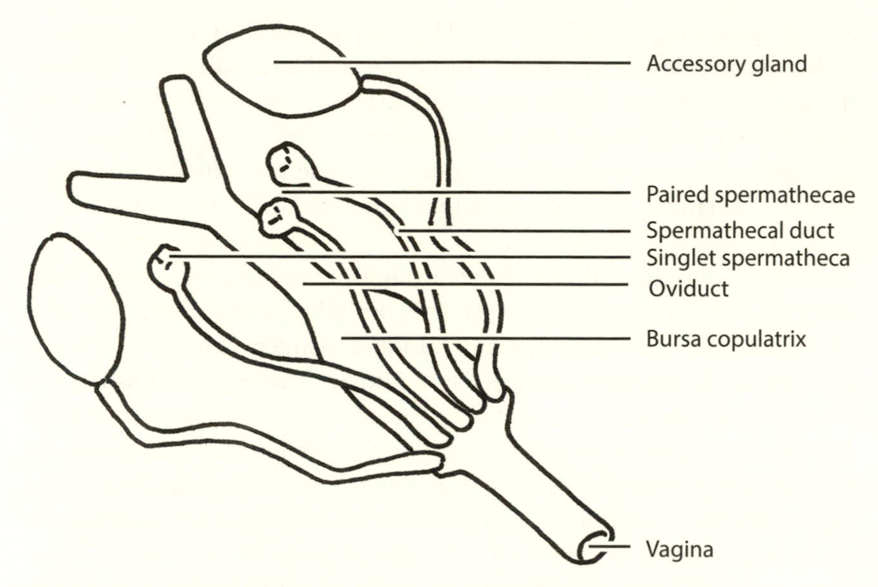

Figure 7.3. Internal genitalia of female yellow dungfly (after Ward 1993).

specimens. Pitnick et al. (1999) showed that spermathecae (fig. 7.3) had been gained or lost 13 times in *Drosophila* and allied genera. They suggested that the spermatheca was primarily recruited for long-term storage of sperm. The length of the short-term storage organ, the seminal receptacle, was highly variable among species and coevolved with sperm length. Similarly, Higginson et al. (2012) recently showed that spermathecal structure coevolved with sperm traits in diving beetles. Genital morphology is thus an exciting system in which to study coevolution of chooser traits with courter phenotypes (chapter 15).

7.3.3 Sperm uptake and rejection: chooser responses to peri-copulatory courtship

Genital morphology sets the stage for choosers to exert behavioral control over fertilization. One consequence of the intricate pathway to fertilization is that insemination is not instantaneous. Accordingly, copulation duration is in general a good predictor of sperm transfer and ultimately fertilization success. This means that choosers can exert preferences during mating by prolonging or interrupting copulation. These preferences are often a function of peri-copulatory courtship (Eberhard 1996). For example, the duration of oral sex immediately before intromission is positively correlated with copulation duration in two fruit bat species (cunnilingus in Indian flying foxes, *Pteropus giganteus*, Maruthupandian & Marimuthu 2013, fig. 7.4a; fellatio in short-nosed fruit bats *Cynopterus sphinx*, Tan et al. 2009, fig. 7.4b). A female kittiwake *Rissa tridactyla* ejects the sperm of her social mate, always after the latter is away from the nest and more often if she interrupted mating by toppling him off (Helfenstein et al. 2003).

Courter traits can similarly influence fertilization bias in a competitive context; for example, Lewis and Austad (1994) found that female flour beetles, mated to one male and then remated, favor the second male more if he produces attractive courtship pheromones.

Correlational studies, like the above on bats, are often the only way forward for studying the intimate acts involved in peri-copulatory courtship, but a growing number of workers have performed experimental manipulation of courter signals or choosers' ability to perceive them. Briceño and Eberhard (2009) used both approaches to evaluate chooser responses to peri-mating courtship in tsetse flies. They experimentally manipulated both signalers by ablating male genital structures hypothesized to function in stimulating females; and receivers by mechanically blocking or ablating female genital structures involved in mechanoreception of courtship signals. Both of these manipulations reduced female ovulation and transfer of sperm into spermathecae. The effects of peri-copulatory courtship on reproductive outcomes have long been known in agriculture. An experimental study on beef cattle showed that cows (older females that had borne a calf) were more likely to become pregnant from artificial insemination when it was accompanied by clitoral massage. The same study incidentally showed that clitoral stimulation can serve as a mechanism of sexual selection, since there was a significant effect on pregnancy rate of the individual technician doing the inseminating (Lunstra et al. 1985). Peri-mating courtship is therefore a target of cryptic mate choice, which choosers exercise by varying both the availability of gametes to fertilize and the fate of sperm in the reproductive tract.

Male red flour beetles *Tribolium castaneum* rub a female's elytra during mating. Edvardsson and Arnqvist (2000) were able to directly manipulate chooser perception of peri-mating courtship by shortening the legs; males still performed the movements, but rate of movement no longer had an effect on fertilization rates (fig. 7.5). In this species, female peri-copulatory behavior is also predictive of sperm transfer. Females actively limit the amount of sperm transferred by males, with males transferring twice as much sperm to dead as to living females. Bloch Qazi (2003) showed that sperm transfer was dependent on arrested movement (so-called quiescent

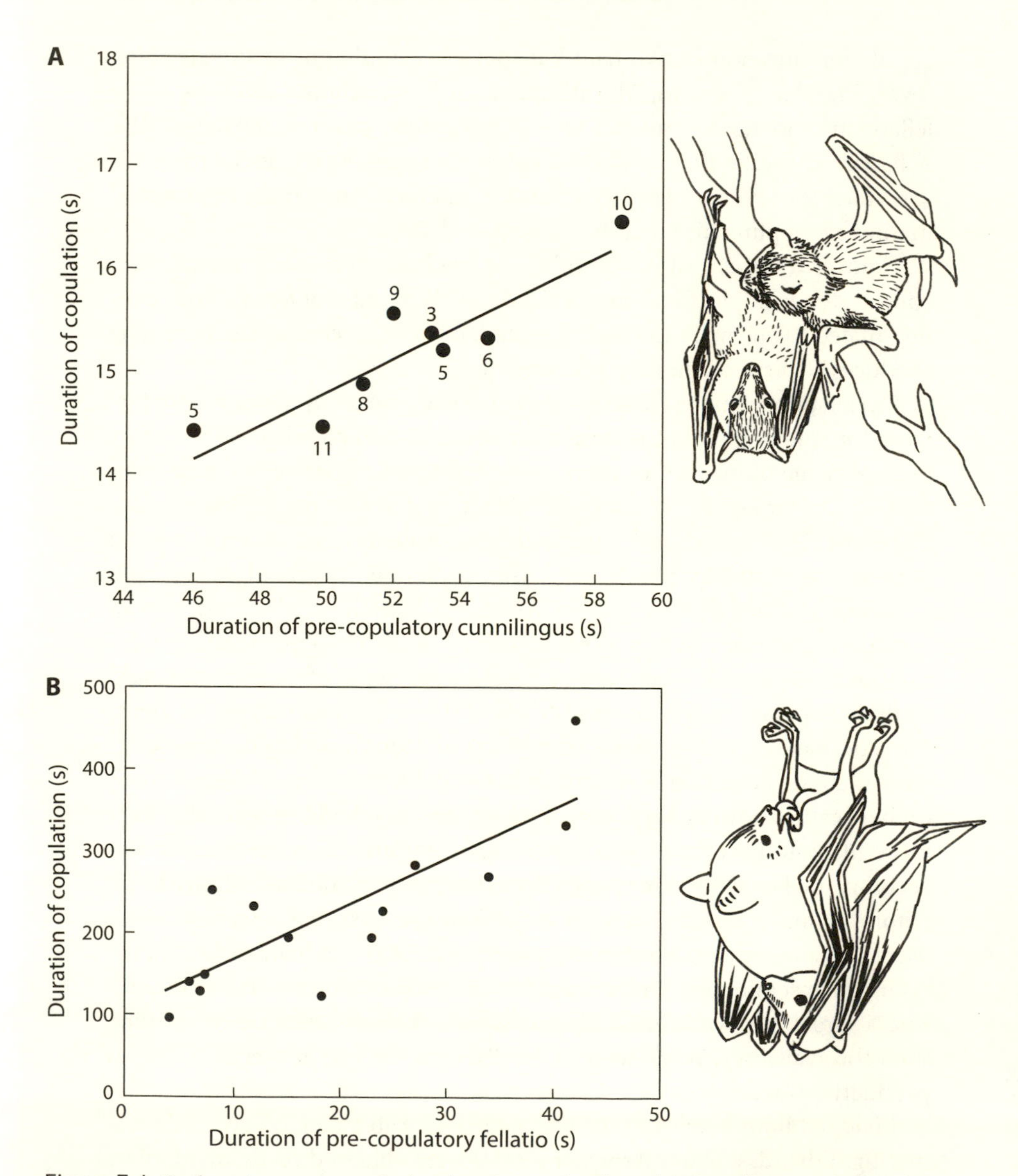

Figure 7.4. Oral sex increases copulation duration in fruit bats. (a) Cunnilingus in Indian flying foxes, *Pteropus giganteus* (after Maruthupandian & Marimuthu 2013); (b) Fellatio in short-nosed fruit bats *Cynopterus sphinx* (after Tan et al. 2009).

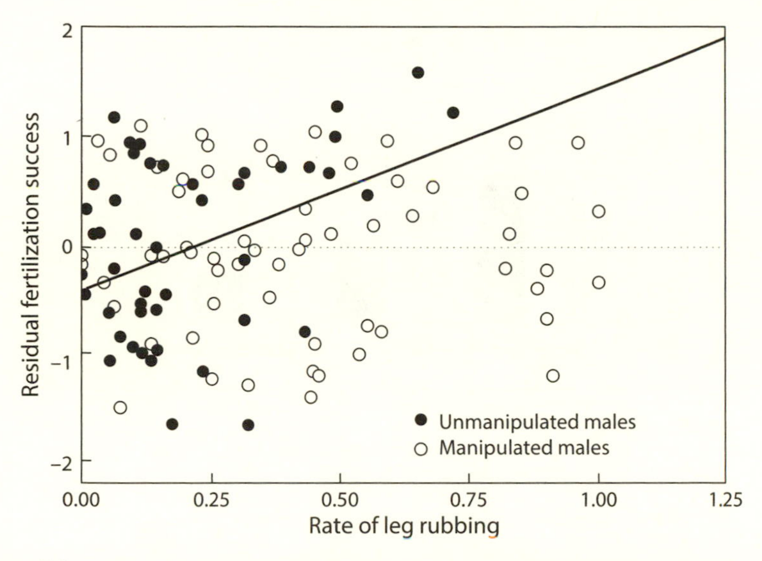

Figure 7.5. Fertilization success as a function of peri-mating courtship in flour beetles *Tribolium castaneum*. Fertilization increases with courtship rate from unmanipulated males, but stays constant when males are morphologically manipulated so that courtship fails to stimulate females. After Edvarsson & Arnqvist (2000).

behavior) during peri-copulatory courtship. Females therefore have a mechanism for active behavioral control of sperm transfer.

Pilastro and colleagues (2004) took advantage of the comparative nature of premating preferences (chapter 6) to design an elegant experimental test of whether premating attractiveness influenced postmating choice in guppies (*Poecilia reticulata*). Subjects were allowed to observe a focal male simultaneously with either a more colorful male or with a drabber male. Females retained more sperm from a focal male when they had evaluated him relative to a drabber male as opposed to a more colorful and thus more attractive male (fig. 7.6).

Postmating processes can also influence fertilization. Female field crickets attend primarily to male song characteristics in premating choice; as a consequence, they mate readily to both siblings and unrelated males. The fitness cost of mating with siblings is mitigated by the fact that unrelated males have a greater share of paternity. Tuni et al. (2013) were able to decouple cryptic choice from other processes by mating female crickets either to siblings or unrelated males, then having them mate-guarded by either a sibling or an unrelated male. Females retained less sperm when guarded by a sibling, irrespective of their actual mate.

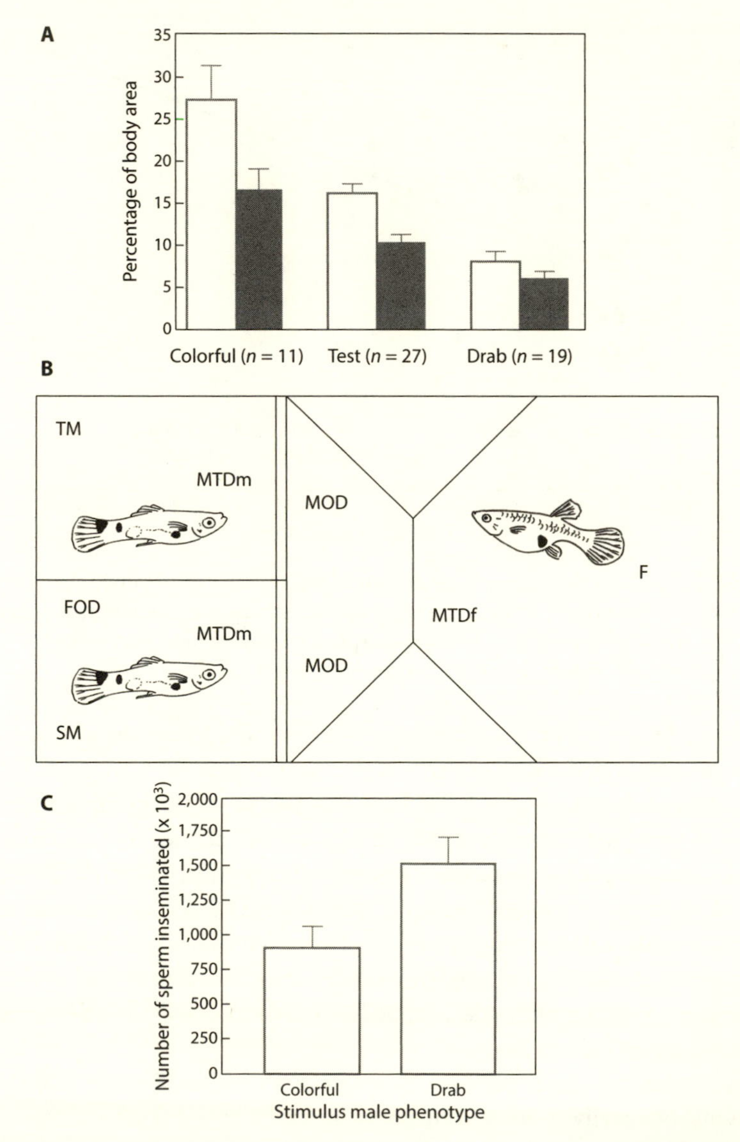

Figure 7.6. Premating preferences and postmating preferences in guppies. (a) Total pigmented area (white) and total carotenoid area (filled) of lateral spots of focal (test) and stimulus (colorful or drab) males. (b) Testing apparatus. A fixed opaque divider ("FOD") prevented interactions between test male (TM) and stimulus male (SM). Females initially sampled both males through a transparent divider (MTDf) which was then removed to allow close-range interactions. An opaque divider (MOD) was then lowered in front of the stimulus male and the test male was released and allowed to mate with the female. (c) Females had more sperm—whether through longer copulation or less sperm ejection—from focal males when these were paired with drab males than when they were paired with attractive males. After Pilastro et al. (2004).

Sperm retention may be a widespread mechanism for minimizing the fitness costs of inbreeding. Bretman and colleagues (2009) showed that female field crickets bias sperm storage toward unrelated males (fig. 7.7). Similarly, female Swedish sand lizards *Lacerta agilis* mated to both siblings and unrelated males skew paternity toward the latter (Olsson et al. 1996).

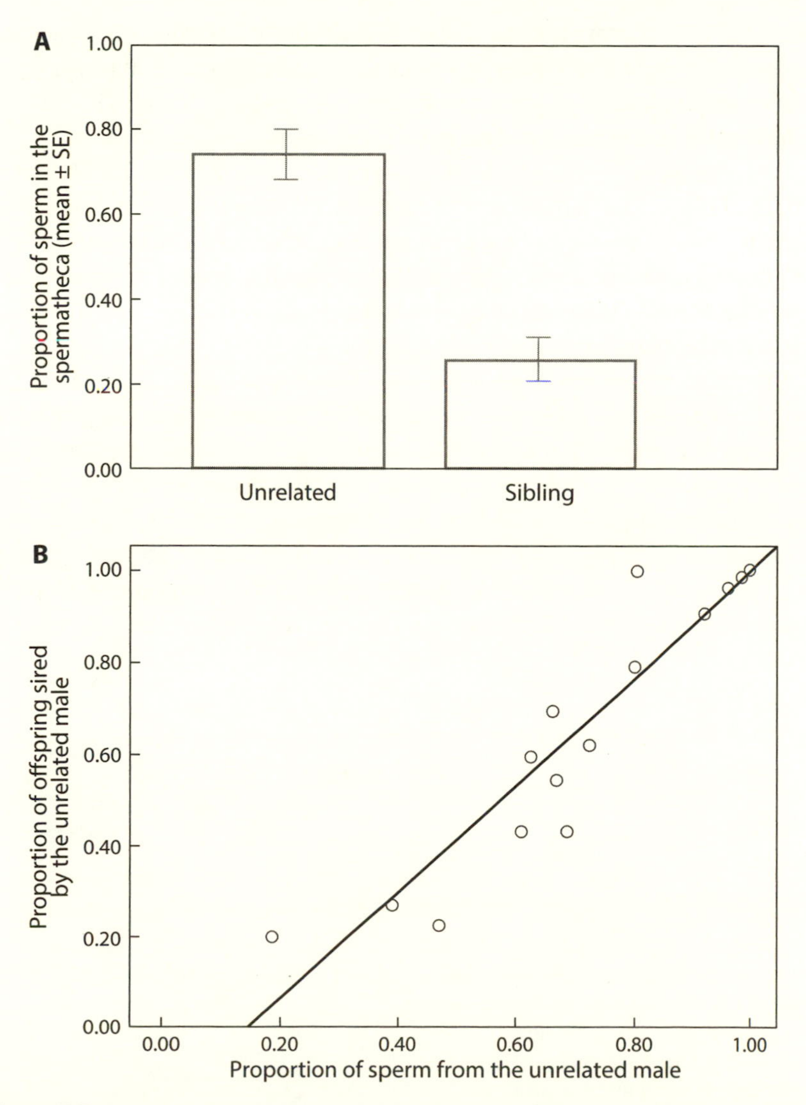

Figure 7.7. (a) Proportion of sperm stored in the spermathecae of female field crickets mated with a sibling and an unrelated male. (b) Sperm storage predicts paternity. After Bretman et al. (2009).

Sperm ejection (or "sperm dumping") is widespread and a powerful mechanism for biasing paternity and exercising postmating choice. In birds, sperm ejection eliminates on average 80% of the sperm from a previous mating (Dean et al. 2011). For example, part of the peri-mating interaction of dunnocks (*Prunella modularis*) involves cloacal pecking by the male, in response to which a female sometimes ejects sperm from her cloaca (Davies 1983; fig. 7.8). Pizzari and Birkhead (2000) showed that female chickens *Gallus gallus* were more likely to eject sperm after coerced matings from subdominant males than from dominant males, and that a male's likelihood of having his sperm ejected changed with manipulated social status.

Peri-mating courtship influences sperm retention; Peretti and Eberhard (2010) showed that female pholcid spiders *Physocyclus globosus* are less likely to discard sperm from a previous mate if he performed more vibratory courtship and genitalic squeezing during mating. Sperm retention plays a major role in the long-standing, lively debate about the function of female orgasm in humans (reviewed in Zuk 2006).

In species where males provide so-called nuptial gifts to females, the properties of the gift influence cryptic choice. In some cases, the gift is left by courters without direct interactions between individuals, becoming an object of premating choice: for example, female springtails (collembolans) *Orchesella cincta* prefer, as nuptial gifts, the spermatophores of males that have interacted with rivals and are therefore somehow altered (Zizzari et al. 2013); in others, spermatophore attachment is accompanied by pericopulatory courtship. Much like duration of copulation, the duration of

Figure 7.8. Peri-mating interactions in dunnocks begin with the male (a) pecking the female's cloaca an average of 28 times, with more pecking if the female has recently mated with another male. (b) Copulation itself lasts only a fraction of a second. After Davies (1983).

attachment often predicts fertilization success (Hall et al. 2010). Cryptic choice can also target the consumption of male tissue itself. In sagebrush crickets (*Cyphoderris strepitans*) females partially consume the hindwings of males during copulation. Hungry females, or females previously mated to males whose wings were experimentally reduced, show a quicker remating rate. This is therefore cryptic choice for (partially) cannibalized males (Johnson et al. 1999).

7.3.4 Reproductive tract physiology and sperm performance

The reproductive tract of females provides yet another mechanism for females to exercise choice after mating. Fewer than 0.001% of the sperm deposited in the human vagina reach the site of fertilization after running a "gauntlet" of antibodies and leucocytes, thereby providing a ready mechanism for choosers to discriminate sperm among and within courters (Zeh & Zeh 1997). In guppies, females use the ovarian fluid to discriminate against the sperm of close relatives; in vitro assays show that sperm of brothers are slower than the sperm of unrelated males (Gasparini & Pilastro 2011). Ovarian fluid also attracts and facilitates swimming for conspecific over heterospecific sperm in two sympatric species of salmonid fishes (Yeates et al. 2013), and two other studies in this externally fertilizing family have shown male x female interactions between ovarian fluid and sperm among conspecifics (Urbach et al. 2005; Rosengrave et al. 2008). Simmons and colleagues (2009) found among-female variation in the effect of egg jelly on sperm motility in an externally fertilizing frog, *Crinia georgiana*. By contrast, Cramer et al. (2014) found no effect of female reproductive tract fluid on conspecific versus heterospecific sperm performance in house sparrows, which may partially account for their propensity to hybridize in nature (chapter 16).

7.3.5 Sperm-egg interactions

Where fertilization is external, biased fertilization after gamete release takes the form of direct interactions between egg and sperm. For organisms that are sessile (permanently attached to a substrate), interactions among gametes are the only way (apart perhaps from chemical communication to synchronize gamete release; Lessios 2011) to exercise mate choice.

One of the most basic kinds of sperm-egg interactions is found in externally fertilizing marine invertebrates. In sea urchins, a sperm-surface protein called bindin attaches to receptors on the surface of the eggs. This is a primary mechanism for conspecific recognition, with eggs failing to recognize and merge with heterospecific bindin (Lessios 2011). In the genus *Echinometra*, urchin gametes fuse assortatively with respect to bindin genotype (Palumbi 1999).

A particularly well-characterized sperm-egg interaction involves the protein lysin and its receptor, vitelline envelope receptor lysin (VERL; reviewed in Lessios 2011). For abalone, lysin is the last gatekeeper in mate choice. Binding of lysin to VERL creates a hole in the vitelline envelope which allows the sperm to reach the cell membrane of the egg. As with bindin, coevolved affinities between lysin and VERL are responsible for recognition of conspecific gametes; by contrast with bindin, however, lysin and VERL are highly conserved within species (Lessios 2011).

Individual gametes can choose their mates. Oliver and Evans (2014) performed simultaneous- and sequential-choice assays, very similar to standard behavioral preference tests (chapter 2), measuring **chemotaxis** of the sperm of mussels *Mytilus galloprovincialis* toward the eggs of individual females. There was a marked effect of individual female identity and some evidence for male x female interactions, suggesting that egg chemoattractants differentially stimulate chemotaxis from sperm. The biases they found in simultaneous-choice trials were strongly correlated with early-stage viability of embryos in sequential-choice trials. Even at the fertilization stage, there is still an opportunity for mate choice, when multiple sperm fuse to an egg (polyspermy). Carré and colleagues (1991) visualized the processes surrounding fertilization in the ctenophore *Beroë ovata* where several sperm penetrate the ovum. In Birkhead's (1998) characterization, the female pronucleus moves between several male pronuclei before "choosing" to fuse with one of them. Carré and colleagues (1991), however, describe the female pronucleus as approaching each male pronucleus in succession, only in an "exceptional situation" returning to a previously sampled site.

Females can therefore choose particular sets of sperm produced by the same courter. This can also be accomplished using sperm dumping, as described above. Female black-legged kittiwakes (*Rissa tridactyla*) are one of the rare species that is genetically monogamous. In repeated matings with the same partner, females reject sperm that were inseminated long before egg-laying, but preferentially retain more recently inseminated sperm; sperm rejection of older sperm predicts hatching success (Wagner et al. 2004).

Sperm-egg interactions allow choosers to discriminate against genetically incompatible mates. They can discriminate generally not only against the sperm of individual courters, like heterospecifics, but also against particular incompatible sperm from individual courters, as when mating with close relatives (Zeh & Zeh 1997). Indeed, a meta-analysis by Slatyer et al. (2012) suggests that selecting against genetic incompatibility (including close relatives or heterospecifics) is a primary benefit of polyandry, and therefore of cryptic mate choice (chapter 14). In mammals, paternity is biased toward sperm with dissimilar MHC haplotypes (Schwensow et al. 2008), with in vitro experiments suggesting that this bias arises from fertilization or blastocyst formation, possibly mediated by MHC antigens expressed on the sperm surface (Wedekind et al. 1996). Consistent with the notion of ova playing a direct role in selection of genetically compatible sperm, female decorated crickets *Gryllodes supplicans* discriminate against sperm of siblings when mated in tandem with unrelated males, despite transferring equivalent amounts of sperm (Stockley 1999). By contrast, in jungle fowl *Gallus gallus*, Løvlie et al. (2013) found that preference against MHC-similar mates occurred in the reproductive tract prior to sperm-egg interactions. Crucially, preference disappeared if females were artificially inseminated, suggesting that females were integrating premating cues into postmating decisions (see below).

7.4 RESOURCE ALLOCATION TO OFFSPRING

7.4.1 Overview

Mating—the act of exchanging gametes—is only part of the theater of mate choice. Adding differential fertilization to the mix accounts for much of the rest. Even in a hypothetical (and unlikely) scenario where mating and fertilization rates for two sets of courters are identical, mate choice involves another dimension that is critically important to the fitness of both courters and choosers, and to the maintenance and breakdown of reproductive isolation between lineages. Namely, choosers distribute resources unevenly among the offspring that they produce with different courters. In section 7.3, I briefly discussed differential abortion and implantation as a function of courter phenotype. In this section, I discuss how choosers bias the distribution of resources they allocate to their offspring, as a function of the phenotype of their mates.

The term **allocation** is often used to specifically refer to the amount of energy that choosers invest in gametes, a decision that usually occurs before fertilization. Allocation can also be construed more broadly to encompass investment by choosers in offspring, including gestation and parental care, and theoretically also the treatment of grandchildren and beyond.

7.4.2 Gamete quantity, size, and nature

Choice can be exercised by favoring the provision of resources to offspring of one courter over another (Sheldon 2000). For animals with varying clutch sizes, a basic decision is how many gametes to allocate to an individual mating (a decision related to fertilization bias, above). For example, Reyer et al. (1999) found that female *Rana lessonae* laid fewer eggs when mated to hybridogenetic *Rana esculenta* than to conspecifics, and thereby suffered a smaller loss of body condition (fig. 7.9). Similarly, Pischedda and colleagues (2011) found that female *D. melanogaster* deposited varying total egg volumes according to male genotype. Female rainbowfish *Melanotaenia australis* deposit more eggs while mating with larger males, reinforcing premating preferences for larger males. Similar to previous-mate effects in premating preferences (chapter 6), females reduced the number of eggs laid with a given male if they had previously been mated to a larger male (Evans et al. 2010).

A host of studies in birds have shown that females vary the size of their eggs, and therefore the amount of resources they provide, as a function of courter phenotype (reviewed in Sheldon 2000; Stiver & Alonzo 2009; Horváthová et al. 2012). For example, female mallards *Anas platyrhynchos* resist

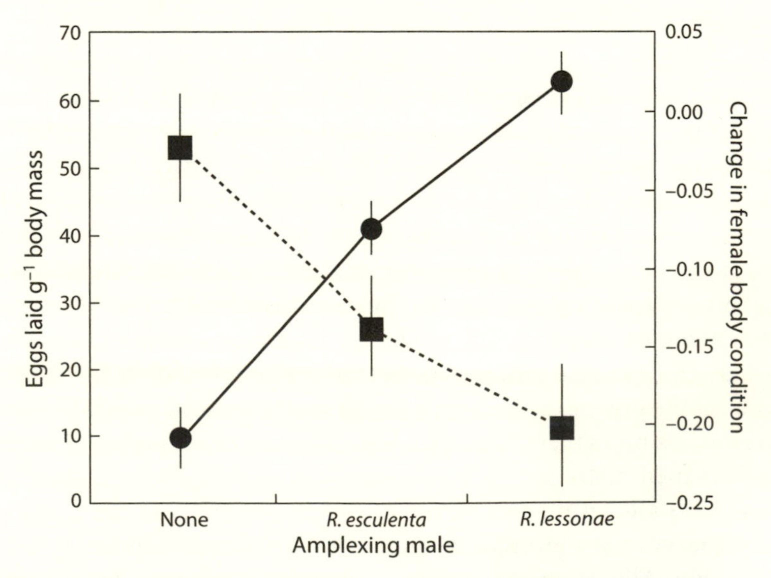

Figure 7.9. Clutch size (circles) and change in body condition after reproduction (squares) of female *Rana lessonae* amplexed conspecifics and to hybridogenetic *R. esculenta* males, and females in isolation. After Reyer et al. (1999).

all extrapair matings (Cunningham 2003), and have mechanical barriers to forced copulation (Brennan et al. 2010), but exert considerable choice after fertilization. Females lay larger eggs after mating with males expressing artificially enhanced carotenoid coloration (Giraudeau et al. 2011). Cunningham and Russell (2000) found that first-year mallard females laid heavier eggs with males who were more attractive according to a consensus of female responses; by contrast, Bluhm and Gowaty (2004) found that second-year females laid heavier eggs with males they had *not* preferred before mating; first-year females showed no effect. These contrasting results within the same species reinforce the need for rigor with catch-all terms like "attractiveness" as well as the importance of age and experience in structuring mating preferences (chapter 9). Female zebra finches (Bolund et al. 2010) also lay heavier eggs with males that are nonpreferred before mating. Grana et al. (2012) found no effect of attractiveness on allocation in male house wrens (*Troglodytes aedon*).

Choosers can also vary the nature of gametes according to courter phenotypes. Female canaries put more testosterone into their eggs after exposure to preferred song (Gil et al. 2004), while female house finches deposit more testosterone into the eggs of less attractive males (Navara et al. 2006). Female flycatchers *Ficedula albicollis* put more testosterone into eggs when mated with younger males but not when mated with attractive males (Michl et al. 2005).

A remarkable instance of allocation occurs when choosers manipulate offspring sex ratio in response to courter phenotype (reviewed in West & Sheldon 2002). As I will examine in chapter 14, they often do so in ways consistent with adaptive predictions (Pryke & Griffith 2009; Calsbeek & Bonneaud 2008; Long & Pischedda 2005). For example, Sheldon et al. (1999) showed that female blue tits mated to males with more ultraviolet chroma in their feathers produced male-skewed sex ratios; the pattern was reversed when UV color was experimentally masked.

Allocation decisions can serve as a mechanism to couple traits and preferences (chapter 15). In zebra finches, females mated to attractive males invest more in eggs, which in turn enhances the attractiveness of male offspring (Tschirren et al. 2012) and the fecundity of female offspring (Gilbert et al. 2012). As I will return to in chapters 11 and 12, the premating preferences of female zebra finches are shaped by social experience in the nest and as adults, and by the physical condition of choosers. Between allocation effects on traits, and social and environmental effects on preferences, nongenetic factors play a dominant role in shaping sexual communication in this system (chapter 18).

7.4.3 Parental care

Choosers can favor particular courters well into the life of their offspring and potentially beyond. In species with parental provisioning and defense, choosers often invest more time and energy into the offspring of mates they find attractive over those they find unattractive (Trivers 1972). And choosers invest in the genetically unrelated progeny of their social mates, including same-sex partners (reviewed in Bailey and Zuk 2009a). Male zebra finches (Burley 1988) and female blue tits (Limbourg et al. 2004) invest more in parental care for the offspring of attractive mates; by contrast, male blue tits invest more in the offspring of unattractive females (Limbourg et al. 2013).

Gale and colleagues (2013) showed in mice that a "Bruce effect" of sorts extends beyond the window for abortion in early pregnancy: females exposed to novel males in late pregnancy invest less in offspring from a previously encountered male during lactation, resulting in measurable differences in mass at weaning.

Caring for one's offspring is perhaps the best-case mate-choice outcome for a courter. It is worth noting that the worst-case outcome can extend well beyond simply taking a courter's nuptial gift and eschewing his sperm. For example, choosers frequently exhibit aggression or attempt cannibalism depending on chooser state and courter signals (Wilder et al. 2009; chapter 6), in the latter case obviating the possibility of any future reproductive activity by the courter.

7.5 MATE CHOICE ACROSS STAGES: PREMATING DECISIONS AND CRYPTIC CHOICE

7.5.1 Contribution of cryptic choice to total mate choice

As the examples above show, mate choice integrates a chain of chooser decisions from the initial detection of chooser signals, often at a considerable distance, through mating, fertilization, and beyond. Zooming out to the whole process of choice, we can ask, first, how important each of these stages is relative to another, and second, whether and how decisions at one stage influence decisions at subsequent stages.

Quantitative estimates of the relative importance of different stages of mate choice are scant and divergent. In section 7.1, I described Thornhill's (1983) finding that 96.5% of female scorpionflies accepted a spermatophore, meaning that cryptic choice accounted for most of the female bias among males. By contrast, Pischedda et al. (2012) found that pre- and postmating processes accounted for similar fractions of total sexual selection on male *D. melanogaster*, but that most of the postmating variation was due to mating order, with the last male to mate having the highest fertilization success. Once mating order was taken into account, postmating selection accounted for only about 2% of the variance in male fertilization success.

7.5.2 Mechanisms linking premating and cryptic stages

Above, I describe several examples where variation in courter signals used in premating courtship predicts peri- and postmating outcomes. We know little about the mechanisms connecting chooser evaluation of courters across stages, but they could involve long-term memory, changes in hormonal state, or perhaps spatial or chemical "labeling" of sets of gametes in the reproductive tract. Correlations (positive or negative) between premating and cryptic attractiveness could be due to such carryover effects within choosers, or they could be due to correlations within courters. For example, in the cricket *Teleogryllus commodus*, there is a positive relationship between premating attractiveness in terms of latency to mate, postmating attractiveness in terms of spermatophore retention time, and postmating fertilization success. This correlation could be due to integrated cryptic choice, or to correlations among sperm competition traits (Hall et al. 2010). By contrast, in fireflies *Photinus greeni*, there is a negative relationship between premating attractiveness and paternity. South and Lewis (2012) independently manipulated a premating signal, flash rate, and spermatophore size, and showed that females preferred higher flash rates and that they were more likely to accept, and gave a larger share of paternity to, males with larger spermatophores; flash rate and spermatophore were negatively correlated across individuals.

In some cases, courter cues are perceived and evaluated by the brain just as happens before mating, but instead of choosers' behavioral and physiological responses affecting the probability of mating, they affect the probability of fertilization or the amount of resource allocation. The often tactile, near-field acoustic, or gustatory cues involved are not trivial to manipulate experimentally, but there is every reason to expect that choosers show preference functions for variation in traits like elytron-rubbing, clitoral massage, and peri-copulatory pheromones, just as they do with traits they experience before mating.

In other cases, perception and evaluation (in the broad sense) are performed directly by the morphology and physiology of the genitalia or by the gametes themselves, although each of these processes presumably respond to brain-mediated neural and hormonal controls. Regardless, each of these processes generates preference functions, since different courter phenotypes will fare better or worse with a given genital morphology, and different chemosignals on the surface of sperm cells will have greater or lower binding affinity to receptors on the egg. Both morphological structures and chemoreceptors can assign positive or negative valence to courter stimuli, and mate sampling can involve accepting or rejecting mates before mating, as emphasized in chapter 6, or accepting or rejecting the gametes of potential sires after mating.

Parker's (1983) distinction between "active" and "passive" traits in a premating context applies equally well to active sperm-dumping processes in response to courtship versus passive constraints set by the length of the seminal receptacle. Along with Eberhard (2000), Telford and Jennions (1998, p. 217) argued that chooser traits "even seemingly 'passive' ones (such as differences in sperm leakage or spermathecae size) that lead to predictable patterns of sperm usage by different females can 'set the rules' under which males compete," in a manner analogous to female preferences operating before mating. A study by Córdoba-Aguilar (1999) illustrates the false dichotomy between "precopulatory, sensory" and "postmating, mechanical." Male damselflies *Calopteryx haemorrhoidalis* use a genital structure to stimulate mechanosensory receptors in the female reproductive tract and thereby induce ejection of the previous male's sperm from the spermatheca. The stimulus-response system is not conceptually different from one where the males were stimulating mechanosensory receptors in the ear at a distance and eliciting positive or negative phonotaxis.

7.6 SYNTHESIS: WHAT IS DIFFERENT ABOUT CRYPTIC CHOICE?

7.6.1 Artificial distinctions between cryptic and premating choice

Prum (2015) articulates a widely shared view of "choice" that emphasizes behavioral decisions—the exercise of "sexual autonomy" by choosers. He argues that "choice" should be restricted to mechanisms involving sensory biology or cognition, drawing a contrast with purely physiological or mechanical structures functioning as "non-choice mechanisms of intersexual selection." The distinction is important, Prum argues, because choice in his narrower sense operates with more independence from coercion or competition among courters. Indeed, choosers can respond to manipulation at the early stages of mate choice by simply ignoring broadcast signals (Rosenthal & Servedio 1999), while later stages require more active engagement with courters.

The previous chapters have called attention to the sophistication and power of the sensory and cognitive mechanisms involved in premating choice. Sometimes, these mechanisms are directly recruited for cryptic choice, as when fertilization bias depends on peri-copulatory courtship. In cases like these, it is difficult and perhaps not very fruitful to distinguish "cryptic choice" as a qualitatively different process. But what about peri- or postmating processes that bypass the nervous system, like genital morphology? In this case, morphology acts as a mechanism to favor some courter-intromittent organs over others. It can evolve in response to selection arising from different outcomes with different courters. This is conceptually similar to a sensory receptor (chapter 3) that favors some courter signals over others and can evolve in response to selection on mate-choice outcomes. The difference, Prum might argue, is that the output of a sensory receptor can be modulated by downstream neural processes. But here the difference is one of degree and not of kind, since the consequences of successful intromission can similarly be modified by differential fertilization or post-fertilization processes. Where and when choice happens is critical to the costs that choosers incur and the extent to which chooser decisions are constrained by those of courters, but the difference is one of degree rather than kind. In chapter 14, I revisit the limits of choice in the context of sexual coercion by courters.

Perhaps the biggest difference between premating and cryptic choice is that methodological constraints inevitably mean that there are different sets of workers studying the two types of mechanisms. Differences in taxa, theory, and authors mean that the literature on cryptic choice stands somewhat apart from that on premating processes. In particular, there has been little theoretical or empirical work on mating decisions (chapter 6) in the specific context of cryptic choice. Eberhard (1996) suggested that cryptic choice involves absolute rather than relative comparisons, but nontransitivity of paternity share (female x male interaction in reproductive success; Wilson et al. 1997; Birkhead et al. 2004; Bjork et al. 2007) is consistent with relative choice, although it could could also arise from early embryo mortality (female x male interactions with respect to viability rather than choice;

Birkhead et al. 2004). Direct visualization of sperm (Manier et al. 2013a, 2013b) may be useful in testing for comparative aspects of cryptic choice.

Work on cryptic choice has arguably been more rigorous than work on premating stages. This is because studies typically combine genetic measures of realized mating outcomes with relevant phenotypic measure of choice like fertilization rate or offspring mass. We who work on premating choice favor convenient behavioral proxies of preference that are difficult to connect to the probability of mating, let alone to subsequent postmating decisions or to chooser fitness. And studies of cryptic choice, with their rich and diverse representation of invertebrate taxa, encompass a fairer taxonomic sample than those of premating choice. I would argue that it is this rigor, perhaps in concert with some inherent features of cryptic choice (next section), that accounts for the relative unpopularity of "good genes" explanations for preference maintenance in the literature on mate choice after mating.

The perceived contrast between premating and cryptic choice is thus in no small part one of culture and taxon bias. There is no bright line between "premating" and "postmating," hence the usefulness of the term "perimating" (Fedina & Lewis 2008). A chooser—and sometimes her gametes—keep sampling after she has mated, and the probability that she will choose to merge one or more of her gametes with those of a courter is a function of the courter's phenotype, or that of his gametes.

7.6.2 Distinctive features of cryptic choice

What, then, are the differences between premating and cryptic choice? The distinctions are mostly in degree rather than in kind. As I will discuss further in chapter 14, cryptic choice imposes different kinds of costs on choosers, making them vulnerable to physical injury, exposure to toxic ejaculates, and sexually transmitted disease (STD), while on the other hand mitigating many of the predation and opportunity costs associated with premating search and evaluation (Eberhard 1996). Counterbalancing these mating-related costs are some distinctive benefits: cryptic mechanisms, particularly sperm-egg interactions, may be most effective at filtering out genetic incompatibilities (Zeh & Zeh 1997; Birkhead & Pizzari 2002; Lessios 2011; chapter 14).

In addition to genetic compatibility, hypotheses about the evolution of cryptic choice have centered on sexual conflict. Relative to premating stages, the intimacy of mating and reproduction may indeed provide greater opportunity for courters to manipulate choosers. This is because harmful premating signals can simply be ignored or not detected, whereas it is much more difficult for choosers to mitigate the cost of direct mechanical or chemical manipulation by courters. Sexual conflict likely plays a bigger role in cryptic processes (Rosenthal & Servedio 1999), but sexual conflict is widespread in premating choice as well (chapter 15).

A major driver of sexual conflict after mating is that chooser decisions become much more entangled with competition among courters. In premating choice, there is at least the possibility that choosers can sample courters independently of each other, such that courters signal without interfering with each other or with choosers. During and after mating, courters are directly competing with each other and directly manipulating the biology of choosers. Much of the early discussion on cryptic choice focused on disentangling it from sperm competition (Birkhead 1998; Andersson & Simmons 2006). As Eberhard (2000, p. 1048) pointed out, "male competition occurs on playing fields whose characteristics are determined by females," which applies equally well to premating interactions where females incite agonistic interactions among males (chapter 6).

7.6.3 Mate choice as an integrated strategy across stages

Cryptic choice is a fundamental part of mate choice whenever choosers mate multiply. We now know at least as much about mechanisms of cryptic choice as we do about premating mechanisms. Where these mechanisms bypass the complexity of the brain, we can sometimes characterize them in a way that meaningfully connects genotypes to choice outcomes. For example, Chow et al. (2010) argued that allelic variation at the SPR (sex-peptide receptor) gene on the *D. melanogaster* X chromosome, which binds to a seminal fluid protein, was responsible for the marked male genotype x female genotype interaction in P_2 (Clark & Begun 1988; Clark et al. 1999). Manier and colleagues (2013a) showed quantitative differences between three sibling *Drosophila* species in the importance of four different mechanisms of cryptic choice (sperm transfer, displacement, ejection, and selection for fertilization); the same group (Manier et al. 2013b) showed that conspecific sperm precedence between *D. simulans* and *D. mauritania* was a function of both differential sperm ejection and differential fertilization bias. We have nowhere near this granular an understanding of the mechanisms underlying behaviorally mediated premating choice.

The artificial separation of cryptic and premating choice sets up a false distinction between these processes from the point of view of choosers, when they in fact form a continuum associated with changes in multiple costs and benefits. It may be useful to think about a chooser's preferences before, during, and after mating as part of an integrated phenotype. Both decision theory (chapter 6) and life-history theory (e.g., Alonzo & Pizzari 2013; Bjork et al. 2007; Kokko & Rankin 2006) can be useful in addressing the question of how an individual should allocate its mate-choice effort across stages. I will return to this topic in chapter 14. In the next chapter, I will relax the distinction between courters and choosers, and consider the dynamics of mate choice when both parties are making decisions about mating, fertilization, and investment.

7.7 ADDITIONAL READING

Birkhead, T. R., and Pizzari, T. 2002. Postcopulatory sexual selection. *Nature Reviews Genetics* 3 (4), 262–273.

Eberhard, W. G. 1996. *Female Control: Sexual Selection by Cryptic Female Choice*. Princeton, NJ: Princeton University Press.

Eberhard, W. G. 2009. Postcopulatory sexual selection: Darwin's omission and its consequences. *Proceedings of the National Academy of Sciences* 106 (Supplement 1):10025–10032. doi: 10.1073/pnas.0901217106

CHAPTER 8

Mutual Mate Choice

8.1 INTRODUCTION

So far, I have treated mate choice as a one-way process, where choosers unilaterally make decisions about which courters to mate with. I have made the implicit assumption that courters will automatically mate if chosen, and that upon mating they will supply choosers with an excess of gametes upon which to exercise cryptic choice. This construct approximates the situation in highly polygynous systems where males contribute few if any resources to the female besides sperm, and in which there is strong competition for mates among males but not females. Indeed, males in some such systems exert no meaningful mate choice among reproductively active females, exhibiting a broad envelope for mate recognition that encompasses dead conspecifics and inanimate objects. As detailed in chapter 6, cognitive constraints and the cost of choice make mating decisions a complex challenge even with these assumptions in place. In this chapter, I introduce another important constraint on mate choice; namely, that choosers are the object of choice themselves.

Mutual mate choice is the rule rather than the exception. Even in the highly polygynous taxa described above, sperm are not in infinite supply (Tang-Martinez & Ryder 2005), and a growing body of evidence has shown that both sexes exert premating and postmating choice (e.g., fruit flies, Byrne & Rice 2006; ungulates, Bro-Jørgensen 2007; toads, Liao & Lu 2009; orb-weaving spiders, Bel-Venner et al. 2008). Mutual mate choice is most easily observed in taxa where both parents invest resources in parental care, in the formation of pair bonds between social partners, and in same-sex pairings. In these cases, both types of partners have a pool of candidates to choose from and a pool of candidates to woo. The evolution of mating systems, and their relationship to mate choice, will be addressed in chapter 14.

In this chapter, I focus on the mechanisms underlying mutual mate choice, notably how preferences are dynamically adjusted in light of the chooser being the chosen. I begin with the simplest form of mutual mate choice, when individuals reciprocally prefer each other's phenotypes. I then turn to the widespread phenomenon of mutual mate choice in systems that are socially promiscuous: both males and females mate multiply, but both males and females express preferences for particular phenotypes in the opposite sex. Mutual mate choice becomes more constrained in social monogamy, where one can only be paired with a single partner at a given time; I discuss how partners reach pairing decisions and how **pair bonds** are maintained and transgressed in both hetero- and homosexual couples. Finally, I briefly cover the interesting case of mate choice in hermaphroditic species.

8.2 RECIPROCAL PREFERENCES

The simplest type of mutual mate choice is when the two parties reciprocally prefer each other's phenotypes (fig. 8.1a–c). Mutual rejection, whereby individuals fail to detect or choose not to respond to each other's sexual cues, is surely the most common form of mutual mate choice. This is because males will reject females who would reject them anyway, like heterospecifics or sexually unreceptive individuals. For example, male goldfish *Carassius auratus* attend to female urinary pheromones, which are preferentially produced during a window of sexual receptivity (Appelt & Sorenson 2007), and male mammals, including humans (Singh & Bronstad 2001), ubiquitously respond preferentially to females displaying the prominent vi-

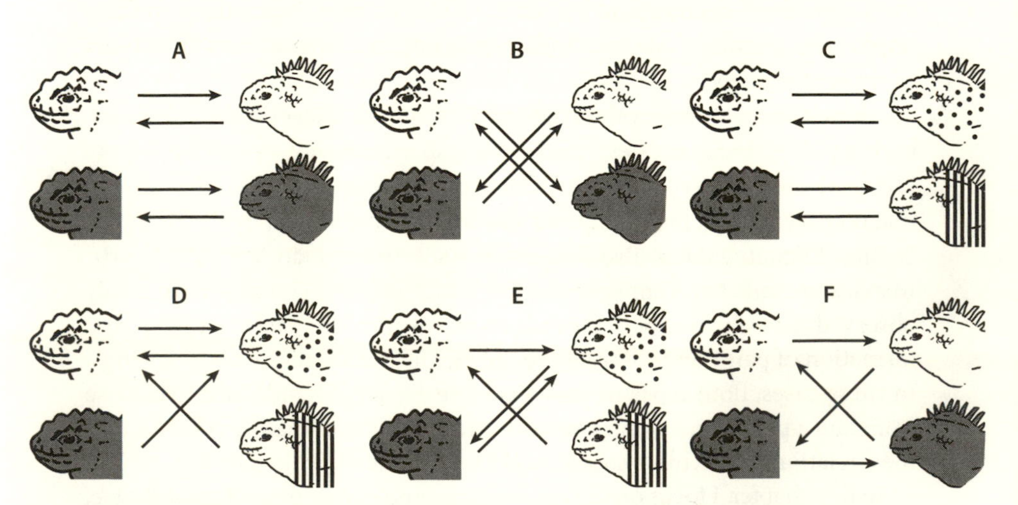

Figure 8.1. Schematic overview of mutual preferences between hypothetical males and females for (a) reciprocal preferences for self-similar traits in males and females; (b) reciprocal preferences for self-dissimilar homologous traits; (c) reciprocal preferences for different traits; (d) one-sided directional or asymmetric preferences for different traits; (e) disassortative preferences in females, directional preferences in males; (f) reciprocally conflicting preferences.

sual and/or olfactory cues associated with receptivity (Clutton-Brock 1989). Reciprocal choice also often involves context-dependent responses to the aversive behavior of a potential mate. For example, male dwarf chameleons (*Bradypodion pumilum*) are smaller than females. Male choice is contingent on female behavior; while males continue to court females after initial rejection, they more often than not terminate courtship after being bitten or receiving high-intensity threat displays (Stuart-Fox & Whiting 2005).

Beyond mutual rejection, reciprocal preferences can have important evolutionary consequences. So-called **homotypic preferences**, where individuals prefer phenotypes similar to their own (fig. 8.1a; Burley 1983), are one of many pathways to **positive assortative mating** and to the formation and maintenance of reproductive isolation (chapter 16). Hamlets (genus *Hypoplectrus*) are small percomorph fishes from the Caribbean and western Atlantic; several species, and genetically distinct morphs between species, coexist in proximity. These fish are distinct among vertebrates in being simultaneously hermaphroditic (section 8.6); peri-mating courtship involves color and postural cues that coordinate the reciprocal release of sperm and eggs. Behavioral observations and genetic evidence indicate that individuals overwhelmingly choose to mate with individuals of their own morphs (Puebla et al. 2012; fig. 8.2). This study also provides an anecdotal answer to a theoretical challenge to the evolution of homotypic preferences, namely that they put rare morphs at an extreme disadvantage due to the difficulties of finding a mate (chapter 16). Contrary to expectations, rare individuals would swim hundreds of meters and meet at established rendezvous points, where they mated as often as or more often than fish from more abundant morphs. Such mutual mate choice may facilitate reproductive isolation (chapter 16).

Hamlets therefore provide an example of choosers reciprocally attending to and preferring a phenotype similar to their own in mate choice. Similarly, in humans there are strong positive correlations between people's self-assessment in four categories (wealth, attractiveness, family commitment, and sexual fidelity) and the importance of that variable in a prospective mate (Buston & Emlen 2003). Humans, or at least a large sample of them from the eastern United States, have homotypic preferences for political affiliation. Alford and colleagues (2011) found that there was a stronger positive correlation between partners for political ideology than for physical traits, many of which are subject to conflicting preferences between partners (see section 8.4). Partnering with socioeconomically similar mates did not account for this correlation, since it persisted even within narrow socioeconomic groups.

The converse of homotypic preferences is **heterotypic** preferences for self-dissimilar phenotypes (fig. 8.1b). These preferences are notably manifest

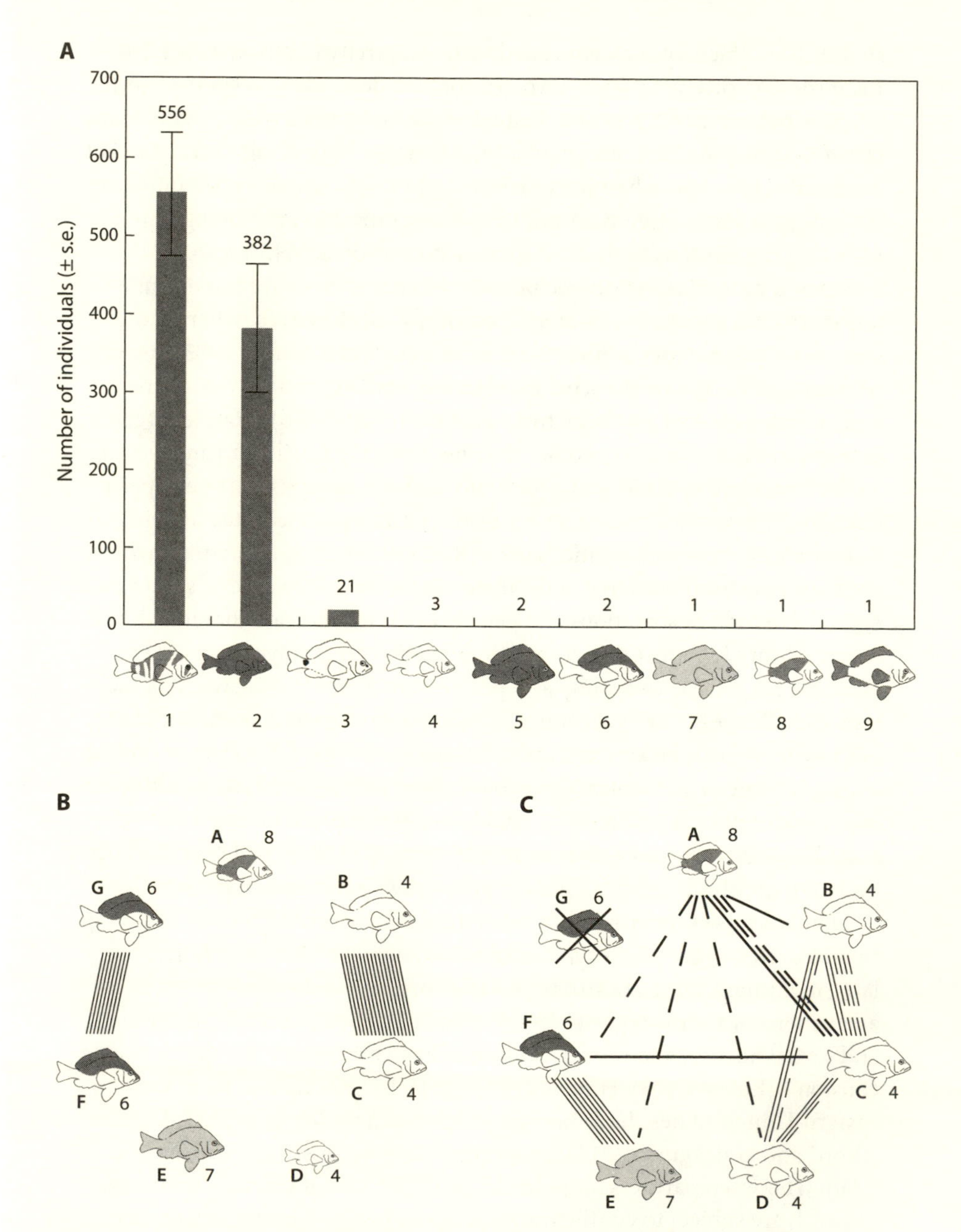

Figure 8.2. (a) Abundances of hamlet morphs (Serranidae: *Hypoplectrus*) on a single reef off Bocas del Toro, Panama. (b, c) Observed courtship (dashed lines) and matings (solid lines) between individuals in September 2009 and April 2010. Bold lines denote heterospecific/heterotypic interactions. One golden hamlet (D) reached sexual maturity and one yellowbelly hamlet (G) disappeared between the observation periods. After Puebla et al. (2012).

in avoidance of kin: for example, male cockroaches *Blattella germanica* preferentially court females other than their sisters, while females are more likely to choose males who are not their brothers (Lihoreau et al. 2008). A number of studies have implicated the highly variable major histocompatibility complex (MHC) both in kin recognition and in avoidance of kin in mate choice. For example, female potbellied seahorses (*Hippocampus abdominalis*) prefer to mate with MHC-dissimilar males (Bahr et al. 2012). In mice, individuals choose mates with dissimilar MHC haplotypes, and MHC haplotype divergence is correlated with the attractiveness of odors in the urine (Potts et al. 1991; reviewed in Penn 2002). Apparent disassortativity based on MHC, however, arises due to correlated differences in major urinary proteins, a class of volatile signal proteins encoded by highly variable genetic loci (Cheetham et al. 2007).

As always, experimentation is required to tease apart the actual cues used in mating decisions from their correlates, which may be more readily (and spuriously) identifiable. This is particularly true when individuals show assortative mating, which can arise from a number of processes other than mutual mate choice, including mate choice by only one sex (Burley 1983; Jiang et al. 2013; chapter 16). Mutual mate choice can also spuriously produce assortative mating (positive or negative) with respect to shared intersexual traits if males and females reciprocally attend to different cues (fig. 8.1c).

If choosers are indeed reciprocally using the same cues—whether to choose self-similar or self-dissimilar individuals—they need a mechanism to incorporate their own phenotype into mating decisions. Such **self-referential phenotype matching** (Mateo & Johnston 2000) can arise through early experience with kin (chapter 12): Penn and Potts (1998) showed that MHC-disassortative mating is reversed by cross-fostering. Contest-mediated assortative mating by body size, as described in section 8.4 below, could also be considered a form of self-referential phenotype matching. More generally, for reciprocal preferences to operate, there must be variation in preferences within or between both males and females; variation in preferences and mating decisions is the focus of the next four chapters.

8.3 SOCIAL PROMISCUITY AND MUTUAL MATE CHOICE

8.3.1 Which sex chooses?

In chapter 1, I argued for the use of *chooser* and *courters* to reflect the fact that females are not automatically the choosy sex. Indeed, **sex-role reversal**, whereby males choose females, is widespread (Amundsen 2000;

Bonduriansky 2001; Edward & Chapman 2011). As I will revisit in chapter 14, males tend to be choosy when males are limiting—as when they provide parental care and/or access to favored territories—and females vary in the benefits they can provide to males. If species are completely sex-role reversed with choosy males and indiscriminate females, the result is unidirectional mate choice like that described in chapter 6. It remains to be seen if there are distinctly male-typical or female-typical approaches to mate choice that hold across taxa. There are few quantitative studies directly comparing the strength of male and female choice. An exception is Rundle and Chenoweth's (2011) study of the Australian fruit fly *Drosophila serrata*, where females and males both make mating decisions based on a multivariate suite of cuticular hydrocarbons (CHCs), volatile pheromones on the exoskeleton. They derived multivariate selection surfaces from simultaneous mating trials of virgin males and females, and found that convex stabilizing selection on female CHC profiles was in fact stronger on females than on males; in other words, males on average had narrower (choosier) unimodal preferences (Rundle & Chenoweth 2011). While this may arise from individual males being choosier, it may also reflect more preference variability among females (chapter 9) or attention to other modalities (chapter 4). Ord et al. (2011) discuss intersexual differences, or lack thereof, in preferences for conspecifics over heterospecifics (chapter 16).

Male-only choice is rarely, if ever observed. Indeed, mate choice is mutual in what is perhaps the most dramatic example of male-limited reproduction: male-pregnant syngnathid fishes (pipefishes and seahorses), where females oviposit into a male brood pouch that in many species acts to transfer oxygen and nutrients to young. The key difference between these fishes and conventional livebearers like poeciliids, goodeids, and mammals is that females are entrusting males with large, nutrient-rich gametes that are costly to produce. As expected, therefore, both males and females can be choosy depending on circumstance (Gwynne & Simmons 1990; Berglund et al. 2005; Forsgren et al. 2004; Aronsen et al. 2013). As I explore further in chapters 11 and 12, environmental and social conditions can have a dramatic impact on choosiness, and this also applies for sex-role reversal. I will return to the evolution of the choosy sex in chapter 15.

8.3.2 Mate availability determines choosiness

Among the proximate conditions that drive which sex chooses, the key variable is the availability of opposite-sex mates. As we saw in chapter 6, the effective distribution of courter signals and the cost of mate searching are important determinants of choosiness. Mattle and Wilson (2009) found in another syngnathid, the potbellied seahorses *H. abdominalis*, that the choosy sex was sensitive to sex ratio. Females normally choose males, but in female-biased populations males were the choosy sex, preferring larger females, while females failed to show a preference. Burley and Calkins (1999) manipulated the adult sex ratio of a captive population of zebra finches. Males invested more in parental care in the male-biased treatment, suggesting that females were making differential allocation decisions (chapter 7). Further, male bill color predicted pairing in the male-biased but not the female-biased treatment, suggesting that females were making pairing decisions based on a male signal only when more males were available. Similarly, Watkins and colleagues (2012) found that women attended more to symmetry cues in male faces when these were presented as part of a male-skewed sample of faces; as with zebra finches, a sex ratio skewed toward the opposite sex had the effect of increasing attention to courter traits.

Theoretical models bolster the intuitive prediction that mate availability drives choosiness. Smaldino and Schank (2012) used an agent-based model incorporating homotypic and invariant preferences to model the effect of mate-searching and spatiotemporal heterogeneity in mate availability in humans; as predicted, search costs and heterogeneity increase when mates are encountered sequentially, because males face uncertainty about whether more attractive mates will be available for mating in the future. Barry and Kokko (2010) reached a similar conclusion with a model inspired by the extreme case of praying mantises *Pseudomantis albofimbriata*. A male will continue to approach a female even after she turns her head toward him, a behavior highly predictive of cannibalism. Imminent death obviously carries an immediate cost to males that counters the benefit of mating. Since the odds of finding another receptive female are scant, however, the model predicts the empirical pattern that a male should persist given the high probability of fertilization from that single mating event relative to the uncertainty of finding subsequent mates. Similarly, redback spiders self-sacrifice to females because the likelihood of surviving to encounter a second mate is small (Andrade 2003). This pattern appears to hold true in systems where males face opportunity costs much less dire than cannibalism. In guppies, Jordan and Brooks (2012) found that males invested more in courting females if they had previously experienced three females in sequence rather than experiencing them simultaneously.

On the other hand, when males interact with many females over the course of a lifetime, they are faced with decisions about how to allocate scarce resources. Beyond the opportunity costs associated with, say, courting a sexually unresponsive female instead of a receptive one, courtship and mating are costly to males in terms of survivorship; this is the case in the mosquito *Sabethes cyaneus*, a polygynous species where both males and females are elaborately ornamented (South et al. 2009; fig. 8.3). A mosquito-inspired model by South and colleagues (2012) showed that male preference can evolve purely as a consequence of female preferences for greater courtship intensity, since selection will favor the strategic allocation of male courtship among females (chapter 15).

The availability of mates in time is also a driver of choice. Female mammals, with a few notable exceptions including humans, are sexually receptive for a discrete estrous period which is typically synchronized among individuals. If the window of estrus is narrow and sex ratios are female-skewed, males may discriminate among receptive females on the basis of these estrus-associated cues (reviewed in Clutton-Brock 2009). Female African topi antelope (*Damaliscus lunatus*) prefer a subset of males on leks. Favored males are therefore sperm-limited, and females behave aggressively toward each other and toward males. Males, however, exert choice by behaving aggressively toward previously mated and aggressive females (Bro-Jørgensen 2007).

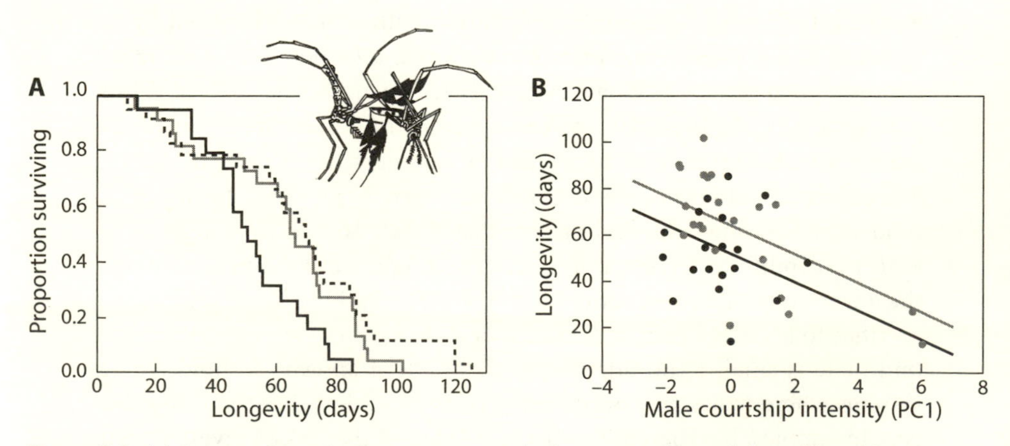

Figure 8.3. (a) Survivorship of male mosquitoes *Sabethes cyaneus* housed alone (dotted line); allowed to court but not mate (gray line); and allowed to court and copulate (black line). (b) Longevity in days as a function of a composite measure of courtship intensity. More intense courtship reduces survivorship in both courtship and courtship plus copulation treatments. After South et al. (2009). Drawing after W. A. Foster, sciencemag.org.

8.3.3 Male preferences and fitness costs for females

A key difference between male and female choice is that male choice is generally more likely to have net fitness costs for females than vice versa. While female choice of males may result in aggression or cannibalism, chosen males experience higher lifetime reproductive success than rejected males. For females, however, too much attention from males can be harmful. Long and colleagues (2009) showed that females preferred by male *D. melanogaster* suffered a disproportionate reduction in lifetime fecundity (fig. 8.4). Since preferred females had higher fitness when access to males was restricted, this had the remarkable evolutionary consequence that male choice acted to counter natural selection favoring more-fecund females (Long et al. 2009; chapter 15). This effect may be exacerbated by an intriguing effect of forced mating on female attractiveness: Dukas and Jongsma (2012) found in the same species that recently force-copulated females were more attractive to males than females that had consensually mated. In polygynous systems, females may therefore suffer more from being chosen by males than by being rejected.

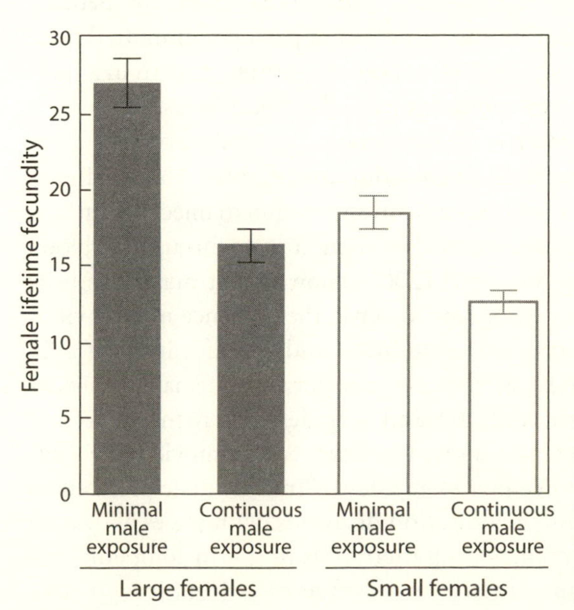

Figure 8.4. Male mate choice for large size erases the fitness advantage of large females. Lifetime fecundity of large females and small females exposed minimally or continuously to males. After Long et al. (2009).

8.3.4 Male choice and intrasexual competition

Choosing males, on the other hand, face two important constraints. The first, which I will discuss in the next section, is that in order for male choice to be realized, a chosen female either has to reciprocate the choice or be susceptible to forced mating. The second is as follows. For females in highly polygynous systems, sharing a preference with other females carries minimal costs and may actually be beneficial in terms of reducing the burden of search and assessment (**mate copying**; chapter 12). Male choice, by contrast, has the interesting dynamic that females that attract more males automatically lead to, on average, the reduced mating success of each courting male, because of sperm competition and the risk of aggressive interference from rivals. This makes it difficult for male preferences to evolve unless these preferences are sensitive to perceived levels of pre- and postmating competition (Jordan et al. 2014; Rowell & Servedio 2009; Servedio & Lande 2006). Critically, males can never evolve preferences for traits that reduce female viability, in stark contrast to female preferences for male traits (chapter 15).

Accordingly, male preferences are intertwined with intrasexual competition for mates. For example, Berglund and colleagues (2005) found that both males and females exercised choice in deep-snouted pipefish *Syngnathus typhle*. The same study also showed that both sexes competed intrasexually for mates; just as with unidirectional mate choice (chapter 6), competition can act to constrain the distribution of potential partners in mutual mate choice. Amundsen and Forsgren (2001) showed a similar pattern in a non-sex-role-reversed fish, the two-spotted goby *Gobiusculus flavescens*.

Male choice is greatly influenced by a male's perception of cues associated with postmating competition (Bonduriansky 2001). Male bedbugs, *Cimex lectularius*, use so-called traumatic insemination to inject sperm into the abdominal cavity of females, where they then migrate through the blood to fertilize eggs. Siva-Jothy and Stutt (2003) showed that males use taste receptors on their intromittent organs to sense the presence of a previous male's sperm, and reduce copulation duration and sperm transfer if a female has previously mated (fig. 8.5). Preferences for virgin females are widespread in insects (Bonduriansky 2001), and in spiders, where they represent the only widespread form of male choice (Andrade & Kasumovic 2005; Zimmer et al. 2014; but see Bel-Venner et al. 2008). Strategic male preferences can also take the form of sperm allocation decisions, with males allocating more sperm—and therefore increasing their odds in sperm competition—when presented with more attractive or novel females (e.g., Joseph et al.

A

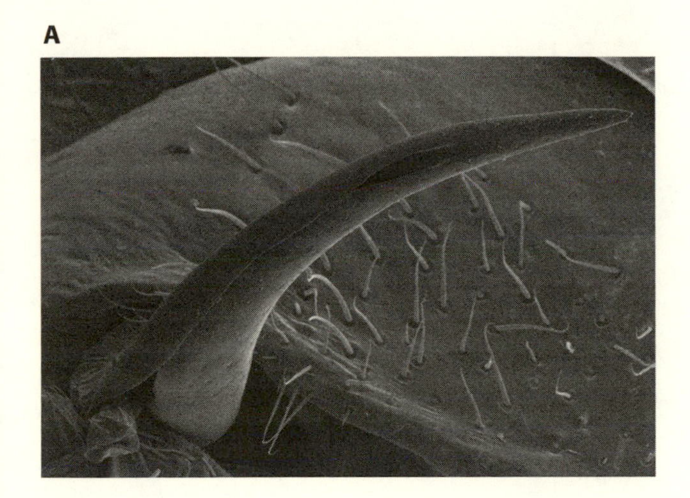

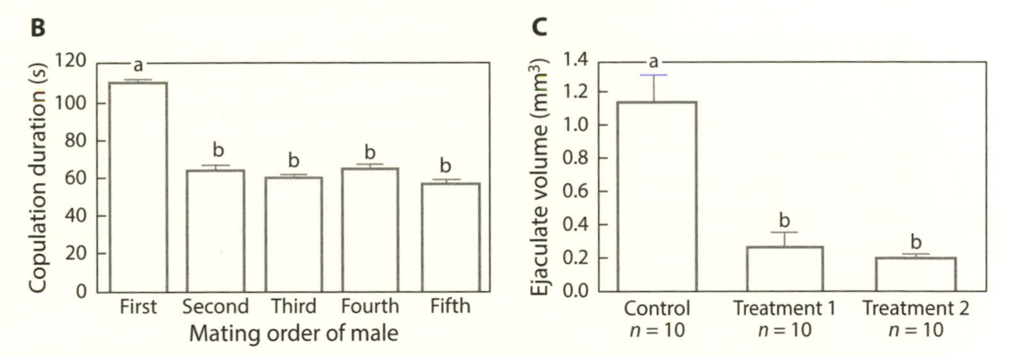

Figure 8.5. (a) Needle-like intromittent organ of a male bedbug (*Cimex lectularius*) used for traumatic insemination of females. (b) Copulation duration, a predictor of sperm transfer, of virgin males as a function of mating order. Copulation duration is highest when males are mated to virgin females. (c) Sperm transfer by virgin males: controls versus intromittent organs painted with ejaculate from a male, and intromittent organs painted with ejaculate removed from a female. After Siva-Jothy & Stutt (2003). Photos by Michael Siva-Jothy.

2012), or when presented with greater risk of sperm competition (Nöbel & Witte 2013).

Intrasexual competition can also directly induce the expression of strategic preferences. For example, male mosquitofish *Gambusia affinis* spent less time with a female if they had previously viewed her associating with another male (Mautz & Jennions 2011). Bel-Venner and colleagues (2008) found that size-assortative pairing in an orb-weaving spider, *Zygiella x-notata*, was stronger under a high-competition regime when male opportunities were

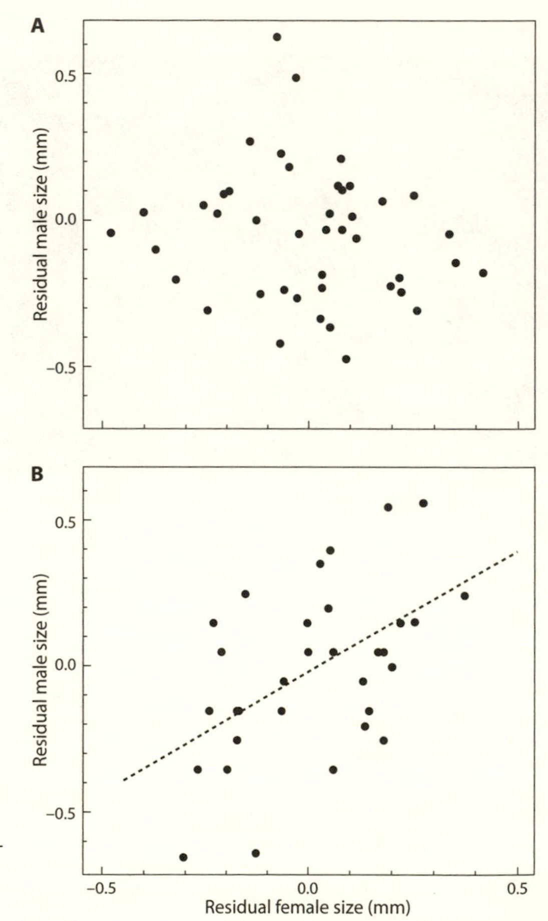

Figure 8.6. Female and male size of mated pairs in the orb-weaving spider, *Zygiella x-notata.* (a) Pairs show no evidence of size-assortativity under a low-male competition treatment. (b) Size-assortative pairing emerges when competition is high between males. After Bel-Venner et al. (2008).

more constrained. This was not simply due to larger males monopolizing larger females, but to small males actively seeking out smaller partners (fig. 8.6). As I will elaborate on later in this chapter, such flexible preferences, modified in response to the likelihood that a partner is attainable, are a hallmark of mutual mate choice.

8.3.5 Male choice depends on female preferences

A study by Gasparini and colleagues (2013) highlights the mutual nature of male mate choice and its sensitivity to competition. As we saw in chapter 6, a courter's attractiveness to choosers is context-dependent, and chooser preferences for a focal courter are stronger if he is encountered after or alongside less attractive courters. Female guppies prefer more colorful males, and exhibit stronger preferences for focal males when these are viewed alongside drabber males. Males are strategic in their preferences, associating more with females that are surrounded by drab males (Gasparini et al. 2013). By contrast, in a species without detectable female premating choice, the eastern mosquitofish *Gambusia holbrooki*, large males showed stronger preferences for larger females, but there was no effect from the presence of other males (Callander et al. 2011). A male's preference therefore appears to incorporate public information about his likely attractiveness to females.

8.4 PAIRING DECISIONS: FINDING A SOCIAL MATE

8.4.1 Social mates and genetic mates

Most birds, some fishes, and a small proportion of mammals and some invertebrate taxa are **socially monogamous**—they associate preferentially with another individual in the context of sexual behavior. In these taxa, reciprocal mate choice results in a pair bond (see below). Socially monogamous pairs almost always engage in biparental care, with butterflyfishes (*Chaetodontidae*; Boyle & Tricas 2014) representing a notable exception.

The offspring that social mates care for are usually, but by no means always, the offspring of both parents. The remaining offspring are the product of **extrapair fertilizations** (and therefore, of extrapair matings). While social monogamy is common in some taxa, genetic monogamy is the exception: a review of bird studies by Griffith and colleagues (2002) found that about 90% of species had extrapair offspring, with an extrapair paternity rate averaging 11%. In the fairy-wrens (genus *Malurus*), extrapair fertilizations account for a *majority* of offspring (Brouwer et al. 2014). In same-sex pairings, any gamete exchange by definition occurs outside the pair bond. To reinforce the point that social and genetic monogamy are very different things, consider DuVal's (2013) finding that 41% of socially polygynous female lance-tailed manakins (*Chiroxiphia lanecolata*) returned to mate with the same male on the lek in successive years.

Accordingly, social and genetic mating can to some extent be decoupled, which enables tests of the relative effects of direct benefits (like provisioning and defending offspring) versus genetic compatibility and "good genes" (Akçay & Roughgarden 2007). Since social and genetic mates provide different sets of costs and benefits to choosers, their expectation is that these should attend to different cues when choosing extrapair versus social mates.

The best-known example of this phenomenon is in humans, where women prefer men with more masculinized faces during the fertile phase of their cycle, and more feminized faces outside the window of ovulation when only social mating is possible (Penton-Voak et al. 1999; chapter 9).

8.4.2 Pairing decisions and preferences

Social pairing can in principle occur in the absence of preferences simply through spatial association, if individuals just mate with an available partner who shares their territory (Mock & Fujioka 1990). For the most part, however, pair bonding is the outcome of a lengthy process of courtship and successive filtering of potential mates. If potential mates have reciprocal preferences for one another (section 8.2), and if there is abundant variation in preferences, pairing proceeds in a straightforward manner. In general, however, there will be a mismatch between chooser preferences and mate availability.

The attractiveness of partners to one another becomes pivotal in the formation of social pairs in monogamous species. In some cases, there may not be much variation among choosers with respect to a particular trait. Here, both unattractive and attractive choosers prefer attractive traits (fig. 8.1d). For example, both female and male pipefish *Syngnathus typhle* prefer larger mates (Berglund et al. 2005). By contrast, the seahorses in Bahr et al.'s (2012) study exhibit the pattern shown in fig. 8.1e: females mate disassortatively with respect to MHC type, while males show an overall preference for larger females. In humans, the pattern is reversed, with men but not women, showing preferences for the faces of MHC-dissimilar women (Lie et al. 2010).

In zebra finches, males and females have preferences of opposite valence (chapter 5) for a trait that is sexually dimorphic. Males have bright red bills, while females' bills are duller. The sexes have opposite directional preferences, with females preferring males with redder bills, and males preferring females with less red bills (fig. 8.1f). Since there is a positive genetic correlation between male and female bill color, Price and Burley (1994) argued that opposing directional preferences by males and females acted to limit the evolution of sexual dichromatism. In chapter 15, I will return to the role of male and female preferences in shaping sexual dimorphism.

Since pair bonding is nearly always associated with biparental care, an important question is how individuals select mates with respect to care functions. In particular, one might expect that individuals might prefer mates who are behaviorally different from themselves and can thus fill a complementary role in parental care. However, a review by Schuett and colleagues (2010) suggests that positive assortativity with respect to personality is more prevalent across species, including humans. Stone et al. (2012) found that human preferences for personality traits were more consistent across choosers than preferences for condition-dependent traits thought to be more relevant to fitness (see chapter 15). I discuss personality as a covariate of preferences again in chapter 9.

8.4.3 Mutual mate choice under conflicting preferences

The cichlid fish *Herichthys tamasopoensis* is endemic to the Río Tamasopo of Mexico's Huasteca Potosina. One of the bigger populations of this species is located in a water park under a high waterfall. Males and females form socially monogamous pairs over the course of the breeding season, aggressively defending their eggs and young from predators. A larger, older partner is beneficial to both males and females; larger females are more fecund, and larger individuals of both sexes are more successful at securing and defending good territories. Yet courtship in these and other substrate-spawning cichlids involves a series of ritualized interactions—parallel swimming, followed by locking jaws and vigorously wrestling—consistent with mutual assessment of body size and physical strength. Mismatched individuals interrupt courtship and fail to pair. Breeding pairs are almost perfectly size-assorted, with larger, older couples occupying prime real estate in the deep, highly oxygenated water under the falls. The smallest couples are relegated to marginal nest sites on concrete staircases leading into the water, subject to being preyed upon by birds and stepped on by humans. The humans doing the stepping are also paired up assortatively: the consensus attractiveness of members of a heterosexual couple (as rated by opposite-sex heterosexuals) is correlated (r =.4 to .6; Simão & Todd 2003). This assortativity occurs even though males and females both prefer mates of greater consensus attractiveness. How does assortativity emerge from directional preferences, and from conflicting preferences more generally?

8.4.4 Pairing by matching

Kalick and Hamilton's (1986) "matching hypothesis" proposed that pairs assess each other sequentially: more attractive individuals form pairs, and the remaining individuals are left with each other. This hypothesis was formalized in a series of economic models (Bergstrom & Real 2000; reviewed in Beckage et al. 2009; Puebla et al. 2011) proposing a matching algorithm similar to that used to match medical residents to hospitals. Here, both sets of individuals rank potential mates according to perceived attractiveness, courting potential mates in rank order until they are no longer rejected. The system stabilizes when no pair forms that can increase the attractiveness for both partners.

In matching algorithms, individuals are not assumed to have any prior information about their own attractiveness. By contrast, another set of models (Todd 1997; Simão & Todd 2002, 2003; Fawcett & Bleay 2008) propose a variant of the adjustable-threshold approach in single-chooser mate searching (chapter 6) where acceptances and rejections shape self-perception of attractiveness. These models posit that individuals go through an "adolescent" or "courtship" period prior to pair-bond formation where they experience acceptance and rejection, and estimate their own overall attractiveness on the basis of the attractiveness of the individuals who accept and reject them. This is essentially a more complex version of the adjustable-threshold models described in chapter 6; individuals adjust their acceptance threshold based not only on the attractiveness of the potential mates they experience, but on the behavior (accept or reject) of those mates toward them. Compared to the "matching" algorithms, a scenario with a courtship period requires substantially fewer bouts of mate sampling to account for observed correlations in attractiveness (Simão & Todd 2003).

Beckage and colleagues (2009) evaluated these alternative models using speed-dating data from heterosexual Berliners in their twenties. In speed dating, individuals engage in brief conversations with potential partners, and then secretly indicate whether or not they wish to subsequently exchange contact information; if both partners concur, information is exchanged at the end of the speed-dating event. Here, third parties also rated the attractiveness of each participant to provide each individual with a global attractiveness score. The offers that choosers made were contingent on their own independently rated attractiveness, consistent with the notion that aspiration levels are set by previous experience, but the pattern of offers did not change over the course of the speed-dating event, suggesting that preferences were robust to short-term experience of acceptance and rejection.

8.4.5 Pairing by market value

An alternative to the matching hypothesis and its adjustable variants is that individuals vary in their preferences. Less attractive individuals are less choosy, settling for less attractive mates (Burley 1983). In line with Miller and Todd's (1998) model of sequential filtering, Burley (1983) proposed that less attractive individuals selected mates based on broad criteria like sex and species, while more attractive ones attended to variation within opposite-sex conspecifics. Biological market theory (Noë & Hammerstein 1995) has provided a framework for evaluating the hypothesis that choosiness covaries with attractiveness in mutual mate choice. Pawłwski and Dunbar (1999) used personal advertisements among heterosexual Britons to estimate the "market value" of men and women as a function of age. Market value was estimated by dividing the number of advertisements seeking opposite-sex partners of a given age cohort (the demand) by the number placed by advertisers specifying that cohort (the supply). Market value peaked in the mid-twenties for women and in the mid-thirties for men (fig. 8.7a). Higher market value should allow choosers to be more selective in a mutual choice task; the personal ads confirmed this prediction, with higher-value individuals of both sexes specifying a larger number of desirable traits in the partner sought (fig. 8.7b).

An important caveat when interpreting these human studies is that personal-advertisement data may not broadly represent the mating pool; Gil-Burmann and colleagues (2002) showed that Spanish personal advertisers were older than average and argued that they had been unsuccessful at conventional approaches to finding mates. Similarly, the average attractiveness (as rated by third parties) of volunteers in Beckage and colleagues' (2009) speed-dating study skewed low (3.98 for women and 2.83 for men on a scale of 1–9).

Focusing on generalized attractiveness elides the fact that choosers are attending to particular traits in mating preferences, and that the strength of preference may differ between sets of partners. This point is illustrated by an analysis of more than 100,000 speed-dating trials, where Stulp et al. (2013) found that conflicting preferences led to formation of pairs that were intermediate between the preference optima of both sexes. Consistent with a previous experimental study on visual preferences by Courtiol, Picq, and colleagues (2010), men preferred women 7 cm shorter than themselves, and women preferred men 26 cm taller. The mean height difference in the sample was 14 cm, but the average for realized pairings was a significantly higher 19 cm. The skew toward the preferred difference for women is because women exhibited stronger preferences with respect to height (i.e., were less likely to say "Yes" to an exchange of contact information). Indeed, Courtiol, Raymond, et al. (2010) found that female preferences for stature and other traits among French couples more closely matched partner traits than did

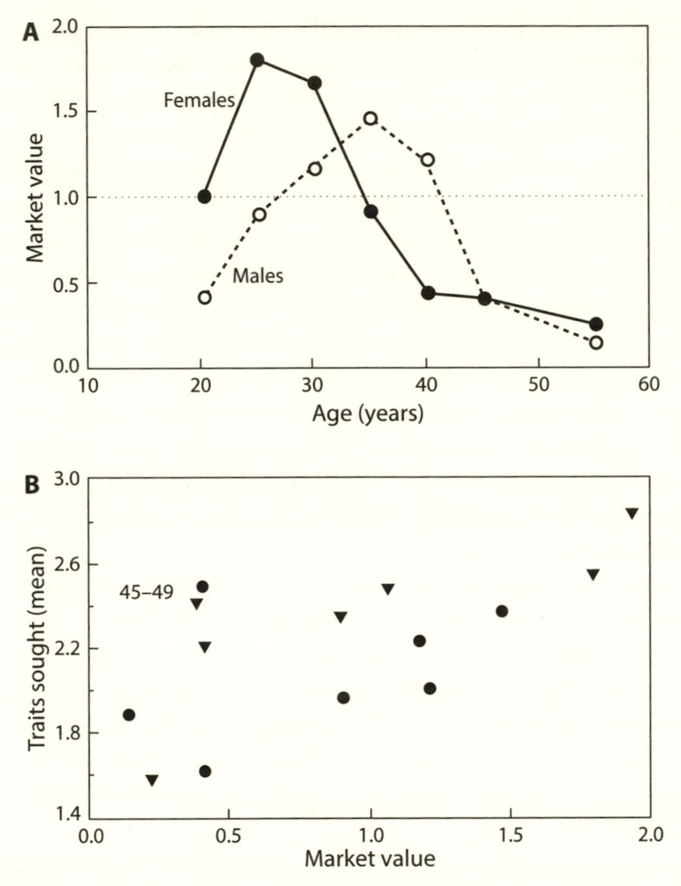

Figure 8.7. (a) Market value (ratio of demand, advertisements seeking an age class to supply, advertisers specifying an age class) of British men and women as a function of age. (b) Across age cohorts, choosiness, in terms of number of traits sought in a potential partner, is positively correlated with market value in both men (circles) and women (triangles).

the preferences of males. The outcome of conflicting preferences is therefore driven by the partner with the stronger preference.

Choosers have some control over their attractiveness to a prospective partner, in that they can modulate the amount of effort they put into providing courtship or nuptial gifts. Patricelli and colleagues (2011) proposed a model of "courtship haggling" where courtship effort is variably invested in response to feedback from prospective partners. In Miller and Todd's (1998) sequential-filtering model, successive aspects of courter phenotypes are more costly for choosers to evaluate. If sequential information-gathering is more costly for both parties, courtship dynamics in mutual mate choice should be sensitive to the distribution of mating opportunities (Patricelli et al. 2011).

A common feature of all models of mutual mate choice is that assortativity at a population level depends on the number of mates sampled; for example, models incorporating self-assessment of attractiveness require fewer assessments for choosers to converge on a partner that approximates the most attractive option available (Simão & Todd 2003). It follows that, just as in single-chooser mate choice (chapter 6) the process of mate searching is an important determinant of mutual mate choice outcomes. Further, choosers express multiple preferences that can shape mutual outcomes in complex ways. Humans have—among many others—directional preferences for global attractiveness, self-referential and unimodal preferences for height, and homotypic preferences for political affiliation. Smaldino and Schank (2012) developed an agent-based model that incorporated features of mate-searching algorithms, and had agents select mates on a similarity rule and on a maximal attractiveness rule. As expected, limited searching decreased interpair matching, and preference rules interacted with spatial structure to produce complex nonlinear outcomes. Humans may mate assortatively with respect to attractiveness, politics, and height, but much of the remaining variation is due to proximity and contingency.

8.4.6 Self-assessment of attractiveness

A critical assumption of most mutual mate choice models is that individuals are capable of assessing their own attractiveness. Royle and Pike (2010) manipulated the attractiveness of male zebra finches by festooning them with green or red leg bands and then subjecting them to two-way interactions with females or one-way interactions where they could see the female but not vice versa. If males had red rings, females performed more proceptive behavior toward them; males that were thus perceived as more attractive in turn performed more courtship behavior toward novel females.

Self-gauging of attractiveness plays a role in dynamic outcomes of mutual mate choice. Humans who rate themselves as highly attractive are choosier (Buston & Emlen 2003). Women appear to be more accurate in their self-assessment and to be more sensitive to their own attractiveness in making mating decisions: men more than women tended to overestimate their own market value when seeking a partner (Pawłwski & Dunbar 1999). Todd and colleagues (2007) found that in speed-dating, men made actual choices based primarily on their partners' physical attractiveness, while women chose men whose overall attractiveness matched their self-perceived physical attractiveness. As I revisit in chapter 11, chooser state, for example as a function of nutrition, disease, or recent social interactions, can have a strong influence on individuals' self-assessment in mutual mating tasks (Holveck & Riebel 2010).

8.5 PAIR BONDING

8.5.1 Forming pair bonds

The outcome of mutual mate choice is very often pair bonding, a social state that typically involves spatial proximity, recurrent mating, and sexual behavior; often, it includes mate-guarding and aggressive responses to other sexually mature individuals (Young et al. 2011; Mock & Fujioka 1990). Individuals exhibit behavioral and hormonal profiles associated with stress when they are forced apart from their partners, and these stress indicators ebb when partners are reunited (e.g., Remage-Healey et al. 2003).

The neurobiology of pair bonding is one of the better-characterized mechanisms of mate choice beyond the sensory periphery. In mammals, pair bonding is triggered by consensual copulation. Pair bonding is dependent on a well-characterized neural circuit that links sensory input (primarily olfactory) with a "social recognition" circuit in the medial amygdala and lateral septum and a circuit associated with affiliative reward (the nucleus accumbens in females and the ventral pallidum in males) that assigns positive hedonic value to familiar social stimuli. This latter circuit is dependent on activation of receptors that bind nonapeptides, neurohormones comprised of nine amino acids—oxytocin for females and arginine vasopressin for males. The network appears to be broadly similar among monogamous mammals (Bales et al. 2007); zebra finches also require nonapeptide receptors for pair bonding (Klatt & Goodson 2012). Targeted overexpression or inhibition of nonapeptide receptors is sufficient to induce or abolish pair bonding in naturally polygynous and monogamous vole (*Microtus*) species, respectively (Young et al. 2011). However, variation in *cis*-regulatory elements of this gene is insufficient to explain interspecific diversity in pair bonding (Fink et al. 2006). Further, intraspecific variation in philandering among male prairie voles (*Microtus ochrogaster*) is not associated with nonapeptide receptor variation in the ventral pallidum, but rather with receptor variation in brain regions involved in spatial memory. Ophir and colleagues (2008) showed that this variation was correlated not only with philandering on the part of males but also with greater space use, arguing that philandering was a consequence of more mating opportunities occasioned by wandering. As I discuss later, pair bonding is perhaps the best-characterized of many mate choice mechanisms that are likely underlain by complex genetic variation and therefore constitute a broad mutational target.

8.5.2 Maintaining pair bonds

Julian Huxley (1938) largely rejected Darwin's idea of premating choice and viewed post-pairing communication as the primary driver of elaborate sexual signals in birds, "acting as a bond to keep members of a pair together." Pair-bonded individuals engage in mating and in intimate mutual courtship, like belly-bumping in grebes (Symes & Price 2015) and kissing in humans (Wlodarski & Dunbar 2013). Huxley's idea has been largely ignored since the late-twentieth-century resurgence of Darwinian views on mate choice (Symes & Price 2015). However, maintaining the pair bond fits tidily into the broader picture of mate choice. Just as mate choice continues after mating (chapter 7), so too does it continue after social pairing. One of the mechanisms of cryptic choice is differential investment in offspring, and choice based on post-pairing behavior plays a role in determining investment decisions (reviewed in Servedio et al. 2013; Templeton et al. 2012). Female Gouldian finches (*Erythrura gouldiae*) have higher levels of circulating stress hormones if paired with a non-preferred partner, which in turn provides a mechanism for stress-mediated differential allocation based on preference (Griffith et al. 2011; chapter 7).

Post-pairing interactions can involve sophisticated cognitive processes. Just as nuptial gifts like spermatophores play an important role in cryptic mate choice (chapter 7), food-sharing is widespread between pair-bonded individuals. Male Eurasian jays *Garrulus glandarius* gauge the "specific satiety" of their mates when food-sharing; that is, they can tell if their mate is tired of a particular food item and give them an alternate item. If they observe their mate eating waxmoth larvae, they are more likely to subsequently feed them mealworm larvae, and vice versa. Importantly, this effect goes away when males are prevented from observing previous feeding by their mate, ruling out differences in the female's food-soliciting behavior as a factor (Ostojić et al. 2013).

8.5.3 Breaking and transgressing pair bonds

Once pair bonded, a chooser may decide to maintain the social relationship for the breeding season or even for life, or he may decide to **divorce** his partner for a new social mate. He also needs to decide how much time and energy to invest in **philandering** (soliciting or accepting matings outside the pair bond) versus directly or indirectly enhancing the fitness of current offspring by engaging in parental care or maintaining the pair bond. Choosers can therefore engage in a variety of mate choice decisions after an initial pairing.

Pair-bonded individuals very frequently engage in **extrapair matings**, more commonly referred to as extrapair copulations (EPCs). Burley et al. (1996) fitted male zebra finches with leg-band colors previously shown to be attractive (red) and unattractive (green) to females on average. Rates of paternal exclusion (chicks at the nest from a father other than the social father) were 16% for red-banded males and 40% for green-banded males. So-called unforced EPCs were highly predictive of exclusion, suggesting that premating choice accounted for most of the extrapair fertilization bias.

The natural history of extrapair mating is intriguing. Social mates often engage in mate guarding and parental care; in order for extrapair mating to occur, one individual (usually the male) has to leave his nest and join the other in a context where he won't be driven off by the social partner. This has the effect of markedly increasing the search costs of mating, and means that choosers may not have the luxury of evaluating a large number of extrapair mates. A study on pied flycatchers (*Ficedula hypoleuca*) by Canal and colleagues (2012) is instructive. In this species, males leave their nests to search for extrapair partners. On average, males philandered with females on nests close to their own, although some individuals bypassed fertile females for others farther away. The timing of philandering was also driven by the reproductive state of a male's mate. Males mate-guarded their fertile female partners before egg-laying, and provided parental care after hatching, concentrating their extrapair matings during incubation.

Divorce occurs when a pair bond is severed. Choudhury (1995) put forth two alternative hypotheses for the proximate cause of divorce in birds: first, that divorce occurs after reproductive failure, with both partners acting to separate and search for more compatible mates; second, that divorce is initiated by one partner when a more attractive social partner is available. Otter and Ratcliffe (1996) tested the latter hypothesis by experimentally removing females from a population of black-capped chickadees (*Parus atricapillus*), thus augmenting the pool of available males. Divorce was initiated by females when a neighboring "widower" was of higher social rank than their current social mate, supporting the hypothesis that choosers divorce to "trade up" to more attractive mates (fig. 8.8). Consistent with the "trade-up" hypothesis, men who are substantially taller than their female partners report less jealousy in interpersonal relationships (Brewer & Riley 2010).

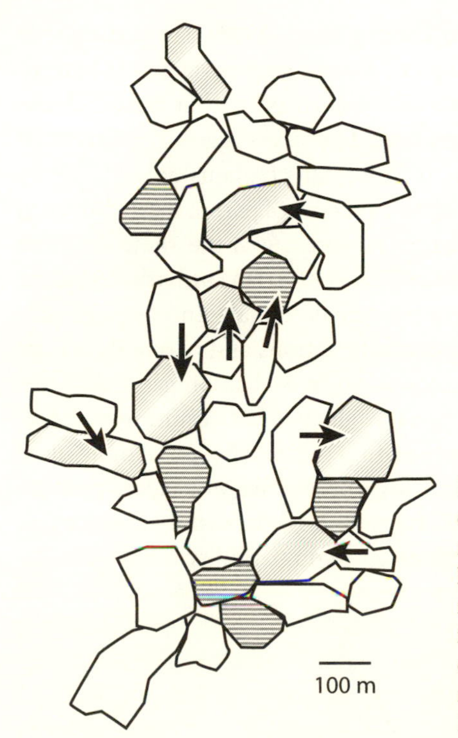

Figure 8.8. Divorce in black-capped chickadees following experimental removal of females from high- (dark hatched) and low-ranked (light hatched) males. Arrows indicate territories of females initiating divorce from their current partner and pairing with the recently widowed male. After Otter & Radcliffe (1996).

By contrast, Galipaud and colleagues (2015) recently found that males of the amphipod *Gammarus pulex* only switch mates when their current mate is smaller and less fecund; the absolute attractiveness of a mate, rather than its attractiveness relative to other options, determines divorce. A meta-analysis of bird studies by Dubois and Cézilly (2002) found a weak but robust effect of breeding failure on divorce, consistent with the first hypothesis that divorce is a consequence of reproductive incompatibility.

8.5.4 Pair bonds between same-sex partners

Female-female pairs in birds with biparental care provide the most clear-cut example of how choice of same-sex partner can have dramatic consequences for chooser fitness. Monogamous seabird species provide some of the best-studied models of the causes and consequences of same-sex pairing (Conover & Hunt 1984). Young and colleagues (2008) studied a Hawaiian population of Laysan albatrosses (*Phoebastria immutabilis*) where 59% of adults are females, and accordingly 31% of breeding pairs are female-female.

All pairs of females were unrelated (Young et al. 2008), suggesting that avoidance of kin in mate choice (chapter 14) may be general across hetero- and homosexual interactions. Remarkably, most pairs that raised chicks in multiple years had offspring from both females, suggesting reciprocal cooperation from season to season. As in humans (Adams & Light 2015), there was no measurable difference in the outcome of parental care between same- and opposite-sex parents; fledgling success was the same for both types of pairings. A study by Elie and colleagues (2011) similarly found no difference in social interactions between female-female and female-male pairs of zebra finches. While within-pair interactions and parental care are the same between homo- and heterosexual pairs, hatching success was markedly lower for female-female pairs of albatrosses, resulting in lower overall fitness of same-sex pairs per breeding season (fig. 8.9; Young et al. 2008). The reasons for reduced hatching success are not clear; perhaps females soliciting extrapair fertilizations were less successful at evaluating genetic mates for cues associated with genetic compatibility (chapter 14).

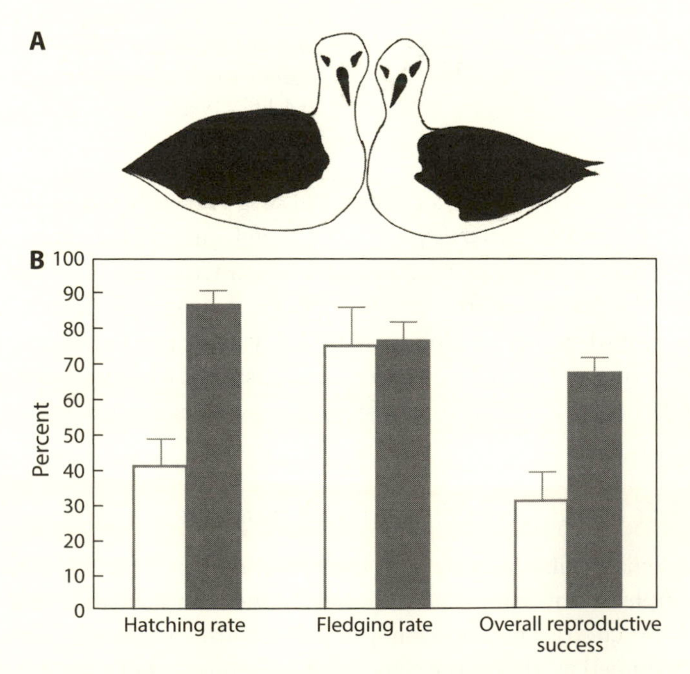

Figure 8.9. (a) Pair-bonded, unrelated female Laysan albatrosses. (b) Fledging rates are similar for chicks raised by same- and opposite-sex partners, but hatching rates are substantially lower, resulting in lower overall reproductive success. After Young et al. (2008). Drawing by Carmen M. Rosenthal Struminger.

8.6 MATE CHOICE IN HERMAPHRODITES

Functional hermaphroditism, whereby the same individual produces both sperm and eggs over the course of its lifetime, is widespread among invertebrates, notably mollusks and nematodes; in vertebrates, it is frequent among marine fishes but otherwise absent. Individuals can be sequential hermaphrodites, changing sex in response to changes in competition and/or mate availability. Simultaneous hermaphrodites, by contrast, can produce both sperm and eggs in the context of a mating interaction, posing an interesting case of mutual mate choice. Despite the lower potential for sexual selection in simultaneous hermaphrodites (Greeff & Michiels 1999; chapter 15), there is growing evidence for pre- and postmating mate choice with respect to fecundity correlates and mating status (reviewed in Leonard 2006 and Schärer & Pen 2013). As mentioned earlier, simultaneous hermaphrodites provide one of the clearest instances of reciprocal homotypic preferences (Puebla et al. 2012; fig. 8.2).

A unique feature of mate choice in simultaneous hermaphrodites is the preference for a particular sex role. Given that a chooser encounters another individual, it is faced with the decision not only of whether to allocate gametes to that individual and how many gametes to allocate, but how to distribute that allocation between sperm and eggs. Two individuals will often prefer the same sex role, leading to tit-for-tat release of sperm and eggs (Leonard 2006). Partners may also choose to transfer different amounts of sperm based on their own (Nakadera et al. 2014) or their partner's (Anthes et al. 2006) size and mating history.

8.7 SYNTHESIS

Mutual mate choice is a dynamic extension of one-way mate choice during and after mating, whereby sampling and evaluation interact with feedback from potential mates and from partners. Mutual mate choice has very different dynamics in different mating systems. In polygynous species, male and female choice ubiquitously coexist but take on very different forms, driven by very different fitness dynamics for the two sexes (chapter 15). Mutual mate choice leading to pairing, by contrast, may include similar preferences in the two partners, including individuals of the same sex. Pairing involves adjusted preferences or negotiations that often end up with both partners having mates that fall far short of their preference maxima, but are the best they can obtain. This kind of mutual mate choice leads not only to mating but also to social pairing, which recruits a distinct set of neural mechanisms and poses new decision points for choosers with regard to post-pairing courtship, philandering, and divorce.

There are probably very few systems in which one of these two kinds of mutual mate choice systems does not operate. The differences between each type of choice system mean that detecting mutual mate choice is an important task for empiricists, since mutual mechanisms will produce different evolutionary constraints and patterns than one-way choice. Detecting mutual mate choice is more difficult than one-way choice, and pertinent perimating and post-pairing signals are challenging to manipulate (chapter 7). Further, a focus on one-way and directional preferences for flamboyant visual and acoustic display traits is likely to inhibit us from looking for preferences involving sexually monomorphic traits. Rundle and Chenoweth (2011) argued that the unimodal preferences they measured in male *D. serrata* for female cuticular hydrocarbon profiles were likely typical of male preferences for fecundity-related traits in females, and suggested that male mate choice may be broadly underestimated because of the prevalence of two-stimulus experimental designs, which fail to distinguish between unimodal and directional preferences (Wagner 1998; chapter 2).

Throughout this chapter, I have largely ignored the multivariate nature of mate choice. Trade-offs between preferences for multiple traits (chapter 6) may particularly complicate mutual mating decisions. These are likely to be widespread in monogamous species, where there may be antagonistic correlations between traits involved in parental care and traits associated with offspring genotype (Gangestad & Simpson 2000; Stiver & Alonzo 2009). Choosers may assign different weights to different traits when weighing these trade-offs. Women in the United States, for example, trade off attractiveness for more resources in a mate, but do not compromise on companionship or age (Waynforth 2001).

An implicit assumption of much of the mutual mate choice literature is that experience with potential mates, and therefore self-assessed attractiveness, is the primary determinant of preference variation in mutual mate choice. Reciprocal preferences resulting in (dis)assortative mating, however, plainly require preferences to vary; and choosers are likely to express preference variation as a result of genotypic differences, environmental influences, and social interactions experienced over a range of timescales. The next four chapters focus on the phenomenon of preference variation.

8.8 ADDITIONAL READING

Bonduriansky, R. 2001. The evolution of male mate choice in insects: a synthesis of ideas and evidence. *Biological Reviews* 76 (3), 305–339.

Simão, J., & Todd, P. M. 2002. Modeling mate choice in monogamous mating systems with courtship." *Adaptive Behavior* 10 (2), 113–136.

Young, K. A., Gobrogge, K. L., Liu, Y., & Wang, Z. 2011. The neurobiology of pair bonding: Insights from a socially monogamous rodent. *Frontiers in Neuroendocrinology* 32, 53–69.

CHAPTER 9

Variation in Preferences and Choices: General Considerations

9.1. INTRODUCTION

The preceding chapters described the mechanisms underlying mating preferences and how they are realized into pre- and postmating choices. From sensory transduction through mate-sampling and evaluation, *any change to any one* of these mechanisms has the potential to change how a given courter stimulus is assessed among species, among individual choosers, and even within the same chooser. It is ultimately this variation—at all of these scales—that makes mate choice interesting. Variation in mating preferences is what makes them important to speciation and diversification, what gives mate choices the power to adapt to changing circumstances, and what drives the variability of courter traits. It is because choosers in different species evolve different preferences, and because choosers are more discerning in some populations than in others, that mate choice is important to evolution (Rodríguez et al. 2013; fig. 9.1). Individual variation in preferences, by contrast, provides both a means for maintaining diversity in courter signals and the raw material upon which selection can act to shape preference evolution (Miller & Svensson 2014).

In this and the next three chapters, I review sources of variation in preferences—genetic, environmental, and social. The questions we ask about preference variation fall into three broad categories:

1. How does *genetic variation* at all scales account for variation in preferences? How much is there, and how many dimensions does it have? Are independent sets of genes associated with different aspects of preference?
2. How much of the *non-genetic variation* can be accounted for by predictable social and ecological processes, like sexual imprinting, condition-dependence, and risk-sensitive choice?
3. How do genes and environment *interact* to produce expressed preferences in natural populations?

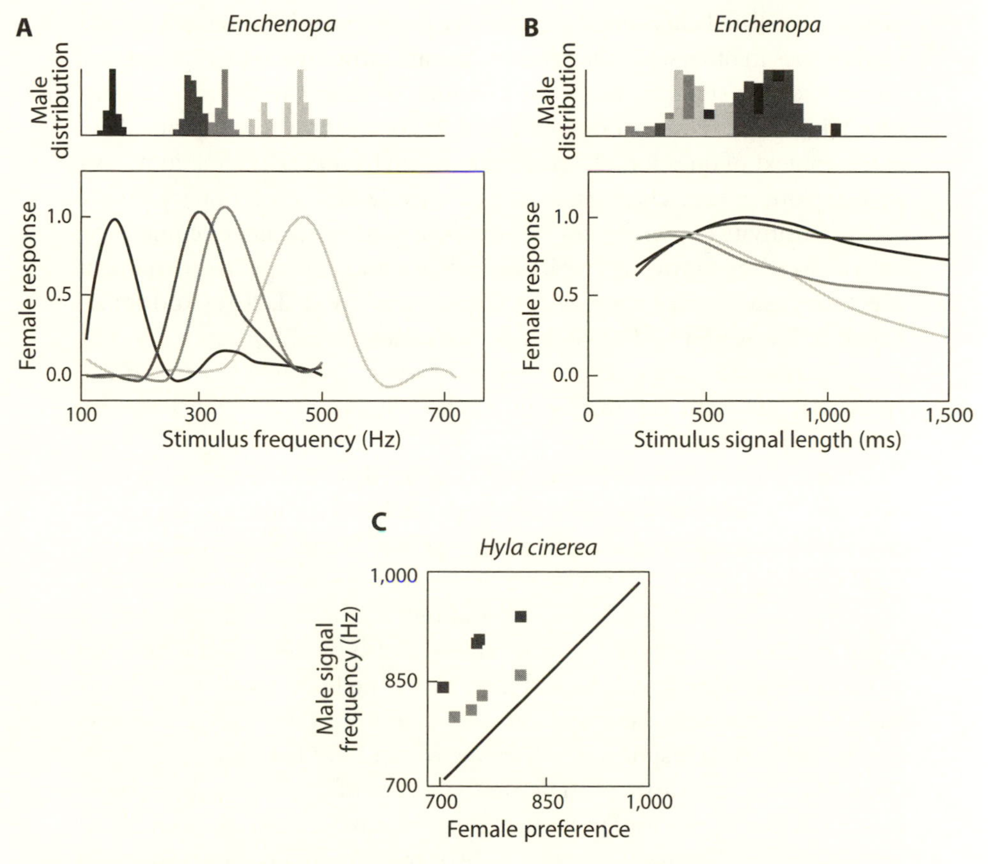

Figure 9.1. Diversity in preference variables and corresponding courter signals. (a) Signal frequency distribution and divergent female preference functions for peak signal frequency in four *Enchenopa* treehopper species associated with different substrates; (b) Overlapping signals and similar, weak preference functions for signal length in the same four species; (c) Peak male signal frequency and peak female preference for signals across eight populations of gray treefrogs *Hyla cinerea*; darker shading indicates "rough" sympatry with the congener *H. gratiosa*. After Rodríguez et al. (2013).

Questions (1) and (2) represent some of the most active areas in contemporary research on mate choice. For several species we have a first-order understanding of the amount of genetic variation in mate choice (reviewed in Prokuda & Roff 2014), and are beginning to gain insight into the genetic architecture of mate choice down to the allelic level. The past few years have also seen an explosion of evidence that preferences are shaped meaningfully by chooser history, condition, and context (e.g., Cotton, Small et al. 2006). These effects can be strikingly variable, spanning the range of preference variation: well-fed choosers show stronger preferences in some species and weaker ones in others; familiar phenotypes are attractive to one species and aversive to its sister species. As we continue to elucidate the mechanisms specifying preferences (questions 1 and 2) we need to think about them in the context of question (3), genotype-by-environment interactions. As I shall explore in later chapters, there is relatively little evolutionary theory on the sensitivity of preferences to individual experience and circumstance, notwithstanding a growing empirical evidence for genotype X environment effects on mate choice (Ingleby et al. 2013; Narraway et al. 2010; Rodriguez, Rebar, & Fowler-Finn 2013; Rebar & Rodriguez 2013). Question (3) also encompasses the relationship of vertical genetic transmission to vertical **epigenetic** and cultural transmission of preferences—so-called inclusive inheritance (Danchin et al. 2011).

Different aspects of preference are associated with different sources of variation. Variation in *responsiveness* (e.g., Ritchie et al. 2005) is often associated with physiological receptivity (see below). Numerous studies document variation in *choosiness*, whether associated with genotype (chapter 10), ecological factors like predation risk and nutritional state (chapter 11), or social experience (chapter 12). Choosiness is also by definition a function of variation in mate searching and mate sampling (chapter 6). It is important to note that responsiveness and choosiness are interrelated: choosers that are entirely unresponsive will by definition have flat preference functions. The failure to distinguish between absence of preference (low choosiness) and lack of motivation to mate (low responsiveness) is a major concern for studies that use association time or other non-specific responses to assay preference (chapter 2).

Processes like speciation and signal diversification depend on choosers having preference functions with different *shapes*—that is, different sets of stimuli are labeled as attractive or unattractive. This is perhaps the aspect of preference variation we are most interested in, and is very often dependent on social experience and sometimes on other environmental or state variables.

This chapter and the next three focus on the many proximate factors contributing to variation in mate preferences, ranging from those that alter the mating decisions of an individual to those that account for differences among broad taxonomic groups. The remainder of this chapter highlights some of the important factors to consider when studying preference variation. Chapter 10 addresses scales and patterns of genetic variation in preferences. Chapter 11 addresses how the broader physical and community environment shapes mating preferences. Finally, chapter 12 focuses on social effects on preference variation, starting with the maternal environment, through sexual imprinting early in life to mate-choice copying.

9.2 SCALES OF VARIATION

Variation in mating preferences exists at all scales. The most obvious differences in mating preferences are those among broad taxonomic groups, where different sensory modalities may be recruited for mate choice—for instance, acoustic cues in crickets versus visual cues in butterflies (chapter 3). Closely related species, by contrast, share the same broad preference mechanisms, but express distinct preferences thanks to differences ranging from sensory tuning to valence assignment; these preference differences constitute a major barrier to genetic exchange (chapter 16). Within species, distinct phenotypic morphs can also vary in preference (Lattanzio et al. 2014).

Much of the interesting variation in preferences occurs below the species level. In a classic review of variation in mating preferences, Jennions and Petrie (1997; and before them, Arnold 1983) lamented the "typological view" of preferences as stereotypic traits conserved within a species. Notwithstanding a large number of studies underscoring the importance of variation below the species level, this view remains subtly pervasive, as evidenced by the neverending flow of papers, including several of my own, titled "Preference for Trait A in Species X." (I have perpetuated this view with numerous sentences in this book that start, *In species X . . .* , but have tried to be more specific where relevant.)

Yet the notion of within-species variation in mate choice has been around since the beginning. Indeed, Darwin (1871) suggested that arbitrary differences in mating preferences among human populations provided the likeliest explanation for the divergent appearances of human racial groups. Recent studies confirm a key prediction of Darwin's suggestion, that human mating preferences are correlated with divergence in sexually dimorphic traits; for example, men from the Hazda hunter-gatherer tribe of Tanzania prefer more protruding buttocks, characteristic of Hazda women, than do American men (Marlowe et al. 2005). In nonhuman animals, interpopulation variation in mating preferences is key evidence for **reinforcement** and **reproductive character displacement** (e.g., Wong et al. 2004; Higgie & Blows 2007; Kronforst et al. 2007; Rundle et al. 2008; chapter 16).

In order for variation among species and populations to evolve, of course, there has to be variation among individuals. Individual variation in behavior has received renewed attention in recent years (Dall et al. 2012), yet relatively few studies have measured individual variation in preference. In large part, the focus on mean preferences is because studies of sexual selection are chiefly interested in measuring aggregate data on preferences (mean and variance) as metrics of sexual selection.

The other reason for focusing on mean preferences is methodological. For some post-copulatory mechanisms, one can quantify morphological mechanisms of choice; for example, yellow dungflies express heritable genetic variation associated with having either three or four spermathecae, which allows more fecund females to mitigate the effects of last-male sperm precedence (Ward 2000). In most cases, however, we are interested in behavioral mechanisms of preference, and it is much more straightforward to assay a sample of individuals to derive aggregate measures of preference. Measuring individual differences, however, is essential in order to quantify the heritability of preferences and in order to understand phenomena like assortative mating and the maintenance of phenotypic variation in courter traits; individual differences in one trait dimension can also modulate shared preferences for other traits (Lee et al. 2014). Individual differences at a given time can also reflect **state-dependent** variation that is important to how preferences are expressed (chapter 11).

9.3 REPEATABILITY

> *La donna è mobile /Qual piuma al vento/Muta d'accento—e di pensiero.*
>
> A woman moves like a feather in the wind, changing in word and thought.
>
> —Piave 1851

Twenty years before Darwin, the librettist for Verdi's *Rigoletto* was thinking about within-individual variation in mating preferences. Within-individual variation sets the upper bound for heritability (Boake 1989) and for stable within-individual suites of behavior, or personalities (see below). Preferences can change predictably over longer timescales with age and reproductive cycle, or they can change in response to immediate circumstances. As we shall see in the next chapters, choosers are flexible, but not windblown, modifying their choices in response to social and environmental inputs in often predictable ways: "fluctuating, but not quite arbitrary."

A key measure for characterizing individual variation in behavior is **repeatability**, which is simply the ratio of among- to within-chooser variation in a trait (Lynch & Walsh 1998; Wolak et al. 2012; chapter 2):

$$r = s^2_A/(s^2_W + s^2_A)$$

where s^2_W is the within-subject variance and s^2_A the among-subject variance. s^2_A is calculated by subtracting mean squares within subjects from mean squares among subjects and adjusting for degrees of freedom, so repeatability will be negative if there is more variation within than among subjects (see Lessells & Boag 1986 for details).

A meta-analysis by Bell and colleagues (2009) found that the repeatability of mating preferences was substantially lower than that for other behavioral traits, particularly in vertebrates. Table 9.1 lists 192 published estimates of repeatability. The highest (0.95) and lowest (–0.37) estimates are from females of the same species, *Gryllus bimaculatus*, to two different acoustic features of male calls (Verburgt et al. 2008), underscoring the point that repeatabilities can vary widely within the same choosers for different courter traits. Consistent with Bell et al.'s analysis, repeatabilities in table 9.1 are qualitatively lower for vertebrates than for insects (fig. 9.2). Caution must be exercised in comparing repeatabilities across studies, in addition to the usual caveats about meta-analysis, since different workers use a variety of approaches to compute them from binary and continuous data. The reader is referred to Nakagawa and Schielzeth (2010) for a detailed treatment of methods for estimating repeatability for discrete and continuous variables.

There are two different goals when we measure repeatability. The first relates to how variation in mating decisions is partitioned in nature. Given all of the external influences on preferences and mate choice, how much of the variation in mating outcomes is due to stable differences among individuals? This is important because repeatability *theoretically* represents the upper bound of heritability (chapter 10), and therefore determines the potential for preferences to coevolve with courter traits and to evolve as reproductive isolating mechanisms. However, repeatability may not provide a meaningful guide to heritability. Dohm (2002) points out several scenarios whereby repeatability is actually lower than heritability. Broadly, repeatability is lowered if we are not measuring the same exact trait at successive times. Given the overwhelming dependence of preference on short- (chapter 6) and long-term experience (chapter 12), we are never dipping twice in the same river when we measure preferences. Widemo and Saether (1999) list multiple factors that can contribute to low repeatability, most of which

Table 9.1. Selected published estimates of repeatability

Source	*Species*	*Sex*	*Courter trait*	*Preference measure*	*Repeatability*
Insects					
Brandt et al. (2005)	*Achroia grisella*	♀	pulse rate	acceptance threshold	0.83 ± 0.109
Shuker & Day (2001)	*Coelopa frigida - strain B22*	pairs	individual identity	copulation/rejection	0.63 ± 0.15
Shuker & Day (2001)	*- strain D51*	pairs	individual identity	copulation/rejection	0.57 ± 0.14
Shuker & Day (2001)	*- strain SMI*	pairs	individual identity	copulation/rejection	0.47 ± 0.13
Reinhold et al. (2002)	*Chorthippus biguttulus*	♀	pause duration	preference (acoustic response)	0.697
Reinhold et al. (2002)	*Chorthippus biguttulus*	♀	pause duration	choosiness (acoustic response)	0.637
Reinhold et al. (2002)	*Chorthippus biguttulus*	♀	pause duration	response rate (acoustic response)	0.662
Verburgt et al. (2008)	*Gryllus bimaculatus*	♀	standard call at trial start	phonotactic precision	0.79 ± 0.13
Verburgt et al. (2008)	*Gryllus bimaculatus*	♀	standard call at trial start	phonotactic precision	0.60 ± 0.23
Verburgt et al. (2008)	*Gryllus bimaculatus*	♀	standard call at trial start	phonotactic precision	0.77 ± 0.12
Verburgt et al. (2008)	*Gryllus bimaculatus*	♀	standard call at trial end	phonotactic precision	0.69 ± 0.18
Verburgt et al. (2008)	*Gryllus bimaculatus*	♀	standard call at trial end	phonotactic precision	0.77 ± 0.14
Verburgt et al. (2008)	*Gryllus bimaculatus*	♀	standard call at trial end	phonotactic precision	0.74 ± 0.14
Verburgt et al. (2008)	*Gryllus bimaculatus*	♀	**BW** spectral bandwidth	polynomial response function (phonotaxis) - **B′**, peak preference	0.79 ± 0.139
Verburgt et al. (2008)	*Gryllus bimaculatus*	♀	**DC** duty cycle	- **B′**, peak preference	0.53 ± 0.259
Verburgt et al. (2008)	*Gryllus bimaculatus*	♀	**FQ** frequency	- **B′**, peak preference	0.69 ± 0.16

Verburgt et al. (2008)	*Gryllus bimaculatus*	♀	**SP** syllable period	- **B′**, peak preference	0.75 ± 0.17
Verburgt et al. (2008)	*Gryllus bimaculatus*	♀	**SP(rev)** syllable period (reverse order)	- **B′**, peak preference	0.78 ± 0.14
Verburgt et al. (2008)	*Gryllus bimaculatus*	♀	**BW** spectral bandwidth	- **Choosiness**	0.84 ± 0.10
Verburgt et al. (2008)	*Gryllus bimaculatus*	♀	DC duty cycle	- **Choosiness**	0.26 ± 0.33
Verburgt et al. (2008)	*Gryllus bimaculatus*	♀	**FQ** frequency	- **Choosiness**	−0.11 ± 0.30
Verburgt et al. (2008)	*Gryllus bimaculatus*	♀	**SP** syllable period	- **Choosiness**	0.67 ± 0.21
Verburgt et al. (2008)	*Gryllus bimaculatus*	♀	**SP(rev)** syllable period (reverse order)	- **Choosiness**	0.71 ± 0.18
Verburgt et al. (2008)	*Gryllus bimaculatus*	♀	**BW** spectral bandwidth	- **R′**, peak response	0.95 ± 0.03
Verburgt et al. (2008)	*Gryllus bimaculatus*	♀	DC duty cycle	- **R′**, peak response	0.80 ± 0.139
Verburgt et al. (2008)	*Gryllus bimaculatus*	♀	**FQ** frequency	- **R′**, peak response	0.067 ± 0.30
Verburgt et al. (2008)	*Gryllus bimaculatus*	♀	**SP** syllable period	- **R′**, peak response	0.76 ± 0.16
Verburgt et al. (2008)	*Gryllus bimaculatus*	♀	**SP(rev)** syllable period (reverse order)	- **R′**, peak response	0.14 ± 0.35
Verburgt et al. (2008)	*Gryllus bimaculatus*	♀	**BW** spectral bandwidth	non-linear response function (phonotaxis) - **B′**, peak preference	0.70 ± 0.18
Verburgt et al. (2008)	*Gryllus bimaculatus*	♀	DC duty cycle	- **B′**, peak preference	0.63 ± 0.21
Verburgt et al. (2008)	*Gryllus bimaculatus*	♀	**FQ** frequency	- **B′**, peak preference	0.47 ± 0.24
Verburgt et al. (2008)	*Gryllus bimaculatus*	♀	**SP** syllable period	- **B′**, peak preference	0.39 ± 0.32
Verburgt et al. (2008)	*Gryllus bimaculatus*	♀	**SP(rev)** syllable period (reverse order)	- **B′**, peak preference	0.32 ± 0.32

Continued on the next page

Table 9.1. *(continued)*

Source	*Species*	*Sex*	*Courter trait*	*Preference measure*	*Repeatability*
Verburgt et al. (2008)	*Gryllus bimaculatus*	♀	**BW** spectral bandwidth	- Choosiness	0.01 ± 0.35
Verburgt et al. (2008)	*Gryllus bimaculatus*	♀	DC duty cycle	- Choosiness	0.52 ± 0.26
Verburgt et al. (2008)	*Gryllus bimaculatus*	♀	**FQ** frequency	- Choosiness	0.64 ± 0.18
Verburgt et al. (2008)	*Gryllus bimaculatus*	♀	**SP** syllable period	- Choosiness	–0.14 ± 0.37
Verburgt et al. (2008)	*Gryllus bimaculatus*	♀	**SP(rev)** syllable period (reverse order)	- Choosiness	0.44 ± 0.29
Verburgt et al. (2008)	*Gryllus bimaculatus*	♀	**BW** spectral bandwidth	- **R′**, peak response	0.57 ± 0.24
Verburgt et al. (2008)	*Gryllus bimaculatus*	♀	DC duty cycle	- **R′**, peak response	0.10 ± 0.35
Verburgt et al. (2008)	*Gryllus bimaculatus*	♀	**FQ** frequency	- **R′**, peak response	0.53 ± 0.22
Verburgt et al. (2008)	*Gryllus bimaculatus*	♀	**SP** syllable period	- **R′**, peak response	–0.02 ± 0.38
Verburgt et al. (2008)	*Gryllus bimaculatus*	♀	**SP(rev)** syllable period (reverse order)	- **R′**, peak response	0.29 ± 0.32
Verburgt et al. (2008)	*Gryllus bimaculatus*	♀	**BW** spectral bandwidth	cubic spline response function (phonotaxis) - **B′**, peak preference	0.83 ± 0.11
Verburgt et al. (2008)	*Gryllus bimaculatus*	♀	DC duty cycle	- **B′**, peak preference	0.53 ± 0.25
Verburgt et al. (2008)	*Gryllus bimaculatus*	♀	**FQ** frequency	- **B′**, peak preference	0.72 ± 0.15
Verburgt et al. (2008)	*Gryllus bimaculatus*	♀	**SP** syllable period	- **B′**, peak preference	0.75 ± 0.17
Verburgt et al. (2008)	*Gryllus bimaculatus*	♀	**SP(rev)** syllable period (reverse order)	- **B′**, peak preference	0.78 ± 0.14

Verburgt et al. (2008)	*Gryllus bimaculatus*	♀	**BW** spectral bandwidth	- **Choosiness**	0.845 ± 0.10
Verburgt et al. (2008)	*Gryllus bimaculatus*	♀	DC duty cycle	- **Choosiness**	0.26 ± 0.33
Verburgt et al. (2008)	*Gryllus bimaculatus*	♀	**FQ** frequency	- **Choosiness**	−0.22 ± 0.29
Verburgt et al. (2008)	*Gryllus bimaculatus*	♀	**SP** syllable period	- **Choosiness**	0.67 ± 0.21
Verburgt et al. (2008)	*Gryllus bimaculatus*	♀	**SP(rev)** syllable period (reverse order)	- **Choosiness**	0.71 ± 0.18
Verburgt et al. (2008)	*Gryllus bimaculatus*	♀	**BW** spectral bandwidth	- **R′**, peak response	0.28 ± 0.33
Verburgt et al. (2008)	*Gryllus bimaculatus*	♀	DC duty cycle	- **R′**, peak response	0.46 ± 0.28
Verburgt et al. (2008)	*Gryllus bimaculatus*	♀	**FQ** frequency	- **R′**, peak response	0.30 ± 0.27
Verburgt et al. (2008)	*Gryllus bimaculatus*	♀	**SP** syllable period	- **R′**, peak response	−0.37 ± 0.33
Verburgt et al. (2008)	*Gryllus bimaculatus*	♀	**SP(rev)** syllable period (reverse order)	- **R′**, peak response	0.15 ± 0.35
Hager & Teale (1994)	*Ips pini*	♀	enantiomer ratio (1 ul)	taxis (trial 1 vs 2)	0.4 ± 0.19
Hager & Teale (1994)	*Ips pini*	♀	enantiomer ratio (5 ul)	taxis (trial 1 vs 2)	0.86 ± 0.06
Hager & Teale (1994)	*Ips pini*	♀	enantiomer ratio (1 ul)	taxis (trial 2 vs 3)	0.20 ± 0.22
Hager & Teale (1994)	*Ips pini*	♀	enantiomer ratio (5 ul)	taxis (trial 2 vs 3)	0.62 ± 0.22
Hager & Teale (1994)	*Ips pini*	♀	enantiomer ratio (1 ul)	taxis (trial 1 vs 3)	0.16 ± 0.22
Hager & Teale (1994)	*Ips pini*	♀	enantiomer ratio (5 ul)	taxis (trial 1 vs 3)	0.47 ± 0.18
Hager & Teale (1994)	*Ips pini*	♀	enantiomer ratio (1 ul)	taxis (3 trials)	0.23 ± 0.08
Hager & Teale (1994)	*Ips pini*	♀	enantiomer ratio (5 ul)	taxis (3 trials)	0.23 ± 0.15
Hager & Teale (1994)	*Ips pini*	♀	enantiomer ratio (5 ul)	within-set responses (trial 1)	0.2 ± 0.08
Hager & Teale (1994)	*Ips pini*	♀	enantiomer ratio (5 ul)	- trial 2	0.109 ± 0.079

Continued on the next page

Table 9.1. (*continued*)

Source	*Species*	*Sex*	*Courter trait*	*Preference measure*	*Repeatability*
Hager & Teale (1994)	*Ips pini*	♀	enantiomer ratio (5 ul)	- trial 3	−0.035 ± 0.061
Wilkinson et al. (1998)	*Cyrtodiopsis dalmanni*	♀	eyespan length	mating	0.59 ± 0.14
Wilkinson et al. (1998)	*Cyrtodiopsis whitei*	♀	eyespan length	mating	0.33 ± 0.17
Isoherranen et al. (1999a)	*Drosophila virilis*	♀	song number	threshold response	0.328 ± 0.236
Fowler-Finn & Rodríguez (2013)	*Enchenopa binotata*	♀	signal frequency	Peak preference (acoustic response)	0.12
Fowler-Finn & Rodríguez (2013)	*Enchenopa binotata*	♀	signal frequency	Responsiveness	0.54
Fowler-Finn & Rodríguez (2013)	*Enchenopa binotata*	♀	signal frequency	Choosiness (“tolerance”)	0.46
Fowler-Finn & Rodríguez (2013)	*Enchenopa binotata*	♀	signal frequency	Choosiness (“strength”)	0.46
Fowler-Finn & Rodríguez (2013)	*Enchenopa binotata*	♀	signal frequency	Choosiness (“selectivity”)	0.51
Greenfield et al. (2004)	*Ephippiger ephippiger*	♀	signal precedence (100–200 ms)	binary preference (age < 50 d; tested in >4 sets)	0.21 ± 0.05
Greenfield et al. (2004)	*Ephippiger ephippiger*	♀	signal precedence (100–200 ms)	binary preference (age >= 50 d; tested in >4 sets)	0.10 ± 0.04

Greenfield et al. (2004)	*Ephippiger ephippiger*	♀	signal precedence (250–350 ms)	binary preference (age < 50 d; tested in >4 sets)	0.12 ± 0.05
Greenfield et al. (2004)	*Ephippiger ephippiger*	♀	signal precedence (250–350 ms)	binary preference (age >= 50 d; tested in >4 sets)	0.15 ± 0.05
Greenfield et al. (2004)	*Ephippiger ephippiger*	♀	signal precedence (100–200 ms)	binary preference (age < 50 d; tested in >9 sets)	0.07 ± 0.05
Greenfield et al. (2004)	*Ephippiger ephippiger*	♀	signal precedence (100–200 ms)	binary preference (age >= 50 d; tested in >9 sets)	0.12 ± 0.06
Greenfield et al. (2004)	*Ephippiger ephippiger*	♀	signal precedence (250–350 ms)	binary preference (age < 50 d; tested in >9 sets)	0.13 ± 0.06
Greenfield et al. (2004)	*Ephippiger ephippiger*	♀	signal precedence (250–350 ms)	binary preference (age >= 50 d; tested in >9 sets)	0.21 ± 0.08
Wagner et al. (1995)	*Gryllus integer*	♀	Pulses per trill	preference slope (taxis)	0.5
Wagner et al. (1995)	*Gryllus integer*	♀	Inter-trill interval	preference slope (taxis)	0.59
Wagner et al. (1995)	*Gryllus integer*	♀	proportion of missing pulses	preference slope (taxis)	–0.02
Boake (1989)	*Tribolium castaneum*	♀	conspecific pheromone vs. blank	binary preference	0
Amphibians					
Michalak (1996)	*Triturus montandoni*	♀	tail-fanning rate	"preference"	–0.097 ± 0.212
Michalak (1996)	*Triturus montandoni*	♀	tail-tip length	"preference"	0.029 ± 0.229
Michalak (1996)	*Triturus montandoni*	♀	individual identity	duration of premating phase	–0.020 ± 0.160
Michalak (1996)	*Triturus montandoni*	♀	individual identity	time to spermatophore transfer	0.045 ± 0.248

Continued on the next page

Table 9.1. *(continued)*

Source	*Species*	*Sex*	*Courter trait*	*Preference measure*	*Repeatability*
Michalak (1996)	*Triturus montandoni*	♀	individual identity	# spermatophores picked up	0.022 ±.222
Michalak (1996)	*Triturus montandoni*	♀	individual identity	duration of sexual responsiveness	−0.041 ± 0.212
Howard & Young (1998)	*Bufo americanus*	♀	size of male (all pairings)	binary preference	0.08 ± 0.07
Howard & Young (1998)	*Bufo americanus*	♀	size of male (female-initiated pairings)	binary preference	0.25 ± 0.30
Howard & Young (1998)	*Bufo americanus*	♀	low dominant frequency call - alternating with high frequency call	binary preference	0.19 ± 0.20
Howard & Young (1998)	*Bufo americanus*	♀	- low frequency precedes high frequency	binary phonotaxis	−.13 ± 0.15
Howard & Young (1998)	*Bufo americanus*	♀	- high frequency precedes low frequency	binary phonotaxis	0.42 ± 0.20
Gerhardt et al. (2000)	*Hyla versicolor*	♀	Call duration	binary phonotaxis	0.69
Jennions et al. (1994)	*Hyperolius marmoratus*	♀	Call frequency	binary preference	0.79
Murphy & Gerhardt (2000)	*Hyla gratiosa*	♀	Call rate (40 vs. 55/min)	binary preference	0.051
Murphy & Gerhardt (2000)	*Hyla gratiosa*	♀	Call rate (45 vs. 65/min)	binary preference	0.031
Murphy & Gerhardt (2000)	*Hyla gratiosa*	♀	Fundamental frequency - 350 vs. 450 Hz	binary preference	0.066

Murphy & Gerhardt (2000)	*Hyla gratiosa*	♀	- 400 vs. 500 Hz	binary preference	0.092
Murphy & Gerhardt (2000)	*Hyla gratiosa*	♀	- 450 vs. 500 Hz	binary preference	0.179
Birds					
Johnsen & Zuk (1996)	*Gallus gallus*	♀	Comb length (March-April)	binary preference	0.19 ± 0.1
Johnsen & Zuk (1996)	*Gallus gallus*	♀	Comb chroma (March-April)	binary preference	0.17 ± 0.1
Johnsen & Zuk (1996)	*Gallus gallus*	♀	Hackle feather chroma (March-April)	binary preference	0.14 ± 0.1
Johnsen & Zuk (1996)	*Gallus gallus*	♀	Comb length (April-May)	binary preference	–0.12 ± 0.1
Johnsen & Zuk (1996)	*Gallus gallus*	♀	Comb chroma (April-May)	binary preference	–0.04 ± 0.11
Johnsen & Zuk (1996)	*Gallus gallus*	♀	Hackle feather chroma (April-May)	binary preference	0.16 ± 0.12
Banbura (1992)	*Hirundo rustica*	♀	Tail length	pair formation	0.15 ± 0.23
Banbura (1992)	*Hirundo rustica*	♀	Wing length	pair formation	0.24 ± 0.18
Møller (1994)	*Hirundo rustica*	♀	Tail length	pair formation	0.18 ± 0.16
Forstmeier (2004)	*Taeniopygia guttata*	♀	individual identity	overall responsiveness	0.63
Forstmeier (2004)	*Taeniopygia guttata*	♀	individual identity	- including nonresponders	0.54
Forstmeier & Birkhead (2004)	*T. guttata*	♀	preference for one of two males	binary preference (proceptive behavior)	0.29

Continued on the next page

Table 9.1. (*continued*)

Source	*Species*	*Sex*	*Courter trait*	*Preference measure*	*Repeatability*
Forstmeier & Birkhead (2004)	*T. guttata*	♀	Body mass	preference slope	–0.08
Forstmeier & Birkhead (2004)	*T. guttata*	♀	Beak color	preference slope	0.17
Forstmeier & Birkhead (2004)	*T. guttata*	♀	avg. song rate in pairwise interactions	preference slope	0.11
Forstmeier & Birkhead (2004)	*T. guttata*	♀	indiv. song rate in pairwise interactions	preference slope	0.24
Forstmeier & Birkhead (2004)	*T. guttata*	♀	song in mixed-sex groups	preference slope	–0.06
Forstmeier & Birkhead (2004)	*T. guttata*	♀	Attractiveness to other females	preference slope	–0.06
Riebel (2000)	*T. guttata*	♀	tutor vs. unfamiliar	operant response (tutored females)	0.71
Riebel (2000)	*T. guttata*	♀	tutor vs. unfamiliar	operant response (naïve females)	0.11
Roulin(1999)	*Tyto alba*	♂	plumage spottiness	pair formation	0.29 ± 0.11
Teleost fishes					
Hoysak & Godin (2007)	*Gambusia holbrooki*	♂	body size (visual cues only)	association time	0.05 ± 0.15
Hoysak & Godin (2007)	*Gambusia holbrooki*	♂	body size (fully interacting)		.36 ± .13

Aspbury & Basolo (2002)	*Heterandria formosa*	♀	body size	association time	0.722 ± 0.133
Cummings & Mollaghan (2006)	*Xiphophorus nigrensis*	♀	body size	association time	0.322 ± 0.087
Cummings & Mollaghan (2006)	*Xiphophorus nigrensis*	♀	body size	glides	−0.088 ± 0.026
Gabor & Aspbury (2008)	*Poecilia latipinna (Alfredo Bonfil)*	♂	species (size-matched)	association time	−.003 ± .169
Gabor & Aspbury (2008)	*Poecilia latipinna (Vicente Guerrero)*	♂	species (size-matched)	association time	.005 ± .169
Gabor & Aspbury (2008)	*Poecilia latipinna (Alfredo Bonfil)*	♂	species (heterospecific larger than conspecific)	association time	0.0003 ± .169
Gabor & Aspbury (2008)	*Poecilia latipinna (Vicente Guerrero)*	♂	species (conspecific larger than heterospecific)	association time	0.03 ± .169
Gabor & Aspbury (2008)	*Poecilia latipinna (Vicente Guerrero)*	♂	species (size-matched)	mating	0.02 ± 0.012
Gabor et al. (2011)	*Poecilia formosa*	♀	body size	time in choice zone	0.092 ± 0.270
Gabor et al. (2011)	*P. latipinna*	♀	body size	time in choice zone	0.700 ± 0.186
Gabor et al. (2011)	*P. mexicana*	♀	body size	time in choice zone	−0.033 ± 0.288
Kodric-Brown & Nicoletto (1997)	*P. reticulata - USA*	♀	ornamented/pale	association time	0.05 ± 0.09
Kodric-Brown & Nicoletto (1997)	*P. reticulata - USA*	♀	ornamented/pale - one way glass	association time	0.4 ± 0.11

Continued on the next page

Table 9.1. (*continued*)

Source	*Species*	*Sex*	*Courter trait*	*Preference measure*	*Repeatability*
Kodric-Brown & Nicoletto (1997)	*P. reticulata - USA*	♀	color saturation - video playback	association time	0.47 ± 0.11
Archard et al. (2006)	*P. reticulata - Lower Tacarigua*	♀	individual identity	**Responsiveness (association time)** - Control – 1 & 2	0.475
Archard et al. (2006)	*P. reticulata - Lower Tacarigua*	♀	individual identity	Control – 1 & 3	0.236
Archard et al. (2006)	*P. reticulata - Lower Tacarigua*	♀	individual identity	Control – 2 & 3	−0.002
Archard et al. (2006)	*P. reticulata - Lower Tacarigua*	♀	individual identity	Control– All trials	0.218
Archard et al. (2006)	*P. reticulata - Lower Tacarigua*	♀	individual identity	Treatment – 1 & 2	0.378
Archard et al. (2006)	*P. reticulata - Lower Tacarigua*	♀	individual identity	Treatment – 1 & 3	0.307
Archard et al. (2006)	*P. reticulata - Lower Tacarigua*	♀	individual identity	Treatment – 2 & 3	0.345
Archard et al. (2006)	*P. reticulata - Lower Tacarigua*	♀	individual identity	All females – 1 & 3	0.266
Archard et al. (2006)	*P. reticulata - Lower Tacarigua*	♀	individual identity	**Discrimination** - C1&2	−0.346
Archard et al. (2006)	*P. reticulata - Lower Tacarigua*	♀	individual identity	Control – 1 & 3	0.248
Archard et al. (2006)	*P. reticulata - Lower Tacarigua*	♀	individual identity	Control – 2& 3	−0.113
Archard et al. (2006)	*P. reticulata - Lower Tacarigua*	♀	individual identity	Control– All trials	−0.066
Archard et al. (2006)	*P. reticulata - Lower Tacarigua*	♀	individual identity	Treatment – 1 & 2	0.344
Archard et al. (2006)	*P. reticulata - Lower Tacarigua*	♀	individual identity	Treatment – 1 & 3	<0.001
Archard et al. (2006)	*P. reticulata - Lower Tacarigua*	♀	individual identity	Treatment – 2& 3	−0.039
Archard et al. (2006)	*P. reticulata - Lower Tacarigua*	♀	individual identity	All females – 1 & 3	0.124

Godin & Dugatkin (1995)	*P. reticulata - Quaré River*	♀	bright/dull	association time	0.58 ± 0.11
Brooks (1996)	*P. reticulata - South Africa*	♀	random pairs - 30 min interval	association time	0.32 ± 0.05
Brooks (1996)	*P. reticulata - South Africa*	♀	random pairs - 24 h interval	association time	0.35 ± 0.06
Brooks (1996)	*P. reticulata - South Africa*	♀	random pairs - 48 h interval	association time	0.35 ± 0.05
Brooks & Endler (2001)	*P. reticulata - Australia*	♀	(overall responsiveness)	association time	0.399 ± .05
Brooks & Endler (2001)	*P. reticulata - Australia*	♀	(overall discrimination)	association time	0.255 ± .06
Brooks & Endler (2001)	*P. reticulata - Australia*	♀	Overall attractiveness	association time	.056 ± .07
Brooks & Endler (2001)	*P. reticulata - Australia*	♀	Predicted attractiveness	association time	−.15 ± .08
Brooks & Endler (2001)	*P. reticulata - Australia*	♀	Body area	association time	.024 ± .07
Brooks & Endler (2001)	*P. reticulata - Australia*	♀	Tail area	association time	.004 ± .07
Brooks & Endler (2001)	*P. reticulata - Australia*	♀	Black area	association time	.271 ± .06
Brooks & Endler (2001)	*P. reticulata - Australia*	♀	Fuzzy black area	association time	.049 ± .07

Continued on the next page

Table 9.1. *(continued)*

Source	*Species*	*Sex*	*Courter trait*	*Preference measure*	*Repeatability*
Brooks & Endler (2001)	*P. reticulata - Australia*	♀	Orange area	association time	–.071 ± .07
Brooks & Endler (2001)	*P. reticulata - Australia*	♀	Orange chroma	association time	–.034 ± .07
Brooks & Endler (2001)	*P. reticulata - Australia*	♀	Orange brightness	association time	.032 ± .07
Brooks & Endler (2001)	*P. reticulata - Australia*	♀	Iridescent area	association time	–.031 ± .07
Brooks & Endler (2001)	*P. reticulata - Australia*	♀	Total spot number	association time	.016 ± .07
Brooks & Endler (2001)	*P. reticulata - Australia*	♀	Mean brightness	association time	.047 ± .07
Brooks & Endler (2001)	*P. reticulata - Australia*	♀	Brightness contrast	association time	.031 ± .07
Brooks & Endler (2001)	*P. reticulata - Australia*	♀	Mean chroma	association time	–.126 ± .07
Brooks & Endler (2001)	*P. reticulata - Australia*	♀	Color contrast	association time	.004 ± .07

Godin & Auld (2013)	*P. reticulata - Upper Aripo*	♂	body size	sexual behavior	0.628 ± 0.099
		♂	body size	association time	0.824 ± 0.053
Morris et al. (2003)	*Xiphophorus cortezi*	♀	vertical bars	association time	0.86 ± 0.006
Morris et al. (2003)	*Xiphophorus cortezi*	♀	vertical bars	association time	0.5 ± 0.028
Howard et al. (1998)	*Oryzias latipes*	♀	body size	mating	0.18 ± 0.11
Lehtonen & Lindstrom (2008)	*Pomatoschistus minutus*	♀	body size	Differential preference	0.613 ± 0.114
Lehtonen & Lindstrom (2008)	*Pomatoschistus minutus*	♀	Nest elaboration	Differential preference	0.602 ± 0.171
Lehtonen & Lindstrom (2008)	*Pomatoschistus minutus*	♀	body size	Absolute preference	0.756 ± 0.078
Lehtonen & Lindstrom (2008)	*Pomatoschistus minutus*	♀	Nest elaboration	Absolute preference	0.66 ± 0.151
Lehtonen & Lindstrom (2008)	*Pomatoschistus minutus*	♀	body size	Relative preference	0.503 ± 0.136
Lehtonen & Lindstrom (2008)	*Pomatoschistus minutus*	♀	Nest elaboration	Relative preference	0.568 ± 0.124

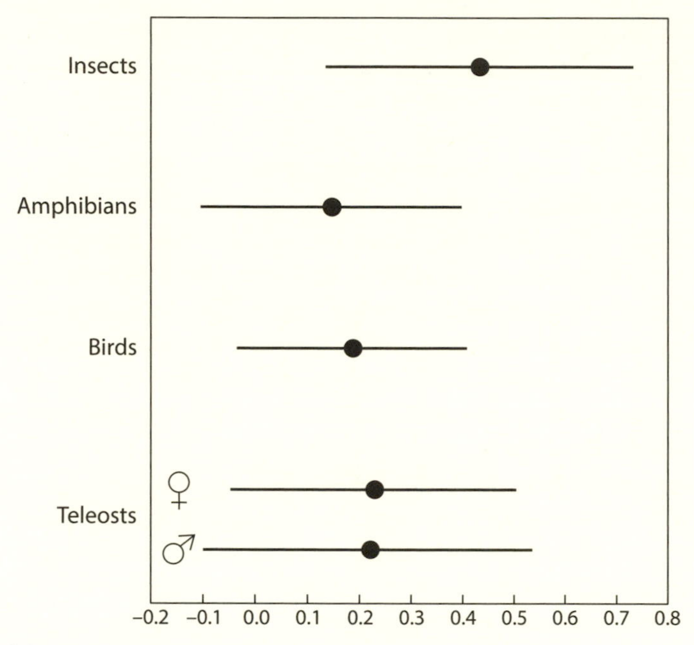

Figure 9.2. Mean and standard deviation of repeatability by taxon for the studies listed in table 9.1. All data on females except for teleost fish.

are detailed here and in the next two chapters; by definition, any shift in individual preferences due to social or environmental inputs will reduce the overall repeatability for that preference.

The second goal relates to minimizing external noise and characterizing the properties of preferences that are inherent to the organism. For example, low repeatability means that gene-mapping studies of mate preference will require large sample sizes to estimate not just genetic effects but the preferences themselves. Here, what we would like to know is: given a fixed set of experiences and conditions, how often would a chooser exhibit a particular preference in nature? Preferences are high-dimensional function-valued traits (Stinchcombe & Kirkpatrick 2012; McGuigan et al. 2008b; Rodríguez, Hallett et al. 2013) since their expression depends on both courter stimulus values and the conditions experienced by choosers. There is therefore an inherent trade-off between the robustness of individual preference estimates and the scope of among- and within-chooser variation that we sample. By design, laboratory studies of the repeatability (and heritability, see below) of mate choice strive to standardize conditions with respect to factors like social experience, nutrition, and reproductive state, thereby artificially increasing repeatability with respect to the wild. By contrast, husbandry conditions and stimulus deficiencies may introduce artifacts of variability into chooser responsiveness. Repeatability is inherently a function of the amount of variation in both chooser preferences and courter stimuli. If all choosers have the same preference, repeatability will be zero even if preferences are strong and individual choosers are entirely consistent in their choices. This is the pattern we would expect if there is strong selection favoring certain preferences (chapter 14). At the other extreme, it is difficult to interpret repeatabilities if preferences are weak or undetected. When Brooks and Endler (2001a) based repeatabilities on female responses weighted by responsiveness, they increased compared to the unweighted estimates presented in table 9.1. Estimated repeatabilities were therefore higher when giving stronger consideration to those individuals actually likely to make mating decisions. In Bosch and Marquez's study (2002; chapter 6) of precedence effects on preferences in female midwife toads *Alytes cisternasii*, repeatability covaried with the strength of preference for the leading call (fig. 9.3).

The range and distribution of courter stimuli assayed also impacts repeatability estimates. For example, Hager and Teale (1994) noted that if a chooser's true peak preference lies between two options in a binary assay, this will have the effect of introducing noise into the observed response as she may arbitrarily select one over the other; this will yield a measured repeatability lower than the true value. Roff and Fairbairn (2015) make a similar point regarding heritability (next chapter). This concern reinforces the desirability of sampling preferences over a range of stimulus values (Wagner 1998).

The very process of assaying preferences changes a chooser's experience and therefore can affect subsequent choices. A widespread limitation on repeatability estimates is that partitioning among- from within-individual variation requires repeated measures of preference from the same individual, which has the inevitable effect of changing an individual's social experience. Negative repeatabilities (table 9.1) point to a role of previous experience with courter signals shaping chooser preferences (chapter 12). If the goal is to identify genetic sources of variation in mating preference, the problem of previous experience can in principle be circumvented by testing naïve, genetically identical individuals (clones or inbred lines) rather than repeatedly testing the same individuals (e.g., Ratterman et al. 2014). This approach, however, creates new problems because inbred lines are less robust to environmental effects (Whitlock & Fowler 1999), and inbreeding depression may generate artifacts of perception and behavior with respect to mate choice.

Even when we account for social and environmental variation and properly sample stimulus space, however, repeatabilities may remain low. This

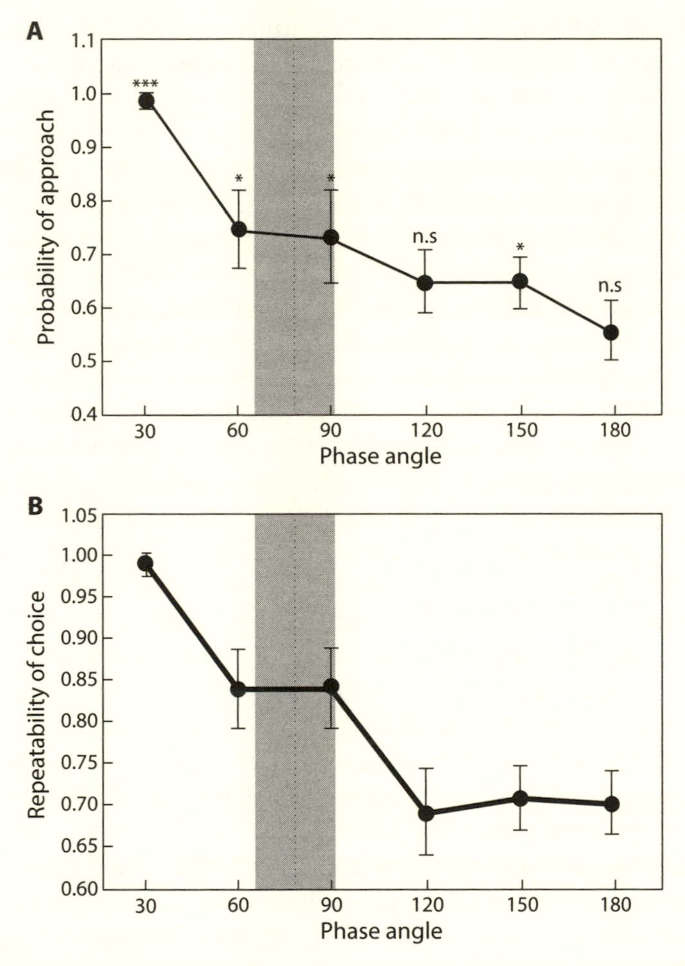

Figure 9.3. (A) Probability of response to the leading call (different from random choice at *P < .05, **P < .001) by Iberian midwife toads as a function of phase angle, with natural phase angles shaded in gray. (B) Repeatability of choice as a function of phase angle. After Bosch & Marquez (2002).

may indicate that mate choice is very sensitive to initial conditions. For example, preferences for visual cues depend heavily on which courter the chooser orients to first (Clark & Uetz 1992). The critical challenge, then, is to tease apart within-individual variation that is meaningful with respect to actual mating decisions in nature, versus spurious variation that arises from the process of assaying choice.

9.4 COVARIATES OF PREFERENCE VARIATION

9.4.1 Covariance among preferences

Most of the studies discussed in the following three chapters address variation that is structured along a single axis of preference. Aspects of preference—notably responsiveness—can vary globally across courter traits, but variation in preferences is often specific to individual courter traits (Lyons, Goedert et al. 2014; Brooks 2002). Further, multiple aspects of preference covary with one another (chapter 4). For example, Bailey (2008) measured individual preference functions for a call-structure parameter in female field crickets (*Teleogryllus oceanicus*) and found that females with quadratic preference functions—i.e., females that preferred an intermediate trait value—were more choosy than females with linear preference functions. Overall choosiness therefore covaried with the shape of the preference function. In pygmy swordtails (*Xiphophorus pygmaeus*), Rosenthal and Ryan (2011a) found that directional preference for body size within species was negatively correlated with avoidance of a sympatric congener. Pierotti and colleagues (2009) applied a compositional data analysis to characterize clusters of individual variation in multivariate preference space of male Lake Victoria cichlids *Neochromis omnicaeruleus* for distinct female color morphs. They found considerable variation in the strength and direction of preference, clustered into distinct preference clusters and a no-preference cluster. Within-individual correlations among different dimensions of preference can therefore represent an important constraint in preference evolution.

Constraints can also arise when preferences covary with aspects of a chooser's phenotype not directly involved in mate choice. Sometimes, these covariates of preference are tied to social and environmental effects, which will be discussed in chapters 11 and 12. Other widespread covariates are discussed below.

9.4.2 Sex

Perhaps the most obvious covariate of preference variation is sex (and in humans, gender). Sex can be genetically, environmentally, or socially determined (Bachtrog et al. 2014); in many aquatic animals and some terrestrial ones, individuals are simultaneously or sequentially hermaphroditic (chapter 8), and it would be interesting to learn whether some aspects of preferences are correlated across sex roles.

Males and females typically have different preferences; at the very least, they typically prefer to mate with the opposite sex (Adkins-Regan 1998). Differences in preference are sometimes apparent in striking morphological differences at the sensory periphery (McClelland et al. 1997; Narins & Capranica 1976; Sison-Mangus et al. 2006; figs. 3.9 and 9.4) and of course in genital anatomy (chapter 7). Such sexual dimorphism in the structures associated with mating can be a useful cue in identifying the modalities and stages (pre-, peri-, or postmating) involved in mate choice.

It is important to recognize, however, that males and females typically share many of the same mechanisms for perceiving courter signals, except that males are evaluating signals of other males in the context of aggression rather than mate choice. In these cases, differences emerge downstream of the sensory periphery. For example, Leonard and Hedrick (2009) found that both female and male field crickets *Gryllus integer* attended to male call differences in simultaneous contexts, but only females showed preferences when calls were presented sequentially. Bernal, Rand,and Ryan (2009) similarly found that while the evoked call responses of male túngara frogs were more permissive than female phonotaxis responses to the same stimuli, these differences disappeared when males were tested on the same task—phonotaxis—as females. Since males and females often share many of the same structures for sensing and integrating signals, differences in evaluative mechanisms and in the execution of choice likely account for much of the intersexual difference in preference. Intersexual correlations among the mechanisms involved in responding to courters pose a potentially important constraint in mate-choice evolution (chapter 14).

Crucially, males and females also exhibit different response functions to external variables that modulate mate choice. For example, females increase and males decrease their mate-searching rate as a function of density in fiddler crabs (*Uca beebei*; deRivera et al. 2003). In female mice, females but not males stop exhibiting proceptive behavior when in moderate pain (Farmer et al. 2014). Such differential responses play an important role in determining how each of the sexes invests in mate choice (chapter 8).

In vertebrates, the differences in male and female sensation, evaluation, and behavior arise largely as a consequence of organizational effects of sex hormones early in development (reviewed in Adkins-Regan 1998); early organizational effects are similarly implicated in the formation of stable same-sex preferences in humans and sheep (Poiani 2010; see below).

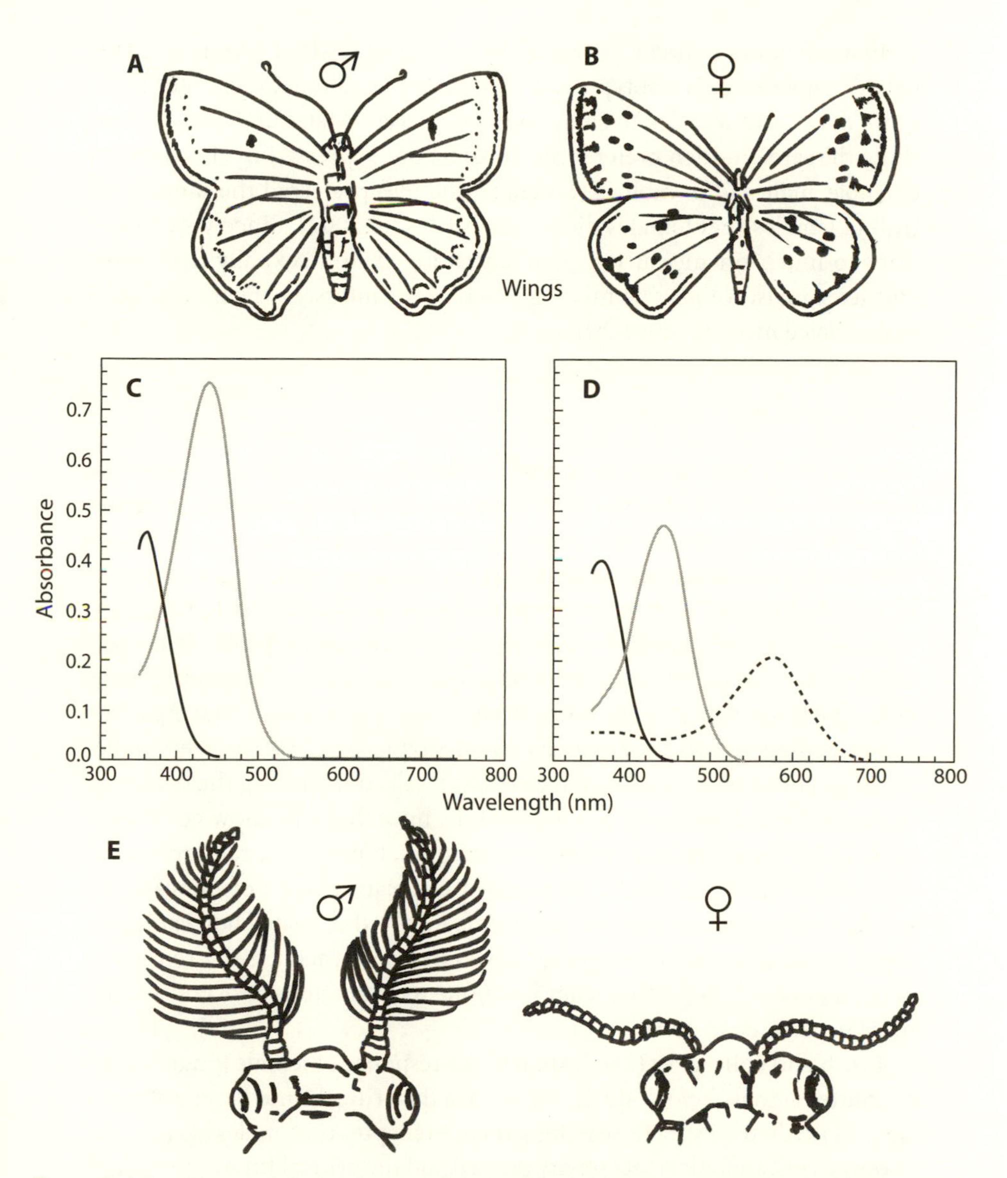

Figure 9.4. (A–D) Sexual dimorphism in the peripheral visual system of the butterfly *Lycaena* rubidus: A–B, wings; C–D, absorbance spectra of visual pigments (after Sison-Mangus et al. 2006); (E) Sexual dimorphism in pectinate antennae of diprionid sawflies (left, male; right, female) (after Keil 1999).

In much of mate choice, however, the subjective value of stimuli is shared between the sexes. Homotypic and heterotypic preferences (chapter 8) require males and females to share the same biases. Males and females may share the same hidden preferences for traits not expressed in choosers; for example, male sailfin mollies (*Poecilia latipinna*) preferred the same novel trait in females that females did in males (Basolo 2002). Since we concentrate so much on measuring preferences for extant, sexually dimorphic courter signals, we may be missing much of the intersexual congruence in mate-choice mechanisms (chapter 8).

9.4.3 Reproductive state

Many animals only behave as choosers when physiologically receptive, that is, when they are capable of exchanging gametes. The hormonal correlates of sexual receptivity are well understood in vertebrates and insects (Ringo 1996); in vertebrates, sexual receptivity is associated with surges in androgens and estrogens in males and females respectively. Physiological receptivity is a universal covariate of mating preferences, and is probably the primary modulator of overall chooser responsiveness to courter stimuli. Many animals exhibit sexual behavior only after sexual maturity or during a discrete breeding season. In mammals, females typically display proceptive behavior, associated with a spike in estrogen levels, only during the ovulatory period of the estrus cycle (Beach 1976). In mice, females show copulation preferences for individual males during estrus, but mate at random otherwise (Zinck & Lima 2013). Changes in circulating sex hormones likely modulate the activity of a host of systems involved in mate choice; in midshipman fish, for example, reproductive hormones increase the sensitivity of the inner ear to high-frequency components characteristic of courter signals (Sisneros et al. 2004).

Lynch and colleagues (2005) studied the response of female túngara frogs to courter signals before, during, and after the critical time for egg deposition. As predicted by mate-searching theory (chapter 6), females became less choosy as physiological receptivity peaked and the critical time approached. Intriguingly, females continued to express preferences even after eggs had been released, despite not being capable of mating.

Indeed, it is by no means the case that mating preferences are only expressed in reproductive choosers. This is particularly evident in species where the ramifications of mating extend beyond the fusion of gametes (e.g., pair-bonding; chapter 8) and in which gametes can be stored for extended periods (chapter 7). In poeciliids, where females use sperm for sires in some cases long after these have died (López-Sepulcre et al. 2013), sexual receptivity and preference are dissociated from endogenous estradiol titers (Ramsey et al. 2010).

In species that are responsive outside the reproductive window, the shape of preference functions can change according to reproductive state. Humans are unusual among mammals in that females are receptive to mating throughout the reproductive cycle. As I mentioned in the previous chapter, preferences throughout the cycle can be recruited to evaluate potential social partners, while genetic mates are evaluated using preferences during the fertile period. In a study that sparked popular attention, Penton-Voak et al. (1999) showed that women preferred more masculinized male faces when ovulating and more feminized faces when menstruating (but see Harris 2011; chapter 18). Preferences for other traits, like body masculinity (Little, Jones, & Burriss 2007) and facial symmetry (Little, Jones, Burt, et al. 2007), also vary across the cycle.

A study by Oinonen and Mazmanian (2006) points out the danger of conflating preference with detectability. Even though preference for facial symmetry is highest during ovulation, women are better at *discriminating* facial symmetry during their menses, when their *preference* for facial symmetry is weakest, but worse at discriminating symmetry among abstract dot patterns (Little, Jones, Burt, et al. 2007).

Reproductive state can also modulate preferences that are arbitrarily learned. For example, the socially constructed preference or antipathy of heterosexual women for beards (Darwin 1871; Janif et al. 2014) varies according to reproductive state, with early-twenty-first-century Finnish women on average showing antipathy for beards when reproductively fertile, and indifference or preference both when menstruating and after menopause (Rantala et al. 2010).

Variation in preference across the menstrual cycle thus appears to be the result at least in part of broad differences in attentional processes and decision making, rather than merely reflecting heightened sensitivity to particular stimuli. Indeed, women self-report stronger commitment to their social partner when fertility is low (Jones et al. 2005) and assign more resources to socially dominant men in the context of a mock job negotiation when ovulating (Senior et al. 2007). A neuroimaging study by van Wingen and colleagues (2008) showed that nasal administration of testosterone, which peaks at ovulation, acted to heighten the encoding and retrieval of male faces in the hippocampus, a brain region associated with automatic memory. Higher-level processes of attention, evaluation, and memory thus act to modulate preferences across the ovarian cycle in humans.

As the latter example underscores, recurrent processes like the estrus cycle—which are themselves modulated by social and environmental stimuli—can also influence both choosiness and the shape of preference functions. Preference variation can be mediated by changes at the sensory periphery; Boulcott and Braithwaite (2007) showed that both female and male sticklebacks increase their sensitivity to red during the breeding season, contradicting earlier results that had shown an effect only in females. Variation can also be mediated by changes in the subjective value of sexual stimuli; for example, Riters and colleagues (2013) found that sexually receptive female starlings exposed to male song exhibited increased opioid activity in brain regions associated with reward.

Behavioral changes across the reproductive cycle can make it problematic to interpret nonspecific assays of preference like association measures. If choosers fail to show an association preference because they are unreceptive, their behavior is irrelevant to mating outcomes, because their "preference" has no impact on mating. Failure to account for unreceptive individuals may thus overestimate permissiveness or variability in preferences. Unreceptive choosers can also generate spurious interpretations of preference shape. For example, Kidd and colleagues (2013) housed female haplochromine cichlids *Astatotilapia burtoni* in a simultaneous-choice test apparatus, flanked by small and large males, for a full reproductive cycle. Females associated primarily with small males (size-assortative shoaling is common in fishes, e.g., Wong & Rosenthal 2005), except on the day of spawning when they reversed their association preference in favor of the large males they actually mated with. Understanding the reproductive physiology of one's study system is therefore especially important when using non-sexual metrics of preference (chapter 2).

Reproductive state, including the timing of sexual maturity and ovulatory cycles, is also strongly influenced by environmentally mediated stress (chapter 11) and by social context (chapter 12). Of particular note is that courters can manipulate chooser receptivity, increasing it during courtship (chapter 6) and reducing it after mating (chapter 7). Physiological receptivity is thus a primary arena for sexual conflict (chapter 15).

9.4.4 Age

Several studies have documented variation as a function of age for precopulatory (e.g., Rantala et al. 2010; Morris et al. 2006; Coleman et al. 2004; Atwell & Wagner 2014) and postmating preferences (Fricke et al. 2013; Engqvist & Sauer 2002). On a shorter timescale, preferences can shift remarkably within a breeding season, in line with mate availability and future prospects for mating (Qvarnström et al. 2000; Passos et al. 2014), I will return to such seasonal shifts in preferences in chapter 11. Unless the same cohorts are tracked over time, it can be difficult to disentangle intrinsic effects of growth and aging from confounding temporal variation in environmental parameters (chapter 11) or social experience (chapter 12; for an exception see Dukas and Baxter 2014, who studied age effects on mate choice in male fruit flies while controlling for social experience). Size, which can covary with peripheral sensitivity (see below), can also scale with age.

One universal effect of age is that older choosers will have experienced more interactions with courters than younger ones (chapter 12). By contrast, two other widespread correlates of age that are probably robust to outside influence are changes in sensory and perceptual abilities (next section), which generally peak when individuals are reproductively active (Ronald et al. 2012), and changes in physiological receptivity (see above).

There has been relatively little work on the development of mating preferences prior to sexual maturity. Two exceptions are ontogenetic studies of preferences for conspecific courter signals in túngara frogs, where preferences are innately specified (Baugh & Ryan 2010), and in white-crowned sparrows, where they depend on an interaction between innate factors and social experience (Whaling et al. 1997). In both studies, choosers-to-be showed behavioral preferences for conspecifics, but were less selective than adults. Fledgling sparrows were more permissive than adults, attending to conspecific songs even if they were played backwards or were simpler than normal. This suggests that preferences for complex acoustic features develop later in ontogeny (Whaling et al. 1997). In túngara frogs, strong preferences for conspecific calls developed gradually but were detected long before sexual maturity (fig. 9.5). Intriguingly, juveniles exhibited phonotaxis behavior (chapter 2) toward calls, even though juveniles neither learn sexual behavior nor mate with courters. Further studies of mate-preference development before maturity will be instructive in understanding how complex preferences come to be expressed (Baugh & Ryan 2010). Importantly, preferences during ontogeny may bias social experience and therefore shape adult mating preferences (chapter 12; Shizuka 2014).

9.4.5 Condition

Condition is a central concept in adaptive thinking about the state of both courters and choosers. There is broad agreement that condition constitutes "the pool from which resources are allocated for use in reproduction,"

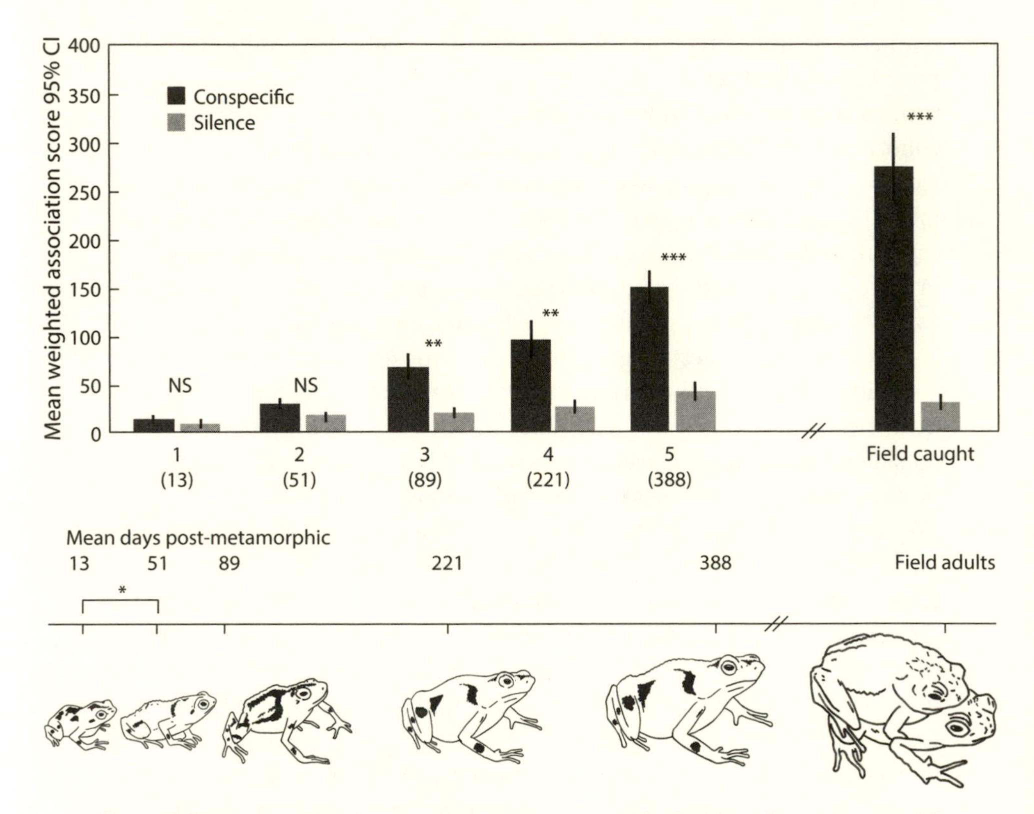

Figure 9.5. Mean weighted association time (+/– 95% CI) of túngara frogs to conspecific versus heterospecific calls. Positive values indicate conspecific preference. Preference emerges by 89 days post-metamorphosis, but is not fully expressed until sexual maturity. Modified from Baugh & Ryan (2010).

including mate choice (Rowe & Houle 1996), but authors have shown great flexibility in operationalizing this concept. The definition of condition has gradually come to be conflated with the supremely fuzzy "Quality" to which I return in chapter 14: for example, Cotton and Pomiankowski (2007) define condition as any "trait showing strong covariance with general viability, such that higher trait values confer greater fitness." Many authors operationalize condition along these lines, using easy-to-measure quantities like body size, tarsus length, or even "the rank order in which birds could be captured by an experimenter" (Birkhead et al. 1998), but fail to demonstrate that the trait covaries with fitness benefits to courters, let alone choosers. I agree with Clancey and Byers (2014, p. 848) that "the definition of condition should be restricted to body fat and total energy reserves, or the term should not be used at all, and authors should simply say what parameters they are measuring or including in models."

Condition is dependent on nutrition, parasite load, and social status, and accordingly is an important mediator of environmental effects on preference (chapter 11; Cotton et al. 2006). Condition is a determinant of reproductive state (see above). As a rule, animals in very poor condition are nonreproductive, but there is substantial condition-dependent variation in preferences among reproductive choosers. As I will revisit later, the costs and benefits of mate choice are often dependent on the condition of the chooser, although not in a consistently predictable direction.

9.4.6 Sensory acuity

Preference can vary with sensory acuity within as well as between the sexes (Ronald et al. 2012). In stalk-eyed flies, *Diasemopsis meigenii*, female choosiness increases with eye-span length, a predictor of photoreceptor density and therefore visual acuity (Cotton et al. 2006). In cricket frogs, *Acris crepitans*, peak frequency sensitivity is negatively correlated with body size, likely because larger animals have thicker auditory tympani (Keddy-Hector et al. 1992); larger females prefer lower-frequency advertisement calls (Ryan, Perrill et al. 1992).

Body size is correlated with choosiness (rather than peak preference) for acoustic cues in painted reed frogs (Jennions et al. 1995). Increased choosiness may be explained by the fact that larger animals have greater interaural distances and are therefore better able to localize sound sources (Heffner & Heffner 1992). Indeed, in field crickets *Gryllus integer*, female choosiness scales with body size for acoustic cues, but not olfactory cues (Hedrick & Kortet 2012). Allometric variation in sensory processes may be a widespread proximate explanation for preference variation; chapter 12 discusses how sensory processes can mediate environmental effects on preferences.

9.4.7 Personality

Stable, correlated suites of behaviors, or personality, may also influence mate choice (Ingley & Johnson 2014; Reinhold & Schielzeth 2014). In particular, many species exhibit variation along the so-called shy/bold axis, with bolder individuals being more likely to engage in exploratory behavior and investigate novel stimuli (Wilson et al. 1994). Bolder individuals may thus engage in more mate sampling (chapter 6) and may respond more positively to unfamiliar signals.

Despite a recent explosion of studies on animal personality (Sih et al. 2004; Carter et al. 2013), work on personality and mate choice has largely been limited to studies of personality assortment in pair-bonded species (chapter 8; reviewed in Schuett et al. 2010). There has been surprisingly little work on personality's relationship to mate-sampling strategies, choosiness, or preference shape. Muraco et al. (2014) found no effect of behavioral type on male mating preferences in sailfin mollies. By contrast, David and Cèzilly (2011) found that high exploratory behavior in female zebra finches was positively correlated with the consistency of individual preferences, but negatively correlated with choosiness and strength of preference. Bierbach and colleagues (2015) showed that bolder males showed stronger audience effects, adjusting courtship effort in the presence of a rival (see chapter 8). Johnson and Sih (2005) showed that female spiders that were more aggressive toward prey were more likely to cannibalize their mates. And in humans, Shackelford and Schmidt (2008) compiled survey data from around the world to show that the tendency to seek short-term sexual partners covaried predictably with personality measures. Personality may represent an important covariate of mate choice.

9.5 SAME-SEX SEXUAL BEHAVIOR

Sexual preference—a bias toward opposite- or same-sex individuals—is a fundamental aspect of mate-choice variation that deserves special mention. When given a choice, choosers as a rule prefer heterosexual partners, but same-sex sexual behavior is ubiquitous in animals, and numerous hypotheses have been put forth to explain its origin and maintenance (reviewed in Bailey & Zuk 2009a; Poiani 2010). While same-sex behavior is widespread, stable sexual *orientation*, whereby an individual preferentially engages in sexual activity with one sex (or gender) more than another, is known only from humans and sheep. There is a consensus that sexual orientation (if not sexual fluidity) is robust to social experience and associated with early organizational effects of sex hormones, although it remains an open question whether these effects are specified by genetics, epigenetics, or the fetal environment (Poiani 2010); I will return to sexual orientation in chapter 18.

In nonhuman animals, same-sex behavior is typically plastic, with choosers engaging in homosexual interactions only when heterosexual partners are unavailable (Poiani 2010; MacFarlane et al. 2010). Accordingly, studies have focused on the proximate drivers and function of homosexual behavior in the general sense, rather than on how same-sex partners are chosen. A review of same-sex behavior in insects by Scharf and Martin (2013) argues that most cases are a consequence of "mistaken identity," or males having a broad preference envelope and mating indiscriminately. In some cases, however, same-sex sexual behavior may be functional. For example, flour beetles *Tribolium castaneum* appear to use same-sex copulation to discard old and less viable spermatophores, which in a small number of cases results in indirect sperm transfer, where a previously mounted male subsequently transfers another male's sperm to his female partner (Levan et al. 2009). In budgerigars, males engage in sexual activity with each other prior to maturity; this does not appear to constitute "practice" for heterosexual courtship, as males that engage in more same-sex activity are less likely to obtain heterosexual mates. Abbassi and Burley (2012) argued that these homosexual interactions served as "social glue" for social interactions distinct from reproduction. In both of these cases, choice of same-sex partner should have measurable fitness consequences for choosers. In the first, selection would favor choice of homosexual mates who are attractive to females, and therefore more likely to transfer sperm indirectly. In the second, selection would favor adolescent interactions with same-sex partners that pay off in terms of more beneficial social relationships later in life.

Phylogenetic comparative analyses by MacFarlane and colleagues (2007, 2010) suggest an inverse relationship, across taxa, between same-sex behavior and the availability of opposite-sex partners. Males were more likely to engage in same-sex behavior if they were in polygynous species, while females were more likely to do so in monogamous species, where female-female pairs engage in biparental care (chapter 8).

9.6 SYNTHESIS

Variation in mate choice and mating preferences can arise from differences within and between individuals in any of the mechanisms used to sense, perceive, and evaluate courter signals before, during, and after mating, and in the mechanisms they use to sample and choose among mates.

Preferences frequently covary with traits not directly involved with mate choice. These patterns of covariation, among preferences and between preferences and other traits, may arise either from properties internal to choosers, like shared underlying mechanisms or statistically associated genetic underpinnings, or from correlations among environmental and social variables that shape preferences. These correlations represent an important constraint on the directions in which preferences can evolve (chapter 15).

Multidimensional preferences can covary in directions that may or may not align with multivariate courter traits, which becomes important when thinking about trait-preference coevolution (chapter 15). Misalignment of traits and preference in multivariate space constrains the effect of mate choice as an agent of sexual selection, and in turn the extent to which courter traits impose indirect selection on preferences (van Homrigh et al. 2007). On the other hand, if individuals have stably idiosyncratic, complex preferences, enough variation may cause them to average out to produce relatively smooth preference surfaces in the aggregate. This phenomenon may account for the weak to nonexistent preferences measured in many multivariate studies (chapter 4).

The next three chapters unpack the sources of variation in mate choice and mating preferences: genetics (chapter 10), the physical and ecological context (chapter 11), and the social environment (chapter 12).

9.7 ADDITIONAL READING

Jennions, M. D., & Petrie, M. 1997. Variation in mate choice and mating preferences: A review of causes and consequences. *Biological Reviews of the Cambridge Philosophical Society, 72*, 283–327.

Poiani, Aldo. 2010. *Animal Homosexuality: A Biosocial Perspective.* Cambridge, UK: Cambridge University Press.

Widemo, F., & S. A. Saether. 1999. Beauty is in the eye of the beholder: causes and consequences of variation in mating preferences. *Trends in Ecology & Evolution, 14* (1):26–31. doi: 10.1016/s0169-5347(98)01531-6

CHAPTER 10

Variation I: Genetics

10.1 INTRODUCTION

Mating decisions, and their dependence on external influences, are the product of networks of genes acting to shape phenotypes. Structural and regulatory differences in these genes are the basis for preference evolution, and genetic variation is therefore a central concern of mate-choice research.

Despite a rapidly growing number of studies on preference genetics, we are just beginning to gain the understanding necessary to test the assumptions and predictions of theory. We begin by asking what we need to know about the genetics of a given aspect of mate choice—say, peak preference for an acoustic signal. The most basic assumption about the evolution of preferences is that preferences are evolvable: we want to know if there is a genetic basis to at least some of the differences in mate choice among species, populations, and individuals. The answer to this question, despite substantial environmental influences on preferences, is confidently "yes."

Second, we want to know how much standing genetic variation there is for preferences in natural populations to provide the raw material for selection (Bakker & Pomiankowski 1995). Third, we want to know the details of the genetic architecture underlying the preference, which can have important consequences for preference evolution (Kirkpatrick et al. 2004): are there many or a few genes involved, are they sex-linked, and how do they interact to produce the phenotype of interest? With respect to these two questions, we now know that both the amount of genetic variance and the underlying genetic architecture vary widely by taxon and by the preference being measured. This likely reflects differences in the evolutionary history of selection, variation in the genetic architecture underlying different types of preferences, and differences in social and environmental influences on the particular preferences being assessed. We are a long way from being able to draw generalizations, however, about the processes underlying this variability.

Finally, we would also like to be able to connect genotype to phenotype by identifying specific allelic variants associated with variation in preference and characterizing their molecular function. There has been some fascinating work in this area, particularly when preference variation relies on differences at the sensory periphery that can be associated with differences in receptor genes, but characterizing the genetic mechanisms underlying perceptual integration, evaluation, and decision making poses a major challenge.

An equally important set of questions concerns the genetic covariance between a given choice mechanism and other traits. Most attention has focused on the critically important question of how **pleiotropy** and **physical linkage** impose covariation on preferences and the traits these preferences act upon (Heisler et al. 1987), and I will return to this in chapter 15. We know very little, by contrast, about genetic covariance between preferences and chooser traits unrelated to mate choice (like age at maturity, sensory acuity, and personality; chapter 9), or among different aspects of preference. And there are only a handful of studies on genotype X environment (G X E) effects on mate choice; these are covered in the next two chapters.

In this chapter, I review the main approaches for characterizing preference genetics and survey the empirical data as they speak to the questions outlined above. Approaches to understanding the genetics underlying preferences (or any other phenotype) take two broad forms. The first approach consists of attempts to identify particular genes or genomic regions associated with preference variation; for preferences, this is typically done using so-called **forward genetics**, whereby variation in phenotype is correlated with variation in genotype. Alternatively, the effects of candidate genes on preference can be characterized using **reverse genetics**, whereby gene structure or function is altered to test its effect on phenotype.

The second approach encompasses **quantitative genetic** studies that assume that the underlying genetic variation is continuous and additive (Bakker 1999). Quantitative genetic models often assume an infinite number of loci each contributing infinitely small positive or negative effect, summing to determine trait value. Classical quantitative genetics has proven a remarkably powerful tool in agricultural breeding programs, since these mathematical assumptions enable the use of continuous functions for modeling phenotypic evolution and estimating selection, and allow one to partition out genotype-by-environment interactions and **indirect genetic effects** (Chenoweth & Blows 2006; Bailey 2012).

10.2 INTERSPECIFIC GENETIC DIFFERENCES

A first step in identifying the genetic basis of preferences is to measure phenotypes in divergent lineages and their hybrids using controlled crosses raised under standard conditions. In addition to establishing that preferences have a genetic basis, statistical techniques can be used to obtain a first-order estimation of the number of genes associated with interspecific differences in preference. For example, female F_1 hybrids between *Drosophila heteroneura* and the sympatric *D. silvestris* are more likely to accept *D. silvestris* males than are *D. heteroneura* females. As expected if preference differences are controlled by a large number of genes, first-generation hybrids showed intermediate preferences between the parental species (Boake et al. 1998). By contrast, measuring the distribution of preferences in second-generation (F_2) hybrids allowed Haesler and Seehausen (2005) to estimate that a minimum of 1.6 genes controlled female preference differences for male phenotypes between two species of *Pundamilia* Rift Lake cichlids (fig. 10.1). Ding and colleagues (2014) used a similar design to determine a similar lower bound for another Rift Lake cichlid, *Maylandia*, with at least two genes interacting non-additively to shape preferences. Such crosses can also reveal that distinct aspects of preference are associated with distinct genetic underpinnings. Isoherranen and colleagues (1999b) found that F1 hybrids between *D. virilis* and *D. montana* resembled *D. virilis* in responsiveness, but *D. montana* in their preference for courtship containing song.

Similar results can be obtained from genetic crosses between naturally and artificially divergent lineages of the same species. For example, populations of the katydid *Ephippiger ephippiger* differ in the shape of their preference functions; where males produce monosyllabic calls, female show directional preferences, increasing antipathy to calls with more syllables (fig. 10.2a). In polysyllabic populations, females show quadratic preference functions with preferences for intermediate syllable numbers (fig.10.2b). First-generation hybrids show intermediate preference functions (fig. 10.2c–d; Ritchie 1996).

Another approach to characterizing preference variation is to compare the phenotypes of inbred or isofemale lines known to differ genetically, to examine the genetic basis of variation among (Hollocher et al. 1997) or within populations (Butlin 1993; Pischedda et al. 2014). Lüpold et al. (2013), following on earlier studies using fixed-chromosome lines by Clark and Begun (1998; Clark et al. 1999), used this approach to show genetic variation in postmating choice in *Drosophila*, showing heritable variation in the

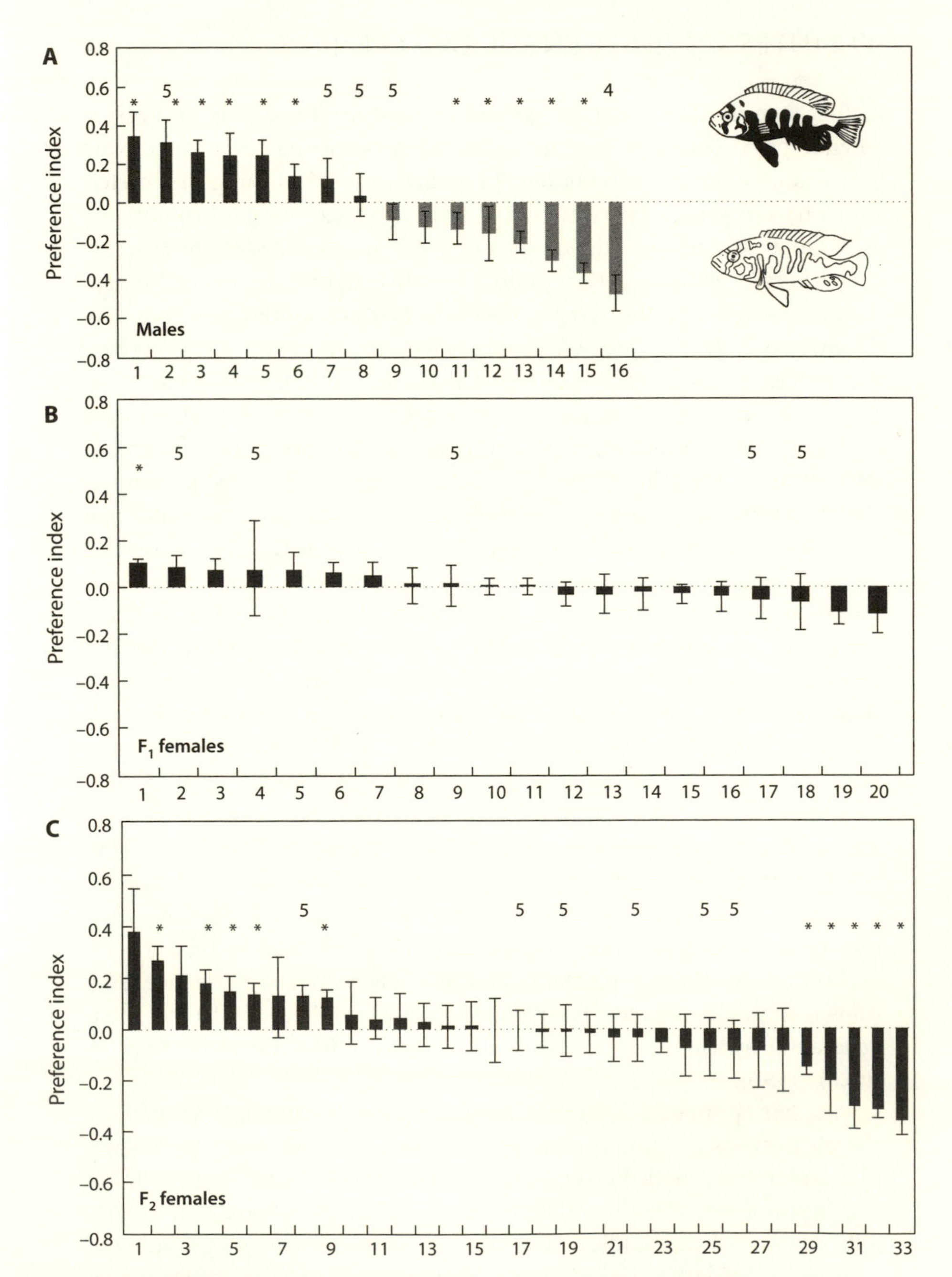

Figure 10.1. (a) Net preferences of individual Lake Victoria cichlids *Pundamilia pundamilia* (top) and *P. nyererei* (bottom) for males of each species. (b) Preferences of F_1 females; (c) preferences of F_2 females. After Haesler & Seehausen (2005).

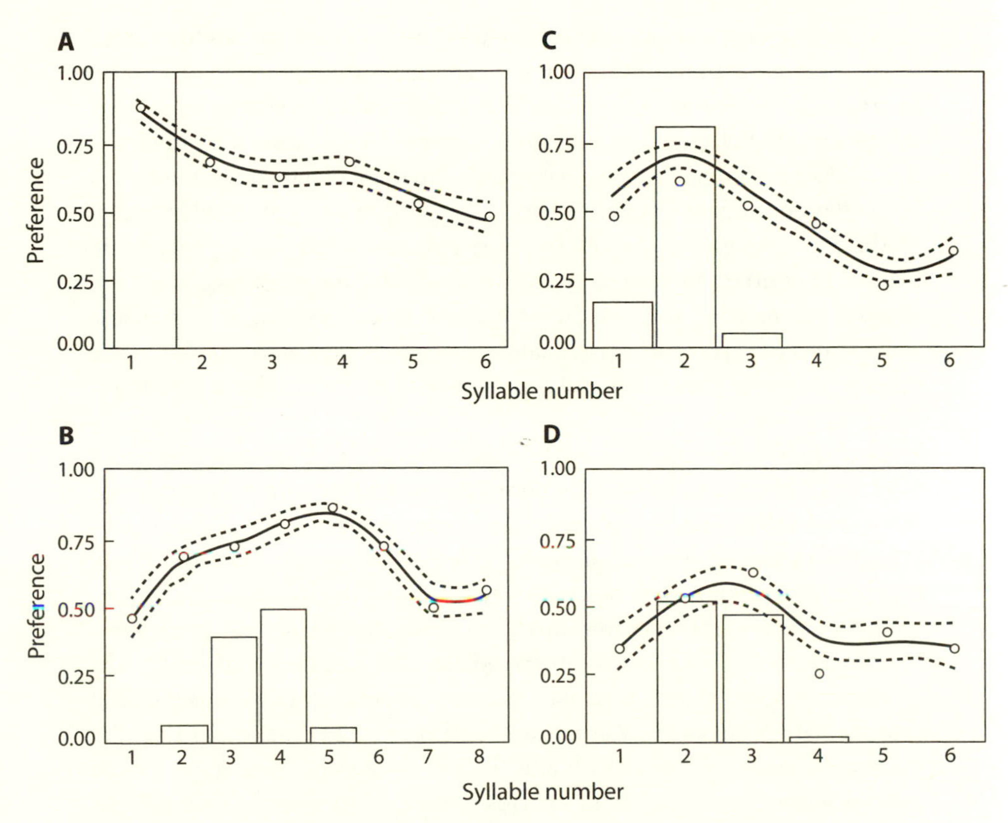

Figure 10.2. (a) Female katydids *Ephippiger ephippiger* from a population with monosyllabic calls show directional antipathy for increased syllable number. Histogram shows male trait distribution. (b) Females from a polysyllabic population show unimodal preferences. (c) First-generation hybrids between monosyllabic females and polysyllabic males and (d) vice versa exhibit intermediate preferences.

timing of sperm ejection, and therefore in the composition of the pool of available sperm, among females having mated with two different males. Intriguingly, two studies on premating preference variation among females from inbred lines of *D. melanogaster* have yielded contrasting results; while Ratterman et al. (2014) found substantial genetic variation in female preference among ten inbred lines using single-choice assays, Foley et al. (2010) found no significant effect of female genotype in two independent lines not used in Ratterman et al.'s (2014) study and tested in a mesocosm population. These differences point either to differences among lines in the amount of preference divergence, or to the importance of context when assaying mating preferences.

A number of studies in insects have used interspecific crosses to characterize the genetic basis of the neural mechanisms underlying mate choice. In crosses between the moths *Heliothis subflexa* and *H. virescens*, behavioral assays showed that preferences for pheromonal blend ratios were intermediate, while preference and antipathy for individual blend components was *H. subflexa* dominant (Vickers 2006a). This pattern was generated by intermediate responses of odorant receptor neurons (ORNs) to pheromone blends. In contrast to vertebrates, where one odorant receptor gene is expressed per neuron, this was likely a result of co-expression of species-specific receptor proteins on the same neurons (Baker et al. 2006) and/or the dominance of the *H. subflexa* genotype with respect to the neural activity of the glomeruli.

In chapter 5, I described how the strain-specific pheromone preferences of E- and Z-strain European corn borers are associated with reciprocal projections of ORNs specific to major and minor components of the pheromone blend onto the medial glomerulus and lateral glomerulus, respectively (Kárpáti et al. 2008). Genetic crosses suggest that the mechanisms underlying the species difference may not be so straightforward. First-generation hybrids showed a preference for intermediate-blend ratios; surprisingly, they consistently showed an *E*-type glomerular morphology despite expressing intermediate preferences. Kárpáti and colleagues (2010) showed that the volume of the medial glomeruli and the electrophysiological responses of antennae were intermediate in hybrids compared to parentals, consistent with the intermediate behavioral preferences.

10.3 GENETIC MAPPING

"Preference genes" are the Philosopher's Stone of preference evolution. An intuitively appealing approach to finding genes responsible for preference differences is **gene mapping**. In quantitative trait locus (QTL) mapping, one conducts crosses as above on species or strains differing in preference, then generates recombinant backcross or second-generation hybrids. With a linkage map, adequate density of genetic marker loci, and sufficient robustness and precision in measuring phenotypes, one can statistically associate phenotypic variation with variation at a particular set of marker genotypes (Chenoweth & Blows 2006). Analogous techniques like admixture mapping and genome-wide association scans (GWAS; Winkler et al. 2010) can in principle be used to map preference genes in natural populations. High-throughput sequencing now makes it straightforward to assay genotypes at hundreds or even thousands of marker loci, even in non-traditional model systems (Schumer, Cui et al. 2014). Genotyping is also becoming increasingly inexpensive, making it easier to achieve the large sample sizes required for robust QTL estimates. When combined with transcriptomic profiling, mapping approaches can identify loci associated with variation in gene expression (eQTL; e.g., Hulse & Cai 2013). Mapping assays thus have the potential to connect variation in gene expression across contexts (Wong et al. 2012) to intra- and interspecific genetic variation in preferences.

Despite these possibilities, genetic mapping of preferences has proved largely frustrating. This is partly because of problems generally inherent to genotype-phenotype mapping of complex traits. At small to moderate sample sizes, mapping only robustly detects genes of large effect. Further, epistatic interactions among genes associated with preferences (section 10.5) can obscure the effect of individual loci (Chenoweth & Blows 2006). These epistatic interactions can be detected by scanning for evidence of an interaction of two genotypes on a phenotype, but this type of analysis is still computationally infeasible in most cases. If preferences have a complex, polygenic basis, gene mapping will only yield a narrow and potentially misleading picture of the underlying genetic architecture. Recombination also imposes a trade-off between the precision of mapping estimates and our power to detect them. In F_2 hybrids, chromosomes are segmented into large blocks inherited from each parental line; at best, we can resolve QTL to one of these blocks, which may contain hundreds of genes; at worst, linked genes of opposite effect can cancel each other out, meaning that genotype-preference associations go undetected. Using later-generation hybrids, as with admixture mapping of natural hybrid populations, gives us smaller ancestry blocks but reduces statistical power because of the need to correct for a greater number of comparisons. These problems are heavily compounded by the issues inherent in obtaining robust, unbiased estimates of individual preferences (chapter 9).

It is therefore not surprising that only a handful of studies have identified preference QTL, primarily in insects (Kronforst et al. 2006; Moehring et al. 2006; McNiven & Moehring 2013). Many of these studies have focused on testing for physical linkage between male traits and female preferences, which should greatly facilitate trait-preference coevolution (chapter 15). For example, none of the five QTL identified between inbred lines of the lesser waxmoth *Achroia grisella* co-localized with QTL for male display traits (Limousin et al. 2012). By contrast, one of the five significant QTL identified by Shaw et al. (2007) between two Hawaiian *Laupala* cricket species co-localized with a QTL for interspecific song differences (Shaw & Lesnick 2009). The five *Laupala* QTL were subsequently confirmed by introgressing these chromosomal regions into parental species and assaying hybrid preferences (Ellison et al. 2011). Many more mapping studies will be needed before we can generalize about the frequency of physical linkage between traits and preferences associated with intra- and interspecific variation.

A series of studies on conspecific mate preference in fruit flies showcases both the potential and limitations of gene-mapping approaches. Ortíz-Barrientos et al. (2004) identified four QTL associated with increased reluctance of female *Drosophila pseudoobscura* to hybridize with *D. persimilis* in sympatric populations. Strikingly, Ortíz-Barrientos and Noor (2005) introgressed sympatric *D. pseudoobscura* alleles at one candidate locus into *D. persimilis* and found that these introgressed females were more reluctant to mate with heterospecifics compared to females with alleles introgressed from the allopatric population. The sympatric allele therefore had the effect of decreasing chooser permissiveness, independent of the species-specific traits females were attending to. Differences in choosiness, rather than differences in peak preference, thus appear to be driving **reproductive character displacement** between these species (chapter 16). It should be noted that Barnwell and Noor (2008) failed to replicate this and another QTL in six additional *D. pseudoobscura* lines, which they suggested may have stemmed from the associated alleles being present at low frequencies in natural populations. Mapping studies in more systems may allow us to generalize about the genetic architecture of within- and among-species differences in the genetic architecture of traits and preferences, but the *Drosophila* results remind us of the limitations of mapping studies for pinpointing the genetic substrates of complex natural variation in preferences.

10.4 GENETIC VARIATION IN NATURAL POPULATIONS

10.4.1 Heritability and genetic variance

Heritable variation is a prerequisite of phenotypic evolution. How much genetic variation is there in mating preferences within a population? This is a critical question if we are interested in how selection operates on preferences. In quantitative genetics, **heritability** is the proportion of phenotypic variation in a population that has a genetic basis:

$$H^2 = V_g / V_p$$

where V_g and V_p are genetic variance and phenotypic variance, respectively. We can further partition genetic variance into **additive genetic variance**

(V_a) arising from linear combinations of effects from each allele, and **non-additive genetic variance** (V_n) which includes effects of epistasis and dominance. An important distinction is between **broad-sense** heritability (H^2), where V_g encompasses V_n as well as vertically transmitted nongenetic effects; and **narrow-sense** heritability (h^2) where heritability is measured as only the contribution of additive genetic variance (V_a/V_p; Visscher et al. 2008). Additive genetic variance becomes important when we consider the response to selection on preferences, and trait-preference coevolution, in chapters 14 and 15. In chapter 11, I will introduce genotype-by-environment interactions; chapter 12 will show that many environmental effects can be modeled as **indirect genetic effects**.

A critical point is that heritability is very different from inheritance. A trait can have a wholly inherited basis, where a mutation at a single locus consistently alters the phenotype independent of external effects, yet heritability is limited by V_g. This leads to the counterintuitive result that if inherited preferences are under strong stabilizing selection, for example in the context of conspecific mate recognition (chapter 16), the underlying alleles will fix. Therefore, there will be little genetic variation and low heritability. The low repeatability of mating preferences (chapter 9), coupled with strong environmental influences (chapters 11 and 12), will also act to reduce heritability. Indeed, a recent meta-analysis of preference heritability by Prokuda and Roff (2014) found heritabilities of preference to be in the same low range as life-history traits also predicted to be under strong stabilizing selection, and markedly lower than those for courter traits. The modal heritability of preference studies is zero (fig. 10.3).

There may be considerable variation among taxa and among preference measures with regard to the heritability of preference, depending on extrinsic and intrinsic factors affecting the robustness and repeatability of preferences. For example, Gray and Cade (1999) estimated the narrow-sense heritability of female preferences for song traits in a natural population of crickets to be 34%. They tested individual females on seven synthetic calls varying solely in the number of pulses per trill. These were second-generation offspring of wild-caught females mated randomly to unrelated males from the same cohort. By design, the researchers standardized factors like social experience and nutrition that would act to increase variation in the wild. They also had the sense to work on insects, which have more repeatable preferences than vertebrates (chapter 9).

Two vertebrate studies estimated heritability from behavioral interactions with live males in the laboratory. Brooks and Endler (2001a) found that of multiple preference variables measured in female guppies attending

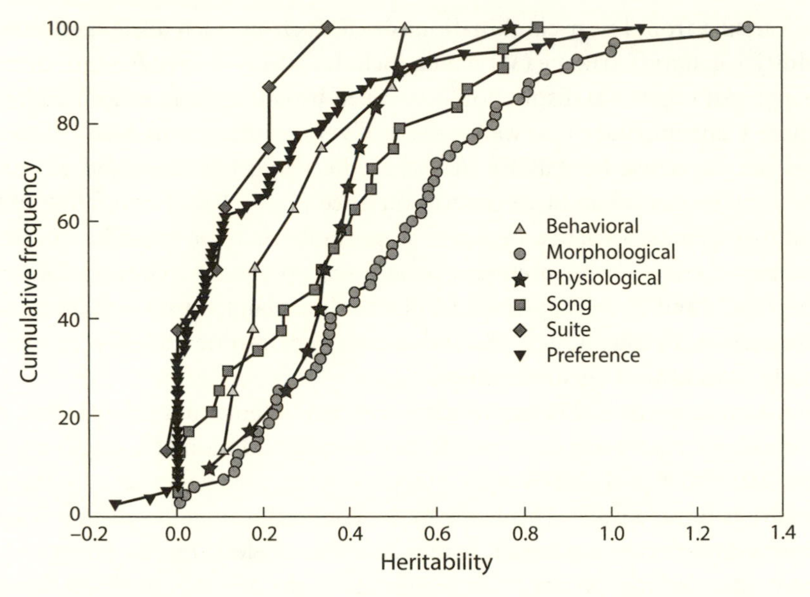

Figure 10.3. Cumulative probability graph of 57 estimates of heritability from 14 species in 20 studies, with curves for courter traits shown for comparison. "Suite" refers to composite traits like attractiveness and multimodal syndromes. After Prokuda & Roff (2014).

to visual traits, only overall responsiveness had significant additive genetic variance. Similarly, Schielzeth and colleagues (2010) cross-fostered sets of unrelated nestling zebra finches in a large experimental population. They then assayed preferences of foster and genetic for eight pairs of males. By testing for correlated preferences between sisters and between nestmates, they were able to decouple genetic effects from effects of rearing (chapter 12). Preference functions were not heritable, but responsiveness and the number of comparisons among males were. They argued that choosiness, responsiveness, and mate-sampling behavior (which are all intertwined), rather than preference function shape might represent the primary source of heritable variation with respect to mate choice in zebra finches. This result is intriguing in light of the extensive work on social determination of preference functions in this species (chapter 12).

Finally, estimates of preference heritability are particularly low when measured as realized mate choice. In socially monogamous birds, social mates can be readily observed, thereby providing a straightforward measure of the preference outcomes. Studies on collared flycatchers (*Ficedula albi-*

collis) provide the only estimates to date of the heritability of realized mate choice. Short-term studies of two populations (Qvarnström et al. 2006; Hegyi et al. 2010) have shown no detectable heritability of choice for male traits; a later analysis of a 25-year data set in a Swedish population did find low but significant heritability for preference for male forehead patch size (h^2 = 0.146, confidence interval 0.042–0.217), along with effects of temperature and rainfall on preference (Robinson et al. 2012; chapter 11). Heritability of mate choice is obviously lowered by greater environmental heterogeneity in the wild. Postma et al. (2006) pointed out that finite mate sampling and mutual mate choice mean that females will often mate with less-preferred males: realized matings will always deviate from latent preferences. Further, peak chooser preferences may be for hidden combinations of traits not represented by current phenotypic variation (van Homrigh et al. 2007; Fisher et al. 2009).

The gamut of studies discussed above speak to the point that which mate-choice variables are measured, and how we measure them, can have major effects on our estimates of heritability. Roff and Fairbairn (2015) draw the distinction between heritability of preference and heritability of choice. Echoing Hager and Teale's (1994; chapter 9) point about repeatability, most estimates of preference are inferred from differential responses to small numbers of individual courter stimuli. The smaller the number of stimuli sampled, the less likely a chooser's "realized" preference in an experimental assay is to reflect her latent preference (fig. 10.4). More generally, if there is less variance in preferences for a trait than in the trait itself, this will bias the heritability of preference downward if preferences are measured using a sample from a natural distribution of courters. Sequential sampling in the wild, whereby choosers mate with one courter from a small sample, will yield the same downward bias, adding to the effects of environmental heterogeneity in reducing heritability (Roff & Fairbairn 2015; cf. Postma et al. 2006).

A study on humans suggests that it is the distinction between realized mate choice and preference, perhaps more than differences in complexity or environmental heterogeneity, that acts to lower estimates of heritability. Zietsch and colleagues (2012) used human mono- and dizygotic twins (wild, in the sense that social and environmental heterogeneity is available to contribute to preference variation) to tease apart the effects of shared environment and shared genotype on mate preferences in surveys of self-reported preferences on an array of variables. The broad-sense heritability of several preference measures was significantly greater than zero, although

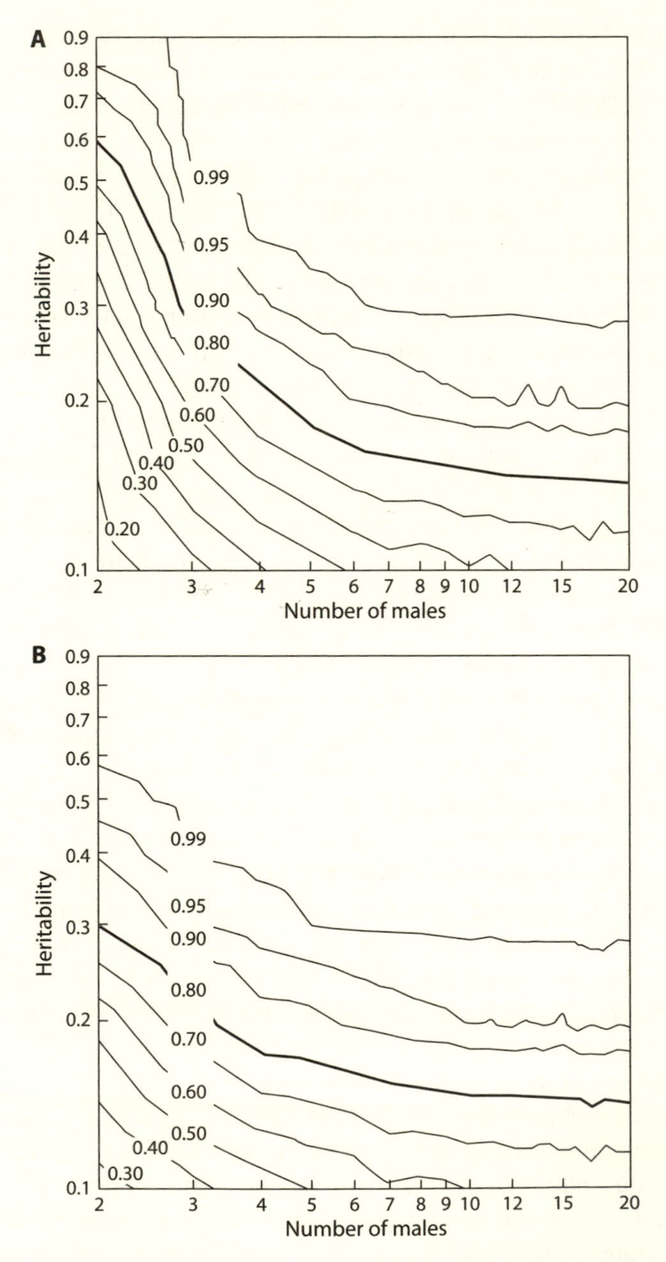

Figure 10.4. Power to detect heritability as a function of the number of courters sampled by choosers. The shaded isoclines show the probability of detecting significant heritability given the number of males sampled and the true heritability. (a) Absolute preferences relative to an internal standard; (b) Relative preferences with respect to the courters sampled. Simulations modeled studies of 50 families with 10 females per family. After Roff & Fairbairn (2015).

it is noteworthy that for no trait was there significant additive genetic variance (table 10.1). Interestingly, the importance of physical attractiveness is heritable in women but not men. A follow-up study by Verweij and colleagues (2012) decomposed physical attractiveness into individual traits. For men, only preference partner's height showed a genotypic effect, while women showed genetic variance in preferences for traits including height, hair color, hair length, and facial hair.

10.4.2 Responses to selection

Responses to selection, in which changes in preference are measured over generations, represent a direct test of the evolvability of mate choice. One of the few studies to directly impose selection on chooser preferences was performed by Sharma and colleagues (2010), who selected on the response of female *Drosophila simulans* to a novel mutant color trait, *ebony*, associated with reduced fitness in males. After five generations of selection, the proportion of females mating with *ebony* males increased substantially (fig. 10.5).

Preferences have also been shown to evolve in response to indirect selection on male traits (chapter 15), although studies have generally not disentangled genetic changes in preference as a consequence of indirect selection from learned preferences for evolved courter trait distributions. Several studies have tested indirect selection on chooser preferences by imposing selection on courter traits (reviewed in Brooks 2002) and are discussed further in chapter 15. In one such study, Brooks and Couldridge (1999) combinatorially imposed artificial selection on orange and black color in male guppies. Females in orange-selected lines increased their preference for orange, and females in black-selected lines increased their preference for black, showing that these two dimensions of preference could evolve independently (fig. 10.6).

10.4.3 Genetic covariance

Brooks and Couldridge's (1999) study addressed an important point if we are to understand the evolution of multiple chooser preferences and multiple ornaments, which is that we want to know the **genetic covariance** among different aspects of preference. If preferences for two courter traits are genetically correlated (for example, because they involve overlapping sets of sensory receptors), preference for one trait could evolve solely as a

Table 10.1. Proportion of variation in mate preferences accounted by additive **(A)** and nonadditive **(D)** genetic influences; **A** is narrow-sense heritability and **D** is broad-sense heritability. **C** is the proportion of variance attributed to a shared environment and **E** is the error term.

	Kind and understanding	*Easygoing*	*Intelligent*	*Healthy*	*Physically attractive*	*Exciting personality*
Females						
A	0.08	0.13	0.16	0.05	0.12	0.00
	(0.00–0.29)	(0.00–0.28)	(0.00–0.36)	(0.00–0.26)	(0.00–0.28)	(0.00–0.26)
D	0.16	—	—	0.16	0.18	0.25
	(0.00–0.31)			(0.00–0.28)	(0.00–0.37)	(0.00–0.32)
A + D	**0.24*****	**0.13**	**0.16**	**0.20*****	**0.30*****	**0.25*****
	(0.15–0.31)	**(0.00–0.28)**	**(0.00–0.36)**	**(0.13–0.28)**	**(0.22–0.37)**	**(0.17–0.33)**
C	—	0.07	0.20*	—	—	—
		(0.00–0.21)	(0.04–0.35)			
E	0.76	0.80	0.64	0.80	0.70	0.75
	(0.69–0.85)	(0.72–0.88)	(0.57–0.71)	(0.72–0.88)	(0.63–0.78)	(0.68–0.83)
Males						
A	0.00	0.17	0.34	0.02	0.11	0.00
	(0.00–0.41)	(0.00–0.40)	(0.00–0.57)	(0.00–0.26)	(0.00–0.35)	(0.00–0.39)
D	—	—	—	0.04	—	0.23
				(0.00–0.28)		(0.00–0.44)
A + D	**0.00**	**0.17**	**0.34**	**0.06**	**0.11**	**0.24**
	(0.00–0.41)	**(0.00–0.40)**	**(0.00–0.57)**	**(0.00–0.28)**	**(0.00–0.35)**	**(0.00–0.45)**
C	0.23	0.06	0.08	—	0.03	—
	(0.00–0.38)	(0.00–0.34)	(0.00–0.48)		(0.00–0.30)	
E	0.77	0.77	0.58	0.94	0.86	0.77
	(0.61–0.95)	(0.59–0.96)	(0.43–0.79)	(0.72–1.00)	(0.65–1.00)	(0.74–0.83)

Source: After Zietsch et al. (2010).
*P < 0.05; **P < 0.01.
Note: Estimates of E are not significance tested, because measurement error is always present.

consequence of selection favoring preference for the other trait (chapter 14). In order to measure genetic correlations among preferences, we need to obtain multiple measures from the same genetic individual. The difficulties of testing the same individual repeatedly on multiple stimuli, coupled with the requirement for very large sample sizes to robustly estimate genetic co-

Wants children	*Good earning capacity*	*Creative and artistic*	*Good housekeeper*	*Good heredity*	*University graduate*	*Religious*
0.14	0.13	0.12	0.05	0.07	0.17	0.14
(0.00–0.28)	(0.00–0.29)	(0.00–0.30)	(0.00–0.27)	(0.00–0.25)	(0.00–0.37)	(0.00–0.35)
—	0.10	—	—	—	—	—
	(0.00–0.31)					
0.14	**0.24*****	**0.12**	**0.05**	**0.07**	**0.17**	**0.14**
(0.00–0.28)	**(0.16–0.31)**	**(0.00–0.30)**	**(0.00–0.27)**	**(0.00–0.25)**	**(0.00–0.37)**	**(0.00–0.35)**
0.07	—	0.10	0.21*	0.10	0.14	0.15
(0.00–0.09)		(0.00–0.24)	(0.03–0.30)	(0.00–0.21)	(0.00–0.28)	(0.00–0.29)
0.80	0.76	0.78	0.75	0.83	0.69	0.71
(0.72–0.88)	(0.69–0.83)	(0.70–0.85)	(0.67–0.819)	(0.75–0.90)	(0.62–0.72)	(0.65–0.76)
0.25	0.06	0.00	0.10	0.00	0.11	0.01
(0.00–0.52)	(0.00–0.37)	(0.00–0.18)	(0.00–0.47)	(0.00–0.32)	(0.00–0.44)	(0.00–0.50)
—	—	—	0.21	0.17	—	0.35
			(0.00–0.49)	(0.00–0.34)		(0.00–0.53)
0.25	**0.06**	**0.00**	**0.31****	**0.17**	**0.11**	**0.36****
(0.00–0.52)	**(0.00–0.37)**	**(0.00–0.18)**	**(0.09–0.49)**	**(0.00–0.35)**	**(0.00–0.44)**	**(0.14–0.53)**
0.09	0.10	0.00	—	—	0.15	—
(0.00–0.42)	(0.00–0.32)	(0.00–0.15)			(0.00–0.40)	
0.66	0.84	1.00	0.69	0.83	0.74	0.64
(0.50–0.73)	(0.63–1.00)	(0.82–1.00)	(0.51–0.91)	(0.66–1.00)	(0.56–0.93)	(0.47–0.76)

variances among traits (Lynch & Walsh 1998), means that we have few estimates of genetic (or, for that matter, phenotypic) correlations among preferences; indeed, multivariate measures are sometimes collapsed into unidimensional indices for the purposes of calculating genetic variance or heritability.

A promising approach to estimating genetic covariances is to model preferences as function-valued traits (FVT),which uses random-regression

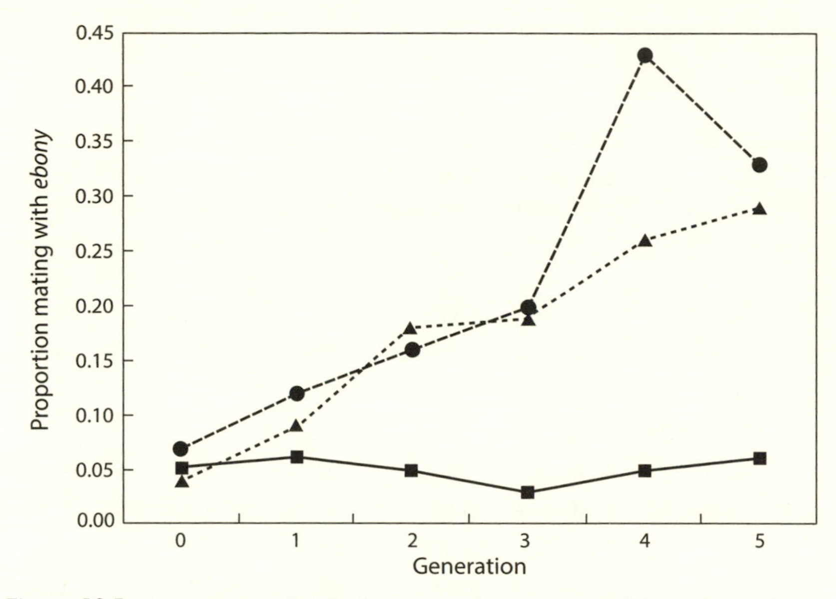

Figure 10.5. Response to artificial selection on *ebony* mutants of *Drosophila melanogaster* for preference for *ebony* males. Dashed lines show two replicate selection lines; solid line is an unselected control. After Sharma et al. (2010).

mixed models to model phenotypes as continuous functions (Meyer & Kirkpatrick 2005). The FVT approach permits the modeling of multiple courter traits as independent variables, with chooser response as the dependent variable. McGuigan and colleagues (2008b) applied an FVT model to estimating genetic variances and covariances in the preferences of *Drosophila bunnanda* for combinations of nine cuticular hydrocarbons produced by males. The FVT approach is well-suited to *Drosophila*, where one can obtain large sample sizes but where playback and other experimental manipulations of courter traits are challenging: McGuigan and colleagues (2008b) used a half-sib mating design of 125 sires mated to 4 dams, where each of the progeny only experienced a single mating encounter with a males whose cuticular hydrocarbon profile was then quantified. Testing females only once eliminated obscuring effects of experience and mating state. They found that about two-thirds of the genetic variation in preference collapsed onto a single significant linear axis (eigenfunction) characterized by negative and positive loadings of individual hydrocarbon components. A later experiment by Delcourt et al. (2010) on *D. serrata* found that a similar proportion of variation was explained by a single axis but that a second axis approached significance, suggesting two independent dimensions of preference variation. It is important to note that this approach only provides a lower bound of the dimensionality of mating preferences, since it is limited

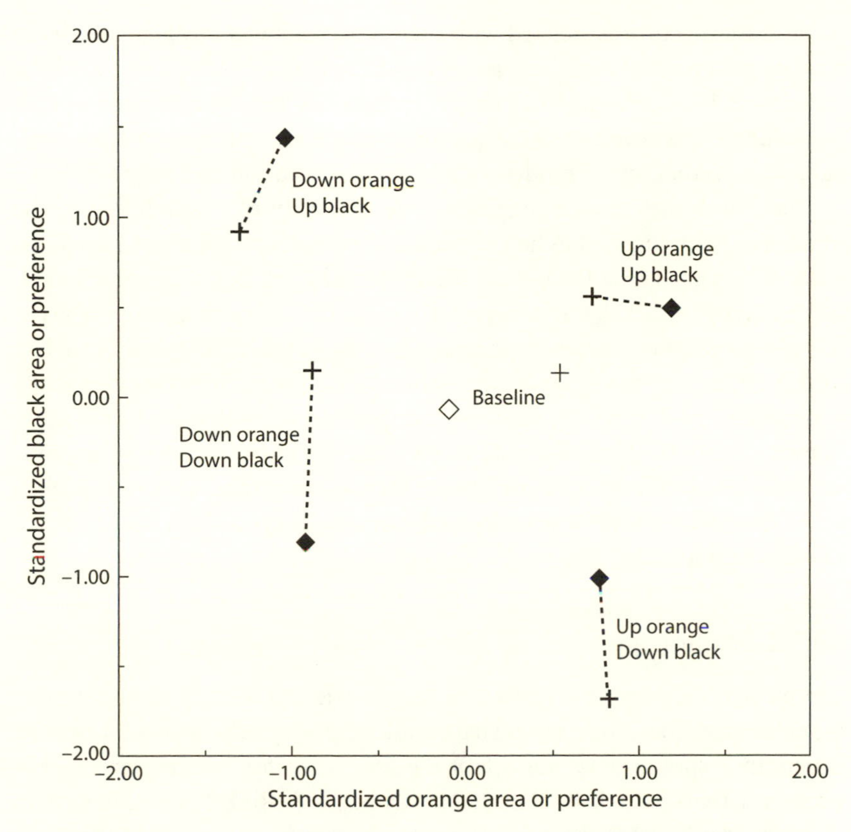

Figure 10.6. Indirect response of female preferences to artificial selection up and down two axes of male trait coloration in guppies. Axes show standardized preference or magnitude of ornament relative to baseline at (0,0). After Brooks & Couldridge (2010).

by the extant (co)variation among courters sampled. In principle, FVT models could be applied to multivariate trials using synthetic calls, as in field crickets (Gray and Cade 1999).

10.5 THE GENETIC ARCHITECTURE OF MATING PREFERENCES

10.5.1 Distribution of allelic effects underlying preference phenotypes

Notwithstanding the difficulties of identifying individual candidate genes (see below), one of the more powerful applications of mapping—and for that matter, classical genetic crosses—is that we can begin to elucidate the genetic architecture associated with preferences. How many genes underlie mating preferences? Are these genes more likely to be located on sex chromosomes? Do different genes shape preferences for different axes of trait variation? The answers to these questions shape assumptions that are critical to the predictions of models of preference evolution and of speciation.

The few studies to date suggest a wide spectrum of possibilities; on the one hand, mapping studies have detected a few loci of large effect associated with preference (Kronforst et al. 2006; Merrill et al. 2011; Limousin et al. 2012; Shaw et al. 2007; Ortíz-Barrientos et al. 2004), but mapping with typical sample sizes is inherently poor at detecting small-effect genes (Lynch & Walsh 1998). On the other hand, assuming polygenic inheritance usually provides a good approximation of the genetics of preferences and other complex traits (section 10.4). We know even less about pleiotropy or physical linkage associated with preference for multiple traits, although Brooks and Couldridge's (1999) selection experiment suggests that these can be readily decoupled.

10.5.2 Sex linkage

The most widely examined aspect of the genetic architecture of preferences is sex linkage. This is because sexual conflict theory predicts that sex-limited inheritance should have important consequences for trait-preference coevolution (Kirkpatrick & Hall 2004; Fairbairn & Roff 2006; reviewed in Dean & Mank 2014; chapter 15), and sex chromosomes are accordingly important candidates for interspecific differences in preference (Albert & Otto 2005; Qvarnström & Bailey 2009; chapter 16). In particular, females from female-heterogametic (ZZ/ZW) species like butterflies and birds should be expected to carry preference alleles on the Z chromosome, since they will thereby be passed on to all sons expressing the preferred traits and will thereby be subject to strong indirect selection.

Iyengar and colleagues (2002) found support for this hypothesis in the arctiid moth *Utetheisa ornatrix*, where the female preference for larger size is patrilineally inherited (fig.10.7). In another female-heterogametic species, Gouldian finches (*Erythrura gouldiae*), females also showed Z-linked inheritance of preferences associated with assortative mating with respect to color morph. Genes associated with color pattern and with reduced fitness of inter-morph hybrids were also Z-linked. Interestingly, there was no effect of cross-fostering with the different morphs or with heterospecifics (Pryke 2010; chapter 12). A similar result was obtained by Saether and colleagues (2007) in collared flycatchers. By contrast, mapping studies of

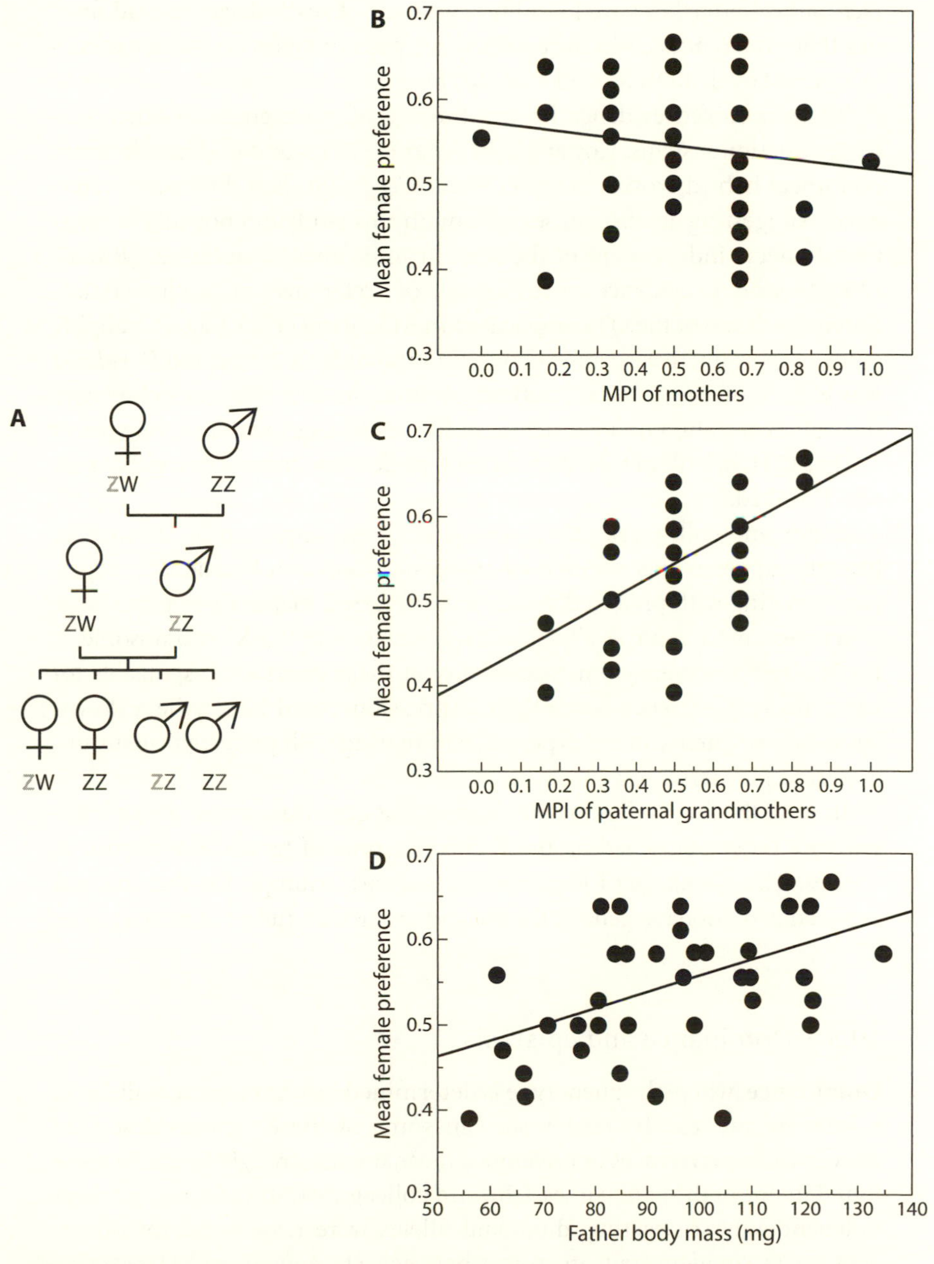

Figure 10.7. (a) Illustration of Z-linked inheritance of preference allele (white). (b) Correlation between body-size preference of female arctiid moths *Utetheisa ornatrix* and their mothers; (c) preference correlation with paternal grandmothers; (d) correlation between preference of females and body size of fathers. After Iyengar et al. (2002).

Heliconius butterflies have found no evidence of sex linkage, instead finding that preferences co-localize with wing-pattern genes on the autosomes (Kronforst et al. 2006; Merrill et al. 2011).

There is mixed evidence for sex linkage of preferences in other sex-determination systems. Hoy and colleagues (1977) found that females from reciprocal hybrid crosses in the cricket *Teleogryllus* had divergent preferences, suggesting an effect of sex, although this study did not rule out maternal effects independent of the sex chromosomes (chapter 12). Ritchie (2000) found no evidence for sex linkage of preferences in another cricket, *Ephippiger.* One of the QTL implicated in Moehring et al.'s (2006) mapping study of reproductive isolation between *Drosophila santomea* and *D. yakuba* was located on the X, but the authors pointed out that this was in line with the null expectation of no enrichment for sex linkage; similarly, one out of the five QTL in Wiley et al.'s study of trait-preference coevolution in *Laupala* was X-linked.

Bailey and colleagues (2011) took a different approach in *Drosophila*, identifying genes that differed in expression between female *D. melanogaster* mating with preferred versus less-preferred males; these genes were overrepresented in physically proximate clusters on the X chromosome. It is not clear, however, to what extent any of these genes are responsible for variation in preference, since their expression could largely be a downstream consequence of the experience of mating with preferred versus unpreferred partners.

In sum, the few studies to date suggest that genes associated with preference are indeed enriched on the Z chromosome of female-heterogametic species. The Z-linkage of loci represents a rare triumph for the dialogue between evolutionary-genetic theory and empirical studies of mate choice.

10.5.3 Dominance and epistasis

Dominance, where the phenotype is determined by one of the two alleles in a heterozygote, can be readily seen in some of the F_1 crosses described above, where preferences or preference mechanisms strongly resemble those found in one of the parent lines. Chu and colleagues (2013) found that three independently segregating, dominant alleles were responsible for preference for *D. simulans* traits in crosses between *D. simulans* and *D. sechellia.* Dominance will become important when we consider the preferences of hybrids (chapter 16).

Epistasis, or nonlinear interactions among genes to produce phenotypes, is a pervasive and theoretically messy consideration when thinking about mate choice. Here, the effect of one allele depends on the so-called *genetic background*—the allelic states at one or more other loci elsewhere in the genome. Genetic compatibility, which is likely to be a major source of selection on preferences (chapter 14), is an epistatic interaction between maternal and paternal alleles with effects on offspring fitness.

Preferences are likely to have complex epistatic bases, as suggested by Zietsch and colleagues' (2012) twin studies, in which human preferences for different attributes were significant only when considering both additive and nonadditive genetic effects. In chapter 4, I argued that nonlinear interactions among preference mechanisms are probably the rule. We should therefore expect epistatic interactions—where the effects of one allele depend on the state of another allele—to be the norm rather than the exception. Consider a simple system where preference depends on the net effect of an excitatory and an inhibitory input, as in European corn borers, above. Recall from chapter 4 that the integration of complex preferences often involves computing contrast between inputs. These inputs are often in turn the product of roughly Gaussian response curves (fig. 3.5). We know, for example with visual pigments (chapter 3) that single amino acid substitutions can shift these response curves in meaningful and variable ways. Figure 10.8 shows the preference of a chooser for one courter versus another differing in a quantitative trait. This preference is determined by a simple integration rule:

$$\text{Preference}_{A|B} \sim (E_A - I_A) - (E_B - I_B)$$

The chooser prefers the stimulus with the greatest net excitatory (E) input relative to inhibitory (I) input. The X axis in figure 10.8 shows the peak of the Gaussian excitatory response, the Y axis the peak of the inhibitory response. The preference for one stimulus value over another changes and reverses as a function of the state of the two alleles, providing for abrupt shifts in preference over a small range of parameter space. Epistatic interactions can thus give rise to complex evolutionary change in preferences, emerging from basic integration of sensory input. Such interactions provide a compelling explanation for why genetic studies using different source strains fail to replicate (Brooks 2002), and should play a pivotal role in shaping the preferences of later-generation hybrids (chapter 16).

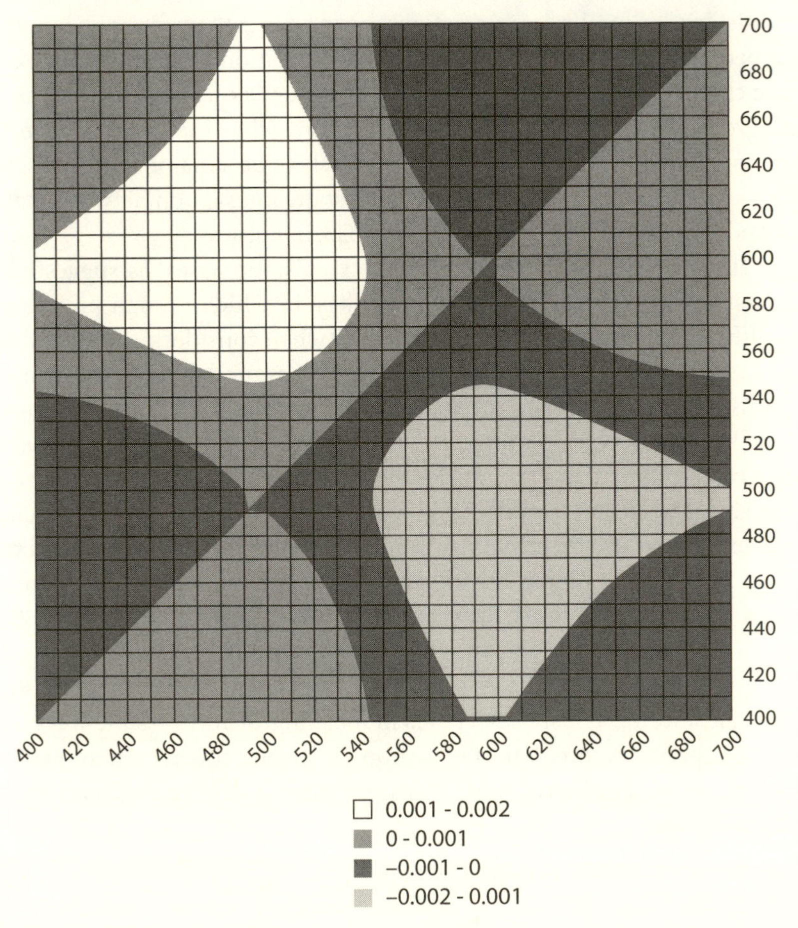

Figure 10.8. Illustration of epistatic interactions between two Gaussian receptor functions responding to a courter signal by computing the net contrast between the two inputs. X and Y axes show the peak sensitivity of each receptor in arbitrary units; responses vary from strong preferences (white) to strong antipathy (light gray) with intermediate tones indicating weaker preferences.

10.6 FUNCTIONAL CHARACTERIZATION OF PREFERENCE GENES

A major goal of evolutionary genetics is to connect structural variation in the genome to functional variation in phenotypes. To date, pheromone receptor proteins in insects (reviewed in chapter 3) provide the only really satisfying way to do this for mate choice in multicellular systems; here, we can connect structural or regulatory changes in a gene to a change in function that directly produces a change in mate choice. We can also do this with the similar system of cell-surface proteins in microorganisms and gametes (chapter 7). Beyond chemoreception and beyond the sensory periphery, however, we know very little about the functional genetics of preference mechanisms.

Reverse-genetic approaches, where we directly manipulate the function of candidate genes, provide a promising way forward, especially in established model systems (Wilkinson et al. 2015). Often, these provide us with insights about the functional mechanisms underlying mate choice, but it is often unclear if these mechanisms play a role in modulating preference variation. For example, *D. melanogaster* females mutant for the brain-expressed transcription factor *datilógrafo* fail to respond to male courtship and always reject males; proper acceptance requires expression of the gene in three distinct brain regions (Schinaman et al. 2014). It is not clear, however, if variation in the expression or structure of *datilógrafo* contributes to variation in preference outcomes.

Monogamy in voles (chapter 8) provides another cautionary tale about mate-choice candidate genes. Regulatory variation in the arginine vasopressin 1a receptor of voles provided a compelling candidate for functional regulation of the tendency to pair-bond and remain sexually faithful to one's partner; monogamous voles became polygynous if vasopressor receptor expression pattern was altered to that of polygynous species, and vice versa. The difference in gene expression is caused by a microsatellite repeat upstream of the receptor gene. However, Ophir and colleagues (2008) showed that intraspecific variation in monogamy was not associated with the same difference in the distribution of receptors (chapter 8), and Fink and colleagues (2006) showed that there was little intra- or interspecific variation in the candidate upstream regulatory region (reviewed in Young & Hammock 2007). Variation in monogamy within and among vole species cannot thus be attributed to this single locus. Similarly, while mutagenesis studies have identified genes in *D. melanogaster* that disrupt mate choice (*fruitless*, chapter 5; *desat1*, Grillet et al. 2006) these have not been associated with preference variation within or among natural populations. A similar caveat applies to studies of differential gene expression (reviewed in Wilkinson et al. 2015). While it is illuminating to identify the suite of genes that are up- or down-regulated in particular mate-choice contexts, differential expression by itself tells us little about heritable differences unless combined with other data, like expression QTL (see above).

10.7 SYNTHESIS

The genetic basis of variation is at the heart of understanding the evolution of mate choice and mating preferences. Characterizing the genetic basis of complex, ecologically interesting traits is always a difficult task, but the peculiarities of mate choice make it a special challenge (Chenoweth & Blows 2006). In addition to the fact that a complex chain of sensory and neural mechanisms are recruited for premating choice, the phenotypes they produce are only expressed with reference to factors outside the organism. By definition, we can only measure preferences with respect to a particular set of external stimuli, and the choice of that set of stimuli has a dispositive effect on basic metrics like the heritability of stimuli. Add to that the fact that preferences have additional dimensions created by environmental and social variation (next chapters) and the task of doing genetics on mate choice seems impossibly daunting.

A detailed focus on proximate mechanisms may be the best way to go about managing this complexity. Understanding the sensory basis of pheromone-based mate choice in moths, for example, allows us to focus on pheromone receptors as a locus of preference evolution in that system, and to characterize their functional properties in isolation and independent of the distribution of courters. As the preceding chapters show, we are beginning to understand a great deal about the sensory, perceptual, and endocrine underpinnings of mate choice, and to correlate gene expression profiles with preference. If variation in preference and choice is reliably correlated with measurable physiological, neuroanatomical, or morphological substrates, we can do genetics in a way that is independent of the phenotypes of courters, natural or engineered. We can then go back and correlate functional variation with variation in realized preferences and choices. The nonneural mechanisms involved in cryptic choice, like the chemistry of the reproductive tract or the evolutionary developmental biology of female genitalia, may represent the most promising entry point here.

Models of preference evolution depend critically on the assumptions they make about genetic architecture and about genetic variance and covariance. The scant data available suggest a range of possibilities for the genetic architecture underlying preferences, but—with the important exception of Z-linkage of preferences in female-heterogametic species—we have far too little data on preference genetics to be able to make broad generalizations. A crucial question, and one not answered by focusing on the relationship between preference and contemporary covariation in courter traits, is the extent to which there is heritable variation in hidden preferences that may respond to novel phenotypes when the distribution of courter traits changes, when communication is altered, or when novel traits emerge.

Measures of genetic variation in preferences run a broad gamut with a mode of zero, but much of this range can be accounted for by variation in what is being measured. The heritability of realized mate choice in the wild will always deviate from the heritability of experimentally measured preferences in the laboratory. Mate sampling, mate availability, and covariance between courter traits will all reduce the heritability of the former relative to the latter. On the other hand, heritable mating outcomes may reflect mutual mate choice (chapter 8) or postmating processes (chapter 7) without heritable behavioral preferences.

Finally, preferences shift according to dynamic environmental and social contexts (e.g., Chaine & Lyon 2008). These have the overall effect of adding to environmental variance and therefore reducing the heritability of preference and choice; as we shall see in the next chapters, however, they often do so in measurable and predictable ways that can be incorporated into models of preference evolution.

10.8 ADDITIONAL READING

Chenoweth, S. F., & Blows, M. W. 2006. Dissecting the complex genetic basis of mate choice. *Nature Reviews Genetics, 7*(9), 681–692.

Roff, D. A., & Fairbairn, D. J. 2015. Bias in the heritability of preference and its potential impact on the evolution of mate choice. *Heredity, 114*(4), 404–412.

Wilkinson, G. S., Breden, F., Mank, J. E., Ritchie, M. G., Higginson, A. D., Radwan, J., Jaquiery, J., Salzburger, W., Arriero, E., Barribeau, S. M., Phillips, P. C., Renn, S.C.P., & Rowe, L. 2015. The locus of sexual selection: moving sexual selection studies into the post-genomics era. *Journal of Evolutionary Biology, 28*(4), 739–755. doi: 10.1111/jeb.12621

CHAPTER 11

Variation II: Biotic and Abiotic Environment

11.1 INTRODUCTION

In this chapter and the next, I focus on how preferences vary within and among choosers as a function of the environment. This chapter focuses on the physical environment and on the biotic community in which choosers live: the resources choosers acquire and the challenges toward which they must allocate those resources. This chapter focuses on how the environment shapes preferences; environmental effects on transmission and perception of courter signals are amply covered elsewhere (Bradbury & Vehrencamp 2011; chapter 3). The effect of the social environment, construed to include closely related species with which mating is possible, is covered in the next chapter.

It is useful to divide environmental effects between those that affect the immediate context of mating decisions and those that change the chooser's phenotype in a way that affects future mating decisions. So-called context-dependent effects fall into two categories. **Context-sensitive** preferences are shaped by immediate circumstances: the sensory environment, predators, other choosers, and most obviously the courters themselves. When preferences are **state-dependent**, a chooser's past history shapes current mating decisions: for example, developmental effects of the sensory environment or social history. Condition (chapter 9), which may vary according to nutritional state, pathogen infection, or social stress, is an important correlate, and likely a driver, of state-dependent preferences.

11.2 CONTEXT-SENSITIVE EFFECTS

In chapter 4, I described how male codling moths respond sexually to a female pheromone cue only if it is paired with an appropriate habitat cue. This is an example of a context-sensitive effect on mate choice. Context-sensitive effects on preferences properly include the consequences of mate sampling and working memory (chapter 6) and mate copying (chapter 12). One of the biggest environmental sources of context-sensitive preference variation is likely the sensory environment (chapter 3). Reception of signals depends on the physics of the transmission medium, the amount and nature of noise, and, for vision, the distribution of incident light energy. For example, non-sexual color preferences in bluefin killifish vary according to diurnal changes in light quality (Johnson et al. 2013). Signal reception and integration also depend on receiver attention and on the spatial arrangement of courters and choosers relative to background (Rosenthal 2007). Increasing the difficulty of perceptual tasks makes receivers less choosy due to reduced detectability or discriminability of stimuli (e.g., Seehausen et al. 1997; Fisher, Wong et al. 2006).

The physical environment can also directly affect receiver properties. In terrestrial ectotherms, body temperature can change markedly according to time of day and microhabitat, resulting in changes in both signal production and chooser preference. In female green treefrogs, *Hyla cinerea*, a 5°C change in testing temperature causes a reversal in preference for frequency characteristics of the lower-frequency component of male calls (components which themselves are shifted higher with increasing temperature). Females at cooler temperatures prefer lower-frequency call components, while warmer females prefer higher-frequency ones (Gerhardt & Mudry 1980).

The physical environment can provide a cue for learned hedonic marking of courter stimuli. Beaulieu and Sockman (2012) played arbitrary sets of male song to female Lincoln's sparrows (*Zonotrichia leucophrys*) at their normal housing temperature (16°C) and under a thermal challenge (1°C) approximating the temperature of dawn choruses under which songs are normally experienced. A day later, females were all tested at 16°C, but showed a marked preference for the songs learned at the colder temperature. The hedonic marking of learned songs as attractive or unattractive (chapter 5), as distinct from the immediate context of stimulus evaluation, was influenced by a physical variable.

As we saw in chapter 6, travel costs play a determining role in modulating choosiness (Booksmythe et al. 2008). One of a few generalizations we can make about environmental effects on mate choice is that increasing the costs of mate searching (chapter 6), like predation risk, almost always make choosers sample fewer males before mating and therefore less choosy (Berglund 1993; Forsgren 1992; Godin and Briggs 1996; Rand et al. 1997; Willis et al. 2012; Kozak & Boughman 2015). Choosers who exhibit strong preferences in a low-risk environment show random preferences in a high-risk environment.

The valence of preference can also be risk-sensitive. Crickets (Hedrick and Dill 1993) and swordtails (Pilakouta & Alonzo 2014) reverse their preferences under increased predation risk, toward less-conspicuous signalers who may incur less immediate risk of predation (Rosenthal et al. 2001; chapter 14).

11.3 STATE-DEPENDENT PREFERENCES

11.3.1 Organizational effects of the environment

Environmental inputs can shape preferences over long timescales. At the other extreme from the context-dependent effects described above, the environment choosers experience during early life can shape their mating decisions as adults. I will return to the early nutritional environment below. As described in chapter 3, the sensory environment can shape peripheral sensitivity during ontogeny, although to date there is scant evidence that environment-induced sensory plasticity directly shapes mate choice (but see Fuller & Noa 2010 for a complex three-way interaction between genotype, rearing environment, and testing environment).

Human-induced alteration of the environment introduces new variables that can affect mate choice, whether early in its development or later in life. In addition to habitat alteration that can affect sensation and mate searching (reviewed in Rosenthal & Stuart-Fox 2012), anthropogenic chemicals can directly affect mate choice; for example, Ward and Blum (2012) showed that exposure to the environmental estrogen bisphenol-A acted to weaken preference for conspecific mates between two species of North American minnows (*Cyprinella*). Aromatase-knockout mice, which cannot produce estrogen, fail to exhibit preferences or proceptive behavior even when treated as adults; this effect is abolished by treatment with estrogen early in life (Brock et al. 2011). Social and environmental effects on developmental hormone profiles may thus contribute substantially to preference variation.

11.3.2 Seasonal and interannual variation

Chooser mating decisions often change over the course of a breeding season (Milner et al. 2010; Qvarnström et al. 2000; Borg et al. 2006; Passos et al. 2014); for example, Myhre and colleagues (2012) found that female two-spotted gobies spent less time mate-sampling, were more likely to initiate courtship, and were less likely to terminate courtship at the end of the breeding season than at the beginning. These changes are often attributed to changes in encounter rate or in the nature of interactions with courters, but there is some evidence that physical variables associated with seasonality can themselves drive changes in preference. Female blue tits (*Cyanistes caeruleus*) are more responsive and more choosy when photoperiod is experimentally increased, expressing a preference for male morphological traits only under long photoperiods (Reparaz et al. 2014).

Recent studies have demonstrated dramatic year-to-year changes in mean chooser preferences in wild populations. Lehtonen and colleagues (2010) found that female sand gobies changed and even reversed their experimentally measured preferences for male body size and nest cover from year to year. Chaine and Lyon (2008) inferred female preferences from the mating success of male lark buntings over successive years, documenting social mate choice from a large sample of observed pairings and characterizing parentage. Multivariate analysis revealed these birds changed and often reversed their preferences across years (fig.11.1); by contrast, traits involved in male-male competition were stable over time in terms of their contribution to male fitness. Environmental variables likely contribute to this year-to-year variation; across species of birds, rates of extrapair paternity and divorce were higher in climatically more variable and unstable habitats, suggesting that preferences may correspondingly be more temporally variable. Botero and Rubenstein (2012) raised the concern that climate change could thereby drive changes to mate choice and pair bonding across bird species. These changes could have far-reaching effects, particularly if they lead to increased rates of interspecific hybridization (chapter 16).

11.3.3 Condition-dependence and nutritional effects

The concept of condition (chapter 9) is important to thinking about state-dependent preferences. The evolution of condition-dependent mate choice and of preferences for condition-dependent chooser traits is discussed in later chapters. Condition is shorthand for the resources an organism has to spare for reproduction, including mate choice; it can be best approximated empirically by measuring fat or other metabolic reserves, but has been applied loosely to a broad range of measures (Clancey & Byers 2014). Condition is positively associated with diet quantity and balance, and negatively associated with immune challenges from disease or parasitism and with physiological or social stress. I will return to these latter effects below in the context of interspecific and social interactions.

A number of studies have manipulated variables that affect condition, notably nutritional state, and shown effects both on condition and female preferences (Hebets et al. 2008; Lerch et al. 2011). An influential review by

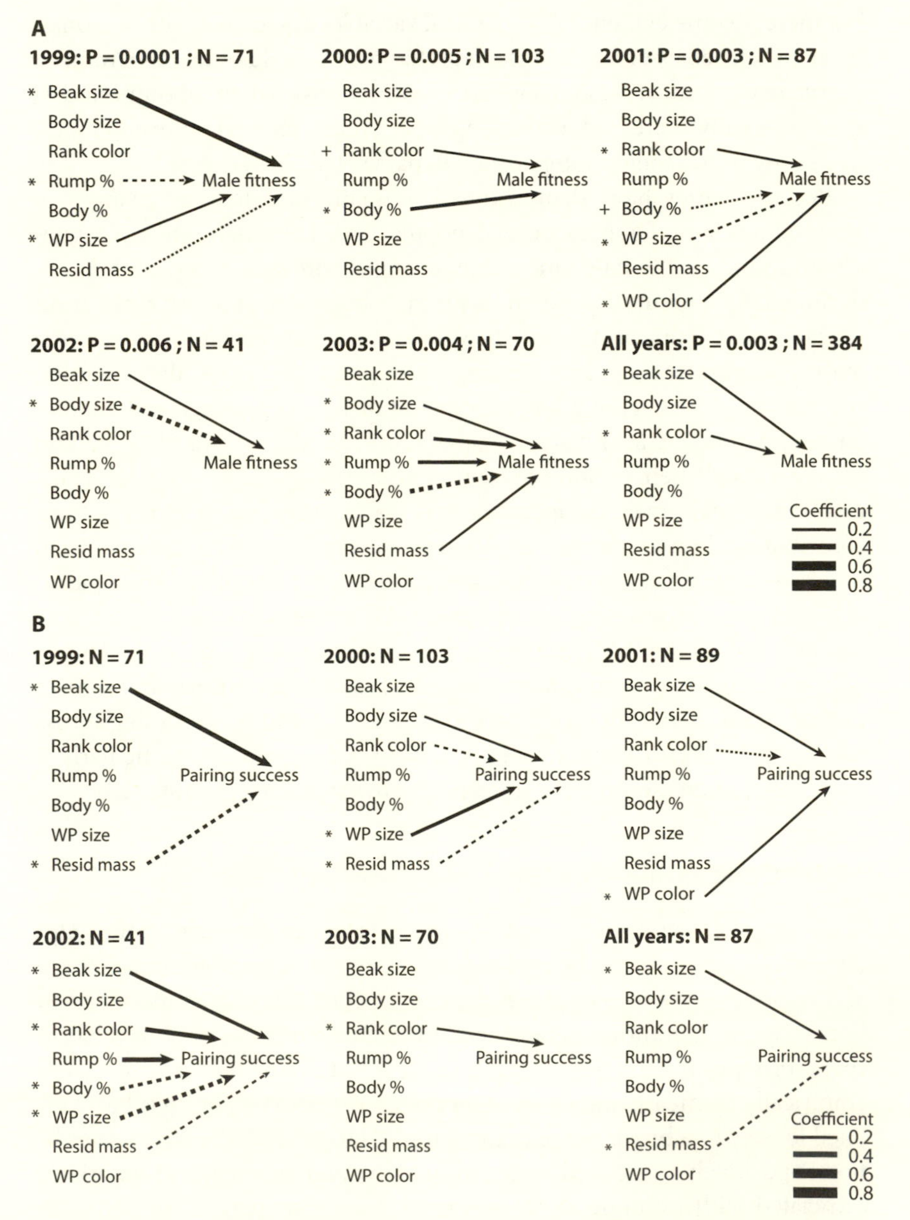

Figure 11.1. (a) Variation across years in male traits predictive of offspring fitness (number of within- and extra-pair chicks fledging) in a wild population of lark buntings (*Calamospiza melanocorys*). Thickness of arrows indicates strength of correlation; dashed arrows show negative relationships. Final panel shows cumulative relationship across years. WP, wing patch; %, percent black feathers. (b) Variation across years in male traits predictive of pairing success, with notation as in (a). After Chaine & Lyon (2008).

Cotton, Rogers, and colleagues (2006) concluded that choosers in poorer condition are less choosy, as expected if mate choice is harder for poor-quality choosers to afford.

In some cases, there are candidate mechanistic links between choosiness and a measure of condition, as with stalk-eyed flies, where better-condition females have larger eye spans and might therefore have better visual acuity to discriminate among courter traits (Cotton, Small, et al. 2006). It is noteworthy that these nutritional effects can be specific to mate choice: Woodgate and colleagues (2010) found that female zebra finches that had been nutritionally stressed during ontogeny were less active during mate-choice trials and made fewer sampling visits to stimulus males, but were equally active in isolation and in dyadic interactions. However, reduced responsiveness did not translate to reduced choosiness: females showed no difference in preferences for more complex song types (Woodgate et al. 2011). Similarly, Judge and colleagues (2014) found no effect of nutritional condition on preferences in field crickets, *Gryllus pennsylvanicus*, but did find that high-condition females took longer to make a choice and were more likely to go through a trial without making a choice; they interpreted this as reflecting greater choosiness, although alternatively, high-condition females may have been less responsive.

The accumulating number of studies demonstrating positive condition-dependence of choosiness has been balanced, particularly since Cotton, Rogers, et al. (2006), by a number of studies showing the opposite: poorer-condition females exhibiting stronger preferences, as expected if choosing is more consequential for poor-condition females (Ortigosa & Rowe 2002; Fisher & Rosenthal 2006a; Eraly et al. 2009; Perry et al. 2009; Griggio & Hoi 2010). In *D. subobscura*, males provide a direct benefit to females by vomiting a drop of a liquid nuptial gift; males in good condition provide more gifts. Preference for good-condition males is strongest in poor-condition females (Immonen et al. 2009).

There are also some plausible low-level mechanisms that could positively couple the motivation to feed with choosiness. For example, hunger—mediated by nutrient titers in the blood—increases the peripheral sensitivity of taste receptors in grasshoppers (Abisgold & Simpson 1988), which could make hungry females more sensitive to non-nutrient cues as well. Hunger could also directly increase sensitivity to food cues produced or proffered by courters, such as edible spermatophores, or to food-mimicking cues (Amcoff et al. 2013).

We would benefit from further studies of the sign (positive or negative) of condition-dependent choosiness in a phylogenetic context, especially as there may be confirmation bias in favor of publishing data that shows a significant relationship to condition. As with "attractiveness" and "quality," to which I will return in chapter 14, "condition" is a loaded and slippery term (Clancey & Byers 2014). In the next section, I will discuss how changes in nutrition and parasite load can change preferences independent of condition.

We must also be mindful of how we measure mate choice, since we may come to very different conclusions depending on what we measure in choosers. Syriatowicz and Brooks (2004) found that food deprivation reduced sexual responsiveness in female guppies, but did not affect preference functions. Female Syrian hamsters (*Mesocricetus auratus*) on a restricted diet performed less vaginal scent marking and spent less time with males (and more time searching for food) compared to *ad libitum* control. Surprisingly, however, they showed the same rates of lordosis (copulation solicitation) and the same learned preferences for test chambers in which mating had previously occurred (Klingerman et al. 2011). As with predation risk, discussed above, sufficiently depressed condition may shut down sexual behavior altogether. In Syrian hamsters, longer-term food deprivation disrupts estrus and shuts down sexual behavior, but estrus and proceptive behavior are restored by administration of the hormone leptin (Schneider et al. 2007). In vertebrates, leptin production and reception might be an important modulator of hunger effects on choosiness.

Choice of assay and stimuli can even determine whether choosiness is positively or negatively associated with condition. Lyons and colleagues (2014) found that female *Xiphophorus multilineatus* raised on high-protein diets had stronger preferences for male vertical bar symmetry, but weaker preferences for male body size. This may be because there is a mechanistic trade-off between assessing body size and assessing symmetry, or, as argued by Lyons and colleagues (2014), because these two preferences are under different selective pressures. Regardless, this study shows that condition can both increase and decrease choosiness within the same individual depending on the trait being evaluated.

Condition-dependent preferences may promote assortative mating (chapter 6). Size-adjusted tarsus length, a frequently (and controversially, Clancey & Byers 2014) used avian proxy for condition, covaries negatively with choosiness for badge size in house sparrows. Below-average females are more choosy, preferring males with averaged-sized over enlarged badges; above-average females have a nonsignificant trend in the opposite direction (Griggio & Hoi 2010). A series of studies has used experimental manipulations to examine condition-dependent preferences of female zebra finches for male songs. Riebel and colleagues (2009) manipulated chooser state by raising birds in randomly assigned large or small broods, with birds from large broods receiving poorer nutrition during ontogeny and consequently having poorer condition. Consistent with the predictions of Cotton, Rogers, and colleagues (2006), preferences were stronger in the small-brood, better-condition females. Holveck and Riebel (2010) then used the same paradigm to manipulate the condition of both females and males. Here, females showed homotypic preferences (chapter 8) based on their own state with poor-condition females preferring the songs of unfamiliar poor-condition males (fig. 11.2); females also showed the same pattern in live, interactive tests (Holveck et al. 2011). Females may be using an assessment of their own internal state to adjust their preferences based on perceived market value (chapter 6). Burley and Foster (2006) experimentally and temporarily reduced flight performance in female zebra finches, and found that females had reduced choosiness for a novel preferred trait, red over green artificial leg bands. Since mutual pairings occurred in a confined setting where flight was limited, the authors argued that this reduction in selectivity was due to self-assessment by females rather than to male assessments of chooser performance (Burley & Foster 2006).

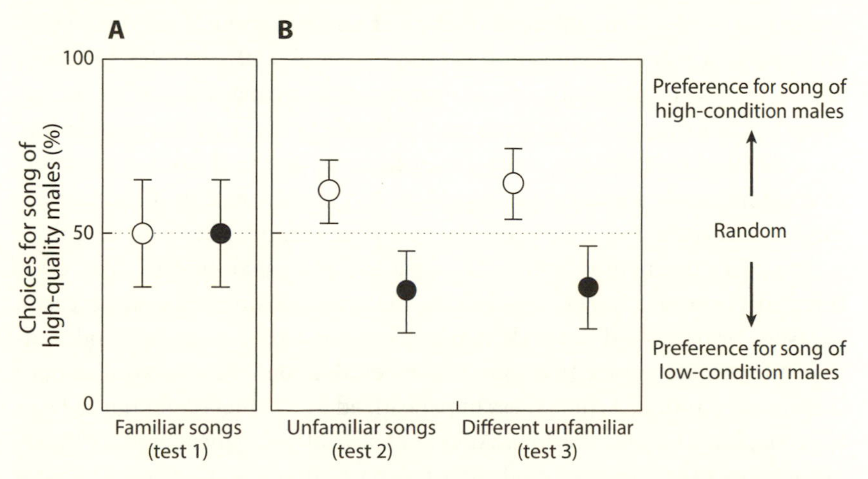

Figure 11.2. Preferences of females raised in large clutches (poor condition, filled circles) and females raised in small clutches (good condition, open circles) for high- and low-condition males. Females showed no difference in baseline preference for familiar songs, but show homotypic preference for condition when presented with unfamiliar songs. After Holbeck & Riebel (2010).

11.3.4 Diet effects beyond condition

Specific nutrients can influence preferences in ways that are distinct from overall condition. Grether et al. (2005) manipulated the concentration of carotenoids in the diet of female guppies, finding that low-carotenoid females have heterotypic preferences for high-carotenoid males (fig.11.3). A pair of studies on swordtail characins (*Corynopoma riisei*) suggests that learning about food can also influence preference independent of condition, if choosers use the same sensory mechanisms for food cues and sexual cues. Swordtail characins experienced with red-colored food items more strongly preferred males with artificially red-colored ornaments than did females that had only experienced standard green-colored food (Amcoff & Kolm 2013; Amcoff et al. 2013; chapter 13).

11.3.5 Parasitoids, parasites, and the microbiome

Parasitoids, pathogens, parasites, and endosymbionts are all part of a continuum of intimate associations between an individual and other species. These commensals, as we can call them for convenience, have different labels depending on how their interests align with one another. They differ in the extent to which the interests of chooser and commensal align, and play an important role in sexual conflict (Hayashi et al. 2007; chapter 15). Mate choice is both a mechanism for commensals to propagate, and a mechanism for choosers to buffer costs and enhance benefits of the association. As with such associations more generally, coevolution between choosers and commensals could be an important force in evolution. Commensal-mediated variation in mate choice has only been addressed very recently.

When a commensal acts to increase the marginal cost of choice for infected choosers, infection tends to reduce choosiness. In these cases, parasite-, or parasitoid-dependent, mate choice is a special case of condition-dependent mate choice (Cotton, Rogers, et al. 2006). Several studies have documented reduced choosiness in parasitized individuals (Pfennig & Tinsley 2002; López 1999; Córdoba-Aguilar et al. 2003; Kavaliers et al. 1998; reviewed in Cotton, Rogers, et al. 2006). A particularly stark example comes from parasitoids, which oviposit into hosts; the larvae consume and kill the host upon hatching. Beckers and Wagner (2013) found that female crickets *Gryllus lineaticeps* infected with fly larvae chose at random; they argued that infected females were under time constraint before being killed by the growing larvae.

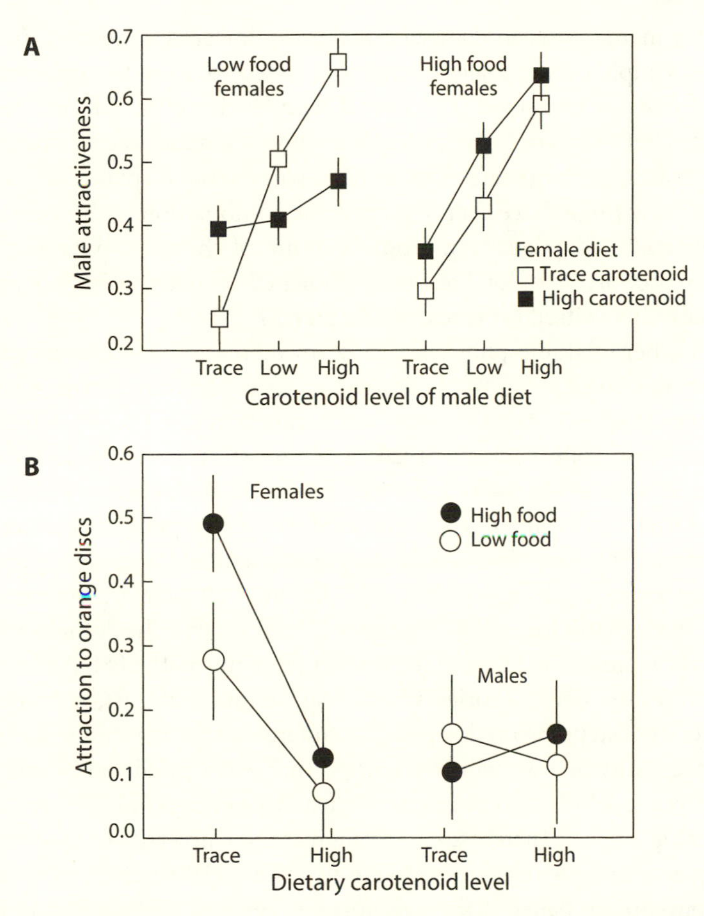

Figure 11.3. Preferences (least-squares means from ANOVA) of females guppies maintained at two carotenoid levels (trace/high) and two levels of food availability (low/high) for males maintained along a gradient of carotenoid levels, with "high" males most colorful. Females on a trace carotenoid diet show stronger preferences at low food concentrations. (b) Females, but not males, show evidence of a sensory bias for orange when raised on a trace, low-food diet. After Grether et al. (2005).

Parasitized individuals are not always less choosy. For example, sexual behavior of female crickets *Gryllus texensis* is not affected by infection with an iridophore virus, while infected males actually increase sexual activity (Adamo et al. 2014). Zuk et al. (1998) found that parasitism did not affect mate choice despite its effects on morphology in jungle fowl. Buchholz (2004) measured the effect of infection on multiple aspects of preference and mate sampling. In wild turkeys, infection with intestinal protozoan parasites incurs costs to choosers, including lowered condition. Infected females sampled more males overall, but failed to show the preference of control females for snood (ornament) length. However, infected females did not differ from controls in latency to choose a mate, in time spent evaluating mates, or in the rate of copulation solicitation displays. A less scrupulous study might have cherry-picked any combination of these results to fit a particular adaptive story. From the point of choosers, parasitism may alter the proportion of total resources allocated to mate choice, not just the way mate choice itself is expressed (chapter 14).

A number of studies, however, show an increase in choosiness parasitized females. In humans, women's preferences for healthy partners, and for cues associated with health, covary with infection risk between countries; DeBruine and colleagues (2010, 2012; Moore et al. 2013) made the argument that choosers at risk of infection should be more choosy with respect to traits that would buffer or minimize infection. Women prefer more masculinized faces in countries with higher mortality and lower life expectancy, pointing to environmental and social influences on facial preferences (DeBruine et al. 2010). Jones and colleagues (2013) assayed the domain-specific disgust of young, white female students at the University of St. Andrews for 21 activities in three categories: sexual disgust, moral disgust, and pathogen disgust (e.g., "stepping on dog poop"). Preferences for masculine faces, and the masculinity of a woman's current partner, were predicted by pathogen disgust but not by sexual or moral disgust. This effect could be modulated by priming subjects (heterosexual women at the University of Queensland) with questionnaires related to disease risk, financial security, or a control questionnaire on belief in the paranormal. Lee and Zietsch (2011) found that subjects preferred more "masculine" traits (which they termed "good genes" traits) in a verbal context when primed for pathogens and more "good-dad" traits (intelligence, creativity) when primed for resource scarcity.

As with nutrition-dependent changes in preference, it would be valuable to study the extent and direction of parasite-mediated changes in choosiness in a phylogenetic context. The situation with commensals is more complicated, however, because the fitness costs of association are very labile; an association can go from symbiotic to parasitic depending on context. Further, commensals can directly manipulate chooser phenotypes. The intracellular bacterium *Wolbachia* illustrates both these points; *Wolbachia* is an endosymbiont within *Drosophila* lineages, but a pathogen in hybrids. *Wolbachia* races therefore mediate reproductive isolation in *Drosophila*. Artificial populations of *Drosophila* mate assortatively to their own *Wolbachia* race, but assortative mating is reduced when flies are put on antibiotics;

crucially, this pattern is maintained for multiple generations, suggesting that absence of the symbiont, not antibiotic treatment, is weakening assortative mating (Koukou et al. 2006; Miller et al. 2010). Koukou et al. (2006) pointed to changes in pheromone profiles as a candidate mechanism, while Miller et al. (2010) speculated that endocellular *Wolbachia* could directly affect responses of the antennal lobe of the brain. A similar result obtains when flies are fed diets that harbor different communities of commensal bacteria, with antibiotic treatment abolishing preference; cruciallly, preferences are restored after reinfection (Sharon et al. 2010; fig. 11.4). Finally, parasites can hijack mate-choice mechanisms for functions unrelated to mating; for example, *Toxoplasma gondii* make their rat hosts sexually attracted to predators (chapter 5).

11.4 GENOTYPE-BY-ENVIRONMENT INTERACTIONS

The responses of preferences to the environment are phenotypes that can evolve. A genetic mutation can enhance, attenuate, eliminate, or reverse the relationship between a preference measure and an environmental variable. Such **genotype-by-environment interactions** (G X E or GEIs) are ubiquitous, and are important targets of selection on preferences (chapter 13); by introducing heterogeneity into mating decisions, they can maintain variation in courter traits (chapter 15; Ingleby et al. 2010; Narraway et al. 2010).

In chapter 10, I introduced the partition of phenotypic variance into genetic and environmental components. A more realistic way to model phenotypic variance, however, is to introduce the variance due to G X E effects:

$$V_p = V_g + V_e + V_{G X E}$$

where V_g is total genetic variance (additive plus nonadditive; chapter 10), V_e is the variance due to environmental factors, and $V_{G X E}$ is variance due to genotype versus environment interactions—in other words, genotypic effects that are different in different environments, which can also be thought of as environmental effects that manifest themselves differently in different genotypes. For details on how to compute components of phenotypic variance, see Lynch and Walsh (1998). We can further decompose V_e and $V_{G X E}$ according to their environmental sources, for example variance due to temperature, variance due to parasite load, etc. For biotic variables, like courter signals, we can use indirect genetic effects (IGEs; Bailey 2012) to model effects of other individuals' phenotypes on the phenotypic variation in courters. Rebar and Rodriguez (2014) showed that interspecific IGEs can

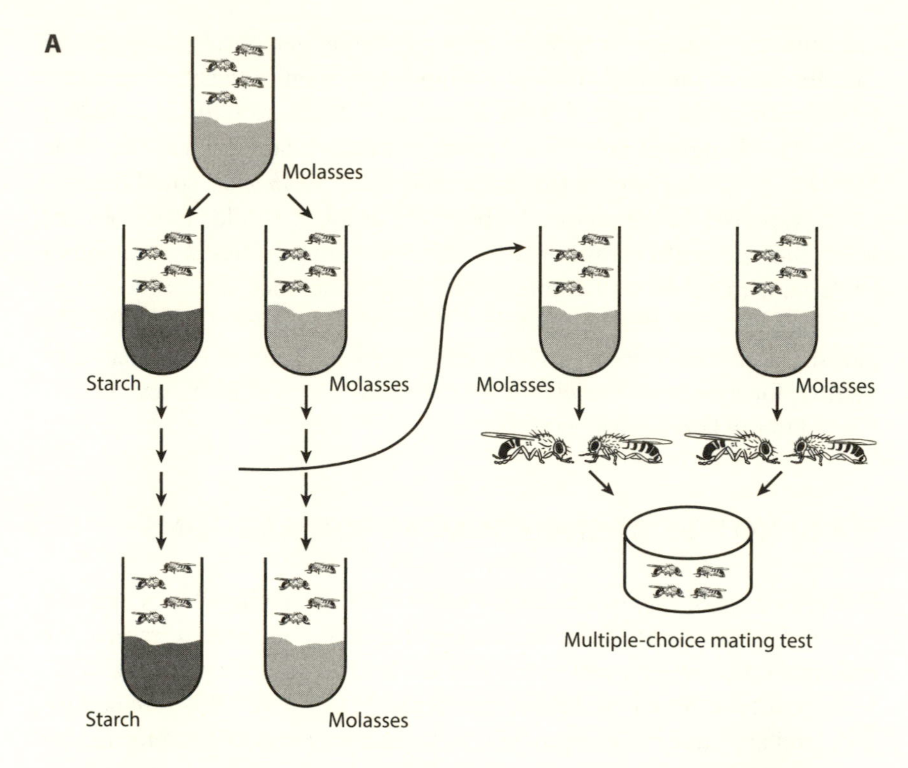

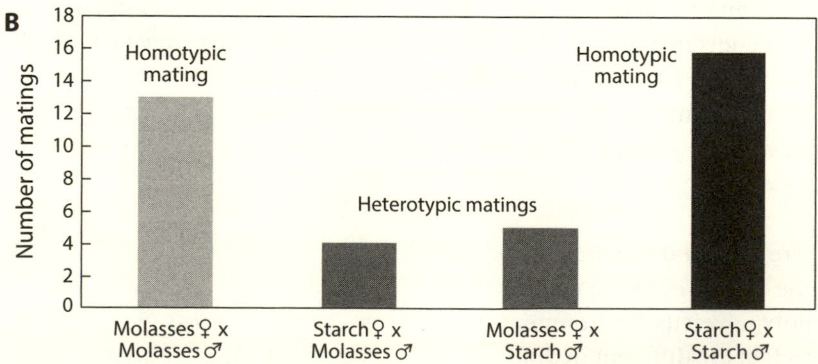

Figure 11.4. (A) Schematic of Sharon and colleagues' (2010) experiment testing the effect of diet-associated microbial meta-communities on mate choice. Flies were reared for 11 generations on either starch or CMY molasses medium, then fed for one generation on molasses to eliminate direct diet effects, then assayed in four-fly groups of one male and one female of each type. (B) Homotypic matings were much more frequent than heterotypic matings; crucially, this effect disappeared after treatment with antibiotics.

markedly affect preference functions in *Enchenopa* treehoppers (fig. 11.5). The IGE approach is particularly relevant to maternal and social effects, and I will return to it in the next chapter.

In the last chapter, I described approaches to modeling preferences as function-valued traits, varying depending on one or more stimulus variables (McGuigan et al. 2008). Here, we add another set of dimensions to preference variation, those associated with environmental variables (Rodríguez, Hallett, et al. 2013). The response of a given genotype over a range of conditions is its **reaction norm**. These conditions can include internal changes like age or reproductive state, as well as external environmental variation. For that matter, the term "reaction norm" can be applied to different chooser responses across a set of courter stimuli. Just as different genotypes can show reversals in preference (chapter 10), reaction norms for preference can undergo so-called *ecological crossover*, with environmental effects acting in opposite directions across genotypes (fig. 11.6). Genotype by environment effects have been detected in several insect studies with respect to temperature (Rodriguez & Greenfield 2003; Ingleby et al. 2013), rearing environment (Narraway et al. 2010; Rodriguez, Hallett, et al. 2013), and social experience (Rebar & Rodríguez 2013; chapter 12). Ecological crossover—reversal of environmental effects with different genotypes—may be quite common. There can therefore be intraspecific genetic variation for the sign of condition-dependent mate choice (see above).

Complementing G X Es is the fact that distinct environmental factors can interact. For example, context and state effects combine to shape the response of mating preferences to predators in guppies. Wild-caught (and therefore predator-experienced) females were less sensitive to predator exposure than lab-reared, predator-naïve females who reduced their choosiness for larger males (Bierbach et al. 2011). Age and diet interact to affect latency to copulate in female wolf spiders *Rabidosa rabisa* (Wilgers & Hebets 2012). Divergent histories of selection can also impact context-dependent choices; for example, male *Brachyrhaphis episcopi* exhibit the widespread preference for novel females, but males from high-predation populations do so only when light levels are low (and perceived predation risk is presumably lower; Simcox et al. 2005). Though not explicitly tested, this pattern is consistent with a genotype (high- versus low-predator lines) by environment (low versus dim light) interaction.

Hybridizing spadefoot toads (*Spea bombifrons* X *S. multiplicata*) from the Sonora Desert of North America exhibit state X context X genotype-dependent preferences for conspecifics over heterospecifics (chapter 16). Female *S. multiplicata* always prefer the calls of conspecific males. Female

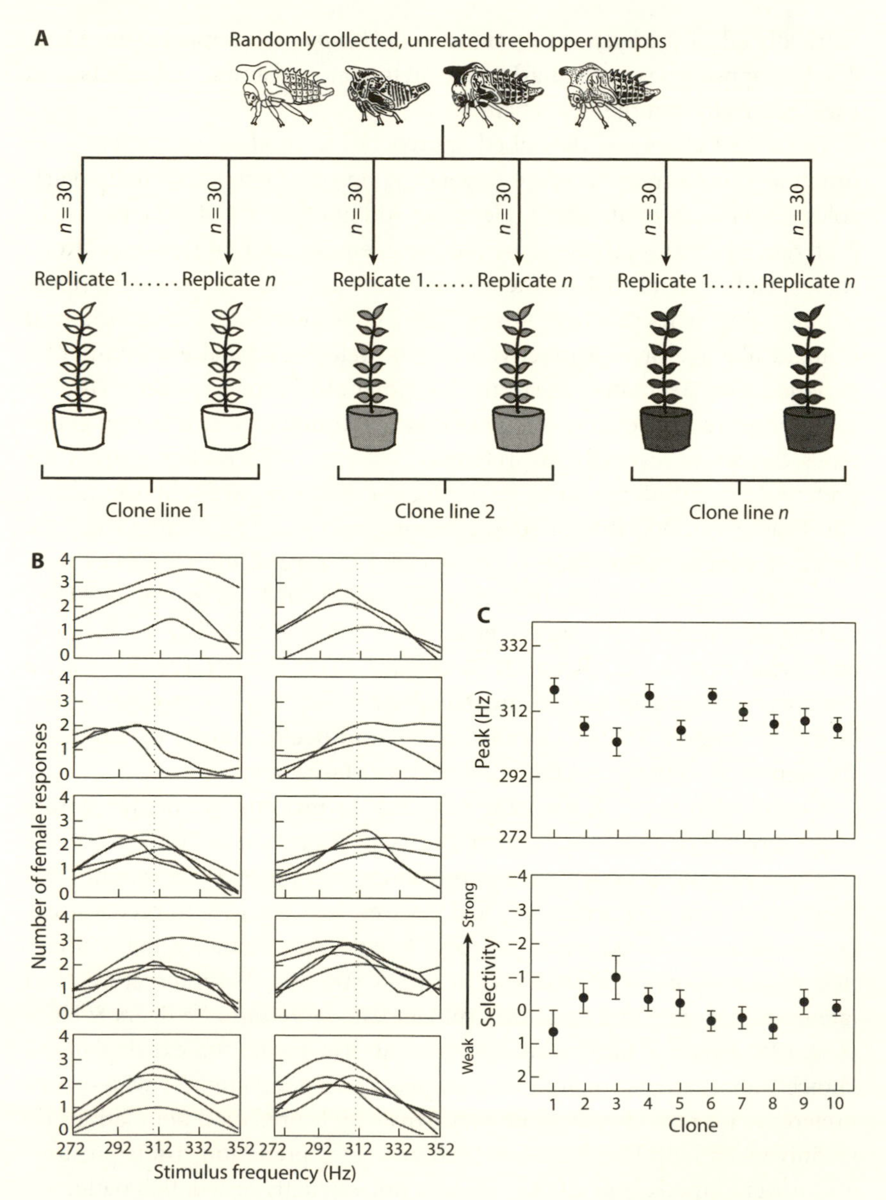

Figure 11.5. (A) Testing indirect genetic effects of plant genotype on preferences in *Enchenopa* treehoppers. Animals were reared on three replicate clones from each host line (wild-collected isolated patches of nannyberry, *Viburnum lentago*). (B) Mean female preference functions (measured by duetting responses) for male peak signal frequency. Panels depict different host lines, individual curves describe mean female preference for each host plant replicate. Dotted line represents mean peak preference. (C) Mean +/− SE for peak preference and choosiness across host plant genotypes. After Rebar & Rodriguez (2014).

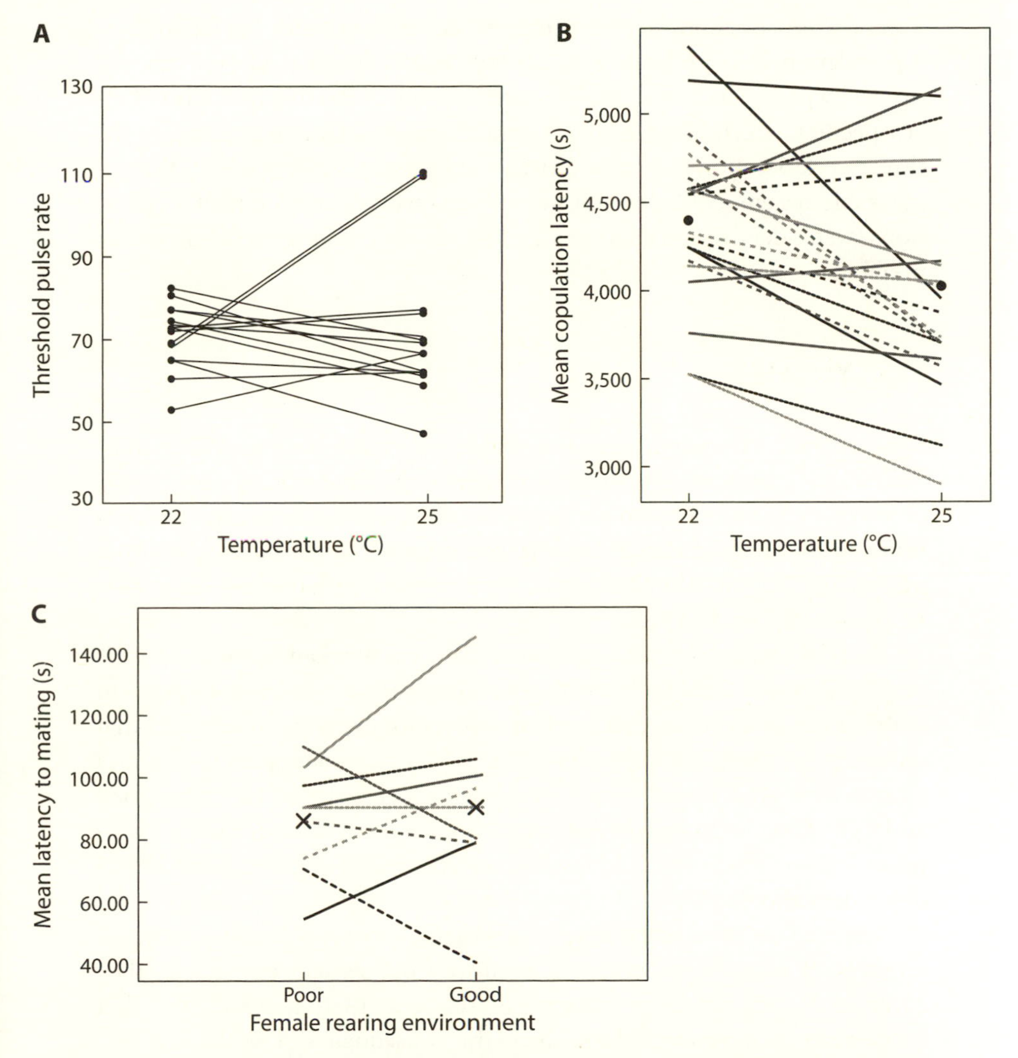

Figure 11.6. Genotype X environment interaction in preferences. (A) Threshold pulse rate for response to a conspecific signal in female lesser waxmoths (*Achroia grisella*). Each line represents mean responses of females from a distinct full-sib family, reared either at 22 or 25 degrees C (Rodriguez & Greenfield 2003). (B) Choosiness (latency to mate) of *Drosophila simulans* isolines as a function of temperature (Ingleby et al. 2013). (C) Latency to mate of *D. melanogaster* isolines as a function of rearing environment; "poor" females were given cold-shocks daily. X's indicate means across genotypes (Narraway et al. 2010).

S. bombifrons, by contrast, prefer conspecifics in deep water, but fail to show a population-level preference in shallow water. This loss of preference is driven by poor-condition females switching and preferring the calls of heterospecifics, which Pfennig (2007) argued was an adaptive response to the threat of desiccation to poorly provisioned eggs in shallow water. This remarkable system highlights how genotype (species), environmental conditions (water depth), and chooser state (condition) can interact to modulate preferences.

11.5 SYNTHESIS

Some of the environment's effects on mate choice are intuitive. When individuals are scared, hungry, or sick, or when they have difficulty perceiving courters—when the immediate task of mate sampling and mate choice is more costly—they are less choosy. It would be worth considering whether personality traits like boldness (chapter 9) covary with context-dependent responses to predation risk.

The relationship of condition to mate choice is more ambiguous. There is empirical evidence for both positively and negatively condition-dependent mate choice, and there are rich opportunities for experimental studies that quantify the condition-dependent fitness costs of mating decisions, and comparative studies that identify ecological correlates of the magnitude and sign of condition on choosiness. In chapter 14, I will return to the question of when selection should favor positively versus negatively condition-dependent choosiness.

A crucial question is the extent to which choosiness is tied to responsiveness. If individuals become less choosy but remain responsive, they continue mating but do so indiscriminately. An environmental effect that weakened choosiness would thus have the consequence of weakening sexual selection and promoting gene flow among divergent lineages. By contrast, if choosers are less choosy because they're less responsive, this would increase sexual selection if courters were competing for a smaller pool of choosers.

In some cases, loss of choosiness is in fact accompanied by a substantial reduction in responsiveness (Rand et al. 1997; Godin & Briggs 1996). By contrast, Berglund (1993) found that predator-exposed pipefish engaged in fewer matings, but transferred more gametes per mating. Manipulation of a hormonal mediator of stress in female green treefrogs also suggests that choosiness is decoupled from responsiveness. Corticosterone-injected females reduced their choosiness for higher call rates, but otherwise showed similar response latencies in phonotaxis experiments as control females (Davis & Leary 2015). In this and perhaps other cases, then, a mechanism exists to modulate the response of choosiness to environmental stressors, independent of responsiveness.

In addition to choosiness, environmental factors can change the shape of preferences, leading to reversals in the relative attractiveness of choosers depending on environment—and "meta-reversals" in the direction of the reversal according to genotype (fig. 11.6). These patterns speak to the importance of considering valence when thinking about mate choice (chapter 5), but they also raise the point that we may be missing subtler environmental effects on preference functions that we lack the power to detect.

Mate choice is a process that connects the broader environment to the exchange of genes between individuals and lineages. We are starting to gain an understanding of how environmental variables influence choosiness and preferences before mating, yet we know very little about environmental effects on cryptic choice. The condition-dependence of pregnancy and parental care outcomes suggest that postmating choice should take different forms depending on chooser condition, as I explore further in chapter 15. Early studies also show promise in elucidating the mechanisms that modulate environmental effects on preferences. Work in these areas will be helpful in predicting environmental impacts on mate choice and their resultant evolutionary consequences.

11.6 ADDITIONAL READING

Cotton, S., Small, J., & Pomiankowski, A. 2006. Sexual selection and condition-dependent mate preferences. *Current Biology, 16*(17), R755–R765.

Narraway, C., Hunt, J., Wedell, N., and Hosken, D. J. 2010. Genotype-by-environment interactions for female preference. *Journal of Evolutionary Biology, 23*(12), 2550–2557.

CHAPTER 12

Variation III: Social Environment and Epigenetics

12.1 INTRODUCTION

Mate choice is by definition a social interaction. In the immediate context of a mating decision, a chooser's behavior is influenced by the phenotype of the courters she experiences (chapter 6). But preferences are shaped by other individuals in a chooser's life, as well. The social processes that influence mating preferences are outlined in fig. 12.1. Even before birth, parents impart epigenetic markings, hormones, and resources that can affect preferences. Prior to sexual maturity, choosers interact with parents and other older individuals in ways that can have dispositive effects on preferences. Throughout adult life, experiences with courters and other choosers can act to influence mating decisions.

Indeed, until the resurrection of Darwin's "taste for the beautiful" in the latter third of the twentieth century, most of the thinking about variation in mate-choice mechanisms focused on their social transmission, a process often viewed, even to this day in some circles, as operating in contrast or to the exclusion of Darwinian evolution (Pinker 2003). The social environment's role in shaping mating decisions was foundational to Sigmund Freud's psychoanalytic theory. In Freud's (1899) view, sexual development incorporated the so-called "Oedipus complex." A key feature of the Oedipus complex, of course, was sexual attraction to the opposite-sex parent (which Freud argued was countered by experience with the same-sex parent later in childhood). Lorenz (1935) later identified the phenomenon of **sexual imprinting** in animals, whereby early experience with a stimulus leads to it being preferred in a sexual context (see below). The ensuing years produced scores of behavioral studies that yielded substantial insight on how preferences develop and how they depend on the nature of social experience, but it is only in the past few decades that this work has become integrated with the thinking on mate choice in behavioral ecology and evolutionary biology. It is now clear that the mechanisms whereby preferences are specified—whether they are purely genetic or epigenetic, whether they are transmitted from parents and siblings or from other individuals as well, and whether choosers learn to be stringent or permissive—can have fundamental effects on speciation and trait diversification (Owens et al. 1999; Price & Irwin 1999; Ihara & Feldman 2003; Verzijden et al. 2012; Bonduriansky & Day 2013; Yeh & Servedio 2015; chapter 16).

This chapter focuses on social interactions, in the broadest sense, as sources of variation in mate choice and mating preferences. These interactions can be divided into three categories corresponding to when they are specified and which individuals are involved (fig.12.1). The first includes effects that are determined before birth and transmitted **vertically** from parents: epigenetic modifications to the genome and the fetal or embryonic environment. The second includes influences between birth and sexual maturity, when the phenotypes of parents and/or other sexually mature, older individuals (**oblique** transmission) direct the development of preferences in choosers. Experience with courters and choosers after sexual maturity, or experience with other juveniles that shapes subsequent preferences, constitutes **peer** (*horizontal*) transmission.

12.2 SOCIAL EFFECTS BEFORE BIRTH: EPIGENETIC AND PARENTAL EFFECTS

12.2.1 Nongenetic effects: overview

Distinctive strings of DNA are only one of many things that parents transmit to their offspring. The rest is lumped collectively under the term **nongenetic inheritance** (Bonduriansky & Day 2009). Nongenetic inheritance provides a mechanism for environmental inputs to have long-term transgenerational consequences, and may represent an important source of variation in preferences and a mechanism for costly preferences to be maintained (Bonduriansky & Day 2009). This section focuses on cellular epigenetics and parental effects before birth, which is a slightly arbitrary dividing line. In particular, effects of the nutritional environment after birth, as through differential allocation (chapter 11), need not be markedly different in their effects on preferences from those before birth. Conversely, the transmission dynamics of nongenetic parental effects are the same as those of vertical cultural transmission (fig.12.1; section 12.3), but I cover social effects after birth in their own section since it is often difficult in practice to disentangle vertical from oblique effects (fig.12.1).

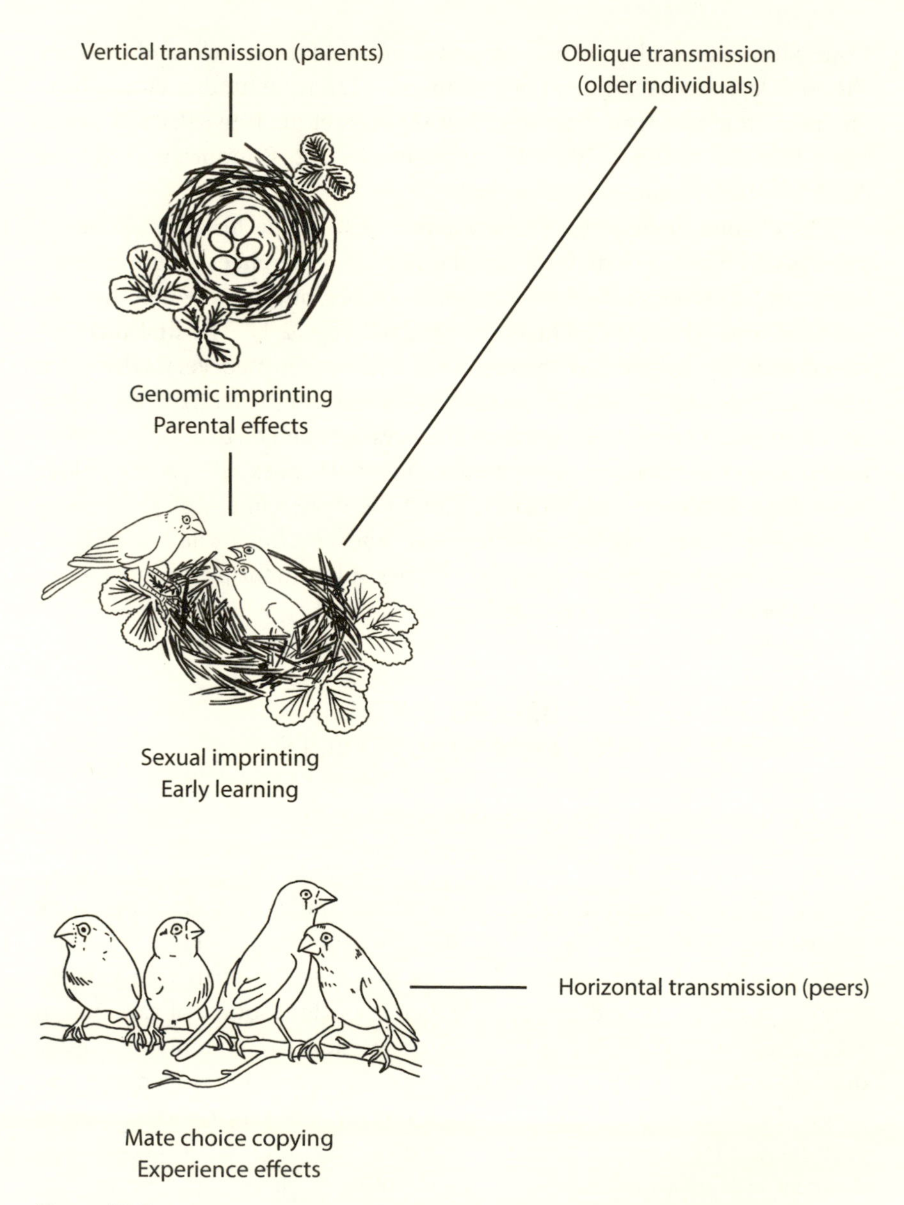

Figure 12.1. Schematic overview of parental and social effects on mating decisions.

12.2.2 Epigenetic marking

The term "**epigenetics**" has narrowed considerably in meaning over the past century. Originally applied to all non-genetic vertical influences on organismal development—which encompass both parental effects and the learned mating preferences discussed in the next section—epigenetics in the contemporary sense refers to cellular changes (or, more narrowly, chromosomal changes) in gene function that do not involve changes in DNA sequence, particularly epigenetic marking through DNA methylation (Bird 2007). Despite widespread evidence for behavioral effects of epigenetic marking, there has been little work on epigenetic effects on mate choice. In rats, exposure to the fungicide vinclozolin causes changes to the sperm epigenome that affect the preferences of female descendants for several generations; epigenetically marked females show stronger preferences for males from an unexposed control lineage. By contrast, there was no epigenetic effect on male traits (Crews et al. 2007).

12.2.3 Other parental effects

Maternal and paternal effects on phenotypes are ubiquitous (Bonduriansky & Day 2009), but only a handful of avian studies provide evidence for parental effects on preference. In chapter 7, we saw that choosers can exercise biases toward particular courters by varying the resources they put into gametes and parental care. These biases interact with the broader environment to shape preference; in the previous chapter, we saw that the stress associated with being in a large clutch acted to modify preferences (Riebel et al. 2009; Holveck & Riebel 2010). Here, social and environmental effects on preference blur together, since it is difficult to tease apart the direct effects of nutritional stress from what is presumably a diminished interaction with parents who have to provide for more mouths, and it is difficult to disentangle nutritional scarcity in the environment from allocation decisions by parents.

A parental effect distinct from nutrition is the allocation of sex hormones to developing embryos. For birds, as we saw in chapter 7, several factors covary with the order in which eggs are laid within a clutch: nutritional resources decline with laying order, as do egg androgen titers. Forstmeier and colleagues (2004) found in zebra finches that female choosiness was negatively correlated with laying order, independent of when the eggs had hatched (fig.12.2a). A contemporaneous study on the same species by Burley

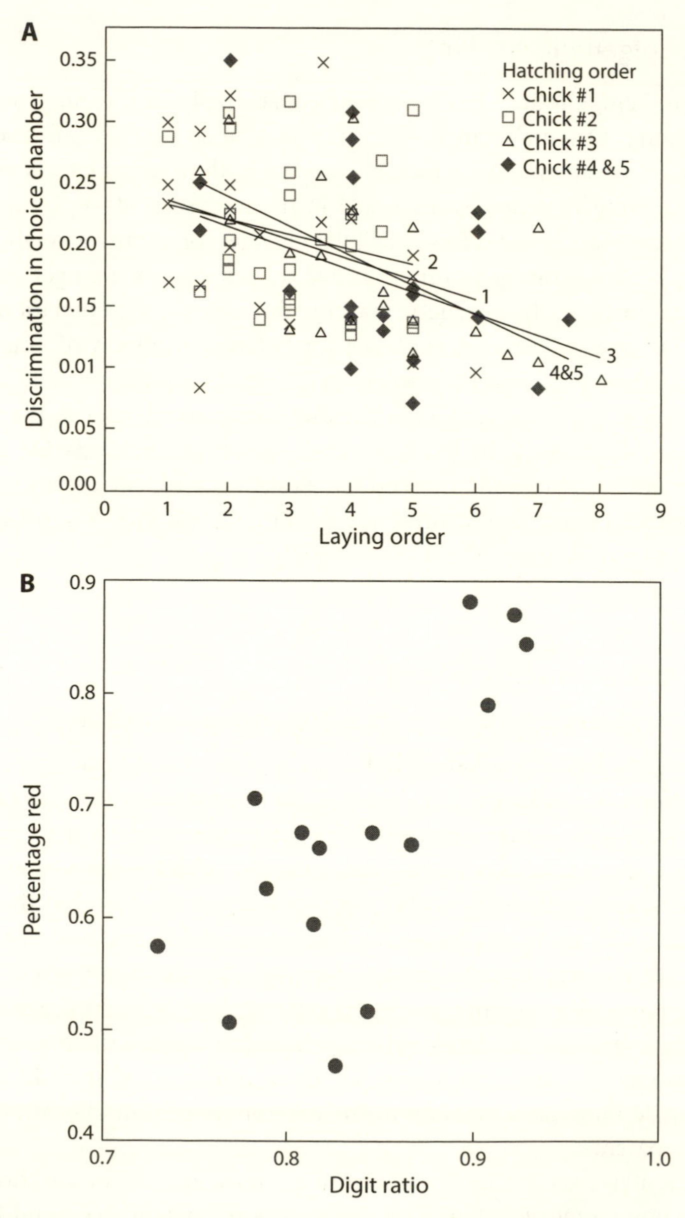

Figure 12.2. Variable effects of laying order on strength of preference in zebra finches. (a) Strength of preference of female zebra finches among males in a four-way preference test decreases with laying order (Forstmeier et al. 2004); (b) strength of preference for red-banded males increases as a function of the length ratio between the second and fourth digits, a positive correlate of laying order (Burley & Foster 2004).

and Foster (2004), however, found an effect in the opposite direction, with putatively more-androgenized females less choosy (fig.12.2b). Experimental manipulation in another bird species corroborated this negative effect of androgenization on choosiness. Bonisoli-Alquati and colleagues (2011) injected testosterone into eggs of ring-necked pheasant (*Phasianus colchicus*), and found that control females copulated more with (more ornamented) testosterone-treated males than did testosterone-treated females, but that both groups copulated at equal rates with control males. As with condition-dependent preferences more broadly (chapter 11), whether choosiness increases or decreases with laying order, or with androgenization, may depend on context, genotype, or other state variables. These parental effects need not be limited to females, since males in some species provide nutrients to embryos (Kvarnemo et al. 2011). Crean and colleagues (2014) showed that the semen of experimentally well-fed male flies *Telostylinus angusticollis* transmitted their condition to the offspring of the females they mated with, including unrelated offspring from those females' subsequent matings with other males.

12.2.4 Nongenetic inheritance of mate choice: summary

The nongenetic inheritance of epigenetic markings, resources, and learned behavior (next section) from one's parents can all contribute to parent-offspring correlations in preference even in the absence of heritable genetic variation. It is important to disentangle these effects, since they lead to different predictions about evolutionary dynamics. In particular, genomic imprinting of germ-line cells affects the offspring's phenotype only, while resource allocation and model behavior affect parent phenotypes as well (Hager et al. 2008). This distinction acts to produce different biases in estimates of quantitative genetic variance components (Santure & Spencer 2006). While parental effects and vertically transmitted culture have similar evolutionary dynamics, social effects after birth are more complex than, say, effects of egg hormone titer, in that they also involve oblique and peer interactions (fig.12.1).

The dynamics of nongenetic inheritance surrounding mate choice are fascinating and would be well worth some theoretical exploration. In particular, nutritional condition (chapter 11) and social stress (see below) modulate the mating preferences of parents. These preferences, in concert with the parent's condition itself, act to shape epigenetic marking and the early social and nutritional environment, which in turn affect the preferences of the next generation. For example, in zebra finches, preference affects rearing environment, with females investing more in the eggs of more attractive males (Tschirren et al. 2012; Gilbert et al. 2011; chapter 7), and postnatal rearing environment affects preference (Riebel et al. 2009; Holveck & Riebel 2010; chapter 11). Such effects may therefore generate considerable variation in preference, particularly choosiness, even in the absence of mate-choice variation, and may help explain some of the considerable year-to-year variation in preference observed in the wild (e.g., Chaine & Lyon 2008).

12.3 SOCIAL STATUS BEFORE AND AFTER MATURITY

Early in life, choosers-to-be develop relationships with other individuals that determine who gets preferential access to resources, including food and mates. At first blush, the effects of social status on preference are more like those described in the previous chapter than they are like the social effects detailed above, since they resemble nutritional and other effects on individual condition. An individual's social status can have marked effects on its growth and condition, with subdominant individuals faring worse (e.g., Wilson & Nussey 2010). Cotton and colleagues (2006a) documented 16 observational studies showing differences in mating outcomes as a function of chooser social status. They argued that status effects were effectively a special case of condition-dependence, with subdominant animals being less choosy (but see chapter 11 for multiple instances of poor-condition individuals being more choosy).

In adulthood, social dominance functions largely in the context of dominant individuals monopolizing reproductive opportunities. In many systems, aggressive interactions act to entirely suppress the reproductive axis of subdominant individuals: "low ranking males are psychophysiologically castrated," in Fricke and Fricke's (1977, p. 830) memorable description of clownfish (*Amphiprion*) social systems. Jockeying for social dominance may therefore involve choosers interfering directly with each other's preferences and choices, which would also act to reduce choosiness. As discussed in the context of male mate choice, agonistic interactions with other choosers act to constrain realized mate choice. Monogamous animals also tend to pair assortatively with respect to social dominance (chapter 9). Social status can therefore have substantial effects on realized preferences both indirectly through its effects on condition (and, for mutual mate choice, on one's own attractiveness) and directly through interference by other, more dominant choosers.

12.4 EARLY LEARNING: IMPACTS ON PREFERENCES

12.4.1 Sexual imprinting as a special case of early learning

The evidence that experience shapes preferences early in life is overwhelming, and indeed, for many of the years since Darwin, provided the default explanation for how the "taste for the beautiful" comes to be. Experience-dependent preferences can be demonstrated through **cross-fostering**, whereby offspring are reared with a novel set of parents. The technique is straightforward in birds, since one can simply swap eggs in a nest before hatching. A comprehensive review by ten Cate and Vos (1999) already documented 101 bird studies showing early learning of preferences. Cross-fostering can also happen spontaneously in wild populations. Rowley and Chapman (1986) conducted detailed observations of sympatric populations of two cockatoos (*Cacatua roseacapilla* and *C. leadbeateri*). In two cases, they documented female *C. roseacapilla* that had been cross-fostered by *C. leadbeateri* pairs after takeover of a nesting site. While all other *roseacapilla* females paired with conspecifics, these females went on to flock and pair with *C. leadbeateri* males. Interspecific interactions in the wild can therefore lead to social and presumably genetic exchange between species.

Cross-fostering studies show convincingly that **early learning** is occurring: adult preference is dependent on early social experience. While several authors use "sexual imprinting" to mean early learning more broadly, Lorenz's (1937) definition of **sexual imprinting** is that it is additionally irreversible and occurs during a defined window of time known as a **critical period** or **sensitive phase** (Clayton 1994). Sexual imprinting encompasses both oblique and vertical transmission of preferences. In other words, if choosers experience certain stimuli during a specific window in early life, those preferences will be robust to subsequent experience as adults. As described in chapter 6 and below, while early experience often has a dispositive hand in shaping preferences, it is rare for preferences to be unmodified by subsequent experience. The distinction between sexual imprinting and early learning more broadly is important because it can affect the opportunity for oblique or peer in addition to vertical transmission, as well as whether preferences are consolidated before dispersal occurs. Indeed, Bischof and Clayton (1991) showed in zebra finches that imprinting can be disrupted if cross-fostering is interrupted, indicating that sustained interactions during the critical period may be required to consolidate preference.

Cross-fostering by itself does not rule out effects of learning later in life or of individuals other than the foster parents, and therefore does not by itself distinguish between imprinting and more generalized learning, or between vertical and oblique transmission. Further, cross-fostering also often includes sibling effects, which can impact mate choice: for example, female pheasants reared in female-biased sex ratios are less choosy (Madden & Whiteside 2013), and female limnetic sticklebacks had stronger preferences for conspecifics if raised with conspecific rather than heterospecific juveniles (Kozak & Boughman 2009).

12.4.2 Learned preferences for conspecifics

Scores of experimental studies have shown that choosers change and often reverse their preferences to favor heterospecifics as a consequence of experience before sexual maturity. Most of these studies have been in birds with biparental care (reviewed in ten Cate & Vos 1999), but a number of studies in the last few decades have shown effects of early learning in a range of other taxa with maternal-only care (sheep and goats, Kendrick et al. 1998; African cichlids, Verzijden & ten Cate 2007; fig. 12.3), male-only care (sticklebacks, Kozak et al. 2011), or no care at all (swordtails, Verzijden and Rosenthal 2011; wolf spiders, Stoffer & Uetz 2015).

Although most early-learning studies have been on vertebrates, a handful of arthropod studies suggest that early experience can influence adult preferences. Early experience with male song made female field crickets *Teleogryllus oceanicus* choosier and less responsive (Bailey & Zuk 2008). In wolf spiders, females require exposure to males during the juvenile stage in order to later express preferences as adults (Hebets & Vink 2007), and favor the phenotypes they were exposed to (Hebets 2003). It is worth noting that there is scant evidence for early learning in taxa (vertebrate or invertebrate) that undergo metamorphosis, and little work on how memory is retained through the process. It would be worth investigating if any courter signals are used by larvae prior to sexual maturity to shape mating decisions.

Sexual imprinting can be directly observed in wild populations. Redhead ducks (*Aythya americana*) sometimes practice brood parasitism, depositing their eggs with closely related canvasback ducks (*A. valisineria*). Experiments by Sorenson and colleagues (2010) showed that redhead males raised with canvasbacks direct more courtship at canvasback ducks (and vice-versa for canvasback males reared by redheads). Misdirecting courtship toward heterospecific females is likely to represent a fitness cost to males reared in parasite nests (Sorenson et al. 2010). It would be fascinating to know the preferences of cross-fostered females, since a preference for other parasitized males could generate assortative mating and a possible first step toward obligate parasitism and speciation (chapter 16).

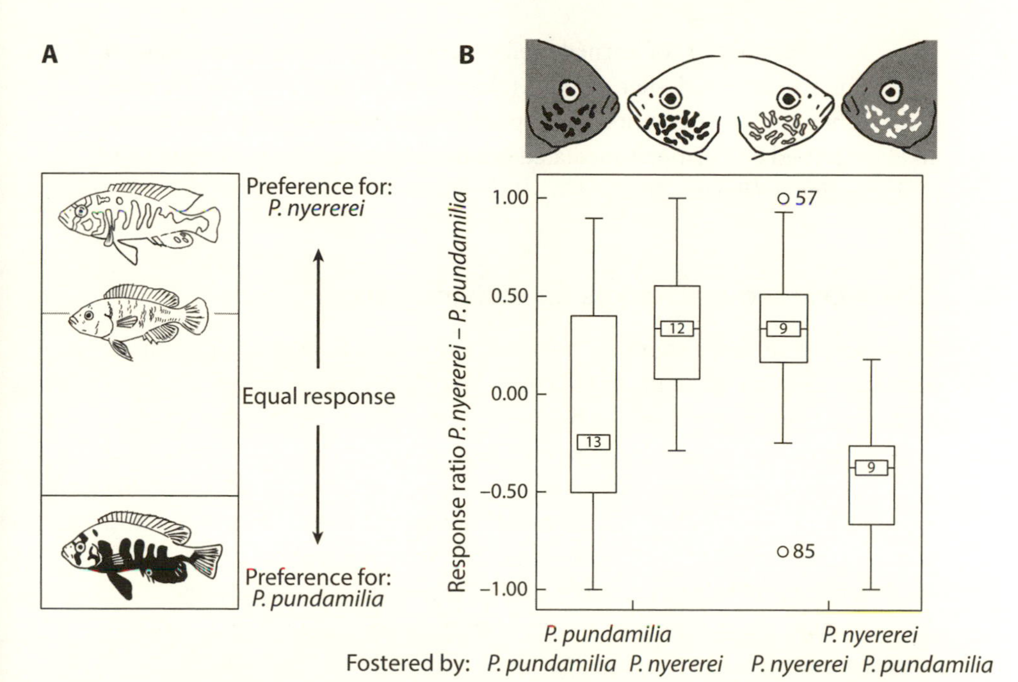

Figure 12.3. Effects of cross-fostering on mating preferences in the African Rift Lake cichlids *Pundamilia pundamilia* and *P. nyererei.* (a) Simultaneous-choice setup for assaying preferences for conspecifics or heterospecifics. Females are smaller and can enter chambers and interact with confined males. (b) Net approaches to *pundamilia* or *nyererei* by females fostered by either species. After Verzijden & ten Cate (2007).

Obligate parasites like cuckoos and cowbirds are faced with a challenge, since courters and choosers can't directly develop conspecific preferences by attending to the individual raising them. In village indigobirds *Vidua chalybeata*, cross-fostering experiments by Payne et al. (2000) showed that cross-fostered females preferred conspecific males singing songs that mimicked the songs of their foster parents; here, host mimicry allows females to use vertical cultural transmission to acquire conspecific mate preferences. In the obligate parasite *Molothrus ater*, the brown-headed cowbird, female mate preferences are shaped by oblique and peer learning later in life; Freeberg et al. (1999) showed that females preferred males from a population with which they'd had adult experience.

12.4.3 Learned preferences for partner sex

Banerjee and Adkins-Regan (2014) found that male zebra finches raised in the absence of females imprinted on males as sexual partners, with a slight majority of males preferring to court other males as adults. To my knowledge, there has been no study of learned preferences in the seabird species known to routinely form female-female pairs (chapter 8). In the two known species with stable sexual orientation—sheep and humans—sexual preference is likely specified by hormone-mediated organizational effects rather than social experience (Poiani 2010).

12.4.4 Learned preferences for individual traits

Experience can shape choosers' preferences for specific traits. For example, zebra finches vary naturally in bill color. Vos (1995) tested the effects of parental bill color experimentally by painting the bills of white zebra finches, with each parent assigned a different color. Females did not attend to the manipulation (see below); males, once mature, preferred females with the manipulated color of their mothers. Males directed courtship at birds with the same manipulated color as their mother, independent of sex (fig. 12.4). Early learning of a trait therefore overrode any innate sexual preference. I will discuss more examples of imprinting on individual traits—notably arbitrary novel traits and song features—in the next few sections.

12.4.5 Preference development robust to social experience

It is important to emphasize that despite abundant and taxonomically widespread evidence for the importance of early learning, many important preference phenotypes are robust to social experience. In treehoppers (*Enchenopa binotata*), choosiness is sensitive to early acoustic experience, but peak preference is not (Fowler-Finn & Rodríguez 2012, 2016). In pied flycatchers, males learn their songs; but the female preferences for these songs are inherited from the paternal Z chromosome, and these preferences are unaffected by cross-fostering (Saether et al. 2007). Z-linked, experience-independent preferences are also found in female Gouldian finches (Pryke 2010; chapter 10).

Such innately specified preferences are likely to be widespread in species that inhabit sharply different sensory environments over ontogeny, or that disperse away from parents before preferences can start developing. For example, fishes living on coral reefs recognize complex spatial patterns as belonging to conspecifics (e.g., Katzir et al. 1981) despite being separated from their parents at fertilization. Innate responses are also present in animals that learn their preferences. Shizuka (2014) showed that six-day-old white-crowned sparrows, *Zonotrichia leucophrys*, display more vocal responses to

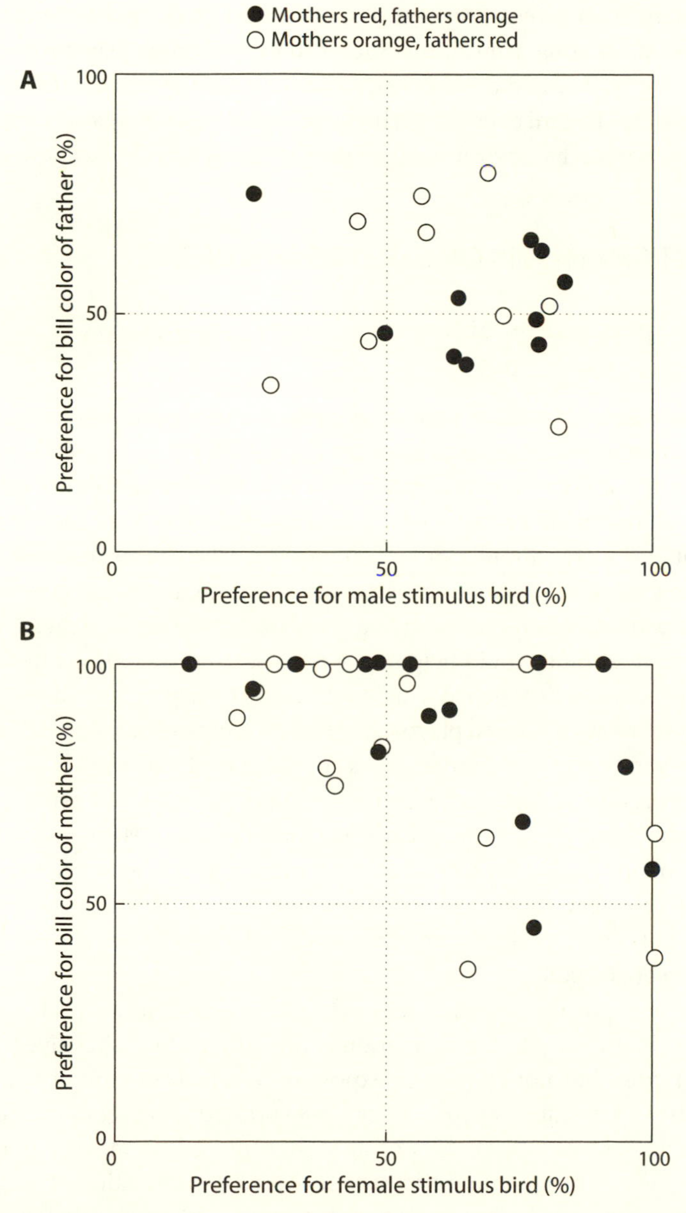

Figure 12.4. Preference ratios of (a) female and (b) male zebra finches for sex and artificial bill color of prospective partners. Open circles, mothers orange, fathers red; filled circles, mothers red, fathers orange. After Vos (1995).

conspecific than heterospecific songs, even though this is before the sensitive period for song acquisition. Such innate responses play an important role in shaping learned preferences, since they can trigger associative learning (Moncho-Begani et al. 2002; chapter 6) and therefore play an important role in genotype-by-environment responses to social cues (section 12.9).

12.5 MECHANISMS OF EARLY LEARNING

12.5.1 Specification of preference mechanisms through experience

In some cases, imprinting may simply involve upregulation of sensory receptors that are responsive to courter stimuli (Nevitt et al. 1994). In mice, the identity of odor receptors directs the functional organization of downstream glomeruli and olfactory bulb circuits (Belluscio et al. 2002), indicating that peripheral stimuli can serve to organize the neural architecture that processes them. Upregulation of sensory receptors and the concomitant organization of downstream wiring provide a mechanism whereby mere exposure to stimuli would heighten sensitivity and thereby create "beauty in the processing experience" (Reber et al. 2004; chapter 5). Indeed, female swamp sparrows preferred playbacks of more typical—i.e., more commonly heard—songs over less familiar types (Lachlan et al. 2014).

Sensitization cannot, however, fully explain the effects of experience on preference (chapter 5). Indeed, early learning can completely reverse the direction of preferences for stimuli that are more detectable on first principles. For example, female green swordtails prefer males with longer sword ornaments, but females raised with short-sworded males prefer shorter swords (Walling et al. 2008).

Passive exposure to stimuli may seldom be sufficient for preferences to develop. In mice, odor discrimination is potentiated by associative learning of odor cues, but not by passive exposure to olfactory stimuli (Abraham et al. 2014). ten Cate and colleagues (1984) raised male zebra finches with conspecifics up to 30–31 days of age, then placed them with Bengalese finches for another month during the so-called consolidation phase. While most birds preferred zebra finch females, consistent with early imprinting, the strength of preference of cross-fostered zebra finch males for Bengalese finches was a positive function of the amount of behavior (both nonaggressive and aggressive) directed toward them by Bengalese finch adults. In cowbirds, West and King (1988) found that male song performance increased as a function of proceptive behavior performed in response by females. Rewarding correct performance with a sexual stimulus should consolidate learning. It would be interesting to unravel how the effects of social experience interact with those of nutritional and social stress during preference acquisition and consolidation (Riebel et al. 2009). Holveck and Riebel (2014) found no effect of developmental stress on the acquisition of preferences for familiar tutors versus unfamiliar males.

Exposure to a mixture of stimuli can generate novel preferences. In a follow-up experiment, ten Cate focused on males with intermediate preferences from the study discussed above (ten Cate et al. 1984). These males "dithered" between both species when presented with a choice, but preferred either species to an unfamiliar white strain of zebra finches or to silverbill finches; these preferences persisted for up to two years (ten Cate 1986). Importantly, ten Cate (1987) went on to show that "ditherers" found hybrids more attractive than either parent species, suggesting that ditherers had a single internal representation—a "mixed standard model" incorporating features of both species. This result has tantalizing implications for **hybrid speciation**, since it provides a mechanism for choosers with mixed social experience to immediately prefer hybrids (chapter 16).

12.5.2 Song learning and sexual imprinting

In songbirds, mating preferences share some important developmental mechanisms with the production of song. Given that cultural transmission plays such an important role in shaping vocalizations in other taxa (e.g., cetaceans, Janik 2014), the shared relationship between transmission and production might offer lessons for how choosers and courters are coupled in other groups of animals. The advertisement calls of oscine passerines certainly awoke Darwin's own "taste for the beautiful," and our terminology for talking about these birds' vocalizations is dangerously anthropomorphic. "Songs" do not consistently use harmonic intervals the way a lot of human music does (Araya-Salas 2012), but they are arranged into distinct, stereotyped acoustic structures called syllables (fig.12.5a). The temporal arrangement of syllables is called the song's "syntax" but does not convey relational meaning the way human syntax does (but see Lachlan & Nowicki 2015). The acquisition, control, and production of song have been much more extensively studied than preference, since it is much easier to record songs and quantify differences as a function of social experience than it is to do the analogous experiment on mating preferences. The cultural

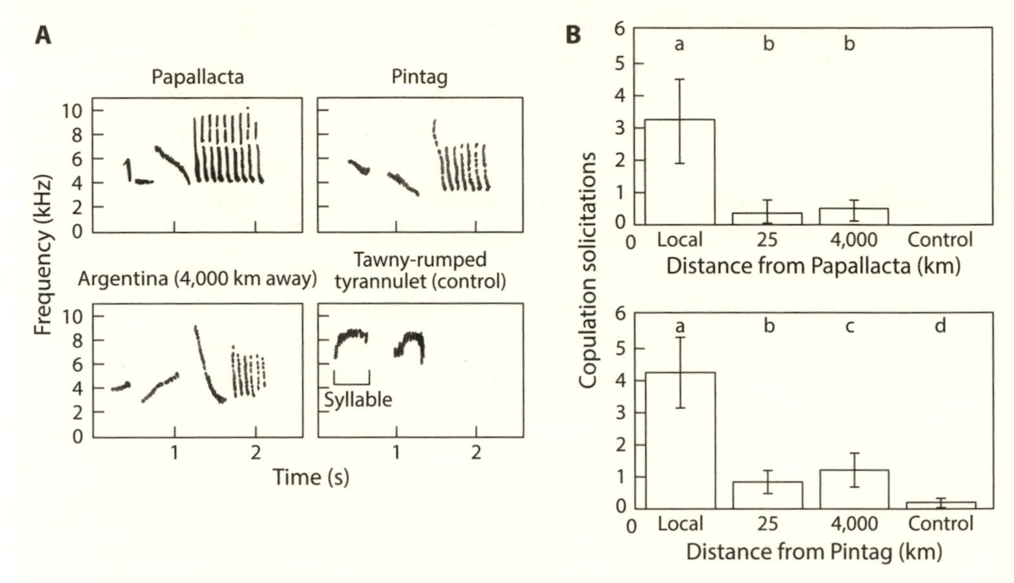

Figure 12.5. (a) Geographic variation in song dialects of rufous-collared sparrows (*Zonotrichia capensis*) in two Andean populations in Ecuador separated by ~25 km, an Argentine population 4000 km away, and a heterospecific used as a control stimulus. (b) Copulation solicitation displays by females from Papallacta and Pintag populations toward local and foreign dialects, and heterospecific control. After Danner et al. (2011).

transmission of song during an early sensitive period means that songs vary not only between species, but also considerably among geographic localities, and even within the same locality over spans of a few years (Derryberry 2009, 2011). These differences are sufficient to elicit strong preferences for contemporary songs in choosers (Derryberry 2007). In particular, urbanization is causing rapid shifts in song structure toward higher frequencies less subject to masking by low-frequency street noise (Slabbekoorn & Peet 2003; Luther & Derryberry 2012).

Clayton (1994) called attention to the likelihood that learned song production and learned song preferences might recruit shared mechanisms. Females and males both show an early bias toward conspecific songs (Shizuka 2014; see below); male variation in song over space and time ("dialects") is the result of cultural transmission. Cross-fostered females in some species prefer songs resembling those of their foster fathers (e.g., Clayton 1990). Females attend to this variation, suggesting that preferences are culturally transmitted along with dialects (Balaban 1988; Danner et al. 2011; fig. 12.5b). In chapter 16, I will address the role of cultural divergence in songs and preferences in driving reproductive isolation.

The mechanisms involved in song preference and song learning, however, may overlap only partially. In oscines, song is uniformly learned during a sensitive period; and while singers might rely on an innate template in the absence of social experience (section 12.9), song formation is invariably sensitive to experience. By contrast, it remains unclear whether there is a sensitive period for shaping song preference (Hernandez et al. 2008); for example, female zebra finches in a study by Holveck and Riebel (2014) were tutored on their foster fathers until 35 days of age and then later on novel males as subadults; previously experienced songs were more attractive irrespective of when they had been experienced. Moreover, several species show no evidence of an effect of auditory experience on preference (reviewed in Hernandez et al. 2008). For example, house finches exposed to songs of local males, songs of "foreign" males, or no song all expressed a preference for the songs of local males (Hernandez & MacDougall-Shackleton 2004).

Experience may, however, shape preferences in more subtle ways. For example, Lauay and colleagues (2004) showed that while female zebra finches had an experience-independent preference for zebra finch over canary songs (see below), they required early exposure to song to express preferences for the song of tutored males over the "abnormal" song of males reared in isolation. Riebel (2000) evaluated the development of preferences of female zebra finches for male song; like most studies of male song learning, females were tutored on recordings of song, in this case of conspecific males unrelated to their fathers. In subsequent tests, they preferred the songs of tutors over unfamiliar songs. Further, females that had been tutored were more consistent in their preferences, showing markedly higher repeatability across trials (table 9.1). Repeatability is therefore contingent on social experience.

12.5.3 Novel traits, generalization, and peak shift

Early learning provides a robust mechanism for the coupling of courter traits and chooser preferences. Female Javan munia *Lonchura leucogastroides* (Witte et al. 2000) and zebra finches (e.g., Burley 2006; Witte & Sawka 2003) imprint on novel, arbitrary traits, meaning that a chooser can both inherit a novel trait allele from her father and exhibit a preference for the same trait. In our old friend the zebra finch, males prefer females with the beak color of their mother and avoid females with the beak color of their father; moreover, they exhibit peak shift, preferring more extreme versions of the maternal color that are further away from the color value for males. Early learning can therefore generate directional preferences that can generate sexual selection and drive interspecific divergence of traits (ten Cate et al. 2006; fig. 5.2). This could either be the result of innate antipathy for a beak color in a paternal context and innate preference for a beak color in a maternal context, or, less plausibly, of positive interactions with mothers and negative interactions with fathers. There may also be **oblique transmission** (fig.12.1) involving explicitly sexual interactions, positive or negative, since courters will often court and even attempt to mate with juveniles.

While there is abundant evidence for early learning of preferences for discrete traits, there is limited evidence for imprinting on continuous courter variation. Testing a sample of 113 zebra finches, Schielzeth et al. (2008a) found no effect of foster father on preference; females neither preferred nor avoided the unfamiliar genetic sons of foster fathers, nor was preference variation between individuals by the behavior of the foster father toward his mate or toward the chooser. In a later study comparing the choices of foster sisters and genetic sisters for individual males, they found no effect of shared rearing environment (nor of shared genetics, see below; Schielzeth et al. 2010). These negative results stand in stark contrast to multiple studies in zebra finches that show learned preferences for conspecifics or heterospecifics, discrete trait variation, novel artificial traits, or same- or opposite sex partners.

12.5.4 Valence of learned stimuli: Oedipus versus Westermarck

Most nonhuman studies have focused on learned preference as a positive function of exposure to a phenotype. Negative imprinting, whereby learned phenotypes are avoided, may be just as important, and may play an important role in the maintenance of variation in chooser preferences and courter traits. Kin recognition based on shared cues and on **self-referential phenotype matching** is ubiquitous in animals (Hepper 2008). Westermarck (1891) noted that humans are as a rule not sexually attracted to individuals they recognize as kin, and an abundance of observational studies have since made the case that this is partly based on shared experience in early life (reviewed in Rantala & Marcinkowska 2011). This **Westermarck effect** constitutes a mechanism for inbreeding avoidance, and one critically dependent on social experience:

> Freud had a wet-nurse, and may not have experienced the early intimacy that would have tipped off his perceptual system that Mrs. Freud was his mother. The Westermarck theory has out-Freuded Freud. (Pinker 1997)

There has been surprisingly little empirical work on negative imprinting in animals since a classic experiment where Bateson (1978) showed that male Japanese quail *Coturnix coturnix* avoided individual females they had been reared with, but preferred the plumage type of the familiar females. He argued that positive and negative imprinting can therefore operate in concert toward "optimal outbreeding" (chapter 14). Subsequent evidence has provided mixed support for experience-mediated avoidance of kin in animals. Schielzeth et al. (2008b) found weak evidence for avoidance of similar genotypes in zebra finches.

Negative imprinting on kin phenotypes may share a basis with more general mechanisms of kin recognition. Major histocompatibility complex (MHC) proteins provide a mechanism for heterotypic preferences (chapter 8), and choosers generally prefer courters with dissimilar MHC haplotypes (Jordan & Bruford 1998; Milinski 2006). However, there is limited support for a role of kin avoidance in mate choice based on MHC-associated cues. In the context of kin recognition, preferential treatment of one's own MHC haplotype can arise through self-referential phenotype matching (Mateo & Johnston 2000) through early learning of familiar phenotypes; Penn and Potts (1998) cross-fostered male house mice (*Mus musculus*) and found that they avoided the MHC haplotype of the foster family. Given the ubiquity of MHC-correlated preferences (chapter 8), early learning of MHC haplotypes is an appealing mechanism for vertically transmitted learned preferences against (or for) familiar phenotypes.

Ironically, some of the clearest evidence for self-referential phenotype matching involves mating preference *for* siblings: male West African cichlids, *Pelvicachromis taeniatus*, isolated from birth prefer their sisters over unrelated females (Thünken et al. 2014). Other studies are consistent with "optimal outbreeding" without specifically invoking kin avoidance. Mating outcomes reflect intermediate levels of dissimilarity in brown trout, *Salmo trutta* (Forsberg et al. 2007). Bonneaud et al. (2006) found that allelic diversity, rather than complementarity, determined genetic pairings in a wild population of house sparrows (*Passer domesticus*). Choosers therefore attend to variation in courters that is correlated with MHC haplotypes, whether or not self-referential phenotype matching or kin discrimination are involved.

Early learning can have positive and negative effects with respect to different traits in the same models. For example, while humans generally avoid kin-associated cues in the context of mating, they are more likely to partner with someone who shares the eye color of their opposite-sex parent (Little et al. 2003). The sensory and perceptual channels that respond to, say, MHC cues and those that respond to color and visual context cues may address distinct brain structures predisposed to respond positively or negatively (section 12.6).

Inputs associated with a particular stimulus type may therefore be hedonically marked based on stimulus properties, as above. An alternative is suggested by ten Cate and colleagues' (1984) experiment described earlier, where more interaction with a heterospecific model elicited a stronger preference for heterospecifics later in life. If associative learning is required to form preferences (Abraham et al. 2014), the outcome may depend on whether the associations are negative or positive. For example, the widespread avoidance of aggressive males by choosy females (Wong & Candolin 2005) may arise from negative interactions with aggressive individuals. An intriguing parallel is found in a recent study by Darden and Watts (2012), who showed that harassment by males led to a disruption of social interactions among females. This may underlie the development of female avoidance of traits associated with aggressive intrasexual interactions (Fisher & Rosenthal 2007), and may account for variation in preferences for aggressive traits across studies (Robinson et al. 2011).

These two factors—hedonic marking of stimuli and subjective value of associations with models—can interact. In humans, there is some suggestion that emotional warmth during childhood may modulate imprinting, but it does so in different ways depending on the cues involved; subjects prefer partners with facial features similar to their opposite-sex parent if that parent was warm to them, but prefer a personality similar to the opposite-sex parent if that parent rejected them during childhood (Gyuris et al. 2010). Preference phenotypes are perhaps a tapestry woven from hedonically marked innate cues interacting with social experience.

12.6 VARIATION IN EARLY LEARNING

12.6.1 Sex differences

Within a brood, the mating preferences of males and females very often respond differently to social experience (ten Cate & Vos 1999). For example, women are influenced by the age of both parents in making judgments of facial attractiveness, preferring "older" faces if their parents are older, while men respond to the age of their mothers but not their fathers (Perrett et al. 2002).

Kendrick et al. (1998) cross-fostered sheep and goats and found that both sexes attended to the maternal foster phenotype. Males retained their preferences throughout their lives, but females reverted to their genetic type within 1–2 years. In these two species, therefore, males are more sensitive than females to social experience. Kozak and Boughman's (2009) study revealed that the effect of learning operates via different mechanisms in males and females of threespine sticklebacks: females appear to learn preferences by positively imprinting on conspecifics, while males do so by negatively imprinting on heterospecifics.

Zebra finch studies have yielded a hodgepodge of results on sex differences. Vos (1995) found that female zebra finches did not attend to bill-color manipulation on either parent, whereas males attended to maternal bill color (fig. 12.4). By contrast, Witte and Sawka (2003) found a learned preference for a novel feather ornament in females, whereas males failed to attend to feather ornaments at all. Interestingly, females did not generalize at all with respect to color, preferring only males with the same color feather that their father had worn.

Between two subspecies of zebra finches, males sing distinct, subspecies-specific songs irrespective of who they're fostered with. By contrast, females of both subspecies prefer the song of their foster father (Clayton 1990). In yet another zebra finch study, both male and female zebra finches attended to subtle details of their mother's, but not their father's phenotype (fig.12.6), and later showed opposite-sex preferences for their mother's ornament (Burley 2006). If there is genotype by environment variation in learning tendencies (see below), heterogeneity of source stocks used by different laboratories may partly account for this variation. Another explanation of sex differences in learning is that males and females have different experiences during their ontogeny or as adults. For instance, parents may behave differently to sons and daughters, such that social experiences differ substantially between sons and daughters. This can generate a sex difference from what might at first glance seem like an identical treatment (ten Cate & Vos 1999; Verzijden et al. 2012).

12.6.2 Interspecific variation in early learning

Since experimental studies on the prevalence of early learning are beginning to accumulate across species (Hernandez & MacDougall-Shackleton 2004), one might apply a meta-analytic approach to predict how courter traits, ecological variables, and phylogeny might predict the existence, timing, and reversibility of early learning. Key variables likely include whether developing choosers experience one or both parents, and how long they associate with them before maturity. *Precocial* birds—hatching at a later

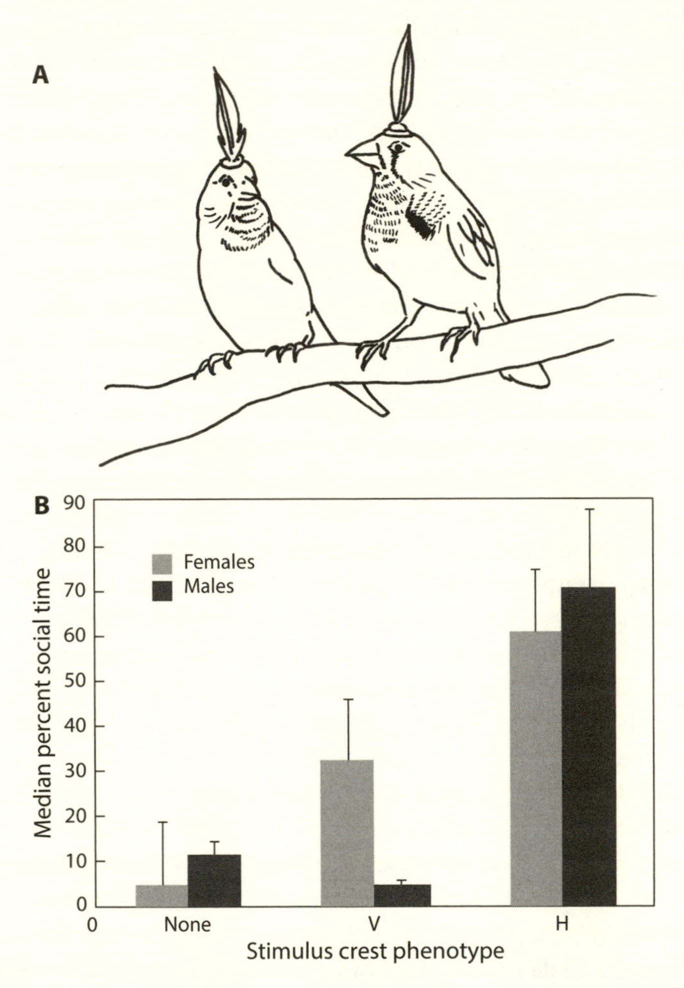

Figure 12.6. (a) Male (left) fitted with a vertical crest and female fitted with a horizontal crest. (b) Heterosexual social preferences of female and male zebra finches reared with a mother with a horizontal crest and a father with a vertical crest, as shown. After Burley (2006).

stage of development—have a period of so-called filial imprinting before the sensitive period, during which they learn to respond preferentially to a familiar phenotype in the context of parental care. The filial period is absent in *altricial* birds that hatch in a less developed state (Bolhuis 1991), and the timing of the sensitive period is also likely to be different between the two types of rearing modes (MacFarlane et al. 2007).

Differences in the effects of social experience can arise between closely related species. Slagsvold and colleagues (2002) performed a large-scale cross-fostering study of wild populations of the genus *Parus* (tits, or chickadees for American Anglophones) in Norway. In particular, there were marked differences of how cross-fostering affected subsequent mating outcomes. Cross-fostering affected blue tit females, but not great tit females; for the cross-fostered blue tit females that chose great tits as social mates, all the offspring fledged were blue tits, indicating extra-pair mating and likely conspecific sperm precedence (chapter 7). This experiment serves as a reminder, of course, that preference behavior does not necessarily translate into mating outcomes.

Between the sister species of swordtails *X. malinche* and *X. birchmanni*, opposite preferences arise from exposure to the same stimuli, both after sexual maturity (chapter 6) and as a consequence of early learning (Verzijden & Rosenthal 2011; Cui et al. in press). This constitutes a genotype X environment interaction on the valence of familiar stimuli, whether because familiarity triggers divergent innate responses or because interactions with adults have different subjective value in the two species. Differences in the canalization of learned stimuli may be a key to understanding preference evolution, as I will discuss in the following section.

12.7 SOCIAL EXPERIENCE AFTER SEXUAL MATURITY

Although most attention has focused on how preferences are shaped by experiences early in life, they can be markedly affected by experiences after sexual maturity. In some cases, early adult experience shapes mating decisions throughout life, as in butterflies *Bicyclus anynana*, where males, but not females, exhibit biased learning for hind-spot patterns; males exposed soon after metamorphosis to females lacking spots have a strong preference for the absence of spots, while naïve males or males exposed to spotted females fail to show a preference (Westerman et al. 2014).

Mating biases can also continue to change over the course of adult life. There is no bright line between immediate-term interactions with courters that shape a mating decision (chapter 6) and the longer- and medium-term patterns discussed here. In some cases, these changes may be related to relatively environment-independent changes in the window of reproductive opportunity (chapter 9), while in others, they may depend on environmental inputs, broadly construed to include sexual interactions with courters.

For example, female-biased sex ratios and increasing age interact to relax choosiness in female sticklebacks (Tinghitella et al. 2013).

The preferences of young and old choosers are often different (chapter 9), and some of these differences may be the effects of social experience. For example, older female bowerbirds prefer males with more intense (higher temporal frequency) courtship displays, while younger females show antipathy to the same displays. A similar pattern is found in Japanese quail (Galef 2008). Patricelli et al. (2002) argued that this age-dependent response reflected a reduced startle response in older females habituated to displaying males.

Different experiences lead to different preferences. This is true if individuals imprint on parental phenotypes during ontogeny, but it is also true when individuals discriminate on the basis of prior encounters as adults. Recent experience with predators, rival choosers, close relatives, or heterospecifics can also influence individual preference decisions. Variation in individual experiences can therefore generate considerable variation in individual preference. Consistency of experience, by contrast, may serve to homogenize preferences, like mate copying below.

12.8 NONINDEPENDENT MATE CHOICE AND COPYING

Nonindependent mate choice occurs when an individual's choices or preferences are contingent on those of other choosers (Westneat et al. 2000). Most attention has focused on **mate copying**, which occurs when choosers modify their mating decisions based on cues from other choosers (Pruett-Jones 1992; Witte et al. 2015). Other forms of nonindependent choice, termed "pseudocopying" by Brooks (1998), include choosers forming associations with one another in a context other than mating, and choosers attending to courter cues, like an increase in courtship rate, brought about by contact with other choosers.

Mate copying is tested experimentally by comparing chooser responses to courters in the presence or absence of another chooser, the so-called model. Much of the early literature on copying focused on improving and validating experimental designs to demonstrate the phenomenon. Pseudocopying can be ruled out by giving courters access to a model chooser, without the focal chooser being able to observe the hidden model (Schlupp et al. 1994). Only a subset of model cues may be required for copying, and some of the most convincing demonstrations of copying come from playback experiments. Freed-Brown and White (2009) showed copying by female brown-headed cowbirds (*Molothrus ater*) for male vocalizations when these were accompanied by a female vocal signal; and Kavaliers et al. (2006) showed that female mice copied on the basis of the presence or absence of estrous urine cues.

Copying has now been robustly established in multiple systems, including a number of manipulative studies that have carefully excluded the possibility of any influence of any stimulus besides cues from another chooser. A recent review by Vakirtzis (2011) documents experimental demonstrations of mate-choice copying in more than two dozen species. Copying may be comparatively rare in monogamous species because of the constraints involved in mutual mate choice (chapter 8), but has nevertheless been shown in zebra finches, where females will prefer males wearing the same leg band color as those of males they previously observed chosen by females (Swaddle et al. 2005).

The nature and extent of copying in humans is controversial. Uller and Johannson (2003) found that, contrary to popular wisdom, a wedding ring on a man's hand did not make him more attractive to heterosexual partners. Waynforth (2007) showed that much like guppies, women attended to the phenotype of the model; a man was more attractive to women when paired with a partner, but only if the latter was rated as attractive. Interestingly enough, social cuing seems to influence social pairing more than it does short-term liaisons; women in Little et al.'s (2008) study copied when rating facial attractiveness in the context of a long-term relationship decision, but not for a one-night stand.

Copying has been demonstrated mostly in females, although some males do copy. As I discussed in chapter 6, females may generally copy more than males, because of the costs males incur from sperm competition (Vakirtzis 2011; Dubois 2014). Indeed, in Japanese quail, copying is positive in females but negative in males, who are less likely to mate with a female observed with another male (White 2004).

Brooks (1996) observed that copying should be more likely when it involves a decision between courters that are relatively similar in attractiveness, but later studies have shown that copying can be a dispositive force in shaping mating preferences, driving them in entirely opposite directions. Female fruit flies reverse their preference for an arbitrary, novel trait (a dusting of green or pink powder; Mery et al. 2009), and female guppies reverse a strong baseline preference for bright orange males (Dugatkin 1998).

They also reverse preferences that are presumed to be under strong selection. Naïve female fruit flies have a preference for males in good nutritional condition (see chapter 11). The presence of a model female does not do more than mere exposure to increase the preference for good-condition males; however, females do show a strong increase in preference for poor-condition males after observing them with a model female (Mery et al. 2009; fig. 12.7). Similarly, naïve female mice perform more appetitive and fewer aversive behaviors in the presence of urine cues of unparasitized males, and perform mostly aversive behaviors around urine cues of parasitized males (Kavaliers et al. 1998). When parasitized-male urine is paired in tandem with that of an estrous female, females reverse their preferences by reducing aversion to the parasitized-male cue (Kavaliers et al. 2006).

Vakirtzis and Roberts (2009) call attention to "mate quality bias," whereby choosers are attending not only to whether a courter is associating with a chooser, but to the phenotype of the model chooser. Female guppies, for example, are more likely to mate with males that they have previously seen interacting with model females than with males they have previously seen alone. The preference is stronger if the model females are older and more experienced (Dugatkin 1993). For example, Hill and Ryan

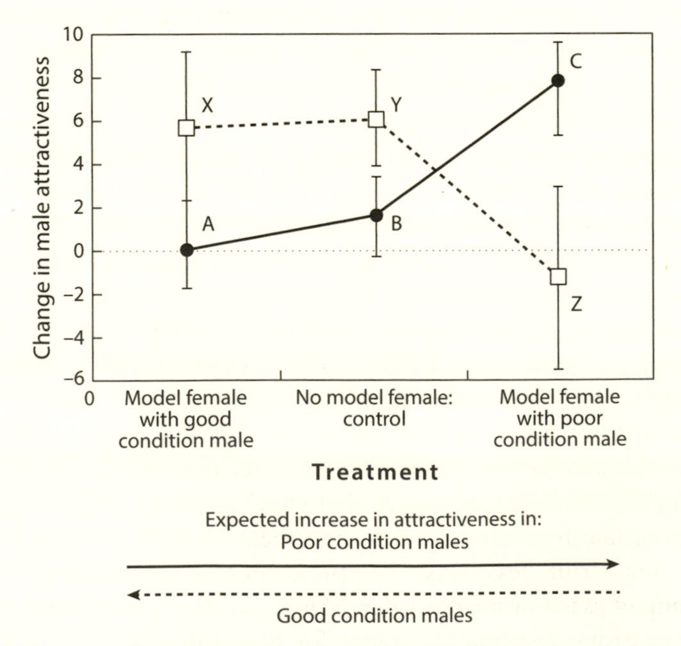

Figure 12.7. Female fruit flies copy the choices of other females. Females initially prefer well-fed males (open circles) over poorly fed males (closed circles). In a second trial, control females maintain this preference, while females who observe another female with the poor-condition male reverse their preference. After Mery et al. (2009).

(2005) showed that choosing female sailfin mollies (*Poecilia latipinna*) were more likely to copy the choices of conspecifics than of a sympatric, asexually reproducing congener. Surprisingly, females of the closely related *P. mexicana* attended to female models and male models equally when making copying decisions (Bierbach et al. 2013). In humans, a woman's perception of the attractiveness of a model female is dispositive in mate copying; women copy the choices of attractive but not unattractive models (Vakirtzis 2011).

Copying may not only be affected by the phenotype of individual choosers, but by the co-distribution of choosers and courters. Drullion and Dubois (2008) provided zebra finch females with video clips of males with two different novel leg band phenotypes. In one set of clips, model females consistently associated with one male phenotype; in the other, they associated with both. Females only reversed their preferences if models consistently associated with the originally unpreferred male.

In Pruett-Jones's (1992) theoretical framework, copying could be either positive or negative: choosers would not only be more likely to mate with an accepted courter, but less likely to mate with a rejected one. Santos et al. (2014) pointed out that since male courtship and copulation attempts are frequently rejected by females (chapter 6) there is abundant public information about rejection. Few studies, however, have demonstrated that choosers copy negative information. Witte and Ueding (2003) showed that female sailfin mollies avoided previously attractive males if they were able to observe rejection by another female. And, consistent with greater copying costs for males, female Japanese quail copy positively, while males copy negatively, avoiding a female they see mating with another male (White 2004). Santos et al. (2014) argued that such a mechanism could facilitate the spread of a novel courter trait (chapter 15).

As with mate preferences more generally, the state and history of the chooser can influence copying. Dugatkin and Godin (1998) manipulated the nutritional state of female guppies and found that only well-fed females copied. Early learning also influences the propensity to copy; Dugatkin (2007) found that female guppies only copied if they had been raised without sexually mature adults.

Copying can involve distinct stages of mate choice for both focal and model individuals. For example, in a wild population of female ocellated wrasse (*Symphodus ocellatus*), mating decisions are positively influenced both by the current presence of a model female in a nest and by the presence of a previous female's eggs (Alonzo 2008). In Japanese quail, the effects of copying manifest in terms of both mating behavior and postmating investment in egg size (reviewed in Galef 2008).

As I will discuss further in chapter 15, positive copying has the effect of increasing reproductive skew among courters; the most preferred individuals will receive the most benefits from copying. Copying can also act to homogenize preference variation among choosers, since it generates a positive correlation among chooser preferences (Brooks 1996). Brooks (1996) pointed out that while copying will generally act to reduce repeatability (since mating decisions will now depend on characteristics of the model chooser, in addition to those of the focal chooser), it will actually increase repeatability if females with similar preferences copy each other more than expected by chance. Preferences learned through copying can be maintained for up to five weeks in sailfin mollies, which may also act to increase repeatability.

Copying based on rejection can theoretically generate selection on a novel trait (Santos et al. 2014). For positive copying to drive courter trait evolution, it has to have a lasting effect that goes beyond an idiosyncratic preference for an individual; it has to result in a generalized preference for some aspect of the courter being observed. In both Japanese quail (White & Galef 2000) and guppies (Godin et al. 2005), choosers do indeed generalize from copying experiences, choosing males with a similar phenotype to that preferred by a model female. Nonindependent learning can therefore have major effects on the consequences of mate choice.

Social experience may also impact the tendency of choosers to copy. This is suggested by a study by Darden and Watts (2012) showing that harassment (copulation attempts) by male guppies resulted in increased aggression among females. Intra- and intersexual interactions may therefore shape the tendency of choosers to rely on public information.

12.9 GENOTYPE BY ENVIRONMENT REVISITED: THE INSTINCT TO LEARN

> I sense from the classical debate between Piaget and Chomsky ... that at least some of us are all too prone to think of learning and instinct as being virtually antithetical. (...)
>
> It is self-evident that this antithesis is false. Just as instincts are products of interactions between genome and environment, even the most extreme case of purely arbitrary, culturally transmitted behavior must, in some sense, be the result of an instinct at work. Functions of instincts may be generalized or highly specialized, but without them learning could not occur. (Marler 1991, p. 37)

Marler's "instinct to learn" cut through a Gordian knot: the false dichotomy of "nature versus nurture." The response to social experience is canalized by genotype. It follows logically that different genotypes will respond differently to social experience. Marler and colleagues' (Marler 1991, 1997; Nelson & Marler 1993) perspective was informed by their pioneering work on vocal learning in song and swamp sparrows. Song sparrows learn swamp sparrow songs in isolation, but reject them if exposed to conspecific songs; birds in isolation sing a crude conspecific song. Swamp sparrows, in contrast, reject heterospecific songs altogether unless they are acoustically modified to incorporate key, species-typical features of a conspecific song. Learning guides song development in these two species, but in a way that is clearly biased toward conspecific signals. Biased learning of songs has been extensively studied in other songbirds (canaries, Leboucher et al. 2012; zebra finches, Riebel 2009), and may be a widespread feature.

Marler's synthesis was paralleled by the emergence of **phenotypic plasticity** as a major area of inquiry in evolutionary ecology (Schlicting & Pigliucci 1998). A fundamental concept in phenotypic plasticity is that genotypes have reaction norms that are functions of an environmental variable or set of variables (chapter 11). Marler was effectively pointing out the converse to scholars used to thinking about learning as a universal process: reaction norms have genotypes.

There is abundant evidence that genotypes canalize the effects of social experience. The first comes from preference tests on naïve individuals showing a bias for features of conspecific stimuli (Shizuka 2014; ter Haar et al. 2014). Just like naïve sparrows in the context of song-learning, naïve zebra finches show an "own-species bias" in the context of conspecific mate preference, even though they fail to attend to intraspecific variation in courter traits (Lauay et al. 2004).

The second line of evidence for G X E effects on social learning comes from divergent preferences among lineages in response to the same social cues. While *D. melanogaster* females copy the choices of other females, the same methods applied to *D. serrata* fail to elicit an effect of copying (Auld et al. 2009). Guppies from different populations also differ in their tendency to copy (Vakirtzis 2011). Early learning is similarly canalized by instinct. For example, Bailey and Zuk (2012) found that exposure to song had highly variable effects among field cricket populations. Rebar and Rodríguez (2013) found strong effects of social environment X family on peak preference and choosiness in treehoppers. We should also expect to see G X E interactions for parental and epigenetic effects (Bonduriansky & Day 2009).

Genotype-by-environment effects may make themselves manifest in subtle and unusual ways. Experience with heterospecifics, interacting with a divergent set of innate factors, may provide an impoverished set of cues to shape subsequent mating decisions. Campbell and Hauber (2009) showed that cross-fostered zebra finches were less choosy with respect to songs of different species, presumably reflecting the conflict between their "instinct to learn" and their social environment. By contrast, when given the opportunity to pair with live partners, the "ditherers" in ten Cate's (1987) study, that had imprinted on both conspecifics and heterospecifics, showed a preference for interspecific hybrids.

Genotype X environment effects on mating preferences are at the heart of what makes mate choice interesting. To address these effects properly, we need to begin understanding the underlying mechanisms that vary among lineages, and we need to incorporate them into genetic models of preference evolution. I will address each of these in turn.

As we saw earlier, the same chooser can show learned preference for one trait and an innate preference for another. The idea of a core subset of biases that are hard-wired and develop independent of experience—so-called **innate releasing mechanisms** (IRMs)—goes back to Konrad Lorenz (1935). These IRMs respond to key features in the environment, so-called sign stimuli. For example, naïve female mice show an innate attraction to nonvolatile contact pheromones produced by males, but naïve females fail to attend to volatile pheromones which can be detected at a distance. The volatile and nonvolatile pheromones are produced by the same males, so females learn to associate these initially neutral volatile cues with the innately preferred nonvolatiles (Moncho-Bogani et al. 2002). Immediate-early-gene studies (chapter 2) suggested that in mice, learning is achieved through integration in the basolateral amygdala of vomeronasal inputs attending to nonvolatile cues, and olfactory inputs attending to volatile cues (Moncho-Bogani et al. 2005). Such associative learning is a generally plausible mechanism for the "instinct to learn": the innate cue can serve as an unconditioned stimulus in a learning paradigm (ten Cate 1994).

There are multiple scenarios whereby mutations would change the hedonic value of social experience, the attention given to different courter traits, or the response functions of IRMs over the course of chooser development. Biased learning need not require an amygdala; for example, the coupling of receptor stimulation to receptor gene expression can be positive, neutral, or negative, thereby biasing learning at the sensory periphery. In either case, the "instinct to learn" may represent a sizable mutational target.

In addition to identifying candidate mechanisms for the evolution of learning, it is helpful to have a biologically realistic framework for thinking about how social effects interact with chooser preferences. In the quantitative-genetic model presented in chapter 11, social effects are part of a myriad of environmental effects. What is distinctive about social effects is that they themselves are organismal phenotypes shaped by genotypic, social, and environmental effects. For example, the mate choice of a model female is a phenotype that affects the phenotype of the focal female by biasing her mate choice. We can therefore think about the focal female not only making a choice due to a direct effect of her own genotype, but also from the **indirect genetic effect** of the model's genotype (Rodríguez, Rebar, and Fowler-Finn. 2013). The effect of the model's phenotype (and the indirect effect of genotype) is scaled by Ψ, the coefficient of social interaction. We can express the phenotype of a focal individual *i* as

$$z_i = a_i + e_i + \Psi z'_j$$

where *a* represents additive genetic variance, *e* represents environmental effects, and z'_j represents the social effect produced by model individual *j*. Ψ represents the scaling of the relationship between the social effect and the focal individual, and can be positive or negative. Another way to think about Ψ is as the hedonic value of social information. Bailey and Zuk (2012) found that Ψ differed substantially, with reversals in sign, between reproductively isolated field cricket populations. Models that incorporate learning in an IGE framework may be invaluable in understanding mate-choice evolution.

12.10 SYNTHESIS

This chapter and the last underscore the importance of the nest—in a metaphorical sense that encompasses the mbuna's mouth and the ewe's udder—in shaping later mating decisions. Scores of studies have experimentally isolated the effects of nutritional stress, social interactions, and other developmental variables in structuring preferences. In nature, these factors will covary within and across generations—for example, food stress presumably will be associated with increased time of parents away from the nest and therefore reduce the amount of social interaction. These effects will shape mating decisions that in turn feed back on the environment experienced by the next generation of offspring (Rodríguez, Rebar, & Fowler-Finn 2013; fig. 12.8). Characterizing ecologically relevant syndromes that shape natural

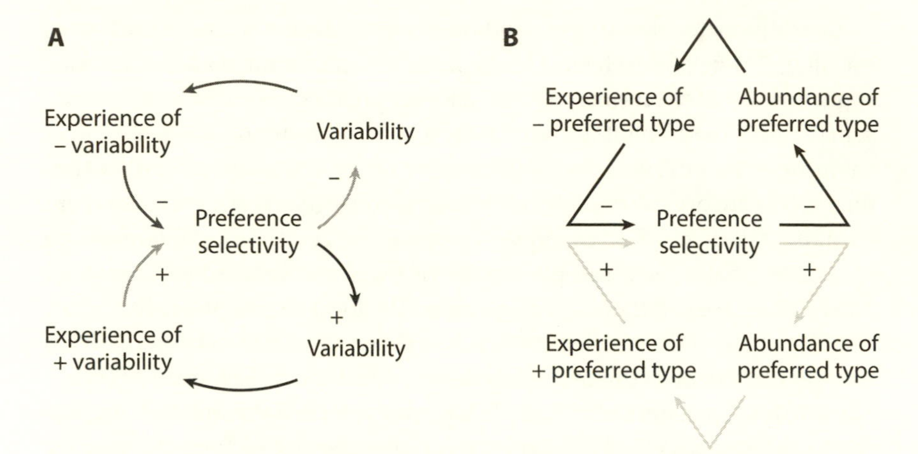

Figure 12.8. Rodriguez and colleagues' (2013a) hypothesis for feedback loops between social experience and mating preferences. (a) Negative feedback loop caused by choosers becoming more selective when exposed to variable stimuli. (b) Positive feedback loop whereby abundance of a particular stimulus type increases its attractiveness, thereby causing it to be favored by selection and ever-more abundant. After Rodriguez et al. (2013a).

variation in preference will be key to understanding its role as an agent of selection.

Another central question is the extent to which aspects of preference are specified early on in life and robust to later influences, whether because of epigenetic or other parental effects in utero or because of social experience during a critical period. This is particularly important in determining mating outcomes if choosers disperse and encounter a distribution of courters that diverges from parental phenotypes.

The "instinct to learn" in mate choice—whether choosers are more sensitive to experience within a critical period, whether innate responses to certain cues trigger associative learning with more complex features, whether flexible learning is biased toward certain stimuli—is, if not ubiquitous, certainly widespread among vertebrate and probably invertebrate taxa. To understand how preferences evolve and operate, we need a synthetic view that encompasses how learning and environmental flexibility interact with genetic variation. This view can be greatly illuminated by an understanding of the proximate mechanisms underlying the integration of social information, since the functional responses of these mechanisms to social cues over development can help us link behavioral variation across contexts with the underlying genetic variation. In mammals, there may be a neuroanatomical locus for the social integration of mate choice in the amygdala. The amygdala may be the locus for associative learning triggered by innate olfactory cues (Moncho-Bogani et al. 2005). The amygdala in female mice responds more strongly to chemosignals of familiar over unfamiliar males (Binns & Brennan 2005). Oxytocin-mediated activity in the amygdala is necessary for individual recognition in female mice (Choleris et al. 2007), and oxytocin-mutant mice fail to copy mating preferences (Kavaliers et al. 2006). Lesions to the amygdala disrupt operant responses to sexual stimuli (Pfaus et al. 2001).

In humans, the amygdala shows functional-neuroanatomical differences between gay and straight men (Savic & Lindstrom 2008), and amygdala size correlates positively with social network size (Bickart et al. 2011). The amygdala is selectively activated by visual mate choice in men (Funayama et al. 2012), and amygdala activity increases for both men and women viewing pornography (Fisher, Aron, & Brown 2006). In fishes, the dorsomedial telencephalon (Dm) is homologous to the basolateral amygdala in mammals. The Dm is one of several brain regions to show elevated expression of the *neuroserpin* gene in female swordtails *X. nigrensis* (Wong et al. 2012); neuroserpin was associated with mate preference in a previous brain-wide microarray expression study (Cummings et al. 2008). Mate choice, associative learning, and individual recognition, not to mention watching pornography, all involve the weighting of distinct but interacting streams of sensory information into a positive, neutral, or negative hedonic evaluation. The amygdala, and corresponding brain regions in the telencephalon of other vertebrates, may play an important role in the social modulation of mate choice.

The neural mechanisms underlying mate copying deserve attention, particularly since they may recruit distinct cognitive processes from the above. Cichlids *Astatotilapia burtoni* perform transitive inference, with a focal male avoiding rival males whom he has observed defeating a male who has previously defeated him (Grosenick et al. 2007). Intriguingly, the Dm was not responsive to a female seeing her preferred male win or lose an agonistic interaction with another male (Desjardins et al. 2010), while regions in the social behavior network showed distinct responses to seeing a preferred courter subsequently win or lose.

Consider a female bird—perhaps one can only imagine zebra finches after reading this chapter, but let's keep it vague—on the cusp of sexual maturity. She enters a phase where she will decide who to partner with, who to mate with, whose sperm to fertilize her eggs with, and how to allocate her effort among those eggs that are fertilized. Much of this decision process will be constrained by the contingencies of where she lives and which courters she happens to encounter, and how other choosers exercise their preferences—not to mention the extent to which those courters and choosers interfere with her and with each other. Notwithstanding these limitations, she has considerable agency in determining who she pairs and reproduces with. Part of that agency is determined when she starts as a zygote, carrying genes on her Z chromosome that will canalize her preference and sex-determining factors that will eventually trigger the hormonal cascade that shapes her brain as female. Another part is determined even before then, when her mother and father's experiences—and their choices regarding each other—are transduced into allocation decisions and epigenetic markings that shape her early development. As our not-zebra-finch hatches and grows, it is her nest—the food she receives, her interactions with parents and siblings—that shapes her mating preferences in ways that will dispose her to be more or less choosy, to be attracted or repulsed by the orange forehead patch on that one male, to mate or not mate with that other male who is so different from her and yet familiar.

These individual decisions, woven from so many strands and connected to the choices of others, are what make sexual selection and reproductive isolation. The rest of this book focuses on how mating preferences come to be, how selection acts on preferences and mating decisions, and on how these decisions shape evolution.

12.11 ADDITIONAL READING

Bonduriansky, R., & Day, T. 2009. Nongenetic inheritance and its evolutionary implications. *Annual Review of Ecology, Evolution, and Systematics, 40*, 103–125.

Riebel, K. 2009. Chapter 6 song and female mate choice in zebra finches: a review. *Advances in the Study of Behavior, 40*, 197–238.

Rodriguez, R. L., Rebar, D., & Fowler-Finn, K. D. 2013. The evolution and evolutionary consequences of social plasticity in mate preferences. *Animal Behaviour,* 85, 1041–1047.

Vakirtzis, A. 2011. Mate choice copying and nonindependent mate choice: a critical review. *Annales Zoologici Fennici, 48*, 91–107.

PART 2

Origins, Evolution, and Consequences

CHAPTER 13

Origins and Histories of Mating Preferences

CHOOSER BIASES

13.1 INTRODUCTION

Where do mating preferences come from? In this chapter, I consider how an organism's biology is recruited to mate choice. The evolution of mate choice has been considered primarily through the lens of how preferences coevolve with the traits they target, which I revisit in chapter 15. But how are preferences expressed in the first place?

Preferences can only exist if there is variation in courter traits; otherwise there is no way a preference can be expressed. But a host of chooser characteristics can potentially function to discriminate among courter phenotypes and therefore bias mate-choice outcomes. When there is no variation in the courter traits targeted by preferences, or when choosers are unable to exercise their choices, mate-choice outcomes do not impose selection on the mechanisms that underlie preferences. While coevolution with courter traits plays a major role in how these mechanisms evolve, accumulating evidence shows that many aspects of preference are conserved whether or not there are traits around to address them. In other words, preference phenotypes very often do not covary with courter traits, and can thus evolve independently of courters at least part of the time.

Preferences can be elicited in choosers with no evolutionary history of a targeted courter trait. Choosers will frequently favor courters with novel traits provided by experimenters. For example, zebra finches prefer mates with artificial colored leg bands, and prefer red over green leg bands. They also prefer mates fitted with artificial crests painted white, but not crests of other colors (Burley & Szymanski 1998). Similarly, female mosquitofish *Gambusia holbrooki*, whose males are unornamented and exhibit entirely coercive mating behavior, show behavioral biases for models of males with novel textures and fin exaggeration (Gould et al. 1999); female túngara frogs prefer conspecific calls adorned with human-made "bells" and "whistles"

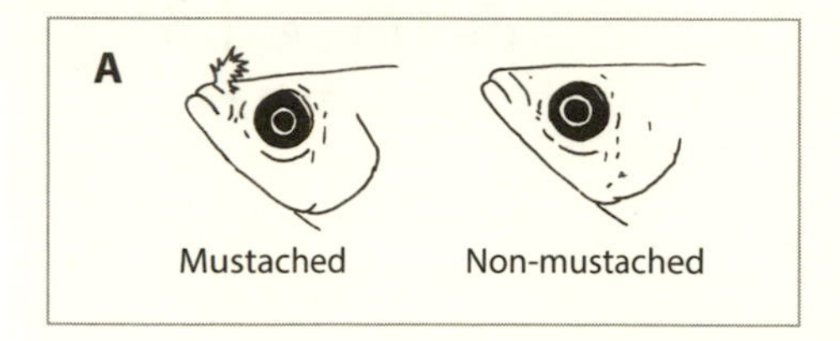

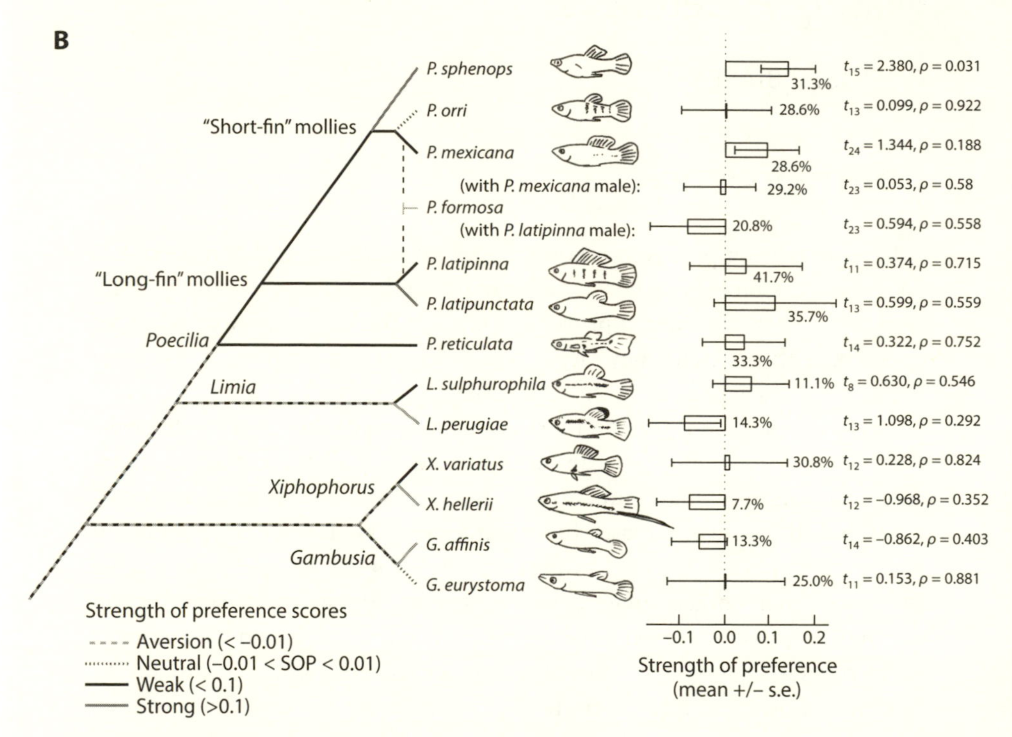

Figure 13.1. Variation in preferences of female poeciliid fish for a novel trait, mustache-like rostral filaments. Bar graph indicates net strength of preference for video-animated conspecific males expressing or lacking mustaches. After McCoy et al. (2003).

(Ryan et al. 2010); and female poeciliids prefer males with possibly ticklish "mustaches" on their snouts (McCoy et al. 2011; fig. 13.1). Amézquita and colleagues (2011) showed in female dendrobatid frogs that the acoustic space that elicits a sexual response from choosers was substantially broader than the signal space of conspecifics. Biases are exposed only by novel opportunities for choices, so these human-made adornments presumably elicit only a tiny subset of the biases that mosquitofish and zebra finches have that could be expressed as mating preferences.

Where do these "hidden preferences" originate? Chooser biases are perhaps most obvious when they involve complex behavioral mechanisms, but they are also evident in the basic molecular constraints on sexual reproduction. Bias is inherent to sex itself: no organism goes around shedding gametes willy-nilly. At its simplest, fertilization is contingent on a chooser (the egg) expressing a membrane-bound receptor that responds to a ligand on the surface of a sperm. The tuning curve of the receptor—the probability that it will bind a ligand as a function of the concentration and chemical structure of that ligand—is, like the tuning curves of the sensory periphery (chapter 3), a multivariate preference function, and one that surely varies across physical and chemical gradients like temperature and pH.

The potential for complex mate choice was therefore present at the creation of sex itself. This minimal requirement for sexual reproduction—molecular recognition between egg and sperm—is a kernel of what mate choice entails in more complex systems: multivariate preference functions dependent on signal dynamics and the external environment, applied as a choice algorithm (in this case, a static threshold). This means that there are also bound to be hidden preferences (chapter 2) outside the current range of courter variation: any novel ligand that elicits a different response at the same concentration, or that responds differently to environment-induced changes, will elicit a different preference.

From a courter's point of view, there are therefore multiple pathways to targeting these **preexisting biases**. Michael Ryan (1990, 1998; Ryan et al. 1990) termed this process **sensory exploitation**: a mutant courter that addressed these hidden preferences would have higher mating success. Courters in different lineages could evolve different traits that elicited the same chooser bias; for example, male frogs in the genus *Physalaemus* have evolved a variety of acoustic adornments that increase call attractiveness (Ryan & Rand 1993a). Continuing with the example of fertilization, a courter could increase its probability of being chosen by producing much more of a less-effective ligand or much less of a more-effective one, or one that is less effective when pH is high but much more effective when pH is low. An even more successful courter could modulate signal structure as a function of pH, and signal concentration as a function of competitor density.

As mate-choice mechanisms become more complex, so do the sensory, perceptual, cognitive, morphological, and physiological components that can be recruited as preferences. This starts to be apparent with internal fertilizers, where it is choosers themselves who determine the chemistry and physics of the environment for gametes. Any mechanism that females use to store sperm, or through which sperm transits, becomes a mechanism for postmating preference. Chooser properties can also become premating choice mechanisms if they are general to the perception and processing of environmental signals. For hearing and vision, the initial detection and attention to a courter signal often depend on broadly tuned, low-level sensory and attentional processes. Choosers prefer courters that incorporate attentional components into their displays (e.g., Christy et al. 2003). As I will discuss below, an even more universal property of sensory systems—habituation to repeated presentations of the same signal—provides a possible explanation for preferences for signals of increasing complexity.

Finally, chooser properties can be recruited for mate choice even if they function in a context entirely different from mating. Nonsexual responses to appetitive stimuli, typically prey, have the effect of increasing proximity and attention to choosers (e.g., Proctor 1991; Macías Garcia & Ramirez 2005; Kolm et al. 2012). And, as noted by Tinbergen (1952), choosers will often respond initially to courters with fear, aggression, or hunger; the differential inhibition of these responses is also a form of mate choice.

All of these very different chooser biases share a key property. They act to generate preference functions; that is, they constitute mechanisms for differentiating among courter phenotypes. But these preference functions exist for reasons that have nothing to do with the current distribution of courter traits; they have not been shaped by selection on chooser mating decisions (Kirkpatrick 1987a; Ryan 1998). The only way we know they exist is because we can present choosers with stimuli they have never encountered before. We can refer to these collectively as chooser biases (cf. "receiver biases," Ryan 1998; "perceptual biases," Ryan & Cummings 2013): they favor some courter phenotypes over others, but their preferences are originally **non-adaptive** with respect to the consequences of mate choice.

When expressed as preferences, these biases may do so in ways that not only benefit choosers but also harm choosers. Once relevant variation emerges in courter phenotypes, these previously hidden biases are now expressed as preferences because they generate differences in mating or fertilization outcomes according to courter phenotype (fig. 13.2). These outcomes can vary in their fitness consequences for choosers, such that aspects of mate choice are subject to direct selection. And they certainly vary in their fitness consequences for courters, which in turn generates indirect selection on preferences. In later chapters, I will address how preferences evolve once they are expressed as such. Here, I survey how biases are shaped prior to their expression as mate preferences and address how we can empirically characterize them.

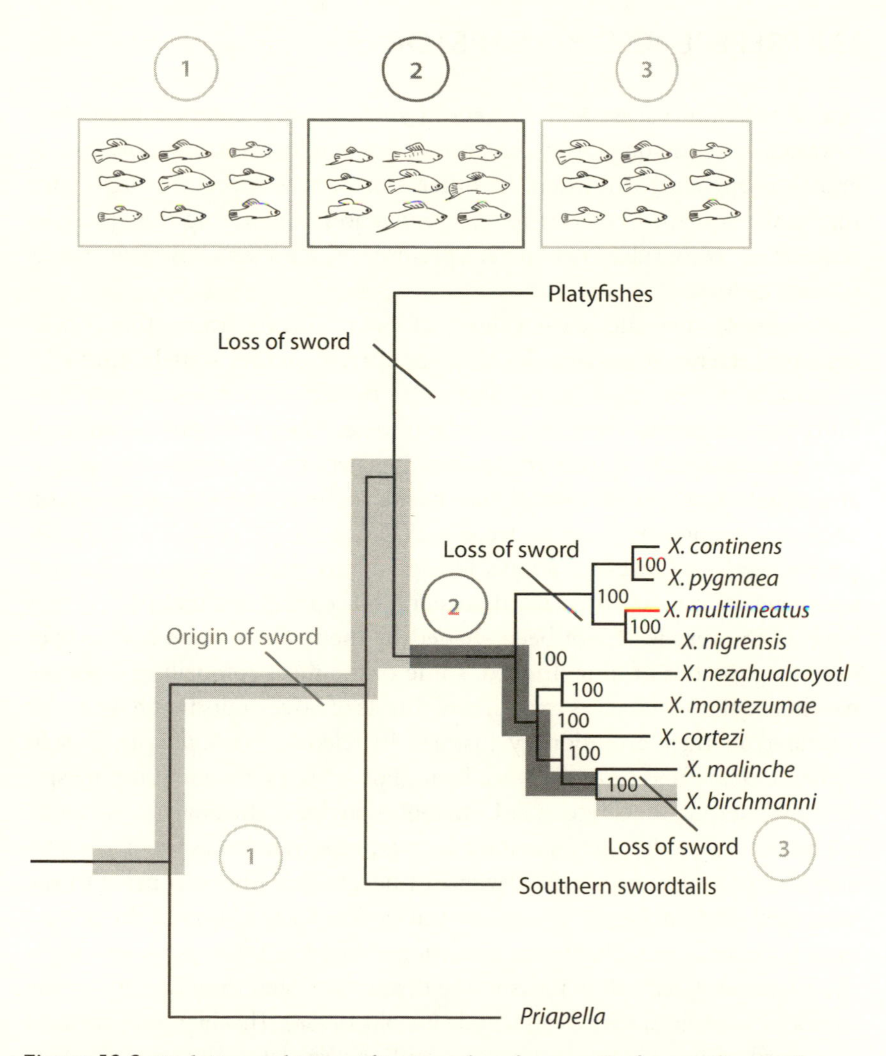

Figure 13.2. Evolutionary history of traits and preferences. Preferences for swords evolved in a common ancestor of *Priapella* and *Xiphophorus* (Basolo 1998a) and were secondarily lost at least once in the northern swordtails (Rosenthal et al. 2002; Wong & Rosenthal 2006). Shaded area shows the opportunity for mate choice with respect to swords in the lineage leading to *X. birchmanni.* Swords evolved in the common ancestor to *Xiphophorus* and were secondarily lost after the split between *malinche* and *birchmanni*. Choosers could not express preferences for males with swords prior to their evolution (1) or after their loss (3). Preferences for swords could only be expressed, and selection could only act on them, during the period encompassed by the darker shading (2).

13.2 PREFERENCES HAVE HISTORIES

A key question that we will address in subsequent chapters is the extent to which selection on mating outcomes plays a role in shaping how a preference is currently expressed. In other words, how much are preference mechanisms shaped by their role *as preferences*? Answering this question requires us to consider how current preferences are shaped by their evolutionary history.

As the examples cited earlier show, we can readily determine that choosers have preferences outside the range of current courter signals simply by measuring preferences for novel traits. But this does not rule out the possibility that these biases reflect a history of mate choice for similar traits that are currently absent in courters. Indeed, it is perhaps impossible to demonstrate that a response to novel signals has *never* been subject to selection on choosers arising from mating decisions (Fuller, Houle, et al. 2005). Phylogenetic analyses however, can yield insights into preference evolution over timescales of millions of years, increasing our confidence that preferences have, at the very least, not been shaped by their role in mate choice over very long periods of time. Indeed, some of the most compelling evidence for chooser biases involves experimental tests of novel traits that have never appeared in their evolutionary history. In teleosts (*Xiphophorus*; Basolo 1990a, 1998a), frogs (*Physalaemus*; Ryan, Fox, et al. 1990), and jumping spiders (*Schizocosa*; McClintock & Uetz 1996) phylogenetic analysis of multiple behavioral studies shows that females express preferences for ornaments characteristic of some species, even in lineages that diverged prior to the ornaments' origin. Figure 13.1 traces the evolutionary history of the preference for swords in *Xiphophorus birchmanni* (fig. 13.2). A bias for swords is the ancestral state, with females in the distantly related *Priapella* preferring males with artificial swords attached (Basolo 1998a). This bias persists until the sword evolves in response, several million years later. The preference for swords in choosers is now coevolving (chapter 15) with the sword ornament in courters. In the lineage leading to *X. birchmanni*, females reverse their preference, now favoring males with no sword. The sword is secondarily lost; since no male *X. birchmanni* have swords, there is no longer any selection on preference for swords in conspecific mate choice. While this suggests that selection is strongest on the mechanisms underlying preference when they are expressed as such, it also suggests that biases may be exposed to selection as preferences for relatively short periods of time. I will return to the relative importance of non-adaptive biases in the final chapter.

There are two important caveats about such phylogenetic studies: first, any interpretation of trait evolution depends on the phylogenetic hypothesis, and different phylogenetic tree topologies can yield different inferences about whether a chooser preference preceded a courter trait (hence the controversy over the preexisting chooser bias for a sword in *Xiphophorus*, reviewed in Cui et al. 2013).

Second, phylogeny-based analyses provide conservative evidence that a preference arose before a current trait, but do not provide very fine resolution. Biases can be very variable across species even in the absence of a trait, often showing reversals in hedonic value (e.g., McCoy et al. 2011; fig. 13.1). Once a chooser bias emerges, it will favor any novel courter trait that addresses it. Phylogeny-based trait reconstruction cannot pinpoint where traits change state between speciation events, so we can only detect preexisting biases when the responses to them are slow and sparse. We will fail to detect a chooser bias that arises between speciation events and is quickly met by courters. For example, Niehuis and colleagues (2013) identified a genetic locus associated with the biosynthesis of a novel pheromone in the wasp *Nasonia vitripennis*. Females were more attracted to pheromone blends containing the novel compound, while females from a sympatric species in the sister lineage were indifferent. They interpreted this as evidence that the preference had evolved after the novel signal had arisen, but this could just as well reflect a bias that evolved somewhere after *N. vitripennis* and its sister lineage diverged, or—just as likely—secondary loss of preference by the sympatric congener (chapter 16). Paradoxically, we can therefore more easily detect chooser biases that have less of an impact on courter trait evolution, and we may therefore be substantially underestimating the importance of preexisting biases to preference evolution.

A compelling case for trait-independent evolution can be made if the bias functions in another context independent of any function in mate choice (Fuller, Houle, et al. 2005). These contexts fall into two categories: structural constraints (including perceptual biases) and non-choice functions. Structural constraints are general, lower-level responses recruited for numerous tasks which may or may not encompass mate choice like orienting to an alerting signal, low-level habituation, or the two-dimensional properties of receptor affinities. Non-choice functions include tasks like feeding responses that are recruited as a mechanism to bias the probability of mating or fertilizing with certain phenotypes. I will address each of these categories in the next section.

13.3 PERCEPTUAL BIASES

13.3.1 Detection and gross stimulation

Perhaps the broadest chooser bias is gross detectability. Ryan and Keddy-Hector's (1992) survey of mate-choice studies at the time showed an overwhelming pattern: all else equal, choosers have directional preferences for courter signals that elicit increased stimulation of the sensory periphery. While studies since then suggest that choosers often reverse the valence of preference (chapter 5), the ability to detect a courter from background is a prerequisite of mate choice (chapter 3). A chooser can't make a decision about courter signals unless they are discernible from background noise.

Sometimes, detection of signal from background involves narrowly tuned responses specific to courter signals in the context of mating; in insects, this is often the case for acoustic (Jain et al. 2014) and chemical signals (Leary et al. 2012). Even though such narrow tuning is prima facie evidence for the coevolution of preferences with courter signals (chapter 15), detection will always be enhanced for courters who produce more intense signals, up to the point of receptor saturation (chapter 3). The directional bias for more intense signals constitutes a chooser bias: more intense signals will always be more readily detected (if by no means always favored), independent of how selection acts on chooser mating decisions. As I review later, it is difficult to disentangle such non-adaptive explanations for the origin of directional preferences from adaptive explanations linking courter investment in signals with benefits to choosers (Kokko et al. 2002), including reduced search costs (Dawkins & Guilford 1996).

As discussed in chapter 3, narrow tuning of the sensory periphery is one extreme of a continuum of domain specificity that varies strikingly among taxa and among modalities. At the other extreme are domain-general aspects of sensation that are common to a broad gamut of ecological tasks, like distinguishing figure from ground in a visual image or localizing the source of a sound. Both these tasks are, as above, going to favor more expensive courter signals (increasing projected area or increasing sound amplitude). Ryan and Keddy-Hector (1992) documented many examples of choosers preferring courters of larger body size, with more high-contrast body markings, or with higher-amplitude calls.

No matter how broad or narrow the sensory process, it is going to discriminate among courters on the basis of how likely courters are to be detected from background noise and how much the sensory periphery is going to be stimulated by courter signals. This stimulation of the sensory process can be explained by universal properties of how sensory modalities work. For example, all eyes are stimulated more strongly with increasing brightness contrast and all ears are stimulated more strongly at a given frequency with increased sound pressure level.

Often, however, detection and gross stimulation depend on the environment in which signaling occurs—the extent to which water-soluble chemicals can propagate through the environment, or low-frequency sounds are scattered, or long wavelengths of light are filtered. John Endler's (1992, 1993; Endler & Basolo 1998) **sensory drive** hypothesis posited that the environment shapes overall sensory function, which in turn drives the design of communication systems including mate choice. Sensory drive provides a particularly powerful explanation for the chromatic tuning of photoreceptors and the wavelength composition of color signals (Cummings & Partridge 2001; Ryan & Cummings 2013).

13.3.2 Attention and habituation

Another set of biases arises from the challenge of filtering sensory inputs in a complex environment. What stimuli should an individual pay attention to? Some of these represent threats like predators or competitors, and some might represent opportunities like shelter, food, or mates. Some of the mechanisms involving selective attention are low-level and broadly shared among taxa. Guilford and Dawkins (1991b) noted that many courtship displays (e.g., the head-bob displays of *Anolis* lizards, Fleishman 1992) began with high-temporal-frequency movements that elicit perceptual attention, like a directed gaze. Courter mating success could be largely a function of how successful courters are in getting choosers to attend to them, in the absence of any structures in choosers that are specialized for mate choice.

Attending to sudden onsets of stimuli is one basic property of receivers; habituating to repeatedly presented stimuli is another. Stimuli that are invariant over space and time quickly become uninteresting, which can be explained by the attenuated response of individual neurons to repeated stimulation. Songbirds have latent biases for more complex songs (zebra finches, Collins 1999; common grackles, Searcy 1992). In zebra finches, repeated presentation of the same song causes attenuated electrophysiological and IEG responses in the auditory forebrain (reviewed in Dong & Clayton 2009). Mate preference for courters producing more complex stimuli could thus emerge from a low-level fundamental property of nervous system function. Habituation could also affect sequential responses to multiple courters, although release from habituation may depend on the attractiveness of the stimulus (Sockman et al. 2002, 2005; chapter 6).

13.3.3 Discriminability and memorability

As detailed in chapter 6, making a comparison among potential mates poses a cognitive challenge. The cognitive constraints on mate sampling can constitute biases that favor some courters over others. For example, Akre and Ryan (2010) showed that call complexity increased female working memory for male signals in túngara frogs. The more complex a call, the more likely a female was to remember its spatial location, and therefore to mate with the courter producing the call. More generally, complex courter ornamentation may generally make it easier for a chooser to remember and distinguish a preferred mate. Complexity, along with novelty, may increase the discriminability and memorability (Guilford & Dawkins 1991a, 1991b) of stimuli: in other words, some stimuli may be inherently easier to remember and distinguish from others, in ways that depend on broad constraints of perceptual systems and are therefore robust to trait-preference coevolution.

13.4 BIASES FROM NON-CHOICE FUNCTIONS

Some of the clearest examples of preferences not coevolving with traits are found when courters use cues that elicit responses from choosers that are clearly distinct from mate choice. These nonsexual responses, typically feeding responses, then increase the probability that a chooser will attend to later stages of courtship and ultimately fertilize her eggs with his sperm.

Sometimes feeding responses are elicited by actual edible substances. Courtship feeding is ubiquitous in birds (Lack 1940; Helfenstein, Wagner, Danchin, & Rossi 2003; Wiggins & Morris 1986), as are nuptial gifts in insects, which are often connected to a spermatophore that inseminates a female (Sakaluk 2000; chapter 7). Once these feeding responses are connected to mating decisions, they become exposed to selection as mating preferences, particularly since the nutritional content of nuptial foods can influence chooser fecundity. As I revisit when discussing sexual conflict in chapter 15, even nuptial gifts are subject to a form of "deceit" by courters in terms of their direct benefits; Warwick and colleagues (2009) showed in crickets *Gryllodes sigillatus* that nuptial gifts were a sort of "candy," flavored with free amino acids but lacking in nutritional value. These gifts were exploiting choosers with sensory cues normally associated with high-nutrient foods.

Often, courters fool hungry choosers into perceiving non-food courter cues as food. For example, female water mites approach males mimicking the temporal pattern of a drowning insect (Proctor 1991). Female sticklebacks in lineages ancestral to the origin of male red nuptial coloration also show a bias for red in a foraging context (Smith et al. 2004). Kolm and colleagues (2012) provide a striking demonstration of how chooser preferences can arise from nonsexual biases. Swordtail characins (*Corynopoma riisei*) feed on terrestrial insects washed into the water column. Males express ornaments that mimic prey items; the shape of the ornament varies according to which insects are more prevalent. Further, recent experience with food items changes the propensity of females to attend to male courtship signals; when females are habituated to flake food or *Drosophila* larvae, they are less likely to attend to the ant-like ornament (fig. 13.3).

Phylogenetic analyses provide further compelling evidence that biases have evolved in a nonsexual context. Conner (1999) argued that sexual

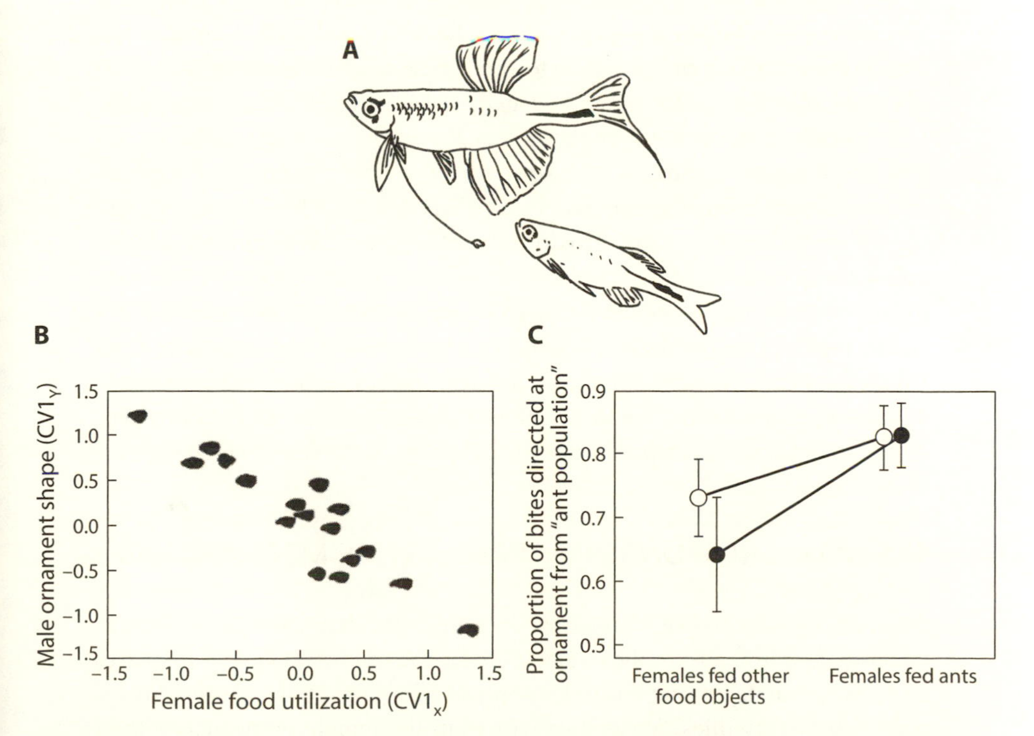

Figure 13.3. (a) Female swordtail characin *Corynopoma riisei* (left) moving in to bite the opercular flag of a male. (b) Ornament shape covaries with female diet, with more ant-like stimuli in populations where females feed heavily on ants. (c) Recent dietary experience changes female responses to male ornaments; ant-fed females direct more bites at ant-like male ornaments than females fed on flake food (filled circles) or *Drosophila* larvae (open circles). After Kolm et al. (2012).

communication in several moth families had arisen several times independently, with male courtship signals exploiting tympanal sound receptors used to avoid bat predation. Males in one lineage of goodeid fishes have "terminal yellow bands" on their caudal fins that resemble insect larvae. Females approach and nip at these bands, ultimately increasing the likelihood that they will mate with a male expressing a band. Critically, artificial bands induce the same response from females in lineages that never had bands (Macías-Garcia & Ramirez 2005; Macías-Garcia & Saldívar Lemus 2012).

Approaching a courter is of course a very early stage of mate choice, and choosers often reject further courtship after approach. Redirecting choosers from attacking prey is another challenge for courters; male goodeids frequently have their fins damaged by female prey-capture attempts (Macías-Garcia & Ramirez 2005). All else equal, however, courters that elicit more approaches are probably going to have higher mating success, independent of any chooser mechanisms specific to mate choice.

Cues associated with predator avoidance may also serve as preference mechanisms. In fiddler crabs, *Uca beebei*, females are drawn to vertical pillars built by males. Christy (1995) termed these pillars a "*sensory trap*" that elicited landmark orientation associated with seeking shelter from predators. Consistent with the hypothesis that the response evolved in the context of predation avoidance, the strength of female preference for pillars increased with increased predation risk (Kim et al. 2009). Female noctuid moths *Spodoptera litura* stop moving when they hear the ultrasonic echolocation calls of bats. Males produce songs that are perceptually indistinguishable from bat songs, inducing the females to stop moving and allow mounting (Nakano et al. 2010). What this tactic has in common with food mimicry is that courters are using particular cues to fool the perceptual systems of choosers in ways that benefit courters, and may or may not benefit choosers. I will return to consider this point in the next two chapters.

13.5 NOVEL RESPONSES OF PREFERENCE MECHANISMS

Mate-choice mechanisms are complex, and a mechanism that evolves in order to discriminate among courters along one trait axis can expose biases that discriminate among courters along another axis. For example, in calopterygid damselflies, sensory structures in the female reproductive tract modulate egg fertilization and oviposition, a mechanism of postcopulatory mate choice (chapter 7). Córdoba-Aguilar (2002) showed that males in one lineage have evolved genital structures that stimulate these structures to induce ejection of sperm from a previous male. A process used in postcopulatory mate choice by females is therefore fooled to postcopulatory mating advantage by males, opening a new front in the evolution of courter traits and chooser preference mechanisms. Females' "ejectability" constitutes a bias, because it exists independent of whether males can take advantage of it.

13.6 BYPRODUCT BIASES: NOVEL BIASES SHAPED BY CURRENT SIGNALS

The chooser biases discussed above all have one thing in common: they were present before courter traits arose to address them. These biases therefore manifest themselves as preferences for novel traits not present among current courters. It is important to distinguish these true chooser biases from other preferences for novel traits which do in fact depend on preferences for current traits, and are therefore linked to selection on realized preferences. These "byproduct biases" evolve in response to selection on mating outcomes (chapter 14) and coevolve with male traits (chapter 15), although their dynamics have not, to my knowledge, been specifically addressed outside the context of how intraspecific mate choice covaries with conspecific mate recognition (chapter 16). They are biases in the sense that they can favor novel traits, but they favor novel traits in a way that is inseparable from preferences that act to produce variation in mating outcomes.

Byproduct biases arise in three main ways: first, learned preferences (chapter 12) based on current trait distributions may cause preferences for novel stimuli via generalization and peak shift. Second, selection for directional preference functions (chapter 3) can result in exaggerated preferences for novel traits. Third, integration rules among preference mechanisms (chapter 4) will often favor unoccupied areas of multivariate trait space. (A fourth, correlated responses to selection on a different aspect of preference, is covered in the next chapter.) In each case, there is a phenotypic correlation between how choosers respond to variation among current traits, and how they respond to variation outside the current range.

The first mechanism for byproduct biases is that experience-dependent preferences, whether short-term (chapter 6) or long-term (chapter 12), will lead choosers to generalize among the stimuli they've experienced, and, perhaps quite often, exhibit peak shift based on previous experience with courter traits (ten Cate & Rowe 2007). Choosers have preferences for traits outside the current range of chooser variation, but these preferences are shaped by experience with current choosers.

The second mechanism is that if preference mechanisms have directional properties that extend outside the current range of courters, they will automatically produce preferences for a level of stimulation outside the current range of variation. The strength of those preferences, however, may depend on how selection acts on preferences within the current range of courter traits. Note that while the strength of preference for supernormal stimuli can evolve in response to selection on current stimuli, this need not be true for the direction of preference, i.e., the preference for gross detectability, above.

Finally, the integration rules for preferences (chapter 4) mean that some novel stimulus *combinations* will be more attractive than extant stimuli (Witte & Curio 1999; Fisher et al. 2009). For example, female swordtails have additive preferences favoring novel males with big bodies and small fins. The preference for large body size, the preference for small fin size, and the rules governing their interaction produce preferences for novel traits, but they are all currently exposed to selection as preferences. I will return to byproduct biases in chapter 16, in the context of the relationship between interspecific and intraspecific mate choice.

13.7 SYNTHESIS

13.7.1 Chooser biases arise independent of courter traits

The first step in answering why preferences exist is to determine whether they exist because they are preferences. Choosers have biases favoring one courter over another. Did these biases evolve because a favored courter was somehow better than another for a chooser, or did they evolve in a context outside of mating? Ryan & Cummings (2013) compiled a conservative set of 117 studies demonstrating a sensory or perceptual basis for a chooser bias (see also Rodríguez & Snedden 2004). Quantitative-genetic studies of traits and preferences show weak correlations between traits and preferences, and phylogenetic studies show frequent mismatches between traits and preferences (chapter 15). In this view, preferences are mostly expressed the way they are because of constraints imposed by evolutionary history and/or current function outside the context of mate choice. Coevolution is relatively unimportant to how mating preferences are currently expressed.

On the other hand, the same phylogenetic studies that show courter traits arising in response to a courter bias show that the trait's origin is followed by a reduction in the strength of preference (chapter 15; Macías-Garcia & Ramirez 2005; Basolo 1998), providing evidence for selection on preferences qua preferences. There may be an intricate interplay among preference mechanisms as selection acts to modify preexisting biases. As chapter 5 argues, a growing number of studies show that greater detectability does not translate into greater preference. Selection in a preference context can favor more sophisticated mechanisms that override low-level perceptual biases (Wong & Rosenthal 2006; chapter 5).

The relationship of sensory biases to mate choice may indeed be more complicated than is commonly thought. In the mechanistically explicit scenario proposed by Madden and Tanner (2003), sensory tuning provides the link between nonsexual and sexual preferences. In this scenario, sensory biases in a foraging context (a consequence of sensory drive, see below) are responsible for chooser preferences in a mating context. A series of studies by Rebecca Fuller and colleagues challenges the hypothesis that sensory biases in the context of mating are likely to arise as a byproduct of selection favoring sensory function in other tasks. A neural network simulation showed that sensory biases could evolve independently with respect to foraging and mating tasks (Fuller 2009). In empirical studies on bluefin killifish *Lucania goodei*, Fuller and colleagues (2010) showed that the relative expression of different cone opsins in the retina—a standard proxy for color sensitivity—accounted for less than 4% of the variation in behavioral preferences for color in a foraging task. This result was consistent with most of the variation in behavioral preferences being modulated by downstream processes (presumably including effects of state-dependent and socially modulated variation, chapters 11 and 12). Fuller and Noa (2010) further found that color preferences in mating tasks were uncorrelated with color preferences in foraging.

A final caveat is that biases in a nonsexual context may be influenced by their use in a sexual context, rather than the other way around. For example, Madden and Tanner (2003) found that female bowerbirds had the same rank-order preferences for artificially colored grapes as males did for decorative objects used in their bowers. While they proposed that foraging biases had driven object preference, they acknowledged that color preferences could have arisen in the context of mating and subsequently biased choices of novel food objects.

Kokko and colleagues (2006) point out that both chooser biases and adaptive coevolution are likely at play in any preference-trait dyad. The question then becomes: how important are non-adaptive biases to mate choice, relative to features that have evolved in the context of optimizing the fitness benefits of mating decisions? Put another way, how important is preference evolution to the way preferences are currently expressed?

13.7.2 Chooser biases are a broad target for novel traits

As the examples in this chapter have shown, biases can occur at every pre-, peri-, and postmating stage of mate choice, and can be capricious in nature. Biases presumably vary among and within individuals, and in response to social and environmental cues, as much as expressed preferences do. As I discuss in chapter 15, a slight bias in a few choosers can kick-start runaway coevolution of traits and preferences. Biases can also be very permissive, which means that there are many routes for them to respond to courters. For example, the chuck in túngara frogs (Kime et al. 1998; Ryan et al. 2010) and the sword in swordtails (Rosenthal & Evans 1998; Haines & Gould 1994) are both no more effective than a wide array of synthetic stimuli. Any novel courter trait that elicited these permissive biases would have an advantage.

13.7.3 Courter traits turn biases into preferences

Until choosers have preferences, natural selection can't act on choosers based on their mating preferences (Kirkpatrick 1987a). Once courters have co-opted a preexisting mechanism in choosers that favors some mates over others, selection can start acting on choosers. In some cases, the courters that produce the most detectable signals might be the ones that maximize direct or indirect benefits; in such cases, selection would favor maintenance of the original bias on the part of choosers, or even an exaggeration of the original response. In other circumstances, selection favors the reduction, loss, or even the reversal of the original preference. For example, female goodeids forage less efficiently after they've been exposed to male ornaments mimicking food items. The chooser bias is no longer a bias, but a preference that affects a chooser's interactions with courters.

Arnqvist and Rowe (2014) echo what is perhaps the prevailing view that one should begin by understanding why preferences exist rather than focusing on the particulars of how they work. But in many if not most cases, at least the rough outline of chooser preference space is broadly constrained by ecology and by the basic nature of sensory function; that is, by functions not directly related to mate choice. How preferences work, in other words, may go a long way toward explaining their existence.

13.8 ADDITIONAL READING

Endler, J. A., & Basolo, A. L. (1998). Sensory ecology, receiver biases and sexual selection. *Trends in Ecology & Evolution, 13*(10), 415–420.

Ryan, Michael J., & Cummings, M. E. (2013). Perceptual biases and mate choice. *Annual Review of Ecology, Evolution, and Systematics, 44*, 437–459.

ten Cate, C., & Rowe, C. (2007). Biases in signal evolution: learning makes a difference. *Trends in Ecology & Evolution, 22*(7), 380–387.

CHAPTER 14

Selection on Mate Choice and Mating Preferences

14.1 INTRODUCTION

The last chapter surveyed the sweep of chooser mechanisms that can serve to discriminate among mates, from the molecular underpinnings of fertilization to feeding behavior. Each of these has the potential to act as a mate-choice mechanism, each has the potential to vary among and within choosers, and each has the potential to respond to selection on mating outcomes. What happens, then, when biases start to have consequences in the context of mating? Once a mechanism operates in the context of mate choice, it becomes subject to selection in that context, as long as mating decisions covary with fitness outcomes for courters (Searcy 1982; Rodriguez & Sneddon 2004).

This chapter focuses on the proximate sources of selection on chooser preferences and mate-choice algorithms. In the next section, I survey the sources of selection on the outcome of choice, some of which I return to in the next chapter on preference-trait coevolution. I then turn to the evolution of mate-searching strategies and choosiness. What is the optimal approach to making a mating decision: how should a chooser sample courters, and how choosy should she be? Next, I examine at what point in the process choosers make choices. Should choosers reject matings from all but the most attractive courters, or should they mate promiscuously and make decisions during or after mating? The rest of the chapter focuses on the evolution of environmental and social reaction norms in mate choice, before turning to the structural and ecological constraints on preference evolution. Mate choice can have positive or negative effects on choosers, and the magnitude and sign of these effects can vary dramatically as a function of ecological and social context over space and time.

14.2 SELECTION ON PREFERENCES FOR COURTER TRAITS

14.2.1 Overview

Mate choice is only good for choosers if there is some payoff to making appropriate decisions, whether in terms of direct benefits to chooser fecundity or survivorship or in terms of indirect benefits in the form of compatible or superior genotypes. We can divide sources of selection along two dimensions (Neff & Pitcher 2005). The first is **direct selection** versus **indirect selection**. Direct selection on mate choice arises when fitness is correlated with variation in mate-choice mechanisms or choice outcomes; in other words, when mate choice affects chooser survival or fecundity, or when the phenotype of courters and choosers affects the fitness of descendants. By contrast, indirect selection occurs when fitness is correlated with a phenotype that in turn is *genetically* correlated with another phenotype; in other words, when mate choice affects the fitness of descendants because of the genotypes they inherit from both parents. A key distinction is that under indirect selection, alleles change in frequency based on the sign and strength of their statistical association, or **linkage disequilibrium**, with alleles at other loci.

The second dimension is the extent to which these benefits are dependent on the phenotype and genotype of the chooser. At one extreme, benefits (or costs) are entirely additive: every chooser maximizes her fitness by mating with one particular Handsome Charlie. At the other extreme, the fitness benefits are entirely idiosyncratic. In other words, the fitness benefits to a chooser are maximized by genotypic or phenotypic compatibility with a given individual (fig. 14.1).

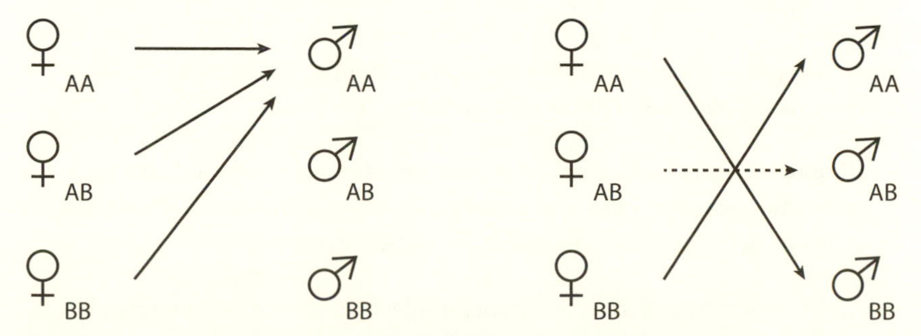

Figure 14.1. Additive and compatibility benefits. With additive benefits, all genotypes of females maximize their fitness with male genotype AA. With compatibility, each female genotype maximizes fitness with a distinct male genotype. After Neff & Pitcher (2005).

14.2.2 On "Quality"

Before we can begin to think about benefits to choosers, it is important to disentangle the benefits that actually accrue to them as opposed to those that *may* accrue to them because of putative attributes of courters. There is a long-standing body of research connecting mate choice to the mechanistic basis of male sexual signals. The overall logic, which I will return to in the next chapter, is that we can gain insight on sexual selection by characterizing the relationship between signal expression and some other, functionally important trait:

> The negative relationship between ornament asymmetry and size suggests that ornament size reliably reflects male quality because the largest secondary sex traits demonstrate the least degree of fluctuating asymmetry. (Møller 1992, p. 238)

> Does morphology indicate immunocompetence, providing females with honest cues to male quality? (Zuk et al. 1995, p. 205)

> If it is both heritable and ecologically valuable, high aerobic capacity would be a direct signal of male genetic quality. (Chappell et al. 1997, p. 511)

> Because call rate is related to hematocrit in Hermann's tortoise males, by responding to fast-rate calls females are also choosing high-quality partners as sires of their offspring. (Galeotti et al. 2005, p. 301)

> A male's original body size, as advertised by [hydroxydanaidal], is therefore an indicator of larval success that is likely to have a genetic component because offspring may inherit genes that enable them to outcompete other larvae. (Kelly et al. 2012, p. 1013)

> ... this pheromone signal informs females about the inbreeding status of their mating partners . (Van Bergen et al. 2013, p. 1)

> We propose that in some species, females select males based on their neuromuscular capabilities and acquired skills and that elaborate steroid-dependent courtship displays evolve to signal these traits. (Fusani et al. 2014, p. 534)

> ... the ornament preferred by females differs between the two populations, but the different ornaments signal similar aspects of male health and genetic quality, specifically information regarding MHC variation and potential indirect genetic benefits to females. (Whittingham et al. 2015, p. 1584)

"Quality" is a seductive shorthand, useful for thought experiments and theoretical models, that allows us to rank courters in terms of their benefits to choosers, without worrying about what those benefits are. The word is ubiquitously used—"mate quality" or "male quality" or "female quality" are used in a staggering 15,000 scholarly publications as of this writing—but is seldom explicitly defined (Wilson & Nussey 2010; Hunt et al. 2004). But Quality is so amorphous and all-encompassing that it does more harm than good. The mate-choice literature reveals a tautological tangle whereby Quality turns out to mean a courter's parasite load, oxygen metabolism, tarsus length, signal amplitude, social dominance, social subordination, attractiveness to other choosers, and so forth. Quite commonly, quality is assumed to be equivalent to "condition," another omnipresent and amorphous term (chapter 8). Many studies simply measure a correlation between chooser preferences and some aspect of the courter phenotype, like physiological performance or residual tarsus length, and take the relationship between these traits and offspring fitness as as axiomatic. As I discuss in chapter 15, there is in fact scant evidence that such courter "quality" measures are generally tied to positive effects on chooser fitness.

A more rigorous definition of quality may be conceptually helpful. Gould and colleagues (1999) define it as "the ability to provide females or their offspring with direct or indirect genetic or material benefits," which is consistent with Wilson and Nussey (2010)'s multivariate quantitative-genetic definition of *phenotypic* quality as "the axis of phenotypic variation that best explains variance in individual fitness." Hunt et al. (2004) define "genetic quality" as the "**breeding value** of an individual for total fitness," that is, the mean fitness of that individual's progeny across all possible mates. In our context, *courter* quality would be the best predictor of *chooser* fitness from observed *courter* trait values.

This is a quantity we can estimate in wild or experimental populations with some effort, but it points to three additional major problems with Quality as a concept. First, the combinations of traits or alleles that maximize fitness vary over space and time. Second, as we shall see below, much of mate choice involves selection of compatible mates, rather than globally superior ones. By focusing on Quality, we assign primacy to (putatively) superior attributes of courters rather than to their compatibility with choosers (Riebel et al. 2010). Third, workers often measure proxies of phenotypic Quality in explicit distinction from attractiveness; to test "good genes," we want to know about that part of total fitness that is associated with traits that are under viability selection, such as oxygen metabolism, parasite load, and so forth. Quality in terms of total fitness is therefore insufficient for teasing apart good genes and nonadaptive models of mate choice. Studies should endeavor to simply state the latent property of courters they're measuring, and to measure empirically not only whether, but how, this property covaries with chooser fitness.

14.2.3 Direct benefits of choosing

It is intuitive that if courters provide a direct benefit to choosers (such as parental care, preferred access to resources, or reduced physical damage from mating), then selection will favor choosers that can accurately pick the courters who will maximize that benefit (Wagner 2011). Such **direct selection** on chooser preferences can also operate in the context of same-sex or other nonreproductive sexual relationships.

Courter phenotypes can have major effects on chooser fitness. Courters may provide nuptial gifts, like spermatophores and food treats. They may also provide access to safer or more fruitful habitats, and they may protect and provide for the chooser's offspring. Courters may also behave aggressively toward choosers and their offspring, or even eat them. Even where mating is the only interaction between internally fertilizing partners, copulation can supply nutritional benefits, toxic chemicals, or infections. Smith and Mueller (2015) recently reviewed the literature on sexually transmitted infections that are beneficial. Any mating (or never mating) incurs net direct costs and benefits.

There is thus always likely to be at least some small variation in the harm or benefit that different courters provide to choosers, and selection will favor choosers that can detect it. The problem, as I alluded to in the previous section, is that most direct benefits cannot be detected directly. Selection favors the coevolution of preferences for traits that are correlated with direct benefit, but these are constrained by the correlation between trait and benefit. I will return to this point in chapter 15.

Sometimes, however, choosers can and do make decisions with stark consequences for their own fecundity or survivorship. For example, male comb-footed spiders, *Anelosimus studiosus*, that mate with rare social females are at lower risk of sexual cannibalism than those that mate with the more common asocial morph (Pruitt & Riechert 2009). Female arctiid moths (*Utetheisa ornatrix*) prefer to mate with larger males, a decision that reaps both direct and indirect benefits (see below). First, large males have larger spermatophores, which contain a greater quantity of defensive alkaloids. These are deposited in eggs, making them less susceptible to predation. Large size is heritable, and larger individuals of both sexes have higher reproductive success (Iyengar & Eisner 1999).

The benefits of mating can be complex and replete with trade-offs. In wild populations of field crickets (*Gryllus campestris*), males guard the periphery of a burrow, allowing females preferential access to the burrow. On the one hand, this acts to inhibit postmating choice, but on the other the male's presence results in dramatically reduced predation risk for the female relative to single females (and increased predation risk for the male; Rodríguez-Muñoz et al. 2011).

Associating with conspicuous courters may put choosers at risk of predation. Preferences for large size have been lost in high-predator populations (Rosenthal & Ryan 2011b), as expected if predation selects indirectly (chapter 15) or directly on preferences for large size (Rosenthal et al. 2002).

One of the most important direct benefits a courter can provide is parental care. In two bird species, the identity of the social father accounts for more of the variance in fledging success than parentage does (blue tits, Hadfield et al. 2006; common yellowthroats *Geothlypis trichas*, Whittingham & Dunn 2005). Alonzo's (2012) model predicted that females should always evolve a preference for males who perform parental care if these can be reliably identified, as long as genetic offspring of males who care have higher fitness than those of males who desert. Female cichlids *Oreochromis mossambicus* appear to directly assay a part of the courter extended phenotype relevant to parental care, preferring males who dig larger pits for fry independent of male body size (Nelson 1995). A robust way of identifying good fathers is the presence of a successful nest, particularly in animals like fishes that face little in the way of trade-offs between number of offspring cared for and survivorship. Indeed, females in a number of fish species prefer males who are already tending eggs previously laid by another female. Lindström and Kangas (1996) showed that while female sand gobies, *Pomatoschistus minutus*, failed to show a preference for novel males with eggs versus ones without eggs, they were more likely to reject males whose eggs were experimentally removed. Females therefore attended to nest failure, presumably recruiting individual recognition and memory on a scale of hours or more.

Sometimes choosers may attend to correlates of parental care. For example, Hill (1991) showed that male house finches (*Carpodacus mexicanus*) with brighter carotenoid coloration spent more time caring for offspring. Here, the carotenoid signal itself may constitute a direct benefit if it facilitates detection or evaluation (section 14.3). I will return to the evolution for preferences of correlates of direct costs and benefits in chapter 15.

A more general direct benefit to choosers, and an underappreciated one, is the efficacy cost of signals used to detect and evaluate mates (Maynard Smith & Harper 2003; Dawkins & Guilford 1993). The costs of seeking and choosing among mates (chapter 6) mean that selection will favor a courter signal that facilitates choosers' perceptual and cognitive tasks (chapter 3). A courter signal that choosers are already sensitive to (chapter 13) can thereby reduce the direct costs of mate choice (Dawkins & Guilford 1993).

It is worth observing that characteristics acquired by a courter, rather than courter genotype, can have major effects on offspring fitness; mates in better condition should be expected to provide better parental care. Courter condition may yield direct benefits even through semen-mediated effects. In the fly *Telostylinus angusticollis*, last-male sperm precedence (chapter 7) means that more recent mates sire the majority of offspring. Remarkably, however, offspring fitness was influenced by the condition of a previous mate, rather than the sire (Crean et al. 2014). Since the heritability of condition tends to be low (chapter 15), choosers may benefit from selecting mates on the basis of traits that are not heritable and therefore carry no indirect genetic benefits.

14.2.4 Additive genetic benefits: "good genes" and offspring attractiveness

The study of mate-choice evolution has disproportionately focused on the genetic benefits of mate choice, and specifically on the additive genetic benefits for offspring viability, or so-called good genes, which I will return to in the next section. Good genes, like Quality, is a conceptually sloppy term in that it means something rather restrictive when rigorously defined, but creeps out to encompass non-additive benefits as well. A casual definition of good genes logically includes the total genetic benefit for fitness that accrues from choosing a particular mate over another. Intuitively, courter genes that are good for a chooser are any alleles that enhance the fitness of her descendants (Boake 1985, 1986), which encompasses the non-additive, compatibility-based benefits that I return to in the next section. However, the evolutionary dynamics of non-additive effects are qualitatively different from additive effects, notably for how they affect the evolution of courter traits. It makes sense to restrict good genes to additive effects, but this further circumscribes their importance relative to the vernacular imagination.

Additive genetic benefits, in turn, can be subdivided into two categories that also involve very different co-evolutionary dynamics (chapter 15). The first is associated with the *attractiveness* of offspring; choosers who mate with attractive courters have higher fitness because they produce offspring that are attractive courters. The second is associated with the *viability* of offspring. Choosers who mate with attractive courters have higher fitness because they produce offspring that are more likely to be favored by natural selection. It is this latter category that is commonly termed good genes, with mates of the highest Quality implicitly defined as those that will deliver the greatest genetic benefit for offspring viability.

As I discuss in the next chapter, good genes is highly overrated as a selective agent on preferences. A meta-analysis of 22 studies by Møller and Alatalo (1999) showed that only about 1.5% of variation in offspring viability was explained by variation in preferred traits. Similarly, Arnqvist and Kirkpatrick (2005) found no evidence of indirect selection on female preferences in the context of extra-pair copulations. An expanded meta-analysis by Prokop and colleagues (2012) similarly failed to detect an effect of courter attractiveness on offspring viability.

Demonstrating good genes, or any genetic effects on mating preferences for that matter, can be challenging. As Hettyey and colleagues (2010) point out, many studies fail to disentangle genetic from nongenetic vertical transmission (epigenetics in the broad sense). Further, it is difficult to disentangle the effects of courter genotype from those of courter-induced investment in offspring by choosers (chapter 15).

Female preferences do confer good genes in multiple systems (train length in peacocks, Petrie 1994; forehead patch area in collared flycatchers, Sheldon et al. 1997; call duration in gray treefrogs, Welch et al. 1998a; mating success in field crickets, Wedell & Tregenza 1999; cuticular hydrocarbon blend in *D. serrata*). Byers and Waits (2006) found that attractive male pronghorn antelope produced substantially more viable offspring than females would obtain if they mated at random (intriguingly, females mated to unattractive males compensated by producing more milk for their offspring). Kokko et al. (2003) make the point that each of these studies only measures components of offspring fitness, some of which trade off against one another; few if any studies have demonstrated an additive genetic benefit on net mean offspring fitness.

By contrast, many studies have failed to find a detectable relationship between courter attractiveness and offspring survivorship (e.g., Boake 1985; Edvarsson & Arnqvist 2006; Rashed & Polak 2009; Sharma et al. 2012; reviewed in Charmantier & Sheldon 2006). Further, there are several studies that demonstrate a preference for genes that are correlated with *deleterious* fitness outcomes. In seed beetles, cryptic choice favors sperm that confer lower fitness on offspring (Bilde et al. 2009). In two poeciliid fishes, mate choice maintains Y chromosomes that carry alleles both for deleterious viability effects and for greater attractiveness (Brooks 2000; Ryan et al. 1992). In northern swordtails, this involves opposing effects of natural and sexual selection on age at maturity, mediated by copy-number variation at the *mc4r* gene (Lampert et al. 2010). Males with fewer *mc4r* copies mature later and at larger size. When mature, they are more attractive to females and have higher mating success, but they are less likely to reach sexual maturity (Ryan et al. 1992).

Similarly, the incompletely sex-linked oncogene *Xmrk* produces a pigment pattern that reduces male viability but is attractive to females in two populations of *X. cortezi*. In a third population, females flip their preferences (chapter 5) and prefer males without the pigment pattern. It is intriguing that this population happens to have higher expression of *Xmrk* in *females* than the other two, suggesting that selection may favor reversal of the preference if the trait's attractiveness in males is counterbalanced by viability costs in both sexes (Fernandez & Morris 2008). It is possible that these "bad-genes" preferences could be maintained by "good genes" if these specific deleterious traits are in turn genetically correlated with greater net viability (see chapter 15), but this has yet to be shown.

Genes can be bad or good according to the environment (Qvarnström 2001; Bussière et al. 2008). Under favorable food and temperature conditions, preferred male waxmoths *Achroia grisella* confer higher fitness on offspring, while less-preferred males produce superior offspring under unfavorable conditions (Jia & Greenfield 1997). Offspring of male grey treefrogs *Hyla versicolor* that produce longer calls were more successful than the offspring of short-call males at low densities, but the pattern was reversed at high densities (Welch 2003). Within-pair offspring of female coal tits (*Parus ater*) have higher fitness earlier in the season and lower fitness later in the season relative to the offspring of extra-pair males (Schmoll et al. 2005; see Schmoll 2011 for a review of bird studies on context-dependent genetic benefits). Whether a female is getting good genes from a male therefore depends on her timing of reproduction.

The genetic benefits of particular mates can also depend on chooser condition, as shown in blue tits (Dreiss et al. 2008; chapter 15). A chooser's environment, in addition to her own phenotype and genotype (see below), is thus dispositive in determining whether or not her partner has good genes for her.

14.2.5 Phenotypic and genetic compatibility

Sometimes the same courter provides greater benefits to some choosers than to others (fig. 14.1). Conspecific males and females of one species usually have higher fitness if they choose to mate with each other than with heterospecifics. This is an extreme example of the *compatibility* between choosers and courters as a major, if little appreciated, selective agent on mate choice. Interfering with mate choice tends to harm choosers (reviewed in Gowaty et al. 2007). In mice (Drickamer et al. 2000; Koeninger Ryan & Altmann 2001) and fruit flies (Partridge 1980; Anderson et al. 2007), both reproductive success and offspring viability were higher when choosers were paired with preferred than with unpreferred partners, independent of the preferred partner's phenotype. Compatibility, rather than absolute Quality, therefore seems to be of primary importance to fitness in these experiments.

The benefits of compatibility can be purely phenotypic. Female Australian frogs *Uperoleia laevigata* maximize fertilization success by mating with preferred males within a similar size range as themselves (Robertson 1990). Depending on how size is inherited, such phenotypic positive assortative mating can play an important role in reproductive isolation (chapter 16).

Personality (chapter 9) may be a primary mediator of phenotypic compatibility. Pruitt and colleagues (2011) showed that the fitness consequences of male mate choice in the spider *Anelosimus studiosus* depended on both the courter's and the chooser's personality; aggressive males mating with aggressive females suffered reduced fitness, including a disproportionate likelihood of being cannibalized. In this system, then, selection favors males who mate disassortatively with respect to personality. This is the reverse of the usual pattern, whereby positive assortative mating by personality is more frequently observed and confers higher fitness on pairs (Schuett et al. 2010; chapter 9). Personality may also covary with other aspects of preference (section 14.6).

The importance of genetic compatibility to fitness is just as self-evident. Sexual reproduction generates combinations of alleles inherited from both parents. The interactions between maternally and paternally derived alleles are prime determinants of offspring fitness. The most obvious of these fitness effects are first, that mating with close relatives can unmask recessive lethal mutations; and second, that mating with heterospecifics can generate lethal genetic incompatibilities. A third is that even between outbreeding individuals in the same population, some matings can also result in recessive or between-locus incompatibilities. A key point is that selection against each of these costs may well favor overlapping sets of mechanisms; the same processes may well function to reject heterospecifics, incompatible conspecifics, and close relatives. We see "inbreeding avoidance" and "conspecific mate recognition" as functional explanations for mate choice, but the objects of our study are neither genetic counselors nor taxonomists.

Inbreeding—reduced genome-wide heterozygosity as a consequence of matings among close relatives—is generally associated with reduced fitness, although, as always, the impacts are variable across taxa and environments (Keller & Waller 2002). Inbreeding avoidance may be a major selective agent favoring mate choice for compatible genotypes (although inbreeding depression is often assumed rather than quantified).

Pusey & Wolf (1996) suggested that mate choice could effect inbreeding avoidance through behavioral mechanisms discriminating against kin (like the Westermarck effect, chapter 12). Tuni and colleagues (2013; chapter 7) showed that female crickets *Teleogryllus oceanicus* biased fertilization toward unrelated males. They did this through interactions during postcopulatory mate guarding, discarding sperm if the mate-guarding male was a relative. Wolak and Reed (2016) recently showed additive genetic variance for the propensity to inbreed in song sparrows (*Melospiza melodia*); interestingly, they showed that the propensity to inbreed shows inbreeding depression, with more-inbred individuals more likely to inbreed themselves. This could arise from assortative pairing as a consequence of conflicting preferences in both males and females for more outbred individuals. Mating preferences can therefore evolve in response to selection against inbreeding.

Zeh and Zeh (1997) suggested that genetic incompatibilities, of which consanguineous mating is a special case, drove the evolution of multiple mating (section 14.4). Indeed, female field crickets *Gryllus bimaculatus* have higher hatching success when they mate with multiple males; Tregenza and Wedell (2002) showed that females suffered no loss of fitness from mating with a sibling, as long as they had mated with another male. However, evidence for choice of unrelated males is mixed. Jennions et al. (2004) found no evidence of postcopulatory mate choice against sibling sperm in black field crickets, *Teleogryllus commodus*. Similarly, inbred female mice were no more likely to mate multiply if they had the opportunity to do so with more genetically diverse males (Thonhauser et al. 2014). Even in the parasitoid wasp *Cotesia glomerata*, where inbreeding depresses fitness by increasing production of less viable diploid males, females fail to discriminate kin from nonkin in mate choice (Ruf et al. 2010). Some of the failure to discriminate kin may stem from proximate constraints on how preferences are specified (chapter 12). For example, Ihle and Forstmeier (2013) found that zebra finches failed to avoid kin as potential mates unless they were in contact with them throughout ontogeny.

Indeed, selection sometimes favors pairs that are more closely related than expected by chance (beyond the obvious case of conspecific matings generally being fitter than heterospecific ones). In humans, both pairing and successful reproduction were associated with higher relatedness than random (Rushton 1988). Similarly, house sparrows mated with more genetically similar individuals than expected, and extrapair mates were more genetically similar to females than their social mates (Bichet et al. 2014). In a handful of examples, choosers prefer siblings over nonrelatives (cichlids, Thünken et al. 2012; fruit flies, Loyau et al. 2012).

Beyond genome-wide relatedness, other interactions may be at play with regard to genetic incompatibility. Rudolfsen et al. (2005) used a crossed in vitro fertilization design to evaluate the effects of male genotype on offspring survival in the externally fertilizing Atlantic cod (*Gadus morhua*). Since males contribute nothing but sperm, this provided an elegant test of genetic effects. They found no effect of male genotype—no one male genotype conferred superior fitness—but a large male X female effect, such that if each female mated with her optimal male she would accrue a 74% increase in offspring survival (fig. 14.2). Wedekind and colleagues (2001) found a similar pattern in Alpine whitefish. Agbali et al. (2010) used an in vitro fertilization design to show that the nonadditive benefits of mate choice correlated with MHC dissimilarity.

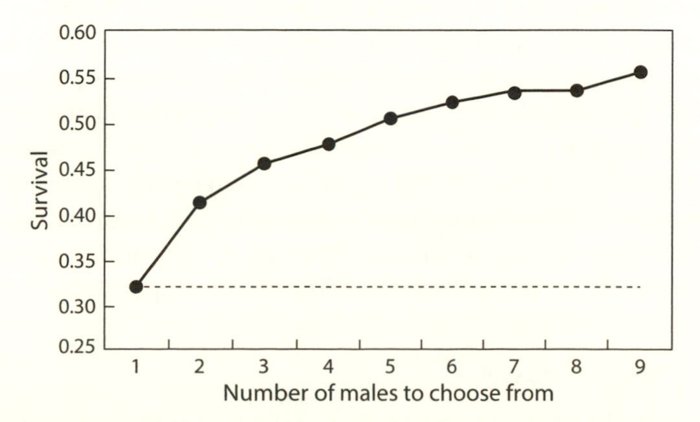

Figure 14.2. Mean offspring survival increases as a function of mate choice for compatible males in Atlantic cod. The hatched line gives the mean outcome of random mating. After Rudolfsen et al. (2005).

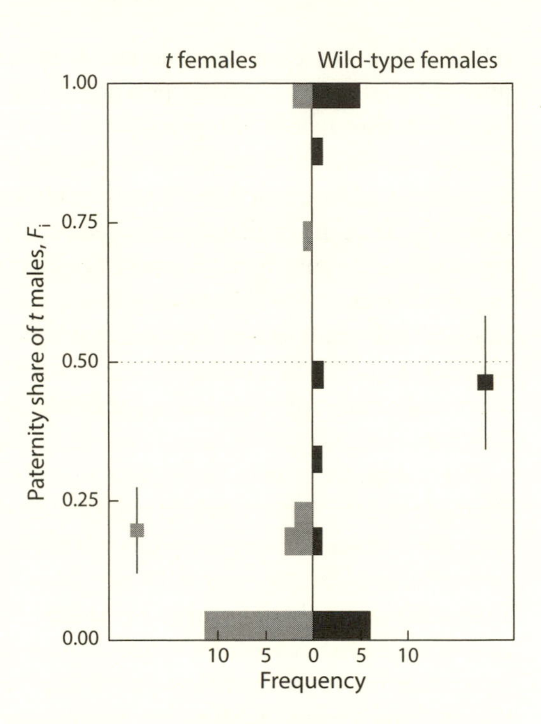

Figure 14.3. Observed distribution of fertilization bias of females carrying *t* allele (white bars) and wild-type females (black bars). *t* females, but not wild-type females, bias paternity toward wild-type males, consistent with selection against matings that produce recessive lethals. Dashed horizontal line = equal paternity share. Squares with whiskers depict mean ± SE of observed values. After Manser et al. (2015).

A striking example of the fitness consequences of (in)compatibility comes from the *t* haplotype in mice. *t* consists of a complex of genes on a large chromosomal inversion that is a recessive lethal in homozygotes, but *t*/+ heterozygotes produce a greater than expected number of *t* offspring through **meiotic drive**. For heterozygous choosers, selection should strongly favor avoidance of other heterozygotes as mates, since this would carry substantial fitness costs. Consistent with this prediction, *t*-heterozygote females, but not wild-type females, strongly favored wild-type males (Manser et al. 2015; fig. 14.3). Zeh and Zeh (1996) suggested that selfish genetic elements like *t* might be a major driver of incompatibility and in turn a major selective agent on mate choice; for a model of how such fertility-driven selection on mate choice interacts with ejaculate allocation, see Engqvist (2012). Intraspecific genetic compatibility is likely to be a major source of selection on mating preferences (Neff & Pitcher 2005, 2009; chapter 15).

14.2.6 Heterozygosity

A final source of selection inhabits a gray area between good genes and compatibility: benefits of genome-wide heterozygosity (Brown 1997; reviewed in Kempenaers 2007). From a good genes standpoint, heterozygosity is a heritable trait (all else equal, parents with higher-than-average heterozygosity will produce offspring that are also more heterozygous than average; Neff & Pitcher 2009) that affects both signal attractiveness and offspring fitness. Choosers select courters that maximize offspring heterozygosity. But like compatibility, the fitness consequences depend on the genotype of both males and females. And, unlike preferences for additive benefits, those for heterozygosity act to maintain rather than erode genetic variation.

Inbred courters are generally less attractive. Pheromones of inbred male, but not female, mealworm beetles (*Tenebrio molitor*) were less attractive to the opposite sex (Pölkki et al. 2012). In both male and female zebra finches, inbreeding made a suite of sexual traits less attractive (Bolund et al. 2010). Ryder and colleagues (2009) showed that choosers preferred heterozygous courters in lekking wire-tailed manakins (*Pipra filicauda*) but found no evidence for an indirect genetic benefit of mating with heterozygotes. By contrast, Neff and Pitcher (2009) showed that preferences for outbred males enhanced mean fitness in a sparrow population by increasing heterozygosity relative to random mating.

Several studies have tested the hypothesis that choosers should maximize dissimilarity in the MHC complex (Penn & Potts 1999; Ziegler et al. 2005; Milinski 2006). Behavioral preferences in experimental populations of pipefish (*Syngnathus typhle*; Roth et al. 2014) and mating patterns in wild populations of Atlantic salmon (*Salmo salar*; Landry et al. 2001), grey mouse lemurs (*Microcebus murinus*; Schwensow et al. 2008), and blue petrels (*Halobaena caerulea*; Strandh et al. 2012), are consistent with mate choice to maximize heterozygosity of the MHC complex. Richardson et al. (2005) found that in Seychelles warblers *Acrocephalus sechellensis*, females showed directional preferences for males with more MHC diversity; they were more likely to solicit extrapair copulations from high-MHC diversity males if their own males had low diversity. In another wild population, Hoffman et al. (2007) found that female fur seals sought to optimize whole-genome heterozygosity together with inbreeding avoidance.

14.2.7 Fixed costs of mate choice

No matter which courter a chooser ends up with, mate choice can carry a **fixed cost** (Otto 2009; Chaffee et al. 2013). A fixed cost is one the chooser pays no matter how little or how much mate choice she engages in. This can be a considerable "ante" associated with the process of mate choice itself, independent of the sampling environment or of mating outcomes. For premating choice, these inherent costs include the cognitive apparatus used in detecting, integrating, and evaluating multiple cues. Even so-called passive choosers, in the idealized case where preference equals detectability, have to allocate resources to the sensory architecture required for detection (although detection of mates may be a pleiotropic consequence of other sensory functions; chapter 13). For peri- and postmating choice, they include the costs of mating described in the next section.

It may be the case that mate choice can sometimes carry a fixed *benefit* by simply slowing down the mating rate. Mating is costly (section 14.4). Gametes are spent, and close physical contact incurs the risk of physical trauma (Hosken & Stockley 2004), attack or consumption by a mate, and disease transmission. And, like any other behavior, mating diverts time, energy, and attention away from other activities. A mating bias that causes some courters to be rejected—on whatever basis—may be an effective mechanism to slow down the mating rate.

14.2.8 Third wheels: exploitation of mate-choice mechanisms by parasites and others

Another factor constraining mate-choice evolution is that mechanisms can be exploited by individuals outside a mating context. A familiar human example, long exploited by intelligence services, is the so-called honey trap whereby men are more susceptible to non-sexual exploitation from women they find attractive (Watabe et al. 2013). Male thynnine wasps *Neozeleboria cryptoides* prefer larger females and females emitting higher concentrations of a pheromone; orchids produce supernormal stimuli mimicking large females and tenfold higher concentrations of chemosignals, promoting pollination for the orchid but increasing mate-searching and mating costs for the wasp, who spends time and sperm mating with a flower instead of with another wasp (Wong & Schiestl 2002; Schiestl 2004; fig. 14.4). Direct exploitation of mate-choice mechanisms may constitute an important constraint on their evolution in many systems.

14.3 EVOLUTION OF CHOOSINESS AND MATE-SAMPLING STRATEGIES

14.3.1 Optimizing choices

In chapter 6, I reviewed the mechanisms choosers use to sample and evaluate potential mates. These are all subject to selection, depending on the cost of making a choice and the benefits to choosers. Below, I summarize models

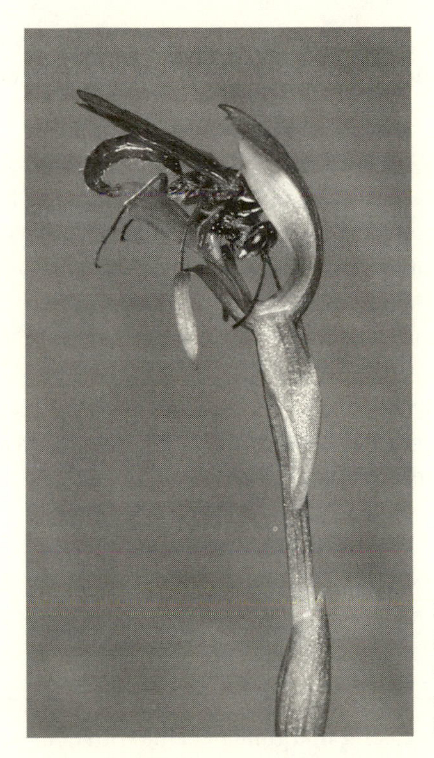

Figure 14.4. Male thynnine wasp (*Neozeleboria cryptoides*) initiating copulation with a sexually deceptive orchid (*Chiloglottis trapeziformis*). Photo courtesy of Rob Peakall.

on optimal search strategies before turning to sources of selection on choosiness. In chapter 15, I discuss how chooser trait distributions coevolve with both of these properties of choosers.

Most current work on mate-sampling traces back to an influential paper by Janetos (1980). He sought to ask how choosers could maximize the relative payoff of mating while minimizing the direct costs of sampling. We need not worry about what those payoffs are, just that they vary among courters and that choosers can discriminate them to some degree. How, then, do different mate-sampling rules optimize the problem of finding the most attractive courter relative to the costs of choice?

Janetos's (1980) model assumed that courters are distributed randomly in space with respect to attractiveness, that sampling costs are an increasing function of the number of choosers sampled, and that a chooser has perfect information about a courter's attractiveness once she encounters him. I will return to these assumptions below. Courters are encountered sequentially within a sampling period, and a chooser can either mate with the nth courter or go on to evaluate courter $n + 1$. The model assumes that choice is a one-way process, with choosers mating only once and courters mating multiply.

The model evaluated four alternatives (chapter 6) to the null strategy of random mating. Comparative evaluation (section 6.6) requires a different and likely more complex model.

Fixed threshold (section 6.4; fig. 14.5a). A chooser will accept the first courter she encounters with a fitness payoff above a certain threshold. If a chooser has limited time to evaluate a series of courters, she will fail to mate unless she encounters a courter whose payoff is above the threshold.

Last-chance option (section 6.5). As above, but if no above-threshold courter is encountered, the chooser will mate with the last courter encountered independent of his payoff.

Adjustable threshold (section 6.5; fig 14.5b). A chooser starts out "very picky indeed," lowering her threshold for acceptance after each encounter with an unacceptable courter, and finally exercising a last-chance option with the last courter encountered.

Best of n (section 6.7; fig. 14.5c). A chooser evaluates a fixed number of courters and selects the most attractive one. Unlike the above three strategies, or the null strategy of random mating, this strategy requires the chooser to remember n courters.

For any strategy besides random mating, the expected payoff for a chooser is always a matter of reconciling decreasing marginal benefits with constant or increasing marginal costs. If the marginal cost of evaluating an additional courter is constant, the cost of sampling is a linear function of n. The chooser's net payoff is maximized when the expected marginal benefit of the nth courter sampled equals the constant cost. It follows that the higher the cost of sampling, the lower the optimal n sampled (chapter 6).

In Janetos's (1980) model, the best-of-n strategy always produced the highest expected payoff from a chooser, particularly for values of $n > 5$. Relaxing the assumptions of the model, however, produces a different outcome. When these search costs are nontrivial (Real 1991), or when a chooser imperfectly assesses a courter's attractiveness (Wiegmann et al. 2010), best-of-n is outperformed by the fixed- or adjustable-threshold strategies (Luttbeg 2002; Wiegmann et al. 2010).

A newer generation of studies has modeled variants of the adjustable-threshold strategy by having choosers perform dynamic Bayesian updating of preferences based on encounters with courters (Luttbeg 1996; Castellano & Cermelli 2011; Castellano 2015). These models perform better than the original Janetos (1980) model under some parameter conditions, and such models may generate predictions we can test by looking at fine-scale chooser responses (e.g., Yorzinski et al. 2013; Clemens et al. 2014).

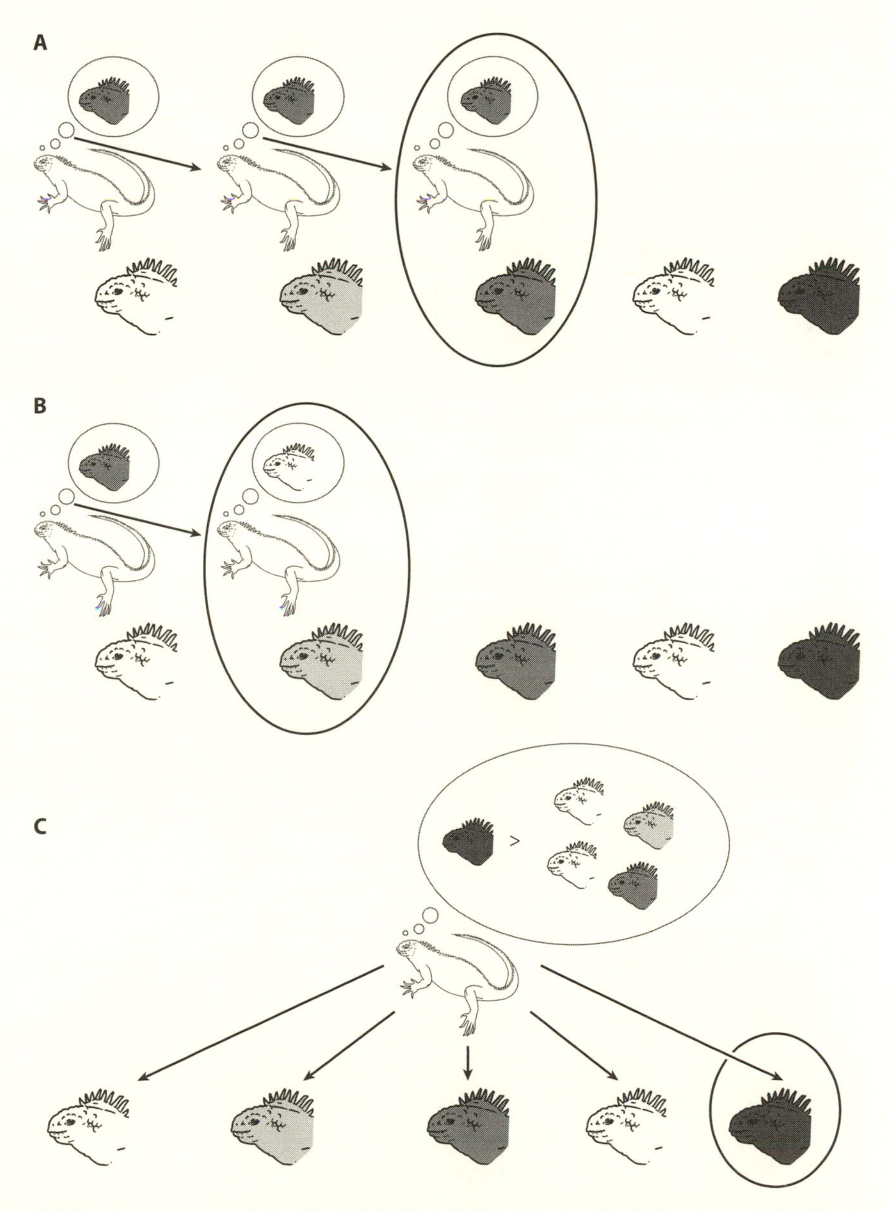

Figure 14.5. Alternatives to the null model of random mating. A focal chooser samples from a population of courters varying in some trait; in this case, choosers prefer courters with darker shading. (a) Under the **fixed-threshold rule**, choosers mate with the first courter they encounter whose trait exceeds a certain value; (b) With an **adjustable-threshold rule**, the threshold value is adjusted dynamically according to the chooser's experience. In this case, the chooser encounters a pale courter first, and lowers her threshold accordingly; (c) Under a **best-of-*n* rule**, a chooser samples *n* courters and selects the one with the highest trait value.

Optimality approaches yield insight into how expected risks and benefits can account for variation in the mate-searching strategies described in chapter 6. Crucial to all these models, however, is how the cost of choice scales with additional courters sampled. This quantity depends critically on the mechanisms that are responsible for additional sampling. It may be inexpensive for choosers sampling broadcast signals, as in a cricket chorus, to ignore $n+1$ courters, but it is usually costly for multiple-mating choosers to mate with $n+1$ partners. This scaling thus depends on costs of choice from multiple sources at different stages of mating, as I will address below.

14.3.2 Direct costs of choosiness

Throughout this book, I have discussed how proximate forces can shape variation in choosiness: how strongly choosers prefer particular courter phenotypes over others. Choosiness depends on a chooser's preferences, her mate-sampling strategy, and the distribution of courters in her midst. It also depends heavily on resources that incur fixed costs: that is, the physical or cognitive architecture she uses to sample males (Chenoweth et al. 2007). Choosiness can be modulated both by adjusting a search strategy (e.g., searching for fewer males under best-of-*n*, or dynamically raising an acceptance threshold) and by adjusting the subjective attractiveness of a distribution of courters (e.g., by changing the slope of a directional preference function or the width of a category boundary). Choosiness can also be modulated independently for different traits, and at different stages of mate choice. This can be accomplished through a variety of mechanisms ranging from changing sensory receptor distribution at the periphery, to changing the sensitivity of sensorimotor brain regions mediating reproductive behavior (Rosenthal 2016).

Choosiness is sensitive both to the direct costs of choice and to the benefits of choosing. All forms of choice carry opportunity costs, that is, the costs of spending time on mate choice that could be spent on other things like foraging, affiliative social interactions, or scanning for predators. Choice carries the risk of being attacked by predators targeting courter signals (e.g., or that choosers will be at increased risk when searching for or attending to predators (Magnhagen 1991; reviewed in Hughes et al. 2012). There may be nontrivial costs associated with copulation itself, which are particularly risky for animals with long copulation times (Siemers et al. 2012; but see Gwynne 1989). This cost of mating may be particularly important when considering the evolution of multiple mating and cryptic choice (see below).

Indeed, one of the few generalizations we can make about mate choice is that it is very sensitive to immediate risk: choosers under threat of predation either do not mate, or mate at random with respect to traits they otherwise attend to (reviewed in Hughes et al. 2012). Even more modest costs, such as having to expend more energy to travel to a preferred mate, can decrease or abolish preferences (e.g., Wong & Jennions 2003). As we will see, selection can favor the context- and state-dependence of choosiness, or the evolutionary loss of preferences for specific traits.

The costs of being choosy are an important feature of coevolutionary models of sexual selection (chapter 15). Dawkins & Guilford (1996) suggested that when courters address preexisting sensory biases (chapter 13), this often acts to reduce search costs, by increasing the detectability (and presumably the memorability and discriminability, Rowe 1999) of signals. Similarly, Wrangham (1980) suggested that search costs could drive the evolution of leks (chapter 15; see also Gibson et al. 1990), since it would be easier for choosers to evaluate courters if these were densely aggregated. Selection on choosers would favor increased attendance at the densest aggregations, which in turn might favor courters that joined these aggregations.

14.3.3 Mate copying

Mate copying (chapter 12) is most favored where choosers are not competing for courters, as among females of highly polygynous species; this is because there is no incentive for "model" females to lie to the female, and because mating with a model does not compromise a courter's value to a chooser (unless it depletes sperm or increases transfer of sexually transmitted pathogens). Among males, however, copying incurs a fitness cost from sperm competition (Vakirtzis 2011). Consistent with this prediction, the sexes copy in opposite directions in Japanese quail, with males avoiding females observed mating with another male (White 2004).

Vakirtzis and Roberts (2010) review a number of mutually reinforcing reasons why copying is predicted to succeed in species with and without biparental care. In birds, the success of biparental care depends on the provisioning efforts of both parents, so that models and choosers are competing. In fishes with paternal care, by contrast, mate-choice copying is favored because larger nests have higher fitness. In threespine stickleback, both sexes copy (Frommen et al. 2009). Female sand gobies gain direct benefits by choosing males with eggs already in the nest (Forsgren, Karlsson, & Kvarnemo 1996). This may be complicated if males abandon nests after a time;

in redlip blennies *Ophioblennius atlanticus*, later-laid batches had half the hatching success of batches laid earlier (Côte and Hunte 1989). Mate copying may thus favor the evolution of male parental care in fishes (Forsgren, Karlsson, & Kvarnemo 1996).

When choosers copy other choosers, they may do so selectively, for example by only attending to the choices of more experienced females ("mate quality bias"; Vakirtzis & Roberts 2009). Further, a chooser's own state—the cost of acquiring personal information about a courter—determines how much weight to place on public information (Nordell & Valone 1998). Dubois et al.'s (2011) game-theoretical model of copying evolution focused on "acquisition of generalized preferences" whereby choosers prefer courters with trait values resembling those of courters observed with models. As in Nordell and Valone's (1998) original model, the chooser's ability to gather personal information played a determining role. The success of copying (for a courter trait predictive of fitness) was negatively frequency-dependent, since copiers rely on the decisions of individuals who use personal information. Certain parameter combinations favored the persistence of copying even if this resulted in suboptimal choices with respect to fitness (Dubois et al. 2011).

14.4 WHEN TO CHOOSE: PRE- VERSUS POSTMATING

14.4.1 Multiple mating

The whole process of mate choice—from initial detection of a long-range courter signal to the provisioning of that courter's great-grandchildren—represents a problem in life-history allocation, as I shall discuss further in the next chapter. Where should the effort of mate choice be invested? Multiple mating is costly (e.g., Franklin et al. 2012). Given that evaluation is cheaper before mating than after, why do choosers ever mate multiply? Cryptic preferences can reap similar indirect and direct genetic benefits (section 14.2) as premating preferences. The sons of female bank voles *Clethrionomys glareolus* mated to two males produce more offspring than those of monogamous mothers, without any detectable trade-offs in terms of offspring size or survivorship (Klemme et al. 2008). Similarly, promiscuous female guppies produce more grand-offspring, also driven by greater reproductive success of sons (Barbosa et al. 2012).

Multiple mating has several additional benefits. First, there may be a direct benefit to multiple mating if it is less costly for choosers to accept matings than to reject mating attempts ("convenience polyandry"; Rowe 1992). Cryptic choice then allows choosers to exercise preferences after forced copulation.

Second, cryptic choice can mitigate some of the costs of mate searching, since females can mate indiscriminately or inaccurately and then select males during or after mating (Parker & Birkhead 2013; Roff & Fairbairn 2015).

Third, cryptic choice might also be more effective at filtering out genetic incompatibilities and avoiding inbreeding, notably if these are manifest in direct sperm-egg interactions (Zeh & Zeh 1997; Birkhead & Pizzari 2002; Lessios 2011; Price & Wedell 2008; chapter 7). Conspecific sperm precedence mitigates the cost of mating with heterospecifics. These mechanisms amount to bet-hedging (Jennions & Petrie 2000)—mating multiply is favored if a subset of males produces viable offspring. Indeed, multiply-mating walking sticks *Timema cristinae* had higher fecundity in the face of a significant male X female genotype effect on fitness (Arbuthnott et al. 2015). Further, female fecundity was higher when paternity skew was strongest, i.e., when a greater proportion of offspring were from one male. Multiple mating thus appears to be a mechanism for females to choose genetically compatible sperm (Arbuthnott et al. 2015). This is consistent with a comparative analysis by Stockley (2003) showing that promiscuous mammal species have lower rates of early reproductive failure.

After fertilization, choosers can make a mid-course correction of sorts if circumstances warrant. Roberts et al. (2012) found that wild geladas made the best of a bad situation by aborting offspring from previous mates, since these were likely to be victims of infanticide if carried to term. I will return to differential allocation after fertilization in chapter 15.

In chapter 7, we saw that some of the same mechanisms can link premating and cryptic choice, for example if choosers use long-distance display traits to bias fertilization. Modeling studies have addressed the interdependence of choice before and after mating. Kokko and Mappes (2005) modeled the evolution of mate choice when females face uncertainty about fertilization; they showed that if there is last-male sperm precedence, females can become effectively choosy even if they are indiscriminate before mating. A model by Lorch and Servedio (2007) suggested that there might be a trade-off between pre-and postmating preference of potential mates, since the evolution of one mechanism of reproductive isolation would act to inhibit the evolution of the other. While they present anecdotal evidence to this effect (e.g., Markow 2002), more systematic surveys across taxa (with careful controls for the validity of negative results) would be valuable, if logistically daunting (see also Kvarnemo & Simmons 2013). Devigili et al. (2015) showed from the courter point of view that male pre- and post-copulatory traits that predict success are positively correlated in guppies. Reid et al. (2014) obtained a similar result in song sparrows.

A critical question concerns the relative importance of cryptic choice to choosers. In chapter 7, I described the wide variance in estimates of the contribution of cryptic choice to total sexual selection. However, when we focus on total sexual selection, the most relevant measure for courter fitness, it may underestimate the importance of cryptic choice to total selection on choosers if cryptic choice involves non-additive compatibility preferences. For example, while there is strong premating selection on call characters in field crickets, cryptic choice is the primary mechanism for inbreeding avoidance (Tuni et al. 2013). Cryptic choice may therefore be more important to the fitness of choosers than we suspect.

14.4.2 Extrapair mating in socially monogamous animals

A special case of multiple mating, **extrapair mating** (EPM, also extrapair copulation or EPC) is common in pair-bonded animals, notably birds. Studies of EPMs have focused on their adaptive value with respect to offspring genotypes. In splendid fairy-wrens, *M. splendens* (Tarvin et al. 2005) and in three shorebird species (Blomqvist et al. 2002), the likelihood of extrapair paternity was higher when social mates were genetically similar to one another, consistent with the hypothesis that choosers were buffering the risk of inbreeding depression.

Forstmeier et al. (2014) review the costs and benefits of extrapair mating in species with social monogamy. They propose that nonadaptive constraints, notably intra- and intersexual genetic correlations with other traits associated with mating, were more likely to maintain observed rates of extrapair mating. This is because any adaptive benefits of extrapair mating are compromised by reduced care on the part of the cuckolded male. Schmoll (2011) noted that the fitness benefits of extrapair mating were environment-dependent. In chapter 15, I will return to differential allocation of resources among courters, including between social and extrapair mates.

14.5 EVOLUTION OF PLASTIC PREFERENCES

14.5.1 Selection on context- and condition-dependence

Both direct and indirect benefits can vary with the environment (Robinson et al. 2012; fig. 14.6), with the social context (Alonzo & Sinervo 2001), and with the state of the chooser (Long et al. 2010). Such context-dependence should favor the evolution of the sorts of plastic preferences we saw in chapters 9–12 (Qvarnström 2001; Badyaev & Qvarnström 2002). Context-dependence may favor the evolution of structures that directly couple mate choice with selection of an appropriate ecological context. The macroglomerular complex of codling moths, which responds to a configural blend of both courter signals and habitat cues (chapter 4), provides an appealing candidate mechanism.

When should chooser preferences depend on chooser state? As described in chapter 11, mating preferences often vary with chooser condition, but they do not do so consistently. There are good arguments to be made for both positive- and negative condition-dependence. One line of argument suggests that the bigger the pool of resources, the more resources choosers can spare to sample and evaluate potential mates (Cotton, Small, &

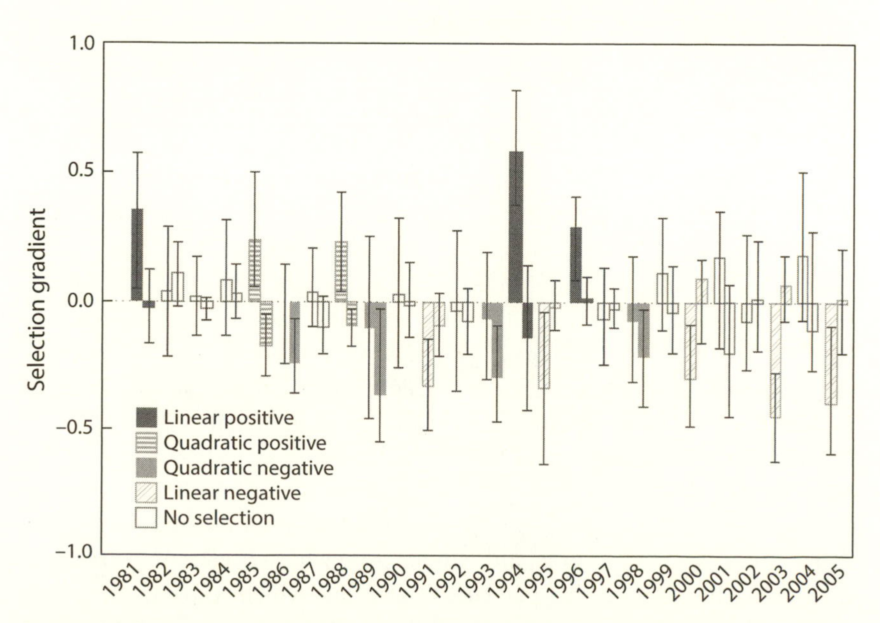

Figure 14.6. Components of selection on female preference for male ornament traits in a long-term study of collared flycatchers. After Robinson et al. (2012).

Pomiankowski 2006; Hunt et al. 2005); alternatively, choosers with more resources may be able to better afford a wasted mating on the wrong courter, or defer choice until after mating (Ortigosa & Rowe 2002; Fisher & Rosenthal 2006a; Eraly et al. 2009; Perry et al. 2009).

Positive condition-dependence, where preferences are expressed more strongly by choosers in better condition, can mitigate the costs of choice and facilitate preference evolution. Tomlinson and O'Donald (1996) presented a model where pleiotropic linkage of strength of preference and fitness facilitated the origin and maintenance of chooser preferences for costly traits. Veen and Otto's (2015) model showed that positive condition-dependence facilitated the evolution of preferences for locally adapted condition-dependent courter traits.

Condition-dependent mate choice can have some important population-level consequences. These effects depend on whether poor-condition choosers mate at random or whether they forgo mating altogether; the net effect can be to weaken sexual selection by introducing non-choosy individuals, or to heighten it by removing parasitized individuals from the mating pool (Poulin & Vickery 1996; Rolff 1998; Vickery & Poulin 1998). This applies critically to interspecific mating and genetic exchange (chapter 16).

14.5.2 Selection on learning rules

Another topic that has focused on trait evolution, rather than the evolution of preference mechanisms, is the mode of transmission of preferences: how are they learned and/or inherited (reviewed in Verzijden et al. 2012; Bonduriansky & Day 2013; Yeh & Servedio 2015)? There has been little theoretical attention paid to the evolution of learning rules in general (Fawcett et al. 2013), let alone in the context of mate choice (Verzijden et al. 2012). Species and populations differ in learning syndromes for both short- (chapter 6) and long-term (chapter 12) learning rules, suggesting that selection may play an important role in shaping how these are specified. An intriguing possibility is that learned preferences might facilitate the evolution of innate biases via the Baldwin effect (summarized in Scheiner 2014).

Both Tramm and Servedio (2008) and Chaffee et al. (2013) modeled the adaptive evolution of imprinting mode (paternal, maternal, oblique, or none at all) as a function of the mean fitness of potential "imprinting sets" (the distribution of traits in model individuals from whom preferences are specified). They both found that paternal imprinting, which restricts models to successful fathers who have survived both natural and sexual selection, was superior to oblique or maternal imprinting, which permitted imprinting on traits of lower mean fitness. Importantly, however, the correlation between the imprinted phenotype and the inherited trait—which would be weakened, say, by the daughter of an extrapair male imprinting on her social father—is key to the success of each strategy; this correlation is highest for daughters imprinting on mothers (Tramm & Servedio 2008). In chapter 16, I discuss the consequences of imprinting mode for speciation. It would be worth giving more theoretical attention to the causes as well as the consequences of how preferences are specified—not only which courters individuals attend to, but also whether they develop preference or antipathy toward the stimuli they experience.

14.6 CONSTRAINTS ON PREFERENCE EVOLUTION

In chapter 13, we saw that courter traits frequently evolve to elicit biases that are originally unrelated to mating; for example, sex pheromones elicit a feeding response in sea hares, *Aplysia* (Blumberg et al. 1998). How important is selection on non-mate choice function? For example, female preference for a food-mimicking ornament covaries with the environment (Kolm et al. 2012). Sensory drive provides a compelling explanation for the diversification of preferences and traits in different habitats, but this very property means that sensory bias is often under strong natural selection; sensory divergence among species is more likely to arise due to direct natural selection on broad sensory abilities, rather than due to selection in the context of mating preferences (Boughman 2002). Selection may thus act on preference changes beyond the sensory periphery, like changes in hedonic value, to modify the behavioral consequences of responding to conserved sensory inputs.

Sensory drive—selection on general sensory function—is important in shaping peripheral preference mechanisms (chapter 13). By analogy, "personality drive" may couple mating preferences to ecological selection. A handful of recent studies have shown that personality traits covary with preferences (chapter 9), but personality effects on preference are likely to be all-pervasive.

A growing body of work suggests that personality should be interdigitated with mate-choice mechanisms. In humans, personality affects the favored mode of associative learning (chapter 6), with impulsive individuals performing worse on learning tasks involving negative reinforcement—punishment—and anxious people performing worse with positive reinforcement (Corr et al. 1995). Correlations with personality may thus constrain how learned preferences are specified.

Indeed, recent studies point to correlations between personality and components of mate choice. Hesse et al. (2016) found positive correlations between levels of intersexual courtship and intrasexual aggression in both male and female cichlids *Pelvicachromis taeniatus*. A heroic study by Forstmeier et al. (2011) on zebra finches used cross-fostering and pedigree information to decouple genetic from rearing effects, and estimated a substantial genetic correlation between male and female extrapair behavior: the tendency of both sexes to philander was genetically correlated, perhaps reflecting underlying personality variation. Han et al. (2015) found that boldness toward predators in a nonsexual context predicted the sensitivity of female, but not male, water striders, to predation risk in the context of mate choice. Pathogen disgust predicts women's preferences for masculine traits (Jones et al. 2013). These correlations mean that ecological selection on behavior can have effects on mate choice. For example, sexual cannibalism might be maladaptive for choosers but be constrained by voracity in the context of predation (Dall et al. 2012). In chapter 16, I will return to the role of personality drive in speciation.

A final point is that mate-choice mechanisms may be shared by males and females, but recruited to evaluate different traits in the two sexes. For example, Forstmeier et al. (2011) showed that the tendency to seek extrapair matings was genetically correlated in male and female zebra finches. They argued that extrapair mating behavior in females might be maintained because it was favored in males. Such **sexual antagonism** arising from different selection optima for male and female values of the same trait may be an unappreciated force in preference evolution.

14.7 COERCION AND CHOICE

Any discussion of the evolution of choice must be mindful that courters are under strong selection to subvert or circumvent mate-choice mechanisms. This especially means that some interactions with non-preferred courters will involve forced matings. Direct selection acts on how choosers respond to coercion. This encompasses both direct resistance to forced matings, and choice after mating has occurred (chapter 7).

Brennan and Prum (2012) make a distinction between the dynamics of "choice" and "display" versus "resistance" and "coercion," drawing from their own research on waterfowl. The breeding biology of ducks is characterized by two very distinct and evolutionarily independent processes, one dominated by choice and the other by coercion. Social pairing occurs early in the breeding season, sometimes before males have regenerated their full penis. In this phase, females evaluate multimodal courtship displays combining elaborate motor patterns, vocalizations, and intricate plumage patterns. Later in the breeding season, females are subject to high rates of forced extrapair matings. Female ducks mitigate the incidence of extrapair fertilizations primarily through specialized genital structures that have coevolved with penis morphology (chapter 7). Premating choice is negated by coercion, but choosers still have agency over fertilization bias. It should be noted that in some systems, like guppies, females subject to forced matings or harassment will physically retaliate, occasionally killing their attackers (Houde 1997; chapter 6). The choice and coercion phases are dramatically different with regard to the basis of the underlying mechanisms of choice, the direct costs and benefits involved, and the ability of choosers to ignore or override courter strategies (Rosenthal & Servedio 1999). Nevertheless, these two very different types of courter-chooser interactions both involve mate choice, in the sense that they recruit mechanisms that affect the differential success of courters.

Clearly, forced mating lies at the opposite end of a continuum from broadcast signals on a lek, which choosers are free to ignore. But the spectrum of sexual interactions runs through the high-intensity, three-dimensional displays of some fishes and birds, in which courters attempt to block avenues of retreat by choosers. Prum (2015) argued that the elaborate structures of bowerbirds, which produce such displays, may have evolved because they provide an elaborate context for mate evaluation and choice while protecting females from sexual attack. Sexual interactions thus vary in the opportunity for choosers to exert choice and in the costs that choosers must incur to reject unwelcome mating attempts. There is the opportunity, subject to constraints, for choosers to exercise "sexual autonomy" (Prum 2015) at multiple stages of choice. We can think about these stages within the same basic framework of trait-preference coevolution, as I do in the next chapter.

14.8 SYNTHESIS

In this chapter, I have endeavored to survey the extrinsic selective forces shaping preferences. These include the environmental pressures that shape the behavioral systems underlying preferences, the constraints on how preferences are specified, and the risks involved in mate sampling.

None of the above are relevant to the evolution of mate choice, however, except in the context of how mate choice—investing in one courter as opposed to another—affects the fitness of choosers. Courters can provide non-genetic and genetic benefits (or costs) at any stage of the process of mate choice. In the next chapter, I will address how traits and preferences coevolve. As discussed earlier, the way we think about mate choice is heavily influenced by traits—flamboyant, exaggerated traits that are targeted by premating preference and assumed to predict additive benefits. While meta-analyses are undoubtedly useful in elucidating general patterns (Jennions, Kokko, & Klug 2012), we may be getting a grossly distorted picture of mate-choice evolution if we only look for the kinds of preferences that select for elaborate plumage. We need to pay at least as much attention to non-adaptive, non-additive, postmating processes as we do to adaptive, additive, premating ones.

To understand preference evolution, we need to at least consider, and where possible actually measure, **total selection on mate choice** in terms of fixed costs, search costs, and the positive or negative consequences of mating outcomes. Importantly, we need to consider the magnitude of total selection on mate choice relative to other organismal function: how important—or unimportant—are mating decisions to total fitness? The Robertsonian framework that Chenoweth and McGuigan (2010) provide for courter traits is useful for mate-choice mechanisms as well (fig. 14.7). In the case of mate choice, nonsexual fitness is determined by direct natural selection on non-sexual functions of mate-choice mechanisms. **Sexual fitness** can in turn be subdivided into fixed costs devoted to mate sampling and evaluation, and costs determined by the outcomes of mate choice. For mate-choice mechanisms, sexual fitness likely depends more on phenotypic and genotypic compatibility than it does on additive genetic benefits from choosers, as I shall detail in the next chapter. Perhaps more often than not, Quality is in the eye of the beholder.

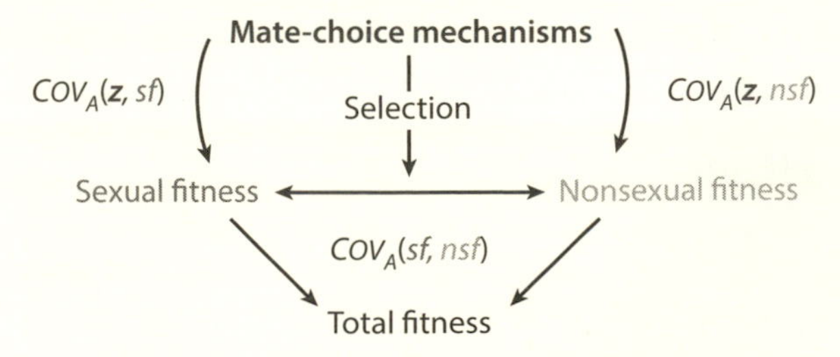

Figure 14.7. A Robertsonian framework for understanding total selection on mate-choice mechanisms, after Chenoweth & McGuigan (2010). Mate-choice phenotypes covary with sexual fitness and nonsexual fitness, which covary with each other and determine total fitness.

14.9 ADDITIONAL READING

Jennions, M. D., Kahn, A. T., Kelly, C. D., & Kokko, H. 2012. Meta-analysis and sexual selection: past studies and future possibilities. *Evolutionary Ecology, 26*(5), 1119–1151.

Neff, B. D., & Pitcher, T. E. 2005. Genetic quality and sexual selection: an integrated framework for good genes and compatible genes. *Molecular Ecology, 14*(1), 19–38.

Wagner, W. E. 2011. Direct benefits and the evolution of female mating preferences: conceptual problems, potential solutions, and a field cricket. *Advances in the Study of Behavior, 43*(273), e319.

CHAPTER 15

Dynamic Evolution of Preferences, Strategies, and Traits

15.1 INTRODUCTION

The last two chapters have shown that preferences and traits don't have to evolve together. Choosers frequently have mechanisms for trait discrimination that can subsequently be recruited for choice; conversely, courter traits evolve in response to latent chooser biases even when these biases are invariant (chapter 13). And the mechanisms underlying preferences are shaped by myriad forces other than heritable properties of courters (chapter 14).

Yet we think about preferences mostly in terms of their interplay with courter traits: that is, how they exert sexual selection on courter traits, and how courter traits exert selection on preferences. Some of this stems from trait-centered thinking; from Darwin (1859) on, the primary interest in mate choice has been as a handmaiden of sexual selection. To this day, sexual selection by mate choice continues to serve as a convenient adaptationist fallback—the ineffable 'taste for the beautiful' provides a handy selective agent driving the evolution of any phenotype that defies easy functional explanation. Therefore, the primary interest in mate-choice evolution has been how preferences coevolve with traits, from the trait's point of view. A long-standing feminist critique of the field of sexual selection is that misogyny and gender stereotyping have led workers to minimize the complexity, power, and autonomy of female agency (reviewed in Perry & Rowe 2012). We are a long way from the days when a female was assumed to passively mate with the winner of male-male aggressive contests, or when the field was a male-only preserve, but most of our intellectual bandwidth in sexual selection is still taken by, or with reference to, the fitness of courter traits.

In defense of courter-centric colleagues across the gender spectrum, there are compelling reasons to frame mate choice largely in terms of sexual selection on courter traits. One is simply practical. Mate choice is only detectable if there is trait variation to elicit it, and courter traits are infinitely easier to characterize from field observations, recordings, and museum specimens than preferences are. The second reason is that some preferences and some traits likely have a long shared evolutionary history; Lockley et al.

(2015) found evidence of courtship arenas in dinosaur taxa closely related to contemporary birds. Along with comparative evidence, this suggests that in birds, some elements of preferences for complex motor displays have been coevolving with their concomitant traits since the Jurassic (Prum 2010).

Studies in a phylogenetic context suggest that a deep history of coevolution between courters and choosers shapes contemporary preferences. Ryan and Rand (1995) reconstructed ancestral calls of túngara frog species and found that female *P. pustulosus* responded more strongly to the hypothesized calls of recent ancestors than to more distantly related taxa; Phelps and colleagues (2001; Phelps & Ryan 2000) went on to show that artificial neural networks developed preferences for current calls if they were taken through a history that mimicked ancestral calls, but not through a mirrored history that took a comparable trajectory through a different area of acoustic space. This showed that current preferences were shaped by a history of interactions with ancestral signals.

The third reason for focusing on trait-preference coevolution is that some degree of coevolution between courter traits and chooser preferences is all but unavoidable, as I address in detail in the next section. The coupling of traits and preferences gives rise to a host of possible evolutionary paths for preferences and the traits they target. Despite the tremendous amount of work on the topic, the conversation for 40 years has not moved on from the importance of nonadaptive processes relative to adaptive processes in the evolution of mating preferences. More specifically, how often can chooser preferences be attributed to additive viability benefits from attractive courters—so-called good genes? As the rest of this chapter suggests, the contribution of good genes, and additive indirect benefits more generally, may be quite small relative to other forms of selection on preferences. I begin by summarizing the simplest scenario for coevolution between traits and preferences, the Fisher-Lande-Kirkpatrick model.

15.2. GENETIC COVARIANCE DRIVES PREFERENCE EVOLUTION: THE FISHER-LANDE-KIRKPATRICK NULL MODEL

15.2.1 Quantitative-genetic models of preference evolution: overview

In this section, I review the major models of trait-preference coevolution in the context of what we know empirically about key model variables. A quantitative-genetic framework, which I have tried to keep as simple as possible, is helpful to understand the key differences between models (fig. 15.1). For a comprehensive treatment of quantitative-genetic models of preference-trait coevolution, see Mead and Arnold (2004).

By definition, there is always *phenotypic* covariance between courter traits and realized preferences for those traits. If there is *genetic* variation in both traits and preferences, choosers mating with preferred courters generates positive **genetic covariance** (C_{tp}; fig.15.1a) between trait and preference. This allows **indirect selection** (not to be confused with *indirect genetic effects*; chapter 12 and below) on traits to cause changes in preference phenotypes (and vice versa). The nature of indirect selection on preferences has been the most studied, most contentious, and most intractable topic in the mate-choice literature for the past four decades. Do preferences change because courter traits benefit choosers, harm choosers, or neither? As I later revisit, the answer to this question is a somewhat unsatisfying "all of the above." I will argue that, in general, the indirect additive effects of traits—particularly good genes—on preference evolution may be smaller than we think.

To think about trait-preference coevolution, it is helpful to use a generalized quantitative-genetic model (Lande 1981; Mead & Arnold 2004; Fuller, Houle, & Travis 2005; Prum 2010; fig. 15.1a). The multivariate phenotype consists of a single courter trait *t*, a single chooser preference *p*, and viability *v* to denote the contribution of so-called good genes ("variation in fecundity not affected by mate quality or number"; Fuller et al. 2005). $\Delta\bar{z}$ describes how phenotypes evolve in response to selection; it is the vector in change of the mean phenotype ($\Delta\bar{z}_t$; $\Delta\bar{z}_p$; $\Delta\bar{z}_v$) over the course of a single generation.

$$\Delta\bar{z} = \mathbf{G}\boldsymbol{\beta} + \mathbf{u}$$

where **G** is the genetic variance-covariance matrix, $\boldsymbol{\beta}$ is the vector of selection gradients (where β_t is selection on traits and so on), and **u** is the vector of mutational effects (fig. 15.1a). The minimal model of preference evolution, which was discussed in chapter 14, is when preferences evolve purely as a direct result of the fitness consequences of choices and there is no genetic covariance between trait and preference, such as when there is no heritable variation in the courter trait (fig. 15.1b). Equally simple are scenarios where preferences are subject to natural selection independent of the identity of the courter (such as fixed costs of choice or non-choice functions of the preference mechanism; fig. 15.1c). Conversely, traits can evolve in response to fixed preferences (chapter 13; fig. 15.1d).

In the last chapter, we saw that preference can evolve if courter traits are predictive of direct benefits to choosers. Here, we consider how preferences

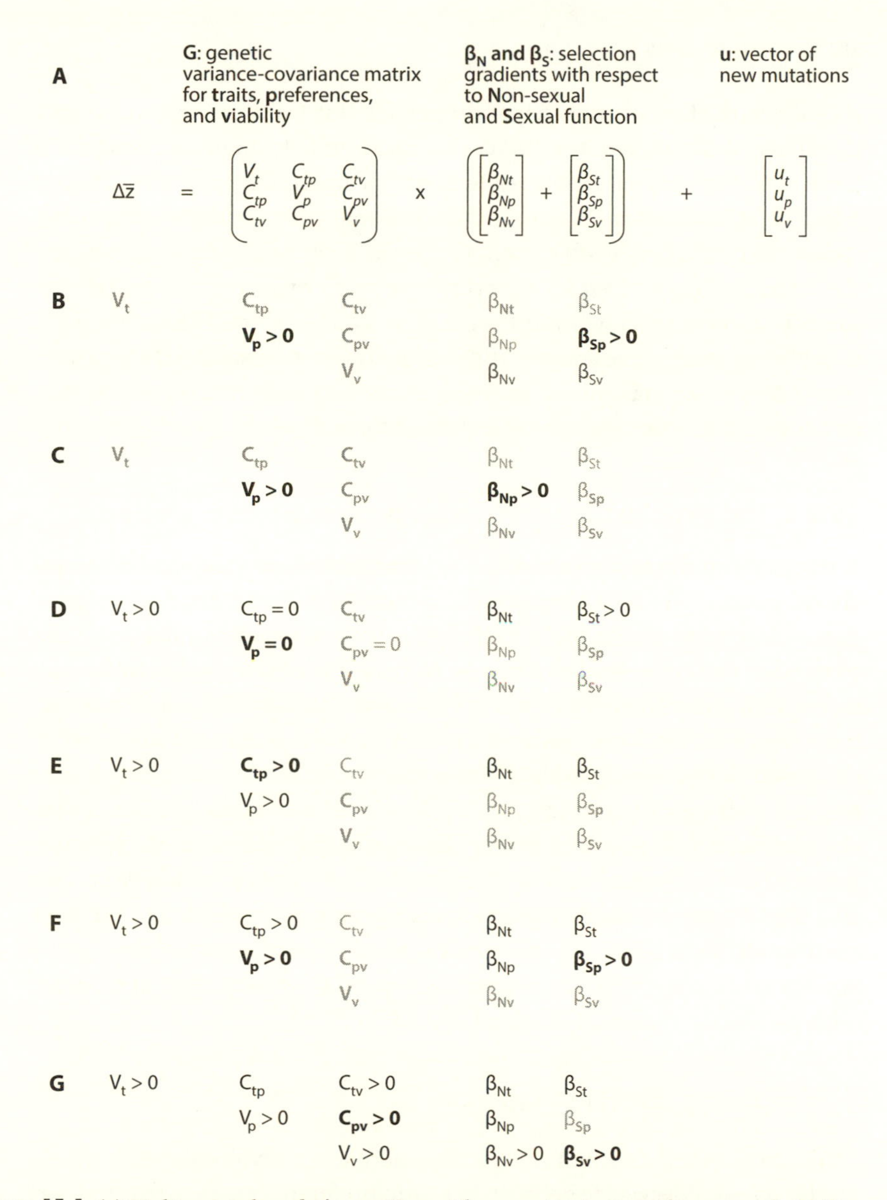

Figure 15.1. (a) Multivariate breeder's equation with respect to **t**rait, **p**reference, and **v**iability, following Fuller et al. (2005). Simplest parameter sets for alternative models of preference and trait evolution. Parameters in light gray have zero values for each minimal model. Values in bold are critical assumptions of each model with respect to choosers. Non-coevolutionary models include (b) preferences evolving in response to non-heritable aspects of partner identity; (c) preferences evolving in response to selection pressures besides partner identity; (d) traits evolving in response to non-heritable aspects of chooser preference. Coevolutionary models include (e) FLK model based on preference-trait genetic covariance alone; (f) direct benefit or sexual conflict model based on covariance between choice and phenotypic payoff; and (g) good genes model based on genetic covariance between preference and viability.

coevolve with chooser traits when preferences for traits are associated with direct and **indirect benefits** and costs. For simplicity, here we assume that the benefits of courter genotypes to courters are entirely additive (chapter 14); that is, all choosers maximize breeding value for attractiveness, viability, or net direct benefits by mating with the same individual courter, and the effects of every courter on every chooser's offspring are exactly the same. In section 15.8 I will address the more complex but ubiquitous case of coevolution when there are epistatic interactions with respect to fitness, such that one chooser maximizes its fitness by mating with one set of courters, and the next chooser does so by mating with another.

15.2.2 The simplest model of trait-preference coevolution

We begin with the case where a courter trait is arbitrary with respect to the chooser's direct benefits and offspring viability; it confers neither good genes nor bad genes. Lande's (1981) quantitative-genetic model and Kirkpatrick's (1982) discrete-locus model of trait-preference evolution formalized a verbal argument by Fisher (1930), as follows: males and females inherit the genes for both courter traits and chooser preferences, but each is expressed in only one sex. Since preference, by definition, causes choosers to mate with courters bearing preferred traits, there will be genetic covariance between trait and preference. This means that selection on traits will have a correlated effect on preferences and vice versa (Lande 1981). This so-called **Fisher-Lande-Kirkpatrick** (FLK) model (fig. 15.1e) is the simplest model to account for trait-preference evolution, since courter traits are not assumed to have any properties beyond how they affect and covary with preference (Prum 2010; Mead & Arnold 2004).

15.2.3 Properties of FLK models

Under FLK, whether or not traits and preferences coevolve depends purely on the genetic covariance between trait and preference, rather than any correlation of the trait with viability. The dynamics of FLK depend critically on C_{tp}/V_t, the ratio of trait-preference genetic covariance to trait genetic variance. If C_{tp}/V_t is lower than a critical value, traits and preferences "walk toward" a line of stable equilibria (fig. 15.2a). Importantly, this means that traits and preferences can coevolve to arbitrary points along the line; any trait value thus has the potential to be the most preferred by choosers. At equilibrium, $\beta_{Nt} = -\beta_{St}$; natural selection against the trait balances sexual selection favoring the trait.

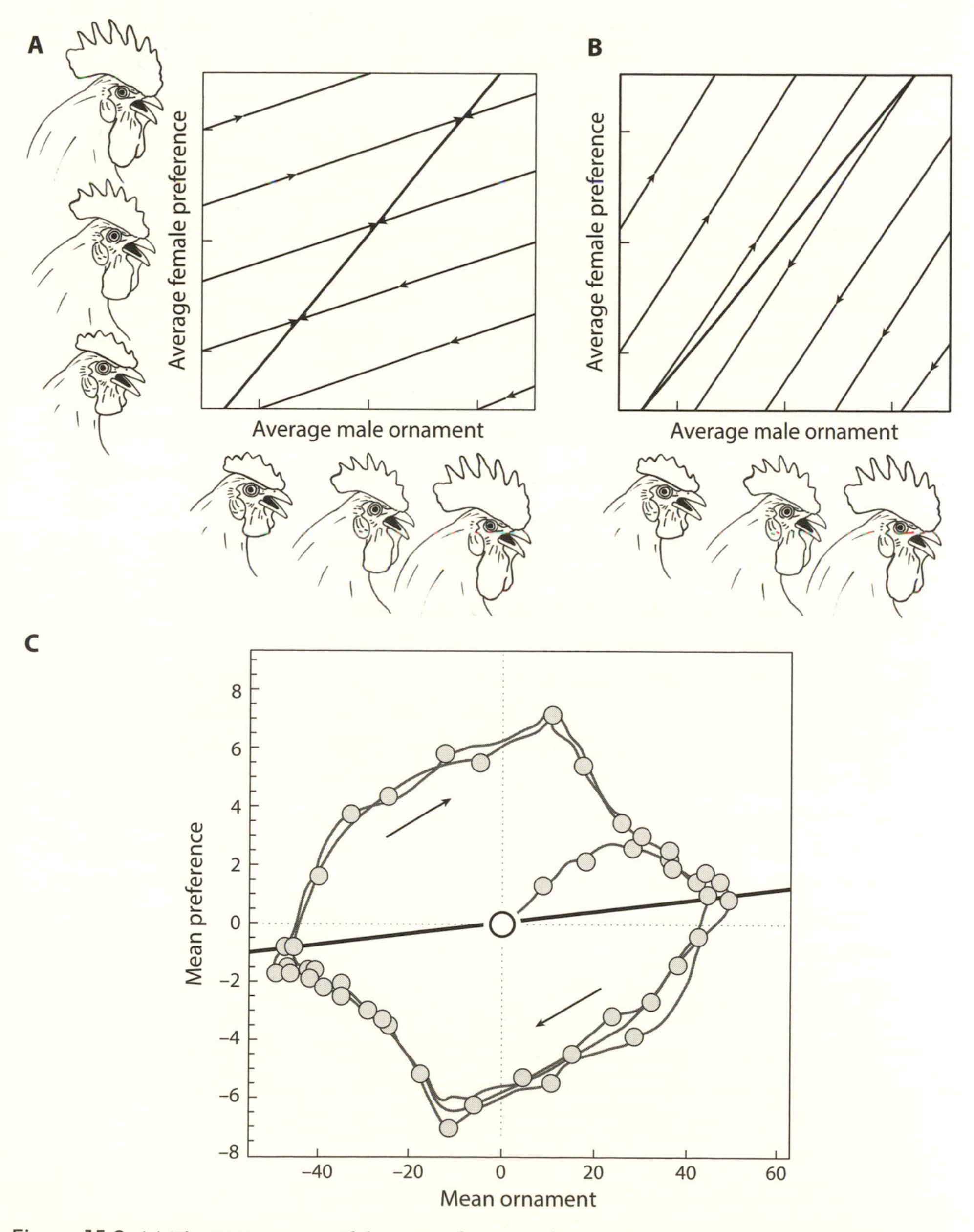

Figure 15.2. (a) The FLK process. If the ratio of trait-preference genetic covariance to trait genetic variance (C_{tp}/V_t) is lower than a critical value, traits and preferences "walk toward" a line of stable equilibria. (b) Higher values of C_{tp}/V_t generate a runaway process, where traits and preference coevolve to extreme values. After Arnold & Mead (2005). (c) simulation showing cyclical evolution of traits and preferences when there is weak stabilizing selection on both. After Kuijper et al. (2012).

Higher values of C_{tp} / V_t generate a "runaway process" (Fisher 1930, where traits and preference coevolve to extreme values (fig. 15.2b). Less attention has been paid to the fact that the FLK model is agnostic about the direction of the runaway; traits and preferences can evolve just as readily down to zero. The model can thus explain gains, losses, and reversals of preferences, as well as preferences for arbitrary trait values that are very far from the optimum with respect to natural selection. The arbitrariness of the FLK process means that it can form the basis for interspecific differences in traits and preferences (Lande 1981, 1982; Pomiankowski & Iwasa 1998) and set the stage for speciation through arbitrary spatial variation in traits and preferences (but see Servedio & Bürger 2014). In chapter 16, I will discuss FLK models in the context of speciation theory.

Traits and preferences can take on arbitrary values in time as well as space. A striking, and surprisingly neglected, property of the basic FLK model is that if it is modified to incorporate certain types of stabilizing selection on both traits and preferences, it predicts perpetual cycling of both courter traits and chooser preferences (Iwasa & Pomiankowski 1995; Mead & Arnold 2004; Kuijper et al. 2012; fig. 15.2). This is also true of models positing adaptive benefits (see below). Empiricists should not assume they are sampling traits and preferences at equilibrium.The arbitrariness of preferences—the extent to which they are indeed free to evolve in any direction with respect to a male trait—is constrained by the mechanisms that underlie them. Preferences shaped by the sensory periphery are likely to be biased in favor of greater sensory stimulation. Lande's (1981) original model analyzed three alternative preferences: unimodal absolute and relative preferences, and a "psychophysical" directional preference based on power-law scaling (chapter 3). This constitutes a preexisting sensory bias for stimuli of greater amplitude (Ryan and Keddy-Hector 1992) whereby a more intense stimulus is always perceived as more attractive than a less intense one. Kirkpatrick and Ryan (1991) argued that such a bias could provide the initial impetus for the FLK process (cf. Fisher 1930).

A model by Tazzyman et al. (2014a) suggests an explanation for why ornaments and preferences are directional in favor of exaggeration (Ryan & Keddy-Hector 1992), when larger signals confer greater *marginal* attractiveness. Power-law scaling and Weber's law (chapter 3), however, suggest that the opposite is likely to be the case; as signals get more extreme, a further unit of exaggeration becomes less efficient at increasing attractiveness. Intriguingly, a handicap model (see below) with similar assumptions predicts that traits of reduced magnitude will be favored as often as those of exaggerated magnitude, but that the latter would be more extreme (Tazzyman et al. 2014b). It would be interesting to explore a range of assumptions about signal efficacy with these models.

15.2.4 Limiting assumptions of FLK

The FLK model assumes that there are no direct marginal costs or benefits to choice—that is, that natural selection acts the same on a chooser expressing a preference for one trait over a chooser expressing a preference for another trait, and that a chooser expressing a preference does not pay any costs relative to a chooser mating at random. The original model fails to produce preference evolution when costs (Bulmer 1989) or genetic drift (Nichols & Butlin 1989; Uyeda et al. 2009) are added. Prum (2010) argues that the costs of choice may be small in nature, and as I argue in preceding chapters, the architecture of choice may emerge as an incidental consequence of other organismal function. Subsequent models show that preferences and traits can in fact coevolve when preferences are costly, under a variety of conditions (biased mutation on traits, Pomiankowski et al. 1991; migration bias, Day 2000; stabilizing natural selection on preferences, Hall et al. 2000; reviewed in Mead & Arnold 2004).

By assuming that preferences are not under direct selection and that choices have no consequences with respect to viability, the FLK model constitutes an evolutionary null model for trait-preference coevolution (Prum 2010). Among coevolutionary models of additive effects, FLK constitutes a more parsimonious model for preference evolution than models that assume viability payoffs depend on courter traits. Prum (2010) takes this one step further and advocates FLK as the null model for hypotheses about ornament and preference evolution more generally, but this perspective does not address all the forces that shape preferences beyond their coevolution with courter traits. As we saw in preceding chapters, these forces include non-additive costs and benefits of preference (like mating with conspecifics and avoiding close relatives), non-heritable properties of courters, biases that exist independent of mate-choice tasks, or biases that are correlated byproducts of preferences under selection.

To summarize, the FLK model makes three key assumptions. The first, that there is no natural selection on preferences, is perhaps the most problematic. As we saw in the preceding chapter, fixed costs and search costs frequently make it so that choosing is more costly than mating at random;

and sensory, perceptual, and cognitive constraints make it so that preferring one trait is less costly than preferring another. The second assumption is that mating decisions are neutral with respect to viability benefits; this assumption is more reasonable, since there is little empirical evidence for courters conferring additive viability benefits. The third assumption is that there is additive genetic variance in the courter trait and in the chooser preference *for the trait*; these both have to be greater than zero in order for them to covary. The next section argues that the opportunity for genetic covariance, while ubiquitous, may be limited in nature.

15.3 CONSTRAINTS ON GENETIC COVARIANCE: (MIS)ALIGNMENT OF PREFERENCES AND TRAITS

15.3.1 Phenotypic and genotypic covariance between traits and preferences

In this section, I review the evidence for genetic covariance between traits and preference. Phenotypic covariance between traits and preferences happens readily whenever choosers vary predictably in their preferences for a certain trait; different choosers will prefer different courters. But this does not mean that the trait and the preference are coevolving. For example, larger female swordtails prefer large males, and the preference is weaker in younger females (Morris et al. 2010). Female size is a function of age and not highly heritable, and male size is a function of repeat variation at a gene on the Y chromosome. Despite the strong phenotypic covariance, there is therefore little to no genetic covariance between the male trait and the female preference.

There are three ways that traits and preferences can be genetically correlated. The first is **pleiotropy**, if the same underlying mechanisms are involved in the production and perception of a trait, such that the same genes affect both. As we shall see later, this applies to vertically imprinted preferences, where the genes that specify the parental phenotype indirectly specify the offspring's preference. The second is **physical linkage**, whereby genes for traits and for preferences are spatially close together on the same chromosome. The third, which happens automatically when chooser genotype determines mating with a particular courter genotype, is **linkage disequilibrium**—a statistical correlation between distinct sets of genes. The greater the correlation between preference genes and trait genes, the greater the genetic covariance between them. The genetic covariance places an upper limit on the extent to which traits and preferences can coevolve, and, as I will discuss further below, is automatically eroded if choosers exhibit preferences associated with additive genetic benefits.

It is important to keep in mind that genotypic covariance is limited by phenotypic covariance. If courters bypass premating choice through coercion, or if choosers are otherwise constrained from expressing their preferences, this will act to weaken the genetic covariance between trait and preference.

15.3.2 Genetic covariance between trait and preference: the data

Genetic covariance between traits and preferences is essential to their coevolution. However, within-population estimates of genetic covariance between traits and preferences are weak, albeit consistently in the expected direction. In a meta-analysis by Greenfield and colleagues (2014), 80% of studies did not find a significant genetic correlation between traits and preferences. This by itself tells us little about the opportunity for coevolution; as Fisher (1930) pointed out, bouts of trait-preference coevolution can be extremely rapid, and at equilibrium there may be very little genetic variation in traits or preferences available to selection.

Indeed, there are strong phylogenetic and among-population signatures of trait-preference coevolution (e.g., Grace & Shaw 2011). In fairy wrens, ultraviolet shifts in sensitivity correlate across species with UV-ward shifts in male plumage (Delhey et al. 2013). Higginson et al. (2012) and Pitnick et al. (1999) showed that female genital traits coevolve with sperm traits in diving beetles and *Drosophila*, respectively. Experimental selection studies provide further evidence of genetic correlations between traits and preferences (Brooks 2002). Miller and Pitnick (2002) artificially selected on the length of the female reproductive tract in *Drosophila*; sperm evolved longer tails, as predicted. Wilkinson and Reillo (1994) manipulated the courter trait, male stalk length, in stalk-eyed flies, and preferences flipped—females preferred short-stalked males—when the courter trait was reduced.

Several studies, however, have failed to detect covariance between trait and preference (e.g., Ritchie et al. 2005; Zhou et al. 2011). In guppies, an early study showed phenotypic covariance between male traits and female preferences (Houde & Endler 1990), while a later study found that traits and preferences were mismatched in two populations (Houde & Hankes 1997). Hall and colleagues (2004) were unable to detect a response to direct selection on male traits or on female preferences in guppies, nor did they detect indirect selection on traits via preferences. Indeed, the genetic covariance between traits and preferences is limited by proximate and evolutionary factors, as I describe next.

15.3.3. Shared and divergent mechanisms underlie traits and preferences

Courter traits and chooser preferences are usually produced by entirely distinct mechanisms, but there is evidence that preferences and traits are coupled in some systems. In such cases, pleiotropy ensures that a genetic change in a preference always produces a genetic change in a trait, and vice versa. Boake's (1991) "genetic coupling" hypothesis argued selection would favor genes that pleiotropically coupled chooser preferences and courter traits (see also Shaw et al. 2011). There are areas of the brain, like the HVC in songbirds and the auditory midbrain in amphibians, associated with both the production of signals by courters and their processing by choosers. Sometimes the same gene can affect both production and perception of a sexual signal; in *D. melanogaster*, mutations to the *desat1* gene affect both production and perception of a cuticular pheromone (Marcillac et al. 2005; Bousquet et al. 2012). Shaw and Lesnick (2009) found that a QTL for song differences between two Hawaiian *Laupala* cricket species co-localized with a QTL for preferences.

A very important, and perhaps widespread, form of pleiotropy comes from learned preferences as an indirect effect of courter phenotypes (Bailey & Moore 2012). This is because when a chooser learns her preferences from the phenotype of her parents, a mutation that affects the parental phenotype indirectly shapes the preference (Immelmann 1975; Verzijden et al. 2012). Bailey and Moore (2012) recently showed that runaway dynamics can occur without genetic covariance between traits and preferences, if preferences are shaped by early learning. This might reconcile trait-preference coevolution with the rarity of trait-preference genetic correlations (Greenfield et al. 2014). Miller and Moore (2007) make a complementary argument about the indirect genetic effects of maternal allocation on male traits.

Pleiotropy may yet play a major role in the diversification of traits and preferences, but for the most part, there are different genes, different tissues, and different organs involved in production and perception of signals. For example, pigmentation patterns in guppies are controlled by genes that regulate melanophore and iridophore differentiation, while their perception is determined by the protein structure of retinal opsins and by the arrangement of neural networks in the retina and the brain (Deere et al. 2012).

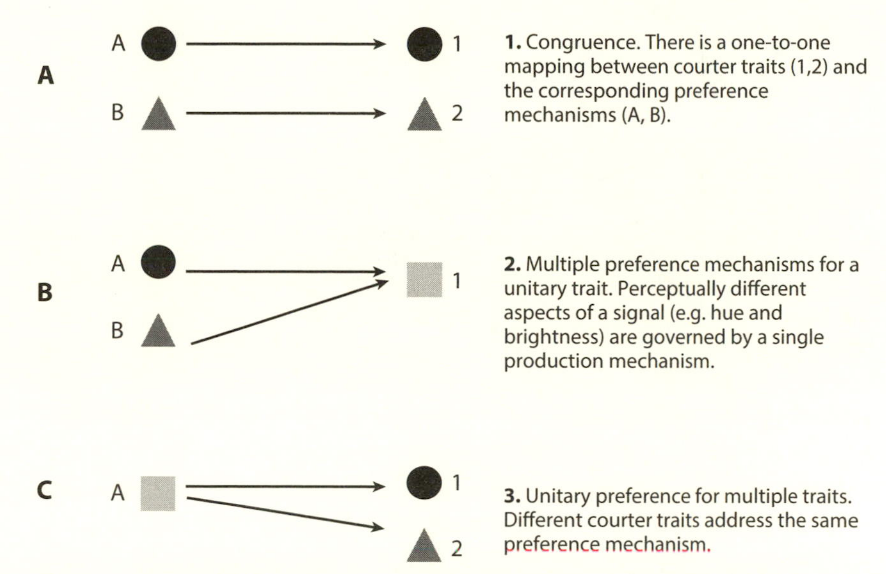

Figure 15.3. Possible relationships among preferences and traits. (a) Independent preferences (1 and 2) for multiple traits (A and B); (b) the same preference mechanisms, or correlated mechanisms, address different traits; (c) uncorrelated preferences for the same trait.

Variation within and among trait mechanisms is therefore very different from variation within and among preference mechanisms. The distinct mechanisms used in choosing and courting means that chooser preferences and courter traits are very unlikely to form a one-to-one mapping (fig. 15.3a), even though some theoretical models have posited that such a mapping should evolve (Tomlinson & O'Donald 1989). Multiple preference mechanisms may address a single courter trait, and vice versa (fig. 15.3b–c).

15.3.4 Trade-offs among traits are different from trade-offs among preferences

Multivariate traits and preferences are subject to different sets of trade-offs, and there is no tidy mapping between the mechanisms that generate traits and the mechanisms that generate preferences (e.g., Fisher et al. 2009; fig. 4.3). Male variable field crickets, *Gryllus lineaticeps*, produce pulsatile calls consisting of repeated "chirps." Females prefer males who both produce more chirps per unit time (a higher chirp rate) and produce longer chirps. In a series of elegant studies, Wagner et al. (2012) showed a negative genetic correlation between chirp rate and chirp duration; this correlation was

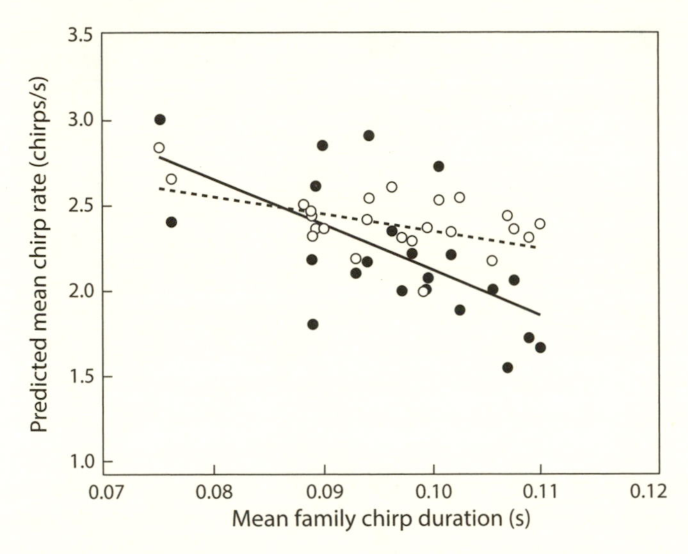

Figure 15.4. Among-family trade-off between chirp duration and chirp rate in male variable field crickets raised in high- (open circles) or low-nutrient (filled circles) environments. Females prefer males with higher chirp rates and longer chirp durations. After Wagner et al. (2012).

stronger in food-limited males, as expected if the trade-off depends on resource limitation (fig. 15.4). Bailey (2008) showed that two components of preference—response probability and response speed—of female crickets, *Gryllus oceanicus*, interacted to shape a complex bivariate preference function that would exert disruptive selection on male traits.

Choosers can similarly face hard trade-offs between perceptual functions. This is the case between brightness sensitivity and spatial resolution, since receptive fields that integrate over a larger area of the retina are more likely to interact with a photon at low light levels. Courters can exploit chooser constraints on spatial resolution by simulating colors via optical smearing of courter color patches that are smaller than the spatial frequency of choosers (Rosenthal 2007).

15.3.5 Genetic variance is depleted in the direction of selection

If choosers prefer a heritable trait in courters, selection via mate choice will deplete genetic variation for that trait. With multiple traits and preferences, genetic variance will be depleted in the direction of selection, i.e., the axis of trait space that is most closely aligned with preference space (fig. 15.5).

Under any indirect selection model, once sexual selection depletes variation in multiple courter traits, the direction of selection via mate choice should be exactly perpendicular to the main axis of multivariate genetic variation among courter traits. The lack of opportunities for multivariate choice then acts to weaken indirect selection on preferences. A study on *Drosophila* by van Homrigh and colleagues (2007) showed that almost all the genetic variance in male cuticular hydrocarbon profiles was in fact oriented perpendicular to the axis of preference; see also McGuigan et al. (2008a).

Mate choice is ultimately frustrating to choosers because the most attractive combinations—strong and sensitive, fun and responsible, free-spirited and reliable—are either extremely rare or physically impossible. This means that courters can make themselves attractive not only by producing signals that are exaggerated along some axis, but also by breaking correlations among traits and expressing novel trait combinations. As we shall see in chapter 16, the extent to which courters succeed in occupying novel, advantageous areas of preference space is a major determinant in the evolutionary success or failure of hybrids between species or populations. On the other hand, the misalignment of traits and preferences may explain why multivariate studies of mate choice in the wild often show fluctuating, weak, or nonexistent preferences by choosers, in contrast to experimental studies that show strong preferences for individually manipulated traits.

Misalignment of traits and preferences may effectively constrain the number of dimensions along which preferences can evolve, despite the fact

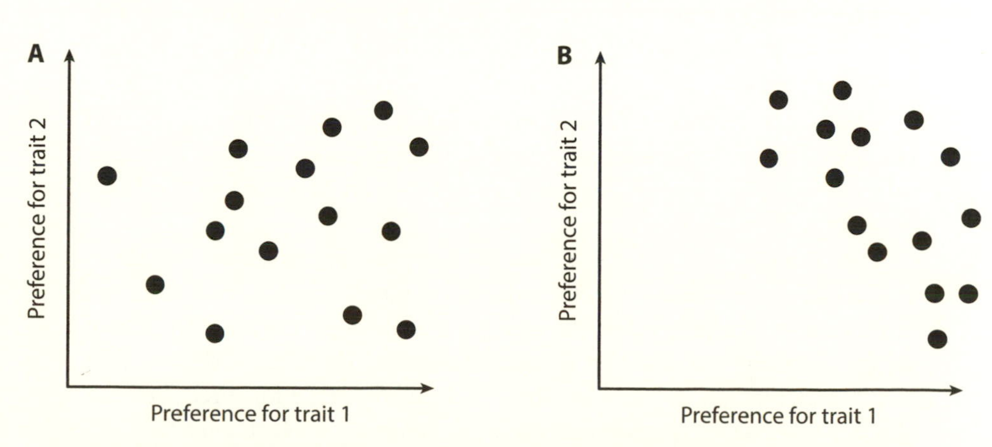

Figure 15.5. Selection via mate choice depletes genetic variance in the direction of selection. (a) Choosers have additive preferences for higher values of both trait 1 and trait 2. (b) Selection favors courters with high values of trait 1 and trait 2, generating misalignment between traits and preferences.

that choosers attend to many different traits. Hohenlohe and Arnold (2010) used preference measures from pairwise tests among multiple species to estimate that interspecific mate choice occupies roughly three genetic dimensions in several lineages (see Nosil & Hohenlohe 2012 for a similar analysis in stick insects). The upper bound of this estimate is constrained, however, by current phenotypic variation the species tested. Generally speaking, multivariate studies of natural mating patterns are limited by the current covariance matrix of traits among courters. Choosers have hidden preferences that are not evident when the only choices are drawn from the individual courters that are available at any one time.

In sum, coevolution of traits and preferences depends on the genetic covariance between them. Evidence suggests that genetic covariance is generally weak, and the highly multivariate, divergent nature of both traits and preferences constrains the potential for a change in one to effect a change in the other.

15.4 ADAPTIVE COEVOLUTION

15.4.1 Indicators of direct benefits

We saw in chapter 14 that preferences for direct benefits evolve readily by direct selection (Heisler et al. 1987; Kirkpatrick & Barton 1997; Wagner 2011; fig. 15.1b). If choosers have higher fecundity from mating with a courter bearing a trait, selection will favor a preference for that trait. This is true even if expressing the preference is costly ($\beta_{Np} < 0$) as long as costs are balanced by the direct benefits of mate choice (β_{Sp}).

Chooser preference in turn generates sexual selection favoring the trait (fig. 15.1f). As with non-heritable direct benefits, above, choosers can maintain costly preferences ($\beta_{Np} < 0$) as long as these are balanced by the direct benefits of mate choice (β_{Sp}).

For example, Alonzo (2012) showed that male parental care, which females can often evaluate directly, can evolve purely under sexual selection from females who prefer parental care, even if males are only slightly related to the offspring they care for. Bonduriansky and Day (2013) showed that costly preferences can be maintained particularly easily for direct benefits that are transmitted epigenetically (e.g., Crean et al. 2014; chapter 12).

The scenario becomes complicated when we consider that some important phenotypic determinants of chooser fitness, like a male bird's ability to care for offspring, cannot be directly assessed prior to mating. Choosers can modify behavior once a courter's parental care abilities are expressed, via differential allocation or reproductive compensation (chapter 7; see below), but only once a substantial investment has been made in offspring. Theory supports the intuitive prediction that selection should favor preferences for *indicator traits* that are reliable predictors of subsequent courter behavior (Price et al. 1993; Wolf et al. 1997; Iwasa & Pomiankowski 1999).

15.4.2 Sexual conflict

The fitness benefits for choosers are compromised, however, because "cheating" courters may compromise indicator reliability, and because there may be trade-offs between producing an indicator and caring for offspring (Kokko 1998). The tension between choosers and cheating courters is an example of **sexual conflict**. Courters' actions are frequently deleterious to chooser fitness. This is self-evident when courters circumvent or ignore choice, as through forced matings or preventing access to preferred individuals, but courters can also subvert mate-choice mechanisms in harmful ways. In Kokko's (1998) example, a courter gains a benefit—a mating—at the expense of a chooser who suffers the cost of missed parental care.

Trivers (1972) characterized male-female interactions as "mutual mistrust for mutual exploitation." If a male and a female each encounter each other only once, the male's fitness is maximized by extracting as much as possible from the female, while the female's is maximized by balancing her current resource allocation against future reproductive events. The reverse is also true, with the female benefiting from more male resources than the male optimum. Selection thus favors courters who can manipulate choosers into investing in a mating, even at a cost to chooser fitness, and favors choosers who can resist such manipulation even at a cost to courter fitness. The most successful courter will be the one who is most adept at manipulating chooser psychology or physiology. Such sexual conflict can drive the rapid coevolution of preferences and traits via direct selection (Chapman et al. 2003; fig. 15.1f). The direct costs incurred as a consequence of sexual conflict should often outweigh any indirect benefits associated with courter traits (Kirkpatrick & Barton 1997; Cameron et al. 2003). A quantitative genetic model by Gavrilets and colleagues (2001) showed that costly mate choice for elaborate traits can evolve through sexual conflict that reduces direct harm to females. However, the model constrains evolutionary outcomes by assuming that choice cannot be lost altogether and varies only by adjusting a fixed threshold for mate acceptance (chapter 6). Like the FLK model, it predicts that trait and preference evolve to a stable equilibrium or enter cyclical evolution.

There are multiple ways for sexual conflict to manifest itself in the context of mating Sexual conflict over the mating rate encapsulates a broad subset of these, and provides a straightforward conceptual example (Holland & Rice 1998). It is in a courter's interest to mate with a chooser, and then to prevent her from mating again with other courters (as well as mitigate parentage from previous matings; chapter 7). Selection favors these strategies in courters, even at the expense of choosers. Three examples of conflict, over remating and over mating too much, are illustrative here. Sakaluk and colleagues (2006) found that feeding on conspecific spermatophores did not reduce remating probabilities in female decorated crickets *Gryllodes sigillatus*. Females of a species that did not provide gifts, *Acheta domesticus*, by contrast, were *less* likely to mate when fed *G. sigillatus* spermatophores. Sakaluk et al. (2006) interpreted this as showing that *G. sigillatus* females had evolved resistance to chemical manipulation from males. Stewart et al. (2005) found that introducing a mutant allele that prevented re-mating acted to increase the fitness of *D. melanogaster* females. The direct benefits of lowered mating rate thus outweighed any indirect benefits of "trading up" through multiple mating (chapter 7). Finally, Maklakov and Arnqvist (2009) manipulated the efficiency of mate choice by resistance in seed beetles by morphologically enhancing or attenuating morphological structures used by females to reject mating attempts. Females that were more resistant had higher fitness, but there was no effect of the manipulation on offspring fitness. This indicated that female resistance had evolved via direct selection on mating outcomes rather than indirect selection.

Sexual conflict is most evident in chemical or mechanical conflict over mating rate, but divergent interests of choosers and courters come into play at every level of mate choice, from coevolving sperm and egg proteins (Civetta 2003) to coevolving ejaculate proteins and female gene expression patterns (Gioti et al. 2012) to coevolving fin elongations and appetitive responses (Macías-Garcia & Valero 2010).

Indeed, Arnqvist (2006) suggested that sexual conflict could manifest itself as courter exploitation of preexisting perceptual biases (chapter 13). If male traits increase direct costs once biases are expressed as preferences, choosers should evolve modified preferences (Macías- Garcia & Saldívar Lemus 2012). This appears to have occurred in goodeid fishes, where females seeking food instead get copulation attempts from males sporting food-mimicking caudal bands (chapter 13). Once food-mimicking bands evolve, females decouple their feeding preferences from their sexual preferences, suggesting a response to selection on the feeding bias (Macías-Garcia & Ramirez 2005). It is intriguing that in swordtails (*Xiphophorus*), preferences for swords became weaker or were reversed once swords evolved (Rosenthal et al. 2002; Wong & Rosenthal 2006). Reduced preferences might be a key phylogenetic signature of sexual conflict (see section 15.12).

The term "exploitation" implies that this "sensory trap" is harmful to choosers. An important caveat, to which I will return later, is that the fitness consequences *to choosers* must be assessed before invoking sexual conflict. Female goodeids experience a cost, but in other systems a feeding bias may provide an inexpensive mechanism for "passive mate choice" sensu Parker (1983), allowing choosers to filter out unacceptable mates without devoting time and attention to mate choice. A similar argument can be made for broad sensory and attentional biases (chapter 13). It is an empirical question as to whether courter traits that elicit these biases are harmful or helpful to choosers. Comparative studies should be instrumental in testing the key prediction that preferences should shift away from deleterious outcomes.

Sexual conflict occurs because a trait benefits a courter at the expense of a chooser, or vice-versa. This is true no matter whether the traits involved are seduction and resistance to seduction, or coercion and resistance to coercion (chapter 14). Coercion is a manifestation of sexual conflict at the premating stages. Peri- and postmating stages offer more sophisticated theaters for sexual conflict. This is because direct chemical and mechanical manipulations of genitalia and reproductive tracts offer greater opportunity for direct harm than sensory manipulations, where choosers can simply stop attending to cues (Rosenthal & Servedio 1999). It is worth observing that sexual conflict plays out along the entire chain of interactions between choosers and courters, from initial responses to signals to differential investment in offspring. Royle et al. (2010) suggested that preferences for stable personality types ("behavioral consistency"; see Schuett et al. 2010) might resolve sexual conflict if they constrained the extent to which courter behavior before mating, say, was predictive of parental care behavior.

Rowe et al. (2005) showed that if courters exploit preference mechanisms under natural selection (like foraging biases), maladaptive preferences can be maintained in an escalating arms race with courter traits. By contrast, choosers should evolve indifference to traits if preferences can be modified without constraint. I will return to sexual conflict in the context of reproductive allocation in section 15.7.

15.4.3 Indirect viability benefits: good genes

"Good genes" is the model of preference evolution that most appeals to our folk eugenic sensibilities. At first blush, preferred courters are healthier and more vigorous than non-preferred courters. If some of this difference is due to heritable factors, preference acts to select courters with good genes (higher breeding value for offspring viability) relative to mating at random. Zahavi (1975) suggested that this could explain the fact that courter traits are typically costly to produce and maintain:

> An individual with a well-developed sexually selected [trait] is an individual which has survived a test. A female which could discriminate between a male possessing a sexually selected character, from one without it, can discriminate between a male which has passed a test and one which has not been tested. Females which selected males with the most developed characters can be sure that they have selected from among the best genotypes of the male population. (Zahavi 1975)

Choosers, the argument goes, are therefore preferring ornamented males because the ornament is a so-called **honest signal** of a courter's genes. Crucially, since preferences confer a benefit, they can be maintained even if choosers pay a cost to expressing them; this is a feature of all models in which choosers gain a benefit by favoring some phenotypes over others (fig. 15.1). Models by Andersson (1986) and Pomiankowski (1988) showed that selection can favor the evolution of preferences for traits that are honest indicators of viability. "Good genes" also predict that choosers should select "amplifiers," cues that are uninformative about a courter's breeding value for fitness but that increase the likelihood that a courter will be chosen (Hasson 1989; Castellano & Cermelli 2010). The feasibility of good genes mechanisms generated a lively debate in the latter decades of the twentieth century (Kirkpatrick 1986; Grafen 1990; Maynard Smith 1991; Kirkpatrick 1992; Kirkpatrick & Barton 1997). It is important to note that good genes in this context is restricted to the additive contribution of genes associated with viability. From a chooser's point of view, additive genetic benefits encompass the courter's total breeding value for fitness, which includes components of attractiveness that evolve under an FLK model (above). Here, we consider whether costly preferences can evolve because females who express their preferences improve the viability of their progeny through additive genetic benefits from their partners.

The minimal good genes model is shown in figure 15.1g. Here, the critical variable is C_{pv}—the genetic covariance between a chooser's preference and the viability benefits her descendants gain from her partner. Here, choosers can maintain costly preferences ($\beta_{Np} < 0$) as long as these are balanced by the viability benefits of mate choice (β_{Sv}).

There are two general problems with good genes scenarios. The first is that any courter signal that is dishonest—i.e., that elicits a preference without incurring a cost—will be strongly favored by both natural and sexual selection. As I will return to shortly, a great deal of research has accordingly focused on the processes that maintain signal honesty. The second is the so-called **lek paradox**, as follows: courters with good genes have higher fitness due to both higher viability and higher attractiveness to choosers. Since certain genotypes have both a viability advantage and a mating advantage, genetic variation with respect to both viability and the preferred trait will quickly erode, and there will no longer be genetic covariance between the trait and viability (Borgia 1979; Kotiaho et al. 2008). Even if courter signals are constrained to be honest, they will no longer be an informative indicator of good genes. Below I discuss mechanisms that maintain genetic variation in traits and preferences.

Good genes can be rescued if and only if (a) signals are mechanistically constrained to be honest; and (b) there is a source of novel genetic variation with respect to viability. The latter can happen via mutation, which I will discuss below, and through changes in the fitness landscape, where what defines genes as "good" changes (i.e., certain genes become more or less adaptive through time). This is what happens as hosts and parasites coevolve. Parasites are constantly evolving strategies for exploiting their hosts, so resistance to parasites continuously requires novel counterstrategies. While there may be fixed optima for, say, locomotor performance or thermoregulation, minimizing the fitness costs of parasitism is a moving target: in Leigh Van Valen's (1973) memorable formulation, host-parasite coevolution is like Alice's Red Queen, who says "it takes all the running you can do to keep in the same place."

Hamilton and Zuk (1982) proposed that parasite-mediated sexual selection could explain female preferences for bright male plumage in many species of birds. The reds, oranges, and yellows of bird feathers are all produced by carotenoid pigments, with more pigment deposition producing more saturated color. These pigments can be metabolically modified to produce an array of plumage colors. Birds, like most animals, can't synthesize carotenoids and have to obtain them from dietary sources. In addition to playing a role in signaling, carotenoids are also used as antioxidants by the immune system. Therefore carotenoids may be mechanistically constrained to be honest indicators of the immune challenge from parasites: if a carotenoid molecule is being used in the immune system, it can't be deposited in a feather. Good genes could work for carotenoid-based plumage in birds as a consistent indicator of the resistance to constantly coevolving parasites. Female preferences for carotenoid colors could therefore be maintained via indirect selection favoring parasite-resistance genes. This honest signaling hypothesis relies, however, on two critical conditions that have not been widely tested: first, that the offspring of parasitized males have lower fitness due to genes they inherit from their fathers; and second, that dietary carotenoids are rare in the diet, and costly to acquire. Nonetheless, preferences for traits dependent on the intensity of carotenoid, like the chroma of a color patch, may be among the better candidates for good genes coevolution (although carotenoids may also function as indicators of direct benefits through parental care; chapter 14). It is useful, however, to bear in mind the rise and fall of fluctuating asymmetry as a good genes candidate, despite its surface appeal (section 5.2.1).

Beginning in chapter 9, I discussed the condition-dependence of chooser preferences; choosers change their preferences depending on the resources they have available for mating. Similarly, courters invest more in ornaments and displays when they are in better condition The condition-dependence of male *traits*, however, partially resolves some of the problems with good genes. Several workers (Rowe & Houle 1996; Bonduriansky 2007a; reviewed in Tomkins et al. 2004) have argued that **condition-dependent traits** provide a reliable target for adaptive mate choice. The logic is that condition provides a large mutational target (so-called genic capture), because a vast array of genes (associated with, e.g., foraging ability, disease and pathogen resistance, social dominance) are involved in determining an individual's overall condition. Traits whose expression depends on condition—that is, traits that are costlier to produce for low-condition than for high-condition individuals (which include carotenoid-based colors in birds) are, by this rationale, constrained to be honest. As with parasite-mediated sexual selection, female preferences can be maintained because they are associated with honest courter signals of maintained genetic variation with respect to viability.

Many studies have suggested that sexually dimorphic traits are disproportionately condition-dependent. Bonduriansky (2007a) found that sexually dimorphic traits were more likely to be condition-dependent in neriid flies *Telostylinus angusticollis*. Parker and Garant (2004) showed that male comb size in jungle fowl was both positively condition-dependent and heritable (see also Parker & Ligon 2007). A variety of mechanisms have been proposed to constrain signal honesty and maintain heritable genetic variation with respect to fitness, notably factors associated with systemic or cellular stress (von Schantz et al. 1999; Buchanan 2000; Hill 2011; Garratt & Brooks 2012; Warren et al. 2013). The relationship between organismal function and trait expression, however, remains murky. A cornerstone of Hamilton and Zuk's (1982) hypothesis, for example, is that parasitized males cannot produce attractive displays. The immunocompetence handicap hypothesis (Folstad & Karter 1992) posits an explicit mechanism whereby testosterone suppresses immune function such that only parasite-free males can afford to produce testosterone-dependent displays. The underlying link between testosterone and immunosuppression, however, is not supported by a meta-analysis (Roberts et al. 2004).

Indeed, current male condition, however defined (chapter 9), is often a poor predictor of offspring viability, making condition-dependent traits relatively uninformative. In general, there is weak evidence for correlations of courter traits with breeding value for fitness (chapter 14; Møller & Alatalo 1999). Part of this variation is underlain by spatiotemporal heterogeneity in fitness optima with respect to natural selection (Day 2000; Holman & Kokko 2014b; Veen & Otto 2015). The fluctuating nature of condition effects on courter fitness mirrors that of genetic benefits to choosers (Robinson et al. 2012).

Delcourt and colleagues (2012) found low genetic covariance between male fitness and multivariate courtship traits in *Drosophila serrata*. There are numerous examples where display traits are not condition-dependent (song and plumage in zebra finches, Bolund et al. 2009), or where courters tactically adjust signal presentation (song rate in zebra finches, David et al. 2012; lateral displays in poeciliids, Řežucha & Reichard 2015). Walker et al. (2013) found that some plumage traits in hihis (*Notymystis cincta*) were sensitive to early nutritional inputs, while others were not. Birkhead et al. (2006) found similarly variable condition dependence of display traits in zebra finches.

The observation that condition and attractiveness are somewhat decoupled speaks to the point that courters engage in a variety of strategies to maximize signal attractiveness, regardless of condition For example, Řežucha and Reichard (2015) recently showed that even in the absence of previous social experience, male *Poecilia wingei* preferentially present their more attractive (more colorful) side to females during courtship displays. The ability of fluctuating asymmetry (chapter 5) or color to serve as an indicator is therefore limited by courters' tactical behavior. Choosers may sometimes discriminate against such compensatory tactics. Kahn et al. (2012) found that female *G. holbrooki* preferred males that grew normally through life as opposed to males that underwent compensatory growth to make up for poor early nutritional condition.

More commonly, however, chooser preferences may not often map on to the condition-dependence of courter traits. Van Homrigh et al. (2007) found that choosers did not attend to condition-dependent axes of multivariate male trait variation. Johnstone et al. (2009) showed that condition-dependence need not coevolve with greater sexual selection; stronger preference did not mean stronger condition-dependence.

A final point is that the condition-dependence of courter traits is by no means a unique prediction of adaptive models of mate choice. Indeed, the distance between the equilibrium favored by mate choice and the natural selection optimum under the FLK model, above, means that good-condition courters will be better at bearing the costs of being attractive (Prum 2010). And if choosers have a preexisting bias for particular trait values, courters with surplus resources will most often be the ones best equipped to express those trait values (Ryan & Cummings 2013). Condition-dependent courter traits may well be the biggest red herring in the study of mate choice and sexual selection.

The ontological problem with good genes models is that the interests of courters and choosers diverge; the optimal mating decision from a chooser's point of view is not the optimal decision from the point of view of a rejected courter. Trait-preference coevolution is, in other words, subject to sexual conflict.

15.5 MODE OF TRANSMISSION AND PREFERENCE-TRAIT COEVOLUTION

15.5.1 Learning

A number of studies have considered the evolutionary effect of various modes of preference inheritance on sexual selection (Bailey & Moore 2012; reviewed in Verzijden et al. 2012). Verzijden et al.'s (2005) model suggests that imprinting (chapter 12) on parents—particularly the widely documented pattern of daughters imprinting on the maternal phenotype—produces mating preferences for an individual's own phenotype and therefore promotes phenotypic assortative mating (chapter 16). A model by Ihara et al. (2003) showed that learned transmission of preferences could facilitate FLK dynamics. Aoki et al. (2001) argued that runaway coevolution of obliquely imprinted preferences on a courter trait might explain continuous trait variation, like that observed for human skin color. Learning becomes particularly important in the context of assortative mating and speciation, as I discuss in chapter 16.

15.5.2 Sex linkage and sexual antagonism

The models discussed above largely assume autosomal, sex-limited inheritance of traits and preferences. Because selection acts differently on each sex in mate choice, sex chromosomes play an outsize role in the coevolution of traits and preferences. Several theoretical studies have shown sex-limited inheritance has important consequences for trait-preference coevolution (Kirkpatrick & Hall 2004; Fairbairn & Roff 2006; reviewed in Dean & Mank 2014).

The evolution of arbitrary traits and arbitrary preferences is constrained by the extent to which each of these is sex-limited. Lande and Arnold (1985) found that preferences can evolve if traits are homologous in males and females, but this can lead to maladaptive outcomes due to sexual antagonism, whereby the same allele has positive fitness effects in one sex and negative effects in the other. Even if a trait allele incurs a fitness cost in choosers, or a preference allele incurs a cost in courters, trait and preference can still increase in frequency. Selection will, however, favor modifier alleles that limit expression to one sex (Kirkpatrick 1982) or sex linkage (Kirkpatrick & Hall 2000).

Theory predicts that in female-heterogametic (ZZ/ZW) species like birds and butterflies, female preferences should be associated with the Z chromosome. This is because preferences will thereby be passed on to all sons expressing the preferred traits and will therefore benefit from strong indirect selection. This prediction is confirmed by a growing number of empirical studies (chapter 10).

Sex linkage determines the dynamics of sexual antagonism. Albert and Otto (2005), building on a model by Seger and Trivers (1986), showed that mate choice could evolve to mitigate sexual antagonism. If sexually antagonistic effects are X-linked or autosomal, females evolve to prefer traits that are beneficial to daughters, whereas in a ZZ/ZW system they prefer traits beneficial to sons. It is worth noting that Y-linked traits do not express themselves in daughters, which may account for the maintenance of preferences associated with deleterious Y alleles (Brooks 2000).

An empirical study by Cox and Calsbeek (2010) on *Anolis sagrei* lizards found that tactical allocation of resources—investing more in sons of attractive males—could maximize indirect costs and mitigate sexual antagonism. Sexual antagonism inherently weakens any benefits of good genes selection if high-Quality males produce lousy daughters. Experiments by Pischedda and Chippindale (2006) show that this acts to completely reverse any positive effects of sexual selection, since high-fitness mothers produce low-fitness sons and vice-versa—a correlation also observed in a wild deer population (Foerster et al. 2007).

15.6 THE LIMITS OF INDIRECT SELECTION

15.6.1 What display traits can't tell us about selection on preferences

Cuando el gallo canta en la madrugada . . . pue´que llueva mucho, que llueva poco o que no llueva nada.

When the rooster crows at dawn . . . it could rain a lot, a little, or not at all.

—El Filósofo de Güemez (apocryphal)

A large part of the conversation about mate choice has focused on the honesty of the signals involved. This is because for good genes to maintain mate choice, display traits have to predict breeding value for viability. Bow-winged grasshoppers *C. biguttulus* provide a satisfying and rare example of the dependence of a display trait on a latent trait predicting offspring fitness. There is a mild negative correlation between genome size and fitness, the consequence of a proliferation of selfish genetic elements. Males with smaller genomes produce more attractive songs (Schielzeth et al. 2014; fig. 15.6). The preference is thus maintained in part by greater fitness returns to females via indirect benefits.

In bow-winged grasshoppers, we can therefore make a clear connection between a courter trait and a genotypic effect on offspring viability. As I discussed in section 15.4, however, there is limited evidence for such predictive power of ornaments across the board. Studies tend to obscure the distinction between a courter's Quality—the latent heritable traits that are thought to confer direct or indirect benefits to choosers—and his breeding value for fitness. The rationale for looking at latent traits like locomotor performance (Husak & Fox 2008; Byers et al. 2010), immunocompetence, social skills (Sih & Bell 2008), and cognition (Boogert et al. 2011), as discussed above, is that they capture a great share of heritable variation with respect to fitness (Rowe & Houle 1996).

These are certainly valuable things to think about in terms of what determines courter allocation to signaling, but they may be somewhat secondary

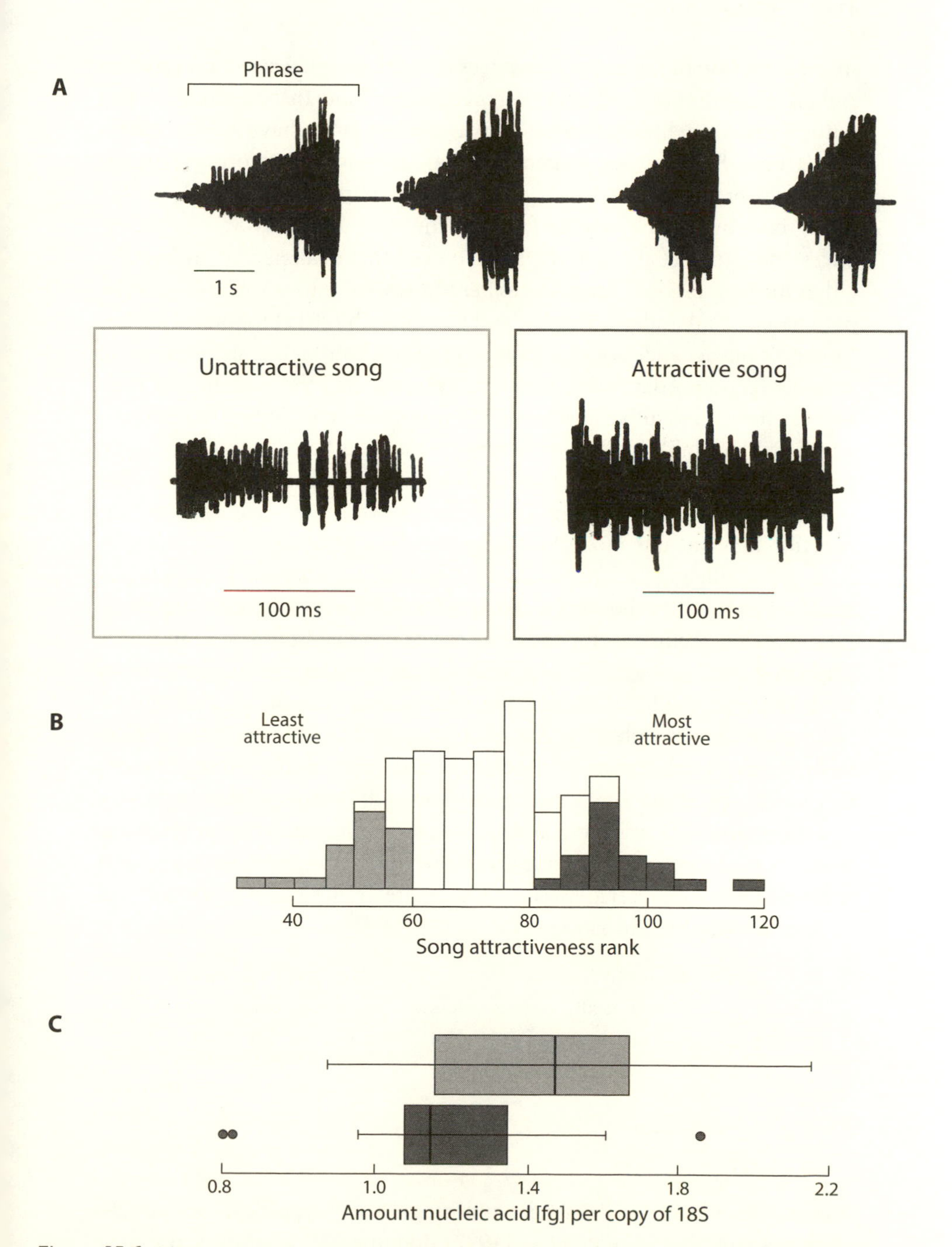

Figure 15.6. (a) Examples of attractive (left) and unattractive (right) songs of male bow-winged grasshoppers *C. biguttulus*. (b) Distribution of song attractiveness across males. (c) Genome size relative to copies of 18S RNA of attractive (dark gray) versus unattractive (light gray) males. After Schielzeth et al. (2014).

to the evolution of the mate preferences that drive sexual selection on these signals (Balmford & Read 1991; Wagner 2011). Traits that are under strong natural selection, like locomotor performance, tend to have low heritabilities (Husak & Fox 2008). When these latent traits are incorporated into models of preference evolution, they reveal dynamics that complicate the interpretation that they should facilitate the evolution of associated preferences. Johnstone et al.'s (2009) model showed that ornaments may become either more or less condition-dependent in response to sexual selection, or not change at all. Indeed, Gosden and Chenoweth (2011) found no change in condition-dependence of cuticular hydrocarbon blends in response to sexual selection. Adamo and Spiteri (2005) modeled the evolution of preference "for" constitutive and inducible immunity, concluding that preference would be unlikely to evolve because these traits failed to track pathogen prevalence and therefore confer reliable fitness benefits to choosers. The indirect fitness gains of mating with males with above-average immune function may be quite modest (Adamo & Spiteri 2009).

Another dilemma of choosing based on latent traits is illustrated by the question of whether choosers should prefer older males. Brooks and Kemp (2001) review arguments for and against why old age should be an indicator of good genes, as suggested by Catchpole (1996) and others. A model by Beck and Powell (2000) showed that the genetic benefits of mating with older males only manifest themselves in populations with low juvenile mortality relative to adult mortality. It also depends on whether old males suffer from reduced fertility or germ-line mutations. It is indisputable that being in good shape with respect to any of these sweeping categories is beneficial to courters. It does not automatically follow that mating with such a courter will be good for choosers. We cannot infer preference history by looking at current courter trait distributions.

> Since mate choice involves communication it involves costs. The existence of a costly ornament is hardly serious evidence that female preferences have evolved to assess a male's genes for survival. (Ryan 1997, p. 193)

15.6.2 Good genes: additive viability benefits don't add up to much

The appeal of "eugenic" (Ryan & Cummings 2013) hypotheses about mate choice is seductive. Alatalo et al. (1997) documented what proved to be a temporary spike (Prokuda & Roff 2014) in the heritabilities of sexually selected traits starting in 1988, the year Pomiankowski's good genes model was published. They argued that the trend was driven by the recent publication of studies with sample sizes too small to obtain robust heritability estimates, as a consequence of "revolutionary" excitement about "good genes" mechanisms. Good genes as typically defined are entirely additive, entirely indirect benefits, and are one of a multitude of sources of selection acting on mate-choice mechanisms. The empirical evidence to date suggests that they are relatively secondary to the evolution of preferences, although they may play a role in maintaining their stability. As detailed earlier, studies vary in the viability effects they detect on preferences. Rodríguez-Muñoz and colleagues (2008) performed controlled crosses of crickets that revealed no good genes but plenty of nonadditive genetic variance, suggestive of genetic diversity maintained by mate choice for compatibility. By contrast, Head and colleagues (2005) found in house crickets that the indirect benefits of mating with attractive males outweighed the direct costs due to reduced survivorship.

The magnitude of indirect selection on preferences has consistently been shown to be weak or undetectable across a host of empirical studies. Møller and Alatalo (1999) found that good genes accounted for only 1.5% of variance in chooser fitness, while Møller and Jennions (2001) found that direct benefits accounted for slightly more. Similarly, Arnqvist and Kirkpatrick (2005) found no evidence of indirect selection on female preferences in the context of extrapair copulations (fig. 15.7; see also Vedder et al. 2013). Slatyer et al.'s (2012) meta-analysis of 40 studies suggested benefits of extrapair copulations consistent with the bet-hedging hypothesis for multiple mating. Females increased their fecundity by mating with extrapair males, but *contra* good genes hypotheses they did not affect the phenotypes of the young.

Prokop and colleagues (2012) recently updated Møller and Alatalo's (1999) study. After filtering out pseudoreplicated studies and designs that did not rule out compatibility preferences, they analyzed 90 studies (40% of which were in Møller and Alatalo's 1999 study). The offspring of attractive courters were highly likely to be attractive themselves, as predicted by the FLK model. By contrast, they did not detect even the modest effect that the previous study had of attractiveness on offspring viability, nor did they detect an effect of attractiveness on offspring performance. A post hoc analysis did, however, reveal an effect of courter phenotype on offspring performance when the analysis was restricted to physiological performance variables, which Prokop et al. (2012) suggested was consistent with the immunocompetence and genic capture hypotheses.

Lande (1987) further observed that good genes preferences should be phylogenetically conserved, with closely related species sharing the same preferences for traits informative about viability. This may be the case for some traits, like body size and carotenoid chroma (both of which happen

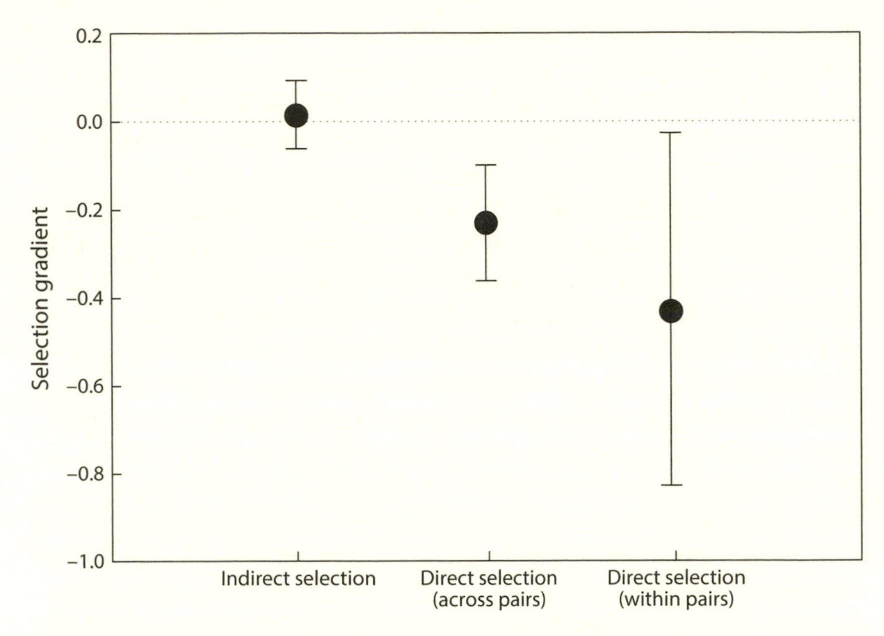

Figure 15.7. Selection gradients (weighted averages ± 95% confidence intervals) on female extrapair mating across bird studies. Direct selection was measured both for different breeding pairs ("across pairs") and with repeated observations of the same pairs ("within pairs"). After Arnqvist & Kirkpatrick (2005).

to be condition-dependent) but good genes likely fails to explain interpopulation trait diversity. Jordan and Ryan (2015) suggested that good genes could be limited by choosers' ability to perceive informative variation associated with breeding values for fitness; they suggest the best candidates for rapidly evolving signals are small molecules directly tied to heritable genetic variation (like major urinary proteins in mammals, Sherborne et al. 2007).

In sum, while good genes are likely to be pervasive selective agents on mate choice, they likely play only a secondary role in the evolution of preferences, and by extension in the evolution of the traits that target them. Good genes is more of a footnote than a default explanation for preference evolution.

15.6.3 Direct costs overwhelm indirect benefits

We saw in chapter 14 that direct selection on chooser biology can shape whether to choose at all, how much effort to devote to choosing, which courter traits to attend to, and whether to choose before, during, and/or after mating. Crucially, these aspects of the chooser phenotype can be under selection independent of courter identity, but still have consequences with respect to courter identity. Despite their importance to preference evolution, direct benefits have received limited attention in the literature, especially relative to good genes hypotheses (Wagner 2011).

Kotiaho and Puurtinen (2007), echoing Kirkpatrick and Barton (1997), argued that indirect selection is likely to be overwhelmed by direct selection. This is because indirect selection will always be limited by both the heritability of fitness and the correlation between chooser decisions and courter breeding values for fitness. The empirical literature suggests that the magnitude of direct selection on preferences may generally be higher than that of indirect selection. A meta-analysis by Møller and Jennions (2001) found that the effects of direct benefits were variable but generally higher than the 1.5% of fitness measured for good genes, depending on the benefit measured (the effect on feeding rates in birds was only 1.3%, while the effect on hatching success in male-guarding ectotherms was 23.6%; fertility and fecundity accounted for 6.3% and 2.3% of variation in fitness, respectively). Hadfield and colleagues (2006) decoupled direct and indirect fitness effects of mate choice in blue tits. Preferred male color traits were weakly heritable and uncorrelated with direct or indirect effects on fitness; paternal genotype had no effect on fitness, while cross-fostering revealed paternal effects on fitness. In socially monogamous zebra finches, Ihle and colleagues (2015) were able to decouple phenotypic and genetic compatibility. Parents had higher fitness when pair-bonded with their chosen partner than when forced to pair-bond with the partner chosen by another individual. This was driven by mortality at the hatchling stage; rates of embryo mortality were no different between treatments.

Sometimes direct and indirect benefits can be in conflict. In the desert-dwelling fruit fly *D. mojavensis*, preferred males confer more desiccation resistance on eggs, but the daughters of preferred males have lower fecundity. Sons of preferred and unpreferred males were similar (Oneal et al. 2007). In variable field crickets, *Gryllus lineaticeps*, different acoustic signal variables are correlated with distinct direct benefits to choosers. Females who choose males with higher chirp rates have higher fecundity, while females who choose males with longer chirp durations suffer lower mortality. Females show preferences for each variable when the other is held constant, but attend only to variation in chirp duration when both are allowed to vary (Wagner & Basolo 2007).

The importance of direct benefits is self-evident where courters supply parental care or nuptial gifts, but is less obvious where individuals interact only to mate, as in lekking systems. Even here, however, the process of sampling mates at different stages is costly in terms of opportunity and risk; and even courters who provide neither care nor gifts vary in their risk of STD transmission, in the toxicity of their ejaculate, or in their fertility. All of these factors impose direct selection on choosers, favoring mechanisms that correctly differentiate prospective partners along these axes.

The behavior of every model of mate-choice evolution hinges on the cost of choice. We know that these costs of choice are spatially and temporally heterogeneous, and that increasing the fixed costs of choice tends to reduce choosiness. We are only beginning, however, to gain an alpha-level understanding of differential costs of preferring one trait over another (Prum 2010), how costs scale with the number of courters being sampled (chapter 6), or of how they scale with the number of traits being sampled (chapter 4). In particular, sampling additional, more conspicuous traits could be *less* costly to choosers if it facilitated detection and localization. Dawkins and Guilford (1996) suggested that highly detectable traits could confer a direct benefit by reducing search costs for choosers. Importantly, the costs of parsing multiple traits are likely to be lower the more correlated they are across choosers (Rowe 1999), which could impose selection on suites of multiple traits.

An important point is that costs, benefits, and constraints vary across the process of mate choice. Direct selection plays a primary role in determining the sensory envelope of the stimuli choosers attend to, and at which stages of mating sampling and choice occur (chapter 14). But these "not quite arbitrary" constraints leave a lot of room for fluctuation. It is probably very inexpensive for choosers to sample broadcast signals at high densities, and in chapter 5 I argued that "flips" in the hedonic value of stimuli might be relatively free of evolutionary constraint. Counterintuitively, FLK processes might be more likely to explain reversals of preferences than their gains or losses. Indirect selection on arbitrary preferences might, as Prum (2010) suggests, accounts for the exuberant diversity in signals that are evaluated at relatively low marginal cost once choosers are attending to courters, like the plumage of lekking birds. Different traits and different preferences in the same species could arise from entirely distinct evolutionary processes (Lande 1987).

15.7 MATE CHOICE IN CONTEXT: SOCIAL AND LIFE HISTORY EVOLUTION

15.7.1 Mate choice and social selection

A major point of this book has been that we need to think about total selection on mate-choice mechanisms. But just as we shouldn't think of preferences as mere agents of sexual selection on courter traits, we shouldn't think of ornaments as mere "courter traits" yoked to mate choice. The evolution of ornaments depends on social forces other than mating, notably parental care and competition over resources. This so-called **social selection** is a unifying term for "all forms of selection due to social interactions, including sexual selection" (Lyon & Montgomerie 2012). As articulated in a landmark paper by West-Eberhard (1983), social selection is an impetus for thinking about complex phenotypes from their own frame of reference. Cornwallis and Uller (2009) offer a similar perspective on sexual traits in an ecological context.

Social selection therefore imposes additional constraints on the traits targeted by chooser preferences. Similarly, the processes that underpin how choosers make mating decisions may often be involved in non-sexual social interactions, like assessing potential rivals or allocating parental care (Rubenstein 2012).

The costs of mate choice are sometimes complex and embedded in a broader social context. Female goodeid fish *Girardinichthys metallicus* discourage sexual harassment by producing a metabolically costly vibratory signal in the presence of males. Interactions with males, however, are not nearly as costly as female-female aggressive interactions. Indeed, the presence of harassing males paradoxically decreases costs for females, since females signal to males instead of fighting each other (Valero et al. 2005). Meanwhile, female guppies also preferentially associate with more attractive females in order to divert sexual harassment from males (Brask et al. 2012).

A more expansive view of social selection views mate-choice mechanisms as functioning primarily in the context of cooperative interactions for physical resources (Roughgarden et al. 2006), mediated by hedonic reward (chapter 5; Roughgarden 2012). This is an attempt to incorporate same-sex and other non-procreative sexual interactions into a comprehensive theory of the evolution of sexual behavior. Rather than invalidating sexual selection theory, as Roughgarden and colleagues claim, their model fits into a well-trodden corpus that views mating decisions from a life-history perspective. As the next section describes, cooperative interactions come into play in the context of conflict over parental care in postmating choice.

15.7.2 Reproductive allocation: sexual conflict after fertilization

As we saw in chapter 7, a key form of cryptic mate choice occurs after fertilization, through differential investment in the offspring of different courters. This dynamic is an interplay between parent-offspring conflict over investment (Uller 2008) and sexual conflict (Ratikainen & Kokko 2009). The same general principles extend to confict over sex roles in hermaphrodites (Charnov 1979; Anthes et al. 2010; Henshaw et al. 2015).

Differential investment is conventionally called "differential allocation" (Burley 1986; Sheldon 2000) if choosers invest more in the offspring of courters with higher-than-average expected offspring fitness (usually termed higher Quality), versus "reproductive compensation" (Gowaty 2008) if they invest more in progeny from worse-than-average courters. In horned dung beetles *Onthophagus taurus*, Kotiaho et al. (2003) showed that females invest more in offspring of large-horned males; this acts to strengthen the phenotypic correlation in horn size between sires and sons, since sons of more attractive males get more resources from their mothers to invest in horn mass.

In chapter 7, I described how choosers can adjust the sex ratio of offspring based on the phenotypes of courters. Vedder and colleagues (2013) evaluated the theoretical prediction that, assuming attractiveness is heritable, females should produce more sons from extrapair offspring, since these will in turn be more successful at attracting extrapair matings. Empirical studies had provided mixed support for this prediction (reviewed in Johnson et al. 2009), which Vedder and colleagues (2013) argued reflected constraints that females might face in determining the sex of individual gametes. In a wild population of blue tits (*Cyanistes caeruleus*), earlier-laid eggs were more likely to be sired by an extrapair male and more likely to be male. Coupled with the lack of evidence for heritability of male attractiveness, they argued that the supposed relationship between sex ratio and extrapair paternity was spurious, offering their result as a cautionary note against the "good genes" bandwagon. Earlier, Arnqvist and Kirkpatrick (2005) came to a similar conclusion, pointing out that there were negative direct costs to EPMs, tied to reduced investment by the social mate, and no indirect benefits. They suggested that EPMs were maintained by sexual conflict.

Outside the context of EPMs, several studies have shown sex-ratio manipulations consistent with adaptive predictions. If male attractiveness is heritable, for example, females should make more males when mated to more attractive males, a prediction borne out by Sheldon et al. (1999) and subsequent studies. The two color morphs of Gouldian finches (*Erythrura gouldiae*) show postzygotic incompatibilities; hybrids between the two morphs have higher mortality than within-morph individuals. Birds (and butterflies, among many others) have female-**heterogametic** sex determination: females have a Z chromosome and a W chromosome, males have two Z chromosomes (mammals and many poeciliids fishes, *inter alia*, are male-heterogametic, with XX females and XY males). Following **Haldane's rule** that hybridization has more deleterious effects on fitness in the heterogametic sex, mortality is 84% higher for daughters but only 40% higher for sons. Females paired with a different morph adjust their sex ratio, producing primarily sons (Pryke & Griffith 2009). Female lizards *Anolis sagrei* produced more sons with larger sires and daughters with smaller sires (Calsbeek & Bonneaud 2008; Cox & Calsbeek 2010), consistent with predictions of sexual conflict theory. Sons of older male *D. melanogaster* perform worse in competitive mating assays; Long and Pischedda (2005) found that females mated to older male *D. melanogaster* initially produced female-skewed broods.

A model by Booksmythe and colleagues (2013) challenged the notion that sex allocation to attractive males acts to erode sexual selection on the trait (Fawcett et al. 2011). They pointed out that if sex allocation is costly, then weakened sexual selection should feed back to weaken or eliminate sex allocation, which in turn produces complex nonlinear patterns of trait-allocation coevolution, with intermediate levels of allocation favored.

Across taxa, choosers use differential allocation, reproductive compensation, or neither (reviewed in Stiver & Alonzo 2009). Gowaty and colleagues (2007) reviewed five studies which all showed nonsignificant trends in favor of reproductive compensation: females constrained to mate with non-preferred courters tended to produce more or heavier eggs. This mirrors findings in a meta-analysis by Harrison et al. (2009) that individuals providing biparental care tend to partially compensate for reduced care by their partners. By contrast, Horváthová and colleagues (2012) used a meta-analysis of bird studies to detect an overall trend in favor of differential allocation; females with female-only parental care laid larger eggs, and females with biparental care laid more, when mated with more attractive males. A model by Harris and Uller (2009) suggested that selection should as a rule favor differential allocation; reproductive compensation evolved only when the marginal impact of chooser investment was low. Their model was also sensitive to chooser age (relative to terminal reproductive investment) and condition (which influenced the relative costs of investments in offspring). Harris and Uller (2009) suggested that these variables may account for the heterogeneity of outcomes across studies of differential investment.

15.7.3 Mating systems and the intensity of sexual selection

Mating systems vary widely among and even within taxa; the evolution of mating systems is a long-standing topic of interest (reviewed in Kokko & Jennions 2008a; Kokko et al. 2014). The relationship of sexual selection to mating-system evolution is of relevance because the mating system forms the biological context of mate choice (chapter 6; Schuster 2009) and because the structure of a mating system is shaped in turn by how choosers express their preferences (Alonzo 2009). The complexity of mating-system evolution is illustrated by a few interesting cases. For example, Opie et al.'s (2013) phylogenetic study showed that social monogamy in primates is driven by the risk of infanticide; selection favors biparental care when fathers can increase their reproductive success by guarding offspring from marauding conspecifics. In social insects, by contrast, genetic monogamy is favored by the dynamics of cooperative behavior among sisters (Boomsma 2013).

Sexual selection, and therefore the opportunity for trait-preference coevolution, is expected to be strongest when courters have to work hard to obtain matings—systems where the operational sex ratio is skewed toward choosers, when choosers have the opportunity to sample multiple courters, and when lifetime reproductive effort is concentrated over a short period (Partridge & Endler 1987; Kokko et al. 2012). A short breeding season also increases the strength of sexual selection (West-Eberhard 1983). Bateman (1948) sought to measure the intensity of sexual selection, assuming "undiscriminating eagerness in males and discriminating passivity in females," as the slope of the number of matings on the number of offspring. For details on conceptualizing and measuring sexual selection, see reviews by Jones and Ratterman (2009), Kokko et al. (2012), and Henshaw et al. (2016).

As expected, sexual selection is stronger on females in species that are sex role reversed, i.e., where males choose and females court (Fritzsche & Arnqvist 2013). A game-theoretical model by Johnstone et al. (1996) suggested that the costs of choice for each sex should be more important in determining relative choosiness than variation in expected benefits. Mac-

Farlane and colleagues (2010) reviewed reports of same-sex behavior in dozens of bird studies and found that homosexual behavior in a given sex varied inversely with that sex's relative amount of parental effort, a proxy for local male competition.

Intensity of sexual selection does not mean intensity of mate choice. The fact that there is weaker sexual selection in less dimorphic species, however, is by no means an indication that mate choice is less consequential or less involved for choosers in these species; indeed, some of the most protracted courtships occur in monogamous species with biparental care (chapter 8; Mayr 1970), and it is in these species where the direct benefits provided by courters are arguably the most important to chooser fitness.

Biparental care provides a poor predictor of sexual selection, let alone mate choice. Trivers (1972, p. 141) said: "What governs the operation of sexual selection is the relative parental investment of the sexes in their offspring." By this logic, competition is stronger among males because females, out of the mating pool providing parental care, are the limiting resource. The fallacy of Trivers's (1972) argument is demonstrated nicely by Kokko and Jennions (2008b) using Bro-Jørgensen's (2008; chapter 8) study of male mate choice in topi, an African ungulate, as an example. Even though female topi invest much more heavily in offspring after fertilization, competition among females arises as a result of male sperm limitation arising from every chooser preferring the same subset of males. The relationship between reward and effort invested in mate choice is evident in McCartney et al.'s (2012) study of searching roles in the bushcricket genus *Poecilimon*, where either males or females search for prospective mates. As predicted, females are more likely to engage in searching when males provide larger nuptial gifts. Clutton-Brock (2007) discusses further examples of male choice in the face of strong female preferences.

15.7.4 Preferences and choices are life-history traits

Preferences may be expressed over an individual's lifetime, and, as we saw in chapter 9, preferences can be expressed at multiple stages before and well after mating. Each of these decisions can be interpreted in the context of reproductive allocation: how does choice of courter X over courter Y impact the current and future reproductive output of the chooser? How does investing in choice itself trade off with allocating resources to other tasks? Economic models from life-history theory may help shed light as to why age, health, and nutritional condition have such large effects on preferences.

These effects are particularly stark in species with large shifts in sex ratio over short breeding seasons. Etienne et al. (2014) modeled choosiness as a function of the "relative searching time," or the proportion of an individual chooser's lifetime devoted to choosing mates. It might not be a stretch to extend this concept to the share of neural, morphological, and physiological resources devoted to mate choice. Indeed, Pérez-Staples et al. (2014) argued that an important mechanism of cryptic mate choice in internal fertilizers, sperm storage, is shaped by life-history constraints on the timing of oviposition. Females with a long-term oviposition window can afford to put sperm in long-term storage, whereas females with a narrow oviposition window should fertilize eggs sooner, thereby limiting the opportunity for post-mating choice. They found support for this prediction in three species of tephritid flies, also finding that within the long-term ovipositing species, females in poor condition diverted more sperm to short-term storage.

As we saw in chapter 6, choosiness is sensitive to the availability of mates in a chooser's environment; the more mates are available for a chooser to evaluate, relative to competition from other choosers, the choosier she can be. But if she's choosy, courters benefit more from advertising than they do from investing in viability, thereby compromising the benefits of choice (Kokko et al. 2002). Such dynamic models, incorporating investment decisions in both courters and choosers, should be helpful in making predictions about how selection on preferences covaries with the benefits of attending to traits.

15.8 COMPATIBILITY AND EPISTASIS

15.8.1 Epistasis

So far, we have considered only additive effects on fitness. In chapter 14, however, I described mate choice based on compatibility, such that one chooser maximizes its fitness by mating with one set of courters, and the next chooser does so by mating with another. In genetic terms, this is epistasis for fitness with respect to mating decisions. The fitness of some preference-trait *combinations* is higher than others. Selection on epistatic interactions encompasses inbreeding avoidance and avoidance of hybridization, as well as other forms of intraspecific compatibility. Epistasis makes it difficult to predict responses to selection, but epistasis is a pervasive feature of preference mechanisms (chapters 4 and 10) as well as a major determinant of the fitness consequences of mate choice. Individual-based simulations (Kuijper et al. 2012) like *Admix'em*, which explicitly incorporates learned and flexible mating preferences (Cui et al. 2016), may prove useful in generating testable evolutionary hypotheses incorporating epistatic effects.

15.8.2 Compatibility and trait-preference coevolution

We saw in chapter 14 that compatibility-based preferences were widespread, with choosers sometimes preferring complementary individuals over globally attractive ones. Selection on compatible and/or outbred genotypes is likely to account for a substantial fraction of total selection on preferences. Multiple aspects of preference are correlated (section 15.11), and compatibility-based selection to avoid close relatives or heterospecifics is likely to generate chooser biases among intraspecific traits (e.g., ten Cate et al. 2006). Some traits will be more characteristic of siblings or of heterospecifics than others, and these will be at a disadvantage with respect to mate choice. Compatibility-based selection may thus indirectly drive additive preference phenotypes and thereby generate sexual selection on courter traits. The partitioning of intraspecific sexual selection from other aspects of mate choice may have limited our thinking in terms of how compatibility-based choice, in the broadest sense, might shape preferences for unrelated conspecifics.

Several studies have addressed preferences for genome-wide heterozygosity, which is heritable since heterozygous parents will on average have more heterozygous offspring. Models by Neff and Pitcher (2008) and Fromhage et al. (2009) showed that, assuming heterozygotes were more fit, directional preferences for heterozygotes evolved if ornament size was correlated with genome-wide heterozygosity. Simulation models showed that these preferences could evolve rapidly and be maintained indefinitely even in the absence of direct selection on preferences (but see Aparicio 2011). It is interesting to note that mutual conflicting preferences (chapter 8) would mean that genetically more homozygous—and therefore more modestly ornamented—individuals would pair with each other. If they did so disassortatively with respect to genotype, this would make their fitness approach that of the heterozygotes.

A model by Colegrave et al. (2002) showed that compatibility preferences acted to mitigate directional good genes preferences, thereby maintaining genetic variation in traits and preferences. Roberts and Gosling (2003) suggested in mice the simultaneous operation of mate choice for MHC dissimilarity (genetic compatibility) and for scent-marking rate (a putative but apparently unvalidated measure of good genes). This study may be limited by pseudoreplication, because females were only allowed to choose among a handful of inbred strains. Mays and Hill (2004) suggested that hierarchical integration (chapter 4) might allow good-genes and compatible-genes selection to act in concert if choosers filtered out the "bad-genes" courters and maximized compatibility among the courters that were left. This hierarchical integration could potentially maintain genetic variance in a population as well given heterogeneity in genetic effects, since the distribution of compatible genotypes a chooser is left after filtering out bad genes can be expected to vary with environmental condition. In a wild population of finches, Oh and Badyaev (2006) found that females relied on complementarity preferences at stages of the season when unrelated mates were readily available. Shifting the traits they attend to within and between sampling bouts may allow the coexistence of complementary and directional preferences, and maintain genetic diversity in preferences.

15.9 MATE CHOICE AS AN AGENT OF TRAIT EVOLUTION

15.9.1 Trait elaboration

The immediate evolutionary consequence of mate choice is sexual selection: the reproductive advantage of some courters versus others (Bateman 1948). The road from differential preference of choosers to differential success of courters is a convoluted one. A behavioral preference for a trait does not mean a mating preference, a mating preference doesn't mean a fertilization preference, a fertilization preference doesn't mean an allocation preference, and courters may be excluded by other courters before they have a chance to compete for the attention of choosers (see below). Nevertheless, chooser agency at multiple stages of mate choice can have the determining say as to which phenotypes reproduce and which don't, even in the face of coercion or courter-courter competition.

Sexual selection via mate choice is a powerful force in the elaboration and diversification of so much we find attractive about the natural world, from the elaborate coloration of West African killifishes to the pulsing displays of fireflies. The reader is referred to Andersson (1994) for a survey of sexually selected courter traits before mating, and to Eberhard (1996) for traits during and after mating. In humans, mate choice has been invoked as an agent of selection in the evolution of cooperation (Farrelly 2011), tooth health (Hendrie & Brewer 2012), vaginal orgasm (Costa et al. 2012), dance (Hugill et al. 2010), and humor (Bressler & Balshine 2006). Mate choice can affect the evolution of the extended phenotype of courters, like the bowers of bowerbirds and pufferfishes, and the evolution of characters above the individual, notably the size and dispersion of leks (Queller 1987). Choosers attend to variation in each of these features, giving mate choice a role to play in their maintenance if not their evolution. It is noteworthy that preferences may also select for general personality traits which may affect other aspects of behavioral function, such as behavioral consistency (Schuett et al. 2010) and boldness (Godin & Dugatkin 1996). Intriguingly, one of the most striking features of sexually selected traits, their complexity of form, varies predictably as a function of body size according to Cope's rule of biological scaling, leading Raia and colleages (2015) to argue that ornament complexity per se is unlikely to be driven by sexual selection.

The FLK and sexual conflict models lead to courter trait diversification among populations, but direct selection can do the same within populations under some circumstances. A model by Thom and Dytham (2012) showed that selection favors individual distinctiveness if choosers have to remember and return to courters during the process of evaluation (chapter 6). Kokko et al. (2007) modeled preference for rare phenotypes and found that low frequencies of rare-preferring choosers could stabilize courter trait evolution, maintaining diversity in both traits and preferences.

15.9.2 Cognition

Since Darwin (1871), workers have suggested that sexual selection via mate choice can favor cognitive sophistication. Sih and Bell (2008) argued that selection favors courters with the "social skills" to adjust to individual variation in chooser behavior (e.g., Patricelli et al. 2002; Ostojić et al. 2013). A more recent idea is that chooser cognitive abilities increase in sophistication in response to their selection as agents of mate choice—the so-called mating mind hypothesis (Miller 2000; Miller & Todd 1998; Cummings & Ramsey 2015). Boogert et al. (2011) suggest cognitive complexity of traits and preferences should coevolve, assuming that complex cognitive traits are costlier for choosers to assess. This assumption awaits testing, as variations in traits produced by complex minds may be comparatively trivial to assess (Madden et al. 2011; Fawcett et al. 2011). For example, drawing a perfect straight line is challenging in terms of cognition and locomotor performance, but a rudimentary visual system can easily detect pattern deviations (Rosenthal 2007). The complex, multistage, dynamic courtship of animals with biparental care may be a more promising candidate for cognitive elaboration.

15.9.3 Intrasexual competition

Direct competition among courters was thought to be the prevailing, if not the only, mode of sexual selection until the late twentieth century, and vigorous debates continue about the relative importance of sperm competition and cryptic female choice (chapter 7). Both before and after mating, intrasexual competition interacts with mate choice. Sperm competition, for example, occurs on a dynamic playing field set by females.

The same courter will often need to signal both to potential mates and to rivals. Fisher (1930) contrasts the evolutionary mechanisms underlying the elaboration of courtship signals and of agonistic (aggressive) signals:

> [I]n the case of attractive ornaments the evolutionary effect upon the female is to fit her to appreciate more and more highly the display offered, while the evolutionary reaction of war paint upon those whom it is intended to impress should be to make them less and less receptive to all impressions save those arising from genuine prowess. Male ornaments acquired in this way might be striking, but could scarcely ever become extravagant.

Signals, however, often function both in courtship and in aggression. Hunt et al. (2009) present an approach for partitioning total sexual selection on courter traits into components from intra- and intersexual competition, much as one can partition pre- and postmating components of mate choice (chapter 7). In some cases, courtship signals can be coopted for intrasexual competition, like vertical bars in swordtails (Morris et al. 2007).

15.10 POPULATION-LEVEL CONSEQUENCES OF MATE CHOICE

15.10.1 Population mean fitness

> Natural selection is apparently defenseless against genotypes that are successful reproducers but do not add to the survival value of the species as a whole. (Mayr 1970, p. 119)

Sexual selection can either increase or decrease the fitness of populations, both by increasing mutational load with respect to natural selection (as with attractive ornaments that are deleterious to survival) and through sexual conflict (Whitlock & Agrawal 2009). The evolution of chooser adaptations and courter counteradaptations via sexual conflict is harmful to fitness (reviewed in Bonduriansky et al. 2009), particularly if sexual selection acts much more strongly in one sex (Connalon et al. 2010). Courter traits are harmful to survival, and coevolutionary dynamics can displace ecologically important traits like parental investment from their naturally selected optima (Wolf et al. 1999) and permit the hitchhiking of deleterious mutations (Brooks 2000). Sexual selection can also reduce population mean fitness by exacerbating the bottleneck effects of genetic drift (Tazzyman & Iwasa 2009). Holman et al. (2015) found that multiple mating (section 14.4) co-evolves with meiotic drive, which under certain conditions can maintain selfish elements at high frequencies.

On the other hand, population mean fitness is increased if mate choice reinforces the success of superior genotypes and prevents unfit mutants from mating (Agrawal 2001) or if mate choice mitigates the costs of sexually antagonistic coevolution (Albert & Otto 2001).

Empirical evidence is scant and equivocal, offering little direct evidence that sexual selection improves population viability (Whitlock & Agrawal 2009). In *D. melanogaster*, experimental evolution by Promislow et al. (1998) found that sexual selection improved fitness metrics, and Hollis and Kawecki (2014) show that males perform worse on cognitive metrics after 100 generations of relaxed sexual selection through enforced monogamy (although this conflates mate choice and intermale competition). By contrast, Holland and Rice (1999) found that monogamy improved mean fitness by removing the load associated with sexual conflict, and Long et al. (2009) found that the fitness of the most fecund females was compromised by the deleterious effects of male preference (chapter 8).

The consequences for population mean fitness depend on the nature of sexual selection via mate choice. Kokko and Rankin's (2006) ecological model showed that populations can persist in the face of sexual conflict if courter and chooser behavior changes with density. In Fierst's (2013) model of gene-network evolution, mate-preference evolution under an FLK model reduced male mutational robustness for viability, while the opposite was true under a good genes scenario. An individual-based model by Lorch et al. (2003) shows that preference for condition-dependent traits accelerates adaptation.

Wedekind (2002), Candolin and Heuschele (2008), and Lumley et al. (2015) address the consequences of sexual selection, including mate choice, for adaptation to environmental change. The models to date suggest that the effect of directional sexual selection should have positive consequences for adaptation if they are associated with direct benefits or good genes, and negative if they evolve through non-adaptive mechanisms (e.g., Servedio & Bürger 2014; chapter 16).

15.10.2 Erosion and maintenance of genetic variation

As noted above, compatible-genes mate choice is a ready mechanism for maintaining genetic variation in traits and preferences (Neff & Pitcher 2005, 2008). When benefits are additive, erosion of genetic variation seems to pose an urgent problem for the maintenance of costly preferences. This is because any advantage of maintaining choice mechanisms inevitably disappears if there is nothing to choose from. In an FLK model, this happens once genetic variation in a courter trait is exhausted. At that point, there may be mate choice based on phenotypic variation in the trait, but there won't be a correlation between preference and genotype. Any indirect advantage therefore disappears. In good genes models, the courter trait can be a phenotypic, non-heritable indicator, but the indirect advantage to choosers goes away when the good genes run out of genetic variation, or the trait and the good genes cease to be phenotypically correlated (fig. 15.6). The so-called lek paradox (Borgia 1979) is that preferences are maintained in the face of persistent directional selection that erodes genetic variation for traits or their correlates, and therefore acts to mitigate any indirect benefits of choice in terms of attractiveness or viability. The problem seems particularly acute when considering that mate-choice copying further acts to homogenize preferences and intensify the effect of sexual selection (Wade & Pruett-Jones 1990; Santos et al. 2014).

As several workers have pointed out, the lek paradox isn't really a paradox (Reynolds & Gross 1990; Ritchie 1996b; Kotiaho et al. 2008), for multiple reasons. The lek paradox is particularly easy to resolve if compatibility-based choice plays a substantial role; indeed, Kotiaho et al. (2008) argue that compatibility sidesteps the lek paradox altogether. If there are hierarchical relationships between preferences whereby compatibility overrides directional preferences, as detailed above, this will act to maintain variation in traits.

The heterogeneous nature of preferences (chapters 9–12) is empirical evidence that preferences are not phenotypically homogenized by selection. The scope of proximate effects on preferences act to lower genetic covariance between traits and preferences, but this in turn also acts to weaken the erosion of genetic variation due to indirect selection. In finite populations, simulations predict substantial variation in preferences among choosers when costs of preference are low (e.g., Uyeda et al. 2009). Engen and Saether (1985), expanding on Lande (1981), showed that genetic variation could be maintained through a combination of sampling and assessment errors (chapter 6) and assortative mating (chapter 8). Bussière et al. (2008) suggest that the lek paradox can be resolved by environment-dependent variation in the magnitude and direction of good gene effects (see chapter 14). And at equilibrium, courter trait expression reflects a balance between natural selection and sexual selection (Lande 1981; Kirkpatrick 1982), and variation in traits can reflect a spectrum of trade-offs between the two (Ryan et al. 1990; Brooks 2000).

Depending on the extent of postmating choice and competition, multiple mating can weaken sexual selection and enhance genetic diversity (Lifjeld et al. 2013). Spatiotemporal variability in the benefits of choice can diversify traits and preferences (Day 2000; Holman & Kokko 2014b; Veen & Otto 2015). The nature and distribution of resources changes the interplay of intrasexual competition and mate choice (Forsgren, Karlsson, & Kvarnemo 1996). Environment-dependent choosiness (chapter 11) should act to weaken sexual selection (and increase the frequency of heterospecific matings under environmental stress; chapter 16). This effect is compounded if unresponsive females remain fertile and subject to forced matings (Syriatowicz & Brooks 2004). When experimental populations of freshwater snails are put on a restricted diet, there is a corresponding reduction in the strength of premating sexual selection, although the contribution thereto of mate choice is unclear (Janicke et al. 2015). Miller and Svensson (2014) provide a general review on fluctuations in sexual selection.

15.11 COEVOLUTION OF MULTIPLE TRAITS AND PREFERENCES

For simplicity, we have focused on the highly improbable situation where choosers use a handful of preference mechanisms to choose courters on the basis of a handful of traits. What explains the extravagantly multivariate nature of both preferences and traits? Pomiankowski and Iwasa (1993) extended the FLK model to multiple preferences for multiple traits. Runaway of multiple ornaments was predicted as long as the joint cost of choice (i.e., the marginal cost of assessing additional ornaments) was relatively low. This quantity is crucial to the behavior of all multiple-preference models; as I will revisit shortly, we are a long way from being able to assert confidently whether sampling multiple traits is more or less costly than sampling a single trait, let alone make statements about how this cost or benefit covaries with courter trait distributions.

The cost of assessing multiple ornaments is likely a function of their phenotypic covariance, which itself may be under selection independent of the effects of chooser preferences. As phenotypic covariance among traits increases, their combined effect approaches that of a single-preference/single-trait model described earlier, and the direct costs of choice are probably reduced for choosers. It is worth incorporating this dynamic cost into coevolutionary models. An important caveat, of course, is that the multivariate covariance between traits and preferences is likely to be complex (fig. 15.3).

A model by Iwasa and Pomiankowski (1994) showed that only one viability indicator evolved alongside multiple traits that evolved via an FLK mechanism. In the original model, multiple indicators were not stable, but later work found that multiple viability indicators could be maintained if they were associated with different components of chooser fitness (the multiple messages hypothesis, Møller & Pomiankowski 1993; van Doorn & Weissing 2004). It is straightforward to imagine how "multiple messages" preferences would evolve in a compatibility context.

Johnstone (1996) suggested that if signals were inexpensive to courters, individual courters should invest heavily in signaling their best latent trait. Van Doorn and Weissing (2006) showed that such sexual conflict over the honesty of viability indicators could maintain stable cycles of preferences for redundant display traits, but that traits and preferences oscillate over time. This may partially account for the general absence of strong good genes effects in natural populations.

These models all hinge on the cost to choosers of assessing multiple traits. Sampling $n + 1$ individuals is generally more costly than sampling n individuals. What about sampling $n + 1$ stimulus features? Does it cost more time, energy, and risk to sample mates more thoroughly? While most models of multiple-trait evolution have assumed that assessing multiple traits increases the costs of choice (Candolin 2003), psychophysical data show that information from multiple channels always facilitates detection, and usually enhances discrimination and memorability of stimuli (Rowe 1999). A familiar example is that visual and acoustic cues reinforce one another in speech ("look at me when I'm talking to you").

Choosers attending to multiple traits will thus generally be better able to localize and choose among potential mates. Multiple mechanisms can, for example, allow for a richer sensory experience; women with four cone opsin alleles rather than the usual three can make finer chromatic comparisons (Jameson et al. 2001). This ability can come at a cost, however. Partan and Marler (2005) point out that processing of multimodal and/or complex signals, like integrating simultaneous visual and auditory cues, may require a sophisticated cognitive architecture which may be costly to produce and maintain, thereby imposing a fixed cost of mate choice on choosers. Indeed, in European tree frogs *Hyla arborea*, females discriminate more consistently between males when given both auditory and visual cues, but take longer to choose than when given auditory cues alone (Gomez et al. 2011). Alternatively, multimodal processing may facilitate mate detection and evaluation in noisy environments. Indeed, Partan (2013) argues for a crucial distinction between multimodal and multicomponent signals, pointing out that multimodality allows for flexible responses to environmental noise. For example, complementary preferences for a visual and an acoustic trait would facilitate mate choice in a dense, noisy forest. More generally, attending to multiple complementary stimuli likely facilitates mate choice in temporally and spatially heterogeneous environments (Muñoz and Blumstein 2012; Bro-Jørgensen 2010).

The costs of assessing multiple cues are dependent on the covariance among cues and the covariance among preferences. In terms of perceptual processing, it is cheap for choosers to pay attention to multiple cues that are correlated with one another among courters—and more expensive to pay attention to multiple uncorrelated cues (Rowe 1999). The cost to choosers is therefore intimately tied to how variation is distributed among courters. Since sexual selection by mate choice directly affects this variation, there is therefore the potential for the cost of choice to change as traits and preferences coevolve.

Another major determinant of the costs of multiple preferences is whether preferences are expressed simultaneously or sequentially. Choosers can evaluate suites of traits in tandem, for example simultaneously attending to multiple acoustic and multiple visual features of a display. Alternatively, they can attend to courter traits sequentially (and therefore hierarchically), using successive sets of traits as filters during different stages in the mate-choice process. Sequential mate choice using different sets of signals is widespread, particularly in species that evaluate long-range broadcast signals in the early stages of mate choice and tactile stimuli in the peri-mating stage. In such cases, the costs of evaluating a chooser at an early stage are almost certainly lower than at later stages, when coercion, injury, and disease transmission come into play.

Increasing the number of traits sampled may also benefit choosers, by mitigating the trade-off between rejection and acceptance (Wiley 1983, 2006; chapter 3); if there are more axes of variation, the smaller the area of overlap between appropriate and inappropriate mates. This should select for configural preferences (chapter 4) where choosers pay attention to a narrow combination of trait values. Preferences can covary, however, even among multiple, distinct courter trait axes. This can be the case, for example, if selection against mating with heterospecifics acts in concert on multiple preference mechanisms, or if the same mechanisms are recruited for evaluating different traits. Within individual female *X. pygmaeus*, directional preferences for large body size and ornamentation are inversely associated with discrimination against a sympatric heterospecific, *X. cortezi*, when compared to size-matched heterospecifics. A female that discriminates against males of a different species is thus more likely to reject large males of her own species (Rosenthal & Ryan 2011a). This places a potentially important trade-off between conspecific mate preference and intrasexual preference. More generally, such correlated responses across contexts may constitute multivariate genetic constraints on preferences (Chenoweth & McGuigan 2010), and may account for the complex selection surfaces that can arise from multivariate preferences (Simmons et al. 2013). We are still gaining a first-order understanding of the costs and constraints of multivariate signal evaluation. How these scale with the number of traits being addressed is surely critical to the dynamics of multivariate trait/preference coevolution.

15.12 SYNTHESIS: A UNIFIED VIEW OF PREFERENCE EVOLUTION

15.12.1 The received view of mate-choice evolution

> Current thinking about sexual selection can be roughly divided into two schools. These are the "good genes" school, which postulates that mate choice evolves under selection for females to mate with ecologically adaptive genotypes, and the "nonadaptive" school, which holds that preferences frequently cause male traits to evolve in ways that are not adaptive with respect to their ecological environment. (Kirkpatrick 1987b, p. 44)

Kirkpatrick's words might as well have been written yesterday. Thirty years later, the "school fight" (Kotiaho & Puurtinen 2007) between proponents of adaptive and nonadaptive mate choice continues unabated. There's still a "good genes" school and a nonadaptive school, and they're still talking about different things. Early good genes work focused on additive genetic benefits as the primary agent of selection on chooser preferences, and early nonadaptive work focused on the balancing effects of natural and sexual selection on courters. Today one school is primarily focused on courters and another is focused on choosers, but their roles have flipped. The good genes school, having failed to turn up much direct evidence of good genes benefits to choosers, has turned its attention to the relationship between Quality and courter signal variation, while the nonadaptive school focuses on proximate constraints on choosers and the arbitrary nature of preference evolution.

The weight of the evidence argues against a primary role for good genes in the evolution of mate choice. Rather, additive genetic benefits tied to viability are omnipresent but generally weak, for sound theoretical reasons. If we can measure and manipulate fitness outcomes for choosers, we can integrate good genes into a comprehensive view of total selection on mate choice.

Good genes is a special case of a preference evolving by indirect selection. Along with others, I have argued that good genes plays a relatively marginal role in preference evolution. What about indirect selection based on attractiveness? Prum (2010) argues that the FLK model should be considered the null model not just for trait-preference coevolution, but for trait evolution and preference evolution more generally. The FLK model rests on two assumptions: first, that there is abundant opportunity for genetic covariance between trait and preference; and second, that the direct costs and benefits of expressing preferences are low.

In section 15.3, I argued that several factors might conspire to limit the opportunity for traits and preferences to covary genetically, once we take into account the full scope of the relationship between courter traits, chooser preferences, and how selection acts upon them. Throughout this book, I have argued that the myriad ways that preference can operate within a chooser collectively represent a broad mutational target, in that a change to any preference mechanism—sensitivity, choosiness, and so forth—can result in different preference outcomes. The importance of coevolutionary scenarios depends on whether there is genetic variation in preferences that aligns with genetic variation in traits. Alternatively, we have seen that courter traits can evolve to address chooser preferences for which there is no genetic variation, and vice versa. A critical empirical question is therefore how often novel mutations for traits and preferences arise and how they are co-distributed.

Prum's (2010) advocacy of FLK as the null model for trait-preference coevolution is part of a broader argument for the primacy of non-adaptive "aesthetic evolution" in shaping preferences for elaborate traits (Prum 2012). The FLK process is one avenue whereby the "taste for the beautiful" can manifest itself, but it is an open empirical question as to whether it is more common for traits and preferences to coevolve in arbitrary directions, or for traits to evolve that address preexisting arbitrary preferences.

The dynamics of trait-preference coevolution are fascinating, but perhaps the focus on indirect selection has shifted our attention away from how mate choice primarily works. Nonadaptive biases (Borgia 2006; Ryan & Cummings 2013), direct benefits (Wagner 2011), and non-additive indirect benefits (Neff & Pitcher 2008) likely contribute more than additive indirect benefits to how mate-choice mechanisms covary with sexual fitness. Roff (2015) further suggested that stochastic variation in mate availability might impose a limiting constraint on mate-choice evolution; there simply isn't enough opportunity for preferences to be expressed as such.

Kokko et al. (2002) and Mead and Arnold (2004) each present coevolutionary models as a continuum, varying only in the extent to which adaptive benefits reinforce the automatic FLK. Further, Kokko et al. (2003) suggest that preexisting biases, genetic covariances between traits and preferences, and genetic covariances between traits and viability are each ubiquitous, pointing out (p. 656) that "indirect selection favors mate choice when females choose males with high breeding values for total fitness, irrespective of the relative contributions of viability, attractiveness, fecundity and other fitness components." I agree with Prum (2010), however, in seeing these relative contributions as critical: the key question is "whether intersexual traits exhibit any meaning or design other than their arbitrarily (co)evolved correspondence with mating preferences" (parentheses mine).

15.12.2 Total selection on mate-choice mechanisms

> Thus, it seems that the only route to a proper understanding of the evolution of female mate choice is empirical studies designed to quantify selection acting on female mating biases (Kirkpatrick 1987).
>
> —Gavrilets et al. 2001

Consider this the latest in 30 years of calls to empirically quantify selection on choice outcomes. A fundamental problem with the field of sexual selection is that our fundamental models predict effects that are indistinguishable from noise given the current way we study them. For example, Blows (1999) showed massively variable selection on the genetic covariance between male and female conspecific mate preferences over generations, consistent with noise (i.e., no effect of mate choice) and with models predicting cyclical evolution of traits and preferences (Iwasa & Pomiankowski 1995). Experimental evolution manipulating sexual selection and measuring multivariate evolution yields similar non-linearities (Bacigalupe et al. 2008). All additive-genetic models of trait-preference coevolution predict a spectrum of outcomes with regard to trait values and preference values. The key differences lie in the form and direction of fitness effects of courter traits on choosers (Mead & Arnold 2004).

Using the Robertsonian framework presented at the end of chapter 14, what proportion of total fitness is a consequence of the sexual fitness on mate choice mechanisms? What constitutes total selection on mate-choice mechanisms? What is the contribution of indirect selection to total selection on preferences? I suggest that we start by partitioning total selection into indirect, direct, additive, and non-additive components, as well as components associated with non-choice functions. Additive, indirect effects on chooser fitness appear to make up a small fraction of total selection on preferences, and good genes a modest share within those.

Some effort is required to disentangle effects of courter genotype, courter phenotype, and differential allocation. Even more effort is required to disentangle the predictions of FLK models from those of viability models, since the former predicts that choice will not improve offspring viability but will increase the number of grand-offspring propagated through the chosen sex. A promising, though laborious, approach is to use an experimental or animal-model design (Wolf et al. 1998; Hunt & Simmons 2002; Bijma et al. 2007a, 2007b) to estimate the direct and indirect genetic effects of courter genotype on offspring phenotype.

15.12.3 The consequences of mate choice

It is hard not to lapse into hyperbole when describing the wondrous and variegated textures, dances, and sounds that have arisen in response to sexual selection via mate choice, not to mention the chemosignals and internal structures we cannot directly perceive. These wonders have been shaped by mate-choice mechanisms, and in turn have shaped the way these mechanisms function. Diversification over space and time is one of the few generalizations we can make about the consequences of mate choice. In the next chapter, I discuss the role that mate choice plays in diversifying, cleaving, and blending evolutionary lineages.

15.13 ADDITIONAL READING

Jones, A. G., & Ratterman, N. L. (2009). Mate choice and sexual selection: what have we learned since Darwin? *Proceedings of the National Academy of Sciences, 106* (Supplement 1), 10001–10008.

Kuijper, B., Pen, I., & Weissing, F. J. (2012). A guide to sexual selection theory. *Annual Review of Ecology, Evolution, and Systematics, 43.*

Prum, R. O. (2010). The Lande–Kirkpatrick mechanism is the null model of evolution by intersexual selection: implications for meaning, honesty, and design in intersexual signals. *Evolution, 64*(11), 3085–3100.

CHAPTER 16

Mate Choice, Speciation, and Hybridization

16.1 INTRODUCTION

For the mid-twentieth-century Modern Evolutionary Synthesis, almost the only use for mate choice—and for that matter, for animal behavior—was as a mechanism to restrict gene flow between species. Dobzhansky (1970), following Fisher (1930), believed that selection against hybridization is the primary force acting on preferences; in fact, his latter-day tome on evolution only mentions mate choice in the context of reproductive isolation. Dobzhansky advocated mate choice as an agent of **reinforcement** (in a very different sense from reinforcement in the context of learning, chapter 6) whereby species could diverge as a result of selection against hybrids.

His contemporary, Ernst Mayr (1970), by contrast, argued that behavioral barriers to mating evolved primarily as a result of geographic isolation, and stressed that such barriers can fail when species come into contact. An expanded version of this view was articulated starting in the 1970s by Hugh Paterson, who saw species as primarily defined by a "specific mate recognition system" (SMRS) that included the full spectrum of mate choice mechanisms from behavioral preferences through fertilization. This system, according to Paterson, involved coadaptation between male and female components, was under stabilizing selection favoring matings with conspecifics, and could evolve to novel optima as a consequence of changing ecological pressures (reviewed in Paterson 1993). But the components of the so-called SMRS are the theater of coevolutionary dynamics that were immediately recognized to drive diversification: the Fisher-Lande-Kirkpatrick (FLK) process (Lande 1981), and sexual conflict (West-Eberhard 1983).

Each of these ideas has, in some form, been synthesized into our contemporary understanding of mate choice and speciation, where there is arguably a more fruitful marriage of empiry and theory than there is for the evolution of intraspecific choice. What can a focus on mate-choice mechanisms tell us about how they work in terms of regulating gene flow between lineages?

In this chapter, I begin by discussing the widespread support for the intuitive predictions that mate choice should promote diversification among geographically isolated species, and that mate choice should evolve to minimize drastic loss of fitness through hybridization. The role of mate choice is more complicated when there is incomplete divergence between lineages; depending on their relationship to other traits under selection, mating preferences can act to accelerate speciation through reinforcement, but they can also act to increase gene flow between divergent lineages. Finally, I address the relationship between individual mating decisions and hybridization between species.

16.2 DIVERGENCE OF PREFERENCES AMONG ISOLATED POPULATIONS

The suggestion that sexual selection via mate choice can generate geographic diversity goes back at least to Darwin (1871), who observed that sexual selection was likely the main cause of the (superficial) differences among human populations (for all his misogyny, Darwin was clear-eyed on the fallacy of "race" as a biological concept). There are several, complementary mechanisms whereby preferences can diverge among populations. As Paterson (1993) pointed out, preferences have different naturally selected optima in different habitats, and thus may be subject to divergent ecological selection with respect to the sensory periphery (Ryan & Cummings 2013; Boughman 2002) or to personality traits associated with preferences or mate-sampling behavior (chapter 13).

Even in the absence of ecological selection, preferences will inevitably diverge among isolated populations due to drift. But two basic coevolutionary mechanisms can drive preferences apart very quickly relative to this neutral expectation. Further, divergence of preferences as a consequence of either Fisher-Lande-Kirkpatrick (FLK) or sexual conflict mechanisms can cause geographically disconnected populations to drift apart more than would be predicted by neutral divergence. The original papers on sexual conflict (West-Eberhard 1983) and FLK (Lande 1981) each suggested that the same processes that drive coevolution of traits and preferences could drive the diversification of traits across species.

Under FLK, trait and preference values can evolve to arbitrary values along an equilibrium line (chapter 15; Lande 1981; Prum 2010). A simulation model by Uyeda and colleagues (2009) showed that populations could become sexually isolated by drifting to different points along the line (fig. 16.1). Pomiankowski and Iwasa (1998) synthesized their original models of ever-cycling preferences (chapter 15; Iwasa & Pomiankowski 1995) and

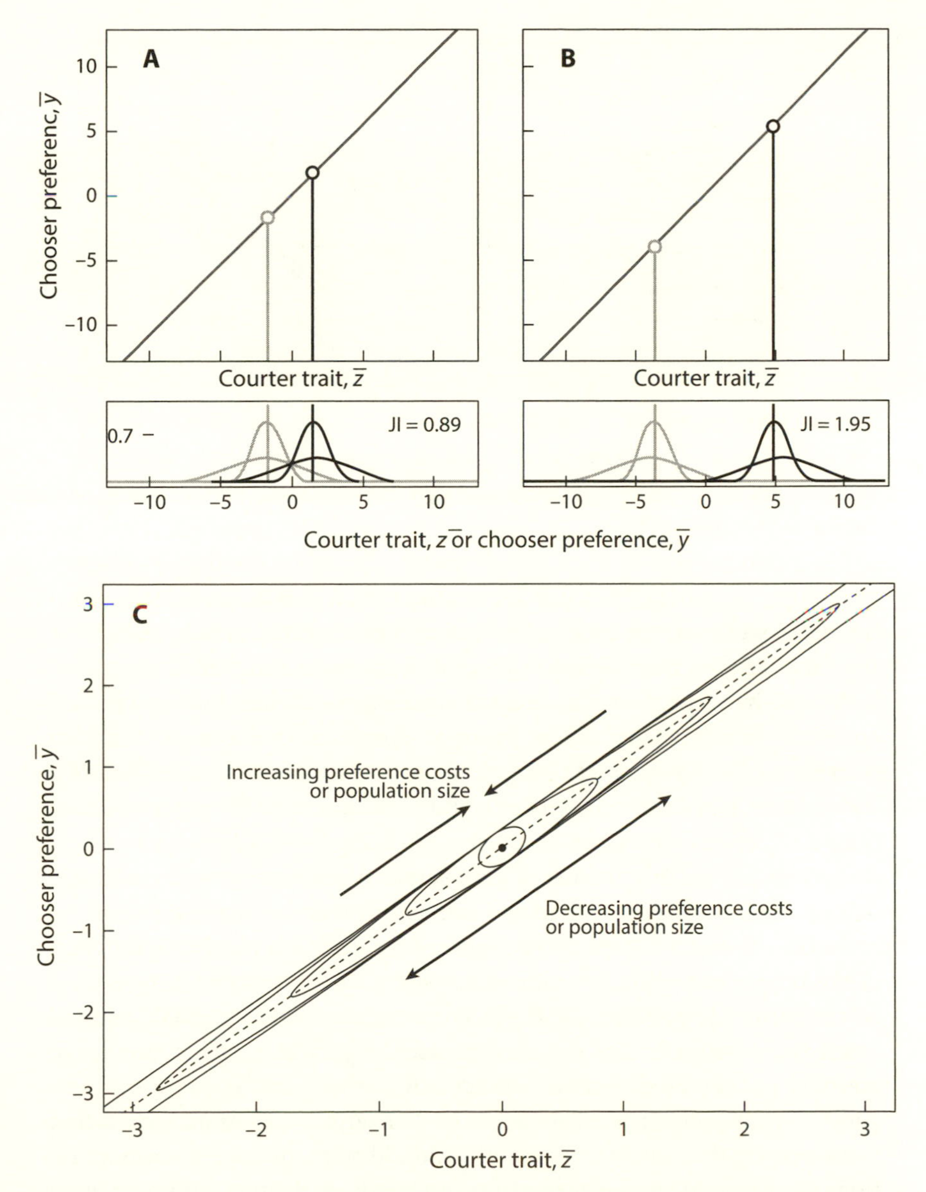

Figure 16.1. Sexual isolation between populations along the equilibrium line predicted by Fisher-Lande-Kirkpatrick dynamics. (a) Traits and preferences overlap between weakly divergent species, resulting in weak isolation. (b) Greater divergence leads to stronger isolation. (c) Preferences diverge more between isolated populations when direct costs of preference are low (chapter 15) and when population sizes are small, allowing for stochastic changes in mean preference due to genetic drift. Ellipses are 95% confidence intervals at equilibrium, with bigger ellipses corresponding to lower costs of preferences as a function of preference at equilibrium. After Uyeda et al. (2009).

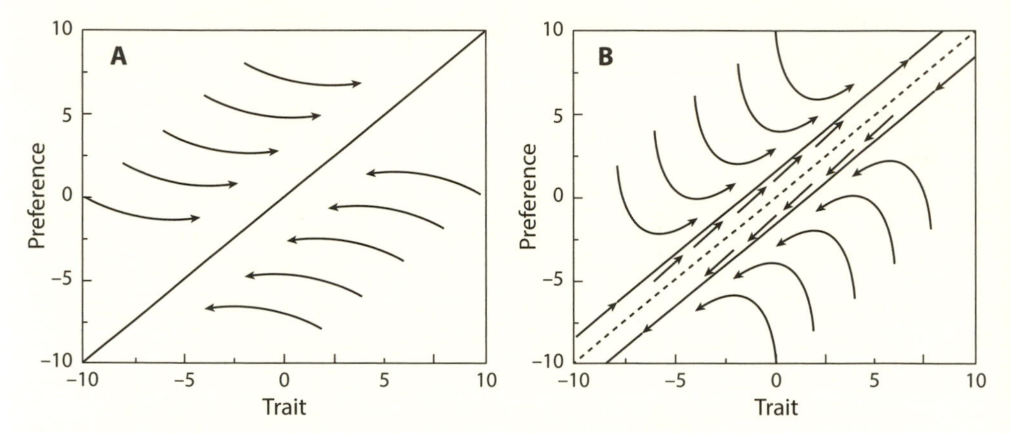

Figure 16.2. Divergence of traits and preferences through sexual conflict. (a) Evolution toward a line of equilibria under weak selection; (b) runaway dynamics when selection is an order of magnitude greater. After Gavrilets (2000).

of multiple ornament evolution (Pomiankowski & Iwasa 1993) into a multivariate extension of Lande's (1981) model predicting runaway-mediated speciation arising from diversification of preferences and traits.

Models by Sergey Gavrilets and colleagues showed that traits and preferences can similarly coevolve in divergent directions due to sexual conflict (Gavrilets 2000; Gavrilets et al. 2001). In particular, Gavrilets's (2000) model showed that traits and preferences could steadily evolve in opposite directions if selection on choosers is sufficiently strong (fig. 16.2). Such a process should be particularly feasible if choosers can easily reverse the hedonic value of their preferences (chapter 5; see also Turner & Burrows 1995).

In Gavrilets's (2000) model, a chooser's fitness was a quadratic function of the proportion of males from whom she rejected mating attempts, and was maximized by an optimal intermediate value. Gavrilets (2000) pointed out that the optimal value should decrease the greater the total number of courters encountered. Sexual conflict might therefore be more likely to generate trait-preference diversification when large populations are isolated from one another, and FLK dynamics should predominate when populations are small due to the effects of drift (Uyeda et al. 2009). If preferences are initially permissive, as suggested by Kaneshiro (1980), this might further facilitate the coevolution of novel traits through an FLK model.

Depending on the history of speciation and isolation, one or the other of these mechanisms may therefore play a more important role in diversification of preferences or traits. An experimental study on dung flies *Sepsis cynipsea* by Martin and Hosken (2003) found support for Gavrilets's (2000) prediction that traits and preferences under sexual conflict should diverge more strongly in larger populations. However, manipulating levels of sexual conflict with similar population sizes of *D. melanogaster* yielded no effect on preference divergence (Wigby & Chapman 2006).

The mechanisms that specify preferences may play a further role in facilitating divergence, independent of how or whether selection acts upon them. Pfennig and Ryan (2007) used neural network simulations to show that selection on preferences for within-population traits could lead to emergent preferences for conspecifics over heterospecifics, even in allopatry. A model by Lachlan and Servedio (2004) showed that if preferences were learned across generations, this could accelerate divergence in allopatry. Within-generational learning might also facilitate interpopulation divergence of traits, particularly if choosers experience positive reinforcement (in the sense of learning, not reduction of gene flow as discussed below) from previous associations with courter phenotypes. This is suggested by a model of **sympatric speciation** with learning by Servedio and Dukas (2013), to which I will return below.

Mate choice thus provides a satisfying explanation for the diversification of ornaments and displays (Seddon et al. 2013; Boul et al. 2007; Rodríguez et al. 2013; Wilkins et al. 2013). M'Gonigle et al.'s (2012) simulation model of speciation with gene flow (see below) showed that mate choice could facilitate the long-term coexistence of ecologically identical species. Wellenreuther and Sánchez-Guillén (2016) provide a compelling argument that the radiations of well-studied European dragonfly genera arose through trait-preference coevolution rather than ecological divergence. Bonduriansky (2011) further suggests that sexual selection and sexual conflict could in turn facilitate ecological adaptation by allowing populations to move more broadly around viability optima in adaptive landscapes.

The lingering question here parallels the one addressed at the end of chapter 15: how much of the variation among species in preferences is due to coevolution from FLK dynamics (which should be more prevalent in smaller populations), how much is due to sexual conflict (more prevalent in larger populations), and how much of it is due to non-coevolutionary divergence of preferences due to drift or ecological selection?

In any event, there are multiple routes whereby mating preferences diverge in allopatry, and trait-preference coevolution can increase genetic differentiation between lineages. It seems a logical next step to say that since preferences accelerate differences between species, they should accelerate speciation through their role as reproductive barriers. As should become apparent in the next section, this is often not the case.

16.3 DIVERGENCE OF PREFERENCES WITH SECONDARY CONTACT

16.3.1 Overview

We now turn to mate choice in the context of how it evolves when divergent lineages come into secondary contact. We start with the extreme case of total hybrid inviability between species. At a minimum, a chooser wastes gametes and time mating with a courter of a different species; furthermore, maternal-fetal incompatibilities increase the risk of maternal mortality (Maddock & Chang 1979). With rare exceptions (see below), there will therefore be strong direct selection on choosers against what Fisher (1930) called "the grossest blunder in sexual preference." Preference displacement is a special case of so-called **reproductive character displacement**, whereby similar traits in sympatric species diverge in response to selection to minimize interference (Pfennig & Pfennig 2009; fig. 16.3). When hybrids are sterile, the evolution of displaced mating preferences is easily explained by direct selection on fecundity (chapter 14) and should favor premating preferences to minimize wasted matings.

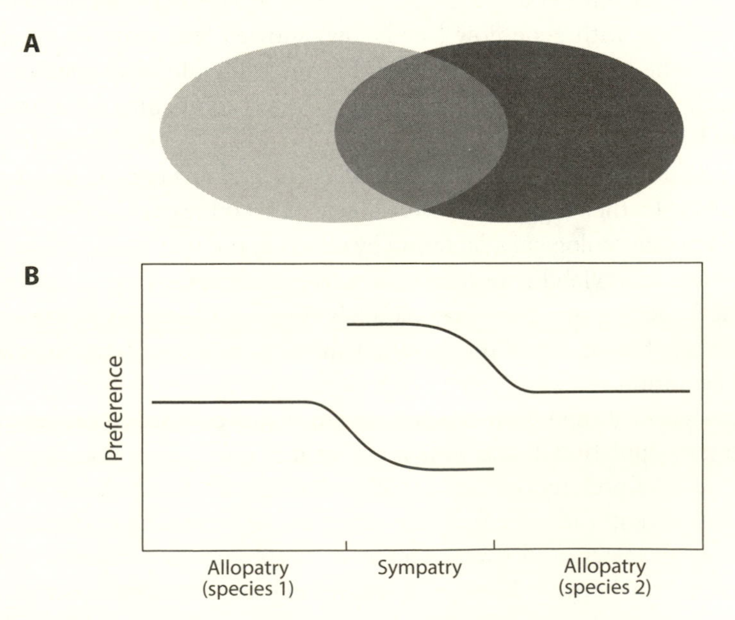

Figure 16.3. Reproductive character displacement. (a) Two species share an area of sympatry. (b) Where species overlap, preferences (and preferred traits) diverge in response to selection against hybridization. After Servedio (2004b).

Indeed, choosers from populations sympatric with another closely related species frequently exhibit stronger preferences for conspecifics and/or antipathy for heterospecifics (e.g., Gabor & Ryan 2001; Wong et al. 2004; Higgie & Blows 2007a; Kronforst et al. 2007; Rundle et al. 2008; Rundle & Dyer 2015; Murphy & Zeyl 2015; reviewed in Ord et al. 2011).

How can choosers avoid heterospecific matings? As discussed in chapter 3, there is a trade-off between being too selective and missing out on mating opportunities, versus being permissive and incurring the costs of mating with a heterospecific (Reeve 1989). For males in polygynous species, opportunities are limited, and preference envelopes are accordingly broad (chapter 6; but see Ord et al. 2011). Such male-driven heterospecific matings are a major cause of interspecific hybridization (Gröning & Hochkirch 2008; fig. 16.4). By contrast, for females in such species, willing conspecific mates are easy to come by. Choosers in sympatry frequently exhibit so-called **peak shift** displacement, preferring conspecific trait values that diverge from the population mean so as to minimize overlap with heterospecific signals (Ryan & Cummings 2013). This can be an important driver of intraspecific sexual selection (section 16.5).

As detailed in the preceding chapters, choosiness is extremely plastic, and heterospecific matings are more likely when choice is more costly. To some

Figure 16.4. Mate choice for heterospecifics by male hylid treefrogs. (a) Male squirrel treefrog (*Hyla squirella*) amplexing a female gray treefrog (*Hyla chrysoscelis*). Photo © Ryan Taylor 2005. (b) Male red-eyed treefrog (*Agalychnis callidryas*) between an amplexed pair of marbled treefrogs (*Trachycephalus venulosus*). Photo © Alexander T. Baugh 2005.

extent this is buffered by the multi-stage, multivariate nature of mate choice. In particular, multiple mating can make up for "blunders" through conspecific sperm precedence (chapter 7; see below).

16.3.2 Preferences are not consistently displaced away from heterospecifics

Conspecific mate preference in sympatry might seem to be what Darwin called "an extremely general result." With exceptions (section 16.5), choosers in general do prefer conspecifics, but a meta-analysis by Ord et al. (2011) found surprisingly weak evidence for further preference displacement in sympatry, although they cautioned about asserting the null from studies suffering from low statistical power and methodological issues (see also Jiang et al. 2013). Ord and colleagues' (2011) analysis suggested that conspecific mating preferences might be driven largely by the direct costs and benefits of individual mating decisions, as I will discuss below.

Perhaps surprisingly, Ord et al.'s (2011) analysis found that males and females varied independently in the magnitude and direction of sexual isolation; in fact, males in two insect species with high direct costs of male mate choice have stronger conspecific mating preferences than females do (Peterson et al. 2005; Svensson et al. 2007). Cryptic female choice for conspecific male sperm (or more generally, conspecific sperm precedence) could also contribute to male choosiness if mating with a heterospecific means a wasted mating (Marshall et al. 2002).

Why don't we always see a signature of preference displacement between sympatric species? The signature may be partly blurred by species that speciated in sympatry (see below), such that divergent preferences were the ancestral state (Servedio 2004a). Another important point is that reproductive character displacement can theoretically evolve through divergence in preference alone (if courters are detectably different to choosers) or through divergence in traits alone (if courters diverge sufficiently to elicit differential responses). Lemmon et al. (2004) showed that, depending on the strength of selection against hybrids along a geographic cline, traits could diverge between species without divergence in preferences (and under a broad range of parameter combinations, neither diverged). Traits were the primary mechanism of reproductive character displacement induced in an experimental study by Higgie et al. (2000), putting *D. serrata* from an allopatric population into experimental sympatry. Cuticular hydrocarbons diverged away from the phenotype of the newly sympatric congener, with the end result that *D. serrata* males were less attractive to conspecific females from the source population. Finally, mate choice is mutual in nearly all cases, and selection on conspecific preferences may also be weak if choosers are already avoided by potential mates of the other species. While males should generally have a broader preference envelope than females, Ord et al.'s (2011) meta-analysis did not detect a sex difference in conspecific mate preference.

Functional constraints may limit the extent to which choosers can respond differentially to conspecific and heterospecific signals. An experimental evolution study by Matute (2015) showed that while *D. yakuba* could evolve behavioral isolation from one other species in sympatry, it failed to do so in a complex community of multiple species. Here, it proves difficult given existing genetic variation to evolve a response mechanism that admits conspecifics and filters out everything else.

Categorical perception, by contrast, might elicit abrupt transitions in preferences (Baugh et al. 2008) away from heterospecific traits, and might be an efficient mechanism for distinguishing conspecifics from heterospecifics on the basis of taxonomically reliable cues. The role of categorical preference in reproductive isolation is worthy of more attention.

16.3.3 Asymmetric preferences

Choosers sometimes fail to prefer conspecifics or even prefer heterospecifics. In particular, choosers in pairs of species frequently have so-called **asymmetric preferences** for heterospecific traits, whereby species share the same biases for traits typical of one species (e.g., Jones & Hunter 1998; Ryan & Wagner 1987; Watanabe & Kawanishi 1979; reviewed in Wirtz 1999). Such asymmetries should be particularly widespread if the arbitrariness of mate choice comes from courter traits arising in response to hidden biases on the part of choosers rather than through tight coevolution with traits (chapter 15).

Kaneshiro (1980) suggested that asymmetries could arise as a result of the biogeography of how new environments were colonized. In small, homogeneous founder populations, Kaneshiro argued, selection would favor reduced choosiness, which would in turn relax selection and facilitate diversification of novel signals. Small populations should also be more likely to have signals diverging as a consequence of FLK dynamics (Uyeda et al. 2009). Tinghitella and Zuk (2009) found support for Kaneshiro's hypothesis in female Polynesian field crickets, *Teleogryllus oceanicus*. Continental Australian crickets were less likely to accept silent males as mates, whereas crickets from Polynesian populations were more permissive; they argued that this had facilitated the parasitoid-driven loss of song in introduced Hawaiian populations. An intriguing parallel is suggested by Lachlan and colleagues' (2013) recent finding that the syntactical structure of birdsong is lost in recently colonized island populations. Both learned and genetically specified elements of preference may similarly be lost following dispersal.

16.3.4. Reproductive isolation across stages of mate choice

Multiple mate-choice mechanisms can favor conspecifics over heterospecifics, and thereby serve as barriers to gene flow; these range from insensitivity to heterospecific broadcast signals, to abandoning hybrid offspring. Selection is predicted to favor mechanisms that minimize the risk of producing heterospecific offspring as early as possible in the decision process, so that choosers don't have to pay the cost of choosing or mating. At the same time, stringent mechanisms that act against heterospecifics later in the process, like incompatibilities between gametes (e.g., Comeault et al. 2016), might permit individuals to be less choosy earlier on.

Consistent with how traits and preferences diverge in allopatry (section 16.2), evidence suggests that premating isolation evolves readily when species diverge. A comparative study by Mendelson (2003) found that premating barriers to mating evolved more quickly than postzygotic barriers in darters, *Etheostoma*, mirroring a classic earlier study on *Drosophila* (Coyne & Orr 1989). Jennings et al. (2014) found premating isolation, sometimes asymmetric, between all pairs of recently diverged *D. montana* populations, while only a subset had evolved postmating, prezygotic isolation.

Several studies have dissected the stages responsible for reproductive isolation in fine detail; in flowering plants, pollinator preferences play a primary role (Ramsey et al. 2003; Lowry et al. 2008). Sánchez-Guillén and colleagues (2012) partitioned stages of reproductive isolation between the damselflies *Ischnura elegans* and *I. graellsi* (fig. 16.5). Although crosses in both directions showed near-total reproductive isolation, the nature of the isolating mechanisms was asymmetric; when *elegans* females mated to *graellsi* males, mating was hampered by mechanical incompatibilities. *graellsi* females mated to *elegans* males, by contrast, were isolated by an accumulation of postmating barriers.

Chooser behavior before mating is a potentially powerful and relatively inexpensive way to reduce the risk of heterospecific matings, but behavior can also contribute to reproductive isolation if hybrid courters are less at-

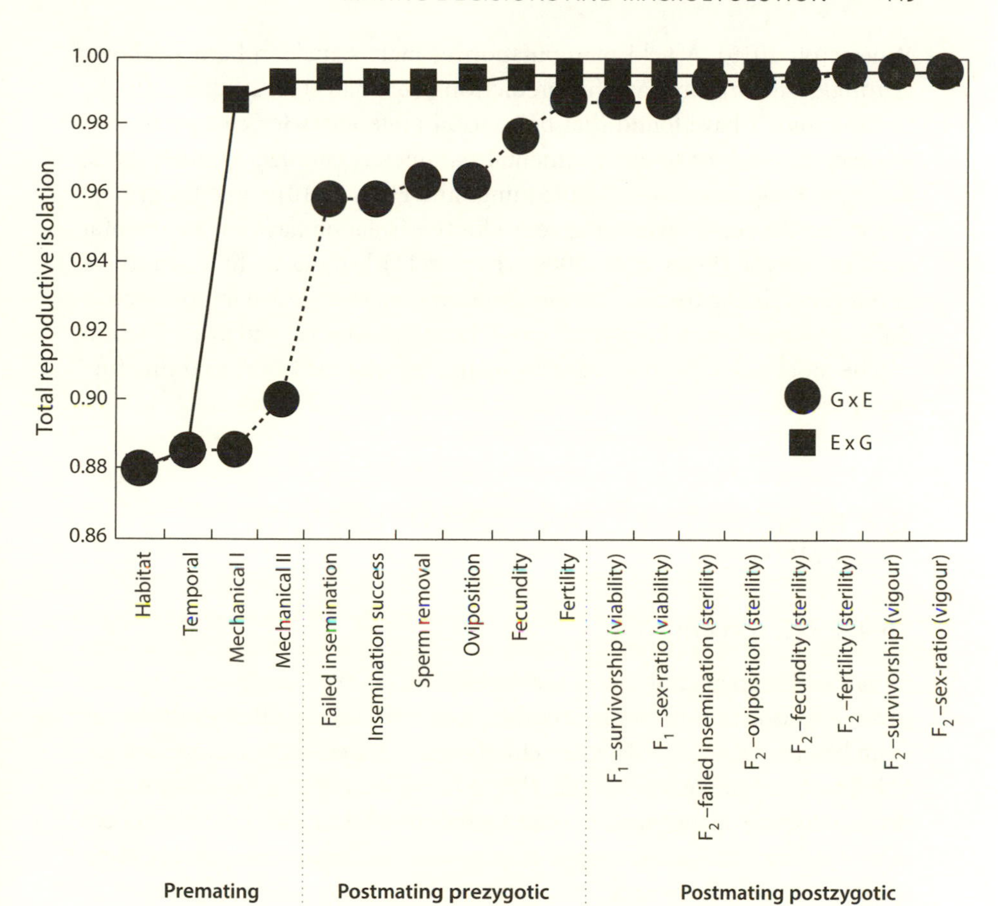

Figure 16.5. Relative contribution to total isolation of reproductive barriers, including mate-choice mechanisms, in reciprocal crosses between damselflies *Ischnura graellsi* females and *I. elegans* males (GxE) and vice versa (ExG). After Sánchez-Guillén et al. (2012).

tractive to parental choosers (e.g., Lemmon & Lemmon 2010). In some cases, the behavioral basis of reproductive isolation can be narrowed down to individual signal molecules and neurons, as in corn borers (chapter 5). Latour et al. (2014) showed sexual selection against natural hybrid mice via odor-based mate choice.

Attending to a small number of cues incurs the danger that choosers will not be able to distinguish conspecifics from heterospecifics, due to signal overlap or environmental noise (e.g., Fisher et al. 2006). Multidimensional preferences (chapter 4) could facilitate choosers discriminating conspecific from heterospecific signals (Proulx & Servedio 2009; Pfennig 1998;

Wheatcroft 2015). A field manipulation of male signals in barn swallows (*Hirundo rustica*) confirms this prediction (Vortman et al. 2013; fig. 16.6).

Two studies have found that behavioral preferences for conspecifics in sympatry are robust to environmental parameters (Berdan & Fuller 2012; Lackey & Boughman 2014), but Jennings and Etges (2010) raised the critical point that the genes involved in reproductive isolation have environmental reaction norms (Etges et al. 2009; chapter 11) leading us to be cautious about generalizing from laboratory data on even the relative importance of different reproductive barriers. Indeed, behavioral barriers to hybridization can be modified or suppressed with surprising ease, as I shall examine further in section 16.6.

16.4. REINFORCEMENT AND SPECIATION WITH GENE FLOW

16.4.1 Reinforcement

So far in this chapter, we have seen that preferences diverge when species diverge in isolation from one another, and preferences often evolve away from heterospecific signals when closely related species come into contact. If hybrids are sterile or inviable, this process occurs as a consequence of direct selection against mating with heterospecifics or hybrids. Differences in mate-choice mechanisms thus emerge as a consequence of the cessation of gene flow among species.

But what about at the early stages of speciation, when hybrids are still interfertile? Consider two populations that are *parapatric*, that is, whose distributions partially overlap. For now, I am assuming that hybridization has net fitness costs and am ignoring the preferences of hybrids; I will revisit these points in section 16.6. Can mate choice act to restrict gene flow when hybridization is costly but hybrids are fertile? If hybrids suffer a fitness cost, selection will favor assortative mating with respect to genotype, which in turn will act to further restrict gene flow. This in turn increases genetic divergence between populations and therefore increases selection on mate-choice mechanisms. This so-called reinforcement driven by selection against less-fit hybrids should ultimately result in the cessation of gene flow between the two populations (fig. 16.7; reviewed in Servedio & Noor 2013; see also Bolnick & Fitzpatrick 2007). The logical extreme of speciation with gene flow is sympatric speciation, whereby lineages diverge from a population where genotypes are initially random with respect to geography.

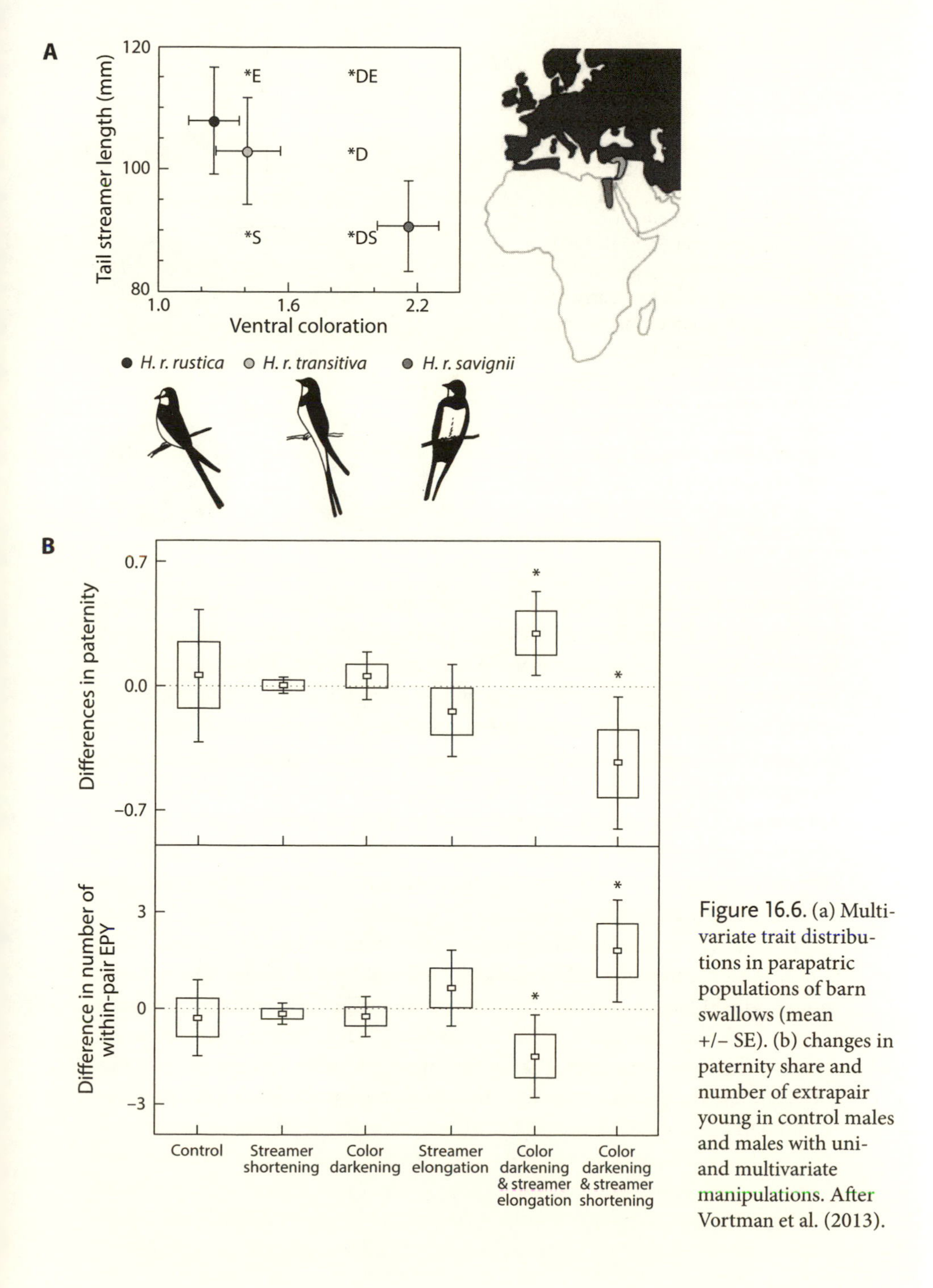

Figure 16.6. (a) Multivariate trait distributions in parapatric populations of barn swallows (mean +/– SE). (b) changes in paternity share and number of extrapair young in control males and males with uni- and multivariate manipulations. After Vortman et al. (2013).

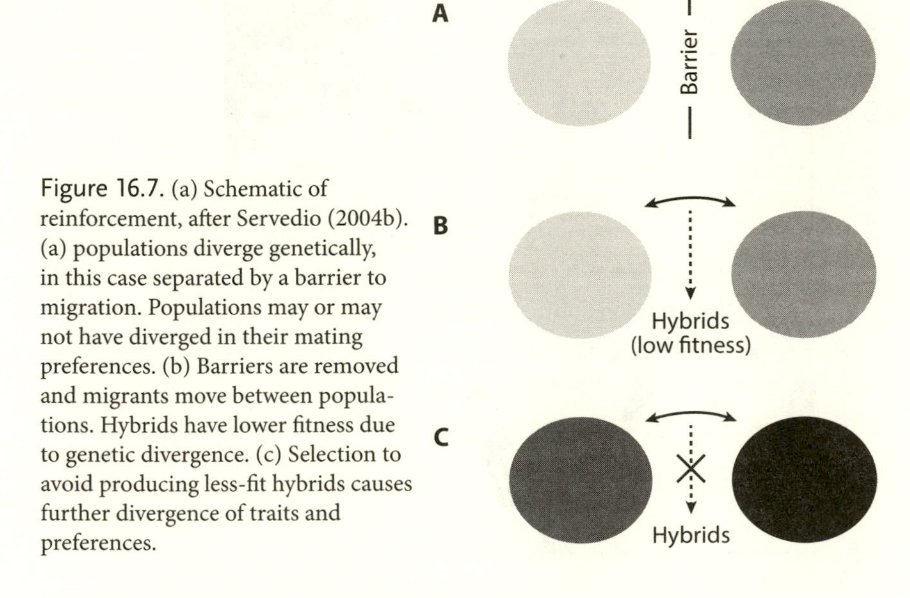

Figure 16.7. (a) Schematic of reinforcement, after Servedio (2004b). (a) populations diverge genetically, in this case separated by a barrier to migration. Populations may or may not have diverged in their mating preferences. (b) Barriers are removed and migrants move between populations. Hybrids have lower fitness due to genetic divergence. (c) Selection to avoid producing less-fit hybrids causes further divergence of traits and preferences.

There is abundant evidence that preferences diverge when gene flow is ongoing (e.g., Butlin & Ritchie 1991; Albert & Schluter 2004; Comeault et al. 2016). Genetic approaches can be recruited to identify the genomic regions underlying preference divergence. In chapter 9, I described experiments by Ortíz-Barrientos and colleagues (2004) identifying QTL involved in conspecific mating preference in *D. pseudoobscura* for conspecifics over the sympatric *D. persimilis*. Ortíz-Barrientos and Noor (2005) introgressed the sympatric allele into *D. persimilis* and found that this increased *conspecific* mate preference in this other species as well. This is therefore a candidate for one-allele assortative mating (see below), one of the strongest routes to reproductive isolation.

Next-generation sequencing can also be used to test for population-genomic signatures of reinforcement in natural hybrid zones. In a house-mouse hybrid zone in eastern Europe, Smadja et al. (2015) found genomic regions including candidate genes for assortative mating via olfactory preferences, showing signatures of selection near where the two species come into contact. These are consistent with stronger assortative odor preferences in sympatry.

16.4.2 Assortative mating

For mate choice to restrict gene flow and ultimately play a role in speciation, it must generate assortative mating: a correlation between homologous trait values across mated pairs (Bolnick & Fitzpatrick 2007; Verzijden et al. 2005). Assortative mating can arise if only one set of partners (e.g., females) expresses preferences, as long as chooser preference for a courter trait covaries with the same trait in choosers themselves (Crespi 1989). Assortative mating includes homotypic assortative mating (chapter 6), but it is important to recognize the distinction between assortative mating and mutual mate choice. Assortative mating can arise for reasons that have nothing to do with mate choice, or even that act in opposition to it; these include spatial or temporal assortment by phenotype or intrasexual competition (see Jiang et al. 2013 for a comprehensive review of assortative mating studies). Conversely, mutual mate choice can generate mating patterns that either play no role in forming population structure, or even—in the case of preferences for heterozygotes—act to oppose it.

Homotypic mating preferences are often favored by direct selection. Assortative mating by personality generally yields higher fitness (Schuett et al. 2010; chapter 9). In a number of anuran species, mechanical constraints limit the size difference between members of a successful pair, with large males unable to grasp small females effectively. For internal fertilizers, some degree of size-assortative mating is almost always imposed by basic biomechanical constraints.

There are numerous studies documenting that individuals prefer to mate with individuals that are similar with respect to a particular phenotype (Martin 2013; reviewed in Jiang et al. 2013). In sticklebacks, for example, females tend to deposit eggs in the nests of males with a similar diet (Snowberg & Bolnick 2008). As an illustrative example, there is a spectrum of ways in which this correlation between the diets of males and females could come about, with major implications for preference divergence and speciation:

1. Individuals are making mating decisions based on direct assessment of diet. In this scenario, diet is a **magic trait** that couples ecological function and a mating cue, as well as the basis for **self-referential phenotype matching** (chapter 6) between partners. Acting in concert, magic trait and phenotype matching provide the most conducive route to speciation (reviewed in Servedio 2016).
2. The situation is somewhat similar when individuals are choosing their mates based on a trait that happens to be correlated with diet. Here,

though, the strength of the correlation between trait and diet places a limit on assortative mating. For example, benthic and limnetic forms of sticklebacks differ in diet but also in both female visual sensitivity and male body coloration. Red-sensitive females prefer red-bellied males based on visual cues, which generates assortative mating with respect to diet (Boughman 2001). In this instance, sensory function is also aligned with ecological performance (see below). Assortative mating with respect to diet will break down, however, if the correlation between diet and either color or visual sensitivity is weakened, for example by disturbance to the visual environment (Wong et al. 2007).

3. In the last scenario, diet is a function of non-heritable variation in condition, with poor-condition individuals competitively excluded from richer food sources. Individuals of each sex would prefer to mate with better-condition individuals, but are constrained to mating with partners as shabby as themselves (chapter 9). Assortative mating on this basis will not contribute to genetic differentiation of subpopulations, since condition is not heritable.

As this series of examples illustrates, understanding both the genetic architecture and the behavioral mechanisms behind assortative mating is critical to predicting its evolutionary consequences. The assortative trait under consideration may or may not be heritable, and the trait itself may or may not be involved directly in mate choice.

Whether assortative mating can restrict gene flow depends on the genetics underlying homotypic mating decisions. Speciation occurs more easily under a so-called one-allele scenario (Felsenstein 1981; Servedio & Noor 2003) whereby the same allele increases assortative mating in both subpopulations (e.g., by increasing choosiness, Ortíz-Barrientos and Noor 2005). As Servedio and Noor (2003) point out, such a one-allele mechanism can be the "instinct to learn"—in this case to imprint on a parental phenotype. The instinct to learn is a one-allele mechanism if both subpopulations evolve a stronger tendency to imprint on parental phenotypes as a consequence of selection against hybridization.

By contrast, if imprinting is the ancestral state and reproductive isolation is a consequence of signal divergence in courters, this constitutes a two-allele mechanism if choosers imprint on the opposite-sex parent. This means that the mode of imprinting can either facilitate or hinder speciation. A model by Verzijden et al. (2005) showed that the evolution of assortative mating depended heavily on how preferences were inherited. Choosers imprinting on the same-sex parent were engaging in a process analogous to self-referential phenotype matching (chapter 12) whereby individuals use their own phenotype as a template for comparison with other individuals.

Self-referential phenotype matching can facilitate genetic divergence. As noted above, assortative mating by body size is widespread; Conte and Schluter (2013) used an elegant experimental design to show that female stickleback used self-referential assessment of body size to mate assortatively with respect to incipient species. Sewall Wright (1921) was probably the first to argue that such "somatic resemblance" could form the basis for assortative mating; individuals would assess their phenotype and then mate preferentially with individuals expressing that same phenotype. Continuing with the diet example above, individuals would assess their own diet (presumably through odorant cues) and select males that matched those odorants. It is important to note that this mechanism could operate whether or not there was genetic variation with respect to diet. Verzijden et al.'s (2005) model also showed that imprinting (chapter 12) on parents—particularly the widely documented pattern of daughters imprinting on the maternal phenotype—would produce mating preferences for an individual's own phenotype and therefore promote assortative mating (see also Servedio et al. 2009).

Mutual mate choice (chapter 9) is a primary driver of assortative mating, as the final example above illustrates. For example, assortative mating can emerge from "prudent choice" in the face of intrasexual aggression, if individuals with better mates are more likely to incur takeovers (Härdling & Kokko 2005).

A final and important point is that the strength of assortative mating, like the strength of sexual selection, is dependent on choosiness (Priklopil et al. 2015), which of course is highly dependent on things like body condition and short-term risk (chapter 11). This becomes important when considering mating decisions that lead to hybridization (section 16.6).

16.4.3 Sympatric speciation: sexual selection alone won't work

Mating preferences can play a powerful role in **sympatric speciation**, as long as they are coupled to a mechanism, like ecological divergence, that exerts natural selection against hybrids (Butlin et al. 2012). Here the distinction between mate choice and sexual selection becomes all-important. While mate choice is often integral to speciation with gene flow, the consequences for speciation of *sexual selection*—i.e., choosers mating differentially among conspecifics—are tenuous and controversial. M'Gonigle et al.'s

(2012) simulation model of speciation with gene flow suggests that sexual selection within sympatric populations can maintain coexistence between species with overlapping ecological niches, provided preferences are sufficiently divergent already.

How likely are mating preferences to diverge enough to produce assortative mating? There is an emerging consensus that additive preferences alone cannot evolve to restrict gene flow within a population, and can even have an inhibitory effect on other speciation mechanisms (see below). A model by Arnegard and Kondrashov (2004) showed that divergence based on directional mating preferences alone was unlikely. Divergent directional preferences can only lead to assortative mating under very restrictive initial conditions. Compatibility-based preferences, by contrast, are definitionally conducive to assortative mating and thereby to the restriction of gene flow.

A recent model by Servedio and Bürger (2014) showed that mate choice can act to *retard* ecological speciation when it operates via a pure FLK process (chapter 15) along with disruptive ecological selection. This is because mating preference alleles introgress faster than alleles for ecologically adapted traits. Ecological divergence is thereby reduced due to the effects of mate choice (fig. 16.8). This process also acts to weaken preference and remove sexual selection via mate choice on the trait.

The nature of selection on preferences greatly influences their role in speciation. Van Doorn et al. (2004) suggested that frequency-dependent selection on chooser preferences was required for sexual selection to lead to sympatric speciation, and detailed how frequency-dependent selection on preferences is implicit in models of sympatric speciation by sexual conflict (Gavrilets & Waxman 2002), mutual mate choice (Almeida & Vistulo de Abreu 2003), and mutually reinforcing ecological and sexual selection (van Doorn & Weissing 2001; see below). Servedio and Dukas (2013) showed that within-generational learning acts to inhibit assortative mating and thereby speciation.

The direct and indirect costs of mate choice may exacerbate the extent to which it inhibits speciation. The direct costs of mating have been largely overlooked (chapter 14), but a model by Kopp and Hermisson (2008) showed that both opportunity costs and viability costs (e.g., predation) associated with choosiness severely restricted the role that mate choice could play in speciation. Natural and sexual selection are directly opposed if preferences are directly or indirectly selected against via predation. An individual-based model by Labonne and Hendry (2010), explicitly simulating the colonization of novel predator regimes by guppies, found that reproductive isolation failed to evolve when natural and sexual selection were thus opposed. It is

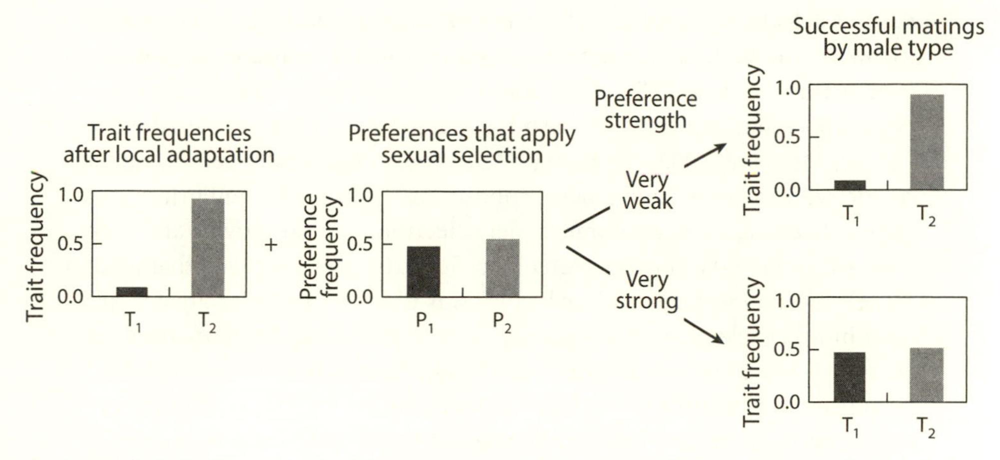

Figure 16.8. The FLK process inhibits local adaptation. From left to right, trait frequencies reflect local adaptation in two populations, subject to sexual selection by both red- and blue-preferring females. Under weak preferences, trait frequencies reflect local adaptation; under strong preferences, trait frequencies mirror the frequencies of chooser preferences in the population. After Servedio & Bürger (2014).

surprising that preferences have diverged relatively little among guppy populations under divergent natural selection regimes. In particular, the direction of preference for ornaments has been largely conserved; females differ primarily in their choosiness.

The comparative conservatism of guppy preferences poses a challenge to Turner and Burrows's (1995) model predicting rapid speciation if a mutation arose such that choosers could reverse their preferences for an elaborate trait costly to courters. It is worth noting that most workers have implicitly or explicitly joined Ritchie (2007) in criticizing such a reversal as implausible. However, as detailed in the preceding chapters, hedonic reversals should be at least as evolvable as subtler changes in preference. Such discontinuous reversals of preference might be worth incorporating into contemporary models of speciation. It is intriguing that, by contrast to guppies, hedonic reversals appear to be common in *Xiphophorus*, another poeciliid fish (e.g., Wong & Rosenthal 2006; Morris 1998a; Morris et al. 2006).

16.4.4 Ecological speciation and mate choice

Where mate choice can play a powerful role is in reducing gene flow in concert with disruptive ecological selection (reviewed in Servedio 2004a). Ecological speciation occurs when populations diverge as a result of divergent natural selection pressures, which favors phenotypic assortative mating (Schluter 2009). Reinforcement can work if targeted traits are under divergent natural selection (Servedio 2000). Physical linkage between preferences and ecological traits can also facilitate the association (Merrill et al. 2011; Saetre et al. 2003). Doebeli (2005) showed that under certain rather restrictive ecological and genetic conditions, linkage disequilibrium can arise between an ecological trait under selection for assortativity, and both a courter trait and a chooser preference. Servedio (2001) argued that natural selection on preferences (see below) was likely to play a far greater role than indirect selection unless mating cues were directly under natural selection. A model by van Doorn et al. (2009) showed that preferences don't even need to evolve for mate choice to facilitate ecological speciation; conserved preferences for condition-dependent traits could cause sexual selection via mate choice to act in concert with ecological divergence (see also Veen & Otto 2015).

The effect of mate choice on speciation thus depends on whether sexual selection via mate choice reinforces or opposes natural selection. Qvarnström et al. (2010) review the sources driving disruptive natural and sexual selection on *Z*-linked preferences in flycatchers. A model by Thibert-Plante and Gavrilets (2013) suggested that traits already under natural selection were more likely to be co-opted as sexual cues and therefore function readily as isolating mechanisms.

Ecological speciation is thus facilitated by genes with pleiotropic effects on both ecological adaptation and mate choice. These genes (or physically linked complexes of genes, Servedio 2009) are associated with so-called magic traits and *magic preferences*, which may play a critical role in coupling mate choice to ecological diversification (Servedio et al. 2011). Just as with preference evolution (chapter 15), Maan and Seehausen (2012) pointed out that focusing on courter traits was insufficient to distinguish between the effects of magic preferences and magic traits (see also Nosil & Schluter 2011; chapter 15).

There is growing evidence for magic traits in the form of courter cues coupled to ecological function. Chung et al. (2014) found that a single fatty acid synthase gene affects both the production of a cuticular hydrocarbon critical for conspecific mate recognition, and the fly's resistance to desiccation. A single gene thus has pleiotropic effects on both a species-typical trait and an ecological trait associated with local adaptation. Merrill et al. (2012) showed that disruptive natural selection on aposematic wing patterns acts in concert with mate choice in *Heliconius* butterflies. Puebla et al. (2007) argued that assortative mating on color patterns in hamlets (chapter 8) was complemented by disruptive selection on color patterns arising from their function in aggressive mimicry.

Correlated responses to ecological divergence can manifest themselves in surprising ways. For example, Oneal and Knowles (2013) found that the genitals of Caribbean crickets *Amphiacusta sanctaecrucis* in populations occupying wet versus dry habitats showed parallel differences in shape, suggesting that genital shape and therefore peri-mating choice mechanisms diverge as an indirect consequence of ecological selection.

16.4.5 Pleiotropy, sensory drive, and personality drive: magic preferences

Preferences can be "magic" if they are coupled to ecological divergence (Servedio & Kopp 2012; Maan & Seehausen 2012). Magic preferences are rarer than magic traits in the literature. Within one chooser, however, so many mechanisms have the potential to influence choice that it seems unlikely that none would be subject to ecological selection. Consider the macroglomerular complex of the antennal lobe of male codling moths (*Cydia pomonella*), which responds to configural combinations of courter and host-plant cues. Choosers only present a mating response if a species-typical odorant is coupled with cues that both chooser and courter are in the appropriate habitat (Trona et al. 2013; fig. 4.7). A shift to feeding on a novel host plant would automatically lead to a novel preference for courters also feeding on that host plant. Lebreton and colleagues (2016) recently showed that one alternative splice variant of a *D. melanogaster* odorant receptor is sensitive to food cues, another to species-typical pheromones. Magic preferences can also arise through purely phenotypic divergence, as when preference is dependent on diet-specific microbiota (Sharon et al. 2010; fig. 11.4). Here, a shift to a novel habitat would directly lead to new, habitat-dependent preferences.

Sensory drive and personality drive are widespread mechanisms that can function as magic traits. This is because different environments will impose different sensory demands on sensation and behavioral function, which in turn will drive divergent mating preferences. Rundle et al. (2005) showed in *Drosophila* that preferences co-diverged with ecological adaptation to different desiccation environments. In *D. melanogaster*, male sexual behavior is modulated by the activity of an odorant receptor neuron sensitive to plant volatiles associated with oviposition sites (Grosjean et al. 2011). There are therefore mechanisms that can directly couple ecology to sexual responses. I will first discuss sensory drive, which has been the topic of decades of research (Boughman 2002), and then the analogous idea of personality drive.

The best-studied systems have involved visual sensitivity at the periphery (chapter 3). Differences in the physical environment—even among microhabitats centimeters apart—favor different sensory abilities. Therefore, different courter signals will be maximally detectable in different sensory environments. The **sensory drive** hypothesis predicts that signals will evolve to optimize detectability to the appropriate receivers (chapter 13; Endler, 1992, 1993). For this reason, sensory biology can play a central role in the formation of new species via so-called **ecological speciation**, because chooser preferences and courter signals will diverge in concert with divergence in ecological niches.

Recently diverged species flocks of freshwater fishes (sticklebacks, Boughman 2001; Lake Victoria cichlids, Hofmann et al. 2009) provide elegant examples of ecological speciation involving both receivers and signalers. In both cases, females occupying shallower or deeper water have different visual sensitivities, and male signals have evolved to maximize detection by females, with red males in shallow water and blue or black males in deep water where red light is unavailable.

Seehausen and colleagues (2008) showed that mate-choice mechanisms allowed sympatric *Pundamilia* along a depth gradient to mate assortatively with respect to sensory function and color pattern. Here, color sensitivity is under natural selection and leads to divergent traits and preferences. Intriguingly, allelic variation at the periphery did not explain the overall pattern of assortative mating, suggesting that this may involve downstream mechanisms of perceptual integration or evaluation.

An underappreciated but appealing candidate for coupling mate choice to ecological function is personality. A swell of empirical studies has called attention to higher-level cognitive and social processes in mate choice. These processes fall under the rubric of personality—they are part of a multivariate network of behavioral mechanisms that is subject to ecological selection, social selection, and sexual selection. Personality measures covary in predictable ways with environmental and social context, and variation in personality in turn covaries with how much is invested in mate choice, how mating preferences are inherited, and which phenotypes are preferred. A substantial majority of studies to date confirm that animals mate assortatively with respect to personality (chapter 8), and there is growing evidence that matching personalities confer direct and indirect benefits to choosers (e.g., Rangassamy et al. 2015). Snowberg and Benkman (2009) directly manipulated feeding rates in male red crossbill (*Loxia curvirostra* complex) and showed that females preferred faster-feeding males. They argued that preferences for faster feeding would be favored if feeding rate was correlated with heritable variation in survivorship or with male provisioning rate. They suggested that preferences for such "information about ecological adaptation" might facilitate ecological speciation via a condition-dependent mechanism similar to that modeled by van Doorn et al. (2009), above.

Personality might thus serve as a magic trait involved in both ecological and mating divergence (Ingley & Johnson 2014). Further, there is likely to be covariance between environment, personality, and mating preferences. For example, the common personality heuristic of anxiety increases in higher risk environments. Heightened anxiety is associated with improved performance in risky contexts, and is likely to shape which display traits individuals are drawn to, how choosy they are, and even how they learn their preferences. Ecological and sexual selection may thus operate in concert to shape divergence of ecological function, personality, and mate choice. Personality drive may therefore facilitate speciation driven by ecological divergence.

16.5 CONSPECIFIC MATE PREFERENCE AND INTRASPECIFIC MATE CHOICE

Recall that intraspecific preference divergence can have effects on interspecific preferences. The reverse is true as well. Intraspecific mate choice can emerge as an incidental consequence of conspecific mate preference (Ryan & Cummings 2013). Intraspecific preferences can thus constitute byproduct biases (chapter 13) whose form did not evolve through intraspecific coevolution.

Preference displacement can have correlated effects on preferences with respect to heterospecifics (Grant & Grant 2010). Several empirical studies suggest conflict between within-species and among-species mate choice (Pfennig 2000; Rosenthal & Ryan 2011a; Collins & Luddem 2002; Higgie & Blows 2007a, 2007b; Pryke & Andersson 2008; Secondi & Théry 2014), whereby choosers compromise on some measured or hypothesized benefits of mate choice to avoid heterospecific mating.

Reinforcement can drive divergence among conspecific populations, if these are subject to different regimes of gene flow with respect to one another (Pfennig & Rice 2014; Comeault & Matute 2016). Pfennig and Pfennig (2009) argued that the trade-off between intra- and interspecific mate choice is why divergent traits in sympatry generally do not spread back into allopatric populations. Kozak et al. (2015) showed that divergence of *Lucania* killifish preferences in sympatry had a correlated consequence of increasing discrimination against conspecific males from different populations, thereby potentially serving to isolate sympatric populations reproductively from allopatric ones.

The conflict between inter- and intraspecific mate choice is alleviated if choosers can select mates along multiple axes. Several workers (Pfennig 1998; Gerhardt & Huber 2002; Hebets & Papaj 2005; Castellano & Cermelli 2006) have suggested that attending to multiple cues can resolve conflicts between different criteria addressed by choice. Multidimensional and multimodal preferences are likely to facilitate conspecific mate recognition more generally (see above).

Divergent preferences can also affect correlated aspects of social communication. Okamoto and Grether's (2013) model showed that aggressive signals of species identity can diverge in response to divergence in mating traits, but not vice-versa. This is because courters minimizing interspecific conflicts suffer reduced attractiveness to conspecific signals. Verrell (1991) raised the possibility that coevolution with eavesdroppers—parasitoids and predators that use mating signals—could similarly drive the divergence of sexual communication in allopatric populations.

16.6 MATE CHOICE AND GENETIC EXCHANGE

16.6.1 Overview

Mate choice plays a complex role in reproductive isolation because it is shaped by an interplay between arbitrary preferences and direct natural selection on mating decisions. And mate choice is always susceptible to error and to deception or coercion by courters. These forces can each lead choosers to mate with courters from divergent lineages, even if hybrids are on average less fit than parentals.

From the point of view of courters, if a mechanism exercises choice, then there are ways to subvert that choice. Heterospecific courters will often mate successfully by eliciting preferences or through coercion (Gröning & Hochkirch 2008). Sometimes, arbitrary preferences will cause choosers to mate with heterospecifics (Servedio & Bürger 2014).

Mate-choice tactics are driven by costs, and mating decisions are sensitive to the environment in which they are exercising choice, as well as to their internal physiological and reproductive strategy. Heterospecific matings may be driven largely by fear, desperation, or blindness, but may nevertheless represent an important mechanism of gene flow between divergent lineages. In this section, I address the role that direct benefits and proximate constraints play in structuring hybridization, as well as how hybridization shapes preferences (Rosenthal 2013).

So far, I have considered hybridization primarily as a negative consequence of mating with heterospecifics and thereby an agent of selection leading to preference displacement and therefore reinforcement. Yet genetic exchange can have important micro- and macroevolutionary consequences. A growing body of literature has called attention to the role that hybridization can play in the evolution of novel phenotypes and in the formation of new species (Abbott et al. 2013). Just as mate-choice mechanisms can both generate and be shaped by divergence between species, so too can they play a primary role in mediating genetic exchange between species.

Much as mate choice can lead to the formation and maintenance of reproductive barriers, mate choice can break down those barriers. One of the most visible signatures of interspecific mating is the asymmetric introgression of courter traits, which is consistent with asymmetric mating preferences (section 16.3). In naturally hybridizing manakins (*Manacus candei X M. vitellinus*), female preference for the golden collar of *M. vitellinus* has favored the introgression of this conspicuous plumage trait into populations of the unornamented *M. candei* (Parsons et al. 1993). A similar pattern is found in red-backed fairy-wrens (*Malurus melanocephalus*) where a red plumage trait is introgressing into the orange subspecies. Baldassarre and Webster (2013) used a plumage manipulation experiment to show that females favored extrapair matings with red males.

The process of hybridization can generate novel mating preferences in hybrids, which in turn can have a determining influence on the evolutionary fate of natural hybrids. Following Arnold (1997), I define **natural hybridization** as a mating that occurs in nature between two individuals from genetically distinguishable populations or species and results in the production of at least some fertile F_1 (first-generation) offspring. This definition excludes the production of sterile or inviable offspring. Natural hybridization has the potential to break down linkage disequilibrium and to rearrange loci involved in producing complex phenotypes (Bell & Travis 2005; Seehausen 2004). There is increasing awareness that hybridization is fundamental to the evolutionary process as a source of adaptive novelty and potentially even as a mechanism of speciation (Mallet et al. 2016).

16.6.2 Proximate causes of hybridization via mate choice

Why do choosers ever commit "the grossest blunder"? We first continue to consider cases in which there is a negative net benefit to choosers of mating with heterospecifics, and then turn to circumstances in which heterospecific matings could be adaptive to choosers. For a comprehensive review, see Willis (2013).

In general, the environmental dependence of choosiness has broad implications for speciation theory and the evolution of reproductive isolation. More specifically, this relationship suggests that a certain proportion of the population will always be non-choosy with respect to preferences. It follows that disturbance events may produce majorities of non-choosers, relaxing sexual selection and facilitating gene flow among species. For example, Pfennig et al. (2013) found that nutritionally stressed female spadefoot toads *S. bombifrons* failed to show a preference for conspecifics in phonotaxis trials, suggesting that hybridization could occur under conditions of ecological stress. It remains to be seen, however, if cryptic choosiness is more robust to environmental variation.

As Mayr (1942) observed, species that are historically allopatric, with no history of sympatry, may come into contact due to range expansion of one or both species; in these cases, mating preferences of one or both species (see asymmetric preferences, above) may result in hybridization. Even if choosers have strong preferences for conspecifics, their preferences could be overridden by coercive mating on the part of (invariably male) heterospecifics (the "satyr effect"; Ribeiro & Spielman 1986, Gröning & Hochkirk 2008; Rohwer et al. 2014). In these cases the fitness consequences of hybridization are highly asymmetric, with males of the coercive species losing relatively little reproductive output to hybridization compared to females of the chooser species.

The communication process involved in mate choice has the potential to be impacted by natural or human-induced changes to the environment. In two cases involving fish, disruption of sensory channels used in mate choice have led to hybridization in the wild. Two species of cichlid fish in Lake Victoria hybridize because females are unable to distinguish red males from blue males under eutrophic conditions (Seehausen et al. 1997). In the swordtail fish *Xiphophorus birchmanni*, females lose their olfactory preference for conspecific males over heterospecific males in polluted water, which may explain hybridization with the congener *X. malinche* (Fisher et al. 2006).

While most interspecific matings are maladaptive for at least one of the parties, there are circumstances whereby direct selection may favor context-dependent mating with heterospecifics (Reyer 2008). If conspecific mates are scarce, or if searching for mates is risky, it may pay a chooser to mate with a heterospecific rather than risk searching for a conspecific (Willis et al. 2011, 2012). Female collared flycatchers accept pied flycatchers as social mates when conspecifics are rare, then solicit extrapair copulations from conspecifics (Veen et al. 2001). Female spadefoot toads also switch from preferring conspecifics to heterospecifics depending on local ecological conditions (Pfennig 2007; chapter 11).

Finally, while vertical transmission of preferences via imprinting can facilitate speciation, it can also cause abrupt transitions in how preferences are expressed. Learning, and preference reversals more generally, could play an important role in determining the extent and direction of hybridization. Grant and Grant (1997) showed that within-species imprinting in Darwin's finches led to females pairing with males who had traits similar to their fathers. A neural network model by Brodin and Haas (2009) showed that sexual imprinting could facilitate the maintenance of preferences for hybrids even if hybrids were unfit. Gee (2003) argued that the direct benefits of pairing early in the season with a social group outweighed the minor genetic costs of hybridization in a wild population of California quail.

In sympatry, two species of flycatcher sing elements of each other's songs. Qvarnström and colleagues (2006) showed that the more overlap between songs, the greater the likelihood of interspecific hybridization. Female collared flycatchers only mate with pied flycatcher males if their songs contain elements of collared flycatcher song. Learned signals thus affect the probability of hybridization. Izzo and Gray (2011) found that previous experience with conspecifics led female *Gryllus rubens* to prefer heterospecific males (in contrast to their sister species, which increased its preference for conspecifics). Finally, Sorenson et al. (2010) found that male ducks cross-fostered through brood-parasitism direct their courtship at heterospecifics (chapter 12).

16.6.3 Effects of hybridization on mate-choice mechanisms

Once hybridization occurs, the fitness of hybrids and their propensity to backcross to parentals depends in no small part on mate-choice mechanisms. The mechanisms underlying preferences in hybrids are shaped by dominance within and epistasis among genes that are functionally divergent between the two species. First-generation hybrid choosers may show a signature consistent with genetic dominance, preferring one or the other parental species (von Helversen & von Helversen 1975). In other cases, they show preferences for signals intermediate to the two parentals, corresponding to intermediate signals in F_1 courters (treefrogs, Doherty & Gerhardt 1983; field crickets, Hoy et al. 1977; tettigonid crickets, Ritchie 1992; Hawaiian crickets, Shaw 2000; butterflies, Melo et al. 2009). Similar parallel shifts in signal and receivers are found in autotriploid treefrog hybrids (Tucker & Gerhardt 2012). Abt and Reyer (1993) and Engeler and Reyer (2001) discuss complex mating patterns in a hybridogenetic complex of frogs. Morgado-Santos et al. (2015) showed that mate-choice dynamics can drive hybrid zone structure in a complex allopolyploid system in fish. In *Ostrinia*, male hybrids overall inherited the paternal preference, but some olfactory neurons responded preferentially to hybrid blends (Anton et al. 1997).

There is a parallel pattern when it comes to the effects of learned mating preferences. In some cases, mixed social experience may in fact generate preferences for hybrids (e.g., ten Cate 1987). This provides an easy mechanism for reproductive isolation of hybrids from parentals, as I discuss later. More commonly, choosers will imprint to one or the other parental species. For example, Rohwer and colleagues (2014) argued that forced heterospecific matings were the primary cause of interspecific hybridization between black-footed (*Phoebastria nigripes*) and Laysan albatrosses (*P. immutabilis*). Directional gene flow from black-footed to Laysan albatrosses, they suggested, arose largely through more instances of black-footed albatrosses coercing Laysan albatrosses. The hybrids would then imprint on the mother's phenotype and preferentially backcross to Laysan albatrosses. Coercion and learning would reinforce asymmetric gene flow between these species.

The albatrosses offer an interesting contrast to Slagsvold et al.'s (2012) field experiment, where cross-fostered females chose heterospecific social mates but sired conspecific offspring. It remains to be seen whether first-generation hybrids would have postmating preferences favoring one or the other species. The effects of hybridization on peri- and postmating mechanisms of mate choice are a ripe subject for study.

Hybridization also has the potential to generate novel preferences through transgressive segregation (Seehausen 2004) and epistatic interactions between genes underlying preference differences between the two species. Recent models (Guillaume & Whitlock 2007), supported by empirical evidence (reviewed in Kalisz & Kramer 2008), suggest that hybridization can influence variation and covariation between traits, altering the evolutionary trajectory of hybrid populations. Hybridization may break down co-adapted gene complexes responsible for both integrated male phenotypes and integrated female preferences for those multitrait phenotypes. Sandkam et al.

(2013) hypothesized that recombinant and heterozygous sensory receptor phenotypes could lead to increased sensory acuity, but it is equally plausible that they could lead to loss of recognition. A study of responses to fruit odor—not a mating signal—in flies is instructive. Dambroski et al. (2005) found that F_1 crosses between different host races of *Rhagoletis* flies failed to navigate to odor sources, while F_2 hybrids mostly recovered the odor preferences of one of the parent species (biased toward the preference of their maternal grandmother, suggesting maternally transmitted epigenetic effects, chapter 12).

A final reason for permissiveness in hybrids is if preferences are positively condition-dependent, such that less-fit hybrid females are less choosy. Given that mate choice is sometimes negatively condition-dependent, this may not be a general phenomenon.

Mate choice in hybrid populations is therefore in general predicted to be more variable, more permissive, and therefore a weaker agent of sexual selection, than in parental populations. This should facilitate gene flow between parental species since hybrid choosers will backcross to parentals and perhaps express novel preferences within parentals. In addition, hybridization may under some circumstances generate courter signals that are more attractive to parentals, whether hybrid courters are more or less fit. This is partly because hybridization can relieve trade-offs by breaking conflicting correlations among courter traits and chooser preferences (Seehausen 2004; chapter 15). The traits of hybrids may thereby be more attractive to choosers from the parental species (Fisher et al. 2009). Taken together, these two factors suggest that once hybridization occurs, backcrossing should be facilitated, often due to the permissiveness of hybrid choosers and sometimes due to the attractiveness of hybrid signals (although hybrid signals can be less attractive to parentals; see above).

16.6.4 Hybrid speciation

Hybrid speciation, whereby new species arise from the fusion of distinct lineages, is most commonly observed in plants, where a common consequence is a change in **ploidy** levels, typically from diploid (two sets of chromosomes) to tetraploid (four). If tetraploid individuals are reproductively incompatible with diploids, this can result in "instant" hybrid speciation since it inhibits backcrossing. Though comparatively rarer in animals, hybrid polyploidy is found in several vertebrates, where it can result in novel preferences (see above), and give rise to unisexual or hybridogenetic lineages.

Homoploid hybrid speciation, unaccompanied by a change in chromosome number, is likely to be less common (Schumer, Rosenthal et al. 2014). As with other forms of sympatric speciation, homoploid hybrid speciation in sympatry requires conditions that generate assortative mating. Abbott and colleagues (2013) describe how this can occur when novel hybrid traits generate a host shift in phytophagous insects, such that hybrids are reproductively isolated from their hosts. Absent such microhabitat shifts, the first step to hybrid speciation is preferences of early-generation hybrids for other hybrids, as described in the examples above. It is premature to make generalizations about how often hybrid choosers prefer hybrid traits, let alone whether these preferences are sufficient to permit persistence of hybrid lineages in the face of backcrossing. If hybrids imprint on their parents and generalize to other hybrids but not to parentals, this should further inhibit gene flow back to parentals (ten Cate 1987; chapter 12). A major lacuna in our understanding of hybrid speciation (and for that matter, of sympatric speciation more generally) is the link between preference mechanisms in earlier-generation hybrids and how they determine the probability of hybrid-hybrid and hybrid-parental matings.

A recent model by Schumer et al. (2015), however, showed that reproductive isolation could evolve through stochastic fixation of incompatibility loci in isolated populations. Seehausen (2004) observed that an immediate consequence of hybridization is to increase variation in traits and preferences, a prerequisite to sexual selection. This should make it more likely for isolated hybrid populations to experience divergence of traits and preferences through FLK or sexual conflict, as outlined in section 16.1. Hybrid swarms formed in separate geographic locations could therefore diverge from parentals, and from each other, through stochastic processes including both fixation of arbitrary incompatibilities and the coevolutionary dynamics of arbitrary traits and preferences, resulting in the formation of new hybrid species in allopatry.

16.6.5 Hybridization as Pandora's box

If transgressive phenotypes are favored by mate choice and/or hybrid mate choice mechanisms are permissive, hybrids can spread even in the absence of recurring matings between the two parent species. This means that short-term perturbations to conspecific mate preference can potentially have long-term evolutionary consequences. If hybridization is rampant, species can collapse altogether into a hybrid swarm (Taylor et al. 2006).

A model by Gilman and Behm (2011) suggested that hybridization following short-term bouts of disturbance (e.g., Fisher et al. 2006) can have long-term evolutionary consequences, depending on the duration and severity of disturbance. Counterintuitively, species are more likely to reemerge and avoid collapse after a more severe disturbance. This is because mate choice operates more strongly after a milder disturbance, such that trait loci can cross more readily across lineages (cf. Servedio & Bürger 2014).

Natural hybrids between the swordtails *Xiphophorus malinche* and *X. birchmanni* offer an interesting case study. There is a pervasive signature of hybridization in the genus *Xiphophorus*, including species with high levels of admixture (Cui et al. 2013; Schumer 2016). This pattern occurs because interspecific hybrids show relatively little intrinsic reduction of viability (Schartl 2008; but see Schumer, Cui et al. et al. 2014) and because premating isolating mechanisms are porous, with choosers exhibiting relaxed preferences under adverse conditions (Fisher et al. 2006; Willis et al. 2011, 2012). Mate-choice mechanisms thus favor the production of first-generation hybrids under a broad set of environmental conditions. Yet ongoing hybridization between the two parental species is extremely rare in contemporary populations (Culumber et al. 2011). The breakdown of mate-choice mechanisms was responsible for hybridization, yet reproductive barriers have reemerged between the two species. Hybridization continues through backcrossing and hybrid-hybrid matings (Culumber et al. 2014a), with ecological selection likely maintaining hybrids at intermediate elevations between the highland *X. malinche* and the lowland *X. birchmanni* (Culumber et al. 2012) but with traces of introgression far up- and downstream (Culumber et al. 2011).

Novel reproductive barriers may have emerged among hybrids in at least one population in Mexico's Sierra Madre Oriental. All *Xiphophorus* at Aguazarca are hybrids, but they are distributed into two very distinct *X. malinche* and *X. birchmanni*-like clusters (Schumer, Cui et al. 2014; fig. 16.9a). Both molecular data on mother-offspring pairs (M. Schumer et al., in review) and social interactions in the wild (Culumber et al. 2014; fig. 16.9b) point to near-perfect reproductive isolation before fertilization. While urine-borne pheromones are the primary cue for conspecific mate preference between the parent species, there is no evidence to date for homotypic preferences based on pheromones. It is tempting to speculate that these lineages of hybrid origin could have evolved novel isolating mechanisms.

Mate-choice mechanisms therefore play a determining role in the origin and evolutionary fate of hybrids. This is an emerging field, and it will be

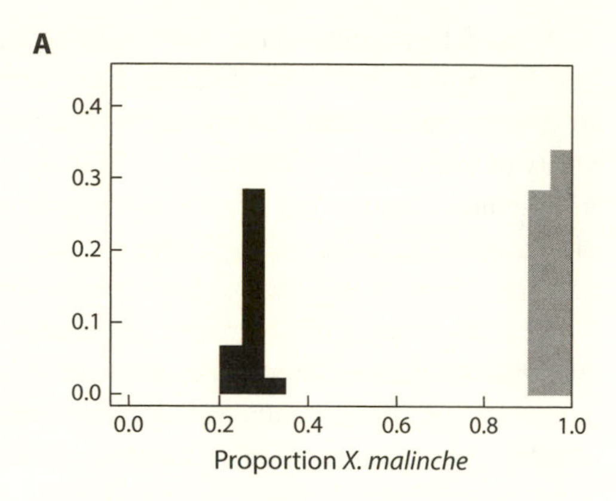

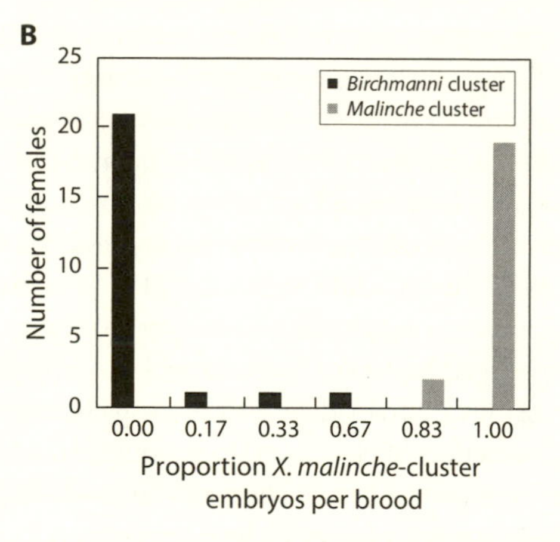

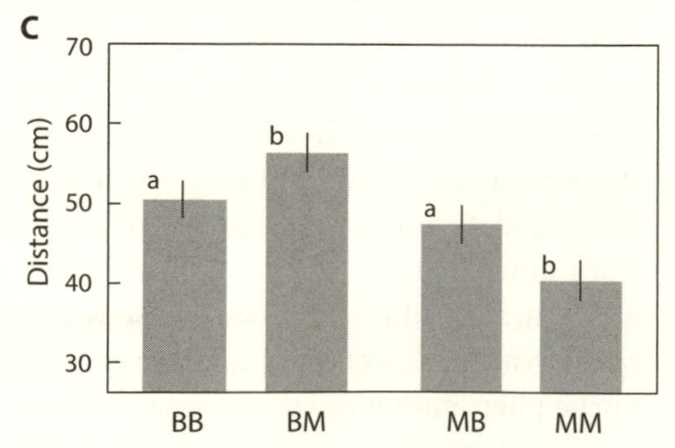

Figure 16.9. Reproductive isolation between sympatric hybrid clusters at Aguazarca. (a) *X. birchmanni*- and *X. malinche*-like hybrids form distinct genotyping clusters (Schumer et al. 2014; courtesy Molly Schumer). (b) Reproductive isolation is in the pre- or peri-mating phase. Pregnant females overwhelmingly have offspring from their own cluster (reanalyzed from Culumber et al. 2014). (c) Females show assortative social preferences for males from their own cluster in artificial mesocosms (BB, distance of *birchmanni*-cluster females to *birchmanni*-cluster males; BM, distance of *birchmanni*-cluster females to *malinche*-cluster males, and so on). Reanalyzed from Culumber et al. 2014, courtesy Zach Culumber.

interesting to see how often novel preferences emerge in hybrid populations, and how often preferences in parent species facilitate versus inhibit hybridization and ultimately the exchange of genomic regions between species.

16.7 SYNTHESIS

A generation ago, this chapter might have been titled "species recognition." I started graduate school when Ryan and Rand (1993b) pointed out that species recognition and sexual selection act on the same underlying mechanisms. Mendelson and Shaw (2012) provide an insightful critique of the term "species recognition" focused on the dependence it places on taxonomic hypotheses (made by humans and, implicitly, by choosers) and the false dichotomy between "compatibility" and "quality" (chapter 14). Implicit in the idea of species recognition is the idea that choosers sequentially or hierarchically assess between-species compatibility before assessing within-species variation. Whether or not this happens—the rules whereby choosers evaluate courters—may be of critical importance to resolving conflicts between species recognition and sexual selection (Schaefer & Ruxton 2015). There is some support for such sequential assessment in a handful of taxa, but this idea has not been systematically tested (Mendelson & Shaw 2012).

Either way, thinking about mate choice in terms of distinct things called sexual selection and species recognition—even as a continuum rather than a dichotomy (Ryan & Rand 1993), gives us an unnecessarily tangled picture of the role of mate choice in speciation. If we shift to thinking about total selection on choosers, things are more straightforward. Ecological selection causes correlated divergence in preference and/or trait mechanisms, whether in sympatry or in allopatry, which in turn inhibits mating when choosers and courters from different lineages interact. Preference evolution through indirect selection causes divergence in allopatry as well, but indirect selection on preferences inhibits gene flow in sympatry.

Direct selection on preferences will also favor interspecific matings under a variety of ecological circumstances. Elucidating the sexual communication systems of natural hybrids will shed light on the role of mate choice in hybrid speciation and in phenotypic diversification through genetic exchange.

Most modern humans bear some trace of Neandertal ancestry (Racimo et al. 2015); we should bear in mind that each time there was a hybridization event between Neandertals and early *Homo sapiens*, it involved mate choice and most probably an element of coercion. The idiosyncracies, excessive consciousness, and infinite variety surrounding human mate choice may reflect our history of hybridization. The unique features and evolution of human mate choice are the focus of the next chapter.

16.8 ADDITIONAL READING

Ritchie, M. G. 2007. Sexual selection and speciation. *Annual Review of Ecology, Evolution, and Systematics*, 79–102.

Schaefer, H. M., & Ruxton, G. D. 2015. Signal diversity, sexual selection, and speciation. *Annual Review of Ecology, Evolution, and Systematics*, 46, 573–592.

Servedio, M. R. 2016. Geography, assortative mating, and the effects of sexual selection on speciation with gene flow. *Evolutionary Applications*. doi:10.1111/eva.12296

CHAPTER 17

Mate Choice and Human Exceptionalism

17.1 INTRODUCTION

17.1.1 Overview: human mate choice in context

Mate choice emerges from an intricate network of complex mechanisms that are exquisitely sensitive to environmental and social context. Like other animals, humans attend to a suite of courter traits, from odorant molecules (chapter 3) to higher-order properties like social status (chapter 4). We assign fluctuating valences to the same stimuli (chapter 5), we evaluate mates sequentially and in comparison, and our choices are all too often circumvented or subverted (chapter 6). Both males and females are somewhat promiscuous, and there is the opportunity for mating biases to be expressed both within the reproductive tract and long after mating (chapter 7). Male choice of short-term sexual partners is surely much more stringent with regard to postzygotic allocation of resources than with regard to the decision to copulate. Humans are variable in their preferences and in their social mating systems, encompassing polygyny, extended or serial monogamy, and—more rarely—polyandry (chapter 8). Our preferences are hyper-variable, within individuals, among individuals, and among populations (chapter 9), and are shaped by genetic (chapter 10), environmental (chapter 11), and social influences (chapter 12). These preferences surely had myriad origins (chapter 13), continue to be subject to direct (chapter 14) and indirect selection (chapter 15), and are shaped by a history of hybridization (chapter 16).

But so much of human mate choice and mating preferences is unique or nearly so, starting with the centrality of things like language, religion, dance, music, and higher-order social interactions to the mate-choice process. These variable, socially modulated, and uniquely human attributes are largely the domain of social and developmental psychologists, and anthropologists, who tend to focus on proximate influences on individual preferences. By contrast, the broadly shared aspects of human sexual behavior, like gender differences in the approach to casual sex and preferences for condition-dependent traits, are the province of evolutionary psychology. With a few notable exceptions, few studies have crossed this divide. Here, I attempt a biological perspective on the diversity and complexity of human mate choice.

17.1.2 Before psychology: Darwin on mate choice and race

The dichotomized approach dates back to the beginning of psychology in the early twentieth century. Before then, Darwin (1871) brought an integrative approach to human sexuality. Darwin attributed the physical differences among human "racial" groups to diversifying sexual selection. He did this by amassing mostly anecdotal evidence about mate preference in human populations. Both this neutralist view of "racial" diversity and the notion of mate choice as a creative force were revolutionary, and have been roundly confirmed in the intervening years. They stem from a heuristic conclusion that the "fluctuating, but not quite arbitrary" nature of mating preferences makes such a process possible. In the *Descent of Man*, Darwin demolishes "race" as a biological concept, marshalling empirical evidence for continuous intergradation and for equal fitnesses of putative "hybrids." In a similar vein, he describes variation among human cultures in their ideals of sexual beauty. He calls attention to their sensitivity to social influences, notably citing examples where colonized people come to prefer the colonizers' standards of beauty. Mate choice tells us that "race" is a consequence solely of arbitrary prejudice.

17.1.3 Social and developmental psychology

Perhaps starting with Freud's (1905) influential *Three Essays on the Theory of Sexuality*, most of psychology has viewed human mating preferences as primarily constructed by social interactions with other people, particularly early in life. Freud's principal ideas about *how* this happens have been thoroughly discredited, but the paradigm remains the same: social forces shape mating preferences, and the social context shapes realized choices. The consequences of mate choice are, in turn, viewed primarily through the lens of their contribution to social and cultural dynamics. A fundamental contribution of this approach concerns how mate choices and reproductive choices are realized in human societies (see below).

17.1.4 Evolutionary psychology and reductionism

Evolutionary psychology arose in reaction to what the field's founders, John Tooby and Leda Cosmides (1992), termed the Standard Social Science Model, or SSSM, which is the aforementioned tendency to understand human behavior as the product of uniquely social forces. The way that evolutionary psychology has conceptualized evolutionary biology, however, bears little resemblance to the real thing. It is not too much of a caricature to summarize Bolhuis et al.'s (2011) excellent critique of the discipline as follows: evolutionary psychology assumes that there was an incredible burst of evolution during the Pleistocene epoch, which produced phenotypes that were perfectly adapted to the social and environmental conditions of the time—what evolutionary psychologists call "the environment of evolutionary adaptation." The serpent in the garden has been the rapid social and technological changes that have left us maladapted to our current conditions, but every one of us bears the imprint of our history, and each and every one of us does whatever would have been optimal in small bands of hunter-gatherers some 30,000 years ago. Even critiques of the evolutionary-psychology approach to mate choice (e.g., Ryan & Jethá 2010) assume that humans have stopped evolving, despite abundant evidence for genetic variation in ecologically important traits and recent adaptation to agriculturally derived diets (Courtiol et al. 2012). The partial decoupling of resources from reproductive success should have particularly interesting consequences for human mate-choice evolution: Nettle and Pollet (2008) found that while British men exhibit the standard pattern of increasing fitness with increasing resources, women exhibit the reverse, perhaps uniquely among species (see also Goodman et al. 2012).

Our Pleistocene forebears, it seems, would have had much to discuss with the early-twentieth-century eugenicists. While there has been some criticism of blanket adaptationism in the context of mate choice (e.g., Skamel 2003), by and large, evolutionary psychology tells us that men want young, nubile women with symmetrical breasts, clear skin, childbearing hips, and red lips that may (Morris 1967) or may not (Johns et al. 2012) remind them of labia majora. Women want square-jawed, testosterone-filled he-men when they're ovulating; and wealthy, compliant protectors when they're not. Mate choice in humans, in this view, is about as complicated as it is in dung beetles. This is perhaps akin to approaching linguistics by focusing on how we push air out of our lungs.

Below I will touch on two points largely overlooked by evolutionary psychologists: that human mate choice leaves comparatively little room for individual agency; and that humans are astonishingly diverse in their preferences.

17.2 SOCIAL INFLUENCES ON HUMAN MATING DECISIONS

Throughout this book, I have focused on mating decisions as an individual process, albeit one influenced by public information (chapter 12) and the broader social context (chapter 15). Beckage et al. (2009) review the literature on single people adjusting their preference thresholds in response to social pressure from married people in their social network. There are two salient social influences that are unique to human mate choice. The first is the role of social systems in regulating how mating preferences are expressed; the second is direct transgenerational involvement in one's choice of social partner.

Contemporary Western society is unusual in the degree to which social pairing decisions are dependent on individual choosers. Parental influence over mating decisions is the norm across traditional societies today, and has likely been the prevalent form of social pairing throughout human history, including Korea and Japan well into the twentieth century (Apostolou 2007, 2010). As of the mid-twentieth century, marriage in most human societies was primarily an economic transaction, involving payments from one family to another; even in societies where individuals choose their spouses, these decisions are often subject to parental approval (Ingoldsby 2006). Today, parental influence over social mating is most prevalent in cultures characterized by collectivism rather than individualism (Buunk et al. 2010; fig. 17.1). Arranged marriage can be viewed with the same lens of conflict and cooperation that characterizes other types of among-individual interactions surrounding mate choice, extending the role of parent-offspring dynamics (chapter 15) into the subsequent generation. The success of arranged marriages is determined by how the interests of choosers and their parents are aligned, ranging from mutual agreement to coercion. Scelza (2011) found that Himba women in Namibia were far more likely to engage in extrapair matings if they had been pressured into marrying their social mates.

In addition to close kin, the broader society has a say in individual mating decisions. Possession may be 9/10ths of the law, but much of the rest is sex. Transgressing sexual laws carries some of the harshest penalties societies impose; there is a registry for sex offenders, but not for violent offenders. Some laws and customs regulating sexual conflict are universal among humans and echo broadly shared mating biases among nonhuman species: these include taboos against mating with close relatives, animals, or preadolescent children.

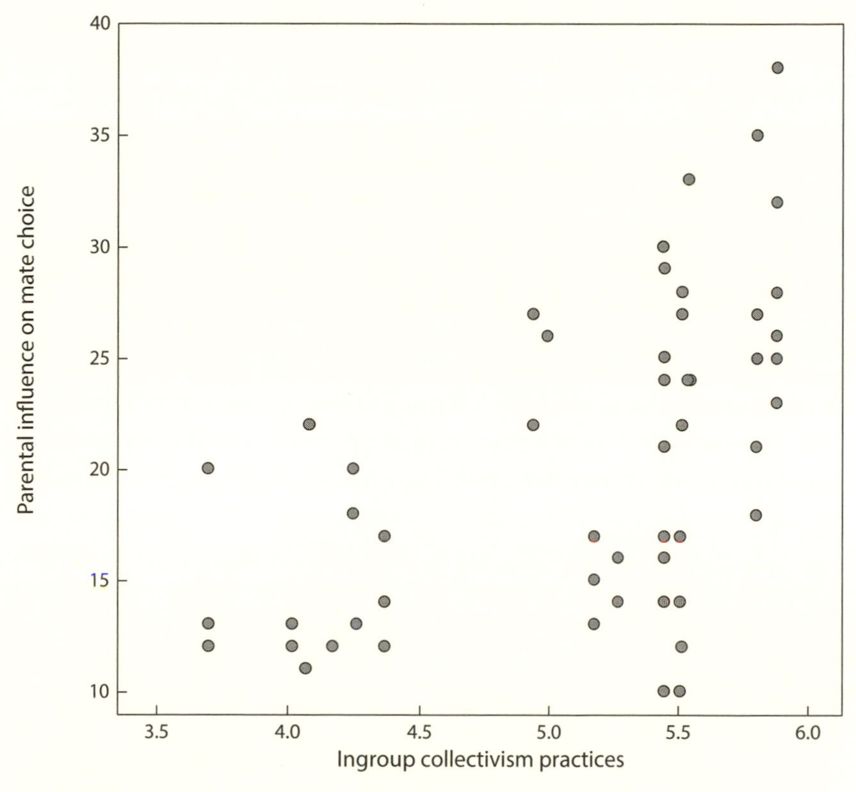

Figure 17.1. Relationship between a cultural measure of collectivism and "parental influence on mate choice" across an ethnically diverse sample of Canadian undergraduates. After Buunk et al. (2010).

Other laws are disconcertingly different across societies and a source of conflict within them. The law generally prohibits sexual coercion, but marital rape was legal in some US states until 1993. Much of the world is still prohibited from divorce, adultery, same-sex behavior, and abortion. Modern democracies are still riven by conflicts over the extent and nature of social control of mate choice. The major social issues of our time revolve around whether or not the state should be allowed to regulate individuals' pre- and postmating decisions.The political history of the United States, perhaps more than other countries, is in no small part about mate choice. The Utah Territory had to ban polygyny in order to gain admission to the Union in 1896; "interracial" marriage was banned in much of the country until 1967, and gay marriage, until 2015.

The societal and legal framework dramatically constrains the scope of mate choice. Constraints on prostitution, for example, affect individuals' abilities to derive direct benefits from short-term sexual encounters, and legal sanctions on promiscuity discourage multiple mating. In order to understand human mating decisions from an evolutionary context, we need to take into account the social context in which they are expressed.

17.3 VARIATION IN HUMAN MATING PREFERENCES

> No healthy person, it appears, can fail to make some addition that might be called perverse to the normal sexual aim; and the universality of this finding is in itself enough to show how inappropriate it is to use the word perversion as a term of reproach. (Freud 1905)

Chapters 9 through 12 detailed how mating preferences vary considerably among and within individuals, and how this variation depends on myriad factors external and internal to choosers. What can a biological perspective bring to our understanding of variation in human mating preferences? The biology of stable sexual orientation, a nearly unique human phenomenon, is covered in detail elsewhere (Gavrilets & Rice 2003; Poiani 2010; LeVay 2011; Gordon & Silva 2015). Here, I focus on an aspect of human preference variation that has been almost entirely overlooked by biologists: fetishes and paraphilias. Ahlers et al. (2011) found that about two-thirds of their sample of German men acknowledged at least one paraphilia. Importantly, these preferences are highly specific and largely hidden; before the arrival of the Internet, someone aroused by stinging nettles was likely unaware that they shared these dispositions with anyone else. As such, we can largely think about them as biases (chapter 13), unfettered by the consequences of choice.

In Freud's *Sexual Aberrations*, "a disposition to perversions is an original and universal disposition." "Perversions"—non-normative sexual preferences, gender identity, and kink—were, according to Freud (1905), entirely produced by (pathological) social influences like sexual imprinting (chapter 12). Havelock Ellis's *Psychology of Sex*, published the same year, deserves more attention from biologists. Ellis's approach is more Darwinian than Freudian, and he makes a compelling case for innate variation in our instinct to learn "perversions."

How much do genetic and epigenetic influences affect our preferences? Some biases might reflect constraints on how we experience sexual desire

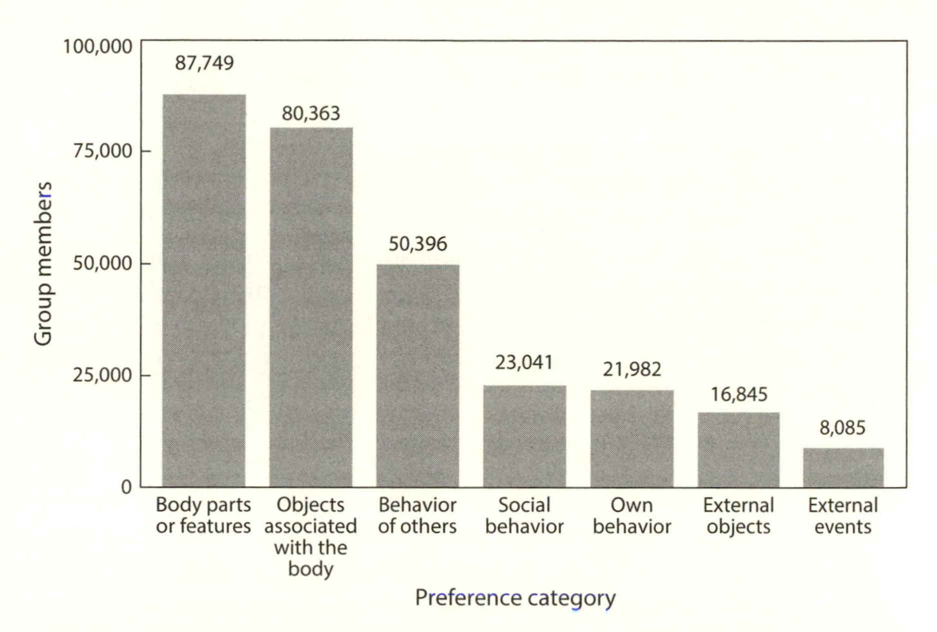

Figure 17.2. Frequency distribution of sexual preferences as estimated from Scorolli et al.'s (2010) analysis of Internet chatroom use.

(Scorolli et al. 2007; fig. 17.2). For example, inputs from the genitals and the lower extremities map to adjacent regions of the somatosensory cortex, which may explain why eroticization of the feet is more common than, say, eroticization of the elbows (Ramachandran et al. 1998). Sexual arousal in response to pedal stimulation may thus be partly explained by genetic and developmental variation in the cross-talk between somatosensory inputs from feet and genitalia. Such mechanistic constraints may broadly influence the form of sexual fetishes. Personality (chapter 9) covaries with sexual fetishes: perhaps unsurprisingly, Hébert and Weaver (2014) found that self-identified sexual dominants differed along several personality dimensions from submissives. Bem (1996) proposed a general mechanism to account for sexual orientation differences among both men and women, based on covariance between personality and mating preferences (chapter 9). Here, personality shapes social experience during adolescence, which in turn can shape generalized sexual feelings toward opposite- or same-sex peers.

Neuroanatomy may explain why some are aroused by foot massages, but it does not fully explain why some are attracted to images of feet. Here, a predisposition to arousal—say, as a result of the organization of the somatosensory cortex—is likely involved in associative learning of parasexual cues.

We saw in chapter 6 that an arbitrary cue—being fitted with a rodent jacket—can elicit a sexual response, and even become necessary for sexual arousal, through association with sexual reward (Pfaus et al. 2012). Similarly, Çetinkaya and Domjan (2006) used sexual conditioning to induce male quail to copulate with a "terrycloth object." About half the subjects developed a fetish for the object. Such a process may explain variation in people's tendency to sexualize clothing and other arbitrary features. Interestingly, the fetishistic quail in Çetinkaya and Domjan's study (2006) were less efficient at copulating with real females in the presence of the object, but fertilized a greater proportion of her eggs as a consequence of the bout, perhaps as a consequence of sperm priming induced by the fetishized stimulus.

Perhaps the most counterintuitive dimension of human sexual preferences is masochism: here, stimuli are sexualized that innately elicit avoidance. Paul Rozin (1999) and colleagues addressed this in the context of a non-sexual preference for painful stimuli, namely the predilection for spicy foods in many human cultures. In addition to several scenarios specific to human psychology, Rozin (1999) posited that downstream products of avoidance responses, like endorphins, might act to enhance pleasure, but noted a lack of empirical evidence that persists to date. Sexual preferences for innately aversive stimuli perhaps represent the ultimate decoupling of sensation and evaluation, as well as reminding us of the highly complex nature of the processes that shape mate choice.

17.4 SYNTHESIS: INTEGRATING EVOLUTIONARY AND SOCIAL-SCIENCE APPROACHES TO HUMAN SEXUALITY

Humans have experienced dramatic evolutionary changes over the past 10,000 years (Courtiol et al. 2012), associated both with adaptation to an agropastoralist lifestyle and with our inordinate fondness for genocide. Mating preferences can change very rapidly owing to the fact that they depend on a large reservoir of standing genetic variation whose effects can be modified and reversed by environmental and social inputs. Brown and colleagues (2012) interpret variation in human mating systems in light of sexual selection theory (chapter 15). Cross-cultural variation (Yu and Shepard 1998; Pisanski & Feinberg 2013) suggests at least the possibility of intraspecific genetic variation in preferences, which would allow us to test Darwin's original hypothesis about the contribution of mating preferences to phenotypic diversity as well as more specific ecological hypotheses about adaptive mating preferences (e.g., DeBruine et al. 2010).

Good genes pan-adaptationism has infected the popular imagination through evolutionary psychology to a degree that we are perhaps only one unscrupulous marketer away from reviving the eugenics of the early twentieth century. Many findings that confirm our normative predictions about sexual behavior in humans thereby receive perhaps more than their fair share of public exposure, and less than their fair share of scrutiny. Our fellow animals, however, tell us that we should take a more heterodox look at human sexual diversity. The influence of social networks and parent/offspring dynamics on realized mate choice is vastly different across societies, and an evolutionary perspective inspired by life-history theory (chapter 15) might be useful here. To the detriment of our understanding of mate choice, approaches focusing on social and cultural effects have been largely divorced from mechanistic and evolutionary studies from a biological perspective.

Like other species, humans harbor tremendous variation in what we find attractive or arousing, and this variation is shaped by environmental, social, and genetic influences. In contemporary global society, rapid technological and societal changes means that individual mating preferences have an unprecedented potential to be revealed and expressed as choices, some with reproductive consequences. Individuals now have more agency than they ever have in human history, with a greater opportunity than ever to sample potential mates. The social and evolutionary consequences are sure to be fascinating.

17.5 ADDITIONAL READING

Bolhuis, J. J., Brown, G. R., Richardson, R. C., & Laland, K. N. 2011. Darwin in mind: new opportunities for evolutionary psychology. *PLoS Biology, 9*(7), e1001109.

Brown, G. R., Laland, K. N., & Mulder, M. B. 2009. Bateman's principles and human sex roles. *Trends in Ecology & Evolution, 24*(6), 297–304.

Rozin, P. 1999. Preadaptation and the puzzles and properties of pleasure. In Edward Diener, Daniel Kahneman, & Norbert Schwarz (Ed.). *Well-Being: The Foundations of Hedonic Psychology* (p. 593). Russell Sage Foundation.

CHAPTER 18

Conclusions: A Mate-Choice View of the World

18.1 THE SWEEP OF MATE CHOICE

Since the dawn of sexual reproduction, since organisms first started swapping genetic material and creating offspring, mate choice has played a critical role in determining what subsequent generations look like and how they behave. There are basic rules described by population- and quantitative-genetic theory that apply to every living thing that has sex. The rules mean that when sex is about producing zygotes, there will be coevolution between the sexes; part of this coevolutionary dynamic will involve mate-choice mechanisms. Despite the ubiquity of coevolution, the contribution of indirect viability benefits to coevolution is much smaller than we commonly think. More generally, choosers can potentially be lied to about anything that provides an adaptive benefit—just one of the ways that sexual conflict affects mate-choice evolution.

Much of what most fascinates us about mate choice has to do with preferences and traits that defy utilitarian explanations. The arbitrary element of mate choice, its capacity to drive traits to flamboyant extremes, and the diversity of possible outcomes emerge from the fundamentals of how coevolution works. But the flamboyant extremes are only possible because an individual chooser has so many different ways to express a preference. Arbitrary preferences evolve most readily by riffing off the pedestrian task of securing a minimally acceptable mate—for the most part, a fertile, unrelated conspecific who won't kill you and all your young. The task of finding a minimal mate generates byproduct biases that can act as preferences whereby choosers prefer arbitrary traits within courters. And choosers contain a host of non-sexual biases that can generate mate choice. Once courters evolve traits that target these biases, they become mechanisms of mate choice, subject to direct and indirect selection.

We should therefore not underestimate the importance of these complex sources of direct selection in shaping mate choice. Mating decisions carry strong downside risks, like producing sterile offspring, never reproducing, or becoming someone's dinner while *in copula*. Finding a minimally acceptable mate is part of a broader context of pairing, mating, and fertilization that determines chooser fitness. Importantly, fitness can be dependent on mating decisions involving sexual behavior, but no exchange of gametes, like those involving same-sex partners.

Since courters and choosers have divergent interests with regard to reproduction, sexual conflict also acts to generate sources of direct selection at all stages of mate choice. Ecological selection also tightly constrains some preference mechanisms, like cone opsin genes. By contrast, others, like narrowly tuned olfactory receptor genes, or assignment of subjective value to configural stimuli, may operate relatively free of ecological constraint. Taken together, these forces guide the shape of mate choice. The sweep of mate choice leaves a lot of room for arbitrariness, shaped both by coevolution and by the evolution of traits that target preexisting chooser biases. This arbitrariness is likely to account for the broad-scale complexity and diversity of courter ornaments and chooser preferences.

Preferences can have major evolutionary consequences whether they never evolve, whether they evolve to equilibria, or whether they cycle in perpetuity. Mate choice can make populations better off or worse off, more diverse or less diverse; it can split species apart and fuse two species into one. The consequences of mate choice depend on how selection acts upon mate-choice mechanisms. It follows that we need to understand how these mechanisms operate and evolve.

18.2 FROM SEXUAL SELECTION TO PREFERENCE EVOLUTION

Darwin sought to understand the train of the peacock, not the mind of the peahen. Up to a point, he appreciated the sophistication of preference mechanisms, but he was perhaps a product of a demographic that thought little about decisions made by females. He also just missed the genetic framework that characterized later insights. Serious thinking about preference evolution among scientists required the confluence of two forces. The first was the mid-twentieth-century interpretation of Darwin's ideas in an explicit genetic context: the Modern Evolutionary Synthesis. The second was the advent of Western feminism, which led to the epiphany that female agency can be important.

But, despite our deepening understanding of preference mechanisms, we still think about them within the frame of reference of courter traits. Models of preference evolution are called "models of sexual selection" because they center on how chooser preferences evolve only insofar as they interact with courter traits.

18.3 HOW WE TALK ABOUT MATE CHOICE

> It is, indeed, fortunate that the law of the equal transmission of characters to both sexes prevails with mammals; otherwise it is probable that man would have become as superior in mental endowment to women, as the peacock is in ornamental plumage to the peahen. (Darwin 1871, p. 565)

As discussed in chapter 17, mate-choice mechanisms in humans are at the center of the norms that govern social interactions. It is therefore unsurprising that the history of mate choice as a topic of study is intertwined with cultural perspectives on sexuality and gender roles. A chooser-centric perspective requires a social context where females actually have agency. Throughout this book, I have endeavored to use the terms "chooser" and "courter" to emphasize the variability of sex roles among and within species. While awkward at times, I believe this has the useful effect of muting the inner voice of barroom wisdom—Bateson's (1983) "unconscious punning"—when we think about mate choice as scientists.

Language also matters in creating artificial distinctions across stages of mate choice, and even within the same processes depending on their fitness outcomes. The literatures on pre- and postmating choice often employ different terms for similar underlying processes, and the way we use terms reflects our prejudices about their putative adaptive value and about the mechanisms that individuals apply to reproductive decisions.

A key point here concerns the word "choice" itself. I have adopted the word in the broadest possible sense, whereas others believe "choice" should require the involvement of the nervous system. When a female túngara frog hops toward a particular calling male in a chorus, we all call that behavior "choice." We say she is making a choice independent of how we believe that behavior affects her fitness. But when we discuss choices made after mating (whether or not they involve behavioral decisions), the literature spawns different terms depending on the fitness consequences of chooser decisions. Some studies of cryptic processes, for example, reserve the term "cryptic choice" for only those interactions with courters after mating that *increase* chooser fitness.[1] And choosers are said to have "resistance" to courter traits if these incur net costs, and "preferences" for courter traits if these incur net benefits. Further, studies of resource allocation draw a distinction between differential allocation, where choosers invest more in the offspring of favored mates, and reproductive compensation, where they invest more in the offspring of less-favored mates.

Assigning different terms based on fitness outcomes has some conceptual utility, but can lead us down a dangerous path, especially if we seldom measure chooser fitness. In particular, distinct terms obscure the fact that direct and indirect fitness effects are highly variable (today's compensation may be yesterday's allocation) and that most pre- and postmating mechanisms can be recruited for dual, opposing roles, either increasing chooser fitness or benefiting courters at the expense of choosers. "Preference" describes the effect of chooser biases independent of historical or current fitness effects of chooser traits. The term "resistance" might be reserved for phenotypes that are costly to choosers but make it more difficult for courters to mate. The etymology of "differential allocation" implies that the allocation can be in either direction and therefore encompass reproductive compensation, but it may be less confusing to refer to "resource allocation" to encompass both positive and negative fitness outcomes.

18.4 HOW WE STUDY MATE CHOICE

The revolution in molecular genetics and bioinformatics means that we can study the consequences of mate choice—from individual mating outcomes to macroevolutionary signatures—with an ease that was unthinkable a few years ago. Combined with techniques like gene-mapping and functional genomics, these approaches are starting to allow us to test evolutionary hypotheses with data from genomic regions associated with traits and preferences. Just as key insights were gained by thinking about mate choice in the light of evolutionary genetics, a new synthesis may emerge from studying genomes.

Phenotypic approaches to studying mate-choice mechanisms, notably behavioral assays, have also been getting increasingly sophisticated. However, the nuanced ways that preferences can change in response to environment

[1] See Karlsson Green & Madjidian (2011) and Perry & Rowe (2012) for critiques of gendered language on sexual conflict, skewing to "active" terms for male processes and "reactive" terms for female ones.

and social context—sometimes over very small but ecologically relevant scales—make preferences inherently challenging to measure. We want to test hypotheses about phenotypes that are function-valued by definition, since preference depends on courter trait value. Most animal behavior studies simply can't muster the sample sizes necessary to test hypotheses about highly multivariate preferences and the complex phenotypic variation that underlies them. Even estimating something as basic and important as the genetic covariance of preference across two environmental treatments is a heroic undertaking.

We lack a culture of reproducibility in animal behavior, and heterogeneity of results is often attributed to factors like variation in genetic stocks, husbandry conditions, and stimulus quality. We don't know how much of the variation we see in mate choice across systems and studies is real, and how much of it is an artefactual consequence of sampling and experimental execution. Too many papers assert the null hypothesis of no effect, without support from power analyses and proper controls. Theory tells us that subtle preferences can have powerful effects on selection and gene flow. Yet many experiments, including most of my own, are designed only to detect preferences of extremely large effect. Methodological rigor, public sharing of stimuli and experimental protocols, and power calculations should be standard prerequisites of publication.

A particular problem in the study of mate choice is the handling of nonsignificant results. Nonsignificant behavioral results, particularly in an experimental context, are—often correctly—dismissed as reflecting deficiencies whereby subjects aren't interested in mating, and require careful controls to assert otherwise. This means, as others have observed, that we open ourselves up to publication bias when the bar for studies that confirm a hypothesis is lower than for ones that fail to provide support. Increasing sample sizes and the thoroughness of experimental design is laborious for empiricists, but may allow patterns to emerge that we aren't currently seeing.

18.5 FOUR OPEN QUESTIONS ABOUT MATE-CHOICE MECHANISMS

18.5.1 What determines variation in how preferences are transmitted?

The way preferences are specified—whether or how they are learned and inherited—helps determine their role in sexual selection and speciation. The next book on mate choice will surely place much more emphasis on cellular epigenetics and maternal effects, and on explicit molecular mechanisms of how genes specify preferences. In the meantime, we know that genetic architecture in the relatively coarse sense—whether preferences are paternally, maternally, or autosomally inherited—has a determining effect on preference evolution. We also know that epigenetics in the broader sense has powerful effects. In some cases, preferences are robust to social experience; in others, choosers prefer familiar phenotypes; and in still others, choosers prefer unfamiliar phenotypes. Little work has been done to explain why different taxa would have these diametrically opposed mechanisms of preference acquisition. Modeling of indirect genetic effects, whereby the phenotypic distribution of courters affects the preference functions of choosers, offers a promising avenue for exploring the effect on preference functions of social experience with courter phenotypes.

18.5.2 What are the costs and benefits of multivariate preferences?

Preferences are usually highly multivariate, reflecting the operation of multiple sensory, perceptual, and other mechanisms. When these operate in sequence, it makes sense to talk about the costs and benefits to mate choice at each stage; theory suggests that the evolution of a mechanism at one stage, like favoring conspecific sperm, can inhibit the evolution of mechanisms at other stages. When they operate simultaneously, a critical parameter is the cost or benefit of multivariate preferences, which depends on how dimensions of preference covary with one another and how multiple traits covary with one another. Understanding how the costs of evaluation scale with signal complexity is a key to understanding the evolution of multiple traits and preferences. A low cost of evaluation, or the benefit of redundant sampling, may play an important role in driving the diversity of traits and preferences. We also need an understanding of how preferences covary with each other—how many axes of mating preferences there are—and how these map on to multivariate distributions of courter traits.

18.5.3 How do mating preferences evolve?

Models of directional preference evolution have typically assumed that preference functions change smoothly, with choosers evolving indifference as an intermediate step before preference reversals. Empirical data on preferences and neural mechanisms, however, suggest that hedonic reversals—

abrupt changes in valence, or direction of preference—may be common, and that interactions among mechanisms may cause unpredictable changes in preferences. Within individuals, the subjective value of the same stimulus can flip based on associative learning or changes in physiological state. Closely related species in sympatry are often repelled and attracted by reciprocal pairs of cues, as are the two sexes. Subjective value may be much more labile than perception. These kinds of changes in the subjective value of stimuli may provide a way for choosers to modify their preferences for sexual stimuli without changing peripheral sensitivities or other biases constrained by natural selection. Along with categorical perception, hedonic reversals suggest that preferences might evolve in a more punctuated manner. Cognitive sophistication may emerge as a way to bypass sensory constraints. It would be desirable to revisit theory with a more expansive view of mate-preference functions and how they may evolve.

18.5.4. How do mutual decisions and social context shape realized outcomes?

Preferences are constrained by the ecological context and by social factors like courter density, coercion, competition among courters, competition among choosers, and mate-choice copying. Approaches based on indirect genetic effects and social network theory should increasingly shed light on these processes, and have the opportunity to be integrated with genetic models. Innovations in computation and molecular biology make it increasingly feasible to track social interactions and mating interactions at large scales and sample sizes (see below). Ambitious theoretical and empirical research programs can address mate choice's role in the evolution of mating systems and sex-role reversal.

18.6 MATE CHOICE AND TOTAL SELECTION

We can arrive to traits and preferences through a garden of forking paths. At a minimum, either a trait has to evolve to match a preexisting preference, or a preference has to evolve to target a variable trait. Traits and preferences can readily coevolve along arbitrary, cyclical, or deterministic trajectories. Each of these processes can occur in conjunction with another or at different points in a preference's evolutionary history.

A tool for discerning regularities about mate-choice evolution is to approach it from the point of view of total selection (fig. 18.1). A first step is

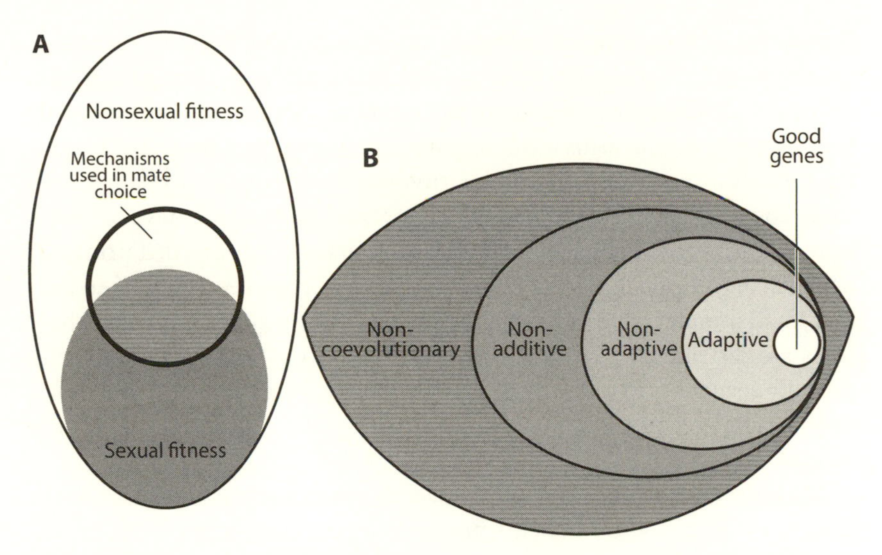

Figure 18.1. Mate choice in the context of total selection. (a) Mate-choice mechanisms impact both sexual fitness and non-sexual fitness. (b) Close-up of determinants of sexual fitness due to mate choice. Good genes selection as a determinant of sexual fitness nested within adaptive coevolution, which in turn is nested within additive coevolution, which in turn are nested within coevolution encompassing compatibility-based choice. The balance of sexual fitness is shaped by selection on mating outcomes or mating behavior independent of the identity of choosers. Relative size of circles is entirely illustrative.

to think about mate-choice mechanisms as potentially determining both sexual fitness and nonsexual fitness (fig. 18.1). Two questions emerge from figure 18.1a. First, how constrained is a preference mechanism by functions that affect nonsexual fitness: how important is sexual fitness versus nonsexual fitness? Second, what proportion of sexual fitness depends on the actions of mate-choice mechanisms (the intersection between the red circle and the tinted area in fig. 18.1a)?

The area where mate-choice mechanisms and sexual fitness intersect may be quite small, but it is the only theater where selection can operate on mate choice as such. Within that area, the contribution of the most widely studied processes of mate-choice evolution may be comparatively small (fig. 18.1b):

> I do not claim that the "Emperor wears no clothes." Rather, I would predict that the "Emperor wears a loincloth." I estimate that the adaptive signaling paradigm covers about the same proportion of the total corpus of intersexual signals as does that humble garment. (Prum 2010, p. 3096)

If good genes is the emperor's loincloth, is coevolution his cargo shorts or his snowsuit? Coevolution with courter traits plays an omnipresent but only partial role in the evolution of mate-choice mechanisms. This is partly because these mechanisms are constrained by non-sexual functions including sensation and personality, and partly because they may carry additional fixed costs independent of choice outcomes. While the fitness of mate-sampling strategies does depend on the courters being sampled, it primarily depends on the direct costs of choice—costs that have been quantified only in a small number of cases. It may be worthwhile to consider the way these costs are distributed. The near-universal loss of choosiness in the face of predation suggests that downside risks, whether from the process of choice or from choice outcomes, may play a primary role in shaping mate-choice evolution.

Mate choice is subject to multiple selective forces beyond variation in the direct or indirect fitness benefits provided by unrelated conspecific males. These include selection against mating with heterospecifics, close relatives, or incompatible conspecifics, and selection on correlated aspects of the preference unrelated to mating.

Reproductive character displacement provides a compelling example. A common outcome is for selection against hybrids to drive courter signals and chooser preferences further apart than they are in allopatric populations. Selection against mating with heterospecifics will inevitably generate a mating bias against conspecifics whose signals too closely resemble the signals of the other species. A similar argument can be made for inbreeding avoidance; choosers will avoid mating with courters that resemble their close relatives, even if these courters are unrelated. Finally, even if current selection on preferences is relaxed—for example, as a result of the local extinction of a closely related species—vestigial biases against mating with heterospecifics could still generate biases among conspecifics. All of these processes will yield mate-choice biases that generate variation in mating success among unrelated, conspecific courters, even in the absence of any selection for choosing among unrelated conspecifics. Mate choice is thus a mechanism of sexual selection, but (at least initially) it functions as such without having evolved in that context. Mate choice is a main agent of sexual selection, but sexual selection is only part of the story of mate choice.

When there is genetic variation in both courters and choosers, preferences can coevolve arbitrarily under a broad range of conditions, encompassing both the Fisher-Lande-Kirkpatrick mechanism and sexual conflict. Additive genetic benefits with respect to viability (so-called good genes) are perhaps widespread, but play a much less important role in shaping preferences than folk wisdom would have us believe. Our view of the preeminence of good genes is based on a paradigm that focuses on courter traits and their putative correlation with chooser fitness. It would likely be more fruitful to focus on empirically quantifying direct and indirect selection on chooser preferences.

Such a chooser-centric perspective is helpful in clarifying the role of mate choice in speciation. The nature of selection on preferences is critical. Preferences can serve to inhibit gene flow if they are subject to direct selection against hybridization, or if they diverge in concert with ecological traits. If direct selection favors hybridization, or if the fitness costs of hybridization are overwhelmed by sexual selection, preferences can have the opposite effect, hindering the formation of new species and facilitating the fusion of existing ones. Therefore, the composition of total selection on preferences plays a determining role in genetic exchange among species and in the formation of new hybrid species.

The list of features that can be recruited as mate-choice mechanisms is staggering. Preferences are multidimensional and nonlinear, and can operate at any level of processing starting with the sensory periphery, through perceptual processing and hedonic evaluation, to morphological and physiological mechanisms of cryptic choice. Each of these mechanisms exists whether or not there is pertinent variation in courter traits, and experiences a markedly different suite of selective forces depending on whether it is expressed as a latent bias or as a mating preference.

The mechanisms underlying preferences are highly labile, which makes mate choice both more interesting and more difficult to study. Preferences are extraordinarily variable within and among individuals, often displaying exquisite sensitivity to environmental and social cues. We can make sense of much of this variation in terms of context-dependent fitness payoffs, but much of this variability—notably whether chooser condition increases or decreases choosiness—awaits more data and perhaps a reassessment under a chooser-centric framework.

The complexity and flexibility with which mate choice is expressed presents an empirical challenge in terms of characterizing the underlying genetics. As we gain a better understanding of the reaction norms guiding preferences, we may be able to design studies where we can manipulate the social and physical environment in meaningful ways that expose heritable variation in preferences. Despite these challenges, we can still perhaps benefit from a shift in conceptual framework, where we think about how preference affects fitness from the chooser's point of view.

18.7 SYNTHESIS: MATE CHOICE AND ITS CONSEQUENCES

Mating decisions are a primary driver of individual fitness. The regularities that emerge around mate choice are of the sobering kind, where the Dionysian exuberance of preferences and ornamentation is checked by Apollonian concerns of fecundity and survivorship. Our intuition, however, tells us that desire is complicated and strange and unpredictable, and this intuition holds true broadly across sexually reproducing organisms. These properties of mate choice are built into the system, from the initial arbitrariness of how traits and preferences originate to the coevolutionary dynamics that drive them in unexpected directions. The seriousness of mate choice, then, is overlain with a mischievous streak that can generate gratuitous beauty and complexity. The regularities and arbitrariness of mate choice each play an integral role in how lineages split apart and come back together. Every creature that came from sex is who they are because somebody made a mating decision. Each of us harbors ancient desires and desires never imagined, desires sensible and wild.

GLOSSARY

Active space The range over which a signal can be detected.

Adaptationism The assumption that an observed trait has been optimized with regard to natural selection.

Adaptive In the evolutionary sense: functionally improved through natural (including sexual) selection. Distinct from **physiological adaptation**.

Additive genetic variance The contribution to total phenotypic variance of allelic effects that add to or subtract from the value of the trait, independent of genetic background

Additive preference A preference that is a function of the weighted sum of two or more variables.

Adjustable-threshold algorithm A **dynamic, sequential sampling** algorithm where a chooser's acceptance threshold depends on successive encounters with courters.

Advertisement A signal produced by a **courter** that is not targeted to a specific **chooser**, and that can be **evaluated** by a chooser without the need for direct interaction.

Allocation In a mate-choice context, provision of resources to zygotes or offspring from a particular chooser.

Amplitude The intensity of a stimulus.

Antipathy A **chooser**'s internal representation of courter traits that elicit an aversive response. See **mate choice** and **preference**.

Appetitive Eliciting positive states like affiliation, appetite, and sexual desire.

Association time A common yet often ambiguous measure of **preference**, measuring time in proximity to a stimulus.

Associative learning Assignment of positive or negative hedonic value to stimuli as a consequence of pairing with hedonically marked stimuli or experience.

Asymmetric preference Typically refers to when preferences are shared between choosers in two species, such that choosers from one species prefer conspecifics and choosers from the other prefer heterospecifics.

Aversive Eliciting negative states and responses such as fear, disgust, and avoidance.

Bimodal preference Preference characterized by two distinct peaks; *multimodal* preferences have multiple peaks.

Breeding value The additive effect of an individual's genes on the average phenotype of its offspring.

Broad-sense heritability Contribution of additive, non-additive, and under some definitions vertically transmitted nongenetic effects to total phenotype.

Brightness contrast Difference in perceived intensity between a visual target and its background.

Categorical preference A **preference** whereby choosers attend only to traits within a certain range of values, but distinguish little within that range.

Ceiling effect A maximal response elicited as an artifact of experimental execution; see **floor effect**.

Chemotaxis Movement toward an odor source (see **phonotaxis**).

Chooser An individual who exercises mate choice among courters. Choosers can be male or female, and the same two individuals can switch between courter and chooser roles.

Choosiness The resources (time, effort, cognitive bandwidth) that a **chooser** is prepared to invest in a mating decision.

Chromatic Relating to color. The chromatic signal is a function of the concentration of photic energy in a stimulus in a particular range of the visual spectrum; chromatic contrast is the difference in chromatic signal between stimuli.

Chromatic adaptation A form of **physiological adaptation** whereby the visual periphery adjusts to the distribution of wavelengths across different environments, helping to maintain **perceptual constancy** with respect to color.

Condition "the pool from which resources are allocated for use in reproduction" (Rowe & Houle 1996).

Condition-dependent preference A preference which is expressed flexibly based on chooser condition.

Condition-dependent trait A courter trait whose expression is correlated with that courter's condition.

Configural preference A preference in which stimuli are only effective in particular combinations with one another, as in face recognition.

Conspecific A member of one's own species.

Conspecific gamete precedence A phenomenon whereby gametes from conspecific courters are preferentially fertilized.

Context-sensitive preference A preference whose expression depends on the environmental or social context.

Coolidge effect Preference of males for unfamiliar females. The term comes from an apocryphal story about US president Calvin Coolidge and his wife touring a chicken farm.

Copying See **mate copying**.

Courter An individual who is chosen or rejected by choosers. Courters can be male or female, and the same two individuals can switch between courter and chooser roles.

Critical period A window during development when choosers are sensitive to social experience for shaping future preferences.

Cross-fostering An experimental procedure whereby young are raised by individuals other than their genetic parents.

Cryptic choice A term collectively referring to the **peri-mating** and **post-mating** phases of choice.

Detectability The probability of detecting a signal against background noise.

Direct selection Selection arising from a correlation between fitness and the phenotype of a chooser and/or of a chosen courter.

Directional preference Preference characterized by a monotonic increasing or decreasing response as a function of courter trait value.

Discrimination Differential responses among a set of alternatives.

Discrimination learning The process whereby individuals learn to differentiate among stimuli. Tested experimentally by training subjects with a negatively rewarded and a positively rewarded stimuli; often induces **peak shift**.

Diversifying selection **Natural selection** favoring different phenotypes or allele states in different lineages. Also called *Darwinian selection*.

Divorce Severing of a pair bond between socially monogamous animals.

Dynamic sampling Mate sampling whereby preferences are modifiable by the number and characteristics of the courters encountered over the course of sampling. See **static sampling**.

Early learning In a mate-choice context, specification of preferences before sexual maturity.

Ecological speciation Formation of reproductive barriers due to divergent selection on function in different environments.

Electroreception Sensation and perception of electric fields.

Epigenetics Historically referred to all nongenetic vertical transmission; in the contemporary sense refers to cellular changes (or, more narrowly, chromosomal changes) in gene function that do not involve changes in DNA sequence, particularly epigenetic marking through DNA methylation.

Epistasis Nonlinear interaction among genotypes at two or more loci.

Evaluation Assignment of **valence** to courter stimuli.

Extrapair mating In socially monogamous animals, a mating outside the pair bond.

Fecundity The number of offspring produced by an individual.

Fisher-Lande-Kirkpatrick (FLK) model The null model for trait-preference coevolution, requiring only genetic covariance between traits and preferences

Fitness The reproductive success of an individual or a **phenotype**.

Fitness function A mathematical expression relating the value of a trait to the expected **fitness** for an individual expressing it.

Fixed cost The costs of producing and maintaining the architecture required for mate choice, independent of the number of courters sampled.

Fixed-threshold algorithm A **static**, **sequential sampling** algorithm where a chooser selects the first courter whose stimulus value exceeds a given value.

Floor effect A minimal response or lack of response as an artifact of experimental execution; see **ceiling effect**.

Forward genetics An experimental approach whereby variation in phenotype is correlated with variation in genotype.

Frequency In acoustics, the inverse of the time interval between successive peaks of a sound wave; more generally, the number of events per unit time.

Gene expression The amount of RNA (ribonucleic acid) of a gene in a cell or tissue at a given point in time. Often used as a proxy for functional activity of a gene.

Gene mapping Encompasses a variety of techniques including quantitative trait locus (QTL) mapping, admixture mapping, and association mapping, all of which involve correlations between **phenotypes** and marker locations on a linkage map.

Generalization A process where responses to learned stimuli are extended to similar stimuli.

Genetic covariance The joint variation among a set of traits due to genetic causes.

Habituation Attenuated response in response to repeated exposure to the same stimuli.

Haldane's rule The observation that hybridization has more deleterious effects on fitness in the **heterogametic** sex.

Hedonic value Association of positive or negative affective states with a stimulus.

Heritability The proportion of phenotypic variation in a population that is due to genetic causes.

Heterogametic In species with chromosomal sex determination, refers to the sex with two distinct chromosomes (for example, XY males in mammals and *Drosophila*, ZW females in birds and butterflies).

Heterospecific A member of a different species.

Heterotypic preference Preference for a phenotypically distinct individual. See **homotypic** preference.

Hierarchical preference A preference in which one stimulus is only effective if values of another stimulus are within a given range.

Homoploid hybrid speciation **Hybrid speciation** without a change in ploidy level (number of sets of chromosomes).

Homotypic preference Preference for a phenotypically similar individual to oneself. See **heterotypic** preference.

Honest signal A signal whose expression is correlated with expected fitness benefits to choosers. Often erroneously taken to mean a signal whose expression is correlated with the fitness of courters. Also called an **indicator trait**.

Hybrid speciation Formation of new species through the fusion of distinct evolutionary lineages.

Immediate-early gene (IEG) A gene whose expression in the brain serves as a marker of short-term changes in neural activity.

Indirect genetic effect Effect on an individual's phenotype as a consequence of the genotype of another individual.

Indirect selection Correlation between fitness and a given phenotype, arising from that phenotype's genetic correlation with another phenotype under direct selection.

Individual recognition Behavior contingent on the unique phenotype of another individual; see also **pair bond**.

Innate releasing mechanism (IRM) A bias that is "hard-wired" and develops independent of social experience.

Intransitivity A violation of **transitivity**. In *strict* intransitivity, a chooser prefers A over B and B over C, but prefers C over A.

Iteroparous Reproducing more than once in the course of a lifetime (cf. *semelparous*).

Just meaningful difference In the context of mate choice, the smallest difference betweeen two stimuli that will elicit a difference in preference from a chooser.

Just noticeable difference The smallest difference between two stimuli that a receiver can detect.

Last-chance option A dynamic modification of a **fixed-threshold algorithm** whereby a **chooser** lowers its **threshold** at the end of a sampling period.

Lek A localized gathering of **courters** for the primary purpose of **advertising** to choosers.

Lek paradox The observation that preferences are maintained in the face of depleted genetic variation in traits and/or their fitness correlates.

Linkage disequilibrium A statistical association between genetic loci.

Maladaptive Harmful to an individual's fitness in the current ecological context.

Magic trait A courter trait or a chooser preference that is correlated with a phenotype under ecological selection sensitivity, whose phenotypic divergence brings about a change in mating patterns.

Mate choice Any aspect of an animal's phenotype that leads to its being more likely to engage in sexual activity with certain individuals than with others.

Mate copying A process whereby choosers modify their mating decisions based on cues from other choosers.

Mating system A population-level construct that describes the array of strategies used by choosers and courters in obtaining mates. Mating systems are generally classified as **monogamous**, **polygynous**, **polyandrous**, or polygynandrous (**promiscuous**). See also **social mate**.

Mechanoreception Sensation and perception of mechanical pressure or distortion.

Meiotic drive A process that causes a gamete carrying a particular allele to be over- or under-represented relative to neutral expectations.

Modality A sensory channel; vision, audition, olfaction, and touch are each different modalities.

Monogamy A genetic **mating system** where both males and females have only one mate. See **social monogamy**.

Multiplicative A preference that is a multiplicative interaction between two or more variables.

Narrow-sense heritability Heritability based strictly on additive genetic effects. Compare **broad-sense** heritability.

Natural hybridization Genetic exchange between lineages in the wild resulting in at least viable first-generation offspring; can arise as a result of human interference.

Natural selection Differential survival and reproductive success as a function of **phenotype**.

Non-adaptive Refers to phenotypes whose structure has not arisen due to direct natural selection.

Non-additive genetic variance Genetic variance arising from dominant and epistatic effects. Compare **additive genetic variance.**

Nongenetic inheritance Refers to effects on phenotypes that are vertically transmitted other than through the genetic material, including preference learning, parental effects, and cellular epigenetics.

Nonindependent mate choice Mate choice contingent on the choices or preferences of other choosers.

Nonlinear compression The principle that the same absolute difference between stimuli becomes less meaningful to receivers as stimulus magnitude increases; see **Weber's law** and **Power law**.

Oblique transmission A mode of preference learning whereby developing choosers shape their preferences in response to the stimuli of older individuals other than their parents.

Ommatidium Individual component of the compound eye of arthropods.

Operational definition In the context of measuring preferences, an explicit and reproducible set of criteria characterizing a response measure.

Operational sex ratio The ratio of sexually responsive individuals (females:males) at a given time and place.

Opsin A seven-pass transmembrane G-protein coupled receptor protein covalently bound to a vitamin A molecule called a chromophore; the amino acid structure of an opsin determines its wavelength sensitivity and its overall probability of responding to a photon.

Order effects An artifact of experimental design whereby responses to a given stimulus or treatment are contingent on previous experiences within the experiment.

P2 The proportion of offspring sired by the second male in a two-mating assay of cryptic choice.

Pair bond A long-term, preferential relationship between two individuals, usually in the context of biparental care.

Peak preference **Preference** at the trait value that elicits the maximum response from a chooser.

Peak shift In **discrimination learning**, displacement of response toward extreme values of the training stimuli, i.e., stimuli that enhance the contrast.

Peer transmission A mode of preference learning whereby developing choosers shape their preferences in response to the stimuli of individuals in their own generation.

Perception Integration of sensory inputs into a neural representation that can be used for decision making.

Perceptual constancy Perception of invariant stimulus attributes like size, shape, and colors despite changes in the sensory environment and in the position and orientation of the receiver relative to the signaler.

Perceptual space A construct describing the extent and dimensionality of the processes that receivers use to perceive differences among stimuli. See ***Umwelt***.

Peri-mating choice **Evaluation** and responses to courter cues when parties are in intimate physical contact; often associated with fertilization decisions.

Permissive Responding to a broad range of stimuli; associated with low **choosiness**.

Phenotype Any measurable feature of an organism external to the genetic material.

Phenotypic plasticity The capacity of a genotype to produce different **phenotypes** in different environments or contexts.

Philandering Soliciting or accepting matings outside a pair bond.

Phonotaxis Movement toward a sound source (see **chemotaxis**).

Physical linkage Association between genes due to their spatial proximity on the same chromosome.

Physiological adaptation In sensory biology, the process whereby an individual's sensory system accommodates short-term changes in the sensory environment. Distinct from evolutionary **adaptation**.

Playback The controlled presentation of natural or synthetic cues.

Pleiotropy Multiple phenotypic effects of the same genetic locus.

Ploidy The number of sets of homologous chromosomes in a genome. Sperm and eggs are haploid, most animals are diploid; **hybrid speciation** can occur when hybridization brings a change in ploidy levels.

Polyandry A genetic **mating system** where females are more likely to mate multiply than males.

Polygyny A genetic **mating system** where males are more likely to mate multiply than females.

Positive assortative mating Mating with a similar phenotype to one's own (see **homotypic preference**).

Positive synergy A **multiplicative preference** whereby the interaction between two variables elicits a more strongly positive response than an additive preference would.

Postmating choice **Evaluation** and responses to **courter** cues after mating; extends to allocation and care decisions after fertilization.

Postzygotic After fertilization.

Preexisting biases Potential mate-choice mechanisms present in the absence of corresponding variation in courter traits.

Power law The principle that receiver response is exponentially related to stimulus value; see **Weber's law** and **nonlinear compression**.

Precedence effect In acoustic communication, a systematic preference for (or against) the first of two temporally overlapping stimuli.

Preference A chooser's internal representation of courter traits that predisposes it to mate with some phenotypes over others. See **mate choice** and **antipathy**.

Preference function A mathematical expression describing the effect of one or more measures of **courter** phenotype on chooser **preference**, usually operationalized as a behavioral or physiological measure of chooser response.

Premating choice The **detection** and **evaluation** of **courter** cues before close physical contact takes place.

Prezygotic Before fertilization.

Private Perceptible only by courters and choosers, and resistant to eavesdropping by unintended receivers.

Proceptive Characteristic of sexual receptivity or responsiveness.

Promiscuity A genetic **mating system** in which both males and females have multiple mates; also called polygynandry.

Pseudoreplication The use of a sample size in a statistical test that is inappropriate to the hypothesis being tested; "testing for treatment effects with an error term inappropriate to the hypothesis being considered" (Hurlbert 1984).

Purifying selection **Natural selection** favoring conserved phenotypes or allele states and acting against novel mutants.

Quality The most fraught word in the study of mate choice. Used in this book to refer to essential or distinctive characteristics of a stimulus, courter, or chooser, rather than in the more widespread (and insidious) sense of a grade of excellence.

Quantitative genetics A body of theory and statistical analysis based on the assumption that phenotypes depend on an infinite number of alleles, each of infinitely small additive effect. Can be used to quantitatively predict responses to selection.

Reaction norm The response of a given genotype over a range of conditions. See **phenotypic plasticity**.

Reinforcement Strengthening or divergence of premating isolation (e.g., mating preferences) in response to selection against hybridization.

Receptor A molecule, usually on the cell membrane, that transduces an exogenous signal (e.g., a photon, mechanical displacement, an odorant molecule) or an endogenous signal (e.g., a neurotransmitter or a steroid hormone) into a cellular response; also used to refer to a whole sensory neuron involved in transducing an external stimulus (*receptor cell*).

Recognition Preference for a courter cue over a null control.

Regulatory evolution Change over time in the timing, tissue-specificity, and amount of gene expression or translation into proteins.

Repeatability The fraction of phenotypic variance due to among-individual rather than within-individual differences.

Reproductive character displacement Divergence of mating traits in sympatric species in response to selection against hybridization; see **reinforcement**.

Reproductive isolation The failure of distinct lineages to produce offspring, whether through **prezygotic** mechanisms including mate choice or through **postzygotic** incompatibilities.

Reproductive success The number of offspring an individual produces that survive to maturity (in some definitions, the number of grand-offspring).

Residual reproductive value An individual's expected number of future offspring.

Resistance Performance of aversive behavior in response to chooser stimuli.

Responsiveness **Chooser** response averaged over the distribution of a **courter** trait in a specified reference population.

Reverse genetics An experimental approach whereby candidate genes are manipulated to discern their effect on phenotypes.

Saturation The point at which additional stimulation fails to elicit additional response; see also **threshold**.

Self-referential phenotype matching Preference for a cue in another individual based on one's own experience of the homologous cue in oneself. See **homotypic preference**.

Sensation The conversion of environmental stimuli into internal neural and chemical responses.

Sensitive phase See **critical period**.

Sensory drive Selection favoring courter signals that are easier to detect and discriminate among in the environment. See also **sensory exploitation**.

Sensory ecology A field focusing on the evolution, diversity, and function of sensory systems in the environment.

Sensory exploitation Use of a signal by courters that addresses a pre-existing bias in the sensory system of choosers. See also **sensory drive**.

Sensory periphery Structures involved in transducing environmental signals, e.g., eyes, ears, noses, and tongues.

Sequential choice tests Assays of **preference** whereby stimuli are presented one at a time to choosers. See also **simultaneous choice tests** and **sequential sampling**.

Sequential sampling Sampling of courters successively over time.

Sex-role reversal Occurs when males primarily choose females and females compete for males, rather than vice-versa.

Sexual antagonism A form of sexual conflict whereby the same allele has positive fitness effects in one sex and negative effects in the other.

Sexual conflict An evolutionary process that occurs when a given **phenotype** has different **fitness** optima for males and females.

Sexual fitness In the context of mate choice, the proportion of fitness determined by mating decisions.

Sexual imprinting A process whereby early experience with a stimulus during a **critical period** results in a strong, often irreversible preference for that stimulus later in life.

Sexual reward The positive hedonic value of sexual activity.

Sexual selection A special case of **natural selection**: differential reproductive success due to the ability to secure matings and/or fertilizations.

Shape The general form of a preference function.

Signal fidelity Any pattern of behavior shown by an individual that leads to its being more likely to engage in sexual activity with certain individuals than with others.

Simultaneous choice tests Assays of **preference** whereby two or more stimuli are presented to a **chooser** within the course of a trial. See also **sequential choice tests** and **simultaneous sampling**.

Simultaneous sampling Sampling of courters at the same time or in overlapping time intervals.

Social mate An individual with whom one preferentially associates and with whom one performs sexual behavior.

Social selection Selection through social interactions encompassing but not limited to mate choice.

Social monogamy Preferential sexual association with a single individual; see **pair bond**.

Spatial acuity The angular resolution of the visual system; determines how small an object an individual can detect at a given distance.

Sperm competition Competition between the sperm of two males to fertilize eggs of a given female.

State-dependent Refers to preferences that are dependent on chooser state, like condition or stress hormone titer. Compare with **context-sensitive**.

Static sampling Mate sampling dependent on the chooser's state when sampling begins, and not modified by the process of sampling. See **dynamic sampling**.

Structural evolution Change over time in the nucleotide sequence of DNA.

Sympatric speciation Formation of new species without geographic isolation.

Threshold The lowest value of a stimulus to elicit a response.

Total selection on mate choice Differential survival and reproductive success of choosers as a consequence of differences in the non-sexual and sexual consequences of mate-choice phenotypes.

Transitivity In the context of mate choice, the logical condition that if a chooser prefers A over B and B over C, then it will prefer A over C. See **intransitivity**.

Umwelt An animal's perceptual universe or "self-centered world," circumscribed by its **sensory** and **perceptual** abilities.

Uncanny valley A range of intermediate values of traits or combination of traits that is **aversive** compared to values outside the range.

Unimodal preference Preference characterized by one distinct peak.

Valence The direction of **preference**, i.e., its sign (positive or negative) along a specified range.

Vertical transmission A mode of preference learning whereby developing choosers shape their preferences in response to the stimuli of their parents. Also refers to other forms of **nongenetic inheritance**.

Weber's law The principle that receivers compare stimulus intensity or quantity by attending to the ratio between stimulus values, rather than the absolute quantity; see **nonlinear compression** and **power law**.

Westermarck effect The principle that individuals develop aversion to sexual relationships with individuals experienced during early development.

Working memory A cognitive system responsible for the transient storage of pertinent information.

LITERATURE CITED

Abbassi, P., & Burley, N. T. (2012). Nice guys finish last: same-sex sexual behavior and pairing success in male budgerigars. *Behavioral Ecology, 23*(4), 775–782.

Abbott, R., Albach, D., Ansell, S., Arntzen, J., Baird, S., Bierne, N., ... Zinner, D. (2013). Hybridization and speciation. *Journal of Evolutionary Biology, 26*(2), 229–246.

Abisgold, J. D., & Simpson, S. J. (1988). The effect of dietary protein levels and haemolymph composition on the sensitivity of the maxillary palp chemoreceptors of locusts. *Journal of Experimental Biology, 135*(1), 215–229.

Abraham, N. M., Vincis, R., Lagier, S., Rodriguez, I., Carleton, A., & Eichenbaum, H. (2014). Long term functional plasticity of sensory inputs mediated by olfactory learning. *eLife, 3*. doi:10.7554/eLife.02109

Abt, G., & Reyer, H.-U. (1993). Mate choice and fitness in a hybrid frog: *Rana esculenta* females prefer *Rana lessonae* males over their own. *Behavioral Ecology and Sociobiology, 32*(4), 221–228.

Ache, B. W., & Restrepo, D. (2000). Olfactory transduction. In T. E. Finger, W. L. Silver, & D. Restrepo (Eds.), *Neurobiology of Taste and Smell* (2nd ed., pp. 159–177). New York: Wiley-Liss.

Adamo, S. A., Kovalko, I., Easy, R. H., & Stoltz, D. (2014). A viral aphrodisiac in the cricket *Gryllus texensis*. *Journal of Experimental Biology, 217*(11), 1970–1976. doi:10.1242/jeb.103408

Adamo, S. A., & Spiteri, R. J. (2005). Female choice for male immunocompetence: when is it worth it? *Behavioral Ecology, 16*(5), 871–879.

———. (2009). He's healthy, but will he survive the plague? Possible constraints on mate choice for disease resistance. *Animal Behaviour, 77*(1), 67–78.

Adams, J., & Light, R. (2015). Scientific consensus, the law, and same sex parenting outcomes. *Social Science Research, 53*, 300–310. doi:http://dx.doi.org/10.1016/j.ssresearch.2015.06.008.

Adkins-Regan, E. (1998). Hormonal mechanisms of mate choice. *American Zoologist, 38*(1), 166–178.

Agbali, M., Reichard, M., Bryjová, A., Bryja, J., & Smith, C. (2010). Mate choice for non-additive genetic benefits correlate with MHC dissimilarity in the rose bitterling (*Rhodeus ocellatus*). *Evolution, 64*(6), 1683–1696.

Agrawal, A. F. (2001). Sexual selection and the maintenance of sexual reproduction. *Nature, 411*(6838), 692–695.

Ahern, T. H., Modi, M. E., Burkett, J. P., & Young, L. J. (2009). Evaluation of two automated metrics for analyzing partner preference tests. *Journal of Neuroscience Methods, 182*(2), 180–188.

Ah-King, M., Barron, A. B., & Herberstein, M. E. (2014). Genital evolution: why are females still understudied? *PLoS Biology, 12*(5), e1001851. doi:10.1371/journal.pbio.1001851

Ahlers, C. J., Schaefer, G. A., Mundt, I. A., Roll, S., Englert, H., Willich, S. N., & Beier, K. M. (2011). How unusual are the contents of paraphilias? Paraphilia-associated sexual arousal patterns in a community-based sample of men. *Journal of Sexual Medicine, 8*(5), 1362–1370.

Aisenberg, A., Barrantes, G., & Eberhard, W. G. (2015). Hairy kisses: tactile cheliceral courtship affects female mating decisions in *Leucauge mariana* (Araneae, Tetragnathidae). *Behavioral Ecology and Sociobiology, 69*(2), 313–323.

Akçay, E., & Roughgarden, J. (2007). Extra-pair paternity in birds: review of the genetic benefits. *Evolutionary Ecology Research, 9*(5), 855–868.

Akre, K. L., Bernal, X., Rand, A. S., & Ryan, M. J. (2014). Harmonic calls and indifferent females: no preference for human consonance in an anuran. *Proceedings of the Royal Society of London B: Biological Sciences, 281*(1789). doi:10.1098/rspb.2014.0986

Akre, K. L., Farris, H. E., Lea, A. M., Page, R. A., & Ryan, M. J. (2011). Signal perception in frogs and bats and the evolution of mating signals. *Science, 333*(6043), 751–752. doi:10.1126/science.1205623

Akre, K. L., & Johnsen, S. (2014). Psychophysics and the evolution of behavior. *Trends in Ecology & Evolution, 29*(5), 291–300. doi:http://dx.doi.org/10.1016/j.tree.2014.03.007

Akre, K. L., & Ryan, M. J. (2010). Complexity increases working memory for mating signals. *Current Biology, 20*(6), 502–505. doi:http://dx.doi.org/10.1016/j.cub.2010.01.021

Alatalo, R. V., Carlson, A., & Lundberg, A. (1988). The search cost in mate choice of the pied flycatcher. *Animal Behaviour, 36*(1), 289–291. doi:http://dx.doi.org/10.1016/S0003-3472(88)80272-0

Alatalo, R. V., Kotiaho, J., Mappes, J., & Parri, S. (1998). Mate choice for offspring performance: major benefits or minor costs? *Proceedings of the Royal Society of London B: Biological Sciences, 265*(1412), 2297–2301.

Alatalo, R. V., Mappes, J., & Elgar, M. A. (1997). Heritabilities and paradigm shifts. *Nature, 385*, 402–403.

Albert, A.Y.K., & Otto, S. P. (2005). Sexual selection can resolve sex-linked sexual antagonism. *Science, 310*(5745), 119–121. doi:10.1126/science.1115328

Albert, A.Y.K., & Schluter, D. (2004). Reproductive character displacement of male stickleback mate preference: reinforcement or direct selection? *Evolution, 58*(5), 1099–1107.

Alexander, W. E. (1891). *The History of Human Marriage*. London: Macmillan.

Alford, J. R., Hatemi, P. K., Hibbing, J. R., Martin, N. G., & Eaves, L. J. (2011). The politics of mate choice. *Journal of Politics, 73*(02), 362–379.

Allison, J. D., & Cardé, R. T. (2008). Male pheromone blend preference function measured in choice and no-choice wind tunnel trials with almond moths, *Cadra cautella*. *Animal Behaviour, 75*, 259–266. doi:10.1016/j.anbehav.2007.04.033

Almeida, C. R., & de Abreu, F. V. (2003). Dynamical instabilities lead to sympatric speciation. *Evolutionary Ecology Research, 5*(5), 739–757.

Alonzo, S. H. (2008). Female mate choice copying affects sexual selection in wild populations of the ocellated wrasse. *Animal Behaviour, 75*, 1715–1723. doi:10.1016/j.anbehav.2007.09.031

———. (2009). Social and coevolutionary feedbacks between mating and parental investment. *Trends in Ecology & Evolution, 25*(2), 99–108.

———. (2012). Sexual selection favours male parental care, when females can choose. *Proceedings of the Royal Society of London B: Biological Sciences, 279*(1734), 1784–1790.

Alonzo, S. H., & Pizzari, T. (2013). Selection on female remating interval is influenced by male sperm competition strategies and ejaculate characteristics. *Philosophical Transactions of the Royal Society of London B: Biological Sciences, 368*(1613), 20120044. doi:10.1098/rstb.2012.0044

Alonzo, S. H., & Sinervo, B. (2001). Mate choice games, context-dependent good genes, and genetic cycles in the side-blotched lizard, *Uta stansburiana. Behavioral Ecology and Sociobiology, 49*(2–3), 176–186.

Amcoff, M., & Kolm, N. (2013). Does female feeding motivation affect the response to a food-mimicking male ornament in the swordtail characin *Corynopoma riisei*? *Journal of Fish Biology, 83*(2), 343–354. doi:10.1111/jfb.12175

Amcoff, M., Lindqvist, C., & Kolm, N. (2013). Sensory exploitation and plasticity in female mate choice in the swordtail characin. *Animal Behaviour, 85*(5), 891–898. doi:http://dx.doi.org/10.1016/j.anbehav.2013.02.001

Amézquita, A., Flechas, S. V., Lima, A. P., Gasser, H., & Hödl, W. (2011). Acoustic interference and recognition space within a complex assemblage of dendrobatid frogs. *Proceedings of the National Academy of Sciences, 108*(41), 17058–17063.

Amstislavskaya, T. G., & Popova, N. K. (2004). Female-induced sexual arousal in male mice and rats: behavioral and testosterone response. *Hormones and Behavior, 46*(5), 544–550. doi:http://dx.doi.org/10.1016/j.yhbeh.2004.05.010

Amundsen, T. (2000). Why are female birds ornamented? *Trends in Ecology & Evolution, 15*(4), 149–155.

Amundsen, T., & Forsgren, E. (2001). Male mate choice selects for female coloration in a fish. *Proceedings of the National Academy of Sciences, 98*(23), 13155–13160.

Amy, M., Monbureau, M., Durand, C., Gomez, D., Théry, M., & Leboucher, G. (2008). Female canary mate preferences: differential use of information from two types of male-male interaction. *Animal Behaviour, 76*(3), 971–982.

Amy, M., Salvin, P., Naguib, M., & Leboucher, G. (2015). Female signalling to male song in the domestic canary, *Serinus canaria. Royal Society Open Science, 2*(1), 140196. doi:10.1098/rsos.140196

Anderson, W. W., Kim, Y.-K., & Gowaty, P. A. (2007). Experimental constraints on mate preferences in *Drosophila pseudoobscura* decrease offspring viability and fitness of mated pairs. *Proceedings of the National Academy of Sciences, 104*(11), 4484–4488.

Andersson, M. (1982). Female choice selects for extreme tail length in a widowbird. *Nature, London, 299*, 818–820.

———. (1986). Evolution of condition-dependent sex ornaments and mating preferences: sexual selection based on viability differences. *Evolution, 40*(4), 804–816.

Andersson, M., & Simmons, L. W. (2006). Sexual selection and mate choice. *Trends in Ecology & Evolution, 21*(6), 296–302. doi:http://dx.doi.org/10.1016/j.tree.2006.03.015

Andersson, M. B. (1994). *Sexual Selection*. Princeton, NJ: Princeton University Press.

Andolfatto, P. (2005). Adaptive evolution of non-coding DNA in *Drosophila. Nature, 437*(7062), 1149–1152.

Andrade, M. (1996). Sexual selection for male sacrifice in redback spiders. *Science, 271*(5245), 70–72.

Andrade, M. C. (2003). Risky mate search and male self-sacrifice in redback spiders. *Behavioral Ecology, 14*(4), 531–538.

Andrade, M. C., & Kasumovic, M. M. (2005). Terminal investment strategies and male mate choice: extreme tests of Bateman. *Integrative and Comparative Biology, 45*(5), 838–847.

Andrés, J. A., & Arnqvist, G. (2001). Genetic divergence of the seminal signal-receptor system in houseflies: the footprints of sexually antagonistic coevolution? Proceedings *of the Royal Society of London B: Biological Sciences, 268*(1465), 399–405. doi:10.1098/rspb.2000.1392.

Anstey, M. L., Rogers, S. M., Ott, S. R., Burrows, M., & Simpson, S. J. (2009). Serotonin mediates behavioral gregarization underlying swarm formation in desert locusts. *Science, 323*(5914), 627–630. doi:10.1126/science.1165939

Anthes, N., David, P., Auld, J. R., Hoffer, J. N., Jarne, P., Koene, J. M., ... Schärer, L. (2010). Bateman gradients in hermaphrodites: an extended approach to quantify sexual selection. *American Naturalist, 176*(3), 249–263.

Anthes, N., Putz, A., & Michiels, N. K. (2006). Hermaphrodite sex role preferences: the role of partner body size, mating history and female fitness in the sea slug *Chelidonura sandrana. Behavioral Ecology and Sociobiology, 60*(3), 359–367.

Anton, S., Löfstedt, C., & Hansson, B. S. (1997). Central nervous processing of sex pheromones in two strains of the European corn borer *Ostrinia nubilalis* (Lepidoptera: Pyralidae). *Journal of Experimental Biology, 200*(7), 1073–1087.

Aoki, K., Feldman, M. W., & Kerr, B. (2001). Models of sexual selection on a quantitative genetic trait when preference is acquired by sexual imprinting. *Evolution, 55*(1), 25–32.

Aparicio, J. M. (2011). The paradox of the resolution of the lek paradox based on mate choice for heterozygosity. *Animal Behaviour, 81*(6), 1271–1279.

Apostolou, M. (2007). Sexual selection under parental choice: the role of parents in the evolution of human mating. *Evolution and Human Behavior, 28*(6), 403–409.

———. (2010). Sexual selection under parental choice in agropastoral societies. *Evolution and Human Behavior, 31*(1), 39–47.

Appelt, C. W., & Sorensen, P. W. (2007). Female goldfish signal spawning readiness by altering when and where they release a urinary pheromone. *Animal Behaviour, 74*(5), 1329–1338.

Araya-Salas, M. (2012). Is birdsong music? Evaluating harmonic intervals in songs of a Neotropical songbird. *Animal Behaviour, 84*(2), 309–313. doi:http://dx.doi.org/10.1016/j.anbehav.2012.04.038

Arbuthnott, D., Crespi, B. J., & Schwander, T. (2015). Female stick insects mate multiply to find compatible mates. *American Naturalist, 186*(4), 519–530.

Archard, G. A., Cuthill, I. C., & Partridge, J. C. (2006). Condition-dependent mate choice in the guppy: a role for short-term food restriction? *Behaviour, 143*(11), 1317–1340. doi:10.1163/156853906778987515

Arianne, Y.K.A., & Otto, S. P. (2005). Sexual selection can resolve sex-linked sexual antagonism. *Science, 310*(5745), 119–121.

Arnegard, M. E., & Kondrashov, A. S. (2004). Sympatric speciation by sexual selection alone is unlikely. *Evolution, 58*(2), 222–237.

Arnegard, M. E., Zwickl, D. J., Lu, Y., & Zakon, H. H. (2010). Old gene duplication facilitates origin and diversification of an innovative communication system—twice. *Proceedings of the National Academy of Sciences.* doi:10.1073/pnas.1011803107

Arnold, M. L. (1997). *Natural Hybridization and Evolution*. Oxford: Oxford University Press.

Arnold, S. J. (1983). Sexual selection: the interface of theory and empiricism. *Mate Choice*, 67–107.

Arnqvist, G. (2004). Sexual conflict and sexual selection: lost in the chase. *Evolution, 58*(6), 1383–1388.

———. (2006). Sensory exploitation and sexual conflict. *Philosophical Transactions of the Royal Society of London B: Biological Sciences, 361*(1466), 375–386.

Arnqvist, G., & Kirkpatrick, M. (2005). The evolution of infidelity in socially monogamous passerines: the strength of direct and indirect selection on extrapair copulation behavior in females. *American Naturalist,165*(S5), S26–S37.

Arnqvist, G., & Rowe, L. (2005). *Sexual Conflict*. Princeton, NJ: Princeton University Press.

———. (2014). The shape of preference functions and what shapes them: a comment on Edwards. *Behavioral Ecology*, aru200.

Aronsen, T., Berglund, A., Mobley, K. B., Ratikainen, I. I., & Rosenqvist, G. (2013). Sex ratio and density affect sexual selection in a sex-role reversed fish. *Evolution, 67*(11), 3243–3257.

Aspbury, A., & Basolo, A. (2002). Repeatable female preferences, mating order and mating success in the poeciliid fish, *Heterandria formosa. Behavioral Ecology and Sociobiology, 51*(3), 238–244. doi:10.1007/s00265-001-0443-1

Atwell, A., & Wagner Jr, W. E. (2014). Female mate choice plasticity is affected by the interaction between male density and female age in a field cricket. *Animal Behaviour, 98*(0), 177–183. doi:http://dx.doi.org/10.1016/j.anbehav.2014.10.007

Auld, H. L., Punzalan, D., Godin, J.-G. J., & Rundle, H. D. (2009). Do female fruit flies (*Drosophila serrata*) copy the mate choice of others? *Behavioural Processes, 82*(1), 78–80. doi:http://dx.doi.org/10.1016/j.beproc.2009.03.004

Avey, M. T., Phillmore, L. S., & MacDougall-Shackleton, S. A. (2005). Immediate early gene expression following exposure to acoustic and visual components of courtship in zebra finches. *Behavioural Brain Research, 165*(2), 247–253. doi:http://dx.doi.org/10.1016/j.bbr.2005.07.002

Azanchi, R., Kaun, K. R., & Heberlein, U. (2013). Competing dopamine neurons drive oviposition choice for ethanol in *Drosophila. Proceedings of the National Academy of Sciences, 110*(52), 21153–21158. doi:10.1073/pnas.1320208110

Bachtrog, D., Mank, J. E., Peichel, C. L., Kirkpatrick, M., Otto, S. P., Ashman, T.-L., ... The Tree of Sex Consortium (2014). Sex determination: why so many ways of doing it? *PLoS Biology, 12*(7), e1001899–e1001899. doi:10.1371/journal.pbio.1001899

Bacigalupe, L. D., Crudgington, H. S., Slate, J., Moore, A. J., & Snook, R. R. (2008). Sexual selection and interacting phenotypes in experimental evolution: a study of *Drosophila pseudoobscura* mating behavior. *Evolution, 62*(7), 1804–1812.

Badyaev, A. V., & Qvarnström, A. (2002). Putting sexual traits into the context of an organism: a life-history perspective in studies of sexual selection. *Auk, 119*(2), 301–310.

Bagemihl, B. (1999). *Biological Exuberance: Animal Homosexuality and Natural Diversity*. New York: Macmillan.

Bahr, A., Sommer, S., Mattle, B., & Wilson, A. B. (2012). Mutual mate choice in the pot-bellied seahorse (*Hippocampus abdominalis*). *Behavioral Ecology*, ars045.

Bailey, N. W. (2008). Love will tear you apart: different components of female choice exert contrasting selection pressures on male field crickets. *Behavioral Ecology, 19*(5), 960–966. doi:10.1093/beheco/arn054

———. (2012). Evolutionary models of extended phenotypes. *Trends in Ecology & Evolution, 27*(10), 561–569. doi:http://dx.doi.org/10.1016/j.tree.2012.05.011

Bailey, N. W., & Moore, A. J. (2012). Runaway sexual selection without genetic correlations: social environments and flexible mate choice initiate and enhance the Fisher process. *Evolution, 66*(9), 2674–2684. doi:10.1111/j.1558-5646.2012.01647.x

Bailey, N. W., & Zuk, M. (2008). Acoustic experience shapes female mate choice in field crickets. *Proceedings of the Royal Society of London B: Biological Sciences, 275*(1651), 2645–2650. doi:10.1098/rspb.2008.0859

———. (2009a). Field crickets change mating preferences using remembered social information. *Biology Letters, 5*(4), 449–451.

———. (2009b). Same-sex sexual behavior and evolution. *Trends in Ecology & Evolution, 24*(8), 439–446.

———. (2012). Socially flexible female choice differs among populations of the pacific field cricket: geographical variation in the interaction coefficient psi (ψ). *Proceedings of the Royal Society of London B: Biological Sciences, 279*(1742), 3589–3596. doi:10.1098/rspb.2012.0631

Bailey, R. I., Innocenti, P., Morrow, E. H., Friberg, U., & Qvarnström, A. (2011). Female *Drosophila melanogaster* gene expression and mate choice: the X chromosome harbours candidate genes underlying sexual isolation. *PLoS ONE, 6*(2), e17358. doi:10.1371/journal.pone.0017358

Bailey, W. J., Cunningham, R. J., & Lebel, L. (1990). Song power, spectral distribution and female phonotaxis in the bushcricket *Requena verticalis* (Tettigoniidae: Orthoptera): active female choice or passive attraction. *Animal Behaviour, 40*, 33–42.

Baker, T. C., Quero, C., Ochieng, S. A., & Vickers, N. J. (2006). Inheritance of olfactory preferences II. Olfactory receptor neuron responses from *Heliothis subflexa* × *Heliothis virescens* hybrid male moths. *Brain, Behavior and Evolution, 68*(2), 75–89.

Bakker, T. C. (1999). The study of intersexual selection using quantitative genetics. *Behaviour, 136*(9), 1237–1266.

Bakker, T.C.M., Kunzler, R., & Mazzi, D. (1999). Sexual selection: condition-related mate choice in sticklebacks. *Nature, 401*(6750), 234–234.

Bakker, T.C.M., & Pomiankowski, A. (1995). The genetic basis of female mate preferences. *Journal of Evolutionary Biology, 8*(2), 129–171. doi:10.1046/j.1420-9101.1995.8020129.x

Balaban, E. (1988). Bird song syntax: learned intraspecific variation is meaningful. *Proceedings of the National Academy of Sciences, 85*(10), 3657–3660.

Baldassarre, D. T., & Webster, M. S. (2013). Experimental evidence that extra-pair mating drives asymmetrical introgression of a sexual trait. *Proceedings of the Royal Society of London B: Biological Sciences, 280*(1771), 20132175.

Baldwin, J., & Johnsen, S. (2009). The importance of color in mate choice of the blue crab *Callinectes sapidus*. *Journal of Experimental Biology, 212*, 3762–3768.

Bales, K. L., Mason, W. A., Catana, C., Cherry, S. R., & Mendoza, S. P. (2007). Neural correlates of pair-bonding in a monogamous primate. *Brain Research, 1184*, 245–253.

Ball, M., & Parker, G. (2003). Sperm competition games: sperm selection by females. *Journal of Theoretical Biology, 224*(1), 27–42.

Balmford, A. (1991). Mate choice on leks. *Trends in Ecology & Evolution, 6*(3), 87–92. doi:http://dx.doi.org/10.1016/0169-5347(91)90181-V

Balmford, A., & Read, A. F. (1991). Testing alternative models of sexual selection through female choice. *Trends in Ecology & Evolution, 6*(9), 274–276.

Bańbura, J. (1992). Mate choice by females of the swallow *Hirundo rustica*: is it repeatable? *Journal für Ornithologie, 133*(2), 125–132. doi:10.1007/BF01639905

Banerjee, S. B., & Adkins-Regan, E. (2014). Same-sex partner preference in adult male zebra finch offspring raised in the absence of maternal care. *Animal Behaviour, 92*, 167–173. doi:http://dx.doi.org/10.1016/j.anbehav.2014.03.030

Barbosa, M., Connolly, S. R., Hisano, M., Dornelas, M., & Magurran, A. E. (2012). Fitness consequences of female multiple mating: a direct test of indirect benefits. *BMC Evolutionary Biology, 12*(1), 185.

Bargmann, C. I. (2006). Comparative chemosensation from receptors to ecology. *Nature, 444*(7117), 295–301.

Bargmann, C. I., & Kaplan, J. M. (1998). Signal transduction in the Caenorhabditis elegans nervous system. *Annual Review of Neuroscience, 21*(1), 279–308.

Barnwell, C. V., & Noor, M.A.F. (2008). Failure to replicate two mate preference QTLs across multiple strains of *Drosophila pseudoobscura*. *Journal of Heredity, 99*(6), 653–656. doi:10.1093/jhered/esn069

Barry, K. L., & Kokko, H. (2010). Male mate choice: why sequential choice can make its evolution difficult. *Animal Behaviour, 80*(1), 163–169.

Basolo, A. L. (1990a). Female preference for male sword length in the green swordtail (Pisces: Poeciliidae). *Animal Behaviour, 40*, 332–338.

———. (1990b). Female preference predates the evolution of the sword in swordtail fish. *Science, 250*, 808–810.

———. (1995). A further examination of a preexisting bias favoring a sword in the genus *Xiphophorus. Animal Behaviour, 50*, 365–375.

———. (1998a). Evolutionary change in a receiver bias: a comparison of female preference functions. *Proceedings of the Royal Society of London B: Biological Sciences, 265*, 2223–2228.

———. (1998b). Shift in investment between sexually-selected traits: tarnishing of the silver spoon. *Animal Behaviour, 55*, 665–671.

———. (2002). Congruence between the sexes in preexisting receiver responses. *Behavioral Ecology, 13*(6), 832–837.

Basolo, A. L., & Trainor, B. C. (2002). The conformation of a female preference for a composite male trait in green swordtails. *Animal Behaviour, 63*(3), 469–474. doi: http://dx.doi.org/10.1006/anbe.2001.1933.

Bateman, A. J. (1948). Intra-sexual selection in Drosophila. *Heredity, 2*(Pt. 3), 349–368.

Bateson, M., & Healy, S. D. (2005). Comparative evaluation and its implications for mate choice. *Trends in Ecology & Evolution, 20*(12), 659–664. doi:http://dx.doi.org/10.1016/j.tree.2005.08.013

Bateson, P. (1978). Sexual imprinting and optimal outbreeding. *Nature, 273*(5664), 659–660.

———. (Ed.) (1983). *Mate Choice*. Cambridge, UK: Cambridge University Press.

Baube, C. L., Rowland, W. J., & Fowler, J. B. (1995). The mechanisms of colour-based mate choice in female threespine sticklebacks: hue, contrast and configurational cues. *Behaviour, 132*(13), 979–996. doi:doi:10.1163/156853995X00405

Baugh, A. T., Akre, K. L., & Ryan, M. J. (2008). Categorical perception of a natural, multivariate signal: mating call recognition in túngara frogs. *Proceedings of the National Academy of Sciences, 105*(26), 8985–8988. doi:10.1073/pnas.0802201105

Baugh, A. T., & Ryan, M. J. (2010). The development of sexual behavior in tungara frogs (*Physalaemus pustulosus*). *J Comp Psychol, 124*(1), 66–80. doi:10.1037/a0017227

Beach, F. A. (1976). Sexual attractivity, proceptivity, and receptivity in female mammals. *Hormones and Behavior, 7*(1), 105–138. doi:http://dx.doi.org/10.1016/0018-506X(76)90008-8

Beaulieu, M., & Sockman, K. W. (2012). Song in the cold is 'hot': memory of and preference for sexual signals perceived under thermal challenge. *Biology Letters, 8*(5), 751–753. doi:10.1098/rsbl.2012.0481

Beck, C., & Powell, L. A. (2000). Evolution of female mate choice based on male age: are older males better mates? *Evolutionary Ecology Research, 2*, 107–118.

Beckage, N., Todd, P. M., Penke, L., & Asendorpf, J. (2009). *Testing sequential patterns in human mate choice using speed dating.* Paper presented at the Proceedings of the 2009 Cognitive Science Conference.

Beckers, O. M., & Wagner Jr., W. E. (2011). Mate sampling strategy in a field cricket: evidence for a fixed threshold strategy with last chance option. *Animal Behaviour, 81*(3), 519–527. doi:http://dx.doi.org/10.1016/j.anbehav.2010.11.022

———. (2013). Parasitoid infestation changes female mating preferences. *Animal Behaviour, 85*(4), 791–796. doi:http://dx.doi.org/10.1016/j.anbehav.2013.01.025

Bell, A. M., Hankison, S. J., & Laskowski, K. L. (2009). The repeatability of behaviour: a meta-analysis. *Animal Behaviour, 77*(4), 771–783. doi:http://dx.doi.org/10.1016/j.anbehav.2008.12.022

Bell, M. A., & Travis, M. P. (2005). Hybridization, transgressive segregation, genetic covariation, and adaptive radiation. *Trends in Ecology & Evolution, 20*(7), 358–361.

Belluck, P. (2012). Health experts dismiss assertions on rape. *New York Times*. Retrieved from http://www.nytimes.com/2012/08/21/us/politics/rape-assertions-are-dismissed-by-health-experts.html

Belluscio, L., Lodovichi, C., Feinstein, P., Mombaerts, P., & Katz, L. C. (2002). Odorant receptors instruct functional circuitry in the mouse olfactory bulb. *Nature, 419*, 296–300.

Bel-Venner, M., Dray, S., Allaine, D., Menu, F., & Venner, S. (2008). Unexpected male choosiness for mates in a spider. *Proceedings of the Royal Society of London B: Biological Sciences, 275*(1630), 77–82.

Bem, D. J. (1996). Exotic becomes erotic: A developmental theory of sexual orientation. *Psychological Review, 103*(2), 320.

Bentsen, C. L., Hunt, J., Jennions, M. D., & Brooks, R. (2006). Complex multivariate sexual selection on male acoustic signaling in a wild population of *Teleogryllus commodus*. *American Naturalist, 167*(4), E102–E116.

Beny, Y., & Kimchi, T. (2014). Innate and learned aspects of pheromone-mediated social behaviours. *Animal Behaviour, 97*(0), 301–311. doi:http://dx.doi.org/10.1016/j.anbehav.2014.09.014

Berdan, E. L., & Fuller, R. C. (2012). A test for environmental effects on behavioral isolation in two species of killifish. *Evolution, 66*(10), 3224–3237.

Bereczkei, T., Gyuris, P., & Weisfeld, G. E. (2004). Sexual imprinting in human mate choice. *Proceedings: Biological Sciences, 271*(1544), 1129–1134. doi:10.1098/rspb.2003.2672

Berglund, A. (1993). Risky sex: male pipefishes mate at random in the presence of a predator. *Animal Behaviour, 46*(1), 169–175. doi:http://dx.doi.org/10.1006/anbe.1993.1172

Berglund, A., Widemo, M. S., & Rosenqvist, G. (2005). Sex-role reversal revisited: choosy females and ornamented, competitive males in a pipefish. *Behavioral Ecology, 16*(3), 649–655.

Bergstrom, C. T., & Real, L. A. (2000). Toward a theory of mutual mate choice: lessons from two-sided matching. Evolutionary Ecology Research, *2*, 493–508.

Bernal, X. E., Rand, A. S., & Ryan, M. J. (2009). Task differences confound sex differences in receiver permissiveness in túngara frogs. *Proceedings of the Royal Societyof London B: Biological Sciences, 276*, 1323–1329. doi:10.1098/rspb.2008.0935

Bernasconi, G., Ashman, T.-L., Birkhead, T., Bishop, J., Grossniklaus, U., Kubli, E., ... Hellriegel, B. (2004). Evolutionary ecology of the prezygotic stage. *Science, 303*(5660), 971–975.

Bichet, C., Penn, D. J., Moodley, Y., Dunoyer, L., Cellier-Holzem, E., Belvalette, M., ... Sorci, G. (2014). Females tend to prefer genetically similar mates in an island population of house sparrows. *BMC Evolutionary Biology, 14*(1), 47.

Bickart, K. C., Wright, C. I., Dautoff, R. J., Dickerson, B. C., & Barrett, L. F. (2011). Amygdala volume and social network size in humans. *Nature Neuroscience, 14*(2), 163–164. doi:10.1038/nn.2724

Bierbach, D., Jung, C. T., Hornung, S., Streit, B., & Plath, M. (2013). Homosexual behaviour increases male attractiveness to females. *Biology Letters, 9*(1). doi:10.1098/rsbl.2012.1038

Bierbach, D., Schulte, M., Herrmann, N., Tobler, M., Stadler, S., Jung, C. T., ... Plath, M. (2011). Predator-induced changes of female mating preferences: innate and experiential effects. *BMC Evolutionary Biology, 11*(1), 190.

Bierbach, D., Sommer-Trembo, C., Hanisch, J., Wolf, M., & Plath, M. (2015). Personality affects mate choice: bolder males show stronger audience effects under high competition. *Behavioral Ecology*, arv079.

Bilde, T., Foged, A., Schilling, N., & Arnqvist, G. (2009). Postmating sexual selection favors males that sire offspring with low fitness. *Science, 324*(5935), 1705–1706.

Billeter, J. C., & Levine, J. D. (2013). Who is he and what is he to you? Recognition in *Drosophila melanogaster*. *Current Opinion in Neurobiology, 23*(1), 17–23. doi:10.1016/j.conb.2012.08.009

Bijma, P., Muir, W. M., Ellen, E. D., Wolf, J. B., & Van Arendonk, J. A. (2007). Multilevel selection 2: estimating the genetic parameters determining inheritance and response to selection. *Genetics, 175*(1), 289–299.

Bijma, P., Muir, W. M., & Van Arendonk, J. A. (2007). Multilevel selection 1: quantitative genetics of inheritance and response to selection. *Genetics, 175*(1), 277–288.

Billeter, J. C., Jagadeesh, S., Stepek, N., Azanchi, R., & Levine, J. D. (2012). *Drosophila melanogaster* females change mating behaviour and offspring production based on social context. *Proceedings of the Royal Society of London B: Biological Sciences, 279*(1737), 2417–2425. doi:10.1098/rspb.2011.2676.

Binns, K. E., & Brennan, P. A. (2005). Changes in electrophysiological activity in the accessory olfactory bulb and medial amygdala associated with mate recognition in mice. *European Journal of Neuroscience, 21*(9), 2529–2537. doi:10.1111/j.1460-9568.2005.04090.x

Bird, A. (2007). Perceptions of epigenetics. *Nature, 447*(7143), 396–398.

Birkhead, T., Fletcher, F., & Pellatt, E. (1998). Sexual selection in the zebra finch *Taeniopygia guttata*: condition, sex traits and immune capacity. *Behavioral Ecology and Sociobiology, 44*(3), 179–191.

Birkhead, T., & Møller, A. (1993). Female control of paternity. *Trends in Ecology & Evolution, 8*(3), 100–104. doi:http://dx.doi.org/10.1016/0169-5347(93)90060-3

Birkhead, T. R. (1998). Cryptic female choice: criteria for establishing female sperm choice. *Evolution, 52*(4), 1212–1218. doi:10.2307/2411251

Birkhead, T. R., Chaline, N., Biggins, J. D., Burke, T., & Pizzari, T. (2004). Nontransitivity of paternity in a bird. *Evolution, 58*(2), 416–420. doi:10.1111/j.0014-3820.2004.tb01656.x

Birkhead, T. R., & Møller, A. P. (1998). *Sperm Competition and Sexual Selection. London*: Academic Press.

Birkhead, T. R., Pellatt, E. J., Matthews, I. M., Roddis, N. J., Hunter, F. M., McPhie, F., . . . Castillo-Juarez, H. (2006). Genic capture and the genetic basis of sexually selected traits in the zebra finch. *Evolution, 60*(11), 2389–2398.

Birkhead, T. R., & Pizzari, T. (2002). Postcopulatory sexual selection. *Nature Reviews Genetics, 3*(4), 262–273.

Bischof, H.-J., & Clayton, N. (1991). Stabilization of sexual preferences by sexual experience in male zebra finches *Taeniopygia guttata castanotis*. *Behaviour, 118*(1), 144–155.

Bjork, A., Starmer, W. T., Higginson, D. M., Rhodes, C. J., & Pitnick, S. (2007). Complex interactions with females and rival males limit the evolution of sperm offence and defence. *Proceedings of the Royal Society of London B: Biological Sciences, 274*(1619), 1779–1788. doi:10.1098/rspb.2007.0293.

Blackwell, A. D., Tamayo, M. A., Beheim, B., Trumble, B. C., Stieglitz, J., Hooper, P. L., . . . Gurven, M. (2015). Helminth infection, fecundity, and age of first pregnancy in women. *Science, 350*(6263), 970–972.

Blais, J., Rico, C., & Bernatichez, L. (2004). Nonlinear effects of female mate choice in wild threespine sticklebacks. *Evolution, 58*(11), 2498–2510.

Bloch Qazi, M. C. (2003). A potential mechanism for cryptic female choice in a flour beetle. *Journal of Evolutionary Biology, 16*(1), 170–176. doi:10.1046/j.1420-9101.2003.00501.x

Blomqvist, D., Andersson, M., Kupper, C., Cuthill, I. C., Kis, J., Lanctot, R. B., . . . Kempenaers, B. (2002). Genetic similarity between mates and extra-pair parentage in three species of shorebirds. *Nature, 419*(6907), 613–615.

Blows, M. W. (1999). Evolution of the genetic covariance between male and female components of mate recognition: an experimental test. *Proceedings of the Royal Society of London B: Biological Sciences, 266*(1434), 2169–2174.

Blows, M. W., Brooks, R., & Kraft, P. G. (2003). Exploring complex fitness surfaces: multiple ornamentation and polymorphism in male guppies. *Evolution, 57*(7), 1622–1630.

Bluhm, C. K., & Gowaty, P. A. (2004). Reproductive compensation for offspring viability deficits by female mallards, *Anas platyrhynchos*. *Animal Behaviour, 68*(5), 985–992. doi:http://dx.doi.org/10.1016/j.anbehav.2004.01.012

Blumberg, S., Haran, T., Botzer, D., Susswein, A., & Teyke, T. (1998). Pheromones linked to sexual behaviors excite the appetitive phase of feeding behavior of *Aplysia fasciata*. I. Modulation and excitation of appetitive behaviors. *Journal of Comparative Physiology A, 182*(6), 777–783.

Blyth, J. E., & Gilburn, A. S. (2011). The function of female behaviours adopted during premating struggles in the seaweed fly, *Coelopa frigida*. *Animal Behaviour, 81*, 77–82. doi:10.1016/j.anbehav.2010.09.013

Boake, C. (1992). Reply from Christine Boake. *Trends in Ecology & Evolution, 7*(1), 30.

Boake, C. B. (1989). Repeatability: its role in evolutionary studies of mating behavior. *Evolutionary Ecology, 3*(2), 173–182. doi:10.1007/BF02270919

Boake, C. R. (1985). Genetic consequences of mate choice: a quantitative genetic method for testing sexual selection theory. *Science, 227*(4690), 1061–1063.

———. (1986). A method for testing adaptive hypotheses of mate choice. *American Naturalist*, 654–666.

———. (1991). Coevolution of senders and receivers of sexual signals: genetic coupling and genetic correlations. *Trends in Ecology & Evolution, 6*(7), 225–227.

Boake, C.R.B. (1989). Repeatability—its role in evolutionary studies of mating behavior. *Evolutionary Ecology, 3*(2), 173–182. doi:10.1007/bf02270919

Boake, C.R.B., Price, D. K., & Andreadis, D. K. (1998). Inheritance of behavioural differences between two interfertile, sympatric species, *Drosophila silvestris* and *D. heteroneura*. *Heredity, 80*, 642–650. doi:10.1046/j.1365-2540.1998.00317.x

Bolhuis, J. J. (1991). Mechanisms of avian imprinting: a review. *Biological Reviews, 66*(4), 303–345.

Bolhuis, J. J., Brown, G. R., Richardson, R. C., & Laland, K. N. (2011). Darwin in mind: new opportunities for evolutionary psychology. *PLoS Biology, 9*(7), e1001109.

Bolnick, D. I., & Fitzpatrick, B. M. (2007). Sympatric speciation: models and empirical evidence. *Annual Review of Ecology, Evolution, and Systematics, 38*(1), 459.

Bolund, E., Martin, K., Kempenaers, B., & Forstmeier, W. (2010). Inbreeding depression of sexually selected traits and attractiveness in the zebra finch. *Animal Behaviour, 79*(4), 947–955.

Bolund, E., Schielzeth, H., & Forstmeier, W. (2009). *Compensatory investment in zebra finches: females lay larger eggs when paired to sexually unattractive males, 276*(1657), 707–715. doi:10.1098/rspb.2008.1251.

———. (2010). No heightened condition dependence of zebra finch ornaments–a quantitative genetic approach. *Journal of Evolutionary Biology, 23*(3), 586–597.

Bonduriansky, R. (2001). The evolution of male mate choice in insects: a synthesis of ideas and evidence. *Biological Reviews, 76*(3), 305–339.

———. (2007a). The evolution of condition-dependent sexual dimorphism. *American Naturalist, 169*(1), 9–19.

———. (2007b). Sexual selection and allometry: a critical reappraisal of the evidence and ideas. *Evolution, 61*(4), 838–849.

———. (2009). Reappraising sexual coevolution and the sex roles. *PLoS Biology, 7*(12), e1000255.

———. (2011). Sexual selection and conflict as engines of ecological diversification. *American Naturalist, 178*(6), 729–745.

Bonduriansky, R., & Day, T. (2009). Nongenetic inheritance and its evolutionary implications. *Annual Review of Ecology, Evolution, and Systematics, 40*, 103–125.

———. (2013). Nongenetic inheritance and the evolution of costly female preference. *Journal of Evolutionary Biology, 26*(1), 76–87. doi:10.1111/jeb.12028

Bonisoli-Alquati, A., Matteo, A., Ambrosini, R., Rubolini, D., Romano, M., Caprioli, M., . . . Saino, N. (2011). Effects of egg testosterone on female mate choice and male sexual behavior in the pheasant. *Hormones and Behavior, 59*(1), 75–82. doi:http://dx.doi.org/10.1016/j.yhbeh.2010.10.013

Bonneaud, C., Chastel, O., Federici, P., Westerdahl, H., & Sorci, G. (2006). Complex *Mhc*-based mate choice in a wild passerine. *Proceedings of the Royal Society of London B: Biological Sciences,* 273, 1111–1116.

Boogert, N. J., Fawcett, T. W., & Lefebvre, L. (2011). Mate choice for cognitive traits: a review of the evidence in nonhuman vertebrates. *Behavioral Ecology, 22*(3), 447–459.

Booksmythe, I., Detto, T., & Backwell, P. R. Y. (2008). Female fiddler crabs settle for less: the travel costs of mate choice. *Animal Behaviour, 76*(6), 1775–1781. doi:http://dx.doi.org/10.1016/j.anbehav.2008.07.022

Booksmythe, I., Schwanz, L. E., & Kokko, H. (2013). The complex interplay of sex allocation and sexual selection. *Evolution, 67*(3), 673–678.

Boomsma, J. J. (2013). Beyond promiscuity: mate-choice commitments in social breeding. *Philosophical Transactions of the Royal Society of London B: Biological Sciences, 368*(1613). doi:10.1098/rstb.2012.0050

Borg, Å. A., Forsgren, E., & Amundsen, T. (2006). Seasonal change in female choice for male size in the two-spotted goby. *Animal Behaviour, 72*(4), 763–771. doi:http://dx.doi.org/10.1016/j.anbehav.2005.11.025

Borg, C., & de Jong, P. J. (2012). Feelings of disgust and disgust-induced avoidance weaken following induced sexual arousal in women. *PLoS ONE, 7*(9).

Borgia, G. (1979). Sexual selection and the evolution of mating systems. In M. Blum & A. Blum (Eds.), *Sexual Selection and Reproductive Competition* (pp. 19–80). Cambridge, MA: Academic Press.

———. (2006). Preexisting male traits are important in the evolution of elaborated male sexual display. *Advances in the Study of Behavior, 36*, 249.

Bosch, J., & Márquez, R. (2002). Female preference function related to precedence effect in an amphibian anuran (*Alytes cisternasii*): tests with non-overlapping calls. *Behavioral Ecology, 13*(2), 149–153. doi:10.1093/beheco/13.2.149

Botero, C. A., & Rubenstein, D. R. (2012). Fluctuating environments, sexual selection and the evolution of flexible mate choice in birds. *PLoS One, 7*(2), e32311–e32311.

Boughman, J. W. (2001). Divergent sexual selection enhances reproductive isolation in sticklebacks. *Nature, 411*(6840), 944–948.

———. (2002). How sensory drive can promote speciation. *Trends in Ecology and Evolution, 17*(12), 571–577.

Boul, K. E., Chris Funk, W., Darst, C. R., Cannatella, D. C., & Ryan, M. J. (2007). Sexual selection drives speciation in an Amazonian frog. *Proceedings of the Royal Society of London B: Biological Sciences, 274*(1608), 399–406. doi:10.1098/rspb.2006.3736

Boulcott, P., & Braithwaite, V. A. (2007). Colour perception in three-spined sticklebacks: sexes are not so different after all. *Evolutionary Ecology, 21*(5), 601–611. doi:10.1007/s10682-006-9138-4

Bousquet, F., Nojima, T., Houot, B., Chauvel, I., Chaudy, S., Dupas, S., ... Ferveur, J.-F. (2012). Expression of a desaturase gene, desat1, in neural and nonneural tissues separately affects perception and emission of sex pheromones in *Drosophila*. *Proceedings of the National Academy of Sciences, 109*(1), 249–254.

Boyle, K. S., & Tricas, T. C. (2014). Discrimination of mates and intruders: visual and olfactory cues for a monogamous territorial coral reef butterflyfish. *Animal Behaviour, 92*, 33–43.

Bradbury, J. W., & Vehrencamp, S. L. (2011). *Principles of Animal Communication* (2nd ed.). Sunderland, MA: Sinauer.

Brandt, L.S.E., Ludwar, B. C., & Greenfield, M. D. (2005). Co-occurrence of preference functions and acceptance thresholds in female choice: mate discrimination in the lesser wax moth. *Ethology, 111*(6), 609–625. doi:10.1111/j.1439-0310.2005.01085.x

Brant, L. J., & Fozard, J. L. (1990). Age changes in pure-tone hearing thresholds in a longitudinal study of normal human aging. *Journal of the Acoustical Society of America, 88*(2), 813–820.

Brask, J. B., Croft, D. P., Thompson, K., Dabelsteen, T., & Darden, S. K. (2012). Social preferences based on sexual attractiveness: a female strategy to reduce male sexual attention. *Proceedings of the Royal Society of London B: Biological Sciences, 279*(1734), 1748–1753.

Brembs, B. (2003). Operant conditioning in invertebrates. *Current Opinion in Neurobiology, 13*, 710–717. doi:10.1016/j.conb.2003.10.002.

Brennan, P.L., Clark, C. J., & Prum, R. O. (2010). Explosive eversion and functional morphology of the duck penis supports sexual conflict in waterfowl genitalia. *Proceedings of the Royal Society of London B: Biological Sciences, 277*(1686), 1309–1314. doi:10.1098/rspb.2009.2139.

Brennan, P. L., & Prum, R. O. (2012). The limits of sexual conflict in the narrow sense: new insights from waterfowl biology. *Philosophical Transactions of the Royal Society of London B: Biological Sciences, 367*(1600), 2324–2338.

Brenowitz, E. A. (1991). Altered perception of species-specific song by female birds after lesions of a forebrain nucleus. *Science, 251*, 303–305.

Bressler, E. R., & Balshine, S. (2006). The influence of humor on desirability. *Evolution and Human Behavior, 27*(1), 29–39.

Bretman, A., Newcombe, D., & Tregenza, T. O. M. (2009). Promiscuous females avoid inbreeding by controlling sperm storage. *Molecular Ecology, 18*(16), 3340–3345. doi: 10.1111/j.1365-294X.2009.04301.x

Brewer, G., & Riley, C. (2010). Sexual dimorphism in stature (SDS), jealousy and mate retention. *Evolutionary Psychology, 8*(4), 147470491000800401

Briceño, R. D., & Eberhard, W. G. (2009). Experimental modifications imply a stimulatory function for male tsetse fly genitalia, supporting cryptic female choice theory. *JournalofEvolutionaryBiology,22*(7),1516–1525.doi:10.1111/j.1420-9101.2009.01761.x

Brewer, G., & Riley, C. (2010). Sexual dimorphism in stature (SDS), jealousy and mate retention. Evolutionary Psychology, 8(4), 147470491000800401

Brock, O., Baum, M. J., & Bakker, J. (2011). The development of female sexual behavior requires prepubertal estradiol. *Journal of Neuroscience, 31*(15), 5574–5578.

Brodie, E. D., 3rd, Moore, A. J., & Janzen, F. J. (1995). Visualizing and quantifying natural selection. *Trends in Ecology and Evolution, 10*(8), 313–318.

Brodin, A., & Haas, F. (2009). Hybrid zone maintenance by non-adaptive mate choice. *Evolutionary Ecology, 23*(1), 17–29.

Bro-Jørgensen, J. (2007). Reversed sexual conflict in a promiscuous antelope. *Current Biology, 17*(24), 2157–2161.

———. (2010). Dynamics of multiple signalling systems: animal communication in a world in flux. *Trends in Ecology & Evolution, 25*(5), 292–300.

Brooks, R. (1996). Copying and the repeatability of mate choice. *Behavioral Ecology and Sociobiology, 39*(5), 323–329. doi:10.1007/s002650050296

———. (1998). The importance of mate copying and cultural inheritance of mating preferences. *Trends In Ecology & Evolution, 13*(2), 45–46.

———. (2000). Negative genetic correlation between male attractiveness and survival. *Nature, 406*, 67–70.

———. (2002). Variation in female mate choice within guppy populations: population divergence, multiple ornaments and the maintenance of polymorphism. *Genetica, 116*(2–3), 343–358.

Brooks, R., & Couldridge, V. (1999). Multiple sexual ornaments coevolve with multiple mating preferences. *American Naturalist, 154*(1), 37–45.

Brooks, R., & Endler, J. A. (2001a). Direct and indirect sexual selection and quantitative genetics of male traits in guppies (*Poecilia reticulata*). *Evolution, 55*(5), 1002–1015.

———. (2001b). Female guppies agree to differ: phenotypic and genetic variation in mate-choice behavior and the consequences for sexual selection. *Evolution, 55*(8), 1644–1655. doi:10.1111/j.0014-3820.2001.tb00684.x

Brooks, R., Hunt, J., Blows, M. W., Smith, M. J., Bussière, L. F., & Jennions, M. D. (2005). Experimental evidence for multivariate stabilizing sexual selection. *Evolution, 59*(4), 871–880.

Brooks, R., & Kemp, D. J. (2001). Can older males deliver the good genes? *Trends in Ecology & Evolution, 16*(6), 308–313.

Brouwer, L., van de Pol, M., & Cockburn, A. (2014). Habitat geometry does not affect levels of extrapair paternity in an extremely unfaithful fairy-wren. *Behavioral Ecology*, aru010.

Brown, G. R., Laland, K. N., & Mulder, M. B. (2009). Bateman's principles and human sex roles. *Trends in Ecology & Evolution, 24*(6), 297–304.

Brown, J. L. (1997). A theory of mate choice based on heterozygosity. *Behavioral Ecology, 8*(1), 60–65.

Bruce, H. M. (1960). A block to pregnancy in the mouse caused by proximity of strange males. *Journal of Reproduction and Fertility, 1*(1), 96–103.

Buchanan, K. L. (2000). Stress and the evolution of condition-dependent signals. *Trends in Ecology & Evolution, 15*(4), 156–160.

Buchholz, R. (2004). Effects of parasitic infection on mate sampling by female wild turkeys (*Meleagris gallopavo*): should infected females be more or less choosy? *Behavioral Ecology, 15*(4), 687–694. doi:10.1093/beheco/arh066

Buckingham, J. N., Wong, B. B., & Rosenthal, G. G. (2007). Shoaling decisions in female swordtails: how do fish gauge group size? *Behaviour, 144*(11), 1333–1346.

Bullock, S., & Cliff, D. (1997). The role of 'hidden preferences' in the artificial coevolution of symmetrical signals. *Proceedings of the Royal Society of London B: Biological Sciences 264.1381*(1997), 505–511.

Bulmer, M. (1989). Structural instability of models of sexual selection. *Theoretical Population Biology, 35*(2), 195–206.

Burley, N. (1983). The meaning of assortative mating. *Ethology and Sociobiology, 4*(4), 191–203.

———. (1986). Sexual selection for aesthetic traits in species with biparental care. *American Naturalist, 127*(4), 415–445. doi:10.2307/2461574

———. (1988). The differential-allocation hypothesis: an experimental test. *American Naturalist, 132*(5), 611–628. doi:10.2307/2461924

Burley, N., & Willson, M. F. (1983). *Mate Choice in Plants: Tactics, Mechanisms, and Consequences*. Princeton, NJ: Princeton University Press.

Burley, N. T. (2006). An eye for detail: selective sexual imprinting in zebra finches. *Evolution, 60*(5), 1076–1085.

Burley, N. T., & Calkins, J. D. (1999). Sex ratios and sexual selection in socially monogamous zebra finches. *Behavioral Ecology, 10*(6), 626–635.

Burley, N. T., & Foster, V. S. (2004). Digit ratio varies with sex, egg order and strength of mate preference in zebra finches. *Proceedings of the Royal Society of London B: Biological Sciences, 271*(1536), 239–244. doi:10.1098/rspb.2003.2562

———. (2006). Variation in female choice of mates: condition influences selectivity. *Animal Behaviour, 72*(3), 713–719. doi:http://dx.doi.org/10.1016/j.anbehav.2006.01.017

Burley, N. T., Parker, P. G., & Lundy, K. (1996). Sexual selection and extrapair fertilization in a socially monogamous passerine, the zebra finch (*Taeniopygia gullata*). *Behavioral Ecology, 7*(2), 218–226.

Burley, N. T., & Szymanski, R. (1998). "A taste for the beautiful": latent aesthetic mate preferences for white crests in two species of Australian grassfinches. *American Naturalist, 152*(6), 792–802.

Bush, S. L., Gerhardt, H. C., & Schul, J. (2002). Pattern recognition and call preferences in treefrogs (Anura: Hylidae): a quantitative analysis using a no-choice paradigm. *Animal Behaviour, 63*, 7–14. doi:10.1006/anbe.2001.1880.

Bushdid, C., Magnasco, M. O., Vosshall, L. B., & Keller, A. (2014). Humans can discriminate more than 1 trillion olfactory stimuli. *Science, 343*(6177), 1370–1372.

Bussière, L. F., Demont, M., Pemberton, A. J., Hall, M. D., & Ward, P. I. (2010). The assessment of insemination success in yellow dung flies using competitive PCR. *Molecular Ecology Resources, 10*(2), 292–303. doi:10.1111/j.1755-0998.2009.02754.x

Bussiere, L. F., Hunt, J., Stölting, K. N., Jennions, M. D., & Brooks, R. (2008). Mate choice for genetic quality when environments vary: suggestions for empirical progress. *Genetica, 134*(1), 69–78.

Buston, P. M., & Emlen, S. T. (2003). Cognitive processes underlying human mate choice: the relationship between self-perception and mate preference in Western society. *Proceedings of the National Academy of Sciences, 100*(15), 8805–8810.

Butkowski, T., Yan, W., Gray, A. M., Cui, R., Verzijden, M. N., & Rosenthal, G. G. (2011). Automated interactive video playback for studies of animal communication. *Journal of Visualized Experiments*(48), e2374. doi:10.3791/2374

Butlin, R., Debelle, A., Kerth, C., Snook, R. R., Beukeboom, L. W., Castillo, C. R., ... Schilthuizen, M. (2012). What do we need to know about speciation? *Trends in Ecology & Evolution, 27*(1), 27–39.

Butlin, R., & Ritchie, M. (1991). Variation in female mate preference across a grasshopper hybrid zone. *Journal of Evolutionary Biology, 4*(2), 227–240.

Butlin, R. K. (1993). The variability of mating signals and preferences in the brown planthopper, *Nilaparvata lugens* (Homoptera: Dephacidae). *Journal of Insect Behavior, 6*(2), 125–140.

Buunk, A. P., Park, J. H., & Duncan, L. A. (2010). Cultural variation in parental influence on mate choice. *Cross-Cultural Research, 44*(1), 23–40.

Byers, B. E., & Kroodsma, D. E. (2009). Female mate choice and songbird song repertoires. *Animal Behaviour, 77*(1), 13–22. doi:http://dx.doi.org/10.1016/j.anbehav.2008.10.003

Byers, J., & Dunn, S. (2012). Bateman in nature: predation on offspring reduces the potential for sexual selection. *Science, 338*(6108), 802–804.

Byers, J., Hebets, E., & Podos, J. (2010). Female mate choice based upon male motor performance. *Animal Behaviour, 79*(4), 771–778.

Byers, J. A., & Waits, L. (2006). Good genes sexual selection in nature. *Proceedings of the National Academy of Sciences, 103*(44), 16343–16345.

Byers, J. A., Wiseman, P. A., Jones, L., & Roffe, T. J. (2005). A large cost of female mate sampling in pronghorn. *American Naturalist, 166*(6), 661–668. doi:10.1086/497401

Byrne, P. G., & Rice, W. R. (2006). Evidence for adaptive male mate choice in the fruit fly *Drosophila melanogaster. Proceedings of the Royal Society of London B: Biological Sciences, 273*(1589), 917–922. doi:10.1098/rspb.2005.3372

Calabrese, G. M., Brady, P. C., Gruev, V., & Cummings, M. E. (2014). Polarization signaling in swordtails alters female mate preference. *Proceedings of the National Academy of Sciences, 111*(37), 13397–13402.

Calkins, J. D., & Burley, N. T. (2003). Mate choice for multiple ornaments in the California quail, *Callipepla californica. Animal Behaviour, 65*(1), 69–81. doi:http://dx.doi.org/10.1006/anbe.2002.2041

Callander, S., Backwell, P. R., & Jennions, M. D. (2011). Context-dependent male mate choice: the effects of competitor presence and competitor size. *Behavioral Ecology*, arr192.

Callander, S., Hayes, C. L., Jennions, M. D., & Backwell, P.R.Y. (2013). Experimental evidence that immediate neighbors affect male attractiveness. *Behavioral Ecology, 24*(3), 730–733. doi:10.1093/beheco/ars208

Calsbeek, R., & Bonneaud, C. (2008). Postcopulatory fertilization bias as a form of cryptic sexual selection. *Evolution, 62*(5), 1137–1148. doi:10.1111/j.1558-5646.2008.00356.x

Calsbeek, R., & Sinervo, B. (2002). Uncoupling direct and indirect components of female choice in the wild. *Proceedings of the National Academy of Sciences, 99*(23), 14897–14902.

Cameron, E., Day, T., & Rowe, L. (2003). Sexual conflict and indirect benefits. *Journal of Evolutionary Biology, 16*(5), 1055–1060.

Campbell, D. M., & Hauber, M. (2009). Cross-fostering diminishes song discrimination in zebra finches (*Taeniopygia guttata*). *Animal Cognition, 12*(3), 481–490. doi:10.1007/s10071-008-0209-5

Canal, D., Jovani, R., & Potti, J. (2012). Male decisions or female accessibility? Spatiotemporal patterns of extra pair paternity in a songbird. *Behavioral Ecology 23(5), 1146–1153.*

Candolin, U. (2003). The use of multiple cues in mate choice. *Biological Reviews, 78*, 575–595.

———. (2004). Why do multiple traits determine mating success? Differential use in female choice and male competition in a water boatman. *Proceedings of the Royal Societyof London B: Biological Sciences, 272*, 47–52.

Candolin, U., & Heuschele, J. (2008). Is sexual selection beneficial during adaptation to environmental change? *Trends in Ecology & Evolution, 23*(8), 446–452.

Carleton, K. L., & Kocher, T. D. (2001). Cone opsin genes of African cichlid fishes: tuning spectral sensitivity by differential gene expression. *Molecular Biology and Evolution, 18*(8), 1540–1550.

Carlson, B. A., & Arnegard, M. E. (2011). Neural innovations and the diversification of African weakly electric fishes. *Communicative & Integrative Biology, 4*(6), 720–725.

Carré, D., Rouvière, C., & Sardet, C. (1991). In vitro fertilization in ctenophores: sperm entry, mitosis, and the establishment of bilateral symmetry in Beroe ovata. *Developmental Biology, 147*(2), 381–391.

Carter, A. J., Feeney, W. E., Marshall, H. H., Cowlishaw, G., & Heinsohn, R. (2013). Animal personality: what are behavioural ecologists measuring? *Biological Reviews, 88*(2), 465–475. doi:10.1111/brv.12007

Castellano, S. (2009a). Towards an information-processing theory of mate choice. *Animal Behaviour, 78*(6), 1493–1497. doi:http://dx.doi.org/10.1016/j.anbehav.2009.10.002

———. (2009b). Unreliable preferences, reliable choice and sexual selection in leks. *Animal Behaviour, 77*(1), 225–232.

———. (2015). Bayes' rule and bias roles in the evolution of decision making. *Behavioral Ecology, 26*(1), 282–292. doi:10.1093/beheco/aru188

Castellano, S., Cadeddu, G., & Cermelli, P. (2012). Computational mate choice: theory and empirical evidence. *Behavioural Processes, 90*, 261–277. doi:10.1016/j.beproc.2012.02.010

Castellano, S., & Cermelli, P. (2006). Reconciling sexual selection to species recognition: a process-based model of mating decision. *Journal of Theoretical Biology, 242*(3), 529–538.

———. (2010). Attractive amplifiers in sexual selection: where efficacy meets honesty. *Evolutionary Ecology, 24*(5), 1187–1197.

———. (2011). Sampling and assessment accuracy in mate choice: a random-walk model of information processing in mating decision. *Journal of Theoretical Biology, 274*(1), 161–169. doi:http://dx.doi.org/10.1016/j.jtbi.2011.01.001

Castellano, S., Zanollo, V., Marconi, V., & Berto, G. (2009). The mechanisms of sexual selection in a lek-breeding anuran, *Hyla intermedia. Animal Behaviour, 77*(1), 213–224.

Catchpole, C. K. (1987). Bird song, sexual selection and female choice. *Trends in Ecology & Evolution, 2*(4), 94–97.

———. (1996). Song and female choice: good genes and big brains? *Trends in Ecology & Evolution, 11*(9), 358–360.

Çetinkaya, H., & Domjan, M. (2006). Sexual fetishism in a quail (Coturnix japonica) model system: Test of reproductive success. *Journal of Comparative Psychology, 120*(4), 427.

Chaffee, D. W., Griffin, H., & Gilman, R. T. (2013). Sexual imprinting: what strategies should we expect to see in nature? *Evolution, 67*(12), 3588–3599.

Chaine, A. S., & Lyon, B. E. (2008). Adaptive plasticity in female mate choice dampens sexual selection on male ornaments in the lark bunting. *Science, 319*(5862), 459–462. doi:10.1126/science.1149167

Chapman, T., Arnqvist, G., Bangham, J., & Rowe, L. (2003). Sexual conflict. *Trends in Ecology & Evolution, 18*(1), 41–47.

Chappell, M. A., Zuk, M., Johnsen, T. S., & Kwan, T. H. (1997). Mate choice and aerobic capacity in red junglefowl. *Behaviour, 134*(7), 511–529.

Charlwood, J. D., & Jones, M. D. R. (1979). Mating behaviour in the mosquito, *Anopheles gambiae* s.1.save. *Physiological Entomology, 4*(2), 111–120. doi:10.1111/j.1365-3032.1979.tb00185.x

Charmantier, A., & Sheldon, B. C. (2006). Testing genetic models of mate choice evolution in the wild. *Trends in Ecology & Evolution, 21*(8), 417–419.
Charnov, E. L. (1979). Simultaneous hermaphroditism and sexual selection. *Proceedings of the National Academy of Sciences, 76*(5), 2480–2484.
Cheetham, S. A., Thom, M. D., Jury, F., Ollier, W. E., Beynon, R. J., & Hurst, J. L. (2007). The genetic basis of individual-recognition signals in the mouse. *Current Biology, 17*(20), 1771–1777.
Chenoweth, S., Petfield, D., Doughty, P., & Blows, M. (2007). Male choice generates stabilizing sexual selection on a female fecundity correlate. *Journal of Evolutionary Biology, 20*(5), 1745–1750.
Chenoweth, S. F., & Blows, M. W. (2006). Dissecting the complex genetic basis of mate choice. *Nature Reviews Genetics, 7*(9), 681–692.
Chenoweth, S. F., & Gosden, T. P. (2015). Variation and selection on preference functions: a comment on Edward. *Behavioral Ecology*. doi:10.1093/beheco/aru233
Chenoweth, S. F., & McGuigan, K. (2010). The genetic basis of sexually selected variation. *Annual Review of Ecology, Evolution, and Systematics, 41*, 81–101.
Chiandetti, C., & Vallortigara, G. (2011). Chicks like consonant music. *Psychological Science*, 22 (10), 1270–1273.doi:10.1177/0956797611418244
Chiou, T.-H., Kleinlogel, S., Cronin, T., Caldwell, R., Loeffler, B., Siddiqi, A., … Marshall, J. (2008). Circular polarization vision in a stomatopod crustacean. *Current Biology, 18*(6), 429–434.
Chittka, L., Skorupski, P., & Raine, N. E. (2009). Speed-accuracy tradeoffs in animal decision making. *Trends in Ecology & Evolution, 24*(7), 400–407. doi:http://dx.doi.org/10.1016/j.tree.2009.02.010
Chivers, M. L., Seto, M. C., Lalumière, M. L., Laan, E., & Grimbos, T. (2010). Agreement of self-reported and genital measures of sexual arousal in men and women: a meta-analysis. *Archives of Sexual Behavior, 39*(1), 5–56. doi:10.1007/s10508-009-9556-9
Choleris, E., Little, S. R., Mong, J. A., Puram, S. V., Langer, R., & Pfaff, D. W. (2007). Microparticle-based delivery of oxytocin receptor antisense DNA in the medial amygdala blocks social recognition in female mice. *Proceedings of the National Academy of Science U S A, 104*(11), 4670–4675. doi:10.1073/pnas.0700670104
Choudhury, S. (1995). Divorce in birds: a review of the hypotheses. *Animal Behaviour, 50*(2), 413–429.
Chouinard-Thuly, L., Gierszewski, S., Rosenthal, G. G., Reader, S. M., Rieucau, G., Woo, K. L., … Stowers, J. R. (2016). Technical and conceptual considerations for using animated stimuli in studies of animal behavior. *Current Zoology*, zow104.
Chow, C. Y., Wolfner, M. F., & Clark, A. G. (2010). The genetic basis for male × female interactions underlying variation in reproductive phenotypes of *Drosophila*. *Genetics, 186*(4), 1355–1365. doi:10.1534/genetics.110.123174
Christensen, T. A., & White, J. (2000). Representation of olfactory information in the brain. In T. E. Finger, W. L. Silver, & D. Restrepo (Eds.), *The Neurobiology of Taste and Smell* (2nd ed., pp. 201–232). New York: Wiley-Liss.
Christy, J. H. (1995). Mimicry, mate choice, and the sensory trap hypothesis. *American Naturalist, 146*(2), 171–181.
Christy, J. H., Backwell, P. R., & Schober, U. (2003). Interspecific attractiveness of structures built by courting male fiddler crabs: experimental evidence of a sensory trap. *Behavioral Ecology and Sociobiology, 53*(2), 84–91.

Chu, Y., Yang, E., Schinaman, J., Chahda, J., & Sousa-Neves, R. (2013). Genetic analysis of mate discrimination in *Drosophila simulans*. *Evolution, 67*(8), 2335–2347.

Chung, H., Loehlin, D. W., Dufour, H. D., Vaccarro, K., Millar, J. G., & Carroll, S. B. (2014). A single gene affects both ecological divergence and mate choice in *Drosophila*. *Science, 343*(6175), 1148–1151.

Civetta, A. (2003). Shall we dance or shall we fight? Using DNA sequence data to untangle controversies surrounding sexual selection. *Genome, 46*(6), 925–929.

Clancey, E., & Byers, J. A. (2014). The definition and measurement of individual condition in evolutionary studies. *Ethology, 120*(9), 845–854.

Clark, A. G., & Begun, D. J. (1998). Female genotypes affect sperm displacement in *Drosophila*. *Genetics, 149*(3), 1487–1493.

Clark, A. G., Begun, D. J., & Prout, T. (1999). Female × male interactions in *Drosophila* sperm competition. *Science, 283*(5399), 217–220. doi:10.1126/science.283.5399.217

Clark, C. J., & Feo, T. J. (2010). Why do Calypte hummingbirds "sing" with both their tail and their syrinx? An apparent example of sexual sensory bias. *American Naturalist, 175*(1), 27–37.

Clark, D. L., Macedonia, J. M., & Rosenthal, G. G. (1997). Testing video playback to lizards in the field. *Copeia, 1997*, 421–423.

Clark, D. L., & Uetz, G. W. (1992). Morph-independent mate selection in a dimorphic jumping spider: demonstration of movement bias in female choice using video-controlled courtship behavior. *Animal Behaviour, 43*, 247–254.

Clark, E., Aronson, L. R., & Gordon, M. (1954). Mating behavior patterns in two sympatric species of xiphophorin fishes: their inheritance and significance in sexual isolation. *Bulletin of the American Museum of Natural History, 103*(2), 135–226.

Clarke, G. M. (1998). Developmental stability and fitness: the evidence is not quite so clear. *American Naturalist, 152*(5), 762–766. doi:10.1086/286207

Clayton, N. S. (1990). Subspecies recognition and song learning in zebra finches. *Animal Behaviour, 40*(6), 1009–1017. doi:10.1016/S0003-3472(05)80169-1

———. (1994). The influence of social interactions on the development of song and sexual preferences in birds. In J. A. Hogan & J. J. Bolhuis (Eds.), *Causal Mechanisms of Behavioural Development* (pp. 98–115). Cambridge, UK: Cambridge University Press.

Clemens, J., Krämer, S., & Ronacher, B. (2014). Asymmetrical integration of sensory information during mating decisions in grasshoppers. *Proceedings of the National Academy of Sciences, 111*(46), 16562–16567. doi:10.1073/pnas.1412741111

Clutton-Brock, T. (1989). Review lecture: mammalian mating systems. *Proceedings of the Royal Society of London B: Biological Sciences, 236*(1285), 339–372.

———. (2007). Sexual selection in males and females. *Science, 318*(5858), 1882–1885.

———. (2009). Sexual selection in females. *Animal Behaviour, 77*(1), 3–11.

Cohen, J. A. (1984). Sexual selection and the psychophysics of female choice. *Zeitschrift für Tierpsychologie, 64*(1), 1–8. doi:10.1111/j.1439-0310.1984.tb00348.x

Cole, G. L., & Endler, J. A. (2015). Variable environmental effects on a multicomponent sexually selected trait. *American Naturalist, 185*(4), 452–468. doi:10.1086/680022

Colegrave, N., Kotiaho, J. S., & Tomkins, J. L. (2002). Mate choice or polyandry: reconciling genetic compatibility and good genes sexual selection. *Evolutionary Ecology Research, 4*(6), 911–917.

Coleman, S. W., Patricelli, G. L., & Borgia, G. (2004). Variable female preferences drive complex male displays. *Nature, 428*, 742–745.

Coleman, S. W., Siani, J., Borgia, G., Patricelli, G. L., & Coyle, B. (2007). Female preferences drive the evolution of mimetic accuracy in male sexual displays. *Biology Letters, 3*(5), 463–466. doi:10.1098/rsbl.2007.0234

Collet, J., Richardson, D. S., Worley, K., & Pizzari, T. (2012). Sexual selection and the differential effect of polyandry. *Proceedings of the National Academy of Sciences, 109*(22), 8641–8645.

Collins, S. (1993). Is there only one type of male handicap? *Proceedings of the Royal Society of London B: Biological Sciences, 252*(1335), 193–197.

Collins, S. A. (1995). The effect of recent experience on female choice in zebra finches. *Animal Behaviour, 49*(2), 479–486. doi:http://dx.doi.org/10.1006/anbe.1995.0062

———. (1999). Is female preference for male repertoires due to sensory bias? *Proceedings of the Royal Society of London B: Biological Sciences, 266*(1435), 2309–2314.

Collins, S. A., & Luddem, S. (2002). Degree of male ornamentation affects female preference for conspecific versus heterospecific males. *Proceedings of the Royal Society of London B: Biological Sciences, 269*(1487), 111–117.

Collins, S. A., & Missing, C. (2003). Vocal and visual attractiveness are related in women. *Animal Behaviour, 65*(5), 997–1004. doi:http://dx.doi.org/10.1006/anbe.2003.2123.

Comeault, A. A., & Matute, D. R. (2016). Reinforcement's incidental effects on reproductive isolation between conspecifics. *Current Zoology, 62*(2), 135–143.

Comeault, A. A., Venkat, A., & Matute, D. R. (2016). *Correlated evolution of male and female reproductive traits drive a cascading effect of reinforcement in Drosophila yakuba.* Paper presented at the Proceedings of the Royal Society B.

Connallon, T., Cox, R. M., & Calsbeek, R. (2010). Fitness consequences of sex-specific selection. *Evolution, 64*(6), 1671–1682.

Conner, W. E. (1999). 'Un chant d'appel amoureux': acoustic communication in moths. *Journal of Experimental Biology, 202*(13), 1711–1723.

Conover, M. R., & Hunt Jr., G. L. (1984). Female-female pairing and sex ratios in gulls: an historical perspective. *Wilson Bulletin*, 619–625.

Conte, G. L., & Schluter, D. (2013). Experimental confirmation that body size determines mate preference via phenotype matching in a stickleback species pair. *Evolution, 67*(5), 1477–1484.

Córdoba-Aguilar, A. (1999). Male copulatory sensory stimulation induces female ejection of rival sperm in a damselfly. *Proceedings of the Royal Society of London B: Biological Sciences, 266*(1421), 779–784. doi:10.1098/rspb.1999.0705

———. (2002). Sensory trap as the mechanism of sexual selection in a damselfly genitalic trait (Insecta: Calopterygidae). *American Naturalist, 160*(5), 594–601.

Córdoba-Aguilar, A., Salamanca-Ocaña, J. C., & Lopezaraiza, M. (2003). Female reproductive decisions and parasite burden in a calopterygid damselfly (Insecta: Odonata). *Animal Behaviour, 66*(1), 81–87. doi:http://dx.doi.org/10.1006/anbe.2003.2198

Coria-Ávila, G. A., Ouimet, A. J., Pacheco, P., Manzo, J., & Pfaus, J. G. (2005). Olfactory conditioned partner preference in the female rat. *Behavioral Neuroscience, 119*(3), 716–725. doi:10.1037/0735-7044.119.3.716

Cornwallis, C. K., & Uller, T. (2009). Towards an evolutionary ecology of sexual traits. *Trends in Ecology & Evolution, 25*(3), 145–152.

Corr, P. J., Pickering, A. D., & Gray, J. A. (1995). Personality and reinforcement in associative and instrumental learning. *Personality and Individual Differences, 19*(1), 47–71.

Costa, R. M., Miller, G. F., & Brody, S. (2012). Women who prefer longer penises are more likely to have vaginal orgasms (but not clitoral orgasms): implications for an evolutionary theory of vaginal orgasm. *Journal of Sexual Medicine, 9*(12), 3079–3088. doi:10.1111/j.1743-6109.2012.02917.x

Cotton, S., & Pomiankowski, A. (2007). Sexual selection: does condition dependence fail to resolve the 'lek paradox'? *Current Biology, 17*(9), R335–R337.

Cotton, S., Rogers, D. W., Small, J., Pomiankowski, A., & Fowler, K. (2006b). Variation in preference for a male ornament is positively associated with female eyespan in the stalk-eyed fly *Diasemopsis meigenii. Proceedings of the Royal Society of London B: Biological Sciences, 273*(1591), 1287–1292. doi:10.1098/rspb.2005.3449

Cotton, S., Small, J., & Pomiankowski, A. (2006a). Sexual selection and condition-dependent mate preferences. *Current Biology, 16*(17), R755–R765. doi:http://dx.doi.org/10.1016/j.cub.2006.08.022

Courtiol, A., Pettay, J. E., Jokela, M., Rotkirch, A., & Lummaa, V. (2012). Natural and sexual selection in a monogamous historical human population. *Proceedings of the National Academy of Sciences, 109*(21), 8044–8049.

Courtiol, A., Picq, S., Godelle, B., Raymond, M., & Ferdy, J.-B. (2010). From preferred to actual mate characteristics: the case of human body shape. *PLoS ONE, 5*(9), e13010.

Courtiol, A., Raymond, M., Godelle, B., & Ferdy, J. B. (2010). Mate choice and human stature: homogamy as a unified framework for understanding mating preferences. *Evolution, 64*(8), 2189–2203.

Cox, R. M., & Calsbeek, R. (2010). Cryptic sex-ratio bias provides indirect genetic benefits despite sexual conflict. *Science, 328*(5974), 92–94.

Coxworth, J. E., Kim, P. S., McQueen, J. S., & Hawkes, K. (2015). Grandmothering life histories and human pair bonding. *Proceedings of the National Academy of Sciences, 112*(38), 11806–11811.

Coyle, B. J., Carleton, K. L., Borgia, G., & Hart, N. S. (2012). Limited variation in visual sensitivity among bowerbird species suggests that there is no link between spectral tuning and variation in display colouration. *Journal of Experimental Biology, 215*(7), 1090–1105. doi:10.1242/jeb.062224

Coyne, J. A., & Orr, H. A. (1989). Patterns of speciation in Drosophila. *Evolution, 43*(2), 362–381.

Cramer, E. A., Laskemoen, T., Eroukhmanoff, F., Haas, F., Hermansen, J., Lifjeld, J., ... Johnsen, A. (2014). Testing a post-copulatory pre-zygotic reproductive barrier in a passerine species pair. *Behavioral Ecology and Sociobiology, 68*(7), 1133–1144. doi:10.1007/s00265-014-1724-9

Crapon de Caprona, M.-D., & Ryan, M. J. (1990). Conspecific mate recognition in swordtails, *Xiphophorus nigrensis* and *X. pygmaeus* (Poeciliidae): olfactory and visual cues. *Animal Behaviour, 39*, 290–296.

Crean, A. J., Kopps, A. M., & Bonduriansky, R. (2014). Revisiting telegony: offspring inherit an acquired characteristic of their mother's previous mate. *Ecology Letters, 17*(12), 1545–1552.

Crespi, B. J. (1989). Causes of assortative mating in arthropods. *Animal Behaviour, 38*(6), 980–1000.

Crews, D., Gore, A. C., Hsu, T. S., Dangleben, N. L., Spinetta, M., Schallert, T., ... Skinner, M. K. (2007). Transgenerational epigenetic imprints on mate preference.

Proceedings of the National Academy of Science U S A, 104(14), 5942. doi:10.1073/pnas.0610410104

Crews, D., Grassman, M., & Lindzey, J. (1986). Behavioral facilitation of reproduction in sexual and unisexual whiptail lizards. *Proceedings of the National Academy of Sciences, 83*(24), 9547–9550.

Cronin, H. (1991). *The Ant and the Peacock.* New York: Cambridge University Press.

Cui, R., Schumer, M., Kruesi, K., Walter, R., Andolfatto, P., & Rosenthal, G. G. (2013). Phylogenomics reveals extensive reticulate evolution in *Xiphophorus* fishes. *Evolution, 67*(8), 2166–2179. doi:10.1111/evo.12099

Cui, R., Schumer, M., & Rosenthal, G. G. (2016). Admix'em: a flexible framework for forward-time simulations of hybrid populations with selection and mate choice. *Bioinformatics, 32*(7), 1103–1105.

Cui, R., Delclós, P. J., Schumer, M., & Rosenthal, G. G. (in press). Early social learning triggers neurogenomic expression changes in a swordtail fish. *Proceedings of the Royal Society of London B: Biological Sciences.*

Culumber, Z. W., Bautista-Hernández, C. E., Monks, S., Arias-Rodriguez, L., & Tobler, M. (2014). Variation in melanism and female preference in proximate but ecologically distinct environments. *Ethology, 120*(11): 1090–1100. doi:10.1111/eth.12282

Culumber, Z. W., Fisher, H. S., Tobler, M., Mateos, M., Barber, P. H., Sorenson, M. D., & Rosenthal, G. G. (2011). Replicated hybrid zones of Xiphophorus swordtails along an elevational gradient. *Molecular Ecology, 20*(2), 342–356. doi:10.1111/j.1365-294X.2010.04949.x

Culumber, Z. W., Ochoa, O. M., & Rosenthal, G. G. (2014). Assortative mating and the maintenance of population structure in a natural hybrid zone. *American Naturalist, 184*(2), 225–232. doi:10.1086/677033

Culumber, Z. W., & Rosenthal, G. G. (2013). Mating preferences do not maintain the tailspot polymorphism in the platyfish, Xiphophorus variatus. *Behavioral Ecology, 24*(6), 1286–1291. doi:10.1093/beheco/art063

Culumber, Z. W., Shepard, D. B., Coleman, S. W., Rosenthal, G. G., & Tobler, M. (2012). Physiological adaptation along environmental gradients and replicated hybrid zone structure in swordtails (Teleostei: Xiphophorus). *Journal of Evolutionary Biology, 25*(9), 1800–1814. doi:10.1111/j.1420-9101.2012.02562.x

Cummings, M., & Mollaghan, D. (2006). Repeatability and consistency of female preference behaviors in a northern swordtail, *Xiphophorus nigrensis. Animal Behaviour, 72,* 217–224.

Cummings, M. E. (2012). Looking for sexual selection in the female brain. *Philosophical Transactions of the Royal Society of London B: Biological Sciences, 367*(1600), 2348–2356. doi:10.1098/rstb.2012.0105.

Cummings, M. E., De León, F. J. G., Mollaghan, D. M., & Ryan, M. J. (2006). Is UV ornamentation an amplifier in swordtails? *Zebrafish, 3*(1), 91–100.

Cummings, M. E., Larkins-Ford, J., Reilly, C.R.L., Wong, R. Y., Ramsey, M., & Hofmann, H. A. (2008). Sexual and social stimuli elicit rapid and contrasting genomic responses. *Proceedings of the Royal Society of London B: Biological Sciences, 275*(1633), 393–402. doi:10.1098/rspb.2007.1454

Cummings, M. E., & Partridge, J. C. (2001). Visual pigments and optical habitats of surfperch (Embiotocidae) in the California kelp forest. *Journal of Comparative Physiology A, 187,* 875–889.

Cummings, M. E., & Ramsey, M. E. (2015). Mate choice as social cognition: predicting female behavioral and neural plasticity as a function of alternative male reproductive tactics. *Current Opinion in Behavioral Sciences, 6*, 125–131.

Cummings, M. E., Rosenthal, G. G., & Ryan, M. J. (2003). A private ultraviolet channel in visual communication. *Proceedings of the Royal Society of London B: Biological Sciences, 270*(1518), 897–904. doi:10.1098/rspb.2003.2334

Cunningham, E.J.A. (2003). Female mate preferences and subsequent resistance to copulation in the mallard. *Behavioral Ecology, 14*(3), 326–333. doi:10.1093/beheco/14.3.326

Cunningham, E.J.A., & Russell, A. F. (2000). Egg investment is influenced by male attractiveness in the mallard. *Nature, 404*(6773), 74–77.

Curtis, J. T. (2010). Female prairie vole mate-choice is affected by the males' birth litter composition. *Physiology & Behavior, 101*(1), 93–100.

Dakin, R., & Montgomerie, R. (2014). Deceptive copulation calls attract female visitors to peacock leks. *American Naturalist, 183*(4), 558–564.

Dall, S. R., Bell, A. M., Bolnick, D. I., & Ratnieks, F. L. (2012). An evolutionary ecology of individual differences. *Ecology Letters, 15*(10), 1189–1198.

Dalziell, A. H., Peters, R. A., Cockburn, A., Dorland, A. D., Maisey, A. C., & Magrath, R. D. (2013). Dance choreography is coordinated with song repertoire in a complex avian display. *Current Biology, 23*(12), 1132–1135. doi:10.1016/j.cub.2013.05.018

Dambroski, H. R., Linn Jr., C., Berlocher, S. H., Forbes, A. A., Roelofs, W., Feder, J. L., & Pellmyr, O. (2005). The genetic basis for fruit odor discrimination in Rhagoletis flies and its significance for sympatric host shifts. *Evolution, 59*(9), 1953–1964.

Danchin, É., Charmantier, A., Champagne, F. A., Mesoudi, A., Pujol, B., & Blanchet, S. (2011). Beyond DNA: integrating inclusive inheritance into an extended theory of evolution. *Nature Reviews Genetics, 12*(7), 475–486.

Danchin, É., Giraldeau, L.-A., Valone, T. J., & Wagner, R. H. (2004). Public information: from nosy neighbors to cultural evolution. *Science, 305*(5683), 487–491.

Dann, S. G., Allison, W. T., Levin, D. B., Taylor, J. S., & Hawryshyn, C. W. (2004). Salmonid opsin sequences undergo positive selection and indicate an alternate evolutionary relationship in *Oncorhynchus*. *Journal of Molecular Evolution, 58*, 400–412.

Danner, J. E., Danner, R. M., Bonier, F., Martin, P. R., Small, T. W., & Moore, I. T. (2011). Female, but not male, tropical sparrows respond more strongly to the local song dialect: implications for population divergence. *American Naturalist, 178*(1), 53–63. doi:10.1086/660283

Darden, S. K., & Watts, L. (2012). Male sexual harassment alters female social behaviour towards other females. *Biology Letters, 8*(2), 186–188. doi:10.1098/rsbl.2011.0807

Darwin, C. (1859). *On the Origins of Species by Means of Natural Selection*. London: Murray, 247.

———. (1871). *The Descent of Man, and Selection in Relation to Sex*. London: Murray.

David, C. T., Kennedy, J. S., & Ludlow, A. R. (1983). Finding of a sex pheromone source by gypsy moths released in the field. *Nature, 303*(5920), 804–806.

David, M., Auclair, Y., Dall, S. R., & Cézilly, F. (2013). Pairing context determines condition-dependence of song rate in a monogamous passerine bird. *Proceedings of the Royal Society of London B: Biological Sciences, 280*(1753), 2012–2177.

David, M., & Cézilly, F. (2011). Personality may confound common measures of mate-choice. *PLoS ONE, 6*(9), e24778. doi:10.1371/journal.pone.0024778

Davidoff, J., & Fagot, J. (2010). Cross-species assessment of the linguistic origins of color categories. *Comparative Cognition & Behavior Reviews 5*, 100–116.

Davies, N. B. (1983). Polyandry, cloaca-pecking and sperm competition in dunnocks. *Nature, 302*(5906), 334–336.

Davis, A. G., & Leary, C. J. (2015). Elevated stress hormone diminishes the strength of female preferences for acoustic signals in the green treefrog. *Hormones and Behavior, 69*, 119–122.

Dawkins, M. S., & Guilford, T. (1995). An exaggerated preference for simple neural network models of signal evolution*Proceedings of the Royal Society of London B: Biological Sciences, 261*(1362), 357–360. doi:10.1098/rspb.1995.0159

———. (1996). Sensory bias and the adaptiveness of female choice. *American Naturalist, 148*(5), 937–942.

Dawkins, R. (1983). *The Extended Phenotype: The Gene as the Unit of Selection*. Oxford: Oxford University Press.

Day, T. (2000). Sexual selection and the evolution of costly female preferences: spatial effects. *Evolution, 54*(3), 715–730.

Dean, R., & Mank, J. E. (2014). The role of sex chromosomes in sexual dimorphism: discordance between molecular and phenotypic data. *Journal of Evolutionary Biology, 27*(7), 1443–1453. doi:10.1111/jeb.12345

Dean, R., Nakagawa, S., & Pizzari, T. (2011). The risk and intensity of sperm ejection in female birds. *American Naturalist, 178*(3), 343–354. doi:10.1086/661244

Dean, R., Zimmer, F., & Mank, J. E. (2014). The potential role of sexual conflict and sexual selection in shaping the genomic distribution of mito-nuclear genes. *Genome Biology and Evolution, 6*(5), 1096–1104.

D'eath, R. B., & Dawkins, M. S. (1996). Laying hens do not discriminate between video images of conspecifics. *Animal Behaviour, 52*, 903–912.

Deb, R., & Balakrishnan, R. (2014). The opportunity for sampling: the ecological context of female mate choice. *Behavioral Ecology, 25*(4), 967–974. doi:10.1093/beheco/aru072.

DeBruine, L. M., Jones, B. C., Crawford, J. R., Welling, L.L.M., & Little, A. C. (2010). The health of a nation predicts their mate preferences: cross-cultural variation in women's preferences for masculinized male faces. *Proceedings of the Royal Society of London B: Biological Sciences, 277*(1692), 2405–2410. doi:10.1098/rspb.2009.2184

DeBruine, L. M., Little, A. C., & Jones, B. C. (2012). Extending parasite-stress theory to variation in human mate preferences. *Behavioral and Brain Sciences, 35*(02), 86–87.

Dechaume-Moncharmont, F.-X., Freychet, M., Motreuil, S., & Cézilly, F. (2013). Female mate choice in convict cichlids is transitive and consistent with a self-referent directional preference. *Frontiers in Zoology, 10*(1), 1–23. doi:10.1186/1742-9994-10-69

Deere, K. A., Grether, G. F., Sun, A., & Sinsheimer, J. S. (2012). Female mate preference explains countergradient variation in the sexual coloration of guppies (*Poecilia reticulata*). *Proceedings of the Royal Society of London B: Biological Sciences, 279*(1734), 1684–1690doi:10.1098/rspb.2011.2132

Dehaene, S. (2003). The neural basis of the Weber-Fechner law: a logarithmic mental number line. *Trends in Cognitive Sciences, 7*, 145–147. doi:10.1016/S1364-6613(03)00055-X

Delcourt, M., Blows, M. W., Aguirre, J. D., & Rundle, H. D. (2012). Evolutionary optimum for male sexual traits characterized using the multivariate Robertson-Price

Identity. *Proceedings of the National Academy of Sciences, 109*(26), 10414–10419. doi:10.1073/pnas.1116828109

Delcourt, M., Blows, M. W., & Rundle, H. D. (2010). Quantitative genetics of female mate preferences in an ancestral and a novel environment. *Evolution, 64*(9), 2758–2766.

Delhey, K., Hall, M., Kingma, S. A., & Peters, A. (2013). Increased conspicuousness can explain the match between visual sensitivities and blue plumage colours in fairy-wrens. *Proceedings of the Royal Society of London B: Biological Sciences, 280*(1750), 20121771.

Del Negro, C., Gahr, M., Leboucher, G., & Kreutzer, M. (1998). The selectivity of sexual responses to song displays: effects of partial chemical lesion of the HVC in female canaries. *Behavioural Brain Research, 96*(1–2), 151–159. doi:http://dx.doi.org/10.1016/S0166-4328(98)00009-6

deRivera, C. E., Backwell, P. R., Christy, J. H., & Vehrencamp, S. L. (2003). Density affects female and male mate searching in the fiddler crab, *Uca beebei*. *Behavioral Ecology and Sociobiology, 53*(2), 72–83. doi:10.1007/s00265-002-0555-2

Derryberry, E. P. (2007). Evolution of bird song affects signal efficacy: an experimental test using historical and current signals. *Evolution, 61*(8), 1938–1945.

———. (2009). Ecology shapes birdsong evolution: variation in morphology and habitat explains variation in white-crowned sparrow song. *American Naturalist, 174*(1), 24–33.

———. (2011). Male response to historical and geographical variation in bird song. *Biology Letters, 7*(1), 57–59. doi:10.1098/rsbl.2010.0519

Desjardins, J. K., Klausner, J. Q., & Fernald, R. D. (2010). Female genomic response to mate information. *Proceedings of the National Academy of Sciences, 107*(49), 21176–21180. doi:10.1073/pnas.1010442107

Devigili, A., Evans, J. P., Di Nisio, A., & Pilastro, A. (2015). Multivariate selection drives concordant patterns of pre- and postcopulatory sexual selection in a livebearing fish. *Nature Communications, 6*, 8291.

Dickson, B. J. (2008). Wired for sex: the neurobiology of *Drosophila* mating decisions. *Science, 322*(5903), 904–909. doi:10.1126/science.1159276

Ding, B., Daugherty, D. W., Husemann, M., Chen, M., Howe, A. E., & Danley, P. D. (2014). Quantitative genetic analyses of male color pattern and female mate choice in a pair of cichlid fishes of Lake Malawi, East Africa. *PLoS ONE, 9*(12), e114798.

Dingemanse, N. J., Both, C., Drent, P. J., & Tinbergen, J. M. (2004). Fitness consequences of avian personalities in a fluctuating environment. *Proceedings of the Royal Society of London B: Biological Sciences, 271*(1541), 847–852.

Dobzhansky, T. (1970). *Genetics of the Evolutionary Process* (Vol. 139). New York: Columbia University Press.

Doebeli, M. (2005). Adaptive speciation when assortative mating is based on female preference for male marker traits. *Journal of Evolutionary Biology, 18*(6), 1587–1600.

Doherty, J. A. (1985). Phonotaxis in the cricket, *Gryllus bimaculatus* DeGeer: comparisons of choice and no-choice paradigms. *Journal of Comparative Physiology A, 157*(3), 279–289. doi:10.1007/BF00618118

Doherty, J. A., & Gerhardt, H. C. (1983). Hybrid tree frogs: vocalizations of males and selective phonotaxis of females. *Science, 220*(4601), 1078–1080.

Dohm, M. R. (2002). Repeatability estimates do not always set an upper limit to heritability. *Functional Ecology, 16*(2), 273–280.

Domingue, M. J., Musto, C. J., Linn Jr., C. E., Roelofs, W. L., & Baker, T. C. (2007). Altered olfactory receptor neuron responsiveness in rare *Ostrinia nubilalis* males attracted to the *O. furnacalis* pheromone blend. *Journal of Insect Physiology, 53*(10), 1063–1071. doi:http://dx.doi.org/10.1016/j.jinsphys.2007.05.013

Dong, S., & Clayton, D. F. (2009). Habituation in songbirds. *Neurobiology of Learning and Memory, 92*, 183–188. doi:10.1016/j.nlm.2008.09.009

Dopman, E. B., Bogdanowicz, S. M., & Harrison, R. G. (2004). Genetic mapping of sexual isolation between E and Z pheromone strains of the European corn borer (*Oshinia nubilalis*). *Genetics, 167*(1), 301–309.

Dougherty, L. R., & Shuker, D. M. (2014). The effect of experimental design on the measurement of mate choice: a meta-analysis. *Behavioral Ecology, 26*(2), 311–319. doi:10.1093/beheco/aru125

Dowling, T. E., & Secor, C. L. (1997). The role of hybridization and introgression in the diversification of animals. *Annual Review of Ecology and Systematics*, 593–619.

Drăgănoiu, T. I., Nagle, L., & Kreutzer, M. (2002). Directional female preference for an exaggerated male trait in canary (*Serinus canaria*) song. *Proceedings of the Royal Society of London B: Biological Sciences, 269*(1509), 2525–2531. doi:10.1098/rspb.2002.2192.

Dreiss, A., Silva, N., Richard, M., Moyen, F., Théry, M., Møller, A., & Danchin, E. (2008). Condition-dependent genetic benefits of extrapair fertilization in female blue tits *Cyanistes caeruleus*. *Journal of Evolutionary Biology, 21*(6), 1814–1822.

Drickamer, L. C., Gowaty, P. A., & Holmes, C. M. (2000). Free female mate choice in house mice affects reproductive success and offspring viability and performance. *Animal Behaviour, 59*(2), 371–378.

Drullion, D., & Dubois, F. (2008). Mate-choice copying by female zebra finches, *Taeniopygia guttata*: what happens when model females provide inconsistent information? *Behavioral Ecology and Sociobiology, 63*(2), 269–276. doi:10.1007/s00265-008-0658-5

———. (2011). Neighbours' breeding success and the sex ratio of their offspring affect the mate preferences of female zebra finches. *PLoS ONE, 6*(12), e29737. doi:10.1371/journal.pone.0029737

Dubois, F. (2014). When being the centre of the attention is detrimental: copiers may favour the use of evasive tactics. *Behavioral Ecology and Sociobiology*, 1–9. doi:10.1007/s00265-014-1831-7

Dubois, F., & Cézilly, F. (2002). Breeding success and mate retention in birds: a meta-analysis. *Behavioral Ecology and Sociobiology, 52*(5), 357–364.

Dubois, F., Drullion, D., & Witte, K. (2011). Social information use may lead to maladaptive decisions: a game theoretic model. *Behavioral Ecology, 23*(1), 225–231.

Dugatkin, L. A. (1998). Genes, copying, and female mate choice: shifting thresholds. *Behavioral Ecology, 9*(4), 323–327. doi:10.1093/beheco/9.4.323

———. (2007). Developmental environment, cultural transmission, and mate choice copying. *Naturwissenschaften, 94*(8), 651–656. doi:10.1007/s00114-007-0238-y

Dugatkin, L. A., & Godin, J.G.J. (1993). Female mate copying in the guppy (*Poecilia reticulata*): age-dependent effects. *Behavioral Ecology, 4*(4), 289–292.

———. (1998). Effects of hunger on mate-choice copying in the guppy. *Ethology, 104*(3), 194–202.

Dukas, R. (2002). Behavioural and ecological consequences of limited attention. *Philosophical Transactions of the Royal Society of London B: Biological Sciences, 357*, 1539–1547.

———. (2008). Learning decreases heterospecific courtship and mating in fruit flies. *Biology Letters, 4*(6), 645–647. doi:10.1098/rsbl.2008.0437

Dukas, R., & Baxter, C. M. (2014). Mate choosiness in young male fruit flies. *Behavioral Ecology*. doi:10.1093/beheco/aru020

Dukas, R., & Jongsma, K. (2012). Effects of forced copulations on female sexual attractiveness in fruit flies. *Animal Behaviour, 84*(6), 1501–1505.

Dumont, R. A., & Gillespie, P. G. (2003). Ion channels: hearing aid. *Nature, 424*(6944), 28–29.

Dunn, S. J., Waits, L. P., & Byers, J. A. (2012). Genetic versus census estimators of the opportunity for sexual selection in the wild. *American Naturalist, 179*(4), 451–462. doi:10.1086/664626

DuVal, E. H. (2013). Female mate fidelity in a lek mating system and its implications for the evolution of cooperative lekking behavior. *American Naturalist, 181*(2), 213–222. doi:10.1086/668830

Easty, L. K., Schwartz, A. K., Gordon, S. P., & Hendry, A. P. (2011). Does sexual selection evolve following introduction to new environments? *Animal Behaviour, 82*(5), 1085–1095.

Eberhard, W. G. (1994). Evidence for widespread courtship during copulation in 131 species of insects and spiders, and implications for cryptic female choice. *Evolution, 48*(3), 711–733.

———. (1996). *Female Control: Sexual Selection by Cryptic Female Choice*. Princeton, NJ: Princeton University Press.

———. (2000). Criteria for demonstrating postcopulatory female choice. *Evolution, 54*(3), 1047–1050. doi:10.1111/j.0014-3820.2000.tb00105.x

———. (2009). Postcopulatory sexual selection: Darwin's omission and its consequences. *Proceedings of the National Academy of Sciences, 106*(Supplement 1), 10025–10032. doi:10.1073/pnas.0901217106

———. (2011). Experiments with genitalia: a commentary. *Trends in Ecology & Evolution, 26*(1), 17–21. doi:http://dx.doi.org/10.1016/j.tree.2010.10.009

Edvardsson, M., & Arnqvist, G. (2000). Copulatory courtship and cryptic female choice in red flour beetles *Tribolium castaneum*. *Proceedings of the Royal Society of London B: Biological Sciences, 267*(1443), 559–563. doi:10.1098/rspb.2000.1037.

———. (2006). No apparent indirect genetic benefits to female red flour beetles preferring males with intense copulatory courtship. *Behavior Genetics, 36*(5), 775–782.

Edward, D. A. (2015). The description of mate choice. *Behavioral Ecology*, 26(2), 301–310. doi:10.1093/beheco/aru142

Edward, D. A., & Chapman, T. (2011). The evolution and significance of male mate choice. *Trends in Ecology & Evolution, 26*(12), 647–654.

Edwards, S. V., & Hedrick, P. W. (1998). Evolution and ecology of MHC molecules: from genomics to sexual selection. *Trends in Ecology & Evolution, 13*(8), 305–311. doi: http://dx.doi.org/10.1016/S0169-5347(98)01416-5

Elias, D. O., Hebets, E. A., Hoy, R. R., & Mason, A. C. (2005). Seismic signals are crucial for male mating success in a visual specialist jumping spider (Araneae: Salticidae). *Animal Behaviour, 69*, 931–938. doi:10.1016/j.anbehav.2004.06.024

Elie, J., Mathevon, N., & Vignal, C. (2011). Same-sex pair-bonds are equivalent to male-female bonds in a life-long socially monogamous songbird. *Behavioral Ecology and Sociobiology, 65*(12), 2197–2208. doi:10.1007/s00265-011-1228-9

Ellis, H. (1905). *Studies in the Psychology of Sex, Volume IV: Sexual Selection in Man.* Philadelphia: F.A. Davis.

Ellison, C. K., Wiley, C., & Shaw, K. L. (2011). The genetics of speciation: genes of small effect underlie sexual isolation in the Hawaiian cricket *Laupala. Journal of Evolutionary Biology, 24*(5), 1110–1119. doi:10.1111/j.1420-9101.2011.02244.x

Emlen, S. T., & Oring, L. W. (1977). Ecology, sexual selection, and the evolution of mating systems. *Science, 197*(4300), 215–223. doi:10.2307/1744497

Endler, J. A. (1992). Signals, signal conditions, and the direction of evolution. *American Naturalist, 139*, S125–S153.

———. (1993). Some general comments on the evolution and design of animal communication systems. *Philosophical Transactions of the Royal Societyof London B: Biological Sciences, 340*, 215–225.

Endler, J. A., & Basolo, A. L. (1998). Sensory ecology, receiver biases and sexual selection. *Trends in Ecology & Evolution, 13*(10), 415–420.

Endler, J. A., & Day, L. B. (2006). Ornament colour selection, visual contrast and the shape of colour preference functions in great bowerbirds, Chlamydera nuchalis. *Animal Behaviour, 72*(6), 1405–1416.

Endler, J. A., Gaburro, J., & Kelley, L. A. (2014). Visual effects in great bowerbird sexual displays and their implications for signal design. *Proceedings of the Royal Society of London B: Biological Sciences, 281*(1783), 20140235. doi:10.1098/rspb.2014.0235

Endler, J. A., & Houde, A. E. (1995). Geographic variation in female preferences for male traits in *Poecilia reticulata. Evolution, 49*(3), 456–468.

Engel, K., Männer, L., Ayasse, M., & Steiger, S. (2015). Acceptance threshold theory can explain occurrence of homosexual behaviour. *Biology Letters, 11*(20140603). doi: http://dx.doi.org/10.1098/rsbl.2014.0603

Engeler, B., & Reyer, H.-U. (2001). Choosy females and indiscriminate males: mate choice in mixed populations of sexual and hybridogenetic water frogs (*Rana lessonae, Rana esculenta*). *Behavioral Ecology, 12*(5), 600–606.

Engen, S., & Bernt-Erik, S. (1985). The evolutionary significance of sexual selection. *Journal of Theoretical Biology, 117*(2), 277–289.

Engen, S., & Saether, B. E. (1985). The evolutionary significance of sexual selection. *Journal of Theoretical Biology, 117*(2), 277–289.

Engqvist, L. (2007). Nuptial gift consumption influences female remating in a scorpionfly: male or female control of mating rate? *Evolutionary Ecology, 21*(1), 49–61. doi: 10.1007/s10682-006-9123-y

———. (2012). Genetic conflicts, intrinsic male fertility, and ejaculate investment. *Evolution, 66*(9), 2685–2696.

Engqvist, L., & Sauer, K. P. (2002). A life-history perspective on strategic mating effort in male scorpionflies. *Behavioral Ecology, 13*(5), 632–636. doi:10.1093/beheco/13.5.632

Enquist, M., & Arak, A. (1994). Symmetry, beauty and evolution. *Nature, 372*(6502), 169–172.

Eraly, D., Hendrickx, F., & Lens, L. (2009). Condition-dependent mate choice and its implications for population differentiation in the wolf spider *Pirata piraticus. Behavioral Ecology, 20*(4), 856–863. doi:10.1093/beheco/arp072

Eriksen, A., Lampe, H. M., & Slagsvold, T. (2009). Interspecific cross-fostering affects song acquisition but not mate choice in pied flycatchers, *Ficedula hypoleuca. Animal Behaviour, 78*(4), 857–863. doi:10.1016/j.anbehav.2009.07.005

Eshel, I. (1979). Sexual selection, population density, and availability of mates. *Theoretical Population Biology, 16*(3), 301–314. doi:http://dx.doi.org/10.1016/0040-5809(79)90019-4

Estrada, C., Schulz, S., Yildizhan, S., & Gilbert, L. E. (2011). Sexual selection drives the evolution of antiaphrodisiac pheromones in butterflies. *Evolution, 65*(10), 2843–2854.

Etges, W. J., de Oliveira, C. C., Gragg, E., Ortíz-Barrientos, D., Mohamed, A.F.N., & Ritchie, M. G. (2007). Genetics of incipient speciation in *Drosophila mojavensis.* I. Male courtship song, mating success, and genotype x environment interactions. *Evolution, 61*(5), 1106. doi:10.1111/j.1558-5646.2007.00104.x

Etges, W. J., de Oliveira, C. C., Ritchie, M. G., & Noor, M. A. (2009). Genetics of incipient speciation in *Drosophila mojavensis*: II. Host plants and mating status influence cuticular hydrocarbon QTL expression and G × E interactions. *Evolution, 63*(7), 1712–1730.

Etienne, L., Rousset, F., Godelle, B., & Courtiol, A. (2014). How choosy should I be? The relative searching time predicts evolution of choosiness under direct sexual selection. *Proceedings of the Royal Society of London B: Biological Sciences, 281*(1785), 20140190.

Evans, J. P., Box, T. M., Brooshooft, P., Tatler, J. R., & Fitzpatrick, J. L. (2010). Females increase egg deposition in favor of large males in the rainbowfish, *Melanotaenia australis. Behavioral Ecology, 21*(3), 465–469. doi:10.1093/beheco/arq006

Evans, J. P., & Garcia-Gonzalez, F. (2016). The total opportunity for sexual selection and the integration of pre- and post-mating episodes of sexual selection in a complex world. *Journal of Evolutionary Biology, 29*(12), 2338–2361. doi:10.1111/jeb.12960

Evans, J. P., Garcia-Gonzalez, F., Almbro, M., Robinson, O., & Fitzpatrick, J. L. (2012). Assessing the potential for egg chemoattractants to mediate sexual selection in a broadcast spawning marine invertebrate. *Proceedings of the Royal Society of London B: Biological Sciences, 279*(1739), 2855–2861. doi:10.1098/rspb.2012.0181

Evans, J. P., Rosengrave, P., Gasparini, C., & Gemmell, N. J. (2013). Delineating the roles of males and females in sperm competition. *Proceedings of the Royal Society of London B: Biological Sciences, 280*(1772), 20132047. doi:10.1098/rspb.2013.2047.

Evans, J. P., & Simmons, L. W. (2008). The genetic basis of traits regulating sperm competition and polyandry: can selection favour the evolution of good- and sexy-sperm? *Genetica, 134*(1), 5–19.

Evans, M. L., Dionne, M., Miller, K. M., & Bernatchez, L. (2011). Mate choice for major histocompatibility complex genetic divergence as a bet-hedging strategy in the Atlantic salmon (*Salmo salar*). *Proceedings of the Royal Society of London B: Biological Sciences*, rspb20110909.

Fairbairn, D. J., & Roff, D. A. (2006). The quantitative genetics of sexual dimorphism: assessing the importance of sex-linkage. *Heredity, 97*(5), 319–328.

Falconer, D. S., & Mackay, T. F. C. (1996). *Introduction to Quantitative Genetics* (4th ed.). London: Pearson.

Farmer, M. A., Leja, A., Foxen-Craft, E., Chan, L., MacIntyre, L. C., Niaki, T., … Mogil, J. S. (2014). Pain reduces sexual motivation in female but not male mice. *Journal of Neuroscience, 34*(17), 5747–5753. doi:10.1523/jneurosci.5337-13.2014

Farrelly, D. (2011). Cooperation as a signal of genetic or phenotypic quality in female mate choice? Evidence from preferences across the menstrual cycle. *British Journal of Psychology, 102*(3), 406–430.

Fawcett, T. W., & Bleay, C. (2008). Previous experiences shape adaptive mate preferences. *Behavioral Ecology, 20*(1), 68–78.

Fawcett, T. W., Boogert, N. J., & Lefebvreb, L. (2011). Female assessment: cheap tricks or costly calculations? *Choice, 4*, 451–460.

Fawcett, T. W., Hamblin, S., & Giraldeau, L.-A. (2013). Exposing the behavioral gambit: the evolution of learning and decision rules. *Behavioral Ecology, 24*(1), 2–11.

Fawcett, T. W., & Johnstone, R. A. (2003). Optimal assessment of multiple cues. *Proceedings of the Royal Society of London B: Biological Sciences, 270*(1524), 1637–1643.

Fawcett, T. W., Kuijper, B., Weissing, F. J., & Pen, I. (2011). Sex-ratio control erodes sexual selection, revealing evolutionary feedback from adaptive plasticity. *Proceedings of the National Academy of Sciences, 108*(38), 15925–15930.

Fechner, G. T. (1860). *Elemente der psychophysik.* Leipzig: Breitkopf und Härtel.

Fedina, T. Y., & Lewis, S. M. (2004). Female influence over offspring paternity in the red flour beetle Tribolium castaneum. *Proceedings of the Royal Society of London B: Biological Sciences, 271*(1546), 1393–1399. doi:10.1098/rspb.2004.2731

———. (2008). An integrative view of sexual selection in *Tribolium* flour beetles. *Biological Reviews, 83*(2), 151–171. doi:10.1111/j.1469-185X.2008.00037.x

Felsenstein, J. (1981). Skepticism towards Santa Rosalia, or why are there so few kinds of animals. *Evolution, 35*(1), 124–138.

Fernandez, A. A., & Morris, M. R. (2008). Mate choice for more melanin as a mechanism to maintain a functional oncogene. *Proceedings of the National Academy of Sciences, 105*(36), 13503–13507.

Ferveur, J.-F. (2010). *Drosophila* female courtship and mating behaviors: sensory signals, genes, neural structures and evolution. *Current Opinion in Neurobiology, 20*, 764–769. doi:10.1016/j.conb.2010.09.007

———. (2005). Cuticular hydrocarbons: their evolution and roles in *Drosophila* pheromonal communication. *Behavioral Genetics, 35*(3), 279–295. doi:10.1007/s10519-005-3220-5

Fessler, D.M.T., & Navarrete, C. D. (2003). Domain-specific variation in disgust sensitivity across the menstrual cycle. *Evolution and Human Behavior, 24*(6), 406–417. doi:http://dx.doi.org/10.1016/S1090-5138(03)00054-0

Feulner, P. G., Plath, M., Engelmann, J., Kirschbaum, F., & Tiedemann, R. (2009). Electrifying love: electric fish use species-specific discharge for mate recognition. *Biology Letters, 5*(2), 225–228. doi:10.1098/rsbl.2008.0566

Fierst, J. L. (2013). Female mating preferences determine system-level evolution in a gene network model. *Genetica, 141*(4–6), 157–170.

Fink, S., Excoffier, L., & Heckel, G. (2006). Mammalian monogamy is not controlled by a single gene. *PNAS, 103*(29), 10956–10960.

Fisher, H. E., Aron, A., & Brown, L. L. (2006). Romantic love: a mammalian brain system for mate choice. *Philosophical Transactions of the Royal Society of London B: Biological Sciences, 361*(1476), 2173–2186. doi:10.1098/rstb.2006.1938.

Fisher, H. S., Mascuch, S., & Rosenthal, G. G. (2009). Multivariate male traits misalign with multivariate female preferences in the swordtail fish, *Xiphophorus birchmanni*. *Animal Behaviour, 78*, 265–269.

Fisher, H. S., & Rosenthal, G. G. (2006). Female swordtail fish use chemical cues to select well-fed mates. *Animal Behaviour, 72*, 721–725. doi:10.1016/j.anbehav.2006.02.009

———. (2006). Hungry females show stronger mating preferences. *Behavioral Ecology, 17*(6), 979–981. doi:10.1093/beheco/arl038

———. (2007). Male swordtails court with an audience in mind. *Biology Letters, 3*(1), 5–7. doi:10.1098/rsbl.2006.0556

Fisher, H. S., & Rosenthal, G. G. (2010). Relative abundance of *Xiphophorus* fishes and its effect on sexual communication. *Ethology, 116*(1), 32–38. doi:10.1111/j.1439-0310.2009.01710.x

Fisher, H. S., Swaisgood, R. R., & Fitch-Snyder, H. (2003). Odor familiarity and female preferences for males in a threatened primate, the pygmy loris *Nycticebus pygmaeus*: applications for genetic management of small populations. *Naturwissenschaften, 90*(11), 509–512. doi:10.1007/s00114-003-0465-9

Fisher, H. S., Wong, B.B.M., & Rosenthal, G. G. (2006). Alteration of the chemical environment disrupts communication in a freshwater fish. *Proceedings of the Royal Society of London B: Biological Sciences, 273*(1591), 1187–1193. doi:10.1098/rspb.2005.3406

Fisher, R. A. (1915). The evolution of sexual preference. *Eugenics Review, 7*(3), 184.

———. (1930). *The Genetical Theory of Natural Selection*. Oxford: Clarendon Press.

Fitzpatrick, J. L., Montgomerie, R., Desjardins, J. K., Stiver, K. A., Kolm, N., & Balshine, S. (2009). Female promiscuity promotes the evolution of faster sperm in cichlid fishes. *Proceedings of the National Academy of Sciences, 106*(4), 1128–1132.

Fleishman, L. J. (1992). The influence of the sensory system and the environment on motion patterns in the visual displays of anoline lizards and other vertebrates. *American Naturalist, 139*, S36–S61.

Fleishman, L. J., McClintock, W. J., D'eath, R. B., Brainard, D. H., & Endler, J. A. (1998). Colour perception and the use of video playback experiments in animal behaviour. *Animal Behaviour, 56*, 1035–1040.

Fletcher, N. H. (2004). A simple frequency-scaling rule for animal communication. *Journal of the Acoustical Society of America, 115*(5 Pt 1), 2334–2338.

Foerster, K., Coulson, T., Sheldon, B. C., Pemberton, J. M., Clutton-Brock, T. H., & Kruuk, L. E. (2007). Sexually antagonistic genetic variation for fitness in red deer. *Nature, 447*(7148), 1107–1110.

Foley, B. R., Genissel, A., Kristy, H. L., & Nuzhdin, S. V. (2010). Does segregating variation in sexual or microhabitat preferences lead to non-random mating within a population of *Drosophila melanogaster? Biology Letters, 6*(1), 102–105. doi:10.1098/rsbl.2009.0608

Folstad, I., & Karter, A. J. (1992). Parasites, bright males, and the immunocompetence handicap. *American Naturalist*, 603–622.

Forister, M. L., & Scholl, C. F. (2012). Use of an exotic host plant affects mate choice in an insect herbivore. *American Naturalist, 179*(6), 805–810.

Forrest, T. G. (1994). From sender to receiver: propagation and environmental effects on acoustic signals. *American Zoologist, 34*(6), 644–654. doi:10.1093/icb/34.6.644

Forsberg, L. A., Dannewitz, J., Peterson, E., & Grahn, M. (2007). Influence of genetic dissimilarity in the reproductive success and mate choice of brown trout—females fishing for optimal MHC dissimilarity. *Journal of Evolutionary Biology, 20*(5), 1859–1869. doi:10.1111/j.1420-9101.2007.01380.x

Forsgren, E. (1992). Predation risk affects mate choice in a gobiid fish. *American Naturalist, 140*(6), 1041–1049. doi:10.2307/2462933

———. (1997). Female sand gobies prefer good fathers over dominant males. *Proceedings of the Royal Society of London B: Biological Sciences, 264*(1386), 1283–1286.

Forsgren, E., Amundsen, T., Borg, Å. A., & Bjelvenmark, J. (2004). Unusually dynamic sex roles in a fish. *Nature, 429*(6991), 551–554.

Forsgren, E., Karlsson, A., & Kvarnemo, C. (1996). Female sand gobies gain direct benefits by choosing males with eggs in their nests. *Behavioral Ecology and Sociobiology, 39*(2), 91–96.

Forsgren, E., Kvarnemo, C., & Lindstrom, K. (1996). Mode of sexual selection determined by resource abundance in two sand goby populations. *Evolution*, 646–654.

Forstmeier, W. (2004). Female resistance to male seduction in zebra finches. *Animal Behaviour, 68*(5), 1005–1015.

Forstmeier, W., & Birkhead, T. R. (2004). Repeatability of mate choice in the zebra finch: consistency within and between females. *Animal Behaviour, 68*, 1017–1028. doi: 10.1016/j.anbehav.2004.02.007

Forstmeier, W., Coltman, D. W., & Birkhead, T. R. (2004). Maternal effects influence the sexual behavior of sons and daughters in the zebra finch. *Evolution, 58*(11), 2574–2583.

Forstmeier, W., Martin, K., Bolund, E., Schielzeth, H., & Kempenaers, B. (2011). Female extrapair mating behavior can evolve via indirect selection on males. *Proceedings of the National Academy of Sciences, 108*(26), 10608–10613.

Forstmeier, W., Nakagawa, S., Griffith, S. C., & Kempenaers, B. (2014). Female extra-pair mating: adaptation or genetic constraint? *Trends in Ecology & Evolution, 29*(8), 456–464.

Fowler-Finn, K. D., & Rodriguez, R. L. (2013). Repeatability of mate preference functions in *Enchenopa* treehoppers (Hemiptera: Membracidae). *Animal Behaviour, 85*(2), 493–499. doi:10.1016/j.anbehav.2012.12.015

Fowler-Finn, K. D., & Rodríguez, R. L. (2012). The evolution of experience-mediated plasticity in mate preferences. *Journal of Evolutionary Biology, 25*(9), 1855–1863. doi:10.1111/j.1420-9101.2012.02573.x

———. (2016). The causes of variation in the presence of genetic covariance between sexual traits and preferences. *Biological Reviews of the Cambridge Philosophical Society*, 91(2), 498–510.

Franklin, A. M., Squires, Z. E., & Stuart-Fox, D. (2012). The energetic cost of mating in a promiscuous cephalopod. *Biology Letters, 8*(5), 754–756. doi:10.1098/rsbl.2012.0556

Freeberg, T. M., Duncan, S. D., Kast, T. L., & Enstrom, D. A. (1999). Cultural influences on female mate choice: an experimental test in cowbirds, Molothrus ater. *Animal Behaviour*, 57(2), 421–426.

Freed-Brown, G., & White, D. J. (2009). Acoustic mate copying: female cowbirds attend to other females' vocalizations to modify their song preferences. *Proceedings of the Royal Society of London B: Biological Sciences, 276*(1671), 3319–3325. doi:10.1098/rspb.2009.0580

Frenzel, H., Bohlender, J., Pinsker, K., Wohlleben, B., Tank, J., Lechner, S. G., ... Lewin, G. R. (2012). A genetic basis for mechanosensory traits in humans. *PLoS Biology, 10*(5), e1001318. doi:10.1371/journal.pbio.1001318

Freud, S. (1905). *Three Essays on the Theory of Sexuality* (J. Strachey, Trans.). New York: Basic Books.

Fricke, C., Green, D., Mills, W. E., & Chapman, T. (2013). Age-dependent female responses to a male ejaculate signal alter demographic opportunities for selection. *Proceedings of the Royal Society of London B: Biological Sciences, 280*(1766). doi:10.1098/rspb.2013.0428

Fricke, C., Perry, J., Chapman, T., & Rowe, L. (2009). The conditional economics of sexual conflict. *Biology Letters, 5*(5), 671–674.

Fricke, H., & Fricke, S. (1977). Monogamy and sex change by aggressive dominance in coral reef fish. *Nature, 266*(5605), 830–832.

Friesen, C. R., Uhrig, E. J., Mason, R. T., & Brennan, P.L.R. (2016). Female behaviour and the interaction of male and female genital traits mediate sperm transfer during mating. *Journal of Evolutionary Biology, 29*(5), 952–964. doi:10.1111/jeb.12836.

Fritzsche, K., & Arnqvist, G. (2013). Homage to Bateman: sex roles predict sex differences in sexual selection. *Evolution, 67*(7), 1926–1936.

Fromberger, P., Jordan, K., von Herder, J., Steinkrauss, H., Nemetschek, R., Stolpmann, G., & Muller, J. L. (2012). Initial orienting towards sexually relevant stimuli: preliminary evidence from eye movement measures. *Archives of Sexual Behavior, 41*(4), 919–928. doi:10.1007/s10508-011-9816-3

Fromhage, L., Kokko, H., & Reid, J. M. (2009). Evolution of mate choice for genome-wide heterozygosity. *Evolution, 63*(3), 684–694.

Frommen, J., Rahn, A., Schroth, S., Waltschyk, N., & Bakker, T. M. (2009). Mate-choice copying when both sexes face high costs of reproduction. *Evolutionary Ecology, 23*(3), 435–446. doi:10.1007/s10682-008-9243-7

Fujii, T., Fujii, T., Namiki, S., Abe, H., Sakurai, T., Ohnuma, A., … Shimada, T. (2011). Sex-linked transcription factor involved in a shift of sex-pheromone preference in the silkmoth *Bombyx mori*. *Proceedings of the National Academy of Sciences, 108*(44), 18038–18043. doi:10.1073/pnas.1107282108

Fuller, R. C. (2009). A test of the critical assumption of the sensory bias model for the evolution of female mating preference using neural networks. *Evolution, 63*(7), 1697–1711. doi:10.1111/j.1558-5646.2009.00659.x

Fuller, R. C., Carleton, K. L., Fadool, J. M., Spady, T. C., & Travis, J. (2004). Population variation in opsin expression in the bluefin killifish, *Lucania goodei*: a real-time PCR study. *Journal of Comparative Physiology A, 190*, 147–154.

———. (2005). Genetic and environmental variation in the visual properties of bluefin killifish, *Lucania goodei*. *Journal of Evolutionary Biology, 18*, 516–523.

Fuller, R. C., Houle, D., Travis, J., Associate Editor: Trevor, P., & Editor: Jonathan, B. L. (2005). Sensory bias as an explanation for the evolution of mate preferences. *American Naturalist, 166*(4), 437–446. doi:10.1086/444443

Fuller, R. C., & Noa, L. A. (2010). Female mating preferences, lighting environment, and a test of the sensory bias hypothesis in the bluefin killifish. *Animal Behaviour, 80*(1), 23–35. doi:http://dx.doi.org/10.1016/j.anbehav.2010.03.017

Fuller, R. C., Noa, L. A., & Strellner, R. S. (2010). Teasing apart the many effects of lighting environment on opsin expression and foraging preference in bluefin killifish. *American Naturalist, 176*(1), 1–13. doi:10.1086/652994

Funayama, R., Sugiura, M., Sassa, Y., Jeong, H., Wakusawa, K., Kawashima, R., … Kawashima, R. (2012). Neural bases of human mate choice: multiple value dimensions, sex difference, and self-assessment system. *Social Neuroscience, 7*(1), 59–73. doi:10.1080/17470919.2011.580120

Funk, D. J., Nosil, P., & Etges, W. J. (2006). Ecological divergence exhibits consistently positive associations with reproductive isolation across disparate taxa. *Proceedings of the National Academy of Sciences of the United States of America, 103*(9), 3209–3213. doi:10.1073/pnas.0508653103

Fusani, L., Barske, J., Day, L. D., Fuxjager, M. J., & Schlinger, B. A. (2014). Physiological control of elaborate male courtship: female choice for neuromuscular systems. *Neuroscience & Biobehavioral Reviews, 46*, 534–546.

Gabor, C. R., & Aspbury, A. S. (2008). Non-repeatable mate choice by male sailfin mollies, *Poecilia latipinna*, in a unisexual-bisexual mating complex. *Behavioral Ecology, 19*(4), 871–878. doi:10.1093/beheco/arn043

Gabor, C. R., Parmley, M. H., & Aspbury, A. S. (2011). Repeatability of female preferences in a unisexual-bisexual mating system. *Evolutionary Ecology Research, 13*(2), 145–157.

Gabor, C. R., & Ryan, M. J. (2001). Geographical variation in reproductive character displacement in mate choice by male sailfin mollies. *Proceedings of the Royal Society of London B: Biological Sciences, 268*(1471), 1063–1070. doi:10.1098/rspb.2001.1626

Gale, T., Gibson, A. B., Brooks, R. C., & Garratt, M. (2013). Exposure to a novel male during late pregnancy influences subsequent growth of offspring during lactation. *Journal of Evolutionary Biology, 26*(9), 2057–2062. doi:10.1111/jeb.12192

Galef Jr., B. G. (2008). Social influences on the mate choices of male and female Japanese quail. *Comparative Cognition & Behavior Reviews, 3*, 1–12. doi:10.3819/ccbr.2008.30001

Galeotti, P., Sacchi, R., Rosa, D. P., & Fasola, M. (2005). Female preference for fast-rate, high-pitched calls in Hermann's tortoises *Testudo hermanni*. *Behavioral Ecology, 16*(1), 301–308.

Galipaud, M., Bollache, L., Oughadou, A., & Dechaume-Moncharmont, F.-X. (2015). Males do not always switch females when presented with a better reproductive option. *Behavioral Ecology, 26*(2), 359–366.

Gallup, A. C., Hale, J. J., Sumpter, D. J., Garnier, S., Kacelnik, A., Krebs, J. R., & Couzin, I. D. (2012). Visual attention and the acquisition of information in human crowds. *Proceedings of the National Academy of Science U S A, 109*(19), 7245–7250. doi:10.1073/pnas.1116141109

Gangestad, S. W., & Simpson, J. A. (2000). The evolution of human mating: trade-offs and strategic pluralism. *Behavioral and Brain Sciences, 23*(04), 573–644.

Garratt, M., & Brooks, R. C. (2012). Oxidative stress and condition-dependent sexual signals: more than just seeing red. *Proceedings of the Royal Society of London B: Biological Sciences, 279*(1741), 3121–3130.

Gaskett, A., Winnick, C., & Herberstein, M. (2008). Orchid sexual deceit provokes ejaculation. *American Naturalist, 171*(6), E206–E212.

Gasparini, C., & Pilastro, A. (2011). Cryptic female preference for genetically unrelated males is mediated by ovarian fluid in the guppy. *Proceedings of the Royal Society of London B: Biological Sciences, 278*(1717), 2495–2501. doi:10.1098/rspb.2010.2369

Gasparini, C., Serena, G., & Pilastro, A. (2013). Do unattractive friends make you look better? *Proceedings of the Royal Society of London B: Biological Sciences, 280*, 20123072.

Gavrilets, S. (2000). Rapid evolution of reproductive barriers driven by sexual conflict. *Nature, 403*(6772), 886–889.

Gavrilets, S., Arnqvist, G., & Friberg, U. (2001). The evolution of female mate choice by sexual conflict. *Proceedings of the Royal Society of London B: Biological Sciences, 268*(1466), 531–539.

Gavrilets, S., & Boake, C. R. (1998). On the evolution of premating isolation after a founder event. *American Naturalist, 152*(5), 706–716.

Gavrilets, S., & Rice, W. R. (2006). Genetic models of homosexuality: generating testable predictions. *Proceedings of the Royal Society of London B: Biological Sciences, 273*(1605), 3031–3038.

Gavrilets, S., & Waxman, D. (2002). Sympatric speciation by sexual conflict. *Proceedings of the Natural Academy of Science of the U S A, 99*(16), 10533–10538.

Gee, J. M. (2003). How a hybrid zone is maintained: behavioral mechanisms of interbreeding between California and Gambel's quail (*Callipepla californica* and *C. gambelii*). *Evolution, 57*(10), 2407–2415.

Gerhardt, H. C., & Brooks, R. (2009). Experimental analysis of multivariate female choice in gray treefrogs (*Hyla versicolor*): evidence for directional and stabilizing selection. *Evolution, 63*(10), 2504–2512. doi:EVO746 [pii]. 10.1111/j.1558-5646.2009.00746.x

Gerhardt, H. C., & Hobel, G. (2005). Mid-frequency suppression in the green treefrog (*Hyla cinerea*): mechanisms and implications for the evolution of acoustic communication. *Journal of Comparative Physiology A. Neuroethology, Sensory, Neural, and Behavioral Physiology, 191*(8), 707–714. doi:10.1007/s00359-005-0626-8

Gerhardt, H. C., & Huber, F. (2002). *Acoustic Communication in Insects and Anurans: Common Problems and Diverse Solutions.* Chicago: University of Chicago Press.

Gerhardt, H. C., & Mudry, K. (1980). Temperature effects on frequency preferences and mating call frequencies in the green treefrog, *Hyla cinerea* (Anura: Hylidae). *Journal of Comparative Physiology, 137*(1), 1–6. doi:10.1007/BF00656911

Gerhardt, H. C., Tanner, S. D., Corrigan, C. M., & Walton, H. C. (2000). Female preference functions based on call duration in the gray tree frog (*Hyla versicolor*). *Behavioral Ecology, 11*(6), 663–669. doi:10.1093/beheco/11.6.663

Gerlach, N. M., McGlothlin, J. W., Parker, P. G., & Ketterson, E. D. (2012). Reinterpreting Bateman gradients: multiple mating and selection in both sexes of a songbird species. *Behavioral Ecology, 23*(5), 1078–1088. doi:10.1093/beheco/ars077

Gershman, S. N., Mitchell, C., Sakaluk, S. K., & Hunt, J. (2012). Biting off more than you can chew: sexual selection on the free amino acid composition of the spermatophylax in decorated crickets. *Proceedings of the Royal Society of London B: Biological Sciences.* doi:10.1098/rspb.2011.2592

Getty, T. (1999). Chase-away sexual selection as noisy reliable signaling. *Evolution,* 299–302.

Giard, M. H., & Peronnet, F. (1999). Auditory-visual integration during multimodal object recognition in humans: a behavioral and electrophysiological study. *Journal of Cognitive Neuroscience, 11*(5), 473–490. doi:10.1162/089892999563544

Gibson, R. M. (1996). Female choice in sage grouse: the roles of attraction and active comparison. *Behavioral Ecology and Sociobiology, 39*(1), 55–59. doi:10.1007/s002650050266

———. (1989). Field playback of male display attracts females in lek breeding sage grouse. *Behavioral Ecology and Sociobiology, 24*(6), 439–443.

Gibson, R. M., & Bachman, G. C. (1992). The costs of female choice in a lekking bird. *Behavioral Ecology, 3*(4), 300–309.

Gibson, R. M., & Langen, T. A. (1996). How do animals choose their mates? *Trends in Ecology & Evolution, 11*(11), 468–470.

Gibson, R. M., Taylor, C. E., & Jefferson, D. R. (1990). Lek formation by female choice: a simulation study. *Behavioral Ecology, 1*(1), 36–42.

Gierszewski, S., Müller, K., Smielik, I., Hütwohl, J.-M., Kuhnert, K.-D., & Witte, K. (2016). The virtual lover: variable and easily-guided 3D-fish animations as an innovative tool in mate-choice experiments with sailfin mollies. II. Validation. *Current Zoology*, zow108.

Gil, D., Leboucher, G., Lacroix, A., Cue, R., & Kreutzer, M. (2004). Female canaries produce eggs with greater amounts of testosterone when exposed to preferred male song. *Hormones and Behavior, 45*, 64–70. doi:10.1016/j.yhbeh.2003.08.005

Gilbert, L., Williamson, K. A., & Graves, J. A. (2012). Male attractiveness regulates daughter fecundity non-genetically via maternal investment. *Proceedings of the Royal Society of London B: Biological Sciences, 279*(1728), 523–528. doi:10.1098/rspb.2011.0962

Gil-Burmann, C., Peláez, F., & Sánchez, S. (2002). Mate choice differences according to sex and age. *Human Nature, 13*(4), 493–508.

Gilchrist, A. S., & Partridge, L. (2000). Why it is difficult to model sperm displacement in *Drosophila melanogaster*: the relation between sperm transfer and copulation duration. *Evolution, 54*(2), 534–542.

Gillespie, S. R., Tudor, M. S, Moore, A. J., & Miller, C. W. (2014). Sexual selection is influenced by both developmental and adult environments. *Evolution, 68*(12), 3421–3432.

Gilman, R. T., & Behm, J. E. (2011). Hybridization, species collapse, and species reemergence after disturbance to premating mechanisms of reproductive isolation. *Evolution, 65*(9), 2592–2605.

Gioti, A., Wigby, S., Wertheim, B., Schuster, E., Martinez, P., Pennington, C., … Chapman, T. (2012). Sex peptide of *Drosophila melanogaster* males is a global regulator of reproductive processes in females. *Proceedings of the Royal Society of London B: Biological Sciences, 279*(1746), 4423–4432.

Giraudeau, M., Duval, C., Czirjak, G. A., Bretagnolle, V., Eraud, C., McGraw, K. J., & Heeb, P. (2011). Maternal investment of female mallards is influenced by male carotenoid-based coloration. *Proceedings of the Royal Society of London B: Biological Sciences, 278*(1706), 781–788. doi:10.1098/rspb.2010.1115

Godin, J.-G. J., & Auld, H. L. (2013). Covariation and repeatability of male mating effort and mating preferences in a promiscuous fish. *Ecology and Evolution, 3*(7), 2020–2029. doi:10.1002/ece3.607

Godin, J.-G. J., & Briggs, S. E. (1996). Female mate choice under predation risk in the guppy. *Animal Behaviour, 51*, 117–130.

Godin, J.-G. J., & Dugatkin, L. A. (1996). Variability and repeatability of female mating preference in the guppy.*Animal Behaviour, 49*, 1427–1433.

Godin, J.-G. J., Herdman, E.J.E., & Dugatkin, L. A. (2005). Social influences on female mate choice in the guppy, *Poecilia reticulata*: generalized and repeatable trait-copying behaviour. *Animal Behaviour, 69*, 999–1005. doi:10.1016/j.anbehav.2004.07.016

Gomes, C. M., & Boesch, C. (2009). Wild chimpanzees exchange meat for sex on a long-term basis. *PLoS One, 4*(4), e5116.

Gomez, D., Théry, M., Gauthier, A.-L., & Lengagne, T. (2011). Costly help of audiovisual bimodality for female mate choice in a nocturnal anuran (*Hyla arborea*). *Behavioral Ecology, 22*(4), 889–898. doi:10.1093/beheco/arr039

Gomez-Diaz, C., & Benton, R. (2013). The joy of sex pheromones. *EMBO Reports, 14*(10), 874–883. doi:10.1038/embor.2013.140

Goodman, A., Koupil, I., & Lawson, D. W. (2012). Low fertility increases descendant socioeconomic position but reduces long-term fitness in a modern post-industrial society. *Proceedings of the Royal Society of London B: Biological Sciences, 279*(1746), 4342–4351.

Goodson, J. L., & Wang, Y. (2006). Valence-sensitive neurons exhibit divergent functional profiles in gregarious and asocial species. *Proceedings of the National Academy of Sciences, 103*(45), 17013–17017. doi:10.1073/pnas.0606278103

Gordon, L. E., & Silva, T. J. (2015). Inhabiting the sexual landscape: toward an interpretive theory of the development of sexual orientation and identity. *Journal of Homosexuality, 62*(4), 495–530.

Gosden, T., & Chenoweth, S. (2011). On the evolution of heightened condition dependence of male sexual displays. *Journal of Evolutionary Biology, 24*(3), 685–692.

Gould, F., Estock, M., Hillier, N. K., Powell, B., Groot, A. T., Ward, C. M., ... Vickers, N. J. (2010). Sexual isolation of male moths explained by a single pheromone response QTL containing four receptor genes. *Proceedings of the National Academy of Sciences of the U S A, 107*(19), 8660–8665. doi:DOI 10.1073/pnas.0910945107

Gould, J. L., Elliott, S. L., Masters, C. M., & Mukerji, J. (1999). Female preferences in a fish genus without female mate choice. *Current Biology, 9(*9), 497–500.

Gould, S. J. (1994). *Eight Little Piggies: Reflections in Natural History*. New York: W. W. Norton.

Govardovskii, V. I., Fyhrquist, N., Reuter, T., Kuzmin, D. G., & Donner, K. (2000). In search of the visual pigment template. *Visual Neuroscience, 17*, 509–528.

Govic, A., Levay, E. A., Hazi, A., Penman, J., Kent, S., & Paolini, A. G. (2008). Alterations in male sexual behaviour, attractiveness and testosterone levels induced by an adult-onset calorie restriction regimen. *Behavioural Brain Research, 190*(1), 140–146.

Gowaty, P. A. (2008). Reproductive compensation. *Journal of Evolutionary Biology, 21*(5), 1189–1200.

Gowaty, P. A., Anderson, W. W., Bluhm, C. K., Drickamer, L. C., Kim, Y.-K., & Moore, A. J. (2007). The hypothesis of reproductive compensation and its assumptions about mate preferences and offspring viability. *Proceedings of the National Academy of Sciences, 104*(38), 15023–15027.

Gowaty, P. A., Kim, Y.-K., & Anderson, W. W. (2012). No evidence of sexual selection in a repetition of Bateman's classic study of *Drosophila melanogaster*. *Proceedings of the National Academy of Sciences, 109*(29), 11740–11745.

Grace, J. L., & Shaw, K. L. (2011). Coevolution of male mating signal and female preference during early lineage divergence of the Hawaiian cricket, *Laupala cerasina*. *Evolution, 65*(8), 2184–2196.

Grafen, A. (1990). Sexual selection unhandicapped by the Fisher process. *Journal of Theoretical Biology, 144*(4), 473–516.

———. (1991). Modelling in behavioural ecology. In J. R. Krebs & N. B. Davies (Eds.), *Behavioural Ecology* (3rd ed., pp. 5–31). Oxford: Blackwell Scientific Publications.

Grammer, K., Kruck, K. B., & Magnusson, M. S. (1998). The courtship dance: patterns of nonverbal synchronization in opposite-sex encounters. *Journal of Nonverbal Behavior, 22*(1), 3–29.

Grana, S., Sakaluk, S., Bowden, R., Doellman, M., Vogel, L., & Thompson, C. (2012). Reproductive allocation in female house wrens is not influenced by experimentally altered male attractiveness. *Behavioral Ecology and Sociobiology, 66*(9), 1247–1258. doi:10.1007/s00265-012-1378-4

Grant, B. R., & Grant, P. R. (1997). Hybridization, sexual imprinting, and mate choice. *American Naturalist*, 1–28.

———. (2010). Songs of Darwin's finches diverge when a new species enters the community. *Proceedings of the National Academy of Sciences, 107*(47), 20156–20163. doi:10.1073/pnas.1015115107

Graur, D., & Li, W.-H. (2000). *Fundamentals of Molecular Evolution* (2nd ed.). Sunderland, MA: Sinauer.

Gray, D. A., & Cade, W. H. (1999). Quantitative genetics of sexual selection in the field cricket, *Gryllus integer*. *Evolution, 53*(3), 848–854. doi:10.2307/2640724

Greeff, J. M., & Michiels, N. K. (1999). Low potential for sexual selection in simultaneously hermaphroditic animals. *Proceedings of the Royal Society of London B: Biological Sciences, 266*(1429), 1671–1676.

Greenfield, M. D., Alem, S., Limousin, D., & Bailey, N. W. (2014). The dilemma of Fisherian sexual selection: mate choice for indirect benefits despite rarity and overall weakness of trait-preference genetic correlation. *Evolution, 68*(12), 3524–3536.

Greenfield, M. D., & Roizen, I. (1993). Katydid synchronous chorusing is an evolutionarily stable outcome of female choice. *Nature, 364*(6438), 618–620.

Greenfield, M. D., Siegfreid, E., & Snedden, W. A. (2004). Variation and repeatability of female choice in a chorusing katydid, *Ephippiger ephippiger*: an experimental exploration of the precedence effect. *Ethology, 110*(4), 287–299. doi:10.1111/j.1439-0310.2004.00969.x

Gregory, P. G., & Howard, D. J. (1994). A postinsemination barrier to fertilization isolates two closely related ground crickets. *Evolution, 48*(3), 705–710. doi:10.2307/2410480

Grether, G. (2000). Carotenoid limitation and mate preference evolution: a test of the indicator hypothesis in guppies (*Poecilia reticulata*). *Evolution, 54*, 1712–1724.

Grether, G. F. (2010). The evolution of mate preferences, sensory biases, and indicator traits. *Advances in the Study of Behavior, 41*, 35–76.

Grether, G. F., Kolluru, G. R., & Nersissian, K. (2004). Individual colour patches as multicomponent signals. *Biological Reviews, 79*(3), 583–610.

Grether, G. F., Kolluru, G. R., Rodd, F. H., de la Cerda, J., & Shimazaki, K. (2005). Carotenoid availability affects the development of a colour-based mate preference and the sensory bias to which it is genetically linked. *Proceedings of the Royal Society of London B: Biological Sciences, 272*(1577), 2181–2188. doi:10.1098/rspb.2005.3197

Griffith, L. C., & Ejima, A. (2009). Multimodal sensory integration of courtship stimulating cues in *Drosophila melanogaster*. *Annals of the New York Academy of Sciences, 1170*(1), 394–398. doi:10.1111/j.1749-6632.2009.04367.x

Griffith, S. C., Owens, I. P., & Thuman, K. A. (2002). Extra pair paternity in birds: a review of interspecific variation and adaptive function. *Molecular Ecology, 11*(11), 2195–2212.

Griffith, S. C., Owens, I.P.F., & Burke, T. (1999). Female choice and annual reproductive success favour less-ornamented male house sparrows. *Proceedings of the Royal Society of London B: Biological Sciences, 266*(1421), 765–770. doi:10.1098/rspb.1999.0703

Griffith, S. C., Pryke, S. R., & Buttemer, W. A. (2011). Constrained mate choice in social monogamy and the stress of having an unattractive partner. *Proceedings of the Royal Society of London B: Biological Sciences, 278*(1719), 2798–2805.

Griggio, M., Biard, C., Penn, D. J., & Hoi, H. (2011). Female house sparrows "count on" male genes: experimental evidence for MHC-dependent mate preference in birds. *BMC Evolutionary Biology, 11*(1), 44.

Griggio, M., & Hoi, H. (2010). Only females in poor condition display a clear preference and prefer males with an average badge. *BMC Evolutionary Biology, 10*(1), 261.

Grillet, M., Dartevelle, L., & Ferveur, J.-F. (2006). A *Drosophila* male pheromone affects female sexual receptivity. *Proceedings of the Royal Society of London B: Biological Sciences, 273*(1584), 315–323. doi:10.1098/rspb.2005.3332

Gröning, J., & Hochkirch, A. (2008). Reproductive interference between animal species. *Quarterly Review of Biology, 83*(3), 257–282.

Grosenick, L., Clement, T. S., & Fernald, R. D. (2007). Fish can infer social rank by observation alone. *Nature, 445*(7126), 429–432. doi:http://www.nature.com/nature/journal/v445/n7126/suppinfo/nature05511_S1.html

Grosjean, Y., Rytz, R., Farine, J.-P., Abuin, L., Cortot, J., Jefferis, G. S., & Benton, R. (2011). An olfactory receptor for food-derived odours promotes male courtship in Drosophila. *Nature, 478*(7368), 236–240.

Guilford, T., & Dawkins, M. S. (1991a). Receiver psychology and the design of animal signals. *Trends in Neurosciences, 16*, 430–436.

———. (1991b). Receiver psychology and the evolution of animals signals. *Animal Behaviour, 42*(1), 1.

Guillaume, F., & Whitlock, M. C. (2007). Effects of migration on the genetic covariance matrix. *Evolution, 61*(10), 2398–2409.

Gwynne, D., & Rentz, D. (1983). Beetles on the bottle: male buprestids mistake stubbies for females (Coleoptera). *Australian Journal of Entomology, 22*(1), 79–80.

Gwynne, D. T. (1989). Does copulation increase the risk of predation? *Trends in Ecology & Evolution, 4*(2), 54–56.

Gwynne, D. T., & Simmons, L. (1990). Experimental reversal of courtship roles in an insect. *Nature, 346*(6280), 172–174.

Gyuris, P., Járai, R., & Bereczkei, T. (2010). The effect of childhood experiences on mate choice in personality traits: homogamy and sexual imprinting. *Personality and Individual Differences, 49*(5), 467–472. doi:http://dx.doi.org/10.1016/j.paid.2010.04.021

Hadfield, J. D., Burgess, M. D., Lord, A., Phillimore, A. B., Clegg, S. M., & Owens, I. P. (2006). Direct versus indirect sexual selection: genetic basis of colour, size and recruitment in a wild bird. *Proceedings of the Royal Society of London B: Biological Sciences, 273*(1592), 1347–1353.

Haesler, P. M., & Seehausen, O. (2005). Inheritance of female mating preference in a sympatric sibling species pair of Lake Victoria cichlids: implications for speciation. *Proceedings of the Royal Society of London B: Biological Sciences, 272*(1560), 237–245. doi:10.1098/rspb.2004.2946

Hager, B., & Teale, S. (1994). Repeatability of female response to ipsdienol enantiomeric mixtures by pine engraver, *Ips pini* (Coleoptera: Scolytidae). *Journal of Chemical Ecology, 20*(10), 2611–2622. doi:10.1007/BF02036195

Hager, R., Wolf, J. B., & Cheverud, J. M. (2008). Maternal effects as the cause of parent-of-origin effects that mimic genomic imprinting. *Genetics, 178*(3), 1755–1762. doi: 10.1534/genetics.107.080697

Haines, S. E., & Gould, J. L. (1994). Female platys prefer long tails. *Nature, 370*, 512.

Hall, D. W., Kirkpatrick, M., & West, B. (2000). Runaway sexual selection when female preferences are directly selected. *Evolution, 54*(6), 1862–1869.

Hall, M., Lindholm, A. K., & Brooks, R. (2004). Direct selection on male attractiveness and female preference fails to produce a response. *BMC Evolutionary Biology, 4*:1.

Hall, M. D., Bussière, L. F., Demont, M., Ward, P. I., & Brooks, R. C. (2010). Competitive PCR reveals the complexity of postcopulatory sexual selection in *Teleogryllus commodus*. *Molecular Ecology, 19*(3), 610–619. doi:10.1111/j.1365-294X.2009.04496.x

Halliday, T. R. (1983). The study of mate choice. In P. Bateson (Ed.), *Mate Choice* (pp. 3–32). Cambridge, UK: Cambridge University Press.

Hamilton, W. D., & Zuk, M. (1982). Heritable true fitness and bright birds: a role for parasites? *Science, 218*(22 October), 384–387.

Han, C. S., Jablonski, P. G., & Brooks, R. C. (2015). Intimidating courtship and sex differences in predation risk lead to sex-specific behavioural syndromes. *Animal Behaviour, 109*, 177–185.

Hancock, A. M., Witonsky, D. B., Ehler, E., Alkorta-Aranburu, G., Beall, C., Gebremedhin, A., ... Di Rienzo, A. (2010). Human adaptations to diet, subsistence, and ecoregion are due to subtle shifts in allele frequency. *Proceedings of the National Academy of Sciences, 107*(Supplement 2), 8924–8930. doi:10.1073/pnas.0914625107

Haning, R. V., O'Keefe, S. L., Randall, E. J., Kommor, M. J., Baker, E., & Wilson, R. (2007). Intimacy, orgasm likelihood, and conflict predict sexual satisfaction in heterosexual male and female respondents. *Journal of Sex & Marital Therapy, 33*(2), 93–113. doi:10.1080/00926230601098449

Hankison, S. J., & Morris, M. R. (2002). Sexual selection and species recognition in the pygmy swordtail, Xiphophorus pygmaeus: conflicting preferences. *Behavioral Ecology and Sociobiology, 51*(2), 140–145.

———. (2003). Avoiding a compromise between sexual selection and species recognition: female swordtail fish assess multiple species-specific cues. *Behavioral Ecology, 14*(2), 282–287.

Härdling, R., & Kokko, H. (2005). The evolution of prudent choice. *Evolutionary Ecology Research, 7*(5), 697–715.

Hardy, D. F., & DeBold, J. F. (1971). The relationship between levels of exogenous hormones and the display of lordosis by the female rat. *Hormones and Behavior, 2*(4), 287–297. doi:http://dx.doi.org/10.1016/0018-506X(71)90003-1.

Harnad, S. (1987). Psychophysical and cognitive aspects of categorical perception: A critical overview. *In Categorical Perception: The Groundwork of Cognition* (pp. 1–52). Cambridge, UK: Cambridge University Press.

Harris, C. (2011). Menstrual cycle and facial preferences reconsidered. *Sex Roles, 64*(9–10), 669–681. doi:10.1007/s11199-010-9772-8

Harris, W. E., & Uller, T. (2009). Reproductive investment when mate quality varies: differential allocation versus reproductive compensation. *Philosophical Transactions of the Royal Society of London B: Biological Sciences, 364*(1520), 1039–1048.

Harrison, F., Barta, Z., Cuthill, I., & Székely, T. (2009). How is sexual conflict over parental care resolved? A meta-analysis. *Journal of Evolutionary Biology, 22*(9), 1800–1812.

Harrison, P. W., Wright, A. E., Zimmer, F., Dean, R., Montgomery, S. H., Pointer, M. A., & Mank, J. E. (2015). Sexual selection drives evolution and rapid turnover of male gene expression. *Proceedings of the National Academy of Sciences, 112*(14), 4393–4398.

Hart, N. S. (2001). The visual ecology of avian photoreceptors. *Progress in Retinal and Eye Research, 20*(5), 675–703.

Hasson, O. (1989). Amplifiers and the handicap principle in sexual selection: a different emphasis. *Proceedings of the Royal Society of London B: Biological Sciences, 235*, 383–406.

Hayashi, T. I., Marshall, J. L., & Gavrilets, S. (2007). The dynamics of sexual conflict over mating rate with endosymbiont infection that affects reproductive phenotypes. *Journal of Evolutionary Biology, 20*(6), 2154–2164. doi:10.1111/j.1420-9101.2007.01429.x

Head, M. L., Hunt, J., & Brooks, R. (2006). Genetic association between male attractiveness and female differential allocation. *Biology Letters, 2*(3), 341–344.

Head, M. L., Hunt, J., Jennions, M. D., & Brooks, R. (2005). The indirect benefits of mating with attractive males outweigh the direct costs. *PLoS Biology, 3*(2), e33.

Hébert, A., & Weaver, A. (2014). An examination of personality characteristics associated with BDSM orientations. *Canadian Journal of Human Sexuality, 23*(2), 106–115.

Hebets, E. A. (2003). Subadult experience influences adult mate choice in an arthropod: exposed female wolf spiders prefer males of a familiar phenotype. *Proceedings of the National Academy of Science U S A, 100*(23), 13390–13395. doi:10.1073/pnas.2333262100

Hebets, E. A., & Papaj, D. R. (2005). Complex signal function: developing a framework of testable hypotheses. *Behavioral Ecology and Sociobiology, 57*, 197–214.

Hebets, E. A., & Vink, C. J. (2007). Experience leads to preference: experienced females prefer brush-legged males in a population of syntopic wolf spiders. *Behavioral Ecology, 18*(6), 1010–1020. doi:10.1093/beheco/arm070

Hebets, E. A., Wesson, J., & Shamble, P. S. (2008). Diet influences mate choice selectivity in adult female wolf spiders. *Animal Behaviour, 76*(2), 355–363. doi:http://dx.doi.org/10.1016/j.anbehav.2007.12.021

Hedrick, A. V., & Dill, L. M. (1993). Mate choice by female crickets is influenced by predation risk. *Animal Behaviour, 46*, 193–196.

Hedrick, A. V., & Kortet, R. (2012). Effects of body size on selectivity for mating cues in different sensory modalities. *Biological Journal of the Linnean Society, 105*(1), 160–168. doi:10.1111/j.1095-8312.2011.01786.x

Heffner, R., & Heffner, H. (1980). Hearing in the elephant (*Elephas maximus*). *Science, 208*, 518–519.

Heffner, R. S., & Heffner, H. E. (1992). Evolution of sound localization in mammals. In D. B. Webster, R. R. Fay, & A. N. Popper (Eds.), *Evolutionary Biology of Hearing* (pp. 691–715). New York: Springer.

Hegyi, G., Herenyi, M., Wilson, A. J., Garamszegi, L. Z., Rosivall, B., Eens, M., & Torok, J. (2010). Breeding experience and the heritability of female mate choice in collared flycatchers. *PLoS ONE, 5*(11), e13855.

Heisler, I. L. (1994). Quantitative genetic models of the evolution of mating behavior. In C.R.B. Boake (Ed.), *Quantitative Genetic Studies of Behavioral Evolution* (pp. 101–125). Chicago: University of Chicago Press.

Heisler, L., Andersson, M., Arnold, S., Boake, C., Borgia, G., Hausfater, G., . . . O'Donald, P. (1987). The evolution of mating preferences and sexually selected traits. In J. W. Bradbury & M. B. Andersson (Eds.), *Sexual Selection: Testing the Alternatives* (pp. 96–118). Chichester: Wiley.

Hekmat-Scafe, D. S., Dorit, R. L., & Carlson, J. R. (2000). Molecular evolution of odorant-binding protein genes OS-E and OS-F in Drosophila. *Genetics, 155*(1), 117–127.

Helfenstein, F., Wagner, R. H., & Danchin, E. (2003). Sexual conflict over sperm ejection in monogamous pairs of kittiwakes *Rissa tridactyla. Behavioral Ecology and Sociobiology, 54*(4), 370–376.

Helfenstein, F., Wagner, R. H., Danchin, E., & Rossi, J.-M. (2003). Functions of courtship feeding in black-legged kittiwakes: natural and sexual selection. *Animal Behaviour, 65*(5), 1027–1033.

Hendrie, C. A., & Brewer, G. (2012). Evidence to suggest that teeth act as human ornament displays signalling mate quality. *PLoS ONE, 7*(7), e42178.

Henshaw, J. M., Kahn, A. T., & Fritzsche, K. (2016). A rigorous comparison of sexual selection indexes via simulations of diverse mating systems. *Proceedings of the National Academy of Sciences, 113*(3), E300–E308.

Henshaw, J. M., Kokko, H., & Jennions, M. D. (2015). Direct reciprocity stabilizes simultaneous hermaphroditism at high mating rates: a model of sex allocation with egg trading. *Evolution, 69*(8), 2129–2139.

Hernandez, A. M., & MacDougall-Shackleton, S. A. (2004). Effects of early song experience on song preferences and song control and auditory brain regions in female house finches (*Carpodacus mexicanus*). *Journal of Neurobiology, 59*(2), 247–258. doi: 10.1002/neu.10312

Hernandez, A. M., Phillmore, L. S., & MacDougall-Shackleton, S. A. (2008). Effects of learning on song preferences and Zenk expression in female songbirds. *Behavioural Processes, 77*(2), 278–284. doi:http://dx.doi.org/10.1016/j.beproc.2007.11.001

Herz, R., & Cahill, E. (1997). Differential use of sensory information in sexual behavior as a function of gender. *Human Nature, 8*(3), 275–286. doi:10.1007/bf02912495

Hesse, S., Bakker, T. C., Baldauf, S. A., & Thünken, T. (2016). Impact of social environment on inter- and intrasexual selection in a cichlid fish with mutual mate choice. *Animal Behaviour, 111*, 85–92.

Hettyey, A., Hegyi, G., Puurtinen, M., Hoi, H., Török, J., & Penn, D. J. (2010). Mate choice for genetic benefits: time to put the pieces together. *Ethology, 116*(1), 1–9.

Heubel, K. U., Rankin, D. J., & Kokko, H. (2009). How to go extinct by mating too much: population consequences of male mate choice and efficiency in a sexual-asexual species complex. *Oikos, 118*(4), 513–520.

Higgie, M., & Blows, M. W. (2007a). Are traits that experience reinforcement also under sexual selection? *American Naturalist, 170*(3), 409–420.

———. (2007b). The evolution of reproductive character displacement conflicts with how sexual selection operates within a species. *Evolution, 62*(5), 1192–1203.

Higgie, M., Chenoweth, S., & Blows, M. W. (2000). Natural selection and the reinforcement of mate recognition. *Science, 290*(5491), 519–521.

Higginson, D. M., Miller, K. B., Segraves, K. A., & Pitnick, S. (2012). Female reproductive tract form drives the evolution of complex sperm morphology. *Proceedings of the National Academy of Sciences, 109*(12), 4538–4543. doi:10.1073/pnas.1111474109

Hill, G. E. (1991). Plumage coloration is a sexually selected indicator of male quality. *Nature, 350*, 337–339.

———. (2011). Condition-dependent traits as signals of the functionality of vital cellular processes. *Ecology Letters, 14*(7), 625–634.

———. (2015). Sexiness, individual condition, and species identity: the information signaled by ornaments and assessed by choosing females. *Evolutionary Biology, 42*(3), 251–259.

Hill, G. E., & Johnson, J. D. (2013). The mitonuclear compatibility hypothesis of sexual selection. *Proceedings of the Royal Society of London B: Biological Sciences, 280*(1768), 20131314.

Hill, S. E., & Reeve, H. K. (2004). Mating games: the evolution of human mating transactions. *Behavioral Ecology, 15*(5), 748–756.

Hill, S. E., & Ryan, M. J. (2005). The role of model female quality in the mate choice copying behavior of sailfin mollies. *Biology Letters*, 1–3.

Hill, W. F. (1978). Effects of mere exposure on preferences in nonhuman mammals. *Psychological Bulletin, 85*(6), 1177–1198. doi:10.1037/0033-2909.85.6.1177

Hinde, R. A., & Steel, E. (1976). The effect of male song on an estrogen-dependent behavior pattern in the female canary (*Serinus canarius*). *Hormones and Behavior, 7*(3), 293–304. doi:http://dx.doi.org/10.1016/0018-506X(76)90035-0

Hine, E., Lachish, S., Higgie, M., & Blows, M. W. (2002). Positive genetic correlation between female preference and offspring fitness. *Proceedings of the Royal Society of London B: Biological Sciences, 269*(1506), 2215–2219.

Hine, E., McGuigan, K., & Blows, M. W. (2011). Natural selection stops the evolution of male attractiveness. *Proceedings of the National Academy of Sciences, 108*(9), 3659–3664.

Hirtenlehner, S., & Römer, H. (2014). Selective phonotaxis of female crickets under natural outdoor conditions. *Journal of Comparative Physiology A: Neuroethology, Sensory, Neural, and Behavioral Physiology, 200*(3), 239–250. doi:10.1007/s00359-014-0881-7

Hoffman, J., Forcada, J., Trathan, P., & Amos, W. (2007). Female fur seals show active choice for males that are heterozygous and unrelated. *Nature, 445*(7130), 912–914.

Hofmann, C. M., O'Quin, K. E., Marshall, N. J., Cronin, T. W., Seehausen, O., & Carleton, K. L. (2009). The eyes have it: regulatory and structural changes both underlie cichlid visual pigment diversity. *PLoS Biology, 7*(12), e1000266. doi:10.1371/journal.pbio.1000266

Hofmann, C. M., O'Quin, K. E., Smith, A. R., & Carleton, K. L. (2010). Plasticity of opsin gene expression in cichlids from Lake Malawi. *Molecular Ecology, 19*(10), 2064–2074. doi:MEC4621 [pii]. 10.1111/j.1365-294X.2010.04621.x

Höglund, J., & Alatalo, R. V. (1995). *Leks*. Princeton, NJ: Princeton University Press.

Hohenlohe, P. A., & Arnold, S. J. (2010). Dimensionality of mate choice, sexual isolation, and speciation. *Proceedings of the National Academy of Sciences, 107*(38), 16583–16588. doi:10.1073/pnas.1003537107

Hoikkala, A., & Aspi, J. (1993). Criteria of female mate choice in *Drosophila littoralis, D. montana*, and *D. ezoana*. *Evolution, 47*(3), 768–777. doi:10.2307/2410182.

Hoke, K. L., Ryan, M. J., & Wilczynski, W. (2007). Integration of sensory and motor processing underlying social behaviour in tungara frogs. *Proceedings of the Royal Society of London B: Biological Sciences, 274*(1610), 641–649. doi:L332M74L60L17100 [pii]10.1098/rspb.2006.0038

———. (2008). Candidate neural locus for sex differences in reproductive decisions. *Biology Letters, 4*(5), 518–521. doi:M5633L201414486X [pii]. 10.1098/rsbl.2008.0192

———. (2010). Sexually dimorphic sensory gating drives behavioral differences in tungara frogs. *Journal of Experimental Biology, 213*(Pt 20), 3463–3472. doi:213/20/3463 [pii]. 10.1242/jeb.043992

Holland, B., & Rice, W. R. (1998). Perspective: Chase-away sexual selection: antagonistic seduction versus resistance. *Evolution, 52*, 1–7.

Hollis, B., & Kawecki, T. J. (2014). Male cognitive performance declines in the absence of sexual selection. *Proceedings of the Royal Society of London B: Biological Sciences, 281*(1781), 20132873.

Hollocher, H., Ting, C. T., Pollack, F., & Wu, C. I. (1997). Incipient speciation by sexual isolation in *Drosophila melanogaster*: variation in mating preference and correlation between sexes. *Evolution, 51*(4), 1175–1181. doi:10.2307/2411047

Holman, L., & Kokko, H. (2014a). The evolution of genomic imprinting: costs, benefits and long-term consequences. *Biological Reviews, 89*(3), 568–587.

———. (2014b). Local adaptation and the evolution of female choice. *Genotype-by-Environment Interactions and Sexual Selection*. Hoboken, NJ: Wiley-Blackwell.

Holman, L., Price, T. A., Wedell, N., & Kokko, H. (2015). Coevolutionary dynamics of polyandry and sex-linked meiotic drive. *Evolution, 69*(3), 709–720.

Holveck, M.-J., Geberzahn, N., & Riebel, K. (2011). An experimental test of condition-dependent male and female mate choice in zebra finches. *PLoS ONE, 6*(8), e23974.

Holveck, M.-J., & Riebel, K. (2007). Preferred songs predict preferred males: consistency and repeatability of zebra finch females across three test contexts. *Animal Behaviour, 74*(2), 297–309. doi:http://dx.doi.org/10.1016/j.anbehav.2006.08.016

———. (2010). Low-quality females prefer low-quality males when choosing a mate. *Proceedings of the Royal Society of London B: Biological Sciences, 277*, 153–160. doi:10.1098/rspb.2009.1222

———. (2014). Female zebra finches learn to prefer more than one song and from more than one tutor. *Animal Behaviour, 88*, 125–135. doi:http://dx.doi.org/10.1016/j.anbehav.2013.11.023

Horth, L. (2007). Sensory genes and mate choice: evidence that duplications, mutations, and adaptive evolution alter variation in mating cue genes and their receptors. *Genomics, 90*(2), 159–175. doi:10.1016/j.ygeno.2007.03.021

Horváthová, T., Nakagawa, S., & Uller, T. (2011). Strategic female reproductive investment in response to male attractiveness in birds. *Proceedings of the Royal Society of London B: Biological Sciences, 279*(1726), 163–170. doi:10.1098/rspb.2011.0663.

———. (2012). Strategic female reproductive investment in response to male attractiveness in birds. *Proceedings of the Royal Society of London B: Biological Sciences, 279*, 163–170.

Hosken, D. J., & Stockley, P. (2004). Sexual selection and genital evolution. *Trends in Ecology & Evolution, 19*(2), 87–93.

Houde, A. E. (1997). *Sex, Color and Mate Choice in Guppies*. Princeton, NJ: Princeton University Press.

Houde, A. E., & Endler, J. A. (1990). Correlated evolution of female mating preferences and male color patterns in the guppy *Poecilia reticulata*. *Science, 248*, 1405–1408.

Houde, A. E., & Hankes, M. A. (1997). Evolutionary mismatch of mating preferences and male colour patterns in guppies. *Animal Behaviour, 53*, 343–351.

Houle, D., & Kondrashov, A. S. (2002). Coevolution of costly mate choice and condition-dependent display of good genes. *Proceedings of the Royal Society of London B: Biological Sciences, 269*(1486), 97–104.

House, P. K., Vyas, A., & Sapolsky, R. (2011). Predator cat odors activate sexual arousal pathways in brains of *Toxoplasma gondii* infected rats. *PLoS One, 6*(8), e23277. doi:10.1371/journal.pone.0023277

Howard, D. J. (1999). Conspecific sperm and pollen precedence and speciation. *Annual Review of Ecology and Systematics, 30*(1), 109–132. doi:doi:10.1146/annurev.ecolsys.30.1.109

Howard, R. D., Martens, R. S., Innis, S. A., Drnevich, J. M., & Hale, J. (1998). Mate choice and mate competition influence male body size in Japanese medaka. *Animal Behaviour, 55*(5), 1151–1163. doi:http://dx.doi.org/10.1006/anbe.1997.0682

Howard, R. D., & Young, J. R. (1998). Individual variation in male vocal traits and female mating preferences in *Bufo americanus*. *Animal Behaviour, 55*, 1165–1179. doi:10.1006/anbe.1997.0683

Hoy, R. R., Hahn, J., & Paul, R. C. (1977). Hybrid cricket auditory behavior: evidence for genetic coupling in animal communication. *Science, 195*(4273), 82–84.

Hoysak, D. J., & Godin, J.-G. J. (2007). Repeatability of male mate choice in the mosquitofish, *Gambusia holbrooki*. *Ethology, 113*(10), 1007–1018. doi:10.1111/j.1439-0310.2007.01413.x

Huber, B. A. (2005). Sexual selection research on spiders: progress and biases. *Biological Reviews, 80*(03), 363–385. doi:doi:10.1017/S1464793104006700

Hughes, N. K., Kelley, J. L., & Banks, P. B. (2012). Dangerous liaisons: the predation risks of receiving social signals. *Ecology Letters, 15*(11), 1326–1339.

Hugill, N., Fink, B., & Neave, N. (2010). The role of human body movements in mate selection. *Evolutionary Psychology, 8*(1), 66–89.

Hulse, A. M., & Cai, J. J. (2013). Genetic variants contribute to gene expression variability in humans. *Genetics, 193*(1), 95–108. doi:10.1534/genetics.112.146779

Hulse, S. H., Bernard, D. J., & Braaten, R. F. (1995). Auditory discrimination of chord-based spectral structures by European starlings (*Sturnus vulgaris*). *Journal of Experimental Psychology: General, 124*(4), 409–423. doi:10.1037/0096-3445.124.4.409.

Hunt, J., Breuker, C. J., Sadowski, J. A., & Moore, A. J. (2009). Male-male competition, female mate choice and their interaction: determining total sexual selection. *Journal of Evolutionary Biology, 22*(1), 13–26.

Hunt, J., Brooks, R., & Jennions, M. D. (2005). Female mate choice as a condition-dependent life-history trait. *American Naturalist, 166*(1), 79–92.

Hunt, J., Bussiere, L. F., Jennions, M. D., & Brooks, R. (2004). What is genetic quality? *Trends in Ecology & Evolution, 19*(6), 329–333.

Hunt, J., & Simmons, L. W. (2001). Status-dependent selection in the dimorphic beetle Onthophagus taurus. *Proceedings of the Royal Society of London B: Biological Sciences, 268*(1484), 2409–2414.

———. (2002). The genetics of maternal care: Direct and indirect genetic effects on phenotype in the dung beetle Onthophagus taurus. *Proceedings of the National Academy of Sciences, 99*(10), 6828–6832. doi:10.1073/pnas.092676199.

Hunt, J., Snook, R., Mitchell, C., Crudgington, H., & Moore, A. (2012). Sexual selection and experimental evolution of chemical signals in *Drosophila pseudoobscura*. *Journal of Evolutionary Biology, 25*(11), 2232–2241.

Hurlbert, A., & Poggio, T. (1988). Making machines (and artificial intelligence) see. *Daedalus*, 213–239.

Hurlbert, S. H. (1984). Pseudoreplication and the design of ecological field experiments. *Ecological Monographs, 54*(2), 187–211.

Husak, J. F., & Fox, S. F. (2008). Sexual selection on locomotor performance. *Evolutionary Ecology Research, 10*(2), 213–228.

Hutchinson, J.M.C., & Gigerenzer, G. (2005). Simple heuristics and rules of thumb: where psychologists and behavioural biologists might meet. *Behavioural Processes, 69*(2), 97–124. doi:http://dx.doi.org/10.1016/j.beproc.2005.02.019

Hutchinson, J.M.C., & Halupka, K. (2004). Mate choice when males are in patches: optimal strategies and good rules of thumb. *Journal of Theoretical Biology, 231*(1), 129–151. doi:http://dx.doi.org/10.1016/j.jtbi.2004.06.009

Huxley, J. S. (1938). Darwin's theory of sexual selection and the data subsumed by it, in the light of recent research. *American Naturalist*, 416–433.

———. (1966). *Ritualization of behaviour in animals and man.* Paper presented at the International Study on the Main Trends of Research in the Sciences of Man.

Ihara, Y., Aoki, K., & Feldman, M. W. (2003). Runaway sexual selection with paternal transmission of the male trait and gene-culture determination of the female preference. *Theoretical Population Biology, 63*(1), 53–62.

Ihara, Y., & Feldman, M. W. (2003). Evolution of disassortative and assortative mating preferences based on imprinting. *Theoretical Population Biology, 64*(2), 193–200. doi: http://dx.doi.org/10.1016/S0040-5809(03)00099-6

Ihle, M., & Forstmeier, W. (2013). Revisiting the evidence for inbreeding avoidance in zebra finches. *Behavioral Ecology, 24*(6), 1356–1362.

Ihle, M., Kempenaers, B., & Forstmeier, W. (2015). Fitness benefits of mate choice for compatibility in a socially monogamous species. *PLoS Biology, 13*(9), e1002248.

Immelmann, K. (1975). Ecological significance of imprinting and early learning. *Annual Review of Ecology and Systematics*, 15–37.

Immonen, E., Hoikkala, A., Kazem, A., & Ritchie, M. (2009). When are vomiting males attractive? Sexual selection on condition-dependent nuptial feeding in *Drosophila subobscura. Behavioral Ecology*, arp008.

Ingleby, F., Hunt, J., & Hosken, D. (2010). The role of genotype-by-environment interactions in sexual selection. *Journal of Evolutionary Biology, 23*(10), 2031–2045.

Ingleby, F. C., Hunt, J., & Hosken, D. J. (2013). Genotype-by-environment interactions for female mate choice of male cuticular hydrocarbons in *Drosophila simulans. PLoS ONE, 8*(6), e67623.

Ingley, S. J., & Johnson, J. B. (2014). Animal personality as a driver of reproductive isolation. *Trends in Ecology & Evolution, 29*(7), 369–371.

Ingoldsby, B. B. (2006). Mate selection and marriage. *Families in Global and Multicultural Perspective, 2*, 133–146.

Irwin, D. E., & Price, T. (1999). Sexual imprinting, learning and speciation. *Heredity, 82*(4), 347–354. doi:10.1046/j.1365-2540.1999.00527.x

Isoherranen, E., Aspi, J., & Hoikkala, A. (1999a). Inheritance of species differences in female receptivity and song requirement between *Drosophila virilis* and *D. montana. Hereditas, 131*(3), 203–209. doi:10.1111/j.1601-5223.1999.t01-1-00203.x

———. (1999b). Variation and consistency of female preferences for simulated courtship songs in *Drosophila virilis. Animal Behaviour, 57*, 619–625. doi:10.1006/anbe.1998.0981

Ivy, T. M., & Sakaluk, S. K. (2007). Sequential mate choice in decorated crickets: females use a fixed internal threshold in pre- and postcopulatory choice. *Animal Behaviour, 74*(4), 1065–1072.

Iwasa, Y., & Pomiankowski, A. (1994). The evolution of mate preferences for multiple sexual ornaments. *Evolution*, 853–867.

———. (1995). Continual change in mate preferences. *Nature, 377*(6548), 420–422.

———. (1999). Good parent and good genes models of handicap evolution. *Journal of Theoretical Biology, 200*(1), 97–109.

Iyengar, A., Wu, C. F., Imoehl, J., Ueda, A., & Nirschl, J. (2012). Automated quantification of locomotion, social interaction, and mate preference in *Drosophila* mutants. *Journal of Neurogenetics, 26*(3–4), 306–316. doi:10.3109/01677063.2012.729626

Iyengar, V. K., & Eisner, T. (1999). Female choice increases offspring fitness in an arctiid moth (*Utetheisa ornatrix*). *Proceedings of the National Academy of Sciences, 96*(26), 15013–15016.

Iyengar, V. K., Reeve, H. K., & Eisner, T. (2002). Paternal inheritance of a female moth's mating preference. *Nature, 419*(6909), 830–832.

Izzo, A. S., & Gray, D. A. (2011). Heterospecific courtship and sequential mate choice in sister species of field crickets. *Animal Behaviour, 81*(1), 259–264. doi:http://dx.doi.org/10.1016/j.anbehav.2010.10.015

Jacobs, G. H. (1981). *Comparative Color Vision*. New York: Academic Press.

Jacobs, L. F. (2012). From chemotaxis to the cognitive map: the function of olfaction. *Proceedings of the National Academy of Sciences, 109*(Supplement 1), 10693–10700. doi:10.1073/pnas.1201880109

Jain, M., Diwakar, S., Bahuleyan, J., Deb, R., & Balakrishnan, R. (2014). A rain forest dusk chorus: cacophony or sounds of silence? *Evolutionary Ecology, 28*(1), 1–22. doi:10.1007/s10682-013-9658-7

Jameson, K. A., Highnote, S. M., & Wasserman, L. M. (2001). Richer color experience in observers with multiple photopigment opsin genes. *Psychonomic Bulletin and Review, 8*(2), 244–261.

Janetos, A. C. (1980). Strategies of female mate choice: a theoretical analysis. *Behavioral Ecology and Sociobiology, 7*(2), 107–112. doi:10.1007/BF00299515

Jang, Y., & Greenfield, M. D. (1996). Ultrasonic communication and sexual selection in wax moths: female choice based on energy and asynchrony of male signals. *Animal Behaviour, 51*(5), 1095–1106. doi:10.1006/anbe.1996.0111

———. (1998). Absolute versus relative measurements of sexual selection: assessing the contributions of ultrasonic signal characters to mate attraction in lesser wax moths, *Achroia grisella* (Lepidoptera: Pyralidae). *Evolution*, 1383–1393.

Janicke, T., David, P., & Chapuis, E. (2015). Environment-dependent sexual selection: Bateman's parameters under varying levels of food availability. *American Naturalist, 185*(6), 756–768.

Janif, Z. J., Brooks, R. C., & Dixson, B. J. (2014). Negative frequency-dependent preferences and variation in male facial hair. *Biology Letters, 10*(4). doi:10.1098/rsbl.2013.0958

Janik, V. M. (2014). Cetacean vocal learning and communication. *Current Opinion in Neurobiology, 28*, 60–65.

Jennings, J. H., & Etges, W. J. (2010). Species hybrids in the laboratory but not in nature: a reanalysis of premating isolation between *Drosophila arizonae* and *D. mojavensis*. *Evolution, 64*(2), 587–598.

Jennings, J. H., Snook, R. R., & Hoikkala, A. (2014). Reproductive isolation among allopatric Drosophila montana populations. *Evolution, 68*(11), 3095–3108.
Jennions, M., Kokko, H., & Klug, H. (2012). The opportunity to be misled in studies of sexual selection. *Journal of Evolutionary Biology, 25*(3), 591–598.
Jennions, M., & Petrie, M. (1997). Variation in mate choice and mating preferences: a review of causes and consequences. *Biological Reviews of the Cambridge Philosophical Society, 72*, 283–327.
Jennions, M. D., Backwell, P.R.Y., & Passmore, N. I. (1995). Repeatability of mate choice: the effect of size in the African painted reed frog, *Hyperolius marmoratus. Animal Behaviour, 49*(1), 181–186. doi:http://dx.doi.org/10.1016/0003-3472(95)80165-0
Jennions, M. D., Hunt, J., Graham, R., & Brooks, R. (2004). No evidence for inbreeding avoidance through postcopulatory mechanisms in the black field cricket, *Teleogryllus commodus. Evolution, 58*(11), 2472–2477.
Jennions, M. D., Kahn, A. T., Kelly, C. D., & Kokko, H. (2012). Meta-analysis and sexual selection: past studies and future possibilities. *Evolutionary Ecology, 26*(5), 1119–1151.
Jennions, M. D., & Petrie, M. (2000). Why do females mate multiply? A review of the genetic benefits. *Biological Reviews of the Cambridge Philosophical Society, 75*(01), 21–64.
Jia, F.-Y., & Greenfield, M. D. (1997). When are good genes good? Variable outcomes of female choice in wax moths. *Proceedings of the Royal Society of London B: Biological Sciences, 264*(1384), 1057–1063.
Jiang, Y., Bolnick, D. I., & Kirkpatrick, M. (2013). Assortative mating in animals. *American Naturalist, 181*(6), E125–E138.
Johns, S. E., Hargrave, L. A., & Newton-Fisher, N. E. (2012). Red is not a proxy signal for female genitalia in humans. *PLoS One, 7*(4), e34669. doi:10.1371/journal.pone.0034669.
Johnsen, T. S., & Zuk, M. (1996). Repeatability of mate choice in female red jungle fowl. *Behavioral Ecology, 7*(3), 243–246. doi:10.1093/beheco/7.3.243
Johnson, A. M., Stanis, S., & Fuller, R. C. (2013). Diurnal lighting patterns and habitat alter opsin expression and colour preferences in a killifish. *Proceedings of the Royal Society of London B: Biological Sciences, 280*(1763). doi:10.1098/rspb.2013.0796
Johnson, J. C., Ivy, T. M., & Sakaluk, S. K. (1999). Female remating propensity contingent on sexual cannibalism in sagebrush crickets, *Cyphoderris strepitans*: a mechanism of cryptic female choice. *Behavioral Ecology, 10*(3), 227–233. doi:10.1093/beheco/10.3.227
Johnson, J. C., & Sih, A. (2005). Precopulatory sexual cannibalism in fishing spiders (*Dolomedes triton*): a role for behavioral syndromes. *Behavioral Ecology and Sociobiology, 58*(4), 390–396.
Johnson, L. S., Thompson, C. F., Sakaluk, S. K., Neuhauser, M., Johnson, B. G., Soukup, S. S., … Masters, B. S. (2009). Extra-pair young in house wren broods are more likely to be male than female. *Proceedings of the Royal Society of London B: Biological Sciences, 276*(1665), 2285–2289. doi:10.1098/rspb.2009.0283.
Johnstone, R., Rands, S., & Evans, M. (2009). Sexual selection and condition-dependence. *Journal of Evolutionary Biology, 22*(12), 2387–2394.
Johnstone, R. A. (1996). Multiple displays in animal communication: 'backup signals' and 'multiple messages.' *Philosophical Transactions of the Royal Society of London B: Biological Sciences, 351*(1337), 329–338.

Johnstone, R. A., Reynolds, J. D., & Deutsch, J. C. (1996). Mutual mate choice and sex differences in choosiness. *Evolution, 50*(4), 1382–1391.

Jones, A. G., & Ratterman, N. L. (2009). Mate choice and sexual selection: what have we learned since Darwin? *Proceedings of the National Academy of Sciences, 106*(Supplement 1), 10001–10008.

Jones, A. G., Small, C. M., Paczolt, K. A., & Ratterman, N. L. (2010). A practical guide to methods of parentage analysis. *Molecular Ecology Resources, 10*(1), 6–30. doi:10.1111/j.1755-0998.2009.02778.x

Jones, B. C., Feinberg, D. R., Watkins, C. D., Fincher, C. L., Little, A. C., & DeBruine, L. M. (2013). Pathogen disgust predicts women's preferences for masculinity in men's voices, faces, and bodies. *Behavioral Ecology, 24*(2), 373–379. doi:10.1093/beheco/ars173

Jones, B. C., Little, A. C., Boothroyd, L., DeBruine, L. M., Feinberg, D. R., Smith, M.J.L., ... Perrett, D. I. (2005). Commitment to relationships and preferences for femininity and apparent health in faces are strongest on days of the menstrual cycle when progesterone level is high. *Hormones and Behavior, 48*(3), 283–290. doi:http://dx.doi.org/10.1016/j.yhbeh.2005.03.010

Jones, I. L., & Hunter, F. M. (1998). Heterospecific mating preferences for a feather ornament in least auklets. *Behavioral Ecology, 9*(2), 187–192.

Jordan, L. A., & Brooks, R. C. (2012). Recent social history alters male courtship preferences. *Evolution, 66*(1), 280–287. doi:10.1111/j.1558-5646.2011.01421.x

Jordan, L. A., Kokko, H., & Kasumovic, M. (2014). Reproductive foragers: male spiders choose mates by selecting among competitive environments. *American Naturalist, 183*(5), 638–649.

Jordan, L. A., & Ryan, M. J. (2015). The sensory ecology of adaptive landscapes. *Biology Letters, 11*(5), 20141054.

Jordan, W., & Bruford, M. W. (1998). New perspectives on mate choice and the MHC. *Heredity, 81*(2), 127–133.

Joseph, P. N., Sharma, R. K., Agarwal, A., & Sirot, L. K. (2015). Men ejaculate larger volumes of semen, more motile sperm, and more quickly when exposed to images of novel women. *Evolutionary Psychological Science, 1*, 195–200.

Judge, K. A., Ting, J. J., & Gwynne, D. T. (2014). Condition dependence of female choosiness in a field cricket. *Journal of Evolutionary Biology, 27*(11), 2529–2540. doi:10.1111/jeb.12509

Juntti, Scott A., Hilliard, Austin T., Kent, Kai R., Kumar, A., Nguyen, A., Jimenez, Mariana A., ... Fernald, Russell D. (2016). A neural basis for control of cichlid female reproductive behavior by prostaglandin F α. *Current Biology, 26*(7), 943–949. doi:10.1016/j.cub.2016.01.067.

Kacelnik, A., Vasconcelos, M., Monteiro, T., & Aw, J. (2011). Darwin's "tug-of-war" vs. starlings' "horse-racing": how adaptations for sequential encounters drive simultaneous choice. *Behavioral Ecology and Sociobiology, 65*(3), 547–558. doi:10.1007/s00265-010-1101-2

Kahn, A. T., Livingston, J. D., & Jennions, M. D. (2012). Do females preferentially associate with males given a better start in life? *Biology Letters* 8(3), 362–364.

Kalisz, S., & Kramer, E. (2008). Variation and constraint in plant evolution and development. *Heredity, 100*(2), 171–177.

Kaneshiro, K. Y. (1980). Sexual isolation, speciation, and the direction of evolution. Evolution, 34(3), 437–444.

Karlsson Green, K., & Madjidian, J. A. (2011). Active males, reactive females: stereotypic sex roles in sexual conflict research? *Animal Behaviour, 81*(5), 901–907. doi:http://dx.doi.org/10.1016/j.anbehav.2011.01.033

Kárpáti, Z., Dekker, T., & Hansson, B. S. (2008). Reversed functional topology in the antennal lobe of the male European corn borer. *Journal of Experimental Biology, 211*(17), 2841–2848. doi:10.1242/jeb.017319

Kárpáti, Z., Olsson, S., Hansson, B. S., & Dekker, T. (2010). Inheritance of central neuroanatomy and physiology related to pheromone preference in the male European corn borer. *BMC Evolutionary Biology, 10*(1), 286.

Kárpáti, Z., Tasin, M., Cardé, R. T., & Dekker, T. (2013). Early quality assessment lessens pheromone specificity in a moth. *Proceedings of the National Academy of Sciences, 110*(18), 7377–7382. doi:10.1073/pnas.1216145110

Katzir, G. (1981). Visual aspects of species recognition in the damselfish *Dascyllus aruanus L.* (Pisces, Pomacentridae). *Animal Behaviour, 29*(3), 842–849.

Kaupp, U. B. (2010). Olfactory signalling in vertebrates and insects: differences and commonalities. *Nature Reviews Neuroscience, 11*(3), 188–200. doi:http://www.nature.com/nrn/journal/v11/n3/suppinfo/nrn2789_S1.html

Kavaliers, M., & Choleris, E. (2013). Neurobiological correlates of sociality, mate choice and learning. *Trends in Ecology & Evolution, 28*(1), 4–5. doi:http://dx.doi.org/10.1016/j.tree.2012.08.019

Kavaliers, M., Choleris, E., Agmo, A., Braun, W. J., Colwell, D. D., Muglia, L. J., . . . Pfaff, D. W. (2006). Inadvertent social information and the avoidance of parasitized male mice: a role for oxytocin. *Proceedings of the National Academy of Science U S A, 103*(11), 4293–4298. doi:10.1073/pnas.0600410103.

Kavaliers, M., Colwell, D. D., & Choleris, E. (1998). Parasitized female mice display reduced aversive responses to the odours of infected males. *Proceedings of the Royal Society of London B: Biologicval Sciences, 265*(1401), 1111–1118. doi:10.1098/rspb.1998.0406

Kawase, H., Okata, Y., & Ito, K. (2013). Role of huge geometric circular structures in the reproduction of a marine pufferfish. *Scientific Reports, 3, 2106*. doi:10.1038/srep02106

Kazancıoğlu, E., & Alonzo, S. H. (2012). The evolution of optimal female mating rate changes the coevolutionary dynamics of female resistance and male persistence. *Philosophical Transactions of the Royal Society of London B: Biological Sciences, 367*(1600), 2339–2347.

Keddy-Hector, A. C., Wilczynski, W., & Ryan, M. J. (1992). Call patterns and basilar papilla tuning in cricket frogs. II. Intrapopulation variation and allometry. *Brain, Behavior and Evolution, 39*(4), 238–246.

Keil, T. A. (1999). Morphology and development of the peripheral olfactory organs. In B. S. Hansson (Ed.), *Insect Olfaction* (pp. 5–47). Berlin; Heidelberg: Springer.

Keller, I., Wagner, C. E., Greuter, L., Mwaiko, S., Selz, O. M., Sivasundar, A., . . . Seehausen, O. (2013). Population genomic signatures of divergent adaptation, gene flow and hybrid speciation in the rapid radiation of Lake Victoria cichlid fishes. *Molecular Ecology, 22*(11), 2848–2863.

Keller, L. F., & Waller, D. M. (2002). Inbreeding effects in wild populations. *Trends in Ecology & Evolution, 17*(5), 230–241.

Kelley, J. L., Graves, J. A., & Magurran, A. E. (1999). Familiarity breeds contempt in guppies. *Nature, 401*(6754), 661–662.

Kelley, L. A., & Endler, J. A. (2012a). Illusions promote mating success in great bowerbirds. *Science, 335*(6066), 335–338. doi:10.1126/science.1212443
———. (2012b). Male great bowerbirds create forced perspective illusions with consistently different individual quality. *Proceedings of the National Academy of Science U S A, 109*(51), 20980–20985. doi:10.1073/pnas.1208350109
Kelley, L. A., & Kelley, J. L. (2014). Animal visual illusion and confusion: the importance of a perceptual perspective. *Behavioral Ecology, 25*(3), 450–463.
Kelly, C. A., Norbutus, A. J., Lagalante, A. F., & Iyengar, V. K. (2012). Male courtship pheromones as indicators of genetic quality in an arctiid moth (*Utetheisa ornatrix*). *Behavioral Ecology*, ars064.
Kempenaers, B. (2007). Mate choice and genetic quality: a review of the heterozygosity theory. *Advances in the Study of Behavior, 37*, 189–278.
Kendrick, K. M., Hinton, M. R., Atkins, K., Haupt, M. A., & Skinner, J. D. (1998). Mothers determine sexual preferences. *Nature, 395*(6699), 229–230.
Kidd, M. R., O'Connell, L. A., Kidd, C. E., Chen, C. W., Fontenot, M. R., Williams, S. J., & Hofmann, H. A. (2013). Female preference for males depends on reproductive physiology in the African cichlid fish *Astatotilapia burtoni*. *General and Comparative Endocrinology, 180*(0), 56–63. doi:http://dx.doi.org/10.1016/j.ygcen.2012.10.014
Kim, T. W., Christy, J. H., Dennenmoser, S., & Choe, J. C. (2009). The strength of a female mate preference increases with predation risk. *Proceedings of the Royal Society of London B: Biological Sciences, 276*(1657), 775–780. doi:10.1098/rspb.2008.1070
Kime, N., Rand, A., Kapfer, M., & Ryan, M. (1998). Consistency of female choice in the túngara frog: a permissive preference for complex characters. *Animal Behaviour, 55*(3), 641–649.
Kimura, K. I., Ote, M., Tazawa, T., & Yamamoto, D. (2005). Fruitless specifies sexually dimorphic neural circuitry in the *Drosophila* brain. *Nature, 438*(7065), 229–233.
King, S. L. (2015). You talkin' to me? Interactive playback is a powerful yet underused tool in animal communication research. *Biology Letters, 11*(7). doi:10.1098/rsbl.2015.0403.
Kingston, J. J., Rosenthal, G. G., & Ryan, M. J. (2003). The role of sexual selection in maintaining a colour polymorphism in the pygmy swordtail, *Xiphophorus pygmaeus*. *Animal Behaviour, 65*(4), 735–743.
Kirkpatrick, M. (1982). Sexual selection and the evolution of female choice. *Evolution*, 1–12.
———. (1985). Evolution of female choice and male parental investment in polygynous species: the demise of the "sexy son." *American Naturalist*, 788–810.
———. (1986). The handicap mechanism of sexual selection does not work. *American Naturalist, 127*(2), 222–240.
———. (1987a). Evolutionary forces acting on female mating preferences in polygynous animals. In J. W. Bradbury & M. B. Andersson (Eds.), *Sexual Selection: Testing the Alternatives* (pp. 67–82). Chichester: Wiley.
———. (1987b). Sexual selection by female choice in polygynous animals. *Annual Review of Ecology, Evolution, and Systematics, 18*, 43–70.
———. (1992). Direct selection of female mating preferences: comments on Grafen's models. *Journal of Theoretical Biology, 154*(1), 127–129.
Kirkpatrick, M., & Barton, N. (1997). The strength of indirect selection on female mating preferences. *Proceedings of the National Academy of Sciences, 94*(4), 1282–1286.

Kirkpatrick, M., Hall, D. W., & Dunn, P. (2004). Sexual selection and sex linkage. *Evolution, 58*(4), 683–691. doi:10.1554/03-332

Kirkpatrick, M., Rand, A. S., & Ryan, M. J. (2006). Mate choice rules in animals. *Animal Behaviour, 71*, 1215–1225.

Kirkpatrick, M., & Rosenthal, G. G. (1994). Symmetry without fear. *Nature, 372*, 134.

Kirkpatrick, M., & Ryan, M. J. (1991). The evolution of mating preferences and the paradox of the lek. *Nature, 350*(6313), 33–38.

Klatt, J. D., & Goodson, J. L. (2012). Oxytocin-like receptors mediate pair bonding in a socially monogamous songbird. *Proceedings of the Royal Societyof London B: Biological Sciences, 280*, 20122396. doi:10.1098/rspb.2012.2396

Klaudia, W., Hamilton, E. F., Michael, J. R., & Walter, W. (2005). How cricket frog females deal with a noisy world: habitat-related differences in auditory tuning. *Behavioral Ecology, 16*(3), 571–579.

Klemme, I., Ylönen, H., & Eccard, J. A. (2008). Long-term fitness benefits of polyandry in a small mammal, the bank vole *Clethrionomys glareolus*. *Proceedings of the Royal Society of London B: Biological Sciences, 275*(1638), 1095–1100.

Klingerman, C. M., Patel, A., Hedges, V. L., Meisel, R. L., & Schneider, J. E. (2011). Food restriction dissociates sexual motivation, sexual performance, and the rewarding consequences of copulation in female Syrian hamsters. *Behavioural Brain Research, 223*(2), 356–370.

Klug, H., Heuschele, J., Jennions, M., & Kokko, H. (2010). The mismeasurement of sexual selection. *Journal of Evolutionary Biology, 23*(3), 447–462.

Kodric-Brown, A., & Nicoletto, P. F. (2001). Female choice in the guppy (*Poecilia reticulata*): the interaction between male color and display. *American Naturalist, 50*, 346–351.

Koeninger Ryan, K., & Altmann, J. (2001). Selection for male choice based primarily on mate compatibility in the oldfield mouse, *Peromyscus polionotus rhoadsi*. *Behavioral Ecology and Sociobiology, 50*(5), 436–440.

Kohda, M., Heg, D., Makino, Y., Takeyama, T., Shibata, J. Y., Watanabe, K., ... Awata, S. (2009). Living on the wedge: female control of paternity in a cooperatively polyandrous cichlid. *Proceedings of the Royal Societyof London B: Biological Sciences*, 276(1676), 4207–4214. doi:10.1098/rspb.2009.1175

Kohl, J., Ostrovsky, A. D., Frechter, S., & Jefferis, G. S. (2013). A bidirectional circuit switch reroutes pheromone signals in male and female brains. *Cell, 155*(7), 1610–1623.

Kokko, H. (1998). Should advertising parental care be honest? *Proceedings of the Royal Society of London B: Biological Sciences, 265*(1408), 1871–1878.

———. (2001). Fisherian and "good genes" benefits of mate choice: how (not) to distinguish between them. *Ecology Letters, 4*(4), 322–326.

———. (2005). Treat 'em mean, keep 'em (sometimes) keen: evolution of female preferences for dominant and coercive males. *Evolutionary Ecology, 19*(2), 123–135.

Kokko, H., Brooks, R., Jennions, M. D., & Morley, J. (2003). The evolution of mate choice and mating biases. *Proceedings of the Royal Society of London B: Biological Sciences, 270*(1515), 653–664. doi:10.1098/rspb.2002.2235

Kokko, H., Brooks, R., McNamara, J. M., & Houston, A. I. (2002). The sexual selection continuum. *Proceedings of the Royal Society of London B: Biological Sciences, 269* (1498), 1331–1340.

Kokko, H., & Jennions, M. D. (2008a). Parental investment, sexual selection and sex ratios. *Journal of Evolutionary Biology, 21*(4), 919–948.

———. (2008b). Sexual conflict: the battle of the sexes reversed. *Current Biology, 18*(3), R121–R123.

———. (2014). The relationship between sexual selection and sexual conflict. *Cold Spring Harbor Perspectives in Biology, 6*(9), a017517.

———. (2015). Describing mate choice in a biased world: comments on Edward and Dougherty & Shuker. *Behavioral Ecology, 26*(2), 320–321. doi:10.1093/beheco/arv005

Kokko, H., Jennions, M. D., & Houde, A. (2007). Evolution of frequency-dependent mate choice: keeping up with fashion trends. *Proceedings of the Royal Society of London B: Biological Sciences, 274*(1615), 1317–1324.

Kokko, H., & Johnstone, R. A. (2002). Why is mutual mate choice not the norm? Operational sex ratios, sex roles and the evolution of sexually dimorphic and monomorphic signalling. *Philosophical Transactions of the Royal Society of London B: Biological Sciences, 357*(1419), 319–330. doi:10.1098/rstb.2001.0926

Kokko, H., Klug, H., & Jennions, M. D. (2012). Unifying cornerstones of sexual selection: operational sex ratio, Bateman gradient and the scope for competitive investment. *Ecology Letters, 15*(11), 1340–1351.

Kokko, H., Klug, H., & Jennions, M. J. (2014). Chapter 3: mating systems. In D. M. Shuker & L. W. Simmons (Eds.), *The Evolution of Insect Mating Systems* (pp. 42–58). Oxford: Oxford University Press.

Kokko, H., & Mappes, J. (2005). Sexual selection when fertilization is not guaranteed. *Evolution, 59*(9), 1876–1885. doi:10.1111/j.0014-3820.2005.tb01058.x

Kokko, H., & Rankin, D. J. (2006). Lonely hearts or sex in the city? Density-dependent effects in mating systems. *Philosophical Transactions of the Royal Society of London B: Biological Sciences, 361*(1466), 319–334.

Kolm, N. (2002). Male size determines reproductive output in a paternal mouthbrooding fish. *Animal Behaviour, 63*(4), 727–733.

Kolm, N., Amcoff, M., Mann, R. P., & Arnqvist, G. (2012). Diversification of a food-mimicking male ornament via sensory drive. *Current Biology, 22*(15), 1440–1443.

Kopp, M., & Hermisson, J. (2008). Competitive speciation and costs of choosiness. *Journal of Evolutionary Biology, 21*(4), 1005–1023.

Kostarakos, K., Hartbauer, M., & Römer, H. (2008). Matched filters, mate choice and the evolution of sexually selected traits. *PLoS ONE, 3*(8). doi:10.1371/journal.pone.0003005

Kotiaho, J., & Puurtinen, M. (2007). Mate choice for indirect genetic benefits: scrutiny of the current paradigm. *Functional Ecology, 21*(4), 638–644.

Kotiaho, J. S., LeBas, N. R., Puurtinen, M., & Tomkins, J. L. (2008). On the resolution of the lek paradox. *Trends in Ecology & Evolution, 23*(1), 1–3.

Kotiaho, J. S., Simmons, L. W., Hunt, J., & Tomkins, J. L. (2003). Males influence maternal effects that promote sexual selection: a quantitative genetic experiment with dung beetles *Onthophagus taurus*. *American Naturalist, 161*(6), 852–859.

Koukou, K., Pavlikaki, H., Kilias, G., Werren, J. H., Bourtzis, K., & Alahiotis, S. N. (2006). Influence of antibiotic treatment and wolbachia curing on sexual isolation among *Drosophila melanogaster* cage populations. *Evolution, 60*(1), 87–96. doi:10.1111/j.0014-3820.2006.tb01084.x

Kowner, R. (1996). Facial asymmetry and attractiveness judgement in developmental perspective. *Journal of Experimental Psychology: Human Perception and Performance, 22*(3), 662.

Kozak, G. M., & Boughman, J. W. (2009). Learned conspecific mate preference in a species pair of sticklebacks. *Behavioral Ecology, 20*(6), 1282–1288. doi:10.1093/beheco/arp134

———. (2015). Predator experience overrides learned aversion to heterospecifics in stickleback species pairs. *Proceedings of the Royal Society of London B: Biological Sciences, 282*(1805). doi:10.1098/rspb.2014.3066

Kozak, G. M., Head, M. L., & Boughman, J. W. (2011). Sexual imprinting on ecologically divergent traits leads to sexual isolation in sticklebacks. *Proceedings of the Royal Society of London B: Biological Sciences, 278*(1718), 2604–2610. doi:10.1098/rspb.2010.2466

Kozak, G. M., Head, M. L., Lackey, A.C.R., & Boughman, J. W. (2013). Sequential mate choice and sexual isolation in threespine stickleback species. *Journal of Evolutionary Biology, 26*(1), 130–140. doi:10.1111/jeb.12034

Kozak, G. M., Roland, G., Rankhorn, C., Falater, A., Berdan, E. L., & Fuller, R. C. (2015). Behavioral isolation due to cascade reinforcement in *Lucania* killifish. *American Naturalist, 185*(4), 491–506.

Kraaijeveld, K., Kraaijeveld-Smit, F. J., & Komdeur, J. (2007). The evolution of mutual ornamentation. *Animal Behaviour, 74*(4), 657–677.

Krakauer, A., Webster, M., Duval, E., Jones, A., & Shuster, S. (2011). The opportunity for sexual selection: not mismeasured, just misunderstood. *Journal of Evolutionary Biology, 24*(9), 2064–2071.

Kranz, F., & Ishai, A. (2006). Face perception is modulated by sexual preference. *Current Biology, 16*(1), 63–68. doi:http://dx.doi.org/10.1016/j.cub.2005.10.070

Kringelbach, M. L. (2005). The human orbitofrontal cortex: linking reward to hedonic experience. *Nature Reviews Neuroscience, 6*(9), 691–702.

Kringelbach, M. L., & Rolls, E. T. (2004). The functional neuroanatomy of the human orbitofrontal cortex: evidence from neuroimaging and neuropsychology. *Progress in Neurobiology, 72*(5), 341–372. doi:10.1016/j.pneurobio.2004.03.006

Kröger, R.H.H., Bowmaker, J. K., & Wagner, H. J. (1999). Morphological changes in the retina of *Aequidens pulcher* (Cichlidae) after rearing in monochromatic light. *Vision Research, 39*, 2441–2448.

Kronforst, M. R., Young, L. G., & Gilbert, L. E. (2007). Reinforcement of mate preference among hybridizing *Heliconius* butterflies. *Journal of Evolutionary Biology, 20*(1), 278–285. doi:10.1111/j.1420-9101.2006.01198.x

Kronforst, M. R., Young, L. G., Kapan, D. D., McNeely, C., O'Neill, R. J., & Gilbert, L. E. (2006). Linkage of butterfly mate preference and wing color preference cue at the genomic location of wingless. *Proceedings of the National Academy of Sciences, 103*(17), 6575–6580. doi:10.1073/pnas.0509685103

Kroodsma, D. E. (1989). Suggested experimental designs for song playbacks. *Animal Behaviour, 37, Part 4*(0), 600–609. doi:10.1016/0003-3472(89)90039-0

Kuijper, B., Pen, I., & Weissing, F. J. (2012). A guide to sexual selection theory. *Annual Review of Ecology, Evolution, and Systematics, 43*, 287–311.

Kvarnemo, C., Mobley, K. B., Partridge, C., Jones, A. G., & Ahnesjö, I. (2011). Evidence of paternal nutrient provisioning to embryos in broad-nosed pipefish *Syngnathus typhle*. *Journal of Fish Biology, 78*(6), 1725–1737. doi:10.1111/j.1095-8649.2011.02989.x

Kvarnemo, C., & Simmons, L. W. (2013). Polyandry as a mediator of sexual selection before and after mating. *Philosophical Transactions of the Royal Society of London B: Biological Sciences, 368*(1613), 20120042. doi:10.1098/rstb.2012.0042

Labonne, J., & Hendry, A. P. (2010). Natural and sexual selection giveth and taketh away reproductive barriers: models of population divergence in guppies. *American Naturalist, 176*(1), 26–39.

Lachlan, R., & Servedio, M. R. (2004). Song learning accelerates allopatric speciation. *Evolution, 58*(9), 2049–2063.

Lachlan, R. F., Anderson, R. C., Peters, S., Searcy, W. A., & Nowicki, S. (2014). Typical versions of learned swamp sparrow song types are more effective signals than are less typical versions. *Proceedings of the Royal Society B: Biological Sciences, 281*(1785). doi:10.1098/rspb.2014.0252

Lachlan, R. F., & Nowicki, S. (2015). Context-dependent categorical perception in a songbird. *Proceedings of the National Academy of Sciences, 112*(6), 1892–1897. doi:10.1073/pnas.1410844112

Lachlan, R. F., Verzijden, M. N., Bernard, C. S., Jonker, P.-P., Koese, B., Jaarsma, S., ... ten Cate, C. (2013). The progressive loss of syntactical structure in bird song along an island colonization chain. *Current Biology, 23*(19), 1896–1901. doi:10.1016/j.cub.2013.07.057

Lack, D. (1940). Courtship feeding in birds. *Auk*, 169–178.

Lackey, A. C., & Boughman, J. W. (2014). Female discrimination against heterospecific mates does not depend on mating habitat. *Behavioral Ecology*, aru111.

Lampert, K. P., Schmidt, C., Fischer, P., Volff, J.-N., Hoffmann, C., Muck, J., ... Schartl, M. (2010). Determination of onset of sexual maturation and mating behavior by melanocortin receptor 4 polymorphisms. *Current Biology: CB, 20*(19), 1729–1734.

LaMunyon, C. W., & Eisner, T. (1993). Postcopulatory sexual selection in an arctiid moth (*Utetheisa ornatrix*). *Proceedings of the National Academy of Sciences, 90*(10), 4689–4692. doi:10.1073/pnas.90.10.4689

Lande, R. (1981). Models of speciation by sexual selection on polygenic traits. *Proceedings of the National Academy of Sciences, 78*(6), 3721–3725.

———. (1982). A quantitative genetic theory of life history evolution. *Ecology, 63*(3), 607–615.

———. (1987). Genetic correlations between the sexes in the evolution of sexual dimorphism and mating preferences. In J. W. Bradbury & M. B. Andersson (Eds.), Sexual *Selection: Testing the Alternatives* (pp. 83–94). Chichester: Wiley.

Lande, R., & Arnold, S. J. (1985). Evolution of mating preference and sexual dimorphism. *Journal of Theoretical Biology, 117*(4), 651–664.

Landry, C., Garant, D., Duchesne, P., & Bernatchez, L. (2001). 'Good genes as heterozygosity': the major histocompatibility complex and mate choice in Atlantic salmon (*Salmo salar*). *Proceedings of the Royal Society of London B: Biological Sciences, 268*(1473), 1279–1285.

Larhammar, D., Nordstrom, K., & Larsson, T. A. (2009). Evolution of vertebrate rod and cone phototransduction genes. *Philosophical Transactions of the Royal Society of London B: Biological Sciences, 364*(1531), 2867–2880. doi:10.1098/rstb.2009.0077

Latour, Y., Perriat-Sanguinet, M., Caminade, P., Boursot, P., Smadja, C. M., & Ganem, G. (2014). Sexual selection against natural hybrids may contribute to reinforcement in a house mouse hybrid zone. *Proceedings of the Royal Society of London B: Biological Sciences, 281*(1776), 20132733.

Lattanzio, M. S., Metro, K. J., & Miles, D. B. (2014). Preference for male traits differs in two female morphs of the tree lizard, *Urosaurus ornatus*. *PLoS ONE, 9*(7), e101515. doi:10.1371/journal.pone.0101515

Lauay, C., Gerlach, N. M., Adkins-Regan, E., & DeVoogd, T. J. (2004). Female zebra finches require early song exposure to prefer high-quality song as adults. *Animal Behaviour, 68*, 1249–1255. doi:10.1016/j.anbehav.2003.12.025

Lea, A. M., & Ryan, M. J. (2015). Irrationality in mate choice revealed by túngara frogs. *Science, 349*(6251), 964–966.

Lea, J., Dyson, M., & Halliday, T. (2001). Calling by male midwife toads stimulates females to maintain reproductive condition. *Animal Behaviour, 61*(2), 373–377. doi: http://dx.doi.org/10.1006/anbe.2000.1604

Leary, G. P., Allen, J. E., Bunger, P. L., Luginbill, J. B., Linn, C. E., Macallister, I. E., ... Wanner, K. W. (2012). Single mutation to a sex pheromone receptor provides adaptive specificity between closely related moth species. *Proceedings of the National Academy of Sciences, 109*(35), 14081–14086. doi:10.1073/pnas.1204661109

LeBas, N. R. (2006). Female finery is not for males. *Trends in Ecology & Evolution, 21*(4), 170–173.

LeBas, N. R., Hockham, L. R., & Ritchie, M. G. (2004). Sexual selection in the gift-giving dance fly *Rhamphomyia Sulcata*, favors small males carrying small gifts. *Evolution, 58*(8), 1763–1772.

Leboucher, G., & Pallot, K. (2004). Is he all he says he is? Intersexual eavesdropping in the domestic canary, *Serinus canaria*. *Animal Behaviour, 68*(4), 957–963. doi:http://dx.doi.org/10.1016/j.anbehav.2003.12.011

Leboucher, G., Vallet, E., Nagle, L., Béguin, N., Bovet, D., Hallé, F., ... Kreutzer, M. (2012). Studying female reproductive activities in relation to male song: the domestic canary as a model. *Advances in the Study of Behavior, 44*, 183–223. doi:10.1016/B978-0-12-394288-3.00005-8

Lebreton, S., Borrero-Echeverry, F., Gonzalez, F., Solum, M., Wallin, E., Hedenstroem, E., ... Witzgall, P. (2016). The Drosophila pheromone Z4-11Al is encoded together with habitat olfactory cues and mediates species-specific communication. *bioRxiv*. doi: 10.1101/083071

Leclaire, S., Merkling, T., Mulard, H., Lhuillier, E. M., Danchin, E., Raynaud, C., ... Hatch, S. A. (2012). Semiochemical compounds of preen secretion reflect genetic make-up in a seabird species. *Proceedings of the Royal Society of London B: Biological Sciences, 279*(1731), 1185–1193. doi:10.1098/rspb.2011.1611

Lee, A. J., Dubbs, S. L., Von Hippel, W., Brooks, R. C., & Zietsch, B. P. (2014). A multivariate approach to human mate preferences. *Evolution and Human Behavior, 35*(3), 193–203. doi:http://dx.doi.org/10.1016/j.evolhumbehav.2014.01.003

Lee, A. J., & Zietsch, B. P. (2011). Experimental evidence that women's mate preferences are directly influenced by cues of pathogen prevalence and resource scarcity. *Biology Letters, 7*, 892–895.

Lehtonen, T. K., & Lindström, K. (2008). Repeatability of mating preferences in the sand goby. *Animal Behaviour, 75*(1), 55–61. doi:http://dx.doi.org/10.1016/j.anbehav.2007.04.011

Lehtonen, T. K., Wong, B.B.M., & Lindström, K. (2010). Fluctuating mate preferences in a marine fish. *Biology Letters, 6*(1), 21–23. doi:10.1098/rsbl.2009.0558

Leinders-Zufall, T., Brennan, P., Widmayer, P., Maul-Pavicic, A., Jäger, M., Li, X.-H., ... Boehm, T. (2004). MHC class I peptides as chemosensory signals in the vomeronasal organ. *Science, 306*(5698), 1033–1037.

Lemmon, A., Smadja, C., & Kirkpatrick, M. (2004). Reproductive character displacement is not the only possible outcome of reinforcement. *Journal of Evolutionary Biology, 17*(1), 177–183.

Lemmon, E. M., & Lemmon, A. R. (2010). Reinforcement in chorus frogs: lifetime fitness estimates including intrinsic natural selection and sexual selection against hybrids. *Evolution, 64*(6), 1748–1761.

Lenton, A. P., & Francesconi, M. (2011). Too much of a good thing? Variety is confusing in mate choice. *Biology Letters, 7*(4), 528–531. doi:10.1098/rsbl.2011.0098.

Leonard, A. S., & Hedrick, A. V. (2009). Male and female crickets use different decision rules in response to mating signals. *Behavioral Ecology, 20*(6), 1175–1184.

Leonard, J. L. (2006). Sexual selection: lessons from hermaphrodite mating systems. *Integrative and Comparative Biology, 46*(4), 349–367. doi:10.1093/icb/icj041

Lerch, A., Rat-Fischer, L., Gratier, M., & Nagle, L. (2011). Diet quality affects mate choice in domestic female canary *Serinus canaria*. *Ethology, 117*(9), 769–776.

Lessells, C., & Boag, P. T. (1987). Unrepeatable repeatabilities: a common mistake. *The Auk*, 116–121.

Lessios, H. A. (2011). Speciation genes in free-spawning marine invertebrates. *Integrative and Comparative Biology*. doi:10.1093/icb/icr039

Levan, K. E., Fedina, T. Y., & Lewis, S. M. (2009). Testing multiple hypotheses for the maintenance of male homosexual copulatory behaviour in flour beetles. *Journal of Evolutionary Biology, 22*(1), 60–70. doi:10.1111/j.1420-9101.2008.01616.x

LeVay, S. (2011). From mice to men: biological factors in the development of sexuality. *Frontiers in neuroendocrinology, 32*(2), 110–113.

Levine, M. W. (2000). *Levine and Shefner's Fundamentals of Sensation and Perception* (3rd ed.). Oxford: Oxford University Press.

Lewis, S. M., & Austad, S. N. (1994). Sexual selection in flour beetles: the relationship between sperm precedence and male olfactory attractiveness. *Behavioral Ecology, 5*(2), 223–224. doi:10.1093/beheco/5.2.223

Lewontin, R., Kirk, D., & Crow, J. (1968). Selective mating, assortative mating, and inbreeding: Definitions and implications. *Eugenics Quarterly, 15*(2), 141–143. doi:10.1080/19485565.1968.9987764

Li, N., Takeyama, T., Jordan, L. A., & Kohda, M. (2015). Female control of paternity by spawning site choice in a cooperatively polyandrous cichlid. *Behaviour, 152*(2), 231–245. doi:10.1163/1568539X-00003242

Liao, W., & Lu, X. (2009). Male mate choice in the Andrew's toad *Bufo andrewsi*: a preference for larger females. *Journal of Ethology, 27*(3), 413–417. doi:10.1007/s10164-008-0135-7

Liberman, A. M. (1970). Some characteristics of perception in the speech mode. *Research publications—Association for Research in Nervous and Mental Disease, 48*, 238–254.

Lie, H. C., Simmons, L. W., & Rhodes, G. (2010). Genetic dissimilarity, genetic diversity, and mate preferences in humans. *Evolution and Human Behavior, 31*(1), 48–58. doi:http://dx.doi.org/10.1016/j.evolhumbehav.2009.07.001

Lieberman, D., Tooby, J., & Cosmides, L. (2003). The evolution of human incest avoidance mechanisms: an evolutionary psychological approach. In A. Wolf & J. P. Takala (Eds.), *Evolution and the Moral Emotions: Appreciating Edward Westermarck*. Stanford, CA: Stanford University Press.

Lifjeld, J. T., Gohli, J., & Johnsen, A. (2013). Promiscuity, sexual selection, and genetic diversity: a reply to Spurgin. *Evolution, 67*(10), 3073–3074.

Lihoreau, M., Zimmer, C., & Rivault, C. (2008). Mutual mate choice: when it pays both sexes to avoid inbreeding. *PLoS One, 3*(10), e3365. doi:10.1371/journal.pone.0003365

Limbourg, T., Mateman, A. C., Andersson, S., & Lessells, C.K.M. (2004). Female blue tits adjust parental effort to manipulated male UV attractiveness. *Proceedings of the Royal Society of London B: Biological Sciences*, 1–6.

Limbourg, T., Mateman, A. C., & Lessells, C. M. (2013). Opposite differential allocation by males and females of the same species. *Biology Letters, 9*(1), 20120835. doi:10.1098/rsbl.2012.0835.

Limousin, D., Streiff, R., Courtois, B., Dupuy, V., Alem, S., & Greenfield, M. D. (2012). Genetic architecture of sexual selection: QTL mapping of male song and female receiver traits in an acoustic moth. *PLoS One, 7*(9). doi:e44554. 10.1371/journal.pone.0044554

Lin, X. (2009). *Cryptococcus neoformans*: morphogenesis, infection, and evolution. *Infection, Genetics and Evolution, 9*(4), 401–416. doi:10.1016/j.meegid.2009.01.013

Lin, X., Jackson, J. C., Feretzaki, M., Xue, C., & Heitman, J. (2010). Transcription factors Mat2 and Znf2 operate cellular circuits orchestrating opposite- and same-sex mating in *Cryptococcus neoformans. PLoS Genetics, 6*(5), e1000953. doi:10.1371/journal.pgen.1000953

Lindström, K., & Kangas, N. (1996). Egg presence, egg loss, and female mate preferences in the sand goby (*Pomatoschistus minutus*). *Behavioral Ecology, 7*(2), 213–217.

Lindström, K., & Lehtonen, T. K. (2013). Mate sampling and choosiness in the sand goby. *Proceedings of the Royal Society of London B: Biological Sciences, 280*(1765), 20130983. doi:10.1098/rspb.2013.0983

Little, A. C., Burriss, R. P., Jones, B. C., DeBruine, L. M., & Caldwell, C. A. (2008). Social influence in human face preference: men and women are influenced more for long-term than short-term attractiveness decisions. *Evolution and Human Behavior, 29*(2), 140–146. doi:http://dx.doi.org/10.1016/j.evolhumbehav.2007.11.007

Little, A. C., Jones, B. C., & Burriss, R. P. (2007). Preferences for masculinity in male bodies change across the menstrual cycle. *Hormones and Behavior, 51*(5), 633–639. doi:http://dx.doi.org/10.1016/j.yhbeh.2007.03.006

Little, A. C., Jones, B. C., Burt, D. M., & Perrett, D. I. (2007). Preferences for symmetry in faces change across the menstrual cycle. *Biological Psychology, 76*(3), 209–216. doi:http://dx.doi.org/10.1016/j.biopsycho.2007.08.003

Little, A. C., Penton-Voak, I. S., Burt, D. M., & Perrett, D. I. (2003). Investigating an imprinting-like phenomenon in humans: partners and opposite-sex parents have similar hair and eye colour. *Evolution and Human Behavior, 24*(1), 43–51. doi:http://dx.doi.org/10.1016/S1090-5138(02)00119-8

Liu, Y., Rossiter, S. J., Han, X., Cotton, J. A., & Zhang, S. (2010). Cetaceans on a molecular fast track to ultrasonic hearing. *Current Biology, 20*(20), 1834–1839. doi:http://dx.doi.org/10.1016/j.cub.2010.09.008

Locatello, L., Poli, F., & Rasotto, M. B. (2015). Context-dependent evaluation of prospective mates in a fish. *Behavioral Ecology and Sociobiology, 69*(7), 1119–1126. doi:10.1007/s00265-015-1924-y

Lockhart, A. B., Thrall, P. H., & Antonovics, J. (1996). Sexually transmitted diseases in animals: ecological and evolutionary implications. *Biological Reviews, 71*(3), 415–471.

Lockley, M. G., McCrea, R. T., Buckley, L. G., Lim, J. D., Matthews, N. A., Breithaupt, B. H., ... Kim, K. S. (2016). Theropod courtship: large-scale physical evidence of display arenas and avian-like scrape ceremony behaviour by Cretaceous dinosaurs. *Scientific Reports*, 6.

Lombardo, M. P., Thorpe, P. A., & Power, H. W. (1999). The beneficial sexually transmitted microbe hypothesis of avian copulation. *Behavioral Ecology, 10*(3), 333–337.

Long, T. A., & Pischedda, A. (2005). Do female *Drosophila melanogaster* adaptively bias offspring sex ratios in relation to the age of their mate? *Proceedings of the Royal Society of London B: Biological Sciences, 272*(1574), 1781–1787. doi:10.1098/rspb.2005.3165.

Long, T. A., Pischedda, A., & Rice, W. R. (2010). Remating in *Drosophila melanogaster*: are indirect benefits condition dependent? *Evolution, 64*(9), 2767–2774.

Long, T.A.F., Pischedda, A., Stewart, A. D., & Rice, W. R. (2009). A cost of sexual attractiveness to high-fitness females. *PLoS Biology, 7*(12). doi:10.1371/journal.pbio.1000254

López, S. (1999). Parasitized female guppies do not prefer showy males. *Animal Behaviour, 57*(5), 1129–1134. doi:http://dx.doi.org/10.1006/anbe.1998.1064

López-Sepulcre, A., Gordon, S. P., Paterson, I. G., Bentzen, P., & Reznick, D. N. (2013). Beyond lifetime reproductive success: the posthumous reproductive dynamics of male Trinidadian guppies. *Proceedings of the Royal Society of London B: Biological Sciences, 280*(1763). doi:10.1098/rspb.2013.1116

Lorch, P. D., Proulx, S., Rowe, L., & Day, T. (2003). Condition-dependent sexual selection can accelerate adaptation. *Evolutionary Ecology Research, 5*(6), 867–881.

Lorch, P. D., & Servedio, M. R. (2007). The evolution of conspecific gamete precedence and its effect on reinforcement. *Journal of Evolutionary Biology, 20*(3), 937–949. doi:10.1111/j.1420-9101.2007.01306.x

Lorenz, K. Z. (1937). The companion in the bird's world. *The Auk, 54*(3), 245–273.

Løvlie, H., Gillingham, M.A.F., Worley, K., Pizzari, T., & Richardson, D. S. (2013). Cryptic female choice favours sperm from major histocompatibility complex-dissimilar males. *Proceedings of the Royal Society of London B: Biological Sciences, 280*(1769), 20131296. doi:10.1098/rspb.2013.1296

Lowry, D. B., Modliszewski, J. L., Wright, K. M., Wu, C. A., & Willis, J. H. (2008). The strength and genetic basis of reproductive isolating barriers in flowering plants. *Philosophical Transactions of the Royal Society of London B: Biological Sciences, 363*(1506), 3009–3021. doi:10.1098/rstb.2008.0064

Loyau, A., Cornuau, J. H., Clobert, J., & Danchin, É. (2012). Incestuous sisters: mate preference for brothers over unrelated males in Drosophila melanogaster. *PLoS One, 7*(12), e51293.

Lozano, G. (2009). Multiple cues in mate selection: the sexual interference hypothesis. *Bioscience Hypotheses, 2*(1), 37–42.

Lumley, A. J., Michalczyk, Ł., Kitson, J. J., Spurgin, L. G., Morrison, C. A., Godwin, J. L., … Chapman, T. (2015). Sexual selection protects against extinction. *Nature, 522*(7557), 470–473. doi:10.1038/nature14419.

Lunstra, D. D., Hays, W. G., Bellows, R. A., & Laster, D. B. (1985). *Increasing pregnancy rate in beef cattle by clitoral massage during artificial insemination*. Roman L. Hruska U.S. Meat Animal Research Center. Paper 54.

Lüpold, S., Pitnick, S., Berben, K. S., Blengini, C. S., Belote, J. M., & Manier, M. K. (2013). Female mediation of competitive fertilization success in *Drosophila melanogaster*. *Proceedings of the National Academy of Sciences, 110*(26), 10693–10698. doi:10.1073/pnas.1300954110

Luther, D. A., & Derryberry, E. P. (2012). Birdsongs keep pace with city life: changes in song over time in an urban songbird affects communication. *Animal Behaviour, 83*, 1059–1066. doi:10.1016/j.anbehav.2012.01.034

Luttbeg, B. (1996). A comparative Bayes tactic for mate assessment and choice. *Behavioral Ecology, 7*(4), 451–460.

———. (2002). Assessing the robustness and optimality of alternative decision rules with varying assumptions. *Animal Behaviour, 63*(4), 805–814. doi:http://dx.doi.org/10.1006/anbe.2001.1979

Luttbeg, B., Towner, M. C., Wandesforde-Smith, A., Mangel, M., & Foster, S. A. (2001). State-dependent mate-assessment and mate-selection behavior in female threespine sticklebacks (*Gasterosteus aculeatus*, Gasterosteiformes: Gasterosteidae). *Ethology, 107*(6), 545–558. doi:10.1046/j.1439–0310.2001.00694.x

Lynch, K. S., Crews, D., Ryan, M. J., & Wilczynski, W. (2006). Hormonal state influences aspects of female mate choice in the Túngara Frog (*Physalaemus pustulosus*). *Hormones and Behavior, 49*(4), 450–457. doi:http://dx.doi.org/10.1016/j.yhbeh.2005.10.001

Lynch, K. S., Stanely Rand, A., Ryan, M. J., & Wilczynski, W. (2005). Plasticity in female mate choice associated with changing reproductive states. *Animal Behaviour, 69*(3), 689–699. doi:http://dx.doi.org/10.1016/j.anbehav.2004.05.016

Lynch, M., & Walsh, B. (1998). *Genetics and Analysis of Quantitative Traits*. Sunderland, MA: Sinauer Associates.

Lynn, S. K. (2006). Cognition and evolution: learning and the evolution of sex traits. *Current Biology, 16*(11), R421–R423. doi:10.1016/j.cub.2006.05.011.

Lyon, B. E., & Montgomerie, R. (2012). Sexual selection is a form of social selection. *Philosophical Transactions of the Royal Society of London B: Biological Sciences, 367* (1600), 2266–2273.

Lyons, S. M., Beaulieu, M., & Sockman, K. W. (2014). Contrast influences female attraction to performance-based sexual signals in a songbird. *Biology Letters, 10*(10), 20140588. doi:10.1098/rsbl.2014.0588.

Lyons, S. M., Goedert, D., & Morris, M. R. (2014). Male-trait-specific variation in female mate preferences. *Animal Behaviour, 87*, 39–44.

M'Gonigle, L. K., Mazzucco, R., Otto, S. P., & Dieckmann, U. (2012). Sexual selection enables long-term coexistence despite ecological equivalence. *Nature, 484*(7395), 506–509.

Maan, M. E., Hofker, K. D., van Alphen, J.J.M., & Seehausen, O. (2006). Sensory drive in cichlid speciation. *American Naturalist, 167*(6), 947–954.

Maan, M. E., & Seehausen, O. (2011). Ecology, sexual selection and speciation. *Ecology Letters, 14*(6), 591–602.

———. (2012). Magic cues versus magic preferences in speciation. *Evolutionary Ecology Research, 14*(6), 779–785.

MacDorman, K. F., & Chattopadhyay, D. (2016). Reducing consistency in human realism increases the uncanny valley effect; increasing category uncertainty does not. *Cognition, 146*, 190–205. doi:10.1016/j.cognition.2015.09.019.

MacFarlane, G. R., Blomberg, S. P., Kaplan, G., & Rogers, L. J. (2007). Same-sex sexual behavior in birds: expression is related to social mating system and state of development at hatching. *Behavioral Ecology, 18*(1), 21–33. doi:10.1093/beheco/arl065

MacFarlane, G. R., Blomberg, S. P., & Vasey, P. L. (2010). Homosexual behaviour in birds: frequency of expression is related to parental care disparity between the sexes. *Animal Behaviour, 80*(3), 375–390.

Macías Garcia, C. ., & Ramirez, E. (2005). Evidence that sensory traps can evolve into honest signals. *Nature, 434*(7032), 501–505.

Macías Garcia, C. , & Saldívar Lemus, Y. (2012). Foraging costs drive female resistance to a sensory trap. *Proceedings of the Royal Society of London B: Biological Sciences, 279*(1736), 2262–2268.

Macías Garcia, C. , & Valero, A. (2010). Sexual conflict and sexual selection in the Goodeinae, a clade of viviparous fish with effective female mate choice. *Advances in the Study of Behavior, 42*, 1–54.

Mackay, T. F., Stone, E. A., & Ayroles, J. F. (2009). The genetics of quantitative traits: challenges and prospects. *Nature Reviews Genetics, 10*(8), 565–577. doi:10.1038/nrg2612

MacLaren, R. D. (2017). Effects of male apparent length on female preference for absolute body size in Xiphophorus helleri. *acta ethologica, 20*(1), 27–36. doi:10.1007/s10211-016-0245-0

MacLaren, R. D., & Daniska, D. (2008). Female preferences for dorsal fin and body size in Xiphophorus helleri: further investigation of the LPA bias in Poeciliid fishes. *Behaviour, 145*, 897–913.

MacLaren, R. D., & Fontaine, A. (2013). Incongruence between the sexes in preferences for body and dorsal fin size in *Xiphophorus variatus*. *Behavioural Processes, 92*, 99–106.

MacLaren, R. D., Gagnon, J., & He, R. (2011). Female bias for enlarged male body and dorsal fins in *Xiphophorus variatus*. *Behavioural Processes, 87*(2), 197–202.

Madden, J., Isden, J., & Dingle, C. (2011). Commentary on review by Boogert et al.: some problems facing females. *Behavioral Ecology, 22*(3), 461–462.

Madden, J. R., & Tanner, K. (2003). Preferences for coloured bower decorations can be explained in a nonsexual context. *Animal Behaviour, 65*(6), 1077–1083.

Madden, J. R., & Whiteside, M. A. (2013). Variation in female mate choice and mating success is affected by sex ratio experienced during early life. *Animal Behaviour, 86*(1), 139–142. doi:10.1016/j.anbehav.2013.05.003

Maddock, M., & Chang, M. (1979). Reproductive failure and maternal-fetal relationship in a Peromyscus species cross. *Journal of Experimental Zoology, 209*(3), 417–425.

Magnhagen, C. (1991). Predation risk as a cost of reproduction. *Trends in Ecology & Evolution, 6*(6), 183–186.

Magurran, A. E., & Ramnarine, I. W. (2005). Evolution of mate discrimination in a fish. *Current Biology, 15*(21), R867–R868.

Mahady, S. J., & Wolff, J. O. (2002). A field test of the Bruce effect in the monogamous prairie vole (*Microtus ochrogaster*). *Behavioral Ecology and Sociobiology, 52*(1), 31–37.

Maki, S. L., McDaniel, W. F., Boyce, K. C., Brown, C. M., Crane, S. M., Cundey, J., … Marcengill, P. R. (2001). Visual categorical perception by rats with temporal, striate, or sham ablations. *NeuroReport, 12*(16), 3425–3431.

Maklakov, A. A., & Arnqvist, G. (2009). Testing for direct and indirect effects of mate choice by manipulating female choosiness. *Current Biology, 19*(22), 1903–1906.

Maklakov, A. A., Cayetano, L., Brooks, R. C., & Bonduriansky, R. (2009). The roles of life-history selection and sexual selection in the adaptive evolution of mating behavior in a beetle. *Evolution, 64*(5), 1273–1282.

Mallet, J., Besansky, N., & Hahn, M. W. (2016). How reticulated are species? *BioEssays, 38*: 140–149.

Manier, M. K., Belote, J. M., Berben, K. S., Lüpold, S., Ala-Honkola, O., Collins, W. F., & Pitnick, S. (2013). Rapid diversification of sperm precedence traits and processes among three sibling *Drosophila* species. *Evolution, 67*(8), 2348–2362. doi:10.1111/evo.12117

Manier, M. K., Belote, J. M., Berben, K. S., Novikov, D., Stuart, W. T., & Pitnick, S. (2010). Resolving mechanisms of competitive fertilization success in *Drosophila melanogaster. Science, 328*(5976), 354–357. doi:10.1126/science.1187096

Manier, M. K., Lüpold, S., Belote, J. M., Starmer, W. T., Berben, K. S., Ala-Honkola, O., ... Pitnick, S. (2013). Postcopulatory sexual selection generates speciation phenotypes in *Drosophila. Current Biology, 23*(19), 1853–1862. doi:http://dx.doi.org/10.1016/j.cub.2013.07.086

Manser, A., König, B., & Lindholm, A. (2015). Female house mice avoid fertilization by t haplotype incompatible males in a mate choice experiment. *Journal of Evolutionary Biology, 28*(1), 54–64.

Marcillac, F., Grosjean, Y., & Ferveur, J.-F. (2005). A single mutation alters production and discrimination of *Drosophila* sex pheromones. *Proceedings of the Royal Society of London B: Biological Sciences, 272*(1560), 303–309.

Marcinkowska, U. M., & Rantala, M. J. (2012). Sexual imprinting on facial traits of opposite-sex parents in humans. *Evolutionary Psychology: An International Journal of Evolutionary Approaches to Psychology and Behavior, 10*(3), 621–630.

Markow, T. A. (2002). Perspective: female remating, operational sex ratio, and the arena of sexual selection in *Drosophila* species. *Evolution, 56*(9), 1725–1734. doi:10.1111/j.0014-3820.2002.tb00186.x

Marler, P. (1991). The instinct to learn. In S. Carey & R. Gelman (Eds.), *The Epigenesis of Mind; Essays on Biology and Cognition.* (pp. 37–66). Hillsdale, NJ: Lawrence Erlbaum.

———. (1997). Three models of song learning: evidence from behavior. *Journal of Neurobiology, 33*, 501–516.

Marlowe, F., Apicella, C., & Reed, D. (2005). Original article: men's preferences for women's profile waist-to-hip ratio in two societies. *Evolution and Human Behavior, 26*, 458–468. doi:10.1016/j.evolhumbehav.2005.07.005

Márquez, R., Bosch, J., & Eekhout, X. (2008). Intensity of female preference quantified through playback setpoints: call frequency versus call rate in midwife toads. *Animal Behaviour, 75*, 159–166. doi:10.1016/j.anbehav.2007.05.003

Marquis, d. S. (1785). Les 120 journées de Sodome. *Huitième Journée.*

Marshall, J. L., Arnold, M. L., & Howard, D. J. (2002). Reinforcement: the road not taken. *Trends in Ecology & Evolution, 17*(12), 558–563.

Martin, C. H. (2013). Strong assortative mating by diet, color, size, and morphology but limited progress toward sympatric speciation in a classic example: Cameroon crater lake cichlids. *Evolution, 67*(7), 2114–2123. doi:10.1111/evo.12090.

Martin, M. D., & Mendelson, T. C. (2015). The accumulation of reproductive isolation in early stages of divergence supports a role for sexual selection. *Journal of Evolutionary Biology, 29*(4):676–689.

Martin, O. Y., & Hosken, D. J. (2003). The evolution of reproductive isolation through sexual conflict. *Nature, 423*(6943), 979–982.

Martin, P., & Bateson, P. (2007) *Measuring Behaviour: An Introductory Guide.* (3rd ed.). Cambridge, UK: Cambridge University Press.

Martinez-Padilla, J., Vergara, P., Mougeot, F., & Redpath, S. M. (2012). Parasitized mates increase infection risk for partners. *American Naturalist, 179*(6), 811–820.

Maruthupandian, J., & Marimuthu, G. (2013). Cunnilingus apparently increases duration of copulation in the Indian flying fox, *Pteropus giganteus*. *PLoS ONE, 8*(3), e59743. doi:10.1371/journal.pone.0059743

Mateo, J. M., & Johnston, R. E. (2000). Kin recognition and the "armpit effect": evidence of self-referent phenotype matching. *Proceedings of the Royal Society of London B: Biological Sciences, 267*(1444), 695–700.

Matsuda, Y.-T., Okamoto, Y., Ida, M., Okanoya, K., & Myowa-Yamakoshi, M. (2012). Infants prefer the faces of strangers or mothers to morphed faces: an uncanny valley between social novelty and familiarity. *Biology Letters*. doi:10.1098/rsbl.2012.0346

Matsumoto, J. (1965). Studies on fine structure and cytochemical properties of erythrophores in swordtail, *Xiphophorus helleri*, with special reference to their pigment granules (pterinosomes). *Journal of Cell Biology, 27*(3), 493–504. doi:10.1083/jcb.27.3.493

Mattle, B., & Wilson, A. (2009). Body size preferences in the pot-bellied seahorse *Hippocampus abdominalis*: choosy males and indiscriminate females. *Behavioral Ecology and Sociobiology, 63*(10), 1403–1410. doi:10.1007/s00265-009-0804-8

Matute, D. R. (2010). Reinforcement of gametic isolation in Drosophila. *PLoS Biology, 8*(3), e1000341.

———. (2015). Noisy neighbors can hamper the evolution of reproductive isolation by reinforcing selection. *American Naturalist, 185*(2), 253–269.

Maurer, D., Grand, R. L., & Mondloch, C. J. (2002). Review: the many faces of configural processing. *Trends in Cognitive Sciences, 6*, 255–260. doi:10.1016/S1364-6613(02)01903-4

Mautz, B. S., & Jennions, M. D. (2011). The effect of competitor presence and relative competitive ability on male mate choice. *Behavioral Ecology*, arr048.

Mautz, B. S., Wong, B.B.M., Peters, R. A., & Jennions, M. D. (2013). Penis size interacts with body shape and height to influence male attractiveness. *Proceedings of the National Academy of Sciences*. doi:10.1073/pnas.1219361110

Mayr, E. (1942). *Systematics and the Origin of Species*. New York: Columbia University Press.

———. (1970). *Populations, Species, and Evolution: An Abridgment of Animal Species and Evolution*. Cambridge, MA: Harvard University Press.

Mays, H. L., & Hill, G. E. (2004). Choosing mates: good genes versus genes that are a good fit. *Trends in Ecology & Evolution, 19*(10), 554–559.

McCartney, J., Kokko, H., Heller, K.-G., & Gwynne, D. (2012). The evolution of sex differences in mate searching when females benefit: new theory and a comparative test. *Proceedings of the Royal Society of London B: Biological Sciences, 279*(1731), 1225–1232.

McClelland, B. E., Wilczynski, W., & Rand, A. S. (1997). Sexual dimorphism and species differences in the neurophysiology and morphology of the acoustic communication system of two neotropical hylids. *Journal of Comparative Physiology—A Sensory, Neural, and Behavioral Physiology, 180*(5), 451–462. doi:10.1007/s003590050062

McClintock, W. J., & Uetz, G. W. (1996). Female choice and pre-existing bias: visual cues during courtship in two *Schizocosa* wolf spiders (Araneae: Lycosidae). *Animal Behaviour, 52*(1), 167–181.

McCoy, E., Syska, N., Plath, M., Schlupp, I., & Riesch, R. (2011). Mustached males in a tropical poeciliid fish: emerging female preference selects for a novel male trait. *Behavioral Ecology and Sociobiology, 65*(7), 1437–1445.

McDermott, J., & Hauser, M. (2004). Are consonant intervals music to their ears? Spontaneous acoustic preferences in a nonhuman primate. *Cognition, 94*(2), B11–B21. doi:http://dx.doi.org/10.1016/j.cognition.2004.04.004

McDonald, G. C., James, R., Krause, J., & Pizzari, T. (2013). Sexual networks: measuring sexual selection in structured, polyandrous populations. *Philosophical Transactions of the Royal Society of London B: Biological Sciences, 368*(1613), 20120356.

McGlothlin, J. W., Neudorf, D.L.H., Casto, J. M., Nolan, V., & Ketterson, E. D. (2004). Elevated testosterone reduces choosiness in female dark-eyed juncos (*Junco hyemalis*): evidence for a hormonal constraint on sexual selection? *Proceedings of the Royal Society of London B: Biological Sciences, 271*(1546), 1377–1384. doi:10.1098/rspb.2004.2741.

McGraw, K. J. (2002). Environmental predictors of geographic variation in human mating preferences. *Ethology, 108*(4), 303–317.

McGregor, P. K. (2000). Playback experiments: design and analysis. *Acta Ethologica, 3*, 3–8.

McGregor, P. K., Catchpole, C. K., Dabelsteen, T., Falls, J. B., Fusani, L., Gerhardt, H. C., ... Weary, D. M. (1992). Design of playback experiments: the Thornbridge Hall NATO ARW consensus. In P. K. McGregor (Ed.), *Playback and Studies of Animal Communication* (pp. 1–9). New York: Plenum Press.

McGuigan, K., Van Homrigh, A., & Blows, M. W. (2008a). An evolutionary limit to male mating success. *Evolution, 62*(6), 1528–1537.

———. (2008b). Genetic analysis of female preference functions as function-valued traits. *American Naturalist, 172*(2), 194–202. doi:10.1086/588075

McLennan, D. A., & Ryan, M. J. (1997). Responses to conspecific and heterospecific olfactory cues in the swordtail *Xiphophorus cortezi*. *Animal Behaviour, 54*, 1077–1088.

McLennan, D. A., & Ryan, M. J. (1999). Interspecific recognition and discrimination based upon olfactory cues in northern swordtails. *Evolution, 53*(3), 880–888.

McLennan, D. A., & Ryan, M. J. (2008). Female swordtails, Xiphophorus continens, prefer the scent of heterospecific males. *Animal Behaviour, 75*, 1731–1737. doi:10.1016/j.anbehav.2007.10.030

McNiven, V.T.K., & Moehring, A. J. (2013). Identification of genetically linked female preference and male trait. *Evolution, 67*(8), 2155–2165. doi:10.1111/evo.12096

Mead, L. S., & Arnold, S. J. (2004). Quantitative genetic models of sexual selection. *Trends in Ecology & Evolution, 19*(5), 264–271.

Meister, M., & Bonhoeffer, T. (2001). Tuning and topography in an odor map on the rat olfactory bulb. *Journal of Neuroscience, 21*(4), 1351–1360.

Melo, M. C., Salazar, C., Jiggins, C. D., & Linares, M. (2009). Assortative mating preferences among hybrids offers a route to hybrid speciation. *Evolution, 63*(6), 1660–1665.

Mendelson, T. C. (2003). Sexual isolation evolves faster than hybrid inviability in a diverse and sexually dimorphic genus of fish (Percidae: Etheostoma). *Evolution, 57*(2), 317–327.

Mendelson, T. C., & Shaw, K. L. (2012). The (mis) concept of species recognition. *Trends in Ecology & Evolution, 27*(8), 421–427.

Merrill, R. M., Van Schooten, B., Scott, J. A., & Jiggins, C. D. (2011). Pervasive genetic associations between traits causing reproductive isolation in *Heliconius* butterflies. *Proceedings: Biological Sciences, 278*(1705), 511–518.

Merrill, R. M., Wallbank, R. W., Bull, V., Salazar, P. C., Mallet, J., Stevens, M., & Jiggins, C. D. (2012). Disruptive ecological selection on a mating cue. *Proceedings of the Royal Society of London B: Biological Sciences, 279*(1749), 4907–4913.

Mery, F., Varela, S.A.M., Danchin, E., Blanchet, S., Coolen, I., Parejo, D., & Wagner, R. H. (2009). Public versus personal information for mate copying in an invertebrate. *Current Biology, 19*(9), 730–734. doi:10.1016/j.cub.2009.02.064

Meyer, A., & Galizia, C. G. (2012). Elemental and configural olfactory coding by antennal lobe neurons of the honeybee (*Apis mellifera*). *Journal of Comparative Physiology A-Neuroethology Sensory Neural and Behavioral Physiology, 198*(2), 159–171. doi:10.1007/s00359-011-0696-8

Meyer, K., & Kirkpatrick, M. (2005). Up hill, down dale: quantitative genetics of curvaceous traits. *Philosophical Transactions of the Royal Society of London B: Biological Sciences, 360*(1459), 1443–1455.

Meyer, M., Popper, A. N., & Fay, R. R. (2012). Coding of sound direction in the auditory periphery of the lake sturgeon, Acipenser fulvescens. *Journal of Neurophysiology, 107*(2), 658–665. doi:10.1152/jn.00390.2011

Mhatre, N., Bhattacharya, M., Robert, D., & Balakrishnan, R. (2011). Matching sender and receiver: poikilothermy and frequency tuning in a tree cricket. *Journal of Experimental Biology, 214*(15), 2569–2578. doi:10.1242/jeb.057612

Michael, J., Bernal, X. E., & Rand, A. S. (2010). Female mate choice and the potential for ornament evolution in túngara frogs Physalaemus pustulosus. *Current Zoology, 56*(3), 343–357.

Michalak, P. (1996). Repeatability of mating behaviour in Montandon's newt, *Triturus montandoni* (Caudata Salamandridae). *Ethology Ecology & Evolution, 8*, 19–27.

Michl, G., Török, J., Péczely, P., Garamszegi, L. Z., & Schwabl, H. (2005). Female collared flycatchers adjust yolk testosterone to male age, but not to attractiveness. *Behavioral Ecology, 16*(2), 383–388. doi:10.1093/beheco/ari002

Milam, E. L. (2010). *Looking for a Few Good Males: Female Choice in Evolutionary Biology*. Baltimore, MD: Johns Hopkins University Press.

Milinski, M. (2006). The major histocompatibility complex, sexual selection, and mate choice. *Annual Review of Ecology, Evolution, and Systematics*, 159–186.

Milinski, M., & Bakker, T.C.M. (1992). Costs influence sequential mate choice in sticklebacks, *Gasterosteus aculeatus. Proceedings of the Royal Society of London B: Biological Sciences, 250*(1329), 229–233. doi:10.1098/rspb.1992.0153

Milinski, M., Croy, I., Hummel, T., & Boehm, T. (2013). Major histocompatibility complex peptide ligands as olfactory cues in human body odour assessment. *Proceedings of the Royal Society of London B: Biological Sciences, 280*(1755), 20122889.

Miller, C. T., & Bee, M. A. (2012). Receiver psychology turns 20: is it time for a broader approach? *Animal Behaviour, 83*(2), 331–343. doi:http://dx.doi.org/10.1016/j.anbehav.2011.11.025

Miller, C. W., & Moore, A. J. (2007). A potential resolution to the lek paradox through indirect genetic effects. *Proceedings of the Royal Society of London B: Biological Sciences, 274*(1615), 1279–1286.

Miller, C. W., & Svensson, E. I. (2014). Sexual selection in complex environments. *Annual Review of Entomology, 59*, 427–445. doi:10.1146/annurev-ento-011613-162044

Miller, G. (2000). *The Mating Mind: How Sexual Selection Shaped the Evolution of Human Nature*. New York: Doubleday.

Miller, G. F., & Todd, P. M. (1998). Mate choice turns cognitive. *Trends in Cognitive Sciences, 2*(5), 190–198. doi:http://dx.doi.org/10.1016/S1364-6613(98)01169-3

Miller, G. T., & Pitnick, S. (2002). Sperm-female coevolution in *Drosophila. Science, 298*(5596), 1230–1233.

Miller, W. J., Ehrman, L., & Schneider, D. (2010). Infectious speciation revisited: impact of symbiont-depletion on female fitness and mating behavior of *Drosophila paulistorum. PLoS Pathogens, 6*(12), e1001214.

Milner, R.N.C., Detto, T., Jennions, M. D., & Backwell, P.R.Y. (2010). Experimental evidence for a seasonal shift in the strength of a female mating preference. *Behavioral Ecology, 21*(2), 311–316. doi:10.1093/beheco/arp196

Mock, D. W., & Fujioka, M. (1990). Monogamy and long-term pair bonding in vertebrates. *Trends in Ecology & Evolution, 5*(2), 39–43. doi:http://dx.doi.org/10.1016/0169-5347(90)90045-F

Moehring, A. J., Llopart, A., Elwyn, S., Coyne, J. A., & Mackay, T.F.C. (2006). The genetic basis of prezygotic reproductive isolation between *Drosophila santomea* and *D. yakuba* due to mating preference. *Genetics, 173*(1), 215–223.

Moeliker, C. W. (2001). The first case of homosexual necrophilia in the mallard *Anas platyrhynchos* (Aves: Anatidae). *Deinsea, 8*, 243–248.

Møller, A., & Jennions, M. (2001). How important are direct fitness benefits of sexual selection? *Naturwissenschaften, 88*(10), 401–415.

Møller, A., & Pomiankowski, A. (1993). Why have birds got multiple sexual ornaments? *Behavioral Ecology and Sociobiology, 32*(3), 167–176.

Møller, A. P. (1990). Fluctuating asymmetry in male sexual ornaments may reliably reveal male quality. *Animal Behaviour, 40*(6), 1185–1187.

———. (1992). Female swallow preference for symmetrical male sexual ornaments. *Nature, 357*(6375), 238–240.

———. (1994). Repeatability of female choice in a monogamous swallow. *Animal Behaviour, 47*(3), 643–648. doi:http://dx.doi.org/10.1006/anbe.1994.1087

Møller, A. P., & Alatalo, R. V. (1999). Good-genes effects in sexual selection. *Proceedings of the Royal Society of London B: Biological Sciences, 266*(1414), 85–91.

Møller, A. P., & Swaddle, J. P. (1997). *Asymmetry, Developmental Stability and Evolution*. Oxford: Oxford University Press.

Mombaerts, P. (2004). Genes and ligands for odorant, vomeronasal and taste receptors. *Nature Reviews Neuroscience, 5*(4), 263–278.

Moncho-Bogani, J., Lanuza, E., Hernández, A., Novejarque, A., & Martínez-García, F. (2002). Attractive properties of sexual pheromones in mice. Innate or learned? *Physiology & Behavior, 77*, 167–176. doi:10.1016/S0031-9384(02)00842-9

Moncho-Bogani, J., Martínez–García, F., Novejarque, A., & Lanuza, E. (2005). Attraction to sexual pheromones and associated odorants in female mice involves activation of the reward system and basolateral amygdala. *European Journal of Neuroscience, 21*(8), 2186–2198. doi:10.1111/j.1460-9568.2005.04036.x

Moore, A. J. (1994). Genetic evidence for the "good genes" process of sexual selection. *Behavioral Ecology and Sociobiology, 35*(4), 235–241.

Moore, A. J., & Pizzari, T. (2005). Quantitative genetic models of sexual conflict based on interacting phenotypes. *American Naturalist, 165*(S5), S88–S97.

Moore, F. R., Coetzee, V., Contreras-Garduño, J., Debruine, L. M., Kleisner, K., Krams, I., . . . Rantala, M. J. (2013). Cross-cultural variation in women's preferences for cues to sex-and stress-hormones in the male face. *Biology Letters, 9*(3), 20130050.

Moravec, M. L., Striedter, G. F., & Burley, N. T. (2010). 'Virtual parrots' confirm mating preferences of female budgerigars. *Ethology, 116*(10), 961–971. doi:10.1111/j.1439-0310.2010.01809.x

Morgado-Santos, M., Pereira, H. M., Vicente, L., & Collares-Pereira, M. J. (2015). Mate choice drives evolutionary stability in a hybrid complex. *PloS ONE, 10*(7), e0132760.

Mori, M. (1970). Bukimi no tani [The uncanny valley]. *Energy, 7* (4), 33–35. (Translated by Karl F. MacDorman and Takashi Minato in 2005 within Appendix B, for the paper "Androids as an experimental apparatus: why is there an uncanny valley and can we exploit it?") Paper presented at the Proceedings of the CogSci-2005 Workshop: Toward Social Mechanisms of Android Science.

Morris, D. (1967). *The Naked Ape: A Zoologist's Study of the Human Animal.* New York: Random House.

Morris, M. (1998a). Further examination of female preference for vertical bars in swordtails: preference for 'no bars' in a species without bars. *Journal of Fish Biology, 53 (Supplement A)*, 56–63.

Morris, M. R. (1998b). Female preference for trait symmetry in addition to trait size in swordtail fish. *Proceedings of the Royal Society of London B: Biological Sciences, 265* (1399), 907–911.

Morris, M. R., & Casey, K. (1998). Female swordtail fish prefer symmetrical sexual signal. *Animal Behaviour, 55*(1), 33–39.

Morris, M. R., Elias, J. A., & Moretz, J. A. (2001). Defining vertical bars in relation to female preference in the swordtail fish Xiphophorus cortezi (Cyprinodontiformes, Poeciliidae). *Ethology, 107*(9), 827–837.

Morris, M. R., Moretz, J. A., Farley, K., & Nicoletto, P. (2005). The role of sexual selection in the loss of sexually selected traits in the swordtail fish *Xiphophorus continens. Animal Behaviour, 69*, 1415–1424.

Morris, M. R., Mussel, M., & Ryan, M. J. (1995). Vertical bars on male *Xiphophorus multilineatus*: a signal that deters rival males and attracts females. *Behavioral Ecology, 6*, 274–279.

Morris, M. R., Nicoletto, P. F., & Hesselman, E. (2003). A polymorphism in female preference for a polymorphic male trait in the swordtail fish *Xiphophorus cortezi. Animal Behaviour, 65*, 45–52. doi:10.1006/anbe.2002.2042

Morris, M. R., Rios-Cardenas, O., & Brewer, J. (2010). Variation in mating preference within a wild population influences the mating success of alternative mating strategies. *Animal Behaviour, 79*(3), 673–678.

Morris, M. R., Rios-Cardenas, O., & Tudor, M. S. (2006). Larger swordtail females prefer asymmetrical males. *Biology Letters, 2*(1), 8–11. doi:10.1098/rsbl.2005.0387

Morris, M. R., & Ryan, M. J. (1996). Sexual difference in signal-receiver coevolution. *Animal Behaviour, 52*, 1017–1024.

Morris, M. R., Tudor, M. S., & Dubois, N. S. (2007). Sexually selected signal attracted females before deterring aggression in rival males. *Animal Behaviour, 74*(5), 1189–1197.

Morris, M. R., Wagner, W. E., Jr, & Ryan, M. J. (1996). A negative correlation between trait and mate preference in *Xiphophorus pygmaeus. Animal Behaviour, 52*, 1193–1203.

Mowles, S. L., & Ord, T. J. (2012). Repetitive signals and mate choice: insights from contest theory. *Animal Behaviour, 84*(2), 295–304. doi:http://dx.doi.org/10.1016/j.anbehav.2012.05.015

Mudry, K. M., & Capranica, R. R. (1987). Correlation between auditory thalamic area evoked responses and species-specific call characteristics. II. H. *Hyla cinerea* (Anura: Hylidae). *Journal of Comparative Physiology A, 161*(3), 407–416.

Mulard, H., Danchin, E., Talbot, S. L., Ramey, A. M., Hatch, S. A., White, J. F., ... Wagner, R. H. (2009). Evidence that pairing with genetically similar mates is maladaptive in a monogamous bird. *BMC Evolutionary Biology, 9*(1), 147.

Mulder, Raoul A., & Hall, Michelle L. (2013). Animal behaviour: a song and dance about lyrebirds. *Current Biology, 23*(12), R518–R519. doi:http://dx.doi.org/10.1016/j.cub.2013.05.009.

Muller, M. N., Thompson, M. E., & Wrangham, R. W. (2006). Male chimpanzees prefer mating with old females. *Current Biology, 16*(22), 2234–2238.

Muñoz, N. E., & Blumstein, D. T. (2012). Multisensory perception in uncertain environments. *Behavioral Ecology, 23*(3), 457–462. doi:10.1093/beheco/arr220

Muntz, W.R.A., & Mouat, G.S.V. (1984). Annual variations in the visual pigments of brown trout inhabiting lochs providing different light environments. *Vision Research, 24*(11), 1575–1580.

Muraco, J. J., Aspbury, A. S., & Gabor, C. R. (2014). Does male behavioral type correlate with species recognition and stress? *Behavioral Ecology, 25*(1), 200–205.

Murphy, C. G. (2012). Simultaneous mate-sampling by female barking treefrogs (*Hyla gratiosa*). *Behavioral Ecology, 23*(6), 1162–1169. doi:10.1093/beheco/ars093

Murphy, C. G., & Gerhardt, H. C. (2000). Mating preference functions of individual female barking treefrogs, *Hyla gratiosa*, for two properties of male advertisement calls. *Evolution, 54*(2), 660–669.

———. (2002). Mate sampling by female barking treefrogs (*Hyla gratiosa*). *Behavioral Ecology, 13*(4), 472–480. doi:10.1093/beheco/13.4.472

Murphy, H. A., & Zeyl, C. W. (2015). A potential case of reinforcement in a facultatively sexual unicellular eukaryote. *American Naturalist, 186*(2), 312–319.

Musch, J., & Klauer, K. C. (2003). The psychology of evaluation: an introduction. *The Psychology of Evaluation: Affective Processes in Cognition and Emotion* (pp. 1–5). Hillside, NJ: Lawrence Erlbaum Associates.

Myhre, L. C., de Jong, K., Forsgren, E., & Amundsen, T. (2012). Sex roles and mutual mate choice matter during mate sampling. *American Naturalist, 179*(6), 741–755.

Nadrowski, B., & Göpfert, M. C. (2009). Level-dependent auditory tuning transducer-based active processes in hearing and best-frequency shifts. *Communicative and Integrative Biology, 2*(1), 7–10. doi:10.10l6/i.cub.2008.07.095

Naguib, M., Kazek, A., Schaper, S. V., Van Oers, K., & Visser, M. E. (2010). Singing activity reveals personality traits in great tits. *Ethology, 116*(8), 763–769.

Nakadera, Y., Swart, E. M., Maas, J. P. A., Montagne-Wajer, K., Ter Maat, A., & Koene, J. M. (2014). Effects of age, size, and mating history on sex role decision of a simultaneous hermaphrodite. *Behavioral Ecology* 26(1), 232–241. doi:10.1093/beheco/aru184

Nakagawa, S., & Schielzeth, H. (2010). Repeatability for Gaussian and non-Gaussian data: a practical guide for biologists. *Biological Reviews of the Cambridge Philosophical Society, 85*(4), 935–956. doi:10.1111/j.1469-185X.2010.00141.x

Nakano, R., Takanashi, T., Skals, N., Surlykke, A., & Ishikawa, Y. (2010). To females of a noctuid moth, male courtship songs are nothing more than bat echolocation calls. *Biology Letters, 6*(5), 582–584. doi:10.1098/rsbl.2010.0058

Nandi, D., & Balakrishnan, R. (2013). Call intensity is a repeatable and dominant acoustic feature determining male call attractiveness in a field cricket. *Animal Behaviour, 86*(5), 1003–1012.

Narins, P. M., & Capranica, R. R. (1976). Sexual differences in the auditory system of the tree frog *Eleutherodactylus coqui*. *Science, 192*(4237), 378.

———. (1980). Neural adaptations for processing the two-note call of the Puerto Rican treefrog, *Eleutherodactylus coqui*. *Brain, Behavior and Evolution, 17*(1), 48–66.

Narraway, C., Hunt, J., Wedell, N., & Hosken, D. (2010). Genotype-by-environment interactions for female preference. *Journal of Evolutionary Biology, 23*(12), 2550–2557.

Navara, K., Hill, G., & Mendonça, M. (2006). Yolk androgen deposition as a compensatory strategy. *Behavioral Ecology and Sociobiology, 60*(3), 392–398. doi:10.1007/s00265-006-0177-1

Navarrete-Palacios, E., Hudson, R., Reyes-Guerrero, G., & Guevara-Guzmán, R. (2003). Lower olfactory threshold during the ovulatory phase of the menstrual cycle. *Biological Psychology, 63*(3), 269–279. doi:http://dx.doi.org/10.1016/S0301-0511(03)00076-0

Neff, B. D. (2000). Females aren't perfect: maintaining genetic variation and the lek paradox. *Trends in Ecology and Evolution, 15*(10), 395.

Neff, B. D., & Pitcher, T. E. (2005). Genetic quality and sexual selection: an integrated framework for good genes and compatible genes. *Molecular Ecology, 14*(1), 19–38.

———. (2008). Mate choice for non-additive genetic benefits: a resolution to the lek paradox. *Journal of Theoretical Biology, 254*(1), 147–155.

———. (2009). Mate choice for nonadditive genetic benefits and the maintenance of genetic diversity in song sparrows. *Journal of Evolutionary Biology, 22*(2), 424–429.

Nelson, C. M. (1995). Male size, spawning pit size and female mate choice in a lekking cichlid fish. *Animal Behaviour, 50*(6), 1587–1599.

Nelson, D. A., & Marler, P. (1993). Innate recognition of song in white-crowned sparrows: a role in selective vocal learning? *Animal Behaviour, 46*(4), 806–808.

Nelson, D. A., & Marler, P. M. (1989). Categorical perception of a natural stimulus continuum: birdsong. *Science, 244*, 976–978.

Nettle, D., & Pollet, T. V. (2008). Natural selection on male wealth in humans. *American Naturalist, 172*(5), 658–666.

Neuhofer, D., Stemmler, M., & Ronacher, B. (2011). Neuronal precision and the limits for acoustic signal recognition in a small neuronal network. *Journal of Comparative Physiology A, 197*(3), 251–265. doi:10.1007/s00359-010-0606-5

Nevitt, G. A., Dittman, A. H., Quinn, T. P., & Moody, W. J., Jr. (1994). Evidence for a peripheral olfactory memory in imprinted salmon. *Proceedings of the National Academy of Sciences, 91*(10), 4288–4292. doi:10.1073/pnas.91.10.4288

Ng, S. H., Shankar, S., Shikichi, Y., Akasaka, K., Mori, K., & Yew, J. Y. (2014). Pheromone evolution and sexual behavior in *Drosophila* are shaped by male sensory exploitation of other males. *Proceedings of the National Academy of Sciences, 111*(8), 3056–3061.

Nichols, R. A., & Butlin, R. K. (1989). Does runaway sexual selection work in finite populations? *Journal of Evolutionary Biology, 2*(4), 299–313.

Nieder, A., & Miller, E. K. (2003). Coding of cognitive magnitude: compressed scaling of numerical information in the primate prefrontal cortex. *Neuron, 37*(1), 149–157.

Niehuis, O., Buellesbach, J., Gibson, J. D., Pothmann, D., Hanner, C., Mutti, N. S., . . . Schmitt, T. (2013). Behavioural and genetic analyses of *Nasonia* shed light on the evolution of sex pheromones. *Nature, 494*(7437), 345–348.

Ninnes, C. E., Adrion, M., Edelaar, P., Tella, J. L., & Andersson, S. (2015). A receiver bias for red predates the convergent evolution of red color in widowbirds and bishops. *Behavioral Ecology*, arv068.

Nishi, H., Inagi, R., Kato, H., Tanemoto, M., Kojima, I., Son, D., . . . Nangaku, M. (2008). Hemoglobin is expressed by mesangial cells and reduces oxidant stress. *Journal of the American Society of Nephrology, 19*(8), 1500–1508. doi:10.1681/asn.2007101085

Nöbel, S., & Witte, K. (2013). Public information influences sperm transfer to females in sailfin molly males. *PloS ONE, 8*(1), e53865.

Nocke, H. (1972). Physiological aspects of sound communication in crickets (*Gryllus campestris* L.). *Journal of Comparative Physiology, 80*, 141–162.

Noë, R., & Hammerstein, P. (1995). Biological markets. *Trends in Ecology & Evolution, 10*(8), 336–339. doi:http://dx.doi.org/10.1016/S0169-5347(00)89123-5

Nordell, S., & Valone, T. (1998). Mate choice copying as public information. *Ecology Letters, 1*(2), 74–76.

Nosil, P., & Hohenlohe, P. A. (2012). Dimensionality of sexual isolation during reinforcement and ecological speciation in *Timema cristinae* stick insects. *Evolutionary Ecology Research, 14*(4), 467–485.

Nosil, P., & Schluter, D. (2011). The genes underlying the process of speciation. *Trends in Ecology & Evolution, 26*(4), 160–167.

Oakes, E. J., & Barnard, P. (1994). Fluctuating asymmetry and mate choice in paradise whydahs, *Vidua paradisaea*: an experimental manipulation. *Animal Behaviour, 48*(4), 937–943. doi:http://dx.doi.org/10.1006/anbe.1994.1319

Oetting, S., Pröve, E., & Bischof, H.-J. (1995). Sexual imprinting as a two-stage process: mechanisms of information storage and stabilization. *Animal Behaviour, 50*(2), 393–403. doi:http://dx.doi.org/10.1006/anbe.1995.0254

Oh, K. P., & Badyaev, A. V. (2006). Adaptive genetic complementarity in mate choice coexists with selection for elaborate sexual traits. *Proceedings of the Royal Society of London B: Biological Sciences, 273*(1596), 1913–1919.

———. (2010). Structure of social networks in a passerine bird: consequences for sexual selection and the evolution of mating strategies. *American Naturalist, 176*(3), E80–E89.

Oh, K. P., & Shaw, K. L. (2013). Multivariate sexual selection in a rapidly evolving speciation phenotype. *Proceedings of the Royal Society of London B: Biological Sciences, 280*(1761).

Oinonen, K. A., & Mazmanian, D. (2007). Facial symmetry detection ability changes across the menstrual cycle. *Biological Psychology, 75*(2), 136–145.

Okamoto, K. W., & Grether, G. F. (2013). The evolution of species recognition in competitive and mating contexts: the relative efficacy of alternative mechanisms of character displacement. *Ecology Letters, 16*(5), 670–678.

Oliver, M., & Evans, J. P. (2014). Chemically moderated gamete preferences predict offspring. *Proceedings of the Royal Society of London B: Biological Sciences 281*(20140148).

Ollivier, F. J., Samuelson, D. A., Brooks, D. E., Lewis, P. A., Kallberg, M. E., & Komaromy, A. M. (2004). Comparative morphology of the tapetum lucidum (among selected species). *Veterinary Ophthalmology, 7*(1), 11–22.

Olsson, M., Shine, R., Madsen, T., Gullberg, A., & Tegelstrom, H. (1996). Sperm selection by females. *Nature, 383*(6601), 585–585.

Oneal, E., Connallon, T., & Knowles, L. L. (2007). Conflict between direct and indirect benefits of female choice in desert *Drosophila*. *Biology Letters, 3*(1), 29–32.

Oneal, E., & Knowles, L. L. (2013). Ecological selection as the cause and sexual differentiation as the consequence of species divergence? *Proceedings of the Royal Society of London B: Biological Sciences, 280*(1750), 20122236.

Ophir, A. G., Wolff, J. O., & Phelps, S. M. (2008). Variation in neural V1aR predicts sexual fidelity and space use among male prairie voles in semi-natural settings. *Proceedings of the National Academy of Sciences, 105*(4), 1249–1254. doi:10.1073/pnas.0709116105

Opie, C., Atkinson, Q. D., Dunbar, R. I., & Shultz, S. (2013). Male infanticide leads to social monogamy in primates. *Proceedings of the National Academy of Sciences, 110*(33), 13328–13332.

Ord, T. J., King, L., & Young, A. R. (2011). Contrasting theory with the empirical data of species recognition. *Evolution, 65*(9), 2572–2591.

Ortigosa, A., & Rowe, L. (2002). The effect of hunger on mating behaviour and sexual selection for male body size in *Gerris buenoi*. *Animal Behaviour, 64*(3), 369–375. doi:http://dx.doi.org/10.1006/anbe.2002.3065

Ortiz-Barrientos, D., Counterman, B. A., & Noor, M. A. (2004). The genetics of speciation by reinforcement. *PLoS Biology, 2*(12), e416.

Ortíz-Barrientos, D., & Noor, M.A.F. (2005). Evidence for a one-allele assortative mating locus. *Science, 310*(5753), 1467. doi:10.1126/science.1121260

Ostojić, L., Shaw, R. C., Cheke, L. G., & Clayton, N. S. (2013). Evidence suggesting that desire-state attribution may govern food sharing in Eurasian jays. *Proceedings of the National Academy of Sciences, 110*(10), 4123–4128. doi:10.1073/pnas.1209926110

Otter, K., McGregor, P. K., Terry, A.M.R., Burford, F.R.L., Peake, T. M., & Dabelsteen, T. (1999). Do female great tits (*Parus major*) assess males by eavesdropping? A field study using interactive song playback. *Proceedings of the Royal Society of London B: Biological Sciences*, 5.

Otter, K., & Ratcliffe, L. (1996). Female initiated divorce in a monogamous songbird: abandoning mates for males of higher quality. *Proceedings of the Royal Society of London B: Biological Sciences, 263*(1368), 351–355. doi:10.1098/rspb.1996.0054.

Otti, O., Johnston, P. R., Horsburgh, G. J., Galindo, J., & Reinhardt, K. (2014). Female transcriptomic response to male genetic and nongenetic ejaculate variation. *Behavioral Ecology, 26*(3), 681–688. doi:10.1093/beheco/aru209

Otto, S. P. (2009). The evolutionary enigma of sex. *American Naturalist, 174*(S1), S1–S14.

Owens, I. P., Rowe, C., & Thomas, A. L. (1999). Sexual selection, speciation and imprinting: separating the sheep from the goats. *Trends in Ecology & Evolution, 14*(4), 131–132.

Owens, I.P.F. (2006). Where is behavioural ecology going? *Trends in Ecology & Evolution, 21*(7), 356–361.

Paczolt, K. A., & Jones, A. G. (2010). Post-copulatory sexual selection and sexual conflict in the evolution of male pregnancy. *Nature, 464*(7287), 401–404. doi:http://www.nature.com/nature/journal/v464/n7287/suppinfo/nature08861_S1.html

Palmer, A. R. (2000). Quasi-replication and the contract of error: lessons from sex ratios, heritabilities and fluctuating asymmetry. *Annual Review of Ecology and Systematics, 31*(1), 441–480. doi:doi:10.1146/annurev.ecolsys.31.1.441

Palumbi, S. R. (1999). All males are not created equal: fertility differences depend on gamete recognition polymorphisms in sea urchins. *Proceedings of the National Academy of Sciences, 96*(22), 12632–12637.

Panhuis, T. M., Butlin, R., Zuk, M., & Tregenza, T. (2001). Sexual selection and speciation. *Trends in Ecology & Evolution, 16*(7), 364–371.

Parker, G. (2006). Sexual conflict over mating and fertilization: an overview. *Philosophical Transactions of the Royal Society of London B: Biological Sciences, 361*(1466), 235–259.

Parker, G., & Partridge, L. (1998). Sexual conflict and speciation. *Philosophical Transactions of the Royal Society of London B: Biological Sciences, 353*(1366), 261–274.

Parker, G. A. (1970). Sperm competition and its evolutionary consequences in the insects. *Biological Reviews, 45*(4), 525–567. doi:10.1111/j.1469-185X.1970.tb01176.x

———. (1983). Mate quality and mate choice. In P. Bateson (Ed.), *Mate Choice* (pp. 141–166). Cambridge, UK: Cambridge University Press.

Parker, G. A., & Birkhead, T. R. (2013). Polyandry: the history of a revolution. *Philosophical Transactions of the Royal Society of London B: Biological Sciences, 368*(1613), 20120335.

Parker, T., & Garant, D. (2004). Quantitative genetics of sexually dimorphic traits and capture of genetic variance by a sexually-selected condition-dependent ornament in red junglefowl (*Gallus gallus*). *Journal of Evolutionary Biology, 17*(6), 1277–1285.

Parker, T. H., & Ligon, J. D. (2007). Multiple aspects of condition influence a heritable sexual trait: a synthesis of the evidence for capture of genetic variance in red junglefowl. *Biological Journal of the Linnean Society, 92*(4), 651–660.

Parsons, T. J., Olson, S. L., & Braun, M. J. (1993). Unidirectional spread of secondary sexual plumage traits across an avian hybrid zone. *Science, 260*, 1643–1646.

Partan, S. (2004). Multisensory animal communication. In G. Calvert, C. Spence, & B. E. Stein (Eds.), *The Handbook of Multisensory Processes* (pp. 225–240). Cambridge, MA: MIT Press.

———. (2013). Ten unanswered questions in multimodal communication. *Behavioral Ecology and Sociobiology*, 1–17. doi:10.1007/s00265-013-1565-y

Partan, S., & Marler, P. (1999). Communication goes multimodal. *Science, 283*, 1272–1273.

Partan, S. R., & Marler, P. (2005). Issues in the classification of multimodal communication signals. *American Naturalist, 166*, 231–245.

Partridge, L. (1980). Mate choice increases a component of offspring fitness in fruit flies. *Nature, 283*(5744), 290–291.

———. (1987). Life history constraints on sexual selection. In J. W. Bradbury & M. B. Andersson (Eds.), In *Sexual Selection: Testing the Alternatives* (pp. 265–277). Hoboken, NJ: John Wiley & Sons.

———. (1988). The rare-male effect: what is its evolutionary significance? *Philosophical Transactions of the Royal Society of London B: Biological Sciences, 319*(1196), 525–539.

Partridge, L., & Endler, J. A. (1987). Life history constraints on sexual selection. In J. W. Bradbury & M. B. Andersson (Eds.), *Sexual Selection: Testing the Alternatives* (pp. 265–277). Chichester: Wiley.

Partridge, L., & Hill, W. (1984). Mechanisms for frequency-dependent mating success. *Biological Journal of the Linnean Society, 23*(2–3), 113–132.

Passos, C., Tassino, B., Reyes, F., & Rosenthal, G. G. (2014). Seasonal variation in female mate choice and operational sex ratio in wild populations of an annual fish, *Austrolebias reicherti*. *PLoS ONE, 9*(7), e101649. doi: 101610.101371/journal.pone.0101649.

Patel, A. D. (2014). The evolutionary biology of musical rhythm: was Darwin wrong? *PLoS Biology, 12*(3), e1001821. doi:10.1371/journal.pbio.1001821

Paterson, H. E. (1985). The recognition concept of species. In E. S. Vrba (Ed.), Species and speciation, Transvaal Museum Monograph No. 4, Transvaal Museum, Pretoria, South Africa, 21–29.

Paterson, H. (1993). Chapter 8: Variation and the specific–mate recognition system. In P.P.G. Bateson, P. H. Klopfer, & N. S. Thompson (Eds.), *Perspectives in Ethology* (Vol. 10, pp. 209–227). New York: Plenum Press.

Patricelli, G. L., Krakauer, A. H., & McElreath, R. (2011). Assets and tactics in a mating market: economic models of negotiation offer insights into animal courtship dynamics on the lek. *Current Zoology, 57*(2), 225–236.

Patricelli, G. L., Uy, J.A.C., & Borgia, G. (2003). Multiple male traits interact: attractive bower decorations facilitate attractive behavioural displays in satin bowerbirds. *Proceedings of the Royal Society of London B: Biological Sciences, 270*(1531), 2389–2395.

Patricelli, G. L., Uy, J.A.C., Walsh, G., & Borgia, G. (2002). Sexual selection: male displays adjusted to female's response. *Nature, 415*(6869), 279–280.

Patrick, S. C., Chapman, J. R., Dugdale, H. L., Quinn, J. L., & Sheldon, B. C. (2012). Promiscuity, paternity and personality in the great tit. *Proceedings of the Royal Society of London B: Biological Sciences, 279*(1734), 1724–1730.

Pawlisch, B. A., Kelm-Nelson, C. A., Stevenson, S. A., & Riters, L. V. (2012). Behavioral indices of breeding readiness in female European starlings correlate with immunolabeling for catecholamine markers in brain areas involved in sexual motivation. *General and Comparative Endocrinology, 179*(3), 359–368. doi:http://dx.doi.org/10.1016/j.ygcen.2012.09.007

Pawlisch, B. A., & Riters, L. V. (2010). Selective behavioral responses to male song are affected by the dopamine agonist GBR-12909 in female European starlings (*Sturnus vulgaris*). *Brain Research, 1353*, 113–124. doi:10.1016/j.brainres.2010.07.003

Pawlisch, B. A., Stevenson, S. A., & Riters, L. V. (2011). α1-Noradrenegic receptor antagonism disrupts female songbird responses to male song. *Neuroscience Letters, 496*(1), 20–24. doi:http://dx.doi.org/10.1016/j.neulet.2011.03.078

Pawłowski, B., & Dunbar, R. I. (1999). Impact of market value on human mate choice decisions. *Proceedings of the Royal Society of London B: Biological Sciences, 266*(1416), 281–285. doi:10.1098/rspb.1999.0634

Payne, R. B., Payne, L. L., Woods, J. L., & Sorenson, M. D. (2000). Imprinting and the origin of parasite-host species associations in brood-parasitic indigobirds, Vidua chalybeata. *Animal Behaviour, 59*(1), 69–81.

Pelosi, P. (1994). Odorant-binding proteins. *Critical Reviews in Biochemistry and Molecular Biology, 29*(3), 199–228.

Penn, D. J. (2002). The scent of genetic compatibility: sexual selection and the major histocompatibility complex. *Ethology, 108*, 1–21.

Penn, D., & Potts, W. (1998). MHC-disassortative mating preferences reversed by cross-fostering. *Proceedings of the Royal Society of London B: Biological Sciences, 265*(1403), 1299. doi:10.1098/rspb.1998.0433

———. (1998). Chemical signals and parasite-mediated sexual selection. *Trends in Ecology & Evolution, 13*(10), 391–396.

———. (1999). The evolution of mating preferences and major histocompatibility complex genes. *American Naturalist, 153*(2), 145–164.

Penton-Voak, I. S., Perrett, D. I., Castles, D. L., Kobayashi, T., Burt, D. M., Murray, L. K., & Minamisawa, R. (1999). Menstrual cycle alters face preference. *Nature, 399*(6738), 741–742.

Peretti, A., & Eberhard, W. (2010). Cryptic female choice via sperm dumping favours male copulatory courtship in a spider. *Journal of Evolutionary Biology, 23*(2), 271–281.

Pérez-Staples, D., Córdova-García, G., & Aluja, M. (2014). Sperm dynamics and cryptic male choice in tephritid flies. *Animal Behaviour, 89*, 131–139.

Perrett, D., Penton-Voak, I. S., Little, A. C., Tiddeman, B. P., Burt, D. M., Schmidt, N., . . . Barrett, L. (2002). Facial attractiveness judgements reflect learning of parental age characteristics. *Proceedings of the Royal Society of London B: Biological Sciences, 269*(1494), 873–880. doi:10.1098/rspb.2002.1971

Perry, J., & Rowe, L. (2008). Ingested spermatophores accelerate reproduction and increase mating resistance but are not a source of sexual conflict. *Animal Behaviour, 76*(3), 993–1000.

———. (2012). Sex role stereotyping and sexual conflict theory. *Animal Behaviour, 83*(4), e10–e13.

Perry, J. C., Sharpe, D.M.T., & Rowe, L. (2009). Condition-dependent female remating resistance generates sexual selection on male size in a ladybird beetle. *Animal Behaviour, 77*(3), 743–748. doi:http://dx.doi.org/10.1016/j.anbehav.2008.12.013

Peterson, M. A., Honchak, B. M., Locke, S. E., Beeman, T. E., Mendoza, J., Green, J., . . . Monsen, K. J. (2005). Relative abundance and the species-specific reinforcement of male mating preference in the *Chrysochus* (Coleoptera: Chrysomelidae) hybrid zone. *Evolution, 59*(12), 2639–2655.

Petfield, D., Chenoweth, S. F., Rundle, H. D., & Blows, M. W. (2005). Genetic variance in female condition predicts indirect genetic variance in male sexual display traits. *Proceedings of the National Academy of Sciences, 102*(17), 6045–6050.

Petrie, M. (1994). Improved growth and survival of offspring of peacocks with more elaborate trains. *Nature, 371*(6498), 598–599.

Pfaus, J., Kippin, T., Coria-Avila, G., Gelez, H., Afonso, V., Ismail, N., & Parada, M. (2012). Who, what, where, when (and maybe even why)? How the experience of sexual reward connects sexual desire, preference, and performance. *Archives of Sexual Behavior, 41*(1), 31–62. doi:10.1007/s10508-012-9935-5

Pfaus, J. G., Kippin, T. E., & Centeno, S. (2001). Conditioning and sexual behavior: a review. *Hormones and Behavior, 40*(2), 291–321. doi:10.1006/hbeh.2001.1686

Pfennig, K. S. (1998). The evolution of mate choice and the potential for conflict between species and mate-quality recognition. *Proceedings of the Royal Society of London B: Biological Sciences, 265*, 1743–1748.

———. (2000). Female spadefoot toads compromise on mate quality to ensure conspecific matings. *Behavioral Ecology, 11*(2), 220–227.

———. (2007). Facultative mate choice drives adaptive hybridization. *Science, 318*(5852), 965–967. doi:10.1126/science.1146035

———. (2008). Population differences in condition-dependent sexual selection may promote divergence in non-sexual traits. *Evolutionary Ecology Research, 10*(5), 763–773.

Pfennig, K. S., Moncalvo, V.G.R., & Burmeister, S. S. (2013). Diet alters species recognition in juvenile toads. *Biology Letters, 9*(5), 20130599.

Pfennig, K. S., & Pfennig, D. W. (2009). Character displacement: ecological and reproductive responses to a common evolutionary problem. *Quarterly Review of Biology, 84*(3), 253.

Pfennig, K. S., & Rice, A. M. (2014). Reinforcement generates reproductive isolation between neighbouring conspecific populations of spadefoot toads. *Proceedings of the Royal Society of London B: Biological Sciences, 281*(1789), 20140949.

Pfennig, K. S., & Ryan, M. J. (2007). Character displacement and the evolution of mate choice: an artificial neural network approach. *Philosophical Transactions of the Royal Society of London B: Biological Sciences, 362*(1479), 411–419.

Pfennig, K. S., & Tinsley, R. C. (2002). Different mate preferences by parasitized and unparasitized females potentially reduces sexual selection. *Journal of Evolutionary Biology, 15*(3), 399–406. doi:10.1046/j.1420-9101.2002.00406.x

Phelps, S., & Ryan, M. (2000). History influences signal recognition: neural network models of tungara frogs. *Proceedings of the Royal Society of London B: Biological Sciences, 267*(1453), 1633–1639.

Phelps, S., Ryan, M., & Rand, A. (2001). Vestigial preference functions in neural networks and tungara frogs. *Proceedings of the National Academy of Sciences, 98*(23), 13161–13166.

Phelps, S. M., Rand, A. S., & Ryan, M. J. (2006). A cognitive framework for mate choice and species recognition. *American Naturalist, 167*, 28–42.

Piave, F. M. (1851). *Rigoletto*. Venice: Tipografia Gaspari.

Pierotti, M.E.R., Martin-Fernandez, J. A., & Seehausen, O. (2009). Mapping individual variation in male mating preference space: multiple choice in a color polymorphic cichlid fish. *Evolution, 63*(9), 2372–2388. doi:10.1111/j.1558-5646.2009.00716.x

Pignatelli, V., Temple, S. E., Chiou, T.-H., Roberts, N. W., Collin, S. P., & Marshall, N. J. (2011). Behavioural relevance of polarization sensitivity as a target detection mechanism in cephalopods and fishes. *Philosophical Transactions of the Royal Society B: Biological Sciences, 366*(1565), 734–741. doi:10.1098/rstb.2010.0204

Pilakouta, N., & Alonzo, S. H. (2014). Predator exposure leads to a short-term reversal in female mate preferences in the green swordtail, *Xiphophorus helleri*. *Behavioral Ecology, 25*(2), 306–312. doi:10.1093/beheco/art120

Pilastro, A., Simonato, M., Bisazza, A., Evans, J. P., & Pitnick, S. (2004). Cryptic female preference for colorful males in guppies. *Evolution, 58*(3), 665–669.

Pinker, S. (1997). *How the Mind Works*. New York: W. W. Norton.

Pisanski, K., & Feinberg, D. R. (2013). Cross-cultural variation in mate preferences for averageness, symmetry, body size, and masculinity. *Cross-Cultural Research, 47*(2), 162–197.

Pischedda, A., & Chippindale, A. K. (2006). Intralocus sexual conflict diminishes the benefits of sexual selection. *PLoS Biology, 4*(11), e356.

Pischedda, A., & Rice, W. R. (2012). Partitioning sexual selection into its mating success and fertilization success components. *Proceedings of the National Academy of Sciences, 109*(6), 2049–2053.

Pischedda, A., Shahandeh, M. P., Cochrane, W. G., Cochrane, V. A., & Turner, T. L. (2014). Natural variation in the strength and direction of male mating preferences for female pheromones in *Drosophila melanogaster*. *PLoS ONE, 9*(1), e87509. doi:10.1371/journal.pone.0087509

Pischedda, A., Stewart, A. D., Little, M. K., & Rice, W. R. (2011). Male genotype influences female reproductive investment in *Drosophila melanogaster*. *Proceedings of the Royal Society of London B: Biological Sciences, 278*(1715), 2165–2172.

Pitcher, T. E., Neff, B. D., Rodd, F. H., & Rowe, L. (2003). Multiple mating and sequential mate choice in guppies: females trade up. *Proceedings of the Royal Society of London B: Biological Sciences, 270*(1524), 1623–1629.

Pitnick, S., Brown, W. D., & Miller, G. T. (2001). Evolution of female remating behaviour following experimental removal of sexual selection. *Proceedings of the Royal Society of London B: Biological Sciences, 268*(1467), 557–563.

Pitnick, S., Markow, T., & Spicer, G. S. (1999). Evolution of multiple kinds of female sperm-storage organs in *Drosophila. Evolution*, 1804–1822.

Pizzari, T., & Birkhead, T. (2000). Female feral fowl eject sperm of subdominant males. *Nature, 405*(6788), 787–789.

Pizzari, T., & Snook, R. R. (2003). Perspective: sexual conflict and sexual selection: chasing away paradigm shifts. *Evolution, 57*(6), 1223–1236.

Pizzari, T., & Wedell, N. (2013). The polyandry revolution. *Philosophical Transactions of the Royal Society of London B: Biological Sciences, 368*(1613), 20120041.

Plomp, R., & Levelt, W. J. M. (1965). Tonal consonance and critical bandwidth. *Journal of the Acoustical Society of America, 38*(4), 548–560. doi:http://dx.doi.org/10.1121/1.1909741

Poiani, A. (2010). *Animal Homosexuality: A Biosocial Perspective*. Cambridge, UK: Cambridge University Press.

Polak, M. (2008). The developmental instability-sexual selection hypothesis: a general evaluation and case study. *Evolutionary Biology, 35*(3), 208–230.

Pölkki, M., Kortet, R., Hedrick, A., & Rantala, M. J. (2013). Dominance is not always an honest signal of male quality, but females may be able to detect the dishonesty. *Biology Letters, 9*(1), 20121002.

Pölkki, M., Krams, I., Kangassalo, K., & Rantala, M. J. (2012). Inbreeding affects sexual signalling in males but not females of *Tenebrio molitor. Biology Letters, 8*(3), 423–425.

Pomiankowski, A. (1988). *The Evolution of Female Mating Preferences for Male Genetic Quality*. New York: Oxford University Press.

Pomiankowski, A., & Iwasa, Y. (1993). Evolution of multiple sexual preferences by Fisher's runaway process of sexual selection. *Proceedings of the Royal Society of London B: Biological Sciences, 253*(1337), 173–181.

———. (1998). Runaway ornament diversity caused by Fisherian sexual selection. *Proceedings of the National Academy of Sciences, 95*(9), 5106–5111.

Pomiankowski, A., Iwasa, Y., & Nee, S. (1991). The evolution of costly mate preferences I. Fisher and biased mutation. *Evolution, 45*(6), 1422–1430. doi:10.2307/2409889

Poole, K. T. (2005). *Spatial Models of Parliamentary Voting*. Cambridge, UK: Cambridge University Press.

Postma, E., Griffith, S. C., & Brooks, R. (2006). Evolutionary genetics: evolution of mate choice in the wild. *Nature, 444*(7121), E16–E16.

Poulin, R., & Vickery, W. L. (1996). Parasite-mediated sexual selection: just how choosy are parasitized females? *Behavioral Ecology and Sociobiology, 38*(1), 43–49.

Powell, D. L., & Rosenthal, G. G. (2016). What artifice can and cannot tell us about animal behavior. *Current Zoology*, zow091.

Pressey, A. W. (1967). A theory of the Mueller-Lyer illusion. *Perceptual and Motor Skills, 25*(2), 569–572. doi:10.2466/pms.1967.25.2.569

Price, D. K., & Burley, N. T. (1994). Constraints on the evolution of attractive traits: selection in male and female zebra finches. *American Naturalist, 144*(6), 908–934. doi:10.2307/2463135

Price, T., Lewis, Z., Smith, D., Hurst, G., & Wedell, N. (2012). No evidence of mate discrimination against males carrying a sex ratio distorter in *Drosophila pseudoobscura. Behavioral Ecology and Sociobiology, 66*(4), 561–568.

Price, T., Schluter, D., & Heckman, N. E. (1993). Sexual selection when the female directly benefits. *Biological Journal of the Linnean Society, 48*(3), 187–211.

Price, T. A., & Wedell, N. (2008). Selfish genetic elements and sexual selection: their impact on male fertility. *Genetica, 134*(1), 99–111.

Priest, N. K., Galloway, L. F., & Roach, D. A. (2008). Mating frequency and inclusive fitness in *Drosophila melanogaster*. *American Naturalist, 171*(1), 10–21. doi:10.1086/523944

Priklopil, T., Kisdi, E., & Gyllenberg, M. (2015). Evolutionarily stable mating decisions for sequentially searching females and the stability of reproductive isolation by assortative mating. *Evolution, 69*(4), 1015–1026.

Proctor, H. C. (1991). Courtship in the water mite *Neumania papillator*: males capitalize on female adaptations for predation. *Animal Behavior, 42*, 589–598.

Prokop, Z. M., Michalczyk, Ł., Drobniak, S. M., Herdegen, M., & Radwan, J. (2012). Meta-analysis suggests choosy females get sexy sons more than "good genes." *Evolution, 66*(9), 2665–2673.

Prokuda, A. Y., & Roff, D. A. (2014). The quantitative genetics of sexually selected traits, preferred traits and preference: a review and analysis of the data. *Journal of Evolutionary Biology, 27*(11), 2283–2296. doi:10.1111/jeb.12483

Promislow, D. E., Smith, E. A., & Pearse, L. (1998). Adult fitness consequences of sexual selection in *Drosophila melanogaster*. *Proceedings of the National Academy of Sciences, 95*(18), 10687–10692.

Proulx, S. R., & Servedio, M. R. (2009). Dissecting selection on female mating preferences during secondary contact. *Evolution, 63*(8), 2031. doi:10.1111/j.1558-5646.2009.00710.x

Pruett-Jones, S. (1992). Independent versus nonindependent mate choice: do females copy each other? *American Naturalist, 140*(6), 1000–1009. doi:10.1086/285452

Pruitt, J. N., & Riechert, S. E. (2009). Male mating preference is associated with risk of pre-copulatory cannibalism in a socially polymorphic spider. *Behavioral Ecology and Sociobiology, 63*(11), 1573–1580.

Pruitt, J. N., Riechert, S. E., & Harris, D. J. (2011). Reproductive consequences of male body mass and aggressiveness depend on females' behavioral types. *Behavioral Ecology and Sociobiology, 65*(10), 1957–1966.

Prum, R. O. (2010). The Lande-Kirkpatrick mechanism is the null model of evolution by intersexual selection: implications for meaning, honesty, and design in intersexual signals. *Evolution, 64*(11), 3085–3100.

———. (2012). Aesthetic evolution by mate choice: Darwin's really dangerous idea. *Philosophical Transactions of the Royal Society of London B: Biological Sciences, 367*(1600), 2253–2265.

———. (2015). The role of sexual autonomy in evolution by mate choice. In T. Hoquet (Ed.), *Current Perspectives on Sexual Selection* (Vol. History, Philosophy and Theory of the Life Sciences pp. 237–262). New York: Springer Netherlands.

Pryke, S. R. (2010). Sex chromosome linkage of mate preference and color signal maintains assortative mating between interbreeding finch morphs. *Evolution, 64*(5), 1301–1310. doi:10.1111/j.1558-5646.2009.00897.x

Pryke, S. R., & Andersson, S. (2008). Female preferences for long tails constrained by species recognition in short-tailed red bishops. *Behavioral Ecology, 19*(6), 1116–1121.

Pryke, S. R., & Griffith, S. C. (2009). Genetic incompatibility drives sex allocation and maternal investment in a polymorphic finch. *Science, 323*(5921), 1605–1607.

Puebla, O., Bermingham, E., & Guichard, F. (2011). Perspective: matching, mate choice, and speciation. *Integrative and Comparative Biology, 51*(3), 485–491. doi:10.1093/icb/icr025

———. (2012). Pairing dynamics and the origin of species. *Proceedings of the Royal Society of London B: Biological Sciences, 279*(1731), 1085–1092. doi:10.1098/rspb.2011.1549

Puebla, O., Bermingham, E., Guichard, F., & Whiteman, E. (2007). Colour pattern as a single trait driving speciation in *Hypoplectrus* coral reef fishes? *Proceedings of the Royal Society of London B: Biological Sciences, 274*(1615), 1265–1271.

Pusey, A., & Wolf, M. (1996). Inbreeding avoidance in animals. *Trends in Ecology & Evolution, 11*(5), 201–206.

Pynchon, T. (1973). *Gravity's Rainbow*. New York: Penguin.

Queller, D. C. (1987). The evolution of leks through female choice. *Animal Behaviour, 35*(5), 1424–1432.

Qvarnström, A. (2001). Context-dependent genetic benefits from mate choice. *Trends in Ecology & Evolution, 16*(1), 5–7.

Qvarnström, A., & Bailey, R. I. (2009). Speciation through evolution of sex-linked genes. *Heredity, 102*(1), 4–15.

Qvarnström, A., Blomgren, V., Wiley, C., & Svedin, N. (2004). Female collared flycatchers learn to prefer males with an artificial novel ornament. *Behavioral Ecology, 15*(4), 543–548.

Qvarnström, A., Brommer, J. E., & Gustafsson, L. (2006). Testing the genetics underlying the co-evolution of mate choice and ornament in the wild. *Nature, 441*(7089), 84–86.

Qvarnström, A., & Forsgren, E. (1998). Should females prefer dominant males? *Trends in Ecology & Evolution, 13*(12), 498–501.

Qvarnström, A., Haavie, J., Saether, S., Eriksson, D., & Pärt, T. (2006). Song similarity predicts hybridization in flycatchers. *Journal of Evolutionary Biology, 19*(4), 1202–1209.

Qvarnstrom, A., Part, T., & Sheldon, B. C. (2000). Adaptive plasticity in mate preference linked to differences in reproductive effort. *Nature, 405*(6784), 344–347.

Qvarnstrom, A., & Price, T. D. (2001). Maternal effects, paternal effects and sexual selection. *TRENDS in Ecology & Evolution, 16*(2), 95–100. doi:10.1016/s0169-5347(00)02063-2

Qvarnström, A., Rice, A. M., & Ellegren, H. (2010). Speciation in *Ficedula* flycatchers. *Philosophical Transactions of the Royal Society of London B: Biological Sciences, 365*(1547), 1841–1852.

Racimo, F., Sankararaman, S., Nielsen, R., & Huerta-Sanchez, E. (2015). Evidence for archaic adaptive introgression in humans. *Nature Reviews Genetics, 16*(6), 359–371. doi:10.1038/nrg3936

Rado, R., Terkel, J., & Wollberg, Z. (1998). Seismic communication signals in the blind mole-rat (*Spalax ehrenbergi*): electrophysiological and behavioral evidence for their processing by the auditory system. *Journal of Comparative Physiology A, 183*(4), 503–511. doi:10.1007/s003590050275

Raffa, K., Havill, N., & Nordheim, E. (2002). How many choices can your test animal compare effectively? Evaluating a critical assumption of behavioral preference tests. *Oecologia, 133*(3), 422–429. doi:10.1007/s00442-002-1050-1

Raia, P., Passaro, F., Carotenuto, F., Maiorino, L., Piras, P., Teresi, L., ... Baiano, M. A. (2015). Cope's rule and the universal scaling law of ornament complexity. *American Naturalist, 186*(2), 165–175.

Ramachandran, V. S., Blakeslee, S., & Sacks, O. W. (1998). *Phantoms in the Brain: Probing the Mysteries of the Human Mind.* New York: William Morrow.

Ramsey, J., Bradshaw, H., & Schemske, D. W. (2003). Components of reproductive isolation between the monkeyflowers Mimulus lewisii and M. cardinalis (Phrymaceae). *Evolution, 57*(7), 1520–1534.

Ramsey, M. E., Wong, R. Y., & Cummings, M. E. (2010). Estradiol, reproductive cycle and preference behavior in a northern swordtail. *General and Comparative Endocrinology, 170*, 381–390. doi:10.1016/j.ygcen.2010.10.012

Rand, A. S., Bridarolli, M. E., Dries, L., & Ryan, M. J. (1997). Light levels influence female choice in túngara frogs: predation risk assessment? *Copeia, 1997*(2), 447–450. doi:10.2307/1447770

Rangassamy, M., Dalmas, M., Féron, C., Gouat, P., & Rödel, H. G. (2015). Similarity of personalities speeds up reproduction in pairs of a monogamous rodent. *Animal Behaviour, 103*, 7–15.

Rantala, M. J., & Marcinkowska, U. M. (2011). The role of sexual imprinting and the Westermarck effect in mate choice in humans. *Behavioral Ecology and Sociobiology, 65*(5), 859–873. doi:10.1007/s00265-011-1145-y

Rantala, M. J., Pölkki, M., & Rantala, L. M. (2010). Preference for human male body hair changes across the menstrual cycle and menopause. *Behavioral Ecology, 21*(2), 419–423. doi:10.1093/beheco/arp206

Rashed, A., & Polak, M. (2009). Does male secondary sexual trait size reveal fertilization efficiency in Australian *Drosophila bipectinata* Duda (Diptera: Drosophilidae)? *Biological Journal of the Linnean Society, 98*(2), 406–413.

Ratikainen, I. I., & Kokko, H. (2009). Differential allocation and compensation: who deserves the silver spoon? *Behavioral Ecology, 21*(1), 195–200.

Ratterman, N. L., Rosenthal, G. G., Carney, G. E., & Jones, A. G. (2014). Genetic variation and covariation in male attractiveness and female mating preferences in *Drosophila melanogaster. G3 (Bethesda), 4*(1), 79–88. doi:10.1534/g3.113.007468

Raveh, S., Sutalo, S., Thonhauser, K. E., Thoß, M., Hettyey, A., Winkelser, F., & Penn, D. J. (2014). Female partner preferences enhance offspring ability to survive an infection. *BMC Evolutionary Biology, 14*(1), 14.

Real, L. A. (1991). Search theory and mate choice. II. Mutual interaction, assortative mating, and equilibrium variation in male and female fitness. *American Naturalist*, 901–917.

Rebar, D., & Rodríguez, R. L. (2013). Genetic variation in social influence on mate preferences. *Proceedings of the Royal Society of London B: Biological Sciences, 280*(1763). doi:20130803.10.1098/rspb.2013.0803

———. (2014). Genetic variation in host plants influences the mate preferences of a plant-feeding insect. *American Naturalist, 184*(4), 489–499.

Reber, R., Schwarz, N., & Winkielman, P. (2004). Processing fluency and aesthetic pleasure: is beauty in the perceiver's processing experience? *Personality and Social Psychology Review, 8*(4), 364–382. doi:10.1207/s15327957pspr0804_3

Reeve, H. K. (1989). The evolution of conspecific acceptance thresholds. *American Naturalist*, 407–435.

Regier, T., Kay, P., & Khetarpal, N. (2007). Color naming reflects optimal partitions of color space. *Proceedings of the National Academy of Sciences, 104*(4), 1436–1441. doi:10.1073/pnas.0610341104

Reichert, M. S., & Höbel, G. (2015). Modality interactions alter the shape of acoustic mate preference functions in gray treefrogs. *Evolution 69*(9), 2384–2398.

Reid, J., Arcese, P., & Losdat, S. (2014). Genetic covariance between components of male reproductive success: within-pair vs. extra-pair paternity in song sparrows. *Journal of Evolutionary Biology, 27*(10), 2046–2056.

Reid, J. M., & Sardell, R. J. (2011). Indirect selection on female extra-pair reproduction? Comparing the additive genetic value of maternal half-sib extra-pair and within-pair offspring. *Proceedings of the Royal Society of London B: Biological Sciences*, rspb 20112230.

Reid, M. L., & Stamps, J. A. (1997). Female mate choice tactics in a resource-based mating system: field tests of alternative models. *American Naturalist, 150*(1), 98–121. doi:10.1086/286058

Reinhold, K., Reinhold, K., & Jacoby, K. J. (2002). Dissecting the repeatability of female choice in the grasshopper *Chorthippus biguttulus*. *Animal Behaviour, 64*(2), 245–250. doi:http://dx.doi.org/10.1006/anbe.2002.3061

Reinhold, K., & Schielzeth, H. (2014). Choosiness, a neglected aspect of preference functions: a review of methods, challenges and statistical approaches. *Journal of Comparative Physiology A: Neuroethology, Sensory, Neural, and Behavioral Physiology, 201*(1), 171–182. doi:10.1007/s00359-014-0963-6

Remage-Healey, L., Adkins-Regan, E., & Romero, L. M. (2003). Behavioral and adrenocortical responses to mate separation and reunion in the zebra finch. *Hormones and Behavior, 43*(1), 108–114.

Reparaz, L. B., van Oers, K., Naguib, M., Doutrelant, C., Visser, M. E., & Caro, S. P. (2014). Mate preference of female blue tits varies with experimental photoperiod. *PLoS ONE, 9*(3), 1–8. doi:10.1371/journal.pone.0092527

Reyer, H., Frei, G., & Som, C. (1999). Cryptic female choice: frogs reduce clutch size when amplexed by undesired males. *Proceedings of the Royal Society of London B: Biological Sciences, 266*(1433), 2101–2107.

Reyer, H. -U. (2008). Mating with the wrong species can be right. *Trends in Ecology & Evolution, 23*(6), 289–292.

Reynolds, J., & Gross, M. (1990). Costs and benefits of female mate choice: Is there a lek paradox? *American Naturalist, 136*, 230–243.

Režucha, R., & Reichard, M. (2015). Strategic exploitation of fluctuating asymmetry in male Endler's guppy courtship displays is modulated by social environment. *Journal of Evolutionary Biology, 28*(2), 356–367.

Riabinina, O., Albert, J. T., Dai, M., & Duke, T. (2011). Active process mediates species-specific tuning of *Drosophila* ears. *Current Biology, 21*(8), 658–664. doi:10.1016/j.cub.2011.03.001

Ribeiro, J., & Spielman, A. (1986). The satyr effect: a model predicting parapatry and species extinction. *American Naturalist*, 513–528.

Rice, W. R. (1996). Sexually antagonistic male adaptation triggered by experimental arrest of female evolution. *Nature, 381*(6579), 232–234.

Rice, W. R., & Holland, B. (1999). Reply to comments on the chase-away model of sexual selection. *Evolution, 53*(1), 302–306.

Richard, D. (1982). *The Extended Phenotype: The Gene as the Unit of Selection*. Oxford: W. H. Freeman.

Richards, C. L. (2006). Has the evolution of complexity in the amphibian papilla influenced anuran speciation rates? *Journal of Evolutionary Biology, 19*(4), 1222–1230. doi: 10.1111/j.1420-9101.2006.01079.x

Richardson, D. S., Komdeur, J., Burke, T., & Von Schantz, T. (2005). MHC-based patterns of social and extra-pair mate choice in the Seychelles warbler. *Proceedings of the Royal Society of London B: Biological Sciences, 272*(1564), 759–767.

Rick, I. P., Mehlis, M., & Bakker, T.C.M. (2011). Male red ornamentation is associated with female red sensitivity in sticklebacks. *PLoS ONE, 6*(9), e25554. doi:10.1371/journal.pone.0025554

Riebel, K. (2000). Early exposure leads to repeatable preferences for male song in female zebra finches. *Proceedings of the Royal Society of London B: Biological Sciences, 267* (1461), 2553–2558.

———. (2009). Song and female mate choice in zebra finches: a review. *Advances in the Study of Behavior, Volume 40*, 197–238. doi:http://dx.doi.org/10.1016/S0065-3454(09)40006-8

Riebel, K., Holveck, M.-J., Verhulst, S., & Fawcett, T. (2010). Are high-quality mates always attractive?: State-dependent mate preferences in birds and humans. *Communicative & Integrative Biology, 3*(3), 271–273.

Riebel, K., Naguib, M., & Gil, D. (2009). Experimental manipulation of the rearing environment influences adult female zebra finch song preferences. *Animal Behaviour, 78*(6), 1397–1404. doi:http://dx.doi.org/10.1016/j.anbehav.2009.09.011

Riede, K. (1983). Influence of the courtship song of the acridid grasshopper *Gomphocerus rufus* L. on the female. *Behavioral Ecology and Sociobiology, 14*(1), 21–27. doi: 10.1007/bf00366652

Rieke, F., & Rudd, M. E. (2009). The challenges natural images pose for visual adaptation. *Neuron, 64*(5), 605–616. doi:10.1016/j.neuron.2009.11.028

Riley, C. (2010). Sexual dimorphism in stature (SDS), jealousy and mate retention. *Evolutionary Psychology, 8*, 530–544.

Ringo, J. (1996). Sexual receptivity in insects. *Annual Review of Entomology (USA)*, 41, 473–494.

Rintamäki, P., Alatalo, R., Höglund, J., & Lundberg, A. (1995). Mate sampling behaviour of black grouse females (*Tetrao tetrix*). *Behavioral Ecology and Sociobiology, 37*(3), 209–215. doi:10.1007/BF00176719

Rios-Cardenas, O., Darrah, A., & Morris, M. R. (2010). Female mimicry and an enhanced sexually selected trait: what does it take to fool a male? *Behaviour, 147*(11), 1443–1460.

Rios-Cardenas, O., Tudor, M. S., & Morris, M. R. (2007). Female preference variation has implications for the maintenance of an alternative mating strategy in a swordtail fish. *Animal Behaviour, 74*(3), 633–640. doi:10.1016/j.anbehav.2007.01.002.

Ritchie, M. (1992). Behavioral coupling in tettigoniid hybrids (Orthoptera). *Behavior Genetics, 22*(3), 369–379.

———. (1996a). What is 'the paradox of the lek'? *Trends in Ecology & Evolution, 11*(4), 175–176.

———. (1996b). The shape of female mating preferences. *Proceedings of the National Academy of Sciences, 93*(25), 14628–14631.

———. (2000). The inheritance of female preference functions in a mate recognition system. *Proceedings of the Royal Society of London B: Biological Sciences, 267*(1441), 327–332. doi:10.1098/rspb.2000.1004

———. (2007). Sexual selection and speciation. *Annual Review of Ecology, Evolution, and Systematics*, 79–102.

Ritchie, M. G., Saarikettu, M., & Hoikkala, A. (2005). Variation, but no covariance, in female preference functions and male song in a natural population of *Drosophila montana*. *Animal Behaviour, 70*, 849–854. doi:DOI 10.1016/j.anbehav.2005.01.018

Riters, L. V., Ellis, J.M.S., Angyal, C. S., Borkowski, V. J., Cordes, M. A., & Stevenson, S. A. (2013). Research report: links between breeding readiness, opioid immunolabeling, and the affective state induced by hearing male courtship song in female European starlings (*Sturnus vulgaris*). *Behavioural Brain Research, 247*, 117–124. doi: 10.1016/j.bbr.2013.02.041

Riters, L. V., Olesen, K. M., & Auger, C. J. (2007). Evidence that female endocrine state influences catecholamine responses to male courtship song in European starlings. *General and Comparative Endocrinology, 154*(1–3), 137–149. doi:http://dx.doi.org/10.1016/j.ygcen.2007.05.029

Riters, L. V., & Teague, D. P. (2003). The volumes of song control nuclei, HVC and lMAN, relate to differential behavioral responses of female European starlings to male songs produced within and outside of the breeding season. *Brain Research, 978*(1–2), 91–98.

Ritschard, M., Riebel, K., & Brumm, H. (2010). Female zebra finches prefer high-amplitude song. *Animal Behaviour, 79*(4), 877–883. doi:http://dx.doi.org/10.1016/j.anbehav.2009.12.038

Roberts, E. K., Lu, A., Bergman, T. J., & Beehner, J. C. (2012). A Bruce effect in wild geladas. *Science, 335*(6073), 1222–1225. doi:10.1126/science.1213600

Roberts, M. L., Buchanan, K. L., & Evans, M. (2004). Testing the immunocompetence handicap hypothesis: a review of the evidence. *Animal Behaviour, 68*(2), 227–239.

Roberts, S. A., Davidson, A. J., Beynon, R. J., & Hurst, J. L. (2014). Female attraction to male scent and associative learning: the house mouse as a mammalian model. *Animal Behaviour, 97*, 313–321.

Roberts, S. C., & Gosling, L. M. (2003). Genetic similarity and quality interact in mate choice decisions by female mice. *Nature Genetics, 35*(1), 103–106.

Robertson, J. G. (1990). Female choice increases fertilization success in the Australian frog, *Uperoleia laevigata*. *Animal Behaviour, 39*(4), 639–645.

Robinson, D. M., & Morris, M. R. (2010). Unraveling the complexities of variation in female mate preference for vertical bars in the swordtail, Xiphophorus cortezi. *Behavioral Ecology and Sociobiology, 64*(10), 1537–1545. doi:10.1007/s00265-010-0967-3

Robinson, D. M., Tudor, M. S., & Morris, M. R. (2011). Female preference and the evolution of an exaggerated male ornament: the shape of the preference function matters. *Animal Behaviour, 81*, 1015–1021. doi:10.1016/j.anbehav.2011.02.005

Robinson, M. R., Sander van Doorn, G., Gustafsson, L., & Qvarnström, A. (2012). Environment-dependent selection on mate choice in a natural population of birds. *Ecology Letters, 15*(6), 611–618. doi:10.1111/j.1461-0248.2012.01780.x

Robson, L., & Gwynne, D. (2010). Measuring sexual selection on females in sex-role-reversed Mormon crickets (*Anabrus simplex*, Orthoptera: Tettigoniidae). *Journal of Evolutionary Biology, 23*(7), 1528–1537.

Rodd, F. H., Hughes, K. A., Grether, G. F., & Baril, C. T. (2002). A possible non-sexual origin of mate preference: are male guppies mimicking fruit? *Proceedings of the Royal Society of London B: Biological Sciences, 269*, 475–481.

Rodríguez, R., Ramaswamy, K., & Cocroft, R. (2006). Evidence that female preferences have shaped male signal evolution in a clade of specialized plant-feeding insects. *Proceedings of the Royal Society of London B: Biological Sciences, 273*(1601), 2585–2593.

Rodríguez, R. L., Boughman, J. W., Gray, D. A., Hebets, E. A., Höbel, G., Symes, L. B., & Grether, G. (2013). Diversification under sexual selection: the relative roles of mate preference strength and the degree of divergence in mate preferences. *Ecology Letters, 16*(8), 964–974. doi:10.1111/ele.12142

Rodríguez, R. L., & Greenfield, M. D. (2003). Genetic variance and phenotypic plasticity in a component of female mate choice in an ultrasonic moth. *Evolution, 57*(6), 1304–1313. doi:10.1111/j.0014-3820.2003.tb00338.x

Rodríguez, R. L., Hallett, A. C., Kilmer, J. T., & Fowler-Finn, K. D. (2013). Curves as traits: genetic and environmental variation in mate preference functions. *Journal of Evolutionary Biology, 26*(2), 434–442. doi:10.1111/jeb.12061

Rodríguez, R. L., Ramaswamy, K., & Cocroft, R. B. (2006). Evidence that female preferences have shaped male signal evolution in a clade of specialized plant-feeding insects. *Proceedings of the Royal Society of London B: Biological Sciences, 273*(1601), 2585–2593.

Rodríguez, R. L., Rebar, D., & Fowler-Finn, K. D. (2013). The evolution and evolutionary consequences of social plasticity in mate preferences. *Animal Behaviour, 85*(5), 1041–1047. doi:10.1016/j.anbehav.2013.01.006

Rodríguez, R. L., & Snedden, W. A. (2004). On the functional design of mate preferences and receiver biases. *Animal Behaviour, 68*(2), 427–432.

Rodríguez-Muñoz, R., Bretman, A., Hadfield, J. D., & Tregenza, T. (2008). Sexual selection in the cricket *Gryllus bimaculatus*: no good genes? *Genetica, 134*(1), 129–136.

Rodríguez-Muñoz, R., Bretman, A., & Tregenza, T. (2011). Guarding males protect females from predation in a wild insect. *Current Biology, 21*(20), 1716–1719.

Roelofs, W., Glover, T., Tang, X. -H., Sreng, I., Robbins, P., Eckenrode, C., … Bengtsson, B. O. (1987). Sex pheromone production and perception in European corn borer moths is determined by both autosomal and sex-linked genes. *Proceedings of the National Academy of Sciences, 84*(21), 7585–7589.

Roff, D., & Fairbairn, D. (2015). Bias in the heritability of preference and its potential impact on the evolution of mate choice. *Heredity, 114*(4), 404–412.

Roff, D. A. (2015). The evolution of mate choice: a dialogue between theory and experiment. *Annals of the New York Academy of Sciences*, 1–15.

Roff, D. A., & Fairbairn, D. J. (2014). The evolution of phenotypes and genetic parameters under preferential mating. *Ecology and Evolution, 4*(13), 2759–2776.

Rohwer, S., Harris, R. B., & Walsh, H. E. (2014). Rape and the prevalence of hybrids in broadly sympatric species: a case study using albatrosses. *PeerJ, 2*, e409. doi:10.7717/peerj.409

Rolff, J. (1998). Parasite-mediated sexual selection: parasitized non-choosy females do not slow down the process. *Behavioral Ecology and Sociobiology, 44*(1), 73–74.

Römer, H., & Krusch, M. (2000). A gain-control mechanism for processing of chorus sounds in the afferent auditory pathway of the bushcricket *Tettigonia viridissima*

(Orthoptera; Tettigoniidae). *Journal of Comparative Physiology A, 186*(2), 181–191. doi:10.1007/s003590050018

Ronald, K. L., Fernández-Juricic, E., & Lucas, J. R. (2012). Taking the sensory approach: how individual differences in sensory perception can influence mate choice. *Animal Behaviour, 84*(6), 1283–1294. doi:http://dx.doi.org/10.1016/j.anbehav.2012.09.015

Rose, G., & Capranica, R. (1984). Processing amplitude-modulated sounds by the auditory midbrain of two species of toads: matched temporal filters. *Journal of Comparative Physiology A, 154*(2), 211–219. doi:10.1007/bf00604986

Rosenfeld, M. J., & Thomas, R. J. (2012). Searching for a mate: The rise of the Internet as a social intermediary. *American Sociological Review, 77*(4), 523–547.

Rosengrave, P., Gemmell, N. J., Metcalf, V., McBride, K., & Montgomerie, R. (2008). A mechanism for cryptic female choice in chinook salmon. *Behavioral Ecology, 19*(6), 1179–1185.

Rosenqvist, G., & Johansson, K. (1995). Male avoidance of parasitized females explained by direct benefits in a pipefish. *Animal Behaviour, 49*(4), 1039–1045.

Rosenthal, G. (2013). Individual mating decisions and hybridization. *Journal of Evolutionary Biology, 26*(2), 252–255.

Rosenthal, G. G. (2007). Spatiotemporal dimensions of visual signals in animal communication. *Annual Review of Ecology Evolution and Systematics, 38*, 155–178. doi:10.1146/annurev.ecolsys.38.091206.095745

———. (2013). Individual mating decisions and hybridization. *Journal of Evolutionary Biology, 26*(2), 252–255. doi:10.1111/jeb.12004

———. (2016). Mate choice: charting desire's tangled bank. *Current Biology, 26*(7), R294–R296.

Rosenthal, G. G., & Evans, C. S. (1998). Female preference for swords in *Xiphophorus helleri* reflects a bias for large apparent size. *Proceedings of the National Academy of Sciences USA, 95*, 4431–4436.

Rosenthal, G. G., Evans, C. S., & Miller, W. L. (1996). Female preference for dynamic traits in the green swordtail, *Xiphophorus helleri*. *Animal Behaviour, 51*, 811–820. doi:10.1006/anbe.1996.0085

Rosenthal, G. G., Martinez, T.Y.F., de Leon, F.J.G., & Ryan, M. J. (2001). Shared preferences by predators and females for male ornaments in swordtails. *American Naturalist, 158*(2), 146–154. doi:10.1086/321309

Rosenthal, G. G., Rand, A. S., & Ryan, M. J. (2004). The vocal sac as a visual cue in anuran communication: an experimental analysis using video playback. *Animal Behaviour, 68*, 55–58. doi:10.1016/j.anbehav.2003.07.013

Rosenthal, G. G., & Ryan, M. J. (2000). Visual and acoustic communication in non-human animals: a comparison. *Journal of Biosciences, 25*, 285–290.

———. (2005). Assortative preferences for stripes in danios. *Animal Behaviour, 70*, 1063–1066. doi:10.1016/j.anbehav.2005.02.005

———. (2011a). Conflicting preferences within females: sexual selection versus species recognition. *Biology Letters, 7*(4), 525–527. doi:10.1098/rsbl.2011.0027

———. (2011b). Multiple visual cues, receiver psychology, and signal evolution in pygmy swordtails. In M. C. Uribe & H. J. Grier (Eds.), *Viviparous Fishes II*. Homestead, FL: New Life Publications.

Rosenthal, G. G., & Servedio, M. R. (1999). Chase-away sexual selection: resistance to "resistance." *Evolution, 53*(1), 296–299.

Rosenthal, G. G., & Stuart-Fox, D. M. (2012). Environmental disturbance and animal communication. In B.B.M. Wong & U. Candolin (Eds.), *Behavioural Responses to a Changing World: Mechanisms and Consequences* (pp. 16–31). Oxford, UK: Oxford University Press.

Rosenthal, G. G., Wagner, W. E., Jr., & Ryan, M. J. (2002). Secondary reduction of preference for the sword ornament in the pygmy swordtail *Xiphophorus nigrensis* (Pisces: Poeciliidae). *Animal Behaviour, 63*, 37–45.

Roth, O., Sundin, J., Berglund, A., Rosenqvist, G., & Wegner, K. (2014). Male mate choice relies on major histocompatibility complex class I in a sex-role-reversed pipefish. *Journal of Evolutionary Biology, 27*(5), 929–938.

Roughgarden, J. (2012). The social selection alternative to sexual selection. *Philosophical Transactions of the Royal Society of London B: Biological Sciences, 367*(1600), 2294–2303.

Roughgarden, J., Oishi, M., & Akçay, E. (2006). Reproductive social behavior: cooperative games to replace sexual selection. *Science, 311*(5763), 965–969.

Roulin, A. (1999). Nonrandom pairing by male barn owls (*Tyto alba*) with respect to a female plumage trait. *Behavioral Ecology, 10*(6), 688–695. doi:10.1093/beheco/10.6.688

Rowe, C. (1999). Receiver psychology and the evolution of multicomponent signals. *Animal Behaviour, 58*, 921–931.

———. (2013). Receiver psychology: a receiver's perspective. *Animal Behaviour, 85*(3), 517–523.

Rowe, L. (1992). Convenience polyandry in a water strider: foraging conflicts and female control of copulation frequency and guarding duration. *Animal Behaviour, 44*, 189–202.

Rowe, L., & Arnqvist, G. (2014). Two faces of environmental effects on mate choice: a comment on Dougherty and Shuker. *Behavioral Ecology 26*(2), 324. doi:10.1093/beheco/aru199

Rowe, L., Cameron, E., & Day, T. (2003). Detecting sexually antagonistic coevolution with population crosses. *Proceedings of the Royal Society of London B: Biological Sciences, 270*(1528), 2009–2016.

———. (2005). Escalation, retreat, and female indifference as alternative outcomes of sexually antagonistic coevolution. *American Naturalist, 165*(S5), S5–S18.

Rowe, L., & Day, T. (2006). Detecting sexual conflict and sexually antagonistic coevolution. *Philosophical Transactions of the Royal Society of London B: Biological Sciences, 361*(1466), 277–285.

Rowe, L., & Houle, D. (1996). The lek paradox and the capture of genetic variance by condition dependent traits. *Proceedings of the Royal Society of London B: Biological Sciences, 263*(1375), 1415–1421.

Rowell, J.T., & Servedio, M. R. (2009). Gentlemen prefer blondes: the evolution of mate preference among strategically allocated males. *American Naturalist, 173*(1), 12–25. doi:10.1086/593356

Rowley, I., & Chapman, G. (1986). Cross-fostering, imprinting and learning in two sympatric species of cockatoo. *Behaviour, 96*(1), 1–16.

Royle, N., & Pike, T. (2010). Social feedback and attractiveness in zebra finches. *Behavioral Ecology and Sociobiology, 64*(12), 2015–2020. doi:10.1007/s00265-010-1013-1

Royle, N. J., Lindström, J., & Metcalfe, N. B. (2008). Context-dependent mate choice in relation to social composition in green swordtails *Xiphophorus helleri*. *Behavioral Ecology, 19*(5), 998–1005. doi:10.1093/beheco/arn059

Royle, N. J., Schuett, W., & Dall, S. R. (2010). Behavioral consistency and the resolution of sexual conflict over parental investment. *Behavioral Ecology*, arq156.

Rozin, P. (1999). Preadaption and the puzzles and properties of pleasure In E. D. Daniel Kahneman, Norbert Schwarz (Ed.), *Well-Being: The Foundations of Hedonic Psychology* (pp. 593). New York: Russell Sage Foundation.

Rubenstein, D. R. (2012). Sexual and social competition: broadening perspectives by defining female roles. *Philosophical Transactions of the Royal Society of London B: Biological Sciences, 367*(1600), 2248–2252.

Rudolfsen, G., Figenschou, L., Folstad, I., Nordeide, J., & Søreng, E. (2005). Potential fitness benefits from mate selection in the Atlantic cod (*Gadus morhua*). *Journal of Evolutionary Biology, 18*(1), 172–179.

Ruf, D., Mazzi, D., & Dorn, S. (2010). No kin discrimination in female mate choice of a parasitoid with complementary sex determination. *Behavioral Ecology*, arq148.

Rundle, H. D., & Chenoweth, S. F. (2011). Stronger convex (stabilizing) selection on homologous sexual display traits in females than in males: a multipopulation comparison in *Drosophila serrata*. *Evolution, 65*(3), 893–899. doi:10.1111/j.1558-5646.2010.01158.x

Rundle, H. D., Chenoweth, S. F., & Blows, M. W. (2008). Comparing complex fitness surfaces: among-population variation in mutual sexual selection in *Drosophila serrata*. *American Naturalist, 171*(4), 443–454. doi:10.1086/528963

Rundle, H. D., Chenoweth, S. F., Doughty, P., & Blows, M. W. (2005). Divergent selection and the evolution of signal traits and mating preferences. *PLoS Biology, 3*(11), e368.

Rundle, H. D., & Dyer, K. A. (2015). Reproductive character displacement of female mate preferences for male cuticular hydrocarbons in Drosophila subquinaria. *Evolution, 69*(10), 2625–2637.

Rushton, J. P. (1988). Genetic similarity, mate choice, and fecundity in humans. *Ethology and Sociobiology, 9*(6), 329–333.

Ruther, J., Matschke, M., Garbe, L.-A., & Steiner, S. (2009). Quantity matters: male sex pheromone signals mate quality in the parasitic wasp *Nasonia vitripennis*. *Proceedings of the Royal Society B: Biological Sciences, 276*(1671), 3303. doi:10.1098/rspb.2009.0738

Ryan, C., & Jethá, C. (2010). *Sex at Dawn: The Prehistoric Origins of Modern Sexuality*. New York: Harper Collins.

Ryan, M. J. (1986). Neuroanatomy influences speciation rates among anurans. *Proceedings of the National Academy of Sciences, 83*, 1379–1382.

———. (1990). Sensory systems, sexual selection, and sensory exploitation. *Oxford Surveys in Evolutionary Biology, 7*, 157–195.

———. (1997). Sexual selection and mate choice. *Behavioural Ecology: An Evolutionary Approach, 4*, 179–202.

———. (1998). Sexual selection, receiver biases, and the evolution of sex differences. *Science, 281*(5385), 1999–2003.

Ryan, M. J., Akre, K. L., & Kirkpatrick, M. (2009). Cognitive mate choice. In R. Dukas & J. Ratcliffe (Eds.), *Cognitive Ecology II* (pp. 137–155). Chicago: University of Chicago Press.

Ryan, M. J., Bernal, X. E., & Rand, A. S. (2010). Female mate choice and the potential for ornament evolution in túngara frogs Physalaemus pustulosus. Current Zoology, 56(3), 343–357.

Ryan, M. J., & Cummings, M. E. (2013). Perceptual biases and mate choice. *Annual Review of Ecology, Evolution, and Systematics, 44*(1), 437–459. doi:doi:10.1146/annurev-ecolsys-110512-135901

Ryan, M. J., Cummings, M. E., & Futuyma, D. (2005). Animal signals and the overlooked costs of efficacy. *Evolution, 59*(5), 1160–1161.

Ryan, M. J., Fox, J. H., Wilczynski, W., & Rand, A. S. (1990). Sexual selection for sensory exploitation in the frog *Physalaemus pustulosus. Nature, 343*, 66–67.

Ryan, M. J., Hews, D. K., & Wagner, W.E. Jr. (1990). Sexual selection on alleles that determine body size in the swordtail *Xiphophorus nigrensis. Behavioral Ecology and Sociobiology, 26*, 231–237.

Ryan, M. J., & Keddy-Hector, A. (1992). Directional patterns of female mate choice and the role of sensory biases. *American Naturalist, 139*, S4–S35. doi:10.2307/2462426

Ryan, M. J., Pease, C. M., & Morris, M. R. (1992). A genetic polymorphism in the swordtail *Xiphophorus nigrensis*: testing the prediction of equal fitnesses. *American Naturalist, 139*(1), 21–31.

Ryan, M. J., Perrill, S. A., & Wilczynski, W. (1992). Auditory tuning and call frequency predict population-based mating preferences in the cricket frog, *Acris crepitans. American Naturalist, 139*(6), 1370–1383. doi:10.2307/2462346

Ryan, M. J., & Rand, A. S. (1993a). Sexual selection and signal evolution: the ghost of biases past. *Philosophical Transactions of the Royal Society of London B: Biological Sciences, 340*, 187–195.

———. (1993b). Species recognition and sexual selection as a unitary problem in animal communication. *Evolution*, 647–657.

———. (1995). Female responses to ancestral advertisement calls in túngara frogs. *Science, 269*(5222), 390–392.

———. (1999). Phylogenetic influence on mating call preferences in female túngara frogs, *Physalaemus pustulosus. Animal Behaviour, 57*(4), 945–956. doi:10.1006/anbe.1998.1057

———. (2003). Sexual selection in female perceptual space: how female tungara frogs percieve and respond to complex population variation in acoustic mating signals. *Evolution, 57*, 2608–2618.

Ryan, M. J., & Taylor, R. C. (2015). Measures of mate choice: a comment on Dougherty & Shuker. *Behavioral Ecology, 26*(2), 323–324. doi:10.1093/beheco/aru221

Ryan, M. J., & Wagner, W.E. Jr. (1987). Asymmetries in mating preferences between species: female swordtails prefer heterospecific males. *Science, 236*, 595–597.

Ryder, T., Tori, W., Blake, J., Loiselle, B., & Parker, P. (2009). Mate choice for genetic quality: a test of the heterozygosity and compatibility hypotheses in a lek-breeding bird. *Behavioral Ecology, 21*(2), 203–210.

Saether, S. A., Saetre, G.-P., Borge, T., Wiley, C., Svedin, N., Andersson, G., … Qvarnström, A. (2007). Sex chromosome-linked species recognition and evolution of reproductive isolation in flycatchers. *Science, 318*(5847), 95–97. doi:10.1126/science.1141506

Saetre, G. P., Borge, T., Lindroos, K., Haavie, J., Sheldon, B. C., Primmer, C., & Syvänen, A. C. (2003). Sex chromosome evolution and speciation in Ficedula flycatchers. *Proceedings of the Royal Society of London B: Biological Sciences, 270*(1510), 53–59.

Saetre, G. -P., Moum, T., Bures, S., Král, M., Adamjan, M., & Moreno, J. (1997). A sexually selected character displacement in flycatchers reinforces premating isolation. *Nature, 387*(6633), 589–592.

Sakaluk, S. K. (2000). Sensory exploitation as an evolutionary origin to nuptial food gifts in insects. *Proceedings of the Royal Society of London B: Biological Sciences, 267*(1441), 339–343.

Sakaluk, S. K., Avery, R. L., & Weddle, C. B. (2006). Cryptic sexual conflict in gift-giving insects: chasing the chase-away. *American Naturalist, 167*(1), 94–104.

Sakurai, T., Mitsuno, H., Haupt, S. S., Uchino, K., Yokohari, F., Nishioka, T., ... Kanzaki, R. (2011). A single sex pheromone receptor determines chemical response specificity of sexual behavior in the silkmoth *Bombyx mori*. *PLoS Genetics, 7*(6), 1–10. doi:10.1371/journal.pgen.1002115

Saleem, S., Ruggles, P. H., Abbott, W. K., & Carney, G. E. (2014). Sexual experience enhances *Drosophila melanogaster* male mating behavior and success. *PLoS One, 9*(5), e96639. doi:10.1371/journal.pone.0096639

Sanchez-Andrade, G., & Kendrick, K. M. (2009). The main olfactory system and social learning in mammals. *Behavioural Brain Research, 200*(2), 323–335. doi:10.1016/j.bbr.2008.12.021

Sánchez-Guillén, R. A., Wellenreuther, M., & Cordero Rivera, A. (2012). Strong asymmetry in the relative strengths of prezygotic and postzygotic barriers between two damselfly sister species. *Evolution, 66*(3), 690–707.

Sandkam, B. A., Joy, J. B., Watson, C. T., Gonzalez-Bendiksen, P., Gabor, C. R., & Breden, F. (2013). Hybridization leads to sensory repertoire expansion in a gynogenetic fish, the Amazon molly (*Poecilia formosa*): a test of the hybrid-sensory expansion hypothesis. *Evolution, 67*(1), 120–130.

Sandkam, B. A., Young, C. M., & Breden, F. (2015a). Beauty in the eyes of the beholders: colour vision is tuned to mate preference in the Trinidadian guppy (*Poecilia reticulata*). *Molecular Ecology, 24*(3), 596–609.

Sandkam, B. A., Young, C. M., Breden, F. M., Bourne, G. R., & Breden, F. (2015b). Color vision varies more among populations than among species of live-bearing fish from South America. *BMC Evolutionary Biology, 15*(1), 225.

Santos, M., Matos, M., & Varela, S.A.M. (2014). Negative public information in mate choice copying helps the spread of a novel trait. *American Naturalist, 184*(5), 658–672. doi:10.1086/678082

Santure, A. W., & Spencer, H. G. (2006). Influence of mom and dad: quantitative genetic models for maternal effects and genomic imprinting. *Genetics, 173*(4), 2297–2316. doi:10.1534/genetics.105.049494

Sardell, R. J., Arcese, P., Keller, L. F., & Reid, J. M. (2012). Are there indirect fitness benefits of female extra-pair reproduction? Lifetime reproductive success of within-pair and extra-pair offspring. *American Naturalist, 179*(6), 779–793.

Saveer, A. M., Kromann, S. H., Birgersson, G., Bengtsson, M., Lindblom, T., Balkenius, A., ... Ignell, R. (2012). Floral to green: mating switches moth olfactory coding and preference. *Proceedings of the Royal Society of London B: Biological Sciences, 279*(1737), 2314–2322. doi:10.1098/rspb.2011.2710

Savic, I., & Lindström, P. (2008). PET and MRI show differences in cerebral asymmetry and functional connectivity between homo- and heterosexual subjects. *Proceedings of the National Academy of Sciences, 105*(27), 9403–9408. doi:10.1073/pnas.0801566105

Scelza, B. A. (2011). Female choice and extra-pair paternity in a traditional human population. *Biology Letters*, rsbl20110478.

Schacht, R., & Mulder, M. B. (2015). Sex ratio effects on reproductive strategies in humans. *Royal Society Open Science, 2*(1), 140402.

Schaefer, H. M., & Ruxton, G. D. (2015). Signal diversity, sexual selection, and speciation. *Annual Review of Ecology, Evolution, and Systematics, 46*, 573–592.

Schärer, L., & Pen, I. (2013). Sex allocation and investment into pre- and post-copulatory traits in simultaneous hermaphrodites: the role of polyandry and local sperm competition. *Philosophical Transactions of the Royal Society of London B: Biological Sciences, 368*(1613). doi:10.1098/rstb.2012.0052

Scharf, I., & Martin, O. (2013). Same-sex sexual behavior in insects and arachnids: prevalence, causes, and consequences. *Behavioral Ecology & Sociobiology, 67*(11), 1719–1730. doi:10.1007/s00265-013-1610-x

Schartl, M. (2008). Evolution of *Xmrk*: an oncogene, but also a speciation gene? *BioEssays, 30*(9), 822–832.

Scheiner, S. M. (2014). The Baldwin effect: neglected and misunderstood. *American Naturalist, 184*(4), ii–iii.

Schielzeth, H., Bolund, E., & Forstmeier, W. (2010). Heritability of and early environment effects on variation in mating preferences. *Evolution, 64*(4), 998–1006. doi:10.1111/j.1558-5646.2009.00890.x

Schielzeth, H., Burger, C., Bolund, E., & Forstmeier, W. (2008a). Assortative versus disassortative mating preferences of female zebra finches based on self-referent phenotype matching. *Animal Behaviour, 76*, 1927–1934. doi:10.1016/j.anbehav.2008.08.014

———. (2008b). Sexual imprinting on continuous variation: do female zebra finches prefer or avoid unfamiliar sons of their foster parents? *Journal of Evolutionary Biology, 21*(5), 1274–1280. doi:10.1111/j.1420-9101.2008.01568.x

Schielzeth, H., Streitner, C., Lampe, U., Franzke, A., & Reinhold, K. (2014). Genome size variation affects song attractiveness in grasshoppers: evidence for sexual selection against large genomes. *Evolution, 68*(12), 3629–3635.

Schiestl, F. (2004). Floral evolution and pollinator mate choice in a sexually deceptive orchid. *Journal of Evolutionary Biology, 17*(1), 67–75.

Schinaman, J. M., Giesey, R. L., Mizutani, C. M., Lukacsovich, T., & Sousa-Neves, R. (2014). The KRÜPPEL-like transcription factor DATILÓGRAFO is required in specific cholinergic neurons for sexual receptivity in *Drosophila* females. *PLoS Biology, 12*(10), e1001964. doi:10.1371/journal.pbio.1001964

Schlichting, C. D., & Pigliucci, M. (1998). *Phenotypic Evolution: A Reaction Norm Perspective*. Sunderland, MA: Sinauer.

Schlupp, I., Marler, C., & Ryan, M. J. (1994). Benefit to male sailfin mollies of mating with heterospecific females. *Science, 263*, 373–374.

Schluter, D. (1988). Estimating the form of natural selection on a quantitative trait. Evolution, *42*, 849–861.

———. (2009). Evidence for ecological speciation and its alternative. *Science, 323*(5915), 737–741.

Schmidt, A.K.D., Riede, K., & Römer, H. (2011). High background noise shapes selective auditory filters in a tropical cricket. *Journal of Experimental Biology, 214*(10), 1754–1762. doi:10.1242/jeb.053819.

Schmidt, E. M., & Pfennig, K. S. (2015). Hybrid female mate choice as a species isolating mechanism: environment matters. *Journal of Evolutionary Biology 29*(4), 865–869.

Schmoll, T. (2011). A review and perspective on context-dependent genetic effects of extra-pair mating in birds. *Journal of Ornithology, 152*(1), 265–277.

Schmoll, T., Dietrich, V., Winkel, W., Epplen, J. T., Schurr, F., & Lubjuhn, T. (2005). Paternal genetic effects on offspring fitness are context dependent: within the extrapair mating system of a socially: monogamous passerine. *Evolution, 59*(3), 645–657.

Schneider, J. E., Casper, J. F., Barisich, A., Schoengold, C., Cherry, S., Surico, J., ... Rabold, E. (2007). Food deprivation and leptin prioritize ingestive and sex behavior without affecting estrous cycles in Syrian hamsters. *Hormones and Behavior, 51*(3), 413–427. doi:http://dx.doi.org/10.1016/j.yhbeh.2006.12.010

Schuett, W., Godin, J.G.J., & Dall, S. R. (2011). Do female zebra finches, *Taeniopygia guttata*, choose their mates based on their 'personality'? *Ethology, 117*(10), 908–917.

Schuett, W., Tregenza, T., & Dall, S.R.X. (2010). Sexual selection and animal personality. *Biological Reviews, 85*(2), 217–246. doi:10.1111/j.1469-185X.2009.00101.x

Schulte, K. F., Uhl, G., & Schneider, J. M. (2010). Mate choice in males with one-shot genitalia: limited importance of female fecundity. *Animal Behaviour, 80*(4), 699–706. doi:http://dx.doi.org/10.1016/j.anbehav.2010.07.005

Schultzhaus, J. N., Saleem, S., Iftikhar, H., & Carney, G. E. (2017). The role of the Drosophila lateral horn in olfactory information processing and behavioral response. *Journal of Insect Physiology, 98*, 29–37. doi:http://dx.doi.org/10.1016/j.jinsphys.2016.11.007

Schumer, M., Cui, R., Rosenthal, G. G., & Andolfatto, P. (2015). Reproductive isolation of hybrid populations driven by genetic incompatibilities. *PLoS Genetics, 11*(3), e1005041. doi:10.1371/journal.pgen.1005041

Schumer, M., Cui, R., Powell, D. L., Dresner, R., Rosenthal, G. G., & Andolfatto, P. (2014). High-resolution mapping reveals hundreds of genetic incompatibilities in hybridizing fish species. *eLife, 3*, e02535. doi:10.7554/eLife.02535

Schumer, M., Rosenthal, G. G., & Andolfatto, P. (2014). How common is homoploid hybrid speciation? *Evolution, 68*(6), 1553–1560.

Schwagmeyer, P. (1979). The Bruce effect: an evaluation of male/female advantages. *American Naturalist*, 932–938.

Schwartz, J. J., Huth, K., & Hutchin, T. (2004). How long do females really listen? Assessment time for female mate choice in the grey treefrog, *Hyla versicolor*. *Animal Behaviour, 68*(3), 533–540. doi:http://dx.doi.org/10.1016/j.anbehav.2003.09.016

Schwensow, N., Eberle, M., & Sommer, S. (2008). Compatibility counts: MHC-associated mate choice in a wild promiscuous primate. *Proceedings of the Royal Society of London B: Biological Sciences, 275*(1634), 555–564.

Scorolli, C., Ghirlanda, S., Enquist, M., Zattoni, S., & Jannini, E. (2007). Relative prevalence of different fetishes. *International Journal of Impotence Research, 19*(4), 432–437.

Searcy, W. A. (1982). The evolutionary effects of mate selection. *Annual Review of Ecology and Systematics 13*, 57–85.

———. (1992). Song repertoire and mate choice in birds. *American Zoologist, 32*(1), 71–80.

Searcy, W. A., & Marler, P. (1981). A test for responsiveness to song structure and programming in female sparrows. *Science, 213*, 926–928.

Secondi, J., & Théry, M. (2014). An ultraviolet signal generates a conflict between sexual selection and species recognition in a newt. *Behavioral Ecology and Sociobiology, 68*(7), 1049–1058.

Seddon, N., Botero, C. A., Tobias, J. A., Dunn, P. O., MacGregor, H. E., Rubenstein, D. R., … Safran, R. J. (2013). Sexual selection accelerates signal evolution during speciation in birds. *Proceedings of the Royal Society of London B: Biological Sciences, 280*(1766), 20131065.

Seehausen, O. (2004). Hybridization and adaptive radiation. *Trends in Ecology & Evolution, 19*(4), 198–207.

Seehausen, O., & van Alphen, J.J.M. (1998). The effect of male coloration on female mate choice in closely related Lake Victoria cichlids (*Haplochromis nyererei* complex). *Behavioral Ecology and Sociobiology, 42*(1), 1–8. doi:10.1007/s002650050405.

Seehausen, O., Terai, Y., Magalhaes, I. S., Carleton, K. L., Mrosso, H.D.J., Miyagi, R., … Okada, N. (2008). Speciation through sensory drive in cichlid fish. *Nature, 455*(7213), 620–626.

Seehausen, O., van Alphen, J.J.M., & Witte, F. (1997). Cichlid fish diversity threatened by eutrophication that curbs sexual selection. *Science, 277*(5333), 1808–1811.

Seger, J., & Trivers, R. (1986). Asymmetry in the evolution of female mating preferences. *Nature, 319*(6056), 771–773.

Selz, O. M., Pierotti, M. E., Maan, M. E., Schmid, C., & Seehausen, O. (2014). Female preference for male color is necessary and sufficient for assortative mating in 2 cichlid sister species. *Behavioral Ecology, 25*(3), 612–626.

Semmelhack, J. L., & Wang, J. W. (2009). Select *Drosophila* glomeruli mediate innate olfactory attraction and aversion. *Nature, 459*(7244), 218–223. doi:http://www.nature.com/nature/journal/v459/n7244/suppinfo/nature07983_S1.html

Senior, C., Lau, A., & Butler, M.J.R. (2007). The effects of the menstrual cycle on social decision making. *International Journal of Psychophysiology, 63*, 186–191. doi:10.1016/j.ijpsycho.2006.03.009

Servedio, M. (2009). The role of linkage disequilibrium in the evolution of premating isolation. *Heredity, 102*(1), 51–56.

Servedio, M. R. (2000). Reinforcement and the genetics of nonrandom mating. *Evolution, 54*(1), 21–29.

———. (2001). Beyond reinforcement: the evolution of premating isolation by direct selection on preferences and postmating, prezygotic incompatibilities. *Evolution, 55*(10), 1909–1920.

———. (2004a). The evolution of premating isolation: local adaptation and natural and sexual selection against hybrids. *Evolution, 58*(5), 913–924.

———. (2004b). The what and why of research on reinforcement. *PLoS Biology, 2*(12), e420.

———. (2016). Geography, assortative mating, and the effects of sexual selection on speciation with gene flow. *Evolutionary Applications, 9*(1), 91–102.

Servedio, M. R., & Bürger, R. (2014). The counterintuitive role of sexual selection in species maintenance and speciation. *Proceedings of the National Academy of Sciences, 111*(22), 8113–8118. doi:10.1073/pnas.1316484111

Servedio, M. R., & Dukas, R. (2013). Effects on population divergence of within-generational learning about prospective mates. *Evolution, 67*(8), 2363–2375.

Servedio, M. R., & Kopp, M. (2012). Sexual selection and magic traits in speciation with gene flow. *Current Zoology, 58*(3), 510–516.

Servedio, M. R., & Lande, R. (2006). Population genetic models of male and mutual mate choice. *Evolution, 60*(4), 674–685.

Servedio, M. R., & Noor, M. A. (2003). The role of reinforcement in speciation: theory and data. *Annual Review of Ecology, Evolution, and Systematics*, 339–364.

Servedio, M. R., Price, T. D., & Lande, R. (2013). Evolution of displays within the pair bond. *Proceedings of the Royal Society of London B: Biological Sciences, 280*(1757). doi:10.1098/rspb.2012.3020

Servedio, M. R., Saether, S. A., & Saetre, G. -P. (2009). Reinforcement and learning. *Evolutionary Ecology, 23*(1), 109–123.

Servedio, M. R., Van Doorn, G. S., Kopp, M., Frame, A. M., & Nosil, P. (2011). Magic traits in speciation: 'magic' but not rare? *Trends in Ecology & Evolution, 26*(8), 389–397.

Seyfarth, R. M., Silk, J. B., & Cheney, D. L. (2012). Variation in personality and fitness in wild female baboons. *PNAS, 109*(42), 16980–16985.

Shackelford, T. K., & Schmitt, D. P. (2008). Big five traits related to short-term mating: from personality to promiscuity across 46 nations. *Evolutionary Psychology, 6*, 246–282.

Shackleton, M., Jennions, M., & Hunt, J. (2005). Fighting success and attractiveness as predictors of male mating success in the black field cricket, *Teleogryllus commodus*: the effectiveness of no-choice tests. *Behavioral Ecology and Sociobiology, 58*(1), 1–8. doi:10.1007/s00265-004-0907-1

Shapira Lots, I., & Stone, L. (2008). Perception of musical consonance and dissonance: an outcome of neural synchronization. *Journal of the Royal Society Interface, 5*(29), 1429–1434. doi:10.1098/rsif.2008.0143

Sharma, M. D., Griffin, R. M., Hollis, J., Tregenza, T., & Hosken, D. J. (2012). Reinvestigating good genes benefits of mate choice in *Drosophila simulans*. *Biological Journal of the Linnean Society, 106*(2), 295–306.

Sharma, M. D., Tregenza, T., & Hosken, D. J. (2010). Female mate preferences in *Drosophila simulans*: evolution and costs. *Journal of Evolutionary Biology, 23*(8), 1672–1679. doi:10.1111/j.1420-9101.2010.02033.x

Sharon, G., Segal, D., Ringo, J. M., Hefetz, A., Zilber-Rosenberg, I., & Rosenberg, E. (2010). Commensal bacteria play a role in mating preference of *Drosophila melanogaster*. *Proceedings of the National Academy of Sciences, 107*(46), 20051–20056. doi:10.1073/pnas.1009906107

Shaw, K. L. (2000). Interspecific genetics of mate recognition: inheritance of female acoustic preference in Hawaiian crickets. *Evolution, 54*(4), 1303–1312.

Shaw, K. L., Ellison, C. K., Oh, K. P., & Wiley, C. (2011). Pleiotropy,"sexy" traits, and speciation. *Behavioral Ecology, 22*(6), 1154–1155.

Shaw, K. L., & Lesnick, S. C. (2009). Genomic linkage of male song and female acoustic preference QTL underlying a rapid species radiation. *Proceedings of the National Academy of Sciences, 106*(24), 9737–9742. doi:10.1073/pnas.0900229106

Shaw, K. L., Parsons, Y. M., & Lesnick, S. C. (2007). QTL analysis of a rapidly evolving speciation phenotype in the Hawaiian cricket Laupala. *Molecular Ecology, 16*(14), 2879–2892.

Sheldon, B. C. (2000). Differential allocation: tests, mechanisms and implications. *Trends in Ecology & Evolution, 15*(10), 397–402. doi:http://dx.doi.org/10.1016/S0169-5347(00)01953-4

Sheldon, B. C., Andersson, S., Griffith, S. C., Örnborg, J., & Sendecka, J. (1999). Ultraviolet colour variation influences blue tit sex ratios. *Nature, 402*(6764), 874–877.

Sheldon, B. C., Merilö, J., Qvarnström, A., Gustafsson, L., & Ellegren, H. (1997). Paternal genetic contribution to offspring condition predicted by size of male secondary sexual character. *Proceedings of the Royal Society of London B: Biological Sciences, 264*(1380), 297–302.

Shelly, T., & Bailey, W. (1992). Experimental manipulation of mate choice by male katydids: the effect of female encounter rate. *Behavioral Ecology and Sociobiology, 30*(3–4), 277–282. doi:10.1007/BF00166713

Sherborne, A. L., Thom, M. D., Paterson, S., Jury, F., Ollier, W. E., Stockley, P., ... Hurst, J. L. (2007). The genetic basis of inbreeding avoidance in house mice. *Current Biology, 17*(23), 2061–2066.

Shettleworth, S. J. (1999). Female mate choice in swordtails and mollies: symmetry assessment or Weber's law? *Animal Behaviour, 58*(5), 1139–1142. doi:10.1006/anbe.1999.1239

Shizuka, D. (2014). Early song discrimination by nestling sparrows in the wild. *Animal Behaviour, 92*, 19–24. doi:10.1016/j.anbehav.2014.03.021

Shuker, D. M. (2010). Sexual selection: endless forms or tangled bank? *Animal Behaviour, 79*(3), e11–e17.

Shuker, D. M., & Day, T. H. (2001). The repeatability of a sexual conflict over mating. *Animal Behaviour, 61*(4), 755–762. doi:http://dx.doi.org/10.1006/anbe.2000.1645

———. (2002). Mate sampling and the sexual conflict over mating in seaweed flies. *Behavioral Ecology, 13*(1), 83–86. doi:10.1093/beheco/13.1.83

Shuster, S. M. (2009). Sexual selection and mating systems. *Proceedings of the National Academy of Sciences, 106*(Supplement 1), 10009–10016.

Shuster, S. M., & Wade, M. J. (2003). *Mating Systems and Strategies*. Princeton, NJ: Princeton University Press.

Siebeck, U. E., & Marshall, N. J. (2007). Potential ultraviolet vision in pre-settlement larvae and settled reef fish—a comparison across 23 families. *Vision Research, 47*(17), 2337–2352. doi:10.1016/j.visres.2007.05.014

Siemers, B. M., Kriner, E., Kaipf, I., Simon, M., & Greif, S. (2012). Bats eavesdrop on the sound of copulating flies. *Current Biology, 22*(14), R563–R564.

Sih, A., Bell, A., & Johnson, J. C. (2004). Behavioral syndromes: an ecological and evolutionary overview. *Trends in Ecology & Evolution, 19*, 372–378. doi:10.1016/j.tree.2004.04.009

Sih, A., & Bell, A. M. (2008). Insights for behavioral ecology from behavioral syndromes. *Advances in the Study of Behavior, 38*, 227–281.

Simão, J., & Todd, P. M. (2002). Modeling mate choice in monogamous mating systems with courtship. *Adaptive Behavior, 10*(2), 113–136. doi:10.1177/1059712302010002003

———. (2003). Emergent patterns of mate choice in human populations. *Artificial Life, 9*(4), 403–417. doi:10.1162/106454603322694843

Simcox, H., Colegrave, N., Heenan, A., Howard, C., & Braithwaite, V. A. (2005). Context-dependent male mating preferences for unfamiliar females. *Animal Behaviour, 70*(6), 1429–1437. doi:http://dx.doi.org/10.1016/j.anbehav.2005.04.003

Simmons, L. (1986). Female choice in the field cricket *Gryllus bimaculatus* (De Geer). *Animal Behaviour, 34*(5), 1463–1470.

———. (1987). Sperm competition as a mechanism of female choice in the field cricket, *Gryllus bimaculatus*. *Behavioral Ecology and Sociobiology, 21*(3), 197–202.

Simmons, L., Parker, G., & Stockley, P. (1999). Sperm displacement in the yellow dung fly, *Scatophaga stercoraria*: an investigation of male and female processes. *American Naturalist, 153*(3), 302–314.

Simmons, L., Roberts, J., & Dziminski, M. (2009). Egg jelly influences sperm motility in the externally fertilizing frog, *Crinia georgiana*. *Journal of Evolutionary Biology, 22*(1), 225–229.

Simmons, L., Stockley, P., Jackson, R., & Parker, G. (1996). Sperm competition or sperm selection: no evidence for female influence over paternity in yellow dung flies *Scatophaga stercoraria*. *Behavioral Ecology and Sociobiology, 38*(3), 199–206.

Simmons, L. W., House, C. M., Hunt, J., & Garcia-Gonzalez, F. (2009). Evolutionary response to sexual selection in male genital morphology. *Current Biology, 19*(17), 1442–1446.

Simmons, L. W., Thomas, M. L., Simmons, F. W., & Zuk, M. (2013). Female preferences for acoustic and olfactory signals during courtship: male crickets send multiple messages. *Behavioral Ecology*, art036.

Sinervo, B., Clobert, J., Miles, D. B., McAdam, A., & Lancaster, L. T. (2008). The role of pleiotropy vs signaller-receiver gene epistasis in life history trade-offs: dissecting the genomic architecture of organismal design in social systems. *Heredity, 101*(3), 197–211.

Sinervo, B., & Svensson, E. (2002). Correlational selection and the evolution of genomic architecture. *Heredity, 89*(5), 329–338.

Singh, D., & Bronstad, P. M. (2001). Female body odour is a potential cue to ovulation. *Proceedings of the Royal Academy of London B: Biological Sciences, 268*(1469), 797–801. doi:10.1098/rspb.2001.1589

Sirkiä, P. M., & Laaksonen, T. (2009). Distinguishing between male and territory quality: females choose multiple traits in the pied flycatcher. *Animal Behaviour, 78*(5), 1051–1060.

Sisneros, J. A., Forlano, P. M., Deitcher, D. L., & Bass, A. H. (2004). Steroid-dependent auditory plasticity leads to adaptive coupling of sender and receiver. *Science, 305* (5682), 404–407.

Sison-Mangus, M. P., Bernard, G. D., Lampel, J., & Briscoe, A. D. (2006). Beauty in the eye of the beholder: the two blue opsins of lycaenid butterflies and the opsin gene-driven evolution of sexually dimorphic eyes. *Journal of Experimental Biology, 209*, 3079–3090.

Siva-Jothy, M. T., & Stutt, A. D. (2003). A matter of taste: direct detection of female mating status in the bedbug. *Proceedings of the Royal Society of London B: Biological Sciences, 270*(1515), 649–652. doi:10.1098/rspb.2002.2260

Skamel, U. (2003). Beauty and sex appeal: sexual selection of aesthetic preferences. *Evolutionary Aesthetics* (pp. 173–200). New York: Springer.

Slabbekoorn, H., & Peet, M. (2003). Ecology: Birds sing at a higher pitch in urban noise. *Nature, 424*(6946), 267–267.

Slagsvold, T., Hansen, B. T., Johannessen, L. E., & Lifjeld, J. T. (2002). Mate choice and imprinting in birds studied by cross-fostering in the wild. *Proceedings of the Royal Society of London B: Biological Sciences, 269*(1499), 1449–1455.

Slatyer, R. A., Mautz, B. S., Backwell, P.R.Y., & Jennions, M. D. (2012). Estimating genetic benefits of polyandry from experimental studies: a meta-analysis. *Biological Reviews, 87*(1), 1–33. doi:10.1111/j.1469-185X.2011.00182.x

Smadja, C., & Butlin, R. K. (2009). On the scent of speciation: the chemosensory system and its role in premating isolation. *Heredity, 102*(1), 77–97. doi:10.1038/hdy.2008.55

Smadja, C. M., Loire, E., Caminade, P., Thoma, M., Latour, Y., Roux, C., … Boursot, P. (2015). Seeking signatures of reinforcement at the genetic level: a hitchhiking mapping and candidate gene approach in the house mouse. *Molecular Ecology, 24*(16), 4222–4237.

Smaldino, P. E., & Schank, J. C. (2012). Human mate choice is a complex system. *Complexity, 17*(5), 11–22. doi:10.1002/cplx.21382

Smith, C., Barber, I., Wootton, R. J., & Chittka, L. (2004). A receiver bias in the origin of three-spined stickleback mate choice. *Proceedings of the Royal Society of London B: Biological Sciences, 271*(1542), 949–955.

Smith, C. C., & Mueller, U. G. (2015). Sexual transmission of beneficial microbes. *Trends in Ecology & Evolution, 30*(8), 438–440.

Smith, J. M. (1991). Theories of sexual selection. *Trends in Ecology & Evolution, 6*(5), 146–151.

Smith, J. M., & Harper, D. (2003). *Animal Signals*. New York: Oxford University Press.

Snedden, W. A., & Greenfield, M. D. (1998). Females prefer leading males: relative call timing and sexual selection in katydid choruses. *Animal Behaviour, 56*(5), 1091–1098. doi:http://dx.doi.org/10.1006/anbe.1998.0871

Snowberg, L., & Benkman, C. (2009). Mate choice based on a key ecological performance trait. *Journal of Evolutionary Biology, 22*(4), 762–769.

Snowberg, L. K., & Bolnick, D. I. (2008). Assortative mating by diet in a phenotypically unimodal but ecologically variable population of stickleback. *American Naturalist, 172*(5), 733–739. doi:10.1086/591692.

Sockman, K. W. (2007). Neural orchestration of mate-choice plasticity in songbirds. *Journal of Ornithology, 148*(2), 225–230. doi:10.1007/s10336-007-0151-3

Sockman, K. W., Gentner, T. Q., & Ball, G. F. (2002). Recent experience modulates forebrain gene-expression in response to mate-choice cues in European starlings. *Proceedings of the Royal Society of London B: Biological Sciences*, 02PB0616.0611-0602 PB0616.0617.

———. (2005). Complementary neural systems for the experience-dependent integration of mate-choice cues in European starlings. *Journal of Neurobiology, 62*(1), 72–81. doi:10.1002/neu.20068

Sorensen, P. W., Hara, T. J., Stacey, N. E., & Goetz, F. W. (1988). F prostaglandins function as potent olfactory stimulants that comprise the postovulatory female sex pheromone in goldfish. *Biology of Reproduction, 39*(5), 1039–1050. doi:10.1095/biolreprod39.5.1039

Sorenson, M. D., Hauber, M. E., & Derrickson, S. R. (2010). Sexual imprinting misguides species recognition in a facultative interspecific brood parasite. *Proceedings of the Royal Society of London B: Biological Sciences, 277*(1697), 3079–3085. doi:10.1098/rspb.2010.0592

South, A., & Lewis, S. M. (2012). Determinants of reproductive success across sequential episodes of sexual selection in a firefly. *Proceedings of the Royal Society of London B: Biological Sciences, 279*(1741), 3201–3208.

South, S. H., Arnqvist, G., & Servedio, M. R. (2012). Female preference for male courtship effort can drive the evolution of male mate choice. *Evolution, 66*(12), 3722–3735.

South, S. H., Steiner, D., & Arnqvist, G. (2009). Male mating costs in a polygynous mosquito with ornaments expressed in both sexes. *Proceedings of the Royal Society of London B: Biological Sciences*. doi:10.1098/rspb.2009.0991

Springer, S. A., Crespi, B. J., & Swanson, W. J. (2011). Beyond the phenotypic gambit: molecular behavioural ecology and the evolution of genetic architecture. *Molecular Ecology, 20*(11), 2240–2257. doi:10.1111/j.1365-294X.2011.05116.x

Stanik, C. E., & Ellsworth, P. C. (2010). Who cares about marrying a rich man? Intelligence and variation in women's mate preferences. *Human Nature—An Interdisciplinary Biosocial Perspective, 21*(2), 203–217. doi:10.1007/s12110-010-9089-x

Stevens, M. (2014). Confusion and illusion: understanding visual traits and behavior. A comment on Kelley and Kelley. *Behavioral Ecology, 25*(3), 464–465. doi:10.1093/beheco/aru013

Stevens, S. S. (1957). On the psychophysical law. *Psychological Review, 64*(3), 153–181. doi:10.1037/h0046162

Stewart, A. D., Morrow, E. H., & Rice, W. R. (2005). Assessing putative interlocus sexual conflict in Drosophila melanogaster using experimental evolution. *Proceedings of the Royal Society of London B: Biological Sciences, 272*(1576), 2029–2035.

Stinchcombe, J. R., & Kirkpatrick, M. (2012). Genetics and evolution of function-valued traits: understanding environmentally responsive phenotypes. *Trends in Ecology & Evolution, 27*(11), 637–647. doi:http://dx.doi.org/10.1016/j.tree.2012.07.002

Stiver, K. A., & Alonzo, S. H. (2009). Parental and mating effort: is there necessarily a trade-off? *Ethology, 115*(12), 1101–1126.

Stockley, P. (1999). Sperm selection and genetic incompatibility: does relatedness of mates affect male success in sperm competition? *Proceedings of the Royal Society of London B: Biological Sciences, 266*(1429), 1663–1669.

———. (2003). Female multiple mating behaviour, early reproductive failure and litter size variation in mammals. *Proceedings of the Royal Society of London B: Biological Sciences, 270*(1512), 271–278.

Stoffer, B., & Uetz, G. W. (2015). The effects of social experience with varying male availability on female mate preferences in a wolf spider. *Behavioral Ecology and Sociobiology, 69*(6), 927–937.

Stoltz, J. A., & Andrade, M.C.B. (2010). Female's courtship threshold allows intruding males to mate with reduced effort. *Proceedings of the Royal Society of London B: Biological Sciences, 277*(1681), 585–592.

Stone, E. A., Shackelford, T. K., & Buss, D. M. (2012). Is variability in mate choice similar for intelligence and personality traits? Testing a hypothesis about the evolutionary genetics of personality. *Intelligence, 40*(1), 33–37.

Strandh, M., Westerdahl, H., Pontarp, M., Canbäck, B., Dubois, M. -P., Miquel, C., … Bonadonna, F. (2012). Major histocompatibility complex class II compatibility, but not class I, predicts mate choice in a bird with highly developed olfaction. *Proceedings of the Royal Society of London B: Biological Sciences, 279*(1746), 4457–4463.

Stuart-Fox, D. M., & Whiting, M. J. (2005). Male dwarf chameleons assess risk of courting large, aggressive females. *Biology Letters, 1*(2), 231–234. doi:10.1098/rsbl.2005.0299

Stulp, G., Buunk, A. P., Kurzban, R., & Verhulst, S. (2013). The height of choosiness: mutual mate choice for stature results in suboptimal pair formation for both sexes. *Animal Behaviour, 86*(1), 37–46. doi:http://dx.doi.org/10.1016/j.anbehav.2013.03.038

Stumpf, R. M., & Boesch, C. (2006). The efficacy of female choice in chimpanzees of the Taï Forest, Côte d'Ivoire. *Behavioral Ecology and Sociobiology, 60*(6), 749–765.
Sturm, T., Leinders-Zufall, T., Maček, B., Walzer, M., Jung, S., Pömmerl, B., . . . Rammensee, H.-G. (2013). Mouse urinary peptides provide a molecular basis for genotype discrimination by nasal sensory neurons. *Nature Communications, 4*, 1616.
Suschinsky, K. D., Elias, L. J., & Krupp, D. B. (2007). Looking for Ms. Right: allocating attention to facilitate mate choice decisions. *Evolutionary Psychology, 5*(2), 428–441.
Svedin, N., Wiley, C., Veen, T., Gustafsson, L., & Qvarnström, A. (2008). Natural and sexual selection against hybrid flycatchers. *Proceedings of the Royal Society of London B: Biological Sciences, 275*(1635), 735–744.
Svensson, E. I., Eroukhmanoff, F., Karlsson, K., Runemark, A., & Brodin, A. (2010). A role for learning in population divergence of mate preferences. *Evolution, 64*(11), 3101–3113.
Svensson, E. I., Karlsson, K., Friberg, M., & Eroukhmanoff, F. (2007). Gender differences in species recognition and the evolution of asymmetric sexual isolation. *Current Biology, 17*(22), 1943–1947.
Swaddle, J. P., Cathey, M. G., Correll, M., & Hodkinson, B. P. (2005). Socially transmitted mate preferences in a monogamous bird: a non-genetic mechanism of sexual selection. *Proceedings of the Royal Society of London B: Biological Sciences, 272*(1567), 1053. doi:10.1098/rspb.2005.3054
Sweeney, A., Jiggins, C., & Johnsen, S. (2003). Insect communication: Polarized light as a butterfly mating signal. *Nature, 423*(6935), 31–32. doi:10.1038/423031a
Symes, L. B., & Price, T. D. (2015). Sexual stimulation and sexual selection. *American Naturalist, 185*(4), iii–iv. doi:10.1086/680414
Syriatowicz, A., & Brooks, R. (2004). Sexual responsiveness is condition-dependent in female guppies, but preference functions are not. *BMC Ecology, 4*(1), 5.
Számadó, S. (2011). The cost of honesty and the fallacy of the handicap principle. *Animal Behaviour, 81*(1), 3–10.
Tait, C., Batra, S., Ramaswamy, S. S., Feder, J. L., & Olsson, S. B. (2016). Sensory specificity and speciation: a potential neuronal pathway for host fruit odour discrimination in *Rhagoletis pomonella*. *Proceedings of the Royal Society of London B: Biological Sciences, 283*(1845). doi:10.1098/rspb.2016.2101
Tan, C.K.W., Løvlie, H., Greenway, E., Goodwin, S. F., Pizzari, T., & Wigby, S. (2013). Sex-specific responses to sexual familiarity, and the role of olfaction in *Drosophila*. *Proceedings of the Royal Society of London B: Biological Sciences, 280*(1771). doi:10.1098/rspb.2013.1691
Tan, M., Jones, G., Zhu, G., Ye, J., Hong, T., Zhou, S., . . . Zhang, L. (2009). Fellatio by fruit bats prolongs copulation time. *PLoS ONE, 4*(10), e7595.
Tang-Martinez, Z., & Ryder, T. B. (2005). The problem with paradigms: Bateman's worldview as a case study. *Integrative and Comparative Biology, 45*(5), 821–830. doi:10.1093/icb/45.5.821
Tarvin, K. A., Webster, M. S., Tuttle, E. M., & Pruett-Jones, S. (2005). Genetic similarity of social mates predicts the level of extrapair paternity in splendid fairy-wrens. *Animal Behaviour, 70*(4), 945–955. doi:http://dx.doi.org/10.1016/j.anbehav.2005.01.012
Taylor, E., Boughman, J., Groenenboom, M., Sniatynski, M., Schluter, D., & Gow, J. (2006). Speciation in reverse: morphological and genetic evidence of the collapse of a

three-spined stickleback (Gasterosteus aculeatus) species pair. *Molecular Ecology, 15*(2), 343–355.

Taylor, R. C., & Ryan, M. J. (2013). Interactions of multisensory components perceptually rescue túngara frog mating signals. *Science*, 1–4. doi:10.1126/science.1237113

Tazzyman, S. J., & Iwasa, Y. (2010). Sexual selection can increase the effect of random genetic drift—a quantitative genetic model of polymorphism in *Oophaga pumilio*, the strawberry poison-dart frog. *Evolution, 64*(6), 1719–1728.

Tazzyman, S. J., Iwasa, Y., & Pomiankowski, A. (2014a). The handicap process favors exaggerated, rather than reduced, sexual ornaments. *Evolution, 68*(9), 2534–2549.

———. (2014b). Signaling efficacy drives the evolution of larger sexual ornaments by sexual selection. *Evolution, 68*(1), 216–229.

Tazzyman, S. J., Seymour, R. M., & Pomiankowski, A. (2012). Fixed and dilutable benefits: female choice for good genes or fertility. *Proceedings of the Royal Society of London B: Biological Sciences, 279*, 334–340.

Teale, S. A., Hager, B. J., & Webster, F. X. (1994). Pheromone-based assortative mating in a bark beetle. *Animal Behaviour, 48*(3), 569–578.

Telford, S. R., & Jennions, M. D. (1998). Establishing cryptic female choice in animals. *Trends in Ecology & Evolution, 13*(6), 216–218.

Templeton, C. N., Ríos-Chelén, A. A., Quirós-Guerrero, E., Mann, N. I., & Slater, P.J.B. (2013). Female happy wrens select songs to cooperate with their mates rather than confront intruders. *Biology Letters, 9*(1). doi:10.1098/rsbl.2012.0863

Templeton, J. J., Mountjoy, D. J., Pryke, S. R., & Griffith, S. C. (2012). In the eye of the beholder: visual mate choice lateralization in a polymorphic songbird. *Biology Letters, 8*(6), 924–927. doi:10.1098/rsbl.2012.0830

ten Cate, C. (1986). Sexual preferences in zebra finch (*Taeniopygia guttata*) males raised by two species (*Lonchura striata* and *Taeniopygia guttata*): I. A case of double imprinting. *Journal of Comparative Psychology, 100*(3), 248.

———. (1987). Sexual preferences in zebra finch males raised by two species: II. The internal representation resulting from double imprinting. *Animal Behaviour, 35*(2), 321–330.

———. (1994). Perceptual mechanisms in imprinting and song learning. In J. A. Hogan & J. J. Bolhuis (Eds.), *Causal Mechanisms of Behavioural Development* (pp. 116–146). Cambridge, UK: Cambridge University Press.

ten Cate, C., Los, L., & Schilperoord, L. (1984). The influence of differences in social experience on the development of species recognition in zebra finch males. *Animal Behaviour, 32*(3), 852–860. doi:http://dx.doi.org/10.1016/S0003-3472(84)80162-1

ten Cate, C., & Rowe, C. (2007). Biases in signal evolution: learning makes a difference. *Trends in Ecology & Evolution, 22*(7), 380–387.

ten Cate, C., Verzijden, M. N., & Etman, E. (2006). Sexual imprinting can induce sexual preferences for exaggerated parental traits. *Current Biology, 16*(11), 1128–1132. doi: 10.1016/j.cub.2006.03.068

ten Cate, C., & Vos, D. R. (1999). Sexual imprinting and evolutionary processes in birds: a reassessment. *Advances in the Study of Behavior* (Vol. 28, pp. 1–31). San Diego: Academic Press.

ter Haar, S. M., Kaemper, W., Stam, K., Levelt, C. C., & ten Cate, C. (2014). The interplay of within-species perceptual predispositions and experience during song ontogeny in zebra finches (*Taeniopygia guttata*). *Proceedings of the Royal Society of London B: Biological Sciences, 281*(1796). doi:10.1098/rspb.2014.1860

ter Hofstede, Hannah M., Schöneich, S., Robillard, T., & Hedwig, B. Evolution of a communication system by sensory exploitation of startle behavior. *Current Biology, 25*(24), 3245–3252. doi:10.1016/j.cub.2015.10.064

Thibert-Plante, X., & Gavrilets, S. (2013). Evolution of mate choice and the so-called magic traits in ecological speciation. *Ecology Letters, 16*(8), 1004–1013.

Thom, M. D., & Dytham, C. (2012). Female choosiness leads to the evolution of individually distinctive males. *Evolution, 66*(12), 3736–3742.

Thomas, D. A., & Barfield, R. J. (1985). Ultrasonic vocalization of the female rat (*Rattus norvegicus*) during mating. *Animal Behaviour, 33*(3), 720–725.

Thonhauser, K. E., Raveh, S., & Penn, D. J. (2014). Multiple paternity does not depend on male genetic diversity. *Animal Behaviour, 93*, 135–141.

Thornhill, R. (1983). Cryptic female choice and its implications in the scorpionfly *Harpobittacus nigriceps*. *American Naturalist, 122*(6), 765–788.

Thünken, T., Bakker, T. C., & Baldauf, S. A. (2014). "Armpit effect" in an African cichlid fish: self-referent kin recognition in mating decisions of male *Pelvicachromis taeniatus*. *Behavioral Ecology and Sociobiology, 68*(1), 99–104.

Thünken, T., Bakker, T.C.M., Baldauf, S. A., & Kullmann, H. (2007). Direct familiarity does not alter mating preference for sisters in male *Pelvicachromis taeniatus* (Cichlidae). *Ethology*, 113(11), 1107–1112.

Thünken, T., Meuthen, D., Bakker, T.C.M., & Baldauf, S. A. (2012). A sex-specific trade-off between mating preferences for genetic compatibility and body size in a cichlid fish with mutual mate choice. *Proceedings of the Royal Society of London B: Biological Sciences, 79*(1740), 2959–2964. doi:10.1098/rspb.2012.0333

Tinbergen, N. (1952). "Derived" activities; their causation, biological significance, origin, and emancipation during evolution. *Quarterly Review of Biology, 27*(1), 1–32.

Tinghitella, R. M., Weigel, E. G., Head, M., & Boughman, J. W. (2013). Flexible mate choice when mates are rare and time is short. *Ecology and Evolution, 3*(9), 2820–2831. doi:10.1002/ece3.666

Tinghitella, R. M., & Zuk, M. (2009). Asymmetric mating preferences accommodated the rapid evolutionary loss of a sexual signal. *Evolution, 63*(8), 2087–2098.

Todd, P. M. (1997). Searching for the next best mate. In R. Conte, R. Hegselmann, & P. Terna (Eds.), *Simulating Social Phenomena* (Vol. 456, pp. 419–436). Berlin: Springer.

Todd, P. M., Penke, L., Fasolo, B., & Lenton, A. P. (2007). Different cognitive processes underlie human mate choices and mate preferences. *Proceedings of the National Academy of Sciences, 104*(38), 15011–15016. doi:10.1073/pnas.0705290104

Tomkins, J. L., Radwan, J., Kotiaho, J. S., & Tregenza, T. (2004). Genic capture and resolving the lek paradox. *Trends in Ecology & Evolution, 19*(6), 323–328.

Tomkins, J. L., & Simmons, L. W. (1998). Female choice and manipulations of forceps size and symmetry in the earwig *Forficula auricularia* L. *Animal Behaviour, 56*(2), 347–356. doi:10.1006/anbe.1998.0838

Tomlinson, I., & O'Donald, P. (1989). The co-evolution of multiple female mating preferences and preferred male characters: the gene-for-gene hypothesis of sexual selection. *Journal of Theoretical Biology, 139*(2), 219–238.

———. (1996). The influence of female viability differences on the evolution of mate choice. *Heredity, 77*(3), 303–312.

Tooby, J., & Cosmides, L. (1992). The psychological foundations of culture. In J. Barkow, L. Cosmides, & J. Tooby (Eds.), *The Adapted Mind: Evolutionary Psychology and the Generation of Culture* (pp. 19–136). New York: Oxford University Press.

Toomey, M., & McGraw, K. (2012). Mate choice for a male carotenoid-based ornament is linked to female dietary carotenoid intake and accumulation. *BMC Evolutionary Biology, 12*(1), 3.

Trainor, B. C., & Basolo, A. L. (2000). An evaluation of video playback using Xiphophorus helleri. *Animal Behaviour, 59*(1), 83–89. doi:http://dx.doi.org/10.1006/anbe.1999.1289

———. (2006). Location, location, location: stripe position effects on female sword preference. *Animal Behaviour, 71*(1), 135–140. doi:http://dx.doi.org/10.1016/j.anbehav.2005.04.007

Trainor, L. (2008). Science & music: the neural roots of music. *Nature, 453*(7195), 598–599.

Trainor, L. J., Tsang, C. D., & Cheung, V.H.W. (2002). Preference for sensory consonance in 2- and 4-month-old infants. *Music Perception: An Interdisciplinary Journal, 20*(2), 187–194. doi:10.1525/mp.2002.20.2.187

Tramm, N. A., & Servedio, M. R. (2008). Evolution of mate-choice imprinting: competing strategies. *Evolution, 62*(8), 1991–2003.

Tregenza, T., & Wedell, N. (2002). Polyandrous females avoid costs of inbreeding. *Nature, 415*(6867), 71–73.

Trivers, R. L. (1972). Parental investment and sexual selection. In B. Campbell (Ed.), *Sexual Selection and the Descent of Man, 1871–1971* (pp. 136–179). Chicago: Aldine-Atherton.

Trona, F., Anfora, G., Balkenius, A., Bengtsson, M., Tasin, M., Knight, A., … Ignell, R. (2013). Neural coding merges sex and habitat chemosensory signals in an insect herbivore. *Proceedings of the Royal Society of London B: Biological Sciences, 280*(1760). doi:10.1098/rspb.2013.0267

Tschirren, B., Postma, E., Rutstein, A. N., & Griffith, S. C. (2012). When mothers make sons sexy: maternal effects contribute to the increased sexual attractiveness of extra-pair offspring. *Proceedings of the Royal Society of London B: Biological Sciences, 279*(1731), 1233–1240.

Tucker, M. A., & Gerhardt, H. (2012). Parallel changes in mate-attracting calls and female preferences in autotriploid tree frogs. *Proceedings of the Royal Society of London B: Biological Sciences, 279*(1733), 1583–1587.

Tudor, M. S., & Morris, M. R. (2009). Variation in male mate preference for female size in the swordtail Xiphophorus malinche. *Behaviour, 146*, 727–740. doi:10.1163/156853909x446172

———. (2011). Frequencies of alternative mating strategies influence female mate preference in the swordtail Xiphophorus multilineatus. *Animal Behaviour, 82*(6), 1313–1318.

Tuni, C., Beveridge, M., & Simmons, L. (2013). Female crickets assess relatedness during mate guarding and bias storage of sperm towards unrelated males. *Journal of Evolutionary Biology, 26*(6), 1261–1268.

Turner, G. F., & Burrows, M. T. (1995). A model of sympatric speciation by sexual selection. *Proceedings of the Royal Society of London B: Biological Sciences, 260*(1359), 287–292.

Twig, G., Levy, H., & Perlman, I. (2003). Color opponency in horizontal cells of the vertebrate retina. *Progress in Retinal and Eye Research, 22*(1), 31–68. doi:http://dx.doi.org/10.1016/S1350-9462(02)00045-9

Tyack, P., & Whitehead, H. (1983). Male competition in large groups of wintering humpback whales. *Behaviour, 83*(1), 132–154.

Tybur, J. M., & Gangestad, S. W. (2011). Mate preferences and infectious disease: theoretical considerations and evidence in humans. *Philosophical Transactions of the Royal Society of London B: Biological Sciences, 366*(1583), 3375–3388.

Tyler, F., Harrison, X. A., Bretman, A., Veen, T., Rodríguez-Muñoz, R., & Tregenza, T. (2013). Multiple post-mating barriers to hybridization in field crickets. *Molecular Ecology, 22*(6), 1640–1649.

Uetz, G., Roberts, J., Clark, D., Gibson, J., & Gordon, S. (2013). Multimodal signals increase active space of communication by wolf spiders in a complex litter environment. *Behavioral Ecology & Sociobiology, 67*(9), 1471–1482. doi:10.1007/s00265-013-1557-y

Uexküll, J. v. (1957). A stroll through the worlds of animals and men: a picture book of invisible worlds (C. H. Schiller, Trans.). In C. H. Schiller (Ed.), *Instinctive Behavior: The Development of a Modern Concept* New York: International Universities Press.

Uller, T. (2008). Developmental plasticity and the evolution of parental effects. *Trends in Ecology & Evolution, 23*(8), 432–438.

Uller, T., & Johansson, L. C. (2003). Human mate choice and the wedding ring effect. *Human Nature, 14*(3), 267–276. doi:10.1007/s12110-003-1006-0

Urbach, D., Folstad, I., & Rudolfsen, G. (2005). Effects of ovarian fluid on sperm velocity in Arctic charr (*Salvelinus alpinus*). *Behavioral Ecology and Sociobiology, 57*(5), 438–444.

Uy, J.A.C., Patricelli, G., & Borgia, G. (2001). Complex mate searching in the satin bowerbird *Ptilonorhynchus violaceus*. *American Naturalist, 158*(5), 530–542. doi:10.1086/323118.

Uy, J., & Safran, R. (2013). Variation in the temporal and spatial use of signals and its implications for multimodal communication. *Behavioral Ecology & Sociobiology, 67*(9), 1499–1511. doi:10.1007/s00265-013-1492-y

Uyeda, J. C., Arnold, S. J., Hohenlohe, P. A., & Mead, L. S. (2009). Drift promotes speciation by sexual selection. *Evolution, 63*(3), 583–594.

Vakirtzis, A. (2011). Mate choice copying and nonindependent mate choice: a critical review. *Annales Zoologici Fennici, 48*(2), 91–107. doi:10.2307/23737067

Vakirtzis, A., & Roberts, S. C. (2009). Mate choice copying and mate quality bias: different processes, different species. *Behavioral Ecology, 20*(4), 908–911. doi:10.1093/beheco/arp073

———. (2010). Nonindependent mate choice in monogamy. *Behavioral Ecology*, arq092.

Valero, A., Hudson, R., Luna, E. Á., & Garcia, C. M. (2005). A cost worth paying: energetically expensive interactions with males protect females from intrasexual aggression. *Behavioral Ecology and Sociobiology, 59*(2), 262–269.

Vallet, E., & Kreutzer, M. (1995). Female canaries are sexually responsive to special song phrases. *Animal Behaviour, 49*(6), 1603–1610. doi:http://dx.doi.org/10.1016/0003-3472(95)90082-9

Valone, T., Nordell, S., Giraldeau, L. -A., & Templeton, J. (1996). The empirical question of thresholds and mechanisms of mate choice. *Evolutionary Ecology, 10*(4), 447–455. doi:10.1007/BF01237729

van Bergen, E., Brakefield, P. M., Heuskin, S., Zwaan, B. J., & Nieberding, C. M. (2013). The scent of inbreeding: a male sex pheromone betrays inbred males. *Proceedings of the Royal Society of London B: Biological Sciences, 280*(1758), 20130102.

van Doorn, G. S., Dieckmann, U., & Weissing, F. J. (2004). Sympatric speciation by sexual selection: a critical reevaluation. *American Naturalist, 163*(5), 709–725.

van Doorn, G. S., Edelaar, P., & Weissing, F. J. (2009). On the origin of species by natural and sexual selection. *Science, 326*(5960), 1704–1707.

van Doorn, G. S., & Weissing, F. J. (2004). The evolution of female preferences for multiple indicators of quality. *American Naturalist, 164*(2), 173–186.

———. (2006). Sexual conflict and the evolution of female preferences for indicators of male quality. *American Naturalist, 168*(6), 742–757.

Van Gossum, H., De Bruyn, L., & Stoks, R. (2005). Reversible switches between male-male and male-female mating behaviour by male damselflies. *Biology Letters, 1*(3), 268–270. doi:10.1098/rsbl.2005.0315

Van Homrigh, A., Higgie, M., McGuigan, K., & Blows, M. W. (2007). The depletion of genetic variance by sexual selection. *Current Biology, 17*(6), 528–532. doi:10.1016/j.cub.2007.01.055

van Oers, K., Drent, P. J., Dingemanse, N. J., & Kempenaers, B. (2008). Personality is associated with extrapair paternity in great tits, *Parus major*. *Animal Behaviour, 76*(3), 555–563. doi:http://dx.doi.org/10.1016/j.anbehav.2008.03.011

Van Valen, L. (1973). Body size and numbers of plants and animals. *Evolution*, 27–35.

van Wingen, G., Mattern, C., Verkes, R. J., Buitelaar, J., & Fernández, G. (2008). Testosterone biases automatic memory processes in women towards potential mates. *Neuroimage, 43*, 114–120. doi:10.1016/j.neuroimage.2008.07.002

Vedder, O., Magrath, M.J.L., van der Velde, M., & Komdeur, J. (2013). Covariance of paternity and sex with laying order explains male bias in extra-pair offspring in a wild bird population. *Biology Letters, 9*(5). doi:10.1098/rsbl.2013.0616

Vedenina, V. Y., & Pollack, G. S. (2012). Recognition of variable courtship song in the field cricket *Gryllus assimilis*. *Journal of Experimental Biology, 215*(13), 2210–2219. doi:10.1242/jeb.068429

Veen, T., Borge, T., Griffith, S. C., Saetre, G.-P., Bures, S., Gustafsson, L., & Sheldon, B. C. (2001). Hybridization and adaptive mate choice in flycatchers. *Nature, 411*(6833), 45–50.

Veen, T., & Otto, S. (2015). Liking the good guys: amplifying local adaptation via the evolution of condition-dependent mate choice. *Journal of Evolutionary Biology, 28*(10), 1804–1815.

Verburgt, L., Ferguson, J.W.H., & Weber, T. (2008). Phonotactic response of female crickets on the Kramer treadmill: methodology, sensory and behavioural implications. *Journal of Comparative Physiology. A, Neuroethology, Sensory, Neural, and Behavioral Physiology, 194*(1), 79–96. doi:10.1007/s00359-007-0292-0

Verrell, P. A. (1991). Illegitimate exploitation of sexual signalling systems and the origin of species. *Ethology Ecology & Evolution, 3*(4), 273–283.

Verweij, K. J., Burri, A. V., & Zietsch, B. P. (2012). Evidence for genetic variation in human mate preferences for sexually dimorphic physical traits. *PLoS ONE, 7*(11), e49294.

Verzijden, M. N., Culumber, Z. W., & Rosenthal, G. G. (2012). Opposite effects of learning cause asymmetric mate preferences in hybridizing species. *Behavioral Ecology, 23*(5), 1133–1139. doi:10.1093/beheco/ars086

Verzijden, M. N., Etman, E., van Heijningen, C., van der Linden, M., & ten Cate, C. (2007). Song discrimination learning in zebra finches induces highly divergent responses to novel songs. *Proceedings of the Royal Society of London B: Biological Sciences, 274*(1607), 295–301.

Verzijden, M. N., Lachlan, R. F., & Servedio, M. R. (2005). Female mate-choice behavior and sympatric speciation. *Evolution, 59*(10), 2097–2108.

Verzijden, M. N., & Rosenthal, G. G. (2011). Effects of sensory modality on learned mate preferences in female swordtails. *Animal Behaviour, 82*(3), 557–562. doi:10.1016/j.anbehav.2011.06.010.

Verzijden, M. N., & ten Cate, C. (2007). Early learning influences species assortative mating preferences in Lake Victoria cichlid fish. *Biology Letters, 3*(2), 134–136.

Verzijden, M. N., ten Cate, C., Servedio, M. R., Kozak, G. M., Boughman, J. W., & Svensson, E. I. (2012). The impact of learning on sexual selection and speciation. *Trends in Ecology & Evolution, 27*(9), 511–519.

Vickers, N. J. (2006a). Inheritance of olfactory preferences I. Pheromone-mediated behavioral responses of *Heliothis subflexa* × *Heliothis virescens* hybrid male moths. *Brain, Behavior and Evolution, 68*(2), 63–74.

———. (2006b). Inheritance of olfactory preferences III. Processing of pheromonal signals in the antennal lobe of *Heliothis subflexa* × *Heliothis virescens* hybrid male moths. *Brain, Behavior and Evolution, 68*(2), 90–108.

Vickery, W. L., & Poulin, R. (1998). Parasitised non-choosy females do slow down the process: a reply to J. Rolff. *Behavioral Ecology and Sociobiology, 44*(1), 75–76.

Visscher, P. M., Hill, W. G., & Wray, N. R. (2008). Heritability in the genomics era—concepts and misconceptions. *Nature Reviews Genetics, 9*(4), 255–266.

von Helversen, D., & von Helversen, O. (1975). Verhaltens genetische Untersuchungen am akustichen Kommunicationssystem der Feldheuschrecken (Orthoptera, Acrididae) II. Das Lautschema von Artbastarden zwischen *Chorthippus biguttulus* und *Ch. mollis*. *Journal of Comparative Physiology, 104*, 301–323.

von Schantz, T., Bensch, S., Grahn, M., Hasselquist, D., & Wittzell, H. (1999). Good genes, oxidative stress and condition-dependent sexual signals. *Proceedings of the Royal Society of London B: Biological Sciences, 266*(1414), 1–12.

von Schilcher, F. (1976). The function of pulse song and sine song in the courtship of *Drosophila melanogaster*. *Animal Behaviour, 24*(3), 622–625. doi:10.1016/s0003-3472(76)80076-0

von Uexküll, J. (1909). *Umwelt und Innenwelt der Tiere*. Berlin: Springer.

Vortman, Y., Lotem, A., Dor, R., Lovette, I., & Safran, R. J. (2013). Multiple sexual signals and behavioral reproductive isolation in a diverging population. *American Naturalist, 182*(4), 514–523.

Vos, D. R. (1995). The role of sexual imprinting for sex recognition in zebra finches—a difference between males and females. *Animal Behaviour, 50*, 645–653. doi:10.1016/0003-3472(95)80126-x

Vyas, A. (2013). Parasite-augmented mate choice and reduction in innate fear in rats infected by *Toxoplasma gondii*. *Journal of Experimental Biology, 216*(1), 120–126.

Vyas, A., Harding, C., McGowan, J., Snare, R., & Bogdan, D. (2008). Noradrenergic neurotoxin, N-(2-chloroethyl)-N-ethyl-2-bromobenzylamine hydrochloride (DSP-4), treatment eliminates estrogenic effects on song responsiveness in female zebra finches (*Taeniopygia guttata*). *Behavioral Neuroscience, 122*(5), 1148–1157. doi:10.1037/0735-7044.122.5.1148

Wada-Katsumata, A., Silverman, J., & Schal, C. (2013). Changes in taste neurons support the emergence of an adaptive behavior in cockroaches. *Science, 340*(6135), 972–975. doi:10.1126/science.1234854

Wade, M. J., & Pruett-Jones, S. G. (1990). Female copying increases the variance in male mating success. *Proceedings of the National Academy of Sciences, 87*(15), 5749–5753.

Wagner, R. H., Helfenstein, F., & Danchin, E. (2004). Female choice of young sperm in a genetically monogamous bird. *Proceedings of the Royal Society of London B: Biological Sciences, 271*(Suppl. 4), S134–S137.

Wagner, W. E. (1998). Measuring female mating preferences. *Animal Behaviour, 55*(4), 1029–1042. doi:10.1006/anbe.1997.0635

———. (2011). Direct benefits and the evolution of female mating preferences: conceptual problems, potential solutions, and a field cricket. *Advances in the Study of Behavior, 43*(273), e319.

Wagner, W. E., & Basolo, A. L. (2007). The relative importance of different direct benefits in the mate choices of a field cricket. *Evolution, 61*(3), 617–622.

Wagner, W. E., Beckers, O. M., Tolle, A. E., & Basolo, A. L. (2012). Tradeoffs limit the evolution of male traits that are attractive to females. *Proceedings of the Royal Society of London B: Biological Sciences, 279*(1739), 2899–2906. doi:10.1098/rspb.2012.0275

Wagner, W. E., Murray, A. M., & Cade, W. H. (1995). Phenotypic variation in the mating preferences of female field crickets, *Gryllus integer*. *Animal Behaviour, 49*(5), 1269–1281. doi:10.1006/anbe.1995.0159

Wagner, W. E., Smeds, M. R., & Wiegmann, D. D. (2001). Experience affects female responses to male song in the variable field cricket *Gryllus lineaticeps* (Orthoptera, Gryllidae). *Ethology, 107*(9), 769–776. doi:10.1046/j.1439-0310.2001.00700.x

Wakita, M. (2004). Categorical perception of orientation in monkeys. *Behavioural Processes, 67*(2), 263–272.

Walker, L. K., Stevens, M., Karadaş, F., Kilner, R. M., & Ewen, J. G. (2013). A window on the past: male ornamental plumage reveals the quality of their early-life environment. *Proceedings of the Royal Society of London B: Biological Sciences, 280*(1756), 20122852.

Walling, C. A., Royle, N. J., Lindstrom, J., & Metcalfe, N. B. (2008). Experience-induced preference for short-sworded males in the green swordtail, *Xiphophorus helleri*. *Animal Behaviour, 76*, 271–276. doi:10.1016/j.anbehav.2008.03.008

———. (2010). Do female association preferences predict the likelihood of reproduction? *Behavioral Ecology and Sociobiology, 64*(4), 541–548. doi:10.1007/s00265-009-0869-4.

Ward, J. L., & Blum, M. J. (2012). Exposure to an environmental estrogen breaks down sexual isolation between native and invasive species. *Evolutionary Applications, 5*(8), 901–912. doi:10.1111/j.1752-4571.2012.00283.x

Ward, J. L., Love, E. K., Vélez, A., Buerkle, N. P., O'Bryan, L. R., & Bee, M. A. (2013). Multitasking males and multiplicative females: dynamic signalling and receiver preferences in Cope's grey treefrog. *Animal Behaviour, 86*, 231–243. doi:10.1016/j.anbehav.2013.05.016

Ward, P. I. (1993). Females influence sperm storage and use in the yellow dung fly *Scathophaga stercoraria* (L.). *Behavioral Ecology and Sociobiology, 32*(5), 313–319.

———. (2000). Cryptic female choice in the yellow dung fly *Scathophaga stercoraria* (L.). *Evolution, 54*(5), 1680–1686.

Warren, I. A., Gotoh, H., Dworkin, I. M., Emlen, D. J., & Lavine, L. C. (2013). A general mechanism for conditional expression of exaggerated sexually-selected traits. *Bioessays, 35*(10), 889–899.

Warwick, S., Vahed, K., Raubenheimer, D., & Simpson, S. J. (2009). Free amino acids as phagostimulants in cricket nuptial gifts: support for the 'Candymaker' hypothesis. *Biology Letters, 5*(2), 194–196.

Watabe, M., Kato, T. A., Tsuboi, S., Ishikawa, K., Hashiya, K., Monji, A., ... Kanba, S. (2013). Minocycline, a microglial inhibitor, reduces 'honey trap' risk in human economic exchange. *Scientific Reports, 3*, 1685. doi:10.1038/srep01685

Watanabe, T. K., & Kawanishi, M. (1979). Mating preference and the direction of evolution in *Drosophila*. *Science, 205*(4409), 906–907.

Watkins, C. D., Jones, B. C., Little, A. C., DeBruine, L. M., & Feinberg, D. R. (2012). Cues to the sex ratio of the local population influence women's preferences for facial symmetry. *Animal Behaviour, 83*(2), 545–553. doi:http://dx.doi.org/10.1016/j.anbehav.2011.12.002

Watson, N. L., & Simmons, L. W. (2012). Unravelling the effects of differential maternal allocation and male genetic quality on offspring viability in the dung beetle, *Onthophagus sagittarius*. *Evolutionary Ecology, 26*(1), 139–147.

Waynforth, D. (2001). Mate choice trade-offs and women's preference for physically attractive men. *Human Nature, 12*(3), 207–219. doi:10.1007/s12110-001-1007-9

———. (2007). Mate choice copying in humans. *Human Nature, 18*(3), 264–271. doi:10.1007/s12110-007-9004-2

Weddle, C. B., Hunt, J., & Sakaluk, S. K. (2013). Self-referent phenotype matching and its role in female mate choice in arthropods. *Current Zoology, 59*(2), 239–248.

Wedekind, C. (2002). Sexual selection and life-history decisions: implications for supportive breeding and the management of captive populations. *Conservation Biology, 16*(5), 1204–1211.

Wedekind, C., Chapuisat, M., Macas, E., & Rulicke, T. (1996). Non-random fertilization in mice correlates with the MHC and something else. *Heredity, 77*(4), 400–409.

Wedekind, C., Müller, R., & Spicher, H. (2001). Potential genetic benefits of mate selection in whitefish. *Journal of Evolutionary Biology, 14*(6), 980–986.

Wedell, N., Gage, M. J., & Parker, G. A. (2002). Sperm competition, male prudence and sperm-limited females. *Trends in Ecology and Evolution, 17*(7), 313–320.

Wedell, N., & Tregenza, T. (1999). Successful fathers sire successful sons. *Evolution*, 620–625.

Wedell, N., Wiklund, C., & Bergström, J. (2009). Coevolution of non-fertile sperm and female receptivity in a butterfly. *Biology Letters*, rsbl20090452.

Weisman, R., Shackleton, S., Ratcliffe, L., Weary, D., & Boag, P. (1994). Sexual preferences of female zebra finches: imprinting on beak colour. *Behaviour, 128*, 15–24.

Weissing, F. J., Edelaar, P., & van Doorn, G. S. (2011). Adaptive speciation theory: a conceptual review. *Behavioral Ecology and Sociobiology, 65*(3), 461–480. doi:10.1007/s00265-010-1125-7

Welch, A. M. (2003). Genetic benefits of a female mating preference in gray tree frogs are context-dependent. *Evolution, 57*(4), 883–893.

Welch, A. M., Semlitsch, R. D., & Gerhardt, H. C. (1998). Call duration as an indicator of genetic quality in male gray tree frogs. *Science, 280*(5371), 1928–1930.

Wellenreuther, M., & Sánchez-Guillén, R. A. (2016). Nonadaptive radiation in damselflies. *Evolutionary Applications, 9*(1), 103–118.

West, M. J., & King, A. P. (1988). Female visual displays affect the development of male song in the cowbird. *Nature, 334*(6179), 244–246.

West, S. A., & Sheldon, B. C. (2002). Constraints in the evolution of sex ratio adjustment. *Science, 295*(5560), 1685–1688.

West-Eberhard, M. J. (1983). Sexual selection, social competition, and speciation. *Quarterly Review of Biology*, 155–183.

———. (2014). Darwin's forgotten idea: the social essence of sexual selection. *Neuroscience & Biobehavioral Reviews, 46*, 501–508.

Westerman, E. L., Chirathivat, N., Schyling, E., & Monteiro, A. (2014). Mate preference for a phenotypically plastic trait is learned, and may facilitate preference-phenotype matching. *Evolution, 68*(6), 1661–1670. doi:10.1111/evo.12381

Westerman, E. L., Hodgins-Davis, A., Dinwiddie, A., & Monteiro, A. (2012). Biased learning affects mate choice in a butterfly. *Proceedings of the National Academy of Sciences U S A, 109*(27), 10948. doi:10.1073/pnas.1118378109.

Westneat, D. F. (2006). No evidence of current sexual selection on sexually dimorphic traits in a bird with high variance in mating success. *American Naturalist, 167*(6), E171–E189.

Westneat, D. F., Walters, A., McCarthy, T. M., Hatch, M. I., & Hein, W. K. (2000). Review: alternative mechanisms of nonindependent mate choice. *Animal Behaviour, 59*, 467–476. doi:10.1006/anbe.1999.1341

Wey, T. W., Chang, A. T., Fogarty, S., & Sih, A. (2015). Personalities and presence of hyperaggressive males influence male mating exclusivity and effective mating in stream water striders. *Behavioral Ecology and Sociobiology, 69*(1), 27–37.

Whaling, C. S., Solis, M. M., Doupe, A. J., Soha, J. A., & Marler, P. (1997). Acoustic and neural bases for innate recognition of song. *Proceedings of the National Academy of Sciences U S A, 94*(23), 12694–12698.

Wheatcroft, D. (2015). Reproductive interference via display signals: the challenge of multiple receivers. *Population Ecology, 57*(2), 333–337.

White, D. J. (2004). Influences of social learning on mate-choice decisions. *Learning & Behavior, 32*, 105–113.

White, D. J., & Galef, J.B.G. (2000). 'Culture' in quail: social influences on mate choices of female *Coturnix japonica*. *Animal Behaviour, 59*, 975–979. doi:10.1006/anbe.1999.1402

Whitlock, M. C., & Agrawal, A. F. (2009). Purging the genome with sexual selection: reducing mutation load through selection on males. *Evolution, 63*(3), 569–582.

Whitlock, M. C., & Fowler, K. (1999). The changes in genetic and environmental variance with inbreeding in *Drosophila melanogaster*. *Genetics, 152*, 345–353.

Whittingham, L. A., & Dunn, P. O. (2005). Effects of extra-pair and within-pair reproductive success on the opportunity for selection in birds. *Behavioral Ecology, 16*(1), 138–144. doi:10.1093/beheco/arh140

Whittingham, L. A., Freeman-Gallant, C. R., Taff, C. C., & Dunn, P. O. (2015). Different ornaments signal male health and MHC variation in two populations of a warbler. *Molecular Ecology, 24*(7), 1584–1595.

Widemo, F., & Saether, S. A. (1999). Beauty is in the eye of the beholder: causes and consequences of variation in mating preferences. *Trends in Ecology & Evolution, 14*(1), 26–31. doi:10.1016/s0169-5347(98)01531-6

Wiegmann, D. D. (1999). Search behaviour and mate choice by female field crickets, *Gryllus integer. Animal Behaviour, 58*(6), 1293–1298. doi:http://dx.doi.org/10.1006/anbe.1999.1243

Wiegmann, D. D., Angeloni, L. M., Seubert, S. M., & Wade, J. G. (2013). Mate choice decisions by searchers. *Current Zoology, 59*(2), 184–199.

Wiegmann, D. D., Seubert, S. M., & Wade, G. A. (2010). Mate choice and optimal search behavior: fitness returns under the fixed sample and sequential search strategies. *Journal of Theoretical Biology, 262*(4), 596–600.

Wigby, S., & Chapman, T. (2006). No evidence that experimental manipulation of sexual conflict drives premating reproductive isolation in *Drosophila melanogaster. Journal of Evolutionary Biology, 19*(4), 1033–1039.

Wiggins, D. A., & Morris, R. D. (1986). Criteria for female choice of mates: courtship feeding and paternal care in the common tern. *American Naturalist*, 126–129.

Wilczynski, W., Keddy-Hector, A. C., & Ryan, M. J. (1992). Call patterns and basilar papilla tuning in cricket frogs. I. Differences among populations and between sexes. *Brain, Behavior and Evolution, 39*(4), 229–237.

Wilczynski, W., Rand, A. S., & Ryan, M. J. (1995). The processing of spectral cues by the call analysis system of the tungara frog, *Physalaemus pustulosus. Animal Behaviour, 49*, 911–929.

Wilder, S. M., Rypstra, A. L., & Elgar, M. A. (2009). The importance of ecological and phylogenetic conditions for the occurrence and frequency of sexual cannibalism. *Annual Review of Ecology, Evolution, and Systematics, 40*(1), 21–39. doi:doi:10.1146/annurev.ecolsys.110308.120238

Wiley, C., Ellison, C. K., & Shaw, K. L. (2011). Widespread genetic linkage of mating signals and preferences in the Hawaiian cricket *Laupala. Proceedings of the Royal Society of London B: Biological Sciences*, rspb20111740. doi:10.1098/rspb.2011.1740

Wiley, R. H. (1983). The evolution of communication: information and manipulation. In T. R. Halliday & P.J.B. Slater (Eds.), *Animal Behaviour, Vol. 2, Communication* (pp. 156–189). Oxford: Blackwell Scientific Publications.

———. (1994). Errors, exaggeration, and deception in animal communication. *Behavioral Mechanisms in Evolutionary Ecology*, 157–189.

———. (2006). Signal detection and animal communication. In H. J. Brockmann, P.J.B. Slater, C. T. Snowdon, T. J. Roper, M. Naguib, & E.W.E. Katherine (Eds.), *Advances in the Study of Behavior* (Vol. 36, pp. 217–247). Cambridge, MA: Academic Press.

———. (2017). How noise determines the evolution of communication. *Animal Behaviour* 124, 307–313. doi:http://dx.doi.org/10.1016/j.anbehav.2016.07.014

Wiley, R. H., & Poston, J. (1996). Perspective: indirect mate choice, competition for mates, and coevolution of the sexes. *Evolution, 50*(4), 1371–1381. doi:10.2307/2410875

Wilgers, D., & Hebets, E. (2012). Age-related female mating decisions are condition dependent in wolf spiders. *Behavioral Ecology and Sociobiology, 66*(1), 29–38. doi:10.1007/s00265-011-1248-5

Wilkins, M. R., Seddon, N., & Safran, R. J. (2013). Evolutionary divergence in acoustic signals: causes and consequences. *Trends in Ecology & Evolution, 28*(3), 156–166.

Wilkinson, G. S., Breden, F., Mank, J. E., Ritchie, M. G., Higginson, A. D., Radwan, J., ... Rowe, L. (2015). The locus of sexual selection: moving sexual selection studies into the post-genomics era. *Journal of Evolutionary Biology, 28*(4), 739–755. doi:10.1111/jeb.12621

Wilkinson, G. S., Kahler, H., & Baker, R. H. (1998). Evolution of female mating preferences in stalk-eyed flies. *Behavioral Ecology, 9*(5), 525–533. doi:10.1093/beheco/9.5.525

Wilkinson, G. S., & Reillo, P. R. (1994). Female choice response to artificial selection on an exaggerated male trait in a stalk-eyed fly. *Proceedings of the Royal Society of London B: Biological Sciences, 255*, 1–6.

Willi, Y., & Van Buskirk, J. (2005). Genomic compatibility occurs over a wide range of parental genetic similarity in an outcrossing plant. *Proceedings of the Royal Society of London B: Biological Sciences, 272*(1570), 1333–1338.

Williams, H. (2001). Choreography of song, dance and beak movements in the zebra finch (*Taeniopygia guttata*). *Journal of Experimental Biology, 204*(20), 3497–3506.

Willis, P. M. (2013). Why do animals hybridize? *Acta Ethologica, 16*(3), 127–134.

Willis, P. M., Rosenthal, G. G., & Ryan, M. J. (2012). An indirect cue of predation risk counteracts female preference for conspecifics in a naturally hybridizing fish Xiphophorus birchmanni. *PLoS One, 7*(4). doi:10.1371/journal.pone.0034802.

Willis, P. M., Ryan, M. J., & Rosenthal, G. G. (2011). Encounter rates with conspecific males influence female mate choice in a naturally hybridizing fish. *Behavioral Ecology, 22*(6), 1234–1240.

Wilson, A. J., & Nussey, D. H. (2010). What is individual quality? An evolutionary perspective. *Trends in Ecology & Evolution, 25*(4), 207–214.

Wilson, D. R., Bayly, K. L., Nelson, X. J., Gillings, M., & Evans, C. S. (2008). Alarm calling best predicts mating and reproductive success in ornamented male fowl, *Gallus gallus*. *Animal Behaviour, 76*(3), 543–554.

Wilson, N., Tubman, S. C., Eady, P. E., & Robertson, G. W. (1997). Female genotype affects male success in sperm competition. *Proceedings of the Royal Society of London B: Biological Sciences, 264*(1387), 1491–1495.

Wilson, S. D., Clark, A. B., Coleman, K., & Dearstyne, T. (1994). Shyness and boldness in humans and other animals. *Trends in Ecology and Evolution, 9*(11), 442–446. doi:10.1016/0169-5347(94)90134-1

Windmill, J.F.C., Jackson, Joseph C., Tuck, E. J., & Robert, D. (2006). Keeping up with bats: dynamic auditory tuning in a moth. *Current Biology: CB, 16*(24), 2418–2423.

Winkler, C. A., Nelson, G. W., & Smith, M. W. (2010). Admixture mapping comes of age. *Annual Review of Genomics and Human Genetics, 11*(1), 65–89. doi:doi:10.1146/annurev-genom-082509-141523

Winslow, J. T., & Insel, T. R. (2004). Neuroendocrine basis of social recognition. *Current Opinion in Neurobiology, 14*(2), 248–253. doi:http://dx.doi.org/10.1016/j.conb.2004.03.009

Wirtz, P. (1999). Mother species-father species: unidirectional hybridization in animals with female choice. *Animal Behaviour, 58*, 1–12.

Witte, K., & Curio, E. (1999). Sexes of a monomorphic species differ in preference for mates with a novel trait. *Behavioral Ecology, 10*(1), 15–21.

Witte, K., Hirschler, U., & Curio, E. (2000). Sexual imprinting on a novel adornment influences mate preferences in the Javanese mannikin *Lonchura leucogastroides*. *Ethology, 106*(4), 349–363. doi:10.1046/j.1439-0310.2000.00558.x

Witte, K., & Klink, K. B. (2005). No pre-existing bias in sailfin molly females, *Poecilia latipinna*, for a sword in males. *Behaviour, 142*(3), 283–303. doi:10.1163/1568539053778292

Witte, K., Kniel, N., & Kureck, I. M. (2015). Mat-choice copying: status quo and where to go. *Current Zoology, 61*(6), 1073–1081.

Witte, K., & Noltemeier, B. (2002). The role of information in mate-choice copying in female sailfin mollies (*Poecilia latipinna*). *Behavioral Ecology and Sociobiology, 52*(3), 194–202. doi:10.1007/s00265-002-0503-1

Witte, K., Ryan, M. J., & Wilczynski, W. (2001). Changes in the frequency structure of a mating call decrease its attractiveness to females in the cricket frog *Acris crepitans blanchardi*. *Ethology, 107*(8), 685–699. doi:10.1046/j.1439-0310.2001.00715.x

Witte, K., & Sawka, N. (2003). Sexual imprinting on a novel trait in the dimorphic zebra finch: sexes differ. *Animal Behaviour, 65*, 195–203. doi:10.1006/anbe.2002.2009

Witte, K., & Ueding, K. (2003). Sailfin molly females (*Poecilia latipinna*) copy the rejection of a male. *Behavioral Ecology, 14*(3), 389–395. doi:10.1093/beheco/14.3.389

Wlodarski, R., & Dunbar, R. M. (2013). Examining the possible functions of kissing in romantic relationships. *Archives of Sexual Behavior, 42*(8), 1415–1423. doi:10.1007/s10508-013-0190-1

Wolak, M. E., Fairbairn, D. J., & Paulsen, Y. R. (2012). Guidelines for estimating repeatability. *Methods in Ecology and Evolution, 3*(1), 129–137.

Wolak, M. E., & Reid, J. M. (2016). Is pairing with a relative heritable? estimating female and male genetic contributions to the degree of biparental inbreeding in song sparrows (Melospiza melodia). *American Naturalist, 187*(6), 736–752. doi:10.1086/686198

Wolf, J. B., Brodie III, E. D., Cheverud, J. M., Moore, A. J., & Wade, M. J. (1998). Evolutionary consequences of indirect genetic effects. *Trends in Ecology and Evolution, 13*(2), 64–69.

Wolf, J., Brodie, E., & Moore, A. (1999). The role of maternal and paternal effects in the evolution of parental quality by sexual selection. *Journal of Evolutionary Biology, 12*(6), 1157–1167.

Wolf, J. B., Brodie III, E. D., & Moore, A. J. (1997). The evolution of indicator traits for parental quality: the role of maternal and paternal effects. *American Naturalist, 150*(5), 639–649.

Wollerman, L. (1999). Acoustic interference limits call detection in a Neotropical frog *Hyla ebraccata*. *Animal Behaviour, 57*(3), 529–536. doi:10.1006/anbe.1998.1013

Wong, B., & Rosenthal, G. G. (2005). Shoal choice in swordtails when preferences conflict. *Ethology, 111*(2), 179–186.

Wong, B. B., & Jennions, M. D. (2003). Costs influence male mate choice in a freshwater fish. *Proceedings of the Royal Society of London B: Biological Sciences, 270*(Suppl 1), S36–S38.

Wong, B.B.M. (2004). Superior fighters make mediocre fathers in the Pacific blue-eye fish. *Animal Behaviour, 67*, 583–590.

Wong, B.B.M., & Candolin, U. (2005). How is female mate choice affected by male competition? *Biological Reviews, 80*(04), 559–571.

Wong, B.B.M., Candolin, U., & Lindstrom, K. (2007). Environmental deterioration compromises socially enforced signals of male quality in three-spined sticklebacks. *American Naturalist, 170*, 184–189.

Wong, B.B.M., Fisher, H. S., & Rosenthal, G. G. (2005). Species recognition by male swordtails via chemical cues. *Behavioral Ecology, 16*(4), 818–822. doi:10.1093/beheco/ari058

Wong, B.B.M., Keogh, J. S., & Jennions, M. D. (2004). Mate recognition in a freshwater fish: geographical distance, genetic differentiation, and variation in female preference for local over foreign males. *Journal of Evolutionary Biology, 17*(3), 701–708.

Wong, B.B.M., & Kokko, H. (2005). Is science as global as we think? *Trends in Ecology & Evolution, 20*(9), 475–476. doi:10.1016/j.tree.2005.07.003

Wong, B.B.M., & Rosenthal, G. G. (2006). Female disdain for swords in a swordtail fish. *American Naturalist, 167*(1), 136–140. doi:10.1086/498278

Wong, B. B., & Schiestl, F. P. (2002). How an orchid harms its pollinator. *Proceedings of the Royal Society of London B: Biological Sciences, 269*(1500), 1529–1532.

Wong, R. Y., & Cummings, M. E. (2014). Expression patterns of neuroligin-3 and tyrosine hydroxylase across the brain in mate choice contexts in female swordtails. *Brain, Behavior and Evolution, 83*(3), 231–243.

Wong, R. Y., Ramsey, M. E., & Cummings, M. E. (2012). Localizing brain regions associated with female mate preference behavior in a swordtail. *PLoS ONE, 7*(11). doi:10.1371/journal.pone.0050355

Wong, R. Y., So, P., & Cummings, M. E. (2011). How female size and male displays influence mate preference in a swordtail. *Animal Behaviour, 82*, 691–697. doi:10.1016/j.anbehav.2011.06.024

Woo, K., & Rieucau, G. (2011). From dummies to animations: a review of computer-animated stimuli used in animal behavior studies. *Behavioral Ecology and Sociobiology, 65*(9), 1671–1685. doi:10.1007/s00265-011-1226-y

Woodgate, J. L., Bennett, A. T., Leitner, S., Catchpole, C. K., & Buchanan, K. L. (2010). Developmental stress and female mate choice behaviour in the zebra finch. *Animal Behaviour, 79*(6), 1381–1390.

Woodgate, J. L., Leitner, S., Catchpole, C. K., Berg, M. L., Bennett, A.T.D., & Buchanan, K. L. (2011). Developmental stressors that impair song learning in males do not appear to affect female preferences for song complexity in the zebra finch. *Behavioral Ecology, 22*(3), 566–573. doi:10.1093/beheco/arr006

Woolley, S. C., & Doupe, A. J. (2008). Social context-induced song variation affects female behavior and gene expression. *PLoS Biology, 6*(3), e62. doi:10.1371/journal.pbio.0060062

Woolley, S.M.N., Fremouw, T. E., Hsu, A., & Theunissen, F. E. (2005). Tuning for spectro-temporal modulations as a mechanism for auditory discrimination of natural sounds. *Nature Neuroscience, 8*(10), 1371–1379. doi:http://www.nature.com/neuro/journal/v8/n10/suppinfo/nn1536_S1.html

Wrangham, R. W. (1980). Female choice of least costly males; a possible factor in the evolution of leks. *Zeitschrift für Tierpsychologie, 54*(4), 357–367.

Wyttenbach, R. A., May, M. L., & Hoy, R. R. (1996). Categorical perception of sound frequency by crickets. *Science, 273*(5281), 1542.

Xu, M., Cerreta, A. L., Schultz, T. D., & Fincke, O. M. (2014). Selective use of multiple cues by males reflects a decision rule for sex discrimination in a sexually mimetic damselfly. *Animal Behaviour, 92*, 9–18. doi:10.1016/j.anbehav.2014.03.016

Yamamoto, Y., & Jeffery, W. R. (2000). Central role for the lens in cave fish eye degeneration. *Science, 289*, 631–633.

Yeates, S. E., Diamond, S. E., Einum, S., Emerson, B. C., Holt, W. V., & Gage, M. J. (2013). Cryptic choice of conspecific sperm controlled by the impact of ovarian fluid on sperm swimming behavior. *Evolution, 67*(12), 3523–3536.

Yeh, D. J., & Servedio, M. R. (2015). Reproductive isolation with a learned trait in a structured population. *Evolution, 69*(7), 1938–1947. doi:10.1111/evo.12688

Yokoyama, S., & Radlwimmer, F. B. (1998). The "five-sites" rule and the evolution of red and green color vision in mammals. *Molecular Biology and Evolution, 15*(5), 560–567.

Yorzinski, J. L., Patricelli, G. L., Pearson, J. M., Platt, M. L., & Babcock, J. S. (2013). Through their eyes: selective attention in peahens during courtship. *Journal of Experimental Biology, 216*(16), 3035–3046. doi:10.1242/jeb.087338

Young, K. A., Gobrogge, K. L., Liu, Y., & Wang, Z. (2011). The neurobiology of pair bonding: insights from a socially monogamous rodent. *Frontiers in Neuroendocrinology, 32*(1), 53–69. doi:http://dx.doi.org/10.1016/j.yfrne.2010.07.006

Young, L. C., Zaun, B. J., & VanderWerf, E. A. (2008). Successful same-sex pairing in *Laysan albatross. Biology Letters, 4*(4), 323–325. doi:10.1098/rsbl.2008.0191

Young, L. J., & Hammock, E.A.D. (2007). On switches and knobs, microsatellites and monogamy. *Trends in Genetics, 23*(5), 209–212. doi:http://dx.doi.org/10.1016/j.tig.2007.02.010

Young, L. J., & Wang, Z. (2004). The neurobiology of pair bonding. *Nature Neuroscience, 7*(10), 1048–1054.

Yu, D. W., & Shepard, G. H. (1998). Is beauty in the eye of the beholder? *Nature, 396*(6709), 321–322.

Zahavi, A. (1975). Mate selection—a selection for a handicap. *Journal of Theoretical Biology, 53*, 205–214.

Zala, S. M., Bilak, A., Perkins, M., Potts, W. K., & Penn, D. J. (2015). Female house mice initially shun infected males, but do not avoid mating with them. *Behavioral Ecology and Sociobiology, 69*(5), 715–722.

Zeh, D. W., & Zeh, J. A. (1988). Condition-dependent sex ornaments and field tests of sexual-selection theory. *American Naturalist*, 454–459.

Zeh, J. A., & Zeh, D. W. (1996). The evolution of polyandry I: intragenomic conflict and genetic incompatibility. *Proceedings of the Royal Society of London B: Biological Sciences, 263*(1377), 1711–1717.

———. (1997). The evolution of polyandry II: post-copulatory defenses against genetic incompatibility. *Proceedings of the Royal Society of London B: Biological Sciences, 264*(1378), 69–75.

Zeil, J. (2000). Depth cues, behavioural context, and natural illumination: some potential limitations of video playback techniques. *Acta Ethologica, 3*, 39–48.

Zhang, S., Liu, Y., & Rao, Y. (2013). Serotonin signaling in the brain of adult female mice is required for sexual preference. *Proceedings of the National Academy of Sciences*, 110(24), 9968–9973. doi:10.1073/pnas.1220712110

Zhou, M., Guo, J., Cha, J., Chae, M., Chen, S., Barral, J. M., ... Liu, Y. (2013). Non-optimal codon usage affects expression, structure and function of clock protein FRQ. *Nature, 495*(7439), 111–115. doi:10.1038/nature11833

Zhou, Y., Kelly, J. K., & Greenfield, M. D. (2011). Testing the fisherian mechanism: examining the genetic correlation between male song and female response in waxmoths. *Evolutionary Ecology, 25*(2), 307–329.

Ziegler, A., Kentenich, H., & Uchanska-Ziegler, B. (2005). Female choice and the MHC. *Trends in Immunology, 26*(9), 496–502.

Zietsch, B. P., Verweij, K.J.H., & Burri, A. V. (2012). Heretability of preferences for multiple cues of mate quality in humans *Evolution, 66*(6), 1762–1772. doi:10.1111/j.1558-5646.2011.01546.x

Zimmer, S. M., Schneider, J. M., & Herberstein, M. E. (2014). Can males detect the strength of sperm competition and presence of genital plugs during mate choice? *Behavioral Ecology*, aru045.

Zinck, L., & Lima, S. Q. (2013). Mate choice in *Mus musculus* is relative and dependent on the estrous state. *PLoS ONE, 8*(6), e66064.

Zizzari, Z. V., Braakhuis, A., van Straalen, N. M., & Ellers, J. (2009). Female preference and fitness benefits of mate choice in a species with dissociated sperm transfer. *Animal Behaviour, 78*(5), 1261–1267.

Zizzari, Z. V., van Straalen, N. M., & Ellers, J. (2013). Male-male competition leads to less abundant but more attractive sperm. *Biology Letters, 9*(6), 20130762.

Zuk, M. (2006). The case of the female orgasm (review). *Perspectives in Biology and Medicine, 49*(2), 294–298.

Zuk, M., Johnsen, T. S., & Maclarty, T. (1995). Endocrine-immune interactions, ornaments and mate choice in red jungle fowl. *Proceedings of the Royal Society of London B: Biological Sciences, 260*(1358), 205–210.

Zuk, M., Kim, T., Robinson, S. I., & Johnsen, T. S. (1998). Parasites influence social rank and morphology, but not mate choice, in female red junglefowl, *Gallus gallus*. *Animal Behaviour, 56*(2), 493–499. doi:http://dx.doi.org/10.1006/anbe.1998.0807

SUBJECT INDEX

TAXONOMIC INDEX